Pharmacogenomics:
Applications to Patient Care

THIRD EDITION

American College of Clinical Pharmacy
Lenexa, Kansas

Director of Professional Development: Nancy M. Perrin, M.A., CAE
Associate Director of Professional Development: Wafa Y. Dahdal, Pharm.D., BCPS
Publications Project Manager: David Shaw, M.B.A.
Desktop Publisher/Graphic Designer: Mary Ann Kuchta, B.S.
Medical Editor: Kimma Sheldon, Ph.D., M.A.

For order information or questions, contact:
American College of Clinical Pharmacy
13000 W. 87th Street Parkway, Suite 100
Lenexa, KS 66215
(913) 492-3311
(913) 492-0088 (fax)
accp@accp.com

Copyright © 2015 by the American College of Clinical Pharmacy. No part of this publication may be reproduced, stored in a retrieval system, or transmitted, in any form or by any means, electronic or mechanical, including photocopy, without prior written permission of the American College of Clinical Pharmacy.

15 16 17 WPC 10 9 8 7 6 5 4 3 2 1

Printed in the United States of America.

ISBN: 978-1-939862-09-9
Library of Congress Control Number: 2014959868

Pharmacogenomics: Applications to Patient Care
Third Edition

EDITORIAL BOARD

Julie A. Johnson, Pharm.D., FCCP, BCPS
University of Florida
College of Pharmacy
Gainesville, Florida

Vicki L. Ellingrod, Pharm.D., FCCP
University of Michigan
College of Pharmacy
Ann Arbor, Michigan

Deanna L. Kroetz, Ph.D.
University of California—San Francisco
School of Pharmacy
San Francisco, California

Grace M. Kuo, Pharm.D., M.P.H., Ph.D., FCCP
University of California—San Diego
Skaggs School of Pharmacy and Pharmaceutical Sciences
La Jolla, California

CONTRIBUTORS

Peter L. Anderson, Pharm.D.
University of Colorado
School of Pharmacy
Aurora, Colorado

Christina L. Aquilante, Pharm.D.
University of Colorado
Skaggs School of Pharmacy and
Pharmaceutical Sciences
Aurora, Colorado

Jeffery R. Bishop, Pharm.D., M.S., BCPP
University of Minnesota
College of Pharmacy
Minneapolis, Minnesota

Barry E. Bleske, Pharm.D., FCCP
University of Michigan
College of Pharmacy
Ann Arbor, Michigan

Kyle J. Burghardt, Pharm.D.
Wayne State University
Eugene Applebaum College of Pharmacy
and Health Sciences
Detroit, Michigan

William J. Canestaro, MSc
University of Washington
School of Pharmacy
Seattle, Washington

Josh J. Carlson, Ph.D., MPH
University of Washington
School of Pharmacy
Seattle, Washington

Larisa H. Cavallari, Pharm.D., FCCP, BCPS
University of Florida
College of Pharmacy
Gainesville, Florida

Zachary L. Cox, Pharm.D., BCPS
Lipscomb University
College of Pharmacy
Nashville, Tennessee

Courtney V. Fletcher, Pharm.D., FCCP
University of Nebraska Medical Center
College of Pharmacy
Omaha, Nebraska

Reginald F. Frye, Pharm.D., Ph.D., FCCP
University of Florida
School of Pharmacy
Gainesville, Florida

Louis P. Garrison, Jr., Ph.D.
University of Washington
Department of Pharmacy
Seattle, Washington

Anita Gupta, D.O., Pharm.D.
Drexel University
College of Medicine
Philadelphia, Pennsylvania

Susanne B. Haga, Ph.D.
Duke University
School of Medicine
Durham, North Carolina

Issam Hamadeh, Pharm.D.
University of Florida
College of Pharmacy
Gainesville, Florida

Daniel L. Hertz, Pharm.D., Ph.D.
University of Michigan
College of Pharmacy
Ann Arbor, Michigan

Pamala Jacobson, Pharm.D., FCCP
University of Minnesota
College of Pharmacy
Minneapolis, Minnesota

Samuel G. Johnson, Pharm.D., FCCP
Kaiser Permanente Colorado
Department of Clinical Pharmacy Services
Denver, Colorado

Marina Kawaguchi-Suzuki, Pharm.D., BCPS, BCACP
University of Florida
College of Pharmacy
Gainesville, Florida

Jill M. Kolesar, Pharm.D., FCCP, BCPS
University of Wisconsin
School of Pharmacy
Madison, Wisconsin

Taimour Langaee, Ph.D.
University of Florida
College of Pharmacy
Gainesville, Florida

Lisa K. Lee, M.D.
Drexel University
College of Medicine
Philadelphia, Pennsylvania

Ming Ta Michael Lee, Ph.D.
Laboratory for International Alliance on Genomic Medicine
RIKEN Center for Integrative Medical Sciences
Yokohama, Japan

Caitrin W. McDonough, Ph.D.
University of Florida
College of Pharmacy
Gainesville, Florida

David R. Nelson, M.D.
University of Florida
College of Medicine
Gainesville, Florida

Aniwaa Owusu Obeng, Pharm.D.
Icahn School of Medicine at Mount Sinai
The Charles Bronfman Institute for Personalized Medicine
New York, New York

Lori A. Orlando, M.D., MHS
Duke University
Center for Applied Genomics and Precision Medicine,
Department of Medicine
Durham, North Carolina

Michael Pacanowski, Pharm.D., M.P.H.
U.S. Public Health Service
Food and Drug Administration
Office of Clinical Pharmacology
Genomics and Targeted Therapy Group
Silver Spring, Maryland

Anthony J. Perissinotti, Pharm.D., BCOP
University of Michigan Health System
Department of Pharmacy
Ann Arbor, Michigan

Elimika Pfuma, Pharm.D., Ph.D.
Food and Drug Administration
Office of Clinical Pharmacology
Silver Spring, Maryland

Anthony T. Podany, Pharm.D.
University of Nebraska Medical Center
College of Pharmacy
Omaha, Nebraska

Marylyn D. Ritchie, Ph.D.
Pennsylvania State University
Center for Systems Genomics
University Park, Pennsylvania

Hobart Rogers, Pharm.D., Ph.D.
U.S. Public Health Service
Food and Drug Administration
Office of Clinical Pharmacology
Genomics and Targeted Therapy Group
Silver Spring, Maryland

Kinjal Sanghavi, M.Pharm.
University of Minnesota
Experimental and Clinical Pharmacology
Minneapolis, Minnesota

Kimberly K. Scarsi, Pharm.D., M.S., BCPS
University of Nebraska Medical Center
College of Pharmacy
Omaha, Nebraska

Stuart A. Scott, Ph.D.
Icahn School of Medicine at Mount Sinai
Department of Genetics and Genomic Sciences
New York, New York

Sharon M. Seifert, Pharm.D.
University of Colorado—Denver
School of Pharmacy
Aurora, Colorado

Jaekyu Shin, Pharm.D., M.S., BCPS
University of California—San Francisco
Department of Clinical Pharmacy
San Francisco, California

Todd Skaar, Ph.D.
Indiana University
Department of Medicine
Indianapolis, Indiana

James M. Stevenson, Pharm.D., M.S.
University of Pittsburgh
School of Pharmacy
Pittsburgh, Pennsylvania

Alexander G. Vandell, Pharm.D., Ph.D.
Daiichi Sankyo Pharma Development
Edison, New Jersey

Orly Vardeny, Pharm.D., M.S., FCCP, BCACP
University of Wisconsin
School of Pharmacy
Madison, Wisconsin

David L. Veenstra, Pharm.D., Ph.D.
University of Washington
School of Pharmacy
Seattle, Washington

Deepak Voora, M.D.
Duke University
Center for Applied Genomics and Precision Medicine,
Department of Medicine
Durham, North Carolina

Joseph R. Walker, Pharm.D.
Daiichi Sankyo Pharma Development
Edison, New Jersey

Kristin Weitzel, Pharm.D.
University of Florida
College of Pharmacy
Gainesville, Florida

Yanfei Zhang, M.D.
Laboratory for International Alliance on
Genomic Medicine
RIKEN Center for Integrative Medical Sciences
Yokohama, Japan

REVIEWERS

The American College of Clinical Pharmacy, the Editorial Board, and the authors would like to thank the following individuals for their careful review.

Ana Alfirevic, M.D., Ph.D.
University of Liverpool
Department of Molecular and Clinical Pharmacology
Liverpool, United Kingdom

Amber Beitelshees, Pharm.D., MPH, FAHA
University of Maryland—Baltimore
Division of Endocrinology, Diabetes & Nutrition
Baltimore, MD

Kelly E. Caudle, Pharm.D., Ph.D., BCPS
St. Jude Children's Research Hospital
Pharmaceutical Sciences Department
Memphis, Tennessee

Mackenzie L. Cottrell, Pharm.D., M.S., BCPS
University of North Carolina
Eshelman School of Pharmacy
Chapel Hill, North Carolina

Andrea Gaedigk, M.S., Ph.D.
Children's Mercy Hospital
Clinical Pharmacology, Toxicology, and Therapeutic Innovation
Kansas City, Missouri

James M. Hoffman, Pharm.D., MS, BCPS
St. Jude Children's Research Hospital
Pharmaceutical Sciences Department
Memphis, Tennessee

Angela Kashuba, BScPhm, Pharm.D., DABCP
University of North Carolina
Eshelman School of Pharmacy
Chapel Hill, North Carolina

Susan Leckband
VA San Diego Healthcare System
San Diego, California

Nita Limdi, Pharm.D, Ph.D, MSPH
University of Alabama at Birmingham
Department of Neurology
Birmingham, Alabama

Jeannine S. McCune, Pharm.D., FCCP, BCOP
University of Washington
Fred Hutchinson Cancer Research Center
Seattle, Washington

Howard McLeod, Pharm.D.
Moffitt Cancer Center
Division of Population Sciences
Tampa, Florida

Rima Mohammad, Pharm.D., BCPS
University of Michigan
College of Pharmacy
Ann Arbor, Michigan

Jeremiah Momper, Pharm.D., Ph.D.
University of California—San Diego
Skaggs School of Pharmacy and Pharmaceutical Sciences
La Jolla, California

Peter H. O'Donnell, M.D.
University of Chicago, Department of Medicine
University of Chicago Committee on Clinical Pharmacology and Pharmacogenomics
Chicago, Illinois

Jeong M. Park, M.S., Pharm.D., BCPS
University of Michigan
College of Pharmacy
Ann Arbor, Michigan

Mark A. Rothstein, J.D.
University of Louisville
School of Medicine
Louisville, Kentucky

Robert Straka, Pharm.D.
University of Minnesota
College of Pharmacy
Minneapolis, Minnesota

Liewei Wang, M.D., Ph.D.
Mayo Clinic
Department of Molecular Pharmacology and
Experimental Therapeutics
Rochester, Minnesota

Jonathan H. Watanabe, Pharm.D., M.S., Ph.D.
University of California—San Diego
Skaggs School of Pharmacy and
Pharmaceutical Sciences
La Jolla, California

Marc S. Williams
Geisinger Health System
Genomic Medicine Institute
Danville, Pennsylvania

Sook Wah Yee, Ph.D.
University of California—San Francisco
School of Pharmacy
San Francisco, California

CONTENTS

Preface xv

Section 1–Pharmacogenomics Basics

CHAPTER 1

Evolution of Pharmacogenomics and Genetic Variation in the Cytochrome P450 Enzymes 1
Todd Skaar, Ph.D.

CHAPTER 2

Clinical Implementation of Pharmacogenomics—Evaluating the Evidence 19
Samuel G. Johnson, Pharm.D., FCCP; and Christina L. Aquilante, Pharm.D.

CHAPTER 3

Genomic Medicine—Current Advances and Future Projections 32
Lori A. Orlando, M.D., MHS; and Deepak Voora, M.D.

CHAPTER 4

Implementing Pharmacogenetics in the Clinical Setting and Competencies for
Health Care Professionals 45
Kristin Weitzel, Pharm.D.

Section 2–Pharmacogenomics in Clinical Care

CHAPTER 5

Pharmacogenetics of Antiplatelet Drugs 65
Aniwaa Owusu Obeng, Pharm.D.; and Stuart A. Scott, Ph.D.

CHAPTER 6

Pharmacogenetics of Anticoagulant Drugs 79
Larisa H. Cavallari, Pharm.D., FCCP, BCPS; and Jaekyu Shin, Pharm.D., M.S., BCPS

CHAPTER 7

Pharmacogenetics of Lipid-Lowering Drugs with a Focus on SLC01B1 and Statin Therapy .102
*Barry E. Bleske, Pharm.D., FCCP; Zachary L. Cox, Pharm.D., BCPS; and
Orly Vardeny, Pharm.D., M.S., FCCP, BCACP*

CHAPTER 8

Pharmacogenetics in Mental Health 115
*Jeffrey R. Bishop, Pharm.D., M.S., BCPP; James M. Stevenson, Pharm.D., M.S.; and
Kyle J. Burghardt, Pharm.D.*

CHAPTER 9

Genetics in Pain Disorders 135
Anita Gupta, D.O., Pharm.D.; and Lisa K. Lee, M.D.

CHAPTER 10

Pharmacogenetics of Somatic Mutations in Cancer 148
Jill M. Kolesar, Pharm.D., FCCP, BCPS

CHAPTER 11

Germline Pharmacogenetics in Oncology 163
Daniel L. Hertz, Pharm.D., Ph.D.; and Anthony J. Perissinotti, Pharm.D., BCOP

CHAPTER 12

Pharmacogenomics of Immunosuppressants 176
Kinjal Sanghavi, M.Pharm.; and Pamala Jacobson, Pharm.D., FCCP

CHAPTER 13

HLA Pharmacogenetics and Serious Cutaneous Adverse Drug Reactions 196
Ming Ta Michael Lee, Ph.D.; and Yanfei Zhang, M.D.

CHAPTER 14

Pharmacogenetics of Hepatitis C Treatment 216
Marina Kawaguchi-Suzuki, Pharm.D., BCPS, BCACP; David R. Nelson, M.D.; and
Reginald F. Frye, Pharm.D., Ph.D., FCCP

CHAPTER 15

Pharmacogenetics of HIV Treatment 233
Kimberly K. Scarsi, Pharm.D., M.S., BCPS; Sharon Seifert, Pharm.D.; Anthony T. Podany, Pharm.D.;
Peter L. Anderson, Pharm.D.; and Courtney V. Fletcher, Pharm.D., FCCP

Section 3–Pharmacogenomics: Other Issues and Implications

CHAPTER 16

Ethical, Legal, and Social Challenges to Pharmacogenetics 251
Susanne B. Haga, Ph.D.

CHAPTER 17

Cost-Effectiveness, Economic Incentives, and Reimbursement Issues 267
William J. Canestaro, MSc.; Josh J. Carlson, Ph.D., MPH; Louis P. Garrison, Jr., Ph.D.;
and David L. Veenstra, Pharm.D., Ph.D.

CHAPTER 18

The Role of Pharmacogenomics and Targeted Therapeutics in the
FDA Drug Approval Process 288
Hobart Rogers, Pharm.D., Ph.D.; Elimika Pfuma, Pharm.D., Ph.D.; and
Michael Pacanowski, Pharm.D., M.P.H.

CHAPTER 19

Pharmacogenomics in Drug Discovery and Drug Development 298
Alexander G. Vandell, Pharm.D., Ph.D.; and Joseph R. Walker, Pharm.D.

Section 4–Fundamentals of Applied Human Genomics

CHAPTER 20

Principles of Genetics and Genetic Medicine 313
Caitrin W. McDonough, Ph.D.

CHAPTER 21

Applied Molecular and Cellular Biology 335
Taimour Langaee, Ph.D.; and Issam Hamadeh, Pharm.D.

CHAPTER 22

Study Design and Analysis Approaches in Pharmacogenomics Research 369
Marylyn D. Ritchie, Ph.D.

Index 395

PREFACE TO THE THIRD EDITION

Since the conception of the Human Genome Project, the promise of personalized medicine has been a long-term goal for the study of pharmacogenomics. With recent and rapid advances in this field, the implementation of pharmacogenomic tests into clinical practice has become a reality for some therapeutic areas, but not for all. The study of pharmacogenomics is well into its third decade; however, implementation of this science from the laboratory bench to the bedside has been slow in coming because our current treatment protocols for many medications are still in the "one-size-fits-all" model. However, as clinicians become more comfortable with their understanding of this science, which involves understanding the impact of individual patient genetic variation on the body's response to drugs, they can now be more confident in making difficult risk-benefit decisions in the context of a limited health care budget.

Therefore, pharmacogenomics represents both an opportunity and a challenge as this information is incorporated into clinical practice. To meet this challenge, the American College of Clinical Pharmacy (ACCP) served as the springboard for development of the first edition of *Pharmacogenomics: Applications to Patient Care* in 2004. This textbook combined the basics of pharmacogenomics with disease-specific applications to give students and practitioners a solid foundation for understanding the basic science of pharmacogenomics and the skills to integrate pharmacogenomics into their daily clinical practice. The second edition contained enhanced chapters on pharmacogenomics science and clinical management, written by leaders in applied pharmacogenomics. Thus, for this now-current third edition, the overall goal has been to help clinicians implement these pharmacogenomic tests into clinical practice, with clinical cases guiding the implementation process.

Completing the new edition of this textbook has been a bit of a departure from ACCP's development of the previous two editions and has resulted in a frameshift away from an exhaustive literature review for each topic, resulting in work that highlights the genetic variants currently ready to be translated into practice. Therefore, for this third edition, we have combined the skills of individuals from different practice areas, disciplines, and research environments in an effort to provide readers with actionable steps they can use in their practice as we move to the era of personalized medicine. We hope that you will enjoy the results of our efforts and that your ability to provide high-level clinical care to your patients is enriched.

Editorial Board
Pharmacogenomics: Applications to Patient Care, Third Edition
Winter 2014

Section 1
Pharmacogenomics Basics

Chapter 1

EVOLUTION OF PHARMACOGENOMICS AND GENETIC VARIATION IN THE CYTOCHROME P450 ENZYMES

Todd Skaar, Ph.D.

Learning Objectives
1. Recognize the historical beginning of pharmacogenomics.
2. Describe the changing approaches leading to the discoveries in pharmacogenomics.
3. Describe the molecular basis of genetic variations leading to altered functional activity of pharmacogenetically relevant genes.
4. Understand the star allele nomenclature for genetic variation in pharmacogenomic genes.

Keywords: Cytochrome P450, pharmacogenetics, pharmacogenomics, history, polymorphism, activity score, drug-metabolizing enzymes, star alleles

Abbreviations in This Chapter
CPIC	Clinical Pharmacogenetics Implementation Consortium
HLA-B	Human leukocyte antigen B
PharmGKB	Pharmacogenomics Knowledge Base
SNP	Single nucleotide polymorphism
VIP	Very Important Pharmacogenomics (project)

Abstract

Pharmacogenetics has undergone major changes since the early studies and the coining of the term in the 1950s. What started as observations of specific individuals having exaggerated responses to a few drugs has evolved into the implementation of specific genetic-guided therapies. This evolution has been made possible, in part, by the rapid changes in genomic technologies. This chapter briefly highlights the evolution of pharmacogenetics during the past 70 years. It also includes a brief description of the evolution of several different genotyping technologies and Internet resources that have been used and are now available for the study and implementation of pharmacogenetics. Because the genetic variants in the cytochrome P450 (CYP) drug metabolism genes have been on the forefront of pharmacogenetics and described in several later chapters, we have included a description of the major variants in the *CYP* genes, their functional implications, and the star allele and activity nomenclature used for these genes. Detailed gene-drug pairs for many of these genes will be the focus of several later chapters. We anticipate that pharmacogenetics will continue to evolve during the coming years and that it will be driven by advances in genomic technologies, preclinical and clinical gene-drug studies, and the development of clinical implementation guidelines.

Interindividual Variability in Drug Response

Drug therapy is an essential part of modern medicine. Hundreds of drugs are currently used that effectively treat many diseases and symptoms.

However, one major problem that physicians and pharmacists must deal with when using drug therapies is interindividual variability in drug response. This variability affects both the efficacy and toxicity of essentially all drugs. Many factors contribute to this variability. In this chapter, we will describe the current knowledge of how inherited genetics plays a role in many drug responses.

Every year, more than 100,000 deaths result from adverse drug reactions (Lazarou 1998). In hospitalized patients, more than 2 million serious adverse drug reactions occur every year (Lazarou 1998). Adverse drug reactions are the fourth leading cause of death in the United States (FDA 2000). Although several factors contribute to these events, one of the most important elements is the inherited germline genetic variation in genes that regulate drug pharmacokinetics and pharmacodynamics. It is not unusual to see 10-fold or more variation in the exposure of drugs in target populations. For some drugs that have a wide therapeutic index, this may not be a problem. However, for many drugs, the therapeutic index is very narrow, and the result is poor drug efficacy and/or the occurrence of adverse events in individual patients with drug concentrations outside that window.

Elimination of drugs from the body is controlled by various factors. One of the main mechanisms is drug-metabolizing enzymes. Phase I and phase II drug-metabolizing enzymes are two of the most important processes. Phase I enzymes are primarily involved in the oxidation and reduction of drugs. These reactions cause many changes in drug activities. In some cases, they activate the drugs; in others, they inactivate them. Often, these reactions also make them better substrates for the phase II enzymes. The phase II enzymes conjugate the drugs to several large molecules that facilitate the excretion of drugs from the body through the urine and bile. These conjugations include molecules such as glucuronides, sulfates, and glutathione. These phase I and phase II enzymes are located primarily in the liver, but several also exist in peripheral tissues such as the intestines and kidney. Many other enzymes are also involved in drug metabolism, including enzymes such as transferases (e.g., thiopurine methyltransferase) and esterases (e.g., carboxyesterase). Many of these enzymes are located throughout the body in fluids such as the blood.

Because enzymes are encoded by genes and substantial genetic variation exists in most human genes, genetic variation in the genes that encode the proteins controlling drug metabolism results in many different metabolic activities of most of these genes. Thus, these genetic variations greatly influence the pharmacokinetics of drugs and ultimately contribute to many of the clinically observed adverse drug reactions.

Genetic variation in various other genes also contributes to variable drug response and adverse drug reactions. For example, genetic variation in drug targets also contributes to drug sensitivities (e.g., *VKORC1* for warfarin) (Johnson 2011). In addition, specific alleles in the human leukocyte antigen B genes (*HLA-B* genes) confer increased susceptibility to severe adverse reactions to drugs such as abacavir and carbamazepine (Martin 2014; Leckband 2013). These reactions may occur because the proteins encoded by the HLA-B alleles bind to drugs and recognize them as foreign antigens, thus stimulating an immune response.

Early Pharmacogenetic Studies

Succinylcholine
As physicians and investigators became interested in the patients with unusual responses to drugs, the field of pharmacogenetics began. Initial reports of pharmacogenetics date back to some of Werner Kalow's studies in the early 1950s. His initial observations were that some patients had prolonged effects from the muscle relaxant succinylcholine (Kalow 1957). Through his investigations, he determined that the unusually long effect of the drug occurred because of defects in patients taking cholinesterase, the enzyme that destroys succinylcholine. This study then stimulated interest in other drugs that also caused unusual reactions in patient subsets.

Isoniazid
Shortly after that, isoniazid gained similar interest because a subset of the population was experiencing peripheral neuropathies as the result of normal doses of this agent for the treatment of tuberculosis (Kalow 1962). Later, it was determined that the neuropathies were caused by elevated levels of isoniazid. The unusually high isoniazid levels were attributable to reduced *N*-acetyltransferase enzyme activities in the patients with high isoniazid levels and neuropathies. The cause of this impaired *N*-acetyltransferase activity was later discovered to be genetic variations in the gene that encodes the *N*-acetyltransferase enzyme.

Primaquine
A third early observation in the mid-1950s was that some patients treated with primaquine for malaria developed a hemolytic anemia (Luzzatto 2001; Beutler 1993). The cause of this adverse drug reaction was determined to be impaired glucose-6-phosphate dehydrogenase (G6PD). It is now known that G6PD-impaired activity is caused by genetic variation in the gene that encodes the G6PD enzyme. Later, it was shown that African Americans had a relatively

high frequency of this genetic variation, which is likely because it protected its carriers from malaria, an important genetic selection mechanism in Africa.

These early works were summarized in a review article published in 1957 (Motulsky 1957). Because of this early interest, the word *pharmacogenetics* was coined in 1959 (Vogel 1959). Shortly afterward, an additional summary was published in a 1962 book (Kalow 1962). With the realization of genetic contributions to drug response and employment of the term *pharmacogenetics*, the field began to focus on identifying the specific genetic variants responsible for interindividual variability in drug response, and a wide variety of association studies began to identify the role of genetic variants in various other drug responses. These investigations grew into studies such as pharmacokinetics and pharmacodynamics that included many different phenotypes. The field then grew into larger studies focused on testing several genetic variants (more than 1 million per study), which resulted in use of the term *pharmacogenomics*, usually used to refer to studies focused on testing many variants. By contrast, the term *pharmacogenetics* is typically used to refer to studies focused on a limited number of candidate variants that have a rational reason for being thought to contribute to the specific drug response. However, the two terms are sometimes used interchangeably. As these two terms became widespread among several disciplines, they likely contributed to the expanded pharmacogenetic studies because they facilitated discussion on the possibility of the genetic contribution to any drug response. As the field grew, new journals spawned that focused specifically on pharmacogenetics and pharmacogenomics. This has resulted in more studies and the concentration of published studies in discipline-specific journals.

The early studies were followed by pharmacogenetic studies focused on what we now know as the cytochrome P450 (CYP) enzymes. These are the phase I drug-metabolizing enzymes that are primarily expressed in the liver. These enzymes are responsible for transforming many drugs and xenobiotics to facilitate drug removal from the body. There are several CYP families and they have different substrate specificities. Early studies of these enzymes focused on the specific enzymes that metabolize a variety of drugs, including debrisoquine and sparteine (Mikkelsen 2011). In the mid-1970s, studies showed that subpopulations of people had deficiencies in the enzymes that metabolize debrisoquine and sparteine (Bertilsson 1980; Eichelbaum 1979). In large populations, these drugs have multimodal distributions of metabolic ratios (e.g., debrisoquine/4-hydroxydebrisoquine [Steiner 1988]), indicating subpopulations with different metabolic activity. The enzyme involved turned out to be CYP2D6. The gene that encodes CYP2D6 is now known to have a great deal of genetic alterations that cause both abnormally low and abnormally high metabolic capacity (Daly 1996). This discovery spawned several studies of many other genes that encode various drug-metabolizing enzymes known to have an important role in determining drug dosing and the efficacy and toxicity of a variety of drugs. To facilitate the use of this genetic information in understanding genomic structure and using it in clinical trials, the star allele nomenclature was developed. Because the CYPs will be discussed in several chapters later in this book, the next section gives an overview of several aspects of these genes/enzymes and their nomenclature that will be useful in understanding the pharmacogenetics of a wide variety of drugs.

Genetic Variation in Clinically Important *CYP* Alleles

The Star Allele Nomenclature

The genes that encode the *CYP* alleles contain many genetic variations (Human Cytochrome P450 Allele [CYP-allele] Nomenclature Web site [www.cypalleles.ki.se/]). The star allele nomenclature created names for the specific alleles that could be used to classify them by the functional impact of the variations. The nomenclature follows the general rules of human genetics for naming allele families. The gene names start with the root symbol "CYP," denoting the *CYP* gene family. The Arabic numeral indicates the CYP family, followed by a letter indicating the subfamily and then another Arabic numeral indicating the individual gene within the subfamily. An example is the *CYP2D6* gene (Daly 1996). Next, the specific allele is indicated by a *, followed by the allele number. Each allele that is functionally distinct or that has an amino acid change, insertion, or deletion is given a new number. Each allele usually also carries several other variations not thought to be functionally significant (e.g., silent single nucleotide polymorphisms [SNPs], intronic SNPs). These are given a suballele name, which is shown by a letter after the * number. An example is *CYP2D6*4A*. The "wild-type" allele is usually designated the *1* allele (e.g., *CYP2D6*1*). This is usually the allele with normal functional activity. An up-to-date description of the known alleles for the major *CYP* genes is shown on the CYP-allele Web site (www.cypalleles.ki.se). In addition, the Very Important Pharmacogenomes (VIP [project]) descriptions on the Pharmacogenomics Knowledge Base (PharmGKB) Web site (www.pharmgkb.org) provide excellent descriptions of the major variant alleles for these genes. Detailed descriptions of the major clinically important alleles and their population frequencies are also included in the supplementary tables in the gene-drug pair guidelines published by the Clinical Pharmacogenetics Implementation Consortium (CPIC) (Relling 2011). In addition, the major functional alleles for several of the clinically important *CYP* genes are described later in this chapter.

Predicted Metabolizer Phenotypes

Genetics is often used to predict an individual's phenotypes and is broadly classified into four main categories. Individuals with normal activities are classified as "extensive metabolizers." These are usually individuals with either two normal functional alleles or one normal and one partly reduced-function allele. Those with a partly reduced phenotype are called "intermediate metabolizers." These individuals have one normal allele and one nonfunctional allele or, in some cases, two partly reduced-function alleles or one nonfunctional and one partial loss-of-function allele. Which combinations are included in the group classified as intermediate metabolizers depend on the specific gene and the drug of interest. For some drugs, one partly functional allele is sufficient for normal activity, whereas for others, only one partly functional allele significantly impairs the results of the drug. For other agents, even one nonfunctional allele can alter the drug response. In addition, some of the reduced-function alleles appear to be drug-specific because they have reduced function for some drugs but apparently normal activity for others. Individuals with two nonfunctional alleles are called "poor metabolizers." These individuals have no functional capacity to metabolize the drugs through that specific pathway. Finally, some individuals have multiple copies of functional alleles or alleles with apparently more-than-normal activity.

They are called "ultrarapid metabolizers." These patients usually have more-than-normal capacity to metabolize substrate drugs.

Table 1.1 summarizes the typical predicted metabolizer status of each combination of an individual's two alleles. These generic groupings are usually similar across the different genes (e.g., *CYP2D6*, *CYP2C19*). However, for some of the genes (e.g., *CYP2D6*), the classification of some groups as intermediate or extensive metabolizers can be drug-dependent. For example, for codeine, it appears that having a single copy of a normal allele may be sufficient to classify individuals as extensive metabolizers, based on their pain-related outcomes. However, for drugs like dextromethorphan or tamoxifen, individuals with one normal and one null allele may be classified as intermediate metabolizers. Classification may also depend on the phenotype being measured because phenotypes, like drug concentrations or parent drug/metabolite ratios, are often more sensitive measures; thus, it may benefit the study to separate the groups into those with more detailed classifications or even to use the activity scores as described in the next section. Classifications are based on the available literature. As much as possible, deciding which genotypes to call which metabolizer status should be based on published literature for the specific genotype, drug, and phenotype in question. When published literature on which to base the decision is lacking (e.g., because of rare variants or no studies having been done), extrapolation should be made with caution. In addition, as described in the next section, the concurrent medications taken by an individual should be considered potential modifiers of the predicted metabolizer status. Tables with specific genotype combinations and their resulting predicted phenotypes are available in some of the guidelines published by the CPIC investigators, found on the PharmGKB Web site.

Table 1.1. Predicted CYP2D6 Metabolic Phenotype Based on the Individual *CYP2D6* Allele Functions Derived from Genotyping

		Allele 2			
		Null	Reduced	Normal	Multiple Copies
Allele 1	Null	Poor	Intermediate	Intermediate	Extensive
	Reduced	Intermediate	Intermediate	Extensive	Extensive
	Normal	Intermediate	Extensive	Extensive	Ultrarapid
	Multiple copies	Extensive	Extensive	Ultrarapid	Ultrarapid

Predicted Activity Scores

For some genes, the predicted phenotypes are further broken down by assigning a more-specific numeric value to each individual (Gaedigk 2008). This is especially common in research studies of *CYP2D6*. To calculate the activity score, each allele is assigned an activity score, with the scores of both alleles for an individual added together to assign an activity score to that individual. Fully functional alleles are assigned a 1, partly reduced-activity alleles are assigned a 0.5, and nonfunctional alleles are assigned a 0. Thus, the activity scores normally range from 0 to 2. In some cases, individuals inherit more than two alleles. Individuals with more than two copies of functional alleles are assigned an activity score of 3. These scores can be further modified by including the effects of concurrent medications that inhibit the enzyme's metabolic activity (Borges 2010). For potent inhibitors, the individual's activity score is multiplied by 0; thus, all of those taking potent inhibitors would have an activity score of 0. For moderate inhibitors, the activity score is multiplied by 0.5. The specific value of 0.5 does not imply that the activity is assumed to be precisely 50% of the normal alleles; the quantitative effect of the moderate inhibitors and the reduced-function alleles depends on the substrate and the inhibitor drugs.

Currently, use of the *CYP2D6* activity score is limited to research studies. This score provides a more detailed quantitative breakdown of the predicted metabolic status for each study subject, which can be statistically more powerful in association studies. Adding the interacting drugs to the activity score can also be more powerful because it combines both the genetic and drug interaction status into a numeric score. In the future, activity scores will likely be useful in guiding individual therapies. Drugs that require a dose adjustment according to *CYP2D6* status, such as tricyclic antidepressants, are most likely to benefit from a quantitative breakdown of *CYP2D6* status; however, more studies are required to show the clinical value of using activity scores versus the standard predicted metabolizer phenotype described in the previous section.

Pharmacogenetics of Major Drug-Metabolizing Enzyme Genes

Many of the genes encoding the hepatic drug-metabolizing enzymes have been extensively characterized for genetic variants that alter protein function. These include variants that cause complete loss of function, reduced function, and increased function. The reduced-function alleles are also sometimes called dysfunctional alleles because their function can be variable, depending on the drug being considered. Complete loss-of-function alleles include variants that cause changes such as complete loss of the gene, gene rearrangements, premature stop codons, disruptive alternative messenger RNA splicing variants, or altered enzyme structure and function. In general, complete loss-of-function alleles have more dramatic effects on drug response than do reduced-function alleles because of their greater magnitude of effect on enzyme function. Reduced-function, or dysfunctional, alleles typically have effects that are intermediate to the normal and the complete loss-of-function alleles. These alleles are partly evident in pharmacokinetic studies that are typically quite sensitive to variations in enzyme activity. However, because their effects are often moderate given the additional genetic and environmental variability that also contributes to the phenotype being tested, reduced-function alleles are sometimes not statistically different from normal alleles. Increased-function alleles are relatively rare but occasionally have important clinical effects. Increased-function alleles are the result of increased copy numbers of the gene in the genomic or genetic variants that cause increased gene expression. These alleles can also have important effects on ultimate drug response, especially in the metabolism of prodrugs such as codeine to morphine, where increased metabolite production can have amplified pharmacodynamic effects.

The frequency of specific types of variations varies greatly between genes and populations. For example, most whites of European descent have no functional CYP3A5 activity because of a variant that causes aberrant splicing. By contrast, nonfunctional variants in CYP3A4 are extremely rare in any population. The frequency of multiple copies of functional *CYP2D6* alleles is relatively rare in most populations, except in some of the African and Middle East populations, which have relatively high frequencies. Asian populations are known to have very high frequencies of multiple copies of the *CYP2D6*36* allele, but because *CYP2D6*36* is a nonfunctional hybrid, individuals with these alleles actually have a loss-of-function phenotype.

The reason for the intergene variability in the frequency of loss-of-function and gain-of-function alleles is unclear. However, it may be the result of their role in the metabolism of either endogenous or exogenous substrates or their redundancy for specific substrates. Some enzymes are known to be important in the metabolism of endogenous substrates. For example, CYP3A4 metabolizes testosterone. Relatively few variants cause complete loss of function CYP3A4 or gain of function. The variants in CYP3A4 that have a major impact on metabolism may be relatively rare because CYP3A4 is important for endogenous substrates in all populations. Thus, there may be substantial selection pressure to maintain the normal function of this allele in all populations. In comparison, some of the other genes may be more important for the metabolism of exogenous substrates that have common exposures in specific populations. For many of these genes, it is still unclear which endogenous compounds are substrates for those enzymes, though more information is continually being discovered pertaining to this topic.

CYP2D6 Enzyme

CYP2D6 Drugs

Around 12% of drug metabolism is attributable to CYP2D6 (Wienkers 2005). CYP2D6 is well known for its role in metabolizing analgesics, antidepressants, antipsychotics, and β-blockers. Several drugs are also known to inhibit *CYP2D6* activity, resulting in reduced clearance and increased concentrations of substrate drugs. Some of the inhibitors are also substrates of CYP2D6. Several examples of CYP2D6 substrates, inhibitors, and inducers are included in Table 1.2. A more complete list is available at http://medicine.iupui.edu/clinpharm/ddis/main-table/. CYP2D6 is expressed primarily in the liver (Zanger 2004). Some expression may also occur in the brain, lung, and gut; however, the biological significance of the extrahepatic expression is still unclear (Ding 2003). Unlike many other CYP enzymes, CYP2D6 expression does not appear to be very inducible. The only factor currently known to significantly increase CYP2D6 activity is pregnancy (Heikkinen 2003; Wadelius 1997; Hogstedt 1985).

CYP2D6 Genetic Variation

CYP2D6 activity is highly influenced by genetic variation in the *CYP2D6* gene. The genetic variation of CYP2D6 is quite complex and includes SNPs, whole gene deletions, whole gene amplifications, and partial gene hybrids. Currently, 159 different *CYP2D6* haplotypes have been given * allele names. The most common loss-of-function allele is *CYP2D6*4*. The *CYP2D6*5* allele is the deletion of the entire *CYP2D6* allele. Other nonfunctional alleles commonly tested for include *CYP2D6*3, *6, *7*, and **8*. In Asian populations, the *CYP2D6*36* allele is a common nonfunctional hybrid allele that often exists as multiple copies of *CYP2D6*36*, together with a copy of the *CYP2D6*10* allele (Ramamoorthy 2010; Hosono 2009). Several reduced-function alleles are also common. These include *CYP2D6*41, CYP2D6*10* (particularly in Asians), and *CYP2D6*17* and **29* (particularly in African Americans). Some individuals also have more than two of the functional alleles, such as *CYP2D6*1* and *CYP2D6*2*. These alleles are indicated with an "xN" after the allele name (e.g., *CYP2D6*2xN*). These individuals usually have greater-than-normal CYP2D6 activity because of the increased expression of the multiple alleles. Several of the common and functionally important alleles with their corresponding functional information are shown in Table 1.3. Because of the complexities of the *CYP2D6* gene, many of the alleles have multiple variations that result in altered amino acid sequences. In some cases, multiple SNPs need to be genotyped to determine the specific allele. For example, the rs1065852 and rs1135840 SNPs are on the *CYP2D6*10* allele and often on the **4* alleles. Thus, the **10* allele is determined by the presence of these two variations, together with the absence of the rs3892097 SNP. The **4* allele is determined by the presence of the rs3892097 SNP, which causes a splice defect and a truncated and nonfunctional protein, regardless of which other SNP is present. The presence of additional SNPs on the **4* allele determines which **4* suballele it is.

Table 1.4 shows the ethnic differences in minor allele frequencies of several of the most common variant *CYP2D6* alleles. A more extensive breakdown of the ethnicities and additional allele frequencies can be found in the CPIC codeine guideline (Crews 2014).

The genetic variation has consistently been associated with the effects of a variety of CYP2D6 substrate drugs. For example, codeine is metabolized to morphine by CYP2D6. CYP2D6 poor metabolizers make only negligible amounts of morphine (Kirchheiner 2007) and thus receive only minimal analgesia from codeine (Crews

Table 1.2. Examples of CYP2D6 Substrates, Inhibitors, and Inducers

CYP2D6	Examples
Substrates	Codeine, tramadol, amitriptyline, venlafaxine, paroxetine, fluoxetine, haloperidol, risperidone, metoprolol, atomoxetine, donepezil, sparteine, duloxetine
Inhibitors	Bupropion, fluoxetine, paroxetine, quinidine, duloxetine
Inducers	Pregnancy

Table 1.3. Some Common and Functionally Important *CYP2D6* Alleles[a]

CYP2D6 Allele	SNP rs#[b]	Amino Acid Change	Enzymatic Activity	Minor Allele Frequency (%)
*2	rs16947 rs1135840	R296C S486T	Normal	1–30
*3	rs35742686	Frameshift	None	0–1
*4	rs1065852 rs3892097 rs1135840	P34S, splicing defect S486T	None	1–20
*5	None	Gene deletion	None	1–6
*6	rs5030655	Frameshift	None	0–3
*10	rs1065852 rs1135840	P34S S486T	Reduced	1–40
*17	rs28371706 rs16947 rs1135840	T107I R296C S486T	Reduced	1–20
*41	rs16947 rs28371725 rs1135840	R296C, splicing defect S486T	Reduced	0–20
*1xN[c]	No variants	*1 gene duplication	Increased	1–10
*2xN[c]	Same as *2	*2 gene duplication	Increased	0–4
*4xN[c]	Same as *4	*4 gene duplication	None	0–2

[a] Additional information on these and other *CYP2D6* alleles can be found on the PharmGKB Web sites www.pharmgkb.org/search/annotatedGene/cyp2d6/variant.jsp and www.cypalleles.ki.se/cyp2d6.htm.
[b] rs#'s are specific numbers given to each SNP that can be used to easily search for the specific SNP in the public databases, such as the dbSNP (Single Nucleotide Polymorphism Database; www.ncbi.nlm.nih.gov/SNP).
[c] xN represents the number of copies of a gene.

2014). CYP2D6 genetic variation is also associated with the kinetics and effects of several tricyclic antidepressants (Hicks 2013; Kirchheiner 2004). Consequently, CYP2D6 poor metabolizers usually need lower doses of these antidepressants (Hicks 2013).

CYP2C19 Enzyme

CYP2C19 Drugs

Around 12% of drug metabolism is attributable to CYP2C19 (Wienkers 2005). CYP2C19 is well known for its role in metabolizing proton pump inhibitors, antiepileptics, and some antidepressants. In addition, the genetic variation in CYP2C19 has repeatedly been shown to be important for clopidogrel, voriconazole, and cyclophosphamide. CYP2C19 is expressed primarily in the liver. Some tissues, including the kidney and small intestine, appear to have some extrahepatic expression; however, their contribution to drug metabolism is still unclear (Chen 2009; Ding 2003). CYP2C19 expression is induced by several drugs, resulting in increased clearance and reduced circulating concentrations of CYP2C19 substrates. In addition, several of the substrates for CYP2C19, as well as several other drugs, are inhibitors of the CYP2C19 activity that results in reduced clearance and increased concentrations of substrate drugs. Several examples of CYP2C19 substrates, inhibitors, and

Table 1.4. Ethnic Differences in Minor Allele Frequencies for Major *CYP2D6* Alleles[a]

Allele	White	Asian	Black
*CYP2D6*2*	0.27	0.13–0.32	0.14–0.20
*CYP2D6*3*	0.01	0	< 0.01
*CYP2D6*4*	0.19	0.01–0.06	0.03–0.06
*CYP2D6*5*	0.3	0.02–0.05	0.06
*CYP2D6*6*	0.01	< 0.01	0.01–0.03
*CYP2D6*17*	< 0.01	< 0.01	0.19
*CYP2D6*41*	0.09	0.01–0.10	0.10

[a]Additional information on these and additional *CYP2D6* alleles can be found on the PharmGKB Web sites www.pharmgkb.org/search/annotatedGene/cyp2d6/variant.jsp and www.cypalleles.ki.se/cyp2d6.htm.

Table 1.5. CYP2C19 Substrates, Inhibitors, and Inducers

CYP2C19	Examples
Substrates	Omeprazole, diazepam, S-mephenytoin, amitriptyline, citalopram, clopidogrel, cyclophosphamide, voriconazole
Inhibitors	Omeprazole, ketoconazole, voriconazole, fluvoxamine, isoniazid, ticlopidine, cimetidine
Inducers	Carbamazepine, phenobarbital, rifampin, St. John's wort, nevirapine

inducers are included in Table 1.5. A more complete list can be found at http://medicine.iupui.edu/clinpharm/ddis/main-table/.

CYP2C19 Genetic Variation

CYP2C19 activity is highly influenced by genetic variation in the *CYP2C19* gene. Currently, 50 different *CYP2C19* haplotypes have been given * allele names. Several of the common and functionally important alleles with their corresponding functional information are shown in Table 1.6. The most common loss-of-function allele is *CYP2C19*2*. This is particularly true in Asian populations. The *CYP2C19*3–*8* alleles are also loss-of-function alleles, but their frequency in most populations is quite low. The *CYP2C19* gene also has one allele, *CYP2C19*17*, that appears to have more-than-normal activity. The allele frequency of *CYP2C19*17* is variable across different ethnicities, but it is fairly common, ranging upward to around 20%.

The genetic variation in CYP2C19 has been associated with several drugs. For example, clopidogrel is metabolized to an active metabolite by CYP2C19. Thus, CYP2C19 poor metabolizers receive less benefit from clopidogrel. Therefore, in some circumstances, they would likely benefit from a different drug (Hogstedt 1985). CYP2C19 is also associated with the altered kinetics of some antidepressants and thus can be useful in predicting the dose of some antidepressants (Hicks 2013). Table 1.7 shows the ethnic differences in minor allele frequencies of three of the most common variant *CYP2C19* alleles.

CYP2C9 Enzyme

CYP2C9 Drugs

Around 16% of drug metabolism is attributable to CYP2C9 (Wienkers 2005). CYP2C9 is well known for its role in metabolizing NSAIDs (nonsteroidal anti-inflammatory drugs), oral hypoglycemic agents, angiotensin II blockers, and sulfonylureas. In addition, S-warfarin and phenytoin, for which genetic variations in CYP2C9 play important roles in drug dosing, are substrates of CYP2C9. CYP2C9 expression occurs mainly in the liver.

Although it has also been detected in several other tissues, including the small intestine, kidney, and endothelial cells, the contribution of extrahepatic CYP2C9 expression to drug metabolism is still unclear (Chen 2009; Ding 2003). It may also be involved in metabolizing environmental compounds and arachidonic acid (Chen 2009). CYP2C9 expression is induced by several drugs, resulting in increased clearance and reduced circulating concentrations of CYP2C9 substrates. In addition, CYP2C9 activity can be inhibited by several drugs, resulting in reduced clearance and increased concentrations of substrate drugs. Several examples of CYP2C9 substrates, inhibitors, and inducers are included in Table 1.8. A more complete list can be found at http://medicine.iupui.edu/clinpharm/ddis/main-table/.

CYP2C9 Genetic Variation

CYP2C9 activity is highly influenced by genetic variation in the *CYP2C9* gene. Currently, 65 different *CYP2C9* haplotypes have been given * allele names. Several of the common and functionally important alleles with their corresponding functional information are shown in Table 1.9. For example, the two most common alleles with altered activity are *CYP2C9*2* and *CYP2C9*3*. The metabolism of warfarin is reduced about ~30%–40% in *CYP2C9*2* alleles and in 80%–90% in *CYP2C9** alleles. Additional *CYP2C9* alleles with impaired metabolic activity include the *CYP2C9*5, *6, *8*, and *11* alleles. These are less common, but they may have important roles in some ethnicities such as African Americans. Currently, no known copy number variations have been described for the *CYP2C9* gene. Table 1.10 shows the ethnic

Table 1.6. Some Common and Functionally Important *CYP2C19* Alleles[a]

CYP2C19 Allele	SNP rs#	Amino Acid Change	Enzymatic Activity	Minor Allele Frequency (%)
*2	rs4244285	Splicing defect	None	12–61
*3	rs4986893	W212X	None	1–15
*4	rs28399504	M1V	None	< 1
*5	rs56337013	R433W	None	< 1
*6	rs72552267	R132Q	None	< 1
*7	rs72558186	Splicing defect	None	< 1
*8	rs41291556	W120R	None	< 1
*17	rs12248560	Increased expression	Increased	2–21

[a]Additional information on these and additional *CYP2C19* alleles can be found on the PharmGKB Web sites www.pharmgkb.org/search/annotatedGene/cyp2c19/variant.jsp and www.cypalleles.ki.se/cyp2c19.htm.

Table 1.7. Ethnic Differences in Minor Allele Frequencies for Major *CYP2C19* Alleles[a]

Allele	White	Asian	Black
CYP2C19*2	0.12–0.15	0.29–0.35	0.15
CYP2C19*3	< 0.01	0.02–0.09	< 0.01
CYP2C19*17	0.18–0.21	0.27	0.16

[a]Frequencies are from those reported in the CPIC clopidogrel guideline. See Scott SA, Sangkuhl K, Stein CM, et al. Clinical Pharmacogenetics Implementation Consortium guidelines for *CYP2C19* genotype and clopidogrel therapy: 2013 update. Clin Pharmacol Ther 2013;94:317-23.

Table 1.8. Some Common CYP2C9 Substrates, Inhibitors, and Inducers

CYP2C9	Examples
Substrates	Warfarin, phenytoin, ibuprofen, celecoxib, tolbutamide, glipizide, glyburide, losartan
Inhibitors	Fluconazole, voriconazole, fluconazole, amiodarone, efavirenz, lovastatin, paroxetine
Inducers	Carbamazepine, rifampin, phenobarbital, nevirapine, St. John's wort

Table 1.9. Some Common and Functionally Important *CYP2C9* Alleles[a]

CYP2C9 Allele	SNP rs#	Amino Acid Change	Enzymatic Activity	Minor Allele Frequency (%)
*2	rs1799853	R144C	Decreased	0–10
*3	rs1057910	I359L	Decreased	1–13
*5	rs28371686	D360E	Decreased	0–3
*6	rs9332131	273 frameshift	None	0–1
*8	rs7900194	R150H	Decreased	1–10
*11	rs28371685	R335W	Decreased	1–24

[a]Additional information on these and additional *CYP2C9* alleles can be found on the PharmGKB Web site at www.pharmgkb.org/search/annotatedGene/cyp2c9/variant.jsp and www.cypalleles.ki.se/cyp2c9.htm.

Table 1.10. Ethnic Differences in Minor Allele Frequencies for Major *CYP2C9* Alleles[a]

Allele	White	Asian	Black
*CYP2C9*2*	0.13	0	0.03
*CYP2C9*3*	0.07	0.04	0.02

[a]Frequencies are calculated using the genotype information from subjects in the International Warfarin Pharmacogenetics Consortium and modified from the warfarin CPIC guideline. See Johnson JA, Gong L, Whirl-Carrillo M, et al. Clinical Pharmacogenetics Implementation Consortium guidelines for *CYP2C9* and *VKORC1* genotypes and warfarin dosing. Clin Pharmacol Ther 2011;90:625-9.

differences in minor allele frequencies of two of the most common variant *CYP2C9* alleles.

One of the most-studied drugs associated with *CYP2C9* genetic variation is warfarin, which is metabolized to an inactive metabolite; thus, poor metabolizers typically require lower doses of warfarin (Hogstedt 1985).

CYP3A Enzymes

CYP3A Drugs

CYP3A4 and CYP3A5 are similar enzymes that are often grouped together as CYP3A enzymes. They are highly similar in structure, function, and expression. Around 46% of drug metabolism is attributable to CYP3A (Wienkers 2005). The CYP3A enzymes are expressed at high levels in the liver. However, these enzymes are also expressed in significant amounts in the intestinal tract (Ding 2003). This influences the bioavailability of a variety of substrate drugs and results in drug interactions (Pinto 2005; Gorski 1998) with dietary compounds such as grapefruit juice (Paine 2006), which is a CYP3A inhibitor. Inhibition of drug metabolism by grapefruit juice appears to occur primarily through inhibition of intestinal, not hepatic, CYP3A. The CYP3A enzymes are also expressed in other tissues (e.g., lung, colon, esophagus) (Ding 2003); however, the contribution of this expression to drug metabolism is unclear. The list of drugs that are substrates for the 3A enzymes is long; some of the drug classes well known as CYP3A substrates include statins, calcium channel blockers, steroids, macrolide antibiotics, and HIV (human immunodeficiency virus) antivirals. CYP3A expression is induced by several drugs, resulting in substantially increased clearance and reduced circulating concentrations of CYP3A substrates. In addition, several drugs are very effective inhibitors of CYP3A activity, resulting in reduced clearance and increased concentrations of substrate drugs. Both CYP3A inducers and inhibitors are known causes of clinically relevant drug-drug interactions. Several examples of CYP3A substrates, inhibitors, and inducers are included in Table 1.11. A more complete list can be found at http://medicine.iupui.edu/clinpharm/ddis/main-table/.

CYP3A4 Genetic Variation

Although there is much interindividual variability in CYP3A4 activity, there are not a lot of influential genetic variants. Currently, 52 different *CYP3A4* haplotypes have been given * allele names (see Table 1.12). Although no common complete loss-of-function alleles exist for CYP3A4, there is a single report of the *CYP3A4*20* allele, which causes a frameshift. A recent report identified a yet-unnamed allele (c.802C>T) that results in a premature stop codon. Both of these loss-of-function alleles appear to be quite rare; thus, they do not account for much of the interindividual variability in *CYP3A* metabolism. The *CYP3A4*22* allele is more common and, according to several recent studies, appears to have reduced metabolic function. Table 1.13 shows the ethnic differences in minor allele frequencies of two of the variant *CYP3A4* alleles.

CYP3A5 Genetic Variation

In contrast to CYP3A4, CYP3A5 is highly influenced by genetic variation (see Table 1.14). Currently, 31 different

Table 1.11. Examples of CYP3A Substrates, Inhibitors, and Inducers

CYP3A	Examples
Substrates	Testosterone, clarithromycin, erythromycin, alprazolam, midazolam, tacrolimus, cyclosporine, ritonavir, diltiazem, verapamil, simvastatin, atorvastatin, vincristine, astemizole
Inhibitors	Ritonavir, indinavir, nelfinavir, clarithromycin, itraconazole, ketoconazole, grapefruit juice
Inducers	Rifampin, carbamazepine, glucocorticoids, efavirenz, phenobarbital, St. John's wort

Table 1.12. Examples of *CYP3A4* Haplotypes Given * Allele Names[a]

CYP3A4 Allele	SNP rs#	Amino Acid Change	Enzymatic Activity	Minor Allele Frequency (%)
*20	rs67666821	Frameshift	None	< 1
*22	rs35599367	None	Reduced	0–7
(c.802C>T)	Unknown	Stop codon	None	< 1

[a]Additional information on these and additional *CYP3A4* alleles can be found on the PharmGKB Web sites www.pharmgkb.org/search/annotatedGene/cyp3A4/variant.jsp and www.cypalleles.ki.se/cyp3A4.htm.

Table 1.13. Ethnic Differences in Minor Allele Frequencies for Major *CYP3A4* Alleles

Allele	White	Asian	Black
*CYP3A4*20*[a]	< 0.01[a]	Not determined	Not determined
*CYP3A4*22*[b]	0.29	0	0

[a]Frequencies are taken from: Westlind-Johnsson A, Hermann R, Huennemeyer A, et al. Identification and characterization of *CYP3A4*20*, a novel rare *CYP3A4* allele without functional activity. Clin Pharmacol Ther 2006;79:339-49.
[b]Frequencies are taken from: Elens L, van Gelder T, Hesselink DA, et al. *CYP3A4*22*: promising newly identified *CYP3A4* variant allele for personalizing pharmacotherapy. Pharmacogenomics 2013;14:47-62.

CYP3A5 haplotypes have been given * allele names. In fact, in some ethnicities, most people do not express any functional CYP3A5 because of the CYP3A5*3 loss-of-function allele. CYP3A5*3 is the most common allele in most populations and causes a splicing alteration that results in a nonfunctional CYP3A5 protein. CYP3A5*6 and *7 are also loss-of-function alleles, but they are relatively rare in most populations. African American populations appear to have a higher frequency of the *6 and *7 alleles. Table 1.15 shows the ethnic differences in minor allele frequencies of two of the variant CYP3A5 alleles.

Genetic variation in CYP3A5 has been associated with the kinetics and response to multiple CYP3A substrate drugs. For example, genetic variants in CYP3A enzymes are consistently associated with the dose requirements of tacrolimus (Gijsen 2013; Rojas 2013). The increased frequency of the wild-type CYP3A5 allele in African Americans likely accounts for at least part of the increased doses often required in that ethnicity to achieve adequate immunosuppression (Macphee 2002). Consequently, individuals with the CYP3A5*3 genotype usually require reduced doses of tacrolimus.

CYP2A6 Enzyme

CYP2A6 Drugs

Around 2% of drug metabolism is attributable to CYP2A6. CYP2A6 is well known for its role in metabolizing nicotine and is expressed mainly in the liver. Although some CYP2A6 expression has been detected in the respiratory tract, its contribution to drug metabolism is still unclear (Ding 2003). CYP2A6 expression is induced by several drugs, resulting in increased clearance and reduced circulating concentrations of CYP2A6 substrates. In addition, CYP2A6 is inhibited by several drugs, resulting in reduced clearance and increased circulating concentrations of substrate drugs. Several examples of CYP2A6 substrates, inhibitors, and inducers are included in Table 1.16.

CYP2A6 Genetic Variation

CYP2A6 activity is influenced by genetic variation in the CYP2A6 gene (see Table 1.17). The CYP2A6 gene is very complex because of several genetic variations, including SNPs, copy number variations, and hybrid genes. Currently, 89 different CYP2A6 haplotypes have been given * allele names. Among these are many

Table 1.14. CYP3A5 Allele Highly Influenced by Genetic Variation

CYP3A5 Allele	SNP rs#	Amino Acid Change	Enzymatic Activity	Minor Allele Frequency (%)
*3	rs776746	Splicing defect	Severely decreased	15–95
*6	rs10264272	Splicing defect	Severely decreased	1–25
*7	rs41303343	Frameshift	None	0–20

Table 1.15. Ethnic Differences in Minor Allele Frequencies for Major CYP3A4 Alleles[a]

Allele	White	Asian	Black
CYP3A5*3	0.95	0.66–0.74	0.12–0.37
CYP3A5*6	0	0–0.1	0.11–0.16
CYP3A5*7	0	0	0.21

[a]Frequencies are taken from: dbSNP (Single Nucleotide Polymorphism Database; www.ncbi.nlm.nih.gov/SNP).

Table 1.16. Some CYP2A6 Substrates, Inhibitors, and Inducers

CYP2A6	Examples
Substrates	Nicotine, letrozole, coumarin, tegafur
Inhibitors	Pilocarpine, tranylcypromine, selegiline, methoxsalen, menthofuran, grapefruit juice
Inducers	Rifampin, carbamazepine, dexamethasone

reduced-function alleles, including a whole gene deletion (*CYP2A6*4*) and a hybrid with *CYP2A7* (*CYP2A6*12*). Table 1.18 shows the ethnic differences in minor allele frequencies of two of the variant *CYP2A6* alleles.

Genetic variants in the *CYP2A6* gene have been associated with the kinetics of several drugs. One of the most studied is the CYP2A6 association with nicotine metabolism and smoking (Ray 2009). In addition to its association with nicotine kinetics, CYP2A6 variants appear to translate to altered smoking behaviors such as cigarettes per day, nicotine dependence, and quitting success.

CYP2B6 Enzyme

CYP2B6 Expression

Around 2% of drug metabolism is attributable to CYP2B6 (Wienkers 2005). It is well known for its role in metabolizing drugs such as efavirenz, methadone, and bupropion. CYP2B6 is expressed primarily in the liver. However, CYP2B6 expression has also been detected in the brain, kidney, intestine, and respiratory tract (Ding 2003; Gervot 1999). Nevertheless, the contribution of extrahepatic expression to drug metabolism is still unclear. CYP2B6 expression is induced by several drugs, resulting in increased clearance and reduced circulating concentrations of CYP2B6 substrates. In addition, CYP2B6 is inhibited by several drugs, resulting in reduced clearance and increased circulating concentrations of substrate drugs. Several examples of CYP2B6 substrates, inhibitors, and inducers are included in Table 1.19. A more complete list can be found at http://medicine.iupui.edu/clinpharm/ddis/main-table/.

CYP2B6 Genetic Variation

CYP2B6 activity is influenced by genetic variation in the *CYP2B6* gene. Currently, 66 different *CYP2B6* haplotypes have been given * allele names. Two of the common and functionally important alleles with their corresponding functional information are shown in Table 1.20. The *CYP2B6*6* allele is the most common functional alteration in the gene. It encodes for a reduced-function allele because of two altered amino acids. More recent evidence also suggests that *CYP2B6*18*, another reduced-function allele, affects its activity. Table 1.21 shows the ethnic differences in minor allele frequencies of two of the variant *CYP2B6* alleles.

Genetic variants in CYP2B6 cause alterations in the kinetics of several drugs. For example, genetic variants in CYP2B6 are well known to be associated with efavirenz metabolism (Desta 2007). Although additional work needs to be done, the CYP2B6 variants may be useful in guiding the dosing of efavirenz to optimize efficacy and reduce the neuropsychiatric adverse effects of the drug.

Table 1.17. Genetic Variation in the *CYP2A6* Gene[a]

CYP2A6 Allele	SNP rs#	Amino Acid Change	Enzymatic Activity	Minor Allele Frequency (%)
*2	rs1801272	L160H	None	0–10
*4	None	Gene deletion	None	0–2
*7	rs5031016	I471T	Reduced	0–24
*9	rs28399433	TATA box	Reduced	5–25
*10	rs5031016 rs28399468	I471T R485L	Reduced	0–2
*12	CYP2A7 hybrid	10-amino acid substitution	Reduced	0–2
*17	rs28399454	V365M	Reduced	0–15
*20	delAA	Frameshift	None	0–2
*23	rs56256500	R203C	Reduced	0–5
*24	rs72549435 rs143731390	V110L N438Y	Reduced	0–2
*35	rs143731390	N438Y	Reduced	0–8

[a]Additional information on these and additional *CYP2A6* alleles can be found on the PharmGKB Web sites www.pharmgkb.org/search/annotatedGene/cyp2a6/variant.jsp and www.cypalleles.ki.se/cyp2a6.htm.

Online Resources for Pharmacogenomics

Overview

As can be appreciated from the previous descriptions of the genetic variation in each *CYP* gene, the genetic variability in this family of genes is very complex. For example, *CYP2D6* has more than 150 alleles, and many of the other *CYP* genes have well over 50 known alleles. The functional impacts of these alleles are highly variable. Thus, several online resources have been developed that can help investigators and clinicians find and understand the large amount of information provided on each gene. The following sections contain descriptions of several of these online resources.

Pharmacogenomics Knowledge Base

With the increasing amount of pharmacogenomic data generated in the 1990s, the PharmGKB Web site was developed as an online resource that stores and curates large amounts of pharmacogenomic data. Development of the PharmGKB was originally funded through the National Institutes of Health Pharmacogenetics Research Network (NIH-PGRN) to provide a central location for curating the data being generated through the network. However, it rapidly evolved into the major pharmacogenomic Web site that investigators, pharmacists, and clinicians go to for a variety of pharmacogenomic information.

The most frequently visited parts of the site include the VIP project gene summaries and the PharmGKB pathways. The VIP project pages contain information on many of the genes important to pharmacogenomics. In addition, they include information about a wide variety of genetic and pharmacologic aspects of the gene, such as genetic variants, gene and protein structure, and drugs that are related to it by pharmacokinetics or pharmacodynamics. The PharmGKB pathways include drug-centric information about the pharmacokinetics and pharmacodynamics of many drugs. In addition, they include schematic diagrams of drug pharmacokinetics and pharmacodynamics. The diagrams have links to the relevant genes that are important to the drugs. They also include additional summaries about the important pharmacogenomic aspects of the drugs.

The PharmGKB Web site includes the most up-to-date versions of the CPIC guidelines. These guidelines were written to provide clinical guidance for handling genetic information when it is available for making drug prescription recommendations.

CYP-allele Web Site

The CYP-allele Web site is the Human Cytochrome P450 (CYP) Allele Nomenclature Database, which discusses the specific details of the current star (*) alleles for the major *CYP* genes. These descriptions include the specific genetic variants that define each of the star alleles, the functional status of each allele, a link to the specific genetic variant

Table 1.18. Ethnic Differences in Minor Allele Frequencies for Major *CYP2A6* Alleles[a]

Allele	White	Asian	Black
*CYP2A6*2*	0.01–0.03	0–0.01	0–0.01
*CYP2A6*4*	0.01	0.06–0.24	0.02
*CYP2A6*9*	0.05–0.07	0.15–0.22	0.07

[a]Frequencies are taken from: Malaiyandi V, Sellers EM, Tyndale RF. Implications of *CYP2A6* genetic variation for smoking behaviors and nicotine dependence. Clin Pharmacol Ther 2005;77:145-58.

Table 1.19. Some CYP2B6 Substrates, Inhibitors, and Inducers

CYP2B6	Examples
Substrates	Efavirenz, bupropion, methadone, cyclophosphamide, ketamine, propofol, sorafenib
Inhibitors	Voriconazole, ticlopidine, thiotepa, clopidogrel
Inducers	Rifampin, phenobarbital, carbamazepine, efavirenz, phenytoin

Table 1.20. Two Common and Functionally Important *CYP2B6* Alleles[a]

CYP2B6 Allele	SNP rs#	Amino Acid Change	Enzymatic Activity	Minor Allele Frequency (%)
*6	rs3745274	Q172H	Reduced	15–40
	rs2279343	K262R		
*18	rs28399499	I328T	Reduced	1–12

[a]Additional information on these and additional *CYP2B6* alleles can be found on the PharmGKB Web sites www.pharmgkb.org/search/annotatedGene/cyp2b6/variant.jsp and ww.cypalleles.ki.se/cyp2b6.htm.

Table 1.21. Ethnic Differences in Minor Allele Frequencies for Major *CYP2B6* Alleles[a]

Allele	White	Asian	Black
CYP2B6*6	0.25–0.33	0.14–0.19	0.35–0.40
CYP2B6*18	0	0	0.03–0.12

[a]Frequencies are taken from: dbSNP (www.ncbi.nlm.nih.gov/SNP).

database, and a link to the manuscripts that identified the alleles and determined their functional activity.

Genetic Testing Registry

The Genetic Testing Registry Web site (www.ncbi.nlm.nih.gov/gtr) contains information about several different genetic tests that are commercially available. It includes information on the test's purpose, methodology, validity, evidence of usefulness, and laboratory contacts and credentials. Links are given to many providers who run tests for specific genes and variants. The information on this Web site is provided by the test providers.

Genotyping Technologies

Overview

Developing the genotyping methods for the *CYP* genes has been complicated for many reasons. First, there is high homology between several members of the *CYP* genes. This includes both functional genes and pseudogenes. This high homology creates problems when designing polymerase chain reaction (PCR)-based assays because it becomes difficult to find unique sequences around the SNPs of interest that are needed for specific assay design. In some cases, the regions surrounding the SNP are identical for a large distance from the SNPs. Second, *CYP2D6* has several hybrid genes that are combinations of CYP2D6 and CYP2D7, which also complicates the mapping and assay design. Several of the CYP genes also have copy number variations, both complete loss of the gene and gene amplification, resulting in multiple tandem copies on a single chromosome. These structural features also create problems when doing some of the sequencing technologies, such as next-generation sequencing, where the short read lengths (30–100 bp) need to be uniquely mapped back to the genome. For example, for *CYP2D6*, there are large stretches of homology with the pseudogenes *CYP2D7* and *CYP2D8*, making this very difficult with the short reads. Other examples are *CYP3A4* and *CYP3A5*, which also have similar sequences. Collectively, these difficulties have resulted in poor coverage of the *CYP* genes in large genome-wide genotyping arrays and next-generation sequencing applications. Consequently, various assays have been developed that focus specifically on genotyping only one or a few of the genes per assay design. In the section that follows, broad differences are briefly described for the assays of single genes or variants, many variants across the genome, and next-generation sequencing for resequencing large portions of the genome. These methods are used to determine an individual patient's genotype compared with individual assays that test only a single variant and next-generation sequencing that can cover the entire genome. Additional information about assays that can be used to test specific genes or SNPs can be found on the Genetic Testing Registry and in the CPIC guidelines.

Single Genes or Variants

Many different types of assays are available for testing a few variants in individual genes. One of the most common assays is the hydrolysis probe assay. These are often called TaqMan assays, which is the registered trademark name for the assays originally marketed by Applied Biosystems. These relatively simple assays test single genetic variants in each well using a PCR reaction with fluorescent-labeled probes. The probes, which are selectively hydrolyzed, fluoresce if they match the sequence of the sample. They are usually run in plates containing 96 or 384 wells. Recently, newer platforms have become more common that use plates containing 3072 (OpenArray), 2304, or 9216 (Fluidigm arrays) wells. Plates, which can have assays from several different genes, are becoming commonly used in clinical genotyping laboratories.

Other types of assays are focused on the important variants in a specific gene. These are usually array-based hybridization assays. Examples are the AmpliChip® CYP450 (Roche Diagnostics, San Francisco, CA), DMET (Affymetrix, Santa Clara, CA), xTAG (Luminex, Austin, TX), and INFINITI CYP450 2D6I assays (AutoGenomics, Vista, CA). These assays test for most clinically important genetic variants in individual genes. Some of them are FDA label-approved devices and are thus commonly used in clinical genotyping laboratories.

Genome-Wide Assays

Higher-throughput assays are commonly used in research studies to discover genetic variants in genes associated with drug responses. These assays are highly multiplexed, with individual assays often testing more than 1 million variants per sample. The two main types of assays are available from Illumina and Affymetrix. The Affymetrix arrays use oligonucleotides that are synthesized on small chips using photolithography. The DNA is hybridized to the oligonucleotides, and the DNA sequence is determined according to which oligos the DNA binds to. The Illumina arrays use oligonucleotides bound to beads, which are then hybridized to DNA fragments that are created by a DNA sequence–specific ligation before hybridization. Both technologies provide highly accurate genotyping of a great many genetic polymorphisms.

Next-Generation Sequencing

The most comprehensive interrogation of the genome for genetic variants is next-generation sequencing. In this approach, the entire target region is resequenced. The region of interest often includes small regions such as a few individual genes; intermediate regions such as all the exons; and large regions such as the entire genome. Although this is commonly the most comprehensive coverage technology, it is also the most expensive and difficult. In addition, some regions of genes that have highly homologous regions in other parts of the genomes are difficult to analyze with this technology. However, an advantage is that it will cover all the genetic variation within a specific target region without having to design a specific assay for each specific variation. In addition, variants can be discovered that were not previously known to exist.

Evolution of Published Pharmacogenomic Guidelines

As the amount of pharmacogenomic information in clinical trials increased during the 1990s, the FDA created guidelines to provide some guidance for drug manufacturers when voluntarily submitting their pharmacogenomic information to the FDA. Originally drafted in 2011, the guidelines were updated in 2013 (www.fda.gov/downloads/Drugs/GuidanceComplianceRegulatoryInformation/Guidances/UCM337169.pdf). These guidelines include recommendations for drug manufacturers to consider during the exploratory and observational studies intended for use in generating genomic hypotheses that could be tested in phase III studies.

Guidelines for using pharmacogenomics in clinical practice have also evolved substantially during the past decade. As the value of pharmacogenomics was realized, review articles began to be published with recommendations for using pharmacogenomics in clinical practice. For example, recommendations were published for using pharmacogenetics in antidepressant and antipsychotic therapies (Kirchheiner 2004). As more data were published, genotyping tools became available, and interest piqued, experts began forming groups with the specific goal of writing pharmacogenomic guidelines. For example, the Royal Dutch Association for the Advancement of Pharmacy established the Pharmacogenetics Working Group. This group was charged with developing pharmacogenetic-based therapeutic recommendations for a wide variety of drugs for which substantial data existed supporting the use of pharmacogenetics. The resulting guidelines for 53 drugs were published in a single publication (Swen 2011). Additional groups also began publishing individual guidelines for specific gene-drug pairs. For example, the CPNDS (Canadian Pharmacogenomics Network for Drug Safety) clinical recommendation group published guidelines for using *CYP2D6* genotyping for codeine therapy and HLA genotyping for guiding carbamazepine therapy (Madadi 2013). The CPIC investigators have also published several guidelines for clinical recommendations regarding how to use genotype results for tailoring drug therapy (Caudle 2014; Relling 2011).

These include published guidelines for many different gene-drug pairs that are updated every 2 years, which are also available on the PharmGKB Web site. Several of these pairs have already been endorsed by the American Society of Health-System Pharmacists (www.ashp.org/menu/ PracticePolicy/PolicyPositionsGuidelinesBestPractices/ BrowsebyDocumentType/EndorsedDocuments.aspx). Finally, a group from the American Association for Clinical Chemistry published a laboratory medicine practice guideline for use by laboratories conducting pharmacogenomic testing.

Summary

Over the past 70 years or so, the early observations of exaggerated clinical responses to drug therapies have evolved into the use of genetic testing to guide clinical therapies. Through the evolution, our understanding of the role of genetics in drug responses has been greatly improved by the advances in genomics technologies. This has led to more exhaustive interrogation of the genome for biomarkers that predict drug efficacy and toxicity. Although there are still many challenges to overcome for pharmacogenetics to be used in everyday clinical practice, a limited number of genetic biomarkers are already being successfully implemented in the clinic to improve several drug therapies.

References

Bertilsson L, Dengler HJ, Eichelbaum M, et al. Pharmacogenetic covariation of defective N-oxidation of sparteine and 4-hydroxylation of debrisoquine. Eur J Clin Pharmacol 1980;17:153-5.

Beutler E. Study of glucose-6-phosphate dehydrogenase: history and molecular biology. Am J Hematol 1993;42:53-8.

Borges S, Desta Z, Jin Y, et al. Composite functional genetic and comedication CYP2D6 activity score in predicting tamoxifen drug exposure among breast cancer patients. J Clin Pharmacol 2010;50:450-8.

Caudle KE, Klein TE, Hoffman JM, et al. Incorporation of pharmacogenomics into routine clinical practice: the Clinical Pharmacogenetics Implementation Consortium (CPIC) guideline development process. Curr Drug Metab 2014;15:209-17.

Chen Y, Goldstein JA. The transcriptional regulation of the human CYP2C genes. Curr Drug Metab 2009;10:567-78.

Crews KR, Gaedigk A, Dunnenberger HM, et al. Clinical Pharmacogenetics Implementation Consortium guidelines for cytochrome P450 2D6 genotype and codeine therapy: 2014 update. Clin Pharmacol Ther 2014;95:376-82.

Daly AK, Brockmoller J, Broly F, et al. Nomenclature for human CYP2D6 alleles. Pharmacogenetics 1996;6:193-201.

Desta Z, Saussele T, Ward B, et al. Impact of CYP2B6 polymorphism on hepatic efavirenz metabolism in vitro. Pharmacogenomics 2007;8:547-58.

Ding X, Kaminsky LS. Human extrahepatic cytochromes P450: function in xenobiotic metabolism and tissue-selective chemical toxicity in the respiratory and gastrointestinal tracts. Annu Rev Pharmacol Toxicol 2003;43:149-73.

Eichelbaum M, Spannbrucker N, Steincke B, et al. Defective N-oxidation of sparteine in man: a new pharmacogenetic defect. Eur J Clin Pharmacol 1979;16:183-7.

Elens L, van Gelder T, Hesselink DA, et al. CYP3A4*22: promising newly identified CYP3A4 variant allele for personalizing pharmacotherapy. Pharmacogenomics 2013;14:47-62.

Gaedigk A, Simon SD, Pearce RE, et al. The CYP2D6 activity score: translating genotype information into a qualitative measure of phenotype. Clin Pharmacol Ther 2008;83:234-42.

Gervot L, Rochat B, Gautier JC, et al. Human CYP2B6: expression, inducibility and catalytic activities. Pharmacogenetics 1999;9:295-306.

Gijsen VM, van Schaik RH, Elens L, et al. CYP3A4*22 and CYP3A combined genotypes both correlate with tacrolimus disposition in pediatric heart transplant recipients. Pharmacogenomics 2013;14:1027-36.

Gorski JC, Jones DR, Haehner-Daniels BD, et al. The contribution of intestinal and hepatic CYP3A to the interaction between midazolam and clarithromycin. Clin Pharmacol Ther 1998;64:133-43.

Heikkinen T, Ekblad U, Palo P, et al. Pharmacokinetics of fluoxetine and norfluoxetine in pregnancy and lactation. Clin Pharmacol Ther 2003;73:330-7.

Hicks JK, Swen JJ, Thorn CF, et al. Clinical Pharmacogenetics Implementation Consortium guideline for CYP2D6 and CYP2C19 genotypes and dosing of tricyclic antidepressants. Clin Pharmacol Ther 2013;93:402-8.

Hogstedt S, Lindberg B, Peng DR, et al. Pregnancy induced increase in metoprolol metabolism. Clin Pharmacol Ther 1985;37:688-92.

Hosono N, Kato M, Kiyotani K, et al. CYP2D6 genotyping for functional-gene dosage analysis by allele copy number detection. Clin Chem 2009;55:1546-54.

Johnson JA, Gong L, Whirl-Carrillo M, et al. Clinical Pharmacogenetics Implementation Consortium guidelines for CYP2C9 and VKORC1 genotypes and warfarin dosing. Clin Pharmacol Ther 2011;90:625-9.

Kalow W. Pharmacogenetics: Heredity and the Response to Drugs. Philadelphia: W.B. Saunders, 1962.

Kalow W, Gunn DR. The relation between dose of succinylcholine and duration of apnea in man. J Pharmacol Exp Ther 1957;120:203-14.

Kirchheiner J, Nickchen K, Bauer M, et al. Pharmacogenetics of antidepressants and antipsychotics: the contribution of allelic variations to the phenotype of drug response. Mol Psychiatry 2004;9:442-73.

Kirchheiner J, Schmidt H, Tzvetkov M, et al. Pharmacokinetics of codeine and its metabolite morphine in ultra-rapid metabolizers due to *CYP2D6* duplication. Pharmacogenomics J 2007;7:257-65.

Lazarou J, Pomeranz BH, Corey PN. Incidence of adverse drug reactions in hospitalized patients: a meta-analysis of prospective studies. JAMA 1998;279:1200-5.

Leckband SG, Kelsoe JR, Dunnenberger HM, et al. Clinical Pharmacogenetics Implementation Consortium guidelines for *HLA-B* genotype and carbamazepine dosing. Clin Pharmacol Ther 2013;94:324-8.

Luzzatto L, Mehta A, Julliamy T. The Metabolic & Molecular Bases of Inherited Disease. New York: McGraw-Hill, 2001.

Macphee IA, Fredericks S, Tai T, et al. Tacrolimus pharmacogenetics: polymorphisms associated with expression of cytochrome p4503A5 and P-glycoprotein correlate with dose requirement. Transplantation 2002;74:1486-9.

Madadi P, Amstutz U, Rieder M, et al. Clinical practice guideline: *CYP2D6* genotyping for safe and efficacious codeine therapy. J Popul Ther Clin Pharmacol 2013;20:e369-96.

Malaiyandi V, Sellers EM, Tyndale RF. Implications of *CYP2A6* genetic variation for smoking behaviors and nicotine dependence. Clin Pharmacol Ther 2005;77:145-58.

Martin MA, Hoffman JM, Freimuth RR, et al. Clinical pharmacogenetics implementation consortium guidelines for *hla-B* genotype and abacavir dosing: 2014 update. Clin Pharmacol Ther 2014;95:499-500.

Mikkelsen TS, Thorn CF, Yang JJ, et al. PharmGKB summary: methotrexate pathway. Pharmacogenet Genomics 2011;21:679-86.

Motulsky AG. Drug reactions, enzymes, and biochemical genetics. JAMA 1957;165:835-7.

Paine MF, Widmer WW, Hart HL, et al. A furanocoumarin-free grapefruit juice establishes furanocoumarins as the mediators of the grapefruit juice-felodipine interaction. Am J Clin Nutr 2006;83:1097-105.

Pinto AG, Wang YH, Chalasani N, et al. Inhibition of human intestinal wall metabolism by macrolide antibiotics: effect of clarithromycin on cytochrome P450 *3A4/5* activity and expression. Clin Pharmacol Ther 2005;77:178-88.

Ramamoorthy A, Flockhart DA, Hosono N, et al. Differential quantification of *CYP2D6* gene copy number by four different quantitative real-time PCR assays. Pharmacogenet Genomics 2010;20:451-4.

Ray R, Tyndale RF, Lerman C. Nicotine dependence pharmacogenetics: role of genetic variation in nicotine-metabolizing enzymes. J Neurogenet 2009;23:252-61.

Relling MV, Klein TE. CPIC: Clinical Pharmacogenetics Implementation Consortium of the Pharmacogenomics Research Network. Clin Pharmacol Ther 2011;89:464-7.

Rojas LE, Herrero MJ, Boso V, et al. Meta-analysis and systematic review of the effect of the donor and recipient *CYP3A5* 6986A>G genotype on tacrolimus dose requirements in liver transplantation. Pharmacogenet Genomics 2013;23:509-17.

Steiner E, Bertilsson L, Sawe J, et al. Polymorphic debrisoquin hydroxylation in 757 Swedish subjects. Clin Pharmacol Ther 1988;44:431-5.

Swen JJ, Nijenhuis M, de Boer A, et al. Pharmacogenetics: from bench to byte—an update of guidelines. Clin Pharmacol Ther 2011;89:662-73.

U.S. Food and Drug Administration (FDA). Why Learn About Adverse Drug Reactions (ADR)? 2000. Available at www.fda.gov/Drugs/DevelopmentApprovalProcess/DevelopmentResources/DrugInteractionsLabeling/ucm114848.htm. Accessed August 15, 2014.

Vogel F. Moderne probleme der human-genetik. Ergebn Inn Med Kinderhalk 1959;12:52.

Wadelius M, Darj E, Frenne G, et al. Induction of *CYP2D6* in pregnancy. Clin Pharmacol Ther 1997;62:400-7.

Westlind-Johnsson A, Hermann R, Huennemeyer A, et al. Identification and characterization of *CYP3A4*20*, a novel rare *CYP3A4* allele without functional activity. Clin Pharmacol Ther 2006;79:339-49.

Wienkers LC, Heath TG. Predicting in vivo drug interactions from in vitro drug discovery data. Nat Rev Drug Discov 2005;4:825-33.

Zanger UM, Raimundo S, Eichelbaum M. Cytochrome P450 2D6: overview and update on pharmacology, genetics, biochemistry. Naunyn Schmiedebergs Arch Pharmacol 2004;369:23-37.

Chapter 2

CLINICAL IMPLEMENTATION OF PHARMACOGENOMICS—EVALUATING THE EVIDENCE

SAMUEL G. JOHNSON, PHARM.D., FCCP; AND CHRISTINA L. AQUILANTE, PHARM.D.

Learning Objectives

1. Identify the differences between analytical validity, clinical validity, and clinical utility as they pertain to pharmacogenomic tests.
2. Analyze different types of pharmacogenomic studies to understand the relationship between strength of evidence and evidence thresholds required for clinical uptake.
3. Compare and contrast low- versus high-evidence thresholds for the use of pharmacogenomics in clinical practice.
4. Identify challenges associated with producing and evaluating evidence in the pharmacogenomics arena.
5. Compare and contrast systematic versus standardized evidence review processes used in the development of evidence-based clinical guidelines.
6. Estimate the impact of horizon-scanning efforts on the continued evaluation of evidence for pharmacogenomic testing in clinical practice.

Keywords: Evidence threshold, clinical utility, clinical validity, analytic validity, pharmacogenetic, pharmacogenomic

Abbreviations in This Chapter

CDC	Centers for Disease Control and Prevention
CPIC	Clinical Pharmacogenetics Implementation Consortium
EGAPP	Evaluation of Genomic Applications in Practice and Prevention
EWG	EGAPP Working Group
GWAS	Genome-wide association study
PharmGKB	Pharmacogenomics Knowledge Base
RCT	Randomized controlled trial

Abstract

Genetically mediated interindividual variability in drug disposition and response—as first reported in the 1950s—persists today and underscores the need to accurately produce, assess, and translate evidence for pharmacogenomic testing into clinical practice. Despite widespread recognition of the scientific rationale and successful clinical implementation of pharmacogenomics at several major academic medical centers, many clinicians and researchers engaged in this space acknowledge the need to establish evidence thresholds in order to realize the aims and benefits of personalized medicine. The primary objectives of this chapter are to (1) assess the impact of analytic validity, clinical validity, and clinical utility data on clinical pharmacogenetics; (2) discuss the importance of defining evidence thresholds to assist with clinical implementation; and (3) review existing clinical practice guideline resources.

Introduction

The field of pharmacogenomics encompasses a growing knowledge base linking genetic variation to drug disposition and response. Yet even though the U.S. Food and Drug Administration (FDA) began incorporating genetic information into drug labels in 2007, clinical implementation of pharmacogenomics outside research programs or academic medical center settings remains largely absent (Scott 2011). The mapping of the human genome, and subsequent advances in technology, has raised public expectations that predicting patients' responses to drug therapy is now possible in every therapeutic area and that personalized drug therapy will come sooner rather than later. However, debate continues among many stakeholders involved in drug development and clinical decision-making regarding whether pharmacogenomic testing should be routinely used in patient assessment. Realistic application of genomic findings and technologies to clinical practice requires many steps. Broadly speaking, the steps involved in developing and implementing clinical pharmacogenomic testing include (1) discovery and validation of pharmacogenomic markers in well-controlled studies with independent populations; (2) replication of drug-gene associations and demonstration of utility in at-risk patients; (3) development and regulatory approval of companion diagnostic tests or laboratory-developed tests; (4) assessment of the clinical impact and cost-effectiveness of pharmacogenomic testing; and (5) involvement of all stakeholders in clinical implementation (Swen 2007).

Evidence Supporting Clinical Implementation

Four major criteria exist for the evaluation of genetic and genomic tests. These criteria consist of analytic validity (i.e., how accurately and reliably the test measures the genotype of interest), clinical validity (i.e., how consistently and accurately the test detects or predicts the intermediate or final outcomes of interest), clinical utility (i.e., how likely the test is to significantly improve patient outcomes or alter management), and ethical, legal, and social implications. These criteria form a model framework (referred to as analytic validity; clinical validity; clinical utility; and ethical, legal, social issues, or the ACCE framework) for collecting, evaluating, interpreting, and reporting data about DNA (and related) testing for disorders with a genetic component in a format that allows policy-makers to have access to up-to-date and reliable information for decision-making. An important by-product of this framework is the identification of gaps in knowledge that help define future research needs (Haddow 2003).

Analytic Validity

Analytic validity determines how reliably a diagnostic test measures what it is intended to measure, regardless of whether it is an expression pattern, a variant, or a protein. Validating the measurement of genetic biomarkers before their widespread use in diagnostic testing is an important step toward implementing pharmacogenetic testing. This part of the evaluation is concerned with assessing test performance in the laboratory as opposed to the clinic. Analytic validity may be considered in three stages: (1) preanalytic, (2) analytic, and (3) postanalytic. The preanalytic stage includes the processes of obtaining and transporting a sample as well as the cataloging of details on a laboratory database. The analytic stage involves preparing the sample and conducting the analysis. The postanalytic stage includes interpreting the results and reporting them. These stages must be considered thoroughly in order to evaluate the analytic validity of a specific test. Although introducing new tests may lead to improved quality of care, many preanalytic, analytic, and postanalytic factors lead to variations in test results, which affect the test accuracy and ultimately the test outcomes. Because analytic validity data are seldom published in peer-reviewed literature, assessment typically involves a search of gray literature (e.g., abstracts, presentations, or proceedings of national or international meetings). However, because optimal search strategies are lacking for identifying these data from gray literature, their usefulness has not been systematically validated. This may affect evidence-based decision-making for clinical guidelines as well as coverage and reimbursement decisions (AHRQ 2011).

In the United States, there is a separate regulatory oversight process for the approval of pharmacogenetic testing developed as an in-house test by a clinical laboratory (e.g., "laboratory-developed" or "home-grown" tests) versus that for a manufacturer-developed in vitro diagnostic test for a specific drug, which is critical for the safe and effective use of that drug (e.g., companion diagnostic). The quality standards for laboratory-developed tests are governed by the Clinical Laboratory Improvement Amendments (CLIA). In addition, laboratories are certified by the College of American Pathologists, the Joint Commission on Accreditation of Healthcare Organizations, or the health department of each individual state. The CLIA regulations categorize laboratory testing as either "waived" from routine regulatory oversight or "nonwaived," depending on the complexity of the tests. Molecular genetic testing is considered a high-complexity and nonwaived test. Thus, laboratories that perform this type of testing must meet CLIA requirements for nonwaived testing. Briefly, CLIA regulations have measures to ensure the quality of the testing process, which include, but are not limited to, preanalytic, analytic, and postanalytic phases. The preanalytic phase includes test selection and ordering, as well as specimen collection, processing, handling, and delivery to the testing site. The analytic phase includes selecting test methods, performing test procedures, monitoring and verifying the accuracy and reliability of test results, and documenting test findings (Morello 2010). The postanalytic phase includes reporting test results and archiving records, reports, and tested specimens. The analytic phase is perhaps the most laborious and requires performance specifications for accuracy, precision, analytic sensitivity, and analytic specificity; reportable range of test results; reference intervals, or normal values; and other performance characteristics required for test performance. Currently, CLIA regulations do not include proficiency-testing requirements (i.e., rigorous external assessment of laboratory performance) for molecular genetic tests. However, laboratories that do perform molecular genetic testing must verify, at least twice annually, the accuracy of the genetic tests they perform (CDC 2011, 2009).

The FDA has progressively acknowledged the importance of biomarkers, provided new recommendations on companion diagnostic tests (i.e., genetic or biomarker evaluations accompanying specific pharmacotherapies), and required data submissions in addition to the Good Laboratory Practices highlighted previously. These include (1) publication of the FDA Guidance for Pharmacogenomic Data Submission, Guidance on Pharmacogenetic Tests and Genetic Tests for Heritable Markers, and draft guidance for "In Vitro Diagnostic Multivariate Index Assays"; (2) introduction of the Voluntary Data Submission Program; and (3) formation of an Interdisciplinary Pharmacogenomic Review Group to evaluate voluntary submissions as well as to approve and classify different biomarkers (FDA 2010). The FDA's approval of companion diagnostic tests would likely generate more confidence for clinicians, health care facility administrators, and payers, which could enhance use across more clinical settings. Although there is no current formal regulatory process for submitting companion diagnostic tests, the FDA previously ruled that evaluation and approval of the AmpliChip CYP450 test as an in vitro diagnostic device were required (de Leon 2006). Further examples of the FDA's assuming a greater role were the respective companion diagnostic tests approved for crizotinib (*ALK*) and vemurafenib (*BRAF V600E*) (Chapman 2011). With the formation of a personalized medicine group within the Office of In Vitro Diagnostic Devices, Center for Device Evaluation and Radiological Health, more FDA-approved tests will likely be available in the future (FDA 2013).

Clinical Validity

Clinical validity measures the ability of the test to consistently and accurately predict the outcome of interest (e.g., differentiate between responders and nonresponders or identify those at high risk of adverse drug reactions), rather than how reliably a test measures the property or characteristic it is intended to measure.

Establishing Genotype-Phenotype Associations

Although many pharmacogenomic associations have been discovered during the past decade, only a few have been extensively used in clinical practice (Ginsburg 2005). Traditionally, the candidate gene approach has been used to investigate pharmacogenetic traits. This approach involves evaluating the associations between genetic variants within prespecified genes (or a single gene) and a phenotype of interest. The hallmark of this approach is a priori knowledge of mechanisms governing a drug's disposition and response in the body. More recently, the genome-wide approach has been used in pharmacogenomic research. This approach involves evaluating many polymorphisms that canvas variation across the entire genome and determining their associations with a phenotype of interest. In contrast to the candidate gene approach, the genome-wide association approach does not rely on a priori knowledge of drug disposition or response mechanisms (i.e., non-candidate driven). Although a genome-wide association study (GWAS) may identify which variants are associated with a phenotypic outcome, it does not provide information on causality (Lane 2012).

An interesting illustration can be derived from patients with diabetes mellitus receiving metformin therapy. The pharmacokinetic disposition of metformin is primarily mediated by drug transporters in the liver, kidney, and intestine. Candidate gene studies have investigated the impact of polymorphisms within drug transporter genes, primarily organic cation transporters and multidrug and toxin extrusion transporters, on the pharmacokinetics and pharmacodynamics of metformin (Todd 2014). Although some single nucleotide polymorphisms in these genes have been associated with altered metformin pharmacokinetics, these findings have not necessarily been consistent across studies, nor have they reproducibly translated into differences in metformin pharmacodynamic outcomes. More recently, a GWAS in patients with type 2 diabetes mellitus was conducted to elucidate genes and polymorphisms involved in metformin glycemic response (Zhou 2011). The GWAS identified a significant association between metformin glycemic response and a polymorphism located in a locus containing the ataxia-telangiectasia mutated gene (*ATM*) (Zhou 2011). This is a DNA repair gene that is also involved in insulin signaling pathways, beta-cell dysfunction, and AMPK activation. After in vitro experiments, the study concluded that *ATM* acts upstream of AMPK and is required for metformin action

(Zhou 2011). Additional research is under way to further elucidate the mechanism(s) underlying this genetic association. The metformin example earlier shows the different approaches used to elucidate genotype-phenotype associations in the field of pharmacogenomics. The candidate gene approach has traditionally been most commonly used, but it carries significant limitations, namely the possibility of yet-to-be-identified genes that play a role in drug response. As such, the GWAS has emerged as an important tool for making more comprehensive discoveries of previously unknown mediators of drug disposition, response, and toxicity.

Regardless of the approach used to identify and validate genotype-phenotype associations, race/ethnicity and population stratification (i.e., population variation in the prevalence of different genetic variants; e.g., *CYP2D6*, *HLA-B*, *UGT1A1*, and *SLC6A4*), affect the sample size requirements for statistical power in many pharmacogenomic studies. For instance, although the warfarin pharmacogenetic algorithms derived from the work of Gage et al. (Gage 2008) and the International Warfarin Pharmacogenetics Consortium (Klein 2009) are clinically useful, they fail to detect *CYP2C9*8*, which more commonly occurs in African Americans (around 9%) (Scott 2009). In addition, classification of the 10 most common single nucleotide polymorphisms for *VKORC1* into low- or high-dose haplotype groups did not show optimal warfarin dose prediction in Africans compared with Asians and whites (Rieder 2005).

Another example is *HLA-B*15:02*, a strong predictor of carbamazepine-induced severe cutaneous drug reactions in Han Chinese and most Southeast Asians; however, this allelic variant is rare in whites (Man 2007; Roujeau 1995). These race- or population-related variables would likely confound the results of most pharmacogenomic association studies and could complicate the interpretation of study results if not properly accounted for in the statistical analyses. Furthermore, with advances in genome sequencing technologies, there is growing appreciation that interindividual variability in drug disposition, response, and adverse effects is complex, with polygenic and epigenetic interactions. Yet currently, much less is known about the influence of environmental variables and gene-environment interactions on drug disposition and response phenotypes (Dempfle 2008). *Epigenetics* refers to changes in gene expression without nucleotide sequence alteration, which could result in many different phenotypes within a population. In the not-too-distant future, pharmacoepigenetic investigations focused on studying the interaction among drugs, the environment, and genes could provide additional insights into interindividual variability in drug disposition and response, beyond the DNA sequence variations (Powanda 2012).

Thus, clinical validity describes the ability of a pharmacogenetic test to identify the outcome of interest (i.e., altered drug response or predisposition to toxicity caused by a specific variant allele or alleles). This is of course affected not only by the presence of the genetic variant itself, but also by other factors that might affect the penetrance (e.g., the person's environmental exposures or personal behaviors) or by the presence or absence of additional genetic variants. For this reason, the clinical validity of a pharmacogenetic test for a specific variant may vary in different populations.

Clinical Utility

Clinical utility refers to the likelihood that a test result will alter clinical outcomes or management strategies for patients. When applying pharmacogenomics in the clinical setting, clinical utility derives from two main practical concepts: (1) sufficient pretest probability that a pharmacogenetic variant will be identified and (2) identification of a genetic variant that alters therapeutic management with respect to drug selection or dosing. When applying pharmacogenomics broadly to evidence-based practice, clinical utility depends on established evidence thresholds that are based on the risks associated with the test results applied to specific drug-gene pairs.

Challenges Associated with Producing Evidence

Contrasting Clinical Utility for Diagnostic Tests vs. Therapeutics

A growing list of drugs have some reference to pharmacogenomic testing presently included in the labeling (Deverka 2009), either for germline mutations (e.g., CYP metabolizing enzyme gene variations) or for somatic mutations (e.g., *KRAS* expression in colorectal cancer tissue). A few drugs (e.g., trastuzumab for *HER2* [human epidermal growth factor receptor 2]-positive breast cancer or maraviroc for HIV [human immunodeficiency virus]) have been codeveloped with a pharmacogenetic test used to define a clinical trial population, and genetic testing is required before prescribing the medication. However, most labels refer to reported pharmacogenetic associations (e.g., altered drug metabolism or drug-gene interactions) and do not specifically recommend testing before prescribing a drug or provide any recommendation of how prescribing should be altered. The current relative lack of prospective evidence linking pharmacogenetic testing to improved drug use outcomes likely contributes to uncertainty about how to incorporate pharmacogenomics labeling information into clinical practice (Deverka 2008). Whether a pharmacogenetic test reaches the marketplace in the United States—either through the laboratory-developed test pathway or through FDA approval, given that there are no regulatory requirements for developers to provide clinical validity or clinical utility evidence—presently falls to the purview of clinicians and payers (Deverka 2009).

Currently, there is a gap between the translation of pharmacogenetic tests into clinical practice and the prevailing business and development models within the diagnostics industry (Garrison 2006; Phillips 2006). The recent report on pharmacogenomics from the Secretary's Advisory Committee on Genetics, Health, and Society analyzed the translational barriers from the perspective of various stakeholder groups such as industry, the FDA, the Centers for Medicare & Medicaid Services, and other third-party payers and clinical practice guideline developers and came to similar conclusions about the role of payers (NIH 2008). This group ultimately decides which pharmacogenetic tests are readily accessible to patients and clinicians and increasingly relies on clinical utility evidence for that decision. The term *clinical utility* has been defined quite broadly to include not only the impact of the pharmacogenomic tests in routine practice on patient health outcomes, but also the ability of testing to inform clinical decision-making while accounting for the availability of resources and patient preferences. Payers typically support the use of pharmacogenetic tests with proven analytic and clinical validity, but tests that more importantly lead to appropriate clinical decisions resulting in net health benefits are preferred (Deverka 2009).

Evidence Thresholds Vary Depending on Availability of Alternatives or Level of Risk Using the Data

Advocating for the clinical utility of pharmacogenomic testing ensures appropriate, clinically effective, and cost-effective use. Constant debates exist across clinical and scientific communities regarding the required evidence thresholds for the clinical utility of pharmacogenomic tests that are both scientifically appropriate and realistically achievable (Limdi 2010). Although the "gold standard" proof of the clinical utility of a drug is through randomized controlled trials (RCTs), this rigorous assessment of drug efficacy and safety may not adequately measure the benefit of pharmacogenomic testing because of sampling issues, technology limitations (e.g., possibility of discordant results when using different testing platforms), or other challenges (e.g., identification of variants with unknown significance). Because pharmacogenomic testing and therapy selection are so closely intertwined with the concept of personalized medicine, it is difficult to distinguish the clinical utility of testing from that of the specific drug or the drug-test combination. Ultimately, sound clinical judgment is critically important in assessing the risks associated with using pharmacogenomic information to tailor pharmacotherapy regimens for individual patients. For example, routinely obtaining the *CYP2C19* genotype to identify clopidogrel nonresponse may result in the identification of a rare allelic variant (e.g., *CYP2C19* *4, *5, *6, *7, or *8) for which current evidence-based guidelines make no recommendation because of the lack of available evidence supporting its clinical utility (SNPedia 2013). However, because the functional consequence of each of these variants is a nonfunctional CYP2C19 enzyme, it is important to tailor pharmacotherapeutic recommendations according to the probable clinical outcome (i.e., higher risk of stent thrombosis or major adverse cardiac events) associated with the loss of CYP2C19 function, knowing that, based on its rare prevalence, it takes more time and effort to generate evidence in these cases.

Translational Pathway for Pharmacogenomics Includes Different Study Methodologies

In addition to the issues highlighted previously, complex disease etiologies, heterogeneous patient populations, placebo effects, and drug disposition and response variability all underscore the practical issue that large RCTs are not clinically feasible or even necessary for many pharmacogenomic use cases. In contrast to the tenets associated with evidence-based practice for disease management, the emphasis and value of pharmacogenomic testing are on the incremental advantages in safety and effectiveness for individual patients (i.e., for poor metabolizers, ultra-rapid metabolizers, or those susceptible to severe adverse drug reactions) versus standard dosing regimens or traditional dosing regimens for most, if not all, patients. In addition, ethical concerns might preclude conducting RCTs in patients with specific genetic polymorphisms (Young 2008). For example, it would be unethical to prescribe abacavir for patients testing positive for *HLA-B*57:01* because they would have a 6%–8% increased likelihood of a hypersensitivity reaction (Hetherington 2001). Therefore, balancing the scientific rigor and demands of RCTs against the practical value of pharmacogenomic testing for patient care seems appropriate, and plausible alternatives to RCTs (including pragmatic trials, enrichment studies, observational studies, and retrospective analyses of stored samples from RCTs) are increasingly important.

On that basis, many stakeholders advocate for enrichment studies that include patient populations in which the effect of a drug-gene interaction can be most readily shown (Boessen 2013). A clear example of this is a recently approved drug for cystic fibrosis, ivacaftor, which works in a fraction of patients with cystic fibrosis who have particular genetic abnormalities (e.g., *G551D*, *G1244E*, *G1349D*, *G178R*, *G551S*, *S1251N*, *S1255P*, *S549N*, or *S549R*). Because ivacaftor has optimal efficacy for a small percentage of patients with cystic fibrosis, it has a dramatic effect for that population subset. Conducting a study within an unselected population, for which a larger percentage would have little benefit, would probably have failed (Massie 2014; Accurso 2010). Similarly, pragmatic clinical trials designed to evaluate the effectiveness of interventions in real-life practice conditions are increasing in prevalence because they produce results that are more generalizable to different practice settings (Newman 2011). Creation of the National Human Genome Research Institute–funded eMERGE (Electronic Medical Records and Genomics) network reflects this increased prevalence, with nine different health systems contributing to the consortium and individual biorepositories linking DNA specimens to phenotypic data contained in electronic medical records. These records in turn fuel the discovery of genotype-phenotype associations, and these discoveries, once validated, may be introduced back to the electronic medical record to augment clinical care (Gottesman 2013).

Not surprisingly, pharmaceutical manufacturers lack incentives to conduct effort- and time-intensive RCTs, especially for out-of-patent marketed drugs. To facilitate eventual clinical implementation, refocusing the types of studies conducted and/or the quality of study data for evidence of clinical validity and clinical utility is of paramount importance. For example, conducting pragmatic clinical trials in real-world settings has been previously proposed for regulatory decision-making in general or for genomics specifically (Ratner 2013; Staa 2012). A study by Anderson et al. provided comparative effectiveness data for genotype-guided warfarin in 504 patients compared with standard care in 1866 patients and validated the clinical benefit associated with the use of pharmacogenomic biomarkers in a real-world setting (Anderson 2007). Another "grassroots example" comes from the idea that a pragmatic clinical trial could be adopted on a much smaller scale in clinics or stand-alone physician practices. For instance, tolbutamide elimination is 50% and 84% slower in carriers of *CYP2C9*2* and *CYP2C9*3* variants, respectively, than in homozygous carriers of *CYP2C9*1* (Kirchheiner 2005; Shon 2002). However, there is no prospective RCT to show the appropriateness of dose reductions for patients who carry *CYP2C9*2* or **3* alleles. Nevertheless, tolbutamide efficacy can be easily evaluated after these dosage reductions are implemented, and encouraging such efforts in clinical practice (instead of expensive and time-consuming RCTs) enhances the development of clinical utility evidence for *CYP2C9* genotyping for optimizing tolbutamide therapy, together with other drugs that share this metabolic pathway.

Documenting Pharmacogenetic Associations with Important Health Outcomes

Finally, it is imperative to recognize how nuances in evidence-based medicine may either introduce or minimize challenges, depending on the disease or medical specialty. For example, despite evidence that strongly links clopidogrel efficacy to *CYP2C19* polymorphisms (Mega 2010), there is continued debate about whether routine use of *CYP2C19* genotyping is warranted to guide clopidogrel therapy, even with documentation of significantly higher rates of stent thrombosis and major adverse cardiovascular events among carriers of the *CYP2C19*2* loss-of-function allele (Holmes 2011). In 2010, the American College of Cardiology and the American Heart Association recommended against routine *CYP2C19* genotyping because of the lack of outcomes data and issued a joint clinical alert suggesting the need for large, prospective RCTs (Holmes 2010). Even though such trials are planned or under way, the dilemma remains: should patient care in the meantime be sacrificed by waiting for proof of value from RCTs, or should the focus sharpen on clinical implementation to determine cost-effective approaches for genotype-guided antiplatelet therapy? Conversely, in oncology, *KRAS* testing to individualize therapy for patients with advanced or metastatic colorectal cancer is widely used in accordance with National Comprehensive Cancer Network recommendations, despite a similar evidence base for clinical utility (Moorcraft 2013).

Tension Between Availability of Genomic Data and Paucity of Health/Patient Outcomes Data

Comparing Targeted and Preemptive Pharmacogenetic Testing

A natural tension exists between targeted (i.e., just in time) and preemptive (i.e., just in case) approaches to pharmacogenomic testing. Outside large research efforts at academic medical centers, pharmacogenetic testing is typically ordered at the point of care before initiating a treatment. This just-in-time, or targeted, approach has several drawbacks. First, the correct test must be ordered, retrieved, and interpreted by the clinician, after which the patient may need to be recontacted if drug therapy changes are needed. This is especially problematic when treatment cannot be delayed (e.g., clopidogrel). Second, a delay in obtaining genetic results to guide drug selection or dosing may render the information obsolete because interim decisions are required, and providers may not effectively respond to genetic results returned beyond the typical time interval of a clinical encounter. Finally, tests for single genes are expensive relative to the potential benefit obtained in guiding a single therapeutic decision (Johnson 2013; Pulley 2012; Schildcrout 2012; Krews 2011). In light of ongoing clinical research with preemptive pharmacogenomic testing (described in later paragraphs), it is increasingly unlikely that the targeted approach will accumulate sufficient clinical utility evidence.

The alternative is to prospectively collect data for multiple pharmacogenes to store electronically for future use (i.e., preemptive or just-in-case use) (Schildcrout 2012). This enables decreased per-genotype costs and development of advanced clinical decision support tools for use by front-line clinicians (similar to, but more advanced than, drug allergy information). In this paradigm, medication order entry would automatically trigger a search for relevant drug-gene interactions for that patient; if clinically actionable variants were identified, the system would guide the clinician toward an appropriate individualized therapy. A particularly appealing feature of this preemptive approach is that testing can be multiplexed, assaying hundreds to thousands of genetic variants at a single time. This genetic information can be reused as other drugs are prescribed over time and as the knowledge base of drug-gene interactions grows. And, as mentioned in the previous paragraph, clinical utility evidence for this approach is rapidly growing. However, this approach is not without drawbacks. Multiplexed testing may include clinically actionable results in addition to incidental findings or non-clinically actionable information, necessitating separate data storage to ensure that only clinically actionable information (as determined by current evidence for clinical utility) is included in medical records. Separate storage of non-clinically actionable information prevents potential harm associated with the disclosure of results, but future value is retained when new evidence becomes available.

Finally, an important consideration for any type of genetic testing (including targeted tests vs. preemptive tests) is that the process of designating wild-type alleles on the basis of testing methods that focus only on common genetic variation might preclude the identification of previously unknown or undiscovered variants (e.g., a rare loss-of-function allele). Although this is always a possibility, the rare chance that it could lead to inaccurate characterization of a loss-of-function allele as a wild-type allele underscores the need for due diligence when interpreting results.

Cost-Effectiveness and Choice Between Targeted Testing and Preemptive Testing

For a variety of reasons, it is critically important in most health systems to accurately assess whether a test offers significant return on investment. Therefore, the lack of cost-effectiveness data, together with clinical validity and clinical utility, is an additional potential barrier to the clinical implementation of pharmacogenomic testing. Ideally, pharmacogenomic testing would result in cost-effective clinical care improvements for patients who derive benefit from individualized therapy and avoid cost-ineffective treatment in the form of lack of response or increased adverse drug reactions (Daly 2014).

A traditional cost-effectiveness study compares the relative costs and outcomes of two different interventions (Detsky 1990). As previously mentioned, avoiding cost-ineffective treatment is one component of cost-effective clinical care. Antipsychotic drugs provide a useful example of an alternative approach to cost-effectiveness evaluation for pharmacogenomic biomarkers. With an annual cost that is at least 10 times higher, atypical antipsychotic agents are more expensive, yet no more efficacious and less cost-effective, than the typical antipsychotic agents (Perlis 2005). Rather than focusing on using biomarkers to predict efficacy of the more expensive atypical antipsychotic agents (Cacabelos 2011), genotyping for the *Ser9Gly* polymorphism in the dopamine 3 receptor gene (*DRD3*) (Malhotra 1998) might be used to identify patients susceptible to tardive dyskinesia, a highly prevalent adverse drug reaction associated with the use of typical antipsychotic agents. In this way, genetic testing might enable appropriate dose reduction for typical antipsychotic agents and reduce the incidence of adverse drug reactions.

Additional approaches for showing the cost-effectiveness of genotype-guided therapy range from a clinical trial comparing per-patient costs for specific clinical outcomes between genotype-based regimens with standard regimens to a decision model–based study (i.e., one that uses a simulated patient cohort) (Lam 2013). Regardless of the specific approach used, the economic impact and cost-effectiveness of pharmacogenomic screening may be affected by different variables. To illustrate this point, two separate studies used modeling techniques with simulated patient cohorts to evaluate potential clinical and economic outcomes for genotype-guided warfarin dosing. Although the relatively high cost of a *CYP2C9* and *VKORC1* bundled test ($326–$570) resulted in only modest improvements (in quality-adjusted life-years, survival rates, and total adverse rates), the investigators also suggested that improvements in the cost-effectiveness could be achieved in several ways—specifically, by further cost reduction of the genotyping test and using genotype-guided warfarin dosing algorithm in outliers (patients with out-of-range INRs [international normalized ratios] and/or those at high risk of hemorrhage (Eckman 2009; Patrick 2009). Other variables such as different population prevalence of a specific variant and cost of alternative treatment approaches would also affect the economic impact analysis. Ultimately, clinical utility and cost-effectiveness cannot solely determine the relative value of pharmacogenomic testing for optimizing drug therapy for individual patients. Rather, they should be used to supplement the best practices currently in place to achieve optimal drug therapy.

Epigenetic Considerations

Epigenetics refers to heritable non-genetic information (i.e., without change in the DNA sequence) and involves three interacting molecular mechanisms/alterations: (1) DNA methylation, (2) histone modification, and (3) RNA-mediated gene regulation. These alterations potentially offer further explanation for interindividual variations in drug disposition, response, or toxicity that cannot be solely attributed to genetic polymorphisms (Ivanov 2012). Several genes encoding drug-metabolizing enzymes, transporters, and intra- and extra-nuclear receptors are influenced by epigenetic mechanisms (Ingelman-Sundberg 2007). Thus, the role of epigenetics in individual drug disposition, response, and toxicity has just begun to be understood, but its impending influence promises to be important in clinical therapeutics.

Synthesizing Evidence-Based Guidelines

Evidence-based medicine is classically defined as "the conscientious, explicit, and judicious use of current best evidence in making decisions about the care of individual patients" (Sackett 1996). Practicing evidence-based medicine requires integrating clinical expertise with the best available external clinical research evidence. Clinical practice guidelines, which increasingly build on objective analysis of evidence from well-designed studies, have become highly credible sources of information about which forms of care are effective. Therefore, it is critically important to develop thoughtful clinical practice guidelines for pharmacogenomics in order to facilitate clinical implementation. Two guideline development groups will be discussed in the following sections: (1) the Evaluation of Genomic Applications in Practice and Prevention (EGAPP) Working Group (EWG), established in 2005, and (2) the Clinical Pharmacogenetics Implementation Consortium (CPIC), established in 2009.

Comparing Systematic and Standardized Evidence Review

The EWG, an independent panel affiliated with the Centers for Disease Control and Prevention (CDC), is charged with developing systematic processes for evidence-based assessment of genetic tests and other genomic technology applications. Key objectives of the EWG are to develop a transparent, publicly accountable process; minimize conflicts of interest; optimize existing evidence review methods to address the challenges presented by complex and rapidly emerging genomic applications; and provide clear linkage between the scientific evidence and the subsequently developed EWG recommendation statements. The EWG is currently composed of 16 experts in areas such as clinical practice, evidence-based medicine, genomics, public health, laboratory practice, epidemiology, economics, ethics, policy, and health technology assessment (Teutsch 2009). This nonfederal panel is supported by the EGAPP initiative launched in late 2004 by the National Office of Public Health Genomics at the CDC. In addition to supporting the activities of the EWG, EGAPP is developing data collection, synthesis, and review capacity to support the timely and efficient translation of genomic applications to practice; evaluating the products and impact of the EWG pilot phase; and working with the EGAPP Stakeholders Group on topic prioritization, information dissemination, and product feedback (Teutsch 2009). The EWG is not a federal advisory committee; rather, it aims to provide information to clinicians and other key stakeholders on integrating genomics into clinical practice. The EGAPP initiative has no oversight or regulatory authority. Much debate has centered on the definition of a "genetic test." Because of the evolving nature of the tests and technologies, the EWG has adopted the broad view articulated in a recent report of the Secretary's Advisory Committee on Genetics, Health, and Society (Teutsch 2009):

> *A genetic test involves the analysis of chromosomes, deoxyribonucleic acid (DNA), ribonucleic acid (RNA), genes, or gene products (e.g., enzymes and other proteins) to detect heritable or somatic variations related to disease or health. Whether a laboratory method is considered a genetic test also depends on the intended use, claim or purpose of a test.*

Depending on resource limitations, EGAPP focuses on tests that have a wider population application (e.g., higher disorder prevalence, higher frequency of test use), those with the potential to affect clinical and public health practice (e.g., emerging prognostic and pharmacogenomic tests), and those for which there is a significant demand for information. Tests currently eligible for EGAPP review include those used to guide intervention in symptomatic individuals (e.g., diagnosis, prognosis, treatment) or asymptomatic individuals (e.g., disease screening), identify individuals at risk of future disorders (e.g., risk assessment or susceptibility testing), or predict treatment response or adverse events (e.g., *UGT1A1* genotyping for irinotecan, *KRAS* testing for cetuximab, and *CYP2C19/CYP2D6* genotyping for selective serotonin reuptake inhibitors). The EGAPP-commissioned evidence reports and EWG recommendation statements are focused on patients seen in traditional primary or specialty care clinical settings, but they may address other contexts, such as direct Web-based test offerings to consumers without clinician involvement (e.g., direct-to-consumer genetic testing) (EGAPP 2007). The EWG recommendations may vary for different applications of the same test or for different clinical scenarios, and they may address testing algorithms that include preliminary tests (e.g., family history or other laboratory tests that identify high-risk populations). All EWG members, review team members, and consultants disclose potential conflicts of interest for each topic considered. After the scope and outcomes of interest are identified for a systematic review, key questions and an analytic framework are developed by the EWG and later refined by the review team in consultation with a technical expert panel (TEP). The EWG assigns members to serve on the TEP, together with other experts selected by those conducting the review; these members constitute the EWG "topic team" for that review. Because of the multidisciplinary nature of the panel, selection of EWG topic teams aims to include expertise in evidence-based medicine and scientific content.

For five of eight testing applications selected by the EWG to date, CDC-funded systematic evidence reviews have been conducted in partnership with the Agency for Healthcare Research and Quality (AHRQ) Evidence-Based Practice Centers. Because of their expertise in conducting comprehensive, well-documented literature searches and evaluation, these centers represent an important resource for performing comprehensive reviews on applications of genomic technology. However, comprehensive reviews are time- and resource-intensive, and the number of relevant tests is rapidly increasing. Some tests have several applications and require review of more than one clinical scenario. Consequently, the EWG is also investigating alternative strategies to produce shorter, less expensive, but no less rigorous, systematic reviews of the evidence needed to make decisions about immediate usefulness and highlight important gaps in knowledge. A key objective is to develop methods to support "targeted" or "rapid" reviews that are both timely and methodologically sound. Regardless of the source, a primary objective of all evidence reviews is that the final product be a comprehensive evaluation and interpretation of the available evidence, rather than summary descriptions of relevant studies (Teutsch 2009).

The CPIC, a joint effort between the Pharmacogenomics Knowledge Base (PharmGKB: www.pharmgkb.org) and the National Institutes of Health (NIH)-funded Pharmacogenomics Research Network, was created to develop guidelines that would enable the translation of clinically relevant pharmacogenetic test results into actionable therapeutic recommendations (Relling 2011). Members of CPIC are self-nominated and contribute expertise to pharmacogenomic discovery research, successful implementation of pharmacogenomic testing into patient care, and informatics resource development. The underlying assumption of all CPIC guidelines is that clinical high-throughput and preemptive genotyping will eventually become common practice, and clinicians will increasingly have access to patients' genetic profiles before ordering medications (Relling 2011). Therefore, the CPIC guidelines are designed to provide guidance to clinicians regarding how available genetic test results should be used to improve drug therapy, rather than guidance regarding whether a genetic test should or should not be ordered. Examples of the types of evidence reviewed include RCTs with genotype-guided prescribing versus non-genetically based prescribing, preclinical and clinical studies linking drug effects or concentrations to functional pharmacogenetic loci, case studies, and in vivo pharmacokinetic/pharmacodynamics studies. When available, evidence for improved outcomes with genotype-guided prescribing is included. However, for most gene drug pairs, RCTs comparing clinical outcomes with genotype-guided versus conventional therapy selection are not available. Publications identifying a major or strong pharmacogenetic association are considered by the guideline writing committee and scored by individual members (see scheme later in this chapter) according to the cumulative evidence supporting that major finding. The rating scheme uses a scale slightly modified from that of Valdes et al. (Valdes 2010): high evidence includes consistent results from well-designed, well-conducted studies; moderate evidence is sufficient to determine effects, but the strength of the evidence is limited by the number, quality, or consistency of the individual studies, generalizability to routine practice, or indirect nature of the evidence; and weak evidence is insufficient to assess the effects on health outcomes because of limited number or power of studies, important flaws in their design or conduct, gaps in the chain of evidence, or lack of information. It is possible for an evidentiary conclusion based on many papers, each of which may be relatively weak, to be graded as "moderate" or even "high" if there are multiple supportive small case reports or studies without contradictory findings. Primary publications are summarized in an evidence table, which is published in the manuscript supplemental material. The writing committee's evaluation of this evidence provides the basis for the therapeutic recommendations. As of this writing, 13 CPIC guidelines exist for specific drug-gene pairs, including clopidogrel-*CYP2C19* (Scott 2013) and codeine-*CYP2D6* (Crews 2012).

Clinical Practice Guideline Dissemination
The importance of disseminating evidence-based clinical practice guidelines underscores development efforts. At a macro level, ensuring that guidelines are published in many high-impact vehicles (i.e., peer-reviewed journals, Web sites with high levels of traffic, and national clinical guideline sites) facilitates broad communication to most major stakeholders. On a granular level, bioinformatics and clinical decision support tools facilitate the translation of guidelines at the individual health-system or provider level. Several methods are used to distribute and support uptake of the CPIC guidelines after publication. As discussed earlier, all guidelines are posted on the PharmGKB Web site, linked to pages for the gene and the drug that also provide other relevant information. Furthermore, all guidelines are cited in PubMed and classified as a "practice guideline," a method used to organize and index published clinical practice guidelines in PubMed. The final published full-text guidelines are also freely available in the NIH-maintained PubMed Central repository. Moreover, the NIH Genetic Testing Registry, which is a central location for the voluntary submission of genetic test information by providers, gives links to the CPIC guidelines according to condition/phenotype (e.g., thiopurine methyltransferase [TPMT] deficiency; www.ncbi.nlm.nih.gov/gtr/conditions/C0342801/). Recently, the CPIC guidelines were accepted for inclusion into AHRQ-sponsored National Guideline Clearinghouse, and summaries of the guidelines, including the prescribing recommendation, are freely available to the public from this site as well.

Clinical Decision Support and Informatics
To address the growing interest in the informatics aspects of the CPIC guidelines and the clinical implementation of pharmacogenetics, the CPIC created an informatics working group in 2013. The working group's goal is to support clinical adoption of the CPIC guidelines by identifying, and resolving when possible, potential technical barriers to implementing the guidelines within a clinical electronic environment. The first CPIC guideline to include informatics tables was the 2014 update of the guideline for *HLA-B* genotype and abacavir dosing (Martin 2014). Because detailed information translating results from genotype to phenotype and finally to clinical recommendation is needed to implement pharmacogenetics into patient care, the initial focus of the informatics working group is on creating translation tables, which complement similar data gathered by the TPP (Translational Pharmacogenetics Project), a network-wide Pharmacogenomics Research Network venture. Initial versions of these tables have been posted on PharmGKB for *CYP2D6*, *CYP2C19*, and *TPMT* (PharmGKB 2014), and processes to develop and maintain each table as part of the CPIC guideline update process are being developed. Other informatics priorities for CPIC include encouraging vendors that produce clinical electronic references and electronic health records to include the CPIC guidelines and developing recommendations for clinical decision support tools according to the guidelines. Long-term opportunities include working with standards development organizations (e.g., the American Medical Association or the International Health Terminology Standards Development Organization) to create content that supports the CPIC guidelines and pharmacogenetics, such as terms that properly define metabolizer status and other pharmacogenetic phenotypes, and generating machine-readable representations of the CPIC guidelines using formal knowledge representation and standard terminologies. Ultimately, these efforts will position the CPIC guidelines to work effectively in our increasingly electronic health care system.

Importance of Horizon Scanning in Guiding Ongoing Evidence Evaluations
Given the rapidly evolving evidence base for pharmacogenomics, a systematic approach allowing stakeholders to perform horizon scanning (i.e., evaluating emerging health care technologies and innovations in order to better inform patient-centered outcomes) of synthesized evidence sources relevant to the use of a particular genetic test is imperative. A practical example of this is provided by efforts of the Office of Public Health Genomics at the CDC, which recently modified an existing classification system (Dotson 2014) and created a table of evolving genomic tests sorted by level of evidence. The table presents an evidence-based cataloging of "genomic applications" (i.e., use of gene-based tests in specific clinical scenarios to assist with clinical decision-making), rather than an assessment of available diagnostic

tests. The methodologic underpinning of this system is a three-tier classification framework to show how to classify genomic applications:

Tier 1: Genomic applications have a base of synthesized evidence that supports their implementation into practice.

Tier 2: Genomic applications have synthesized evidence that is insufficient to support their implementation into routine practice. Nevertheless, the evidence may be useful for informing selective use strategies (e.g., in clinical trials) through individual clinical, or public health policy, decision-making.

Tier 3: Applications either have (1) synthesized evidence that supports recommendations against or discourages use or (2) no available evidence that is relevant.

This process is intended to inform the critical evaluation of genomic applications until the evidence base becomes more robust and until more comprehensive resources become available. (e.g., ClinGen, funded by the NIH, which aims to catalog medically relevant human gene variants). This horizon-scanning method can also point quickly to existing evidence gaps and/or lack of available guidelines and recommendations.

Summary

Although significant scientific and technological advances enable the identification of gene variants that regulate the disposition and target pathways of drugs, translating the pharmacogenomic findings into clinical practice has been met with continued scientific debates, as well as commercial, economic, educational, ethical, legal, and societal barriers. Despite the well-known potentials of improving drug efficacy and safety, as well as the efficiency of the drug development process, the logistic issues and challenges identified for incorporating pharmacogenomics into clinical practice and drug development can be addressed only with all stakeholders in the field working together and occasionally accepting a paradigm change in their current approach.

References

Accurso FJ, Rowe SM, Clancy JP, et al. Effect of VX-770 in persons with cystic fibrosis and the G551D-CFTR mutation. N Engl J Med 2010;363:1991-2003.

AHRQ.gov [homepage on the Internet]. 2011. Effective Health Care Program. Washington, DC: Agency for Healthcare Research and Quality. Available at http://effectivehealthcare.ahrq.gov/index.cfm/search-for-guides-reviews-and-reports/?productid=315&pageaction=displayproduct. Accessed May 12, 2014.

Anderson JL, Horne BD, Stevens SM, et al. Randomized trial of genotype-guided versus standard warfarin dosing in patients initiating oral anticoagulation. Circulation 2007;116:2563-70.

Boessen R, van der Baan F, Groenwold R, et al. Optimizing trial design in pharmacogenetics research: comparing a fixed parallel group, group sequential, and adaptive selection design on sample size requirements. Pharm Stat 2013;12:366-74.

Cacabelos R, Hashimoto R, Takeda M. Pharmacogenomics of antipsychotics efficacy for schizophrenia. Psychiatry Clin Neurosci 2011;65:3-19.

Centers for Disease Control and Prevention (CDC). 2011. Laboratory Requirements (42 CFR 493). Federal Register. October 1, 2011.

Centers for Disease Control and Prevention (CDC). 2009. Good laboratory practices for molecular genetic testing for heritable diseases and conditions. MMWR 2009;59(RR-06):1-29.

Chapman PB, Hauschild A, Robert C, et al. Improved survival with vemurafenib in melanoma with *BRAF V600E* mutation. N Engl J Med 2011;364:2507-16.

Crews KR, Cross SJ, McCormick JN, et al. Development and implementation of a pharmacist-managed clinical pharmacogenetics service. Am J Health Syst Pharm 2011;68:143-50.

Crews KR, Gaedigk A, Dunnenberger HM, et al. Clinical Pharmacogenetics Implementation Consortium (CPIC) guidelines for codeine therapy in the context of cytochrome P450 2D6 (CYP2D6) genotype. Clin Pharmacol Ther 2012;91:321-6.

Daly AK, Cascorbi I. Opportunities and limitations: the value of pharmacogenetics in clinical practice. Br J Clin Pharmacol 2014;77:583-6.

de Leon J, Susce MT, Murray-Carmichael E. The AmpliChip CYP450 genotyping test: integrating a new clinical tool. Mol Diagn Ther 2006;10:135-51.

Dempfle A, Scherag A, Hein R, et al. Gene-environment interactions for complex traits: definitions, methodological requirements and challenges. Eur J Hum Genet 2008;16:1164-72.

Detsky AS, Naglie IG. A clinician's guide to cost-effectiveness analysis. Ann Intern Med 1990;113:147-54.

Deverka PA. Pharmacogenomics, evidence, and the role of payers. Public Health Genomics 2009;12:149-57.

Deverka PA, McLeod HL. Harnessing economic drivers for successful clinical implementation of pharmacogenetic testing. Clin Pharmacol Ther 2008;84:191-3.

Dotson WD, Douglas MP, Kolor K, et al. Prioritizing genomic applications for action by level of evidence: a horizon-scanning method. Clin Pharmacol Ther 2014;95:394-402.

Eckman MH, Rosand J, Greenberg SM, et al. Cost-effectiveness of using pharmacogenetic information in warfarin dosing for patients with nonvalvular atrial fibrillation. Ann Intern Med 2009;150:73-83.

Evaluation of Genomic Applications in Practice and Prevention (EGAPP) Working Group. Recommendations from the EGAPP Working Group: testing for cytochrome P450 polymorphisms in adults with nonpsychotic depression treated with selective serotonin reuptake inhibitors. Genet Med 2007;9:819-25.

FDA.gov [homepage on the Internet]. 2010. Interdisciplinary Pharmacogenomics Review Group (IPRG). Silver Spring, MD: U.S. Department of Health and Human Services. Available at www.fda.gov/Drugs/ScienceResearch/ResearchAreas/Pharmacogenetics/ucm083889.htm. Accessed May 22, 2014.

FDA.gov [homepage on the Internet]. 2013. Paving the Way for Personalized Medicine: FDA's Role in a New Era of Product Development. Silver Spring, MD: U.S. Food and Drug Administration. Available at www.fda.gov/downloads/scienceresearch/specialtopics/personalizedmedicine/ucm372421.pdf. Accessed February 12, 2014.

Gage BF, Eby C, Johnson JA, et al. Use of pharmacogenetic and clinical factors to predict the therapeutic dose of warfarin. Clin Pharmacol Ther 2008;84:326-31.

Garrison LP Jr, Austin MJ. Linking pharmacogenetics-based diagnostics and drugs for personalized medicine. Health Aff (Millwood) 2006;25:1281-90.

Ginsburg GS, Konstance RP, Allsbrook JS, et al. Implications of pharmacogenomics for drug development and clinical practice. Arch Intern Med 2005;165:2331-6.

Gottesman O, Kuivaniemi H, Tromp G, et al. The Electronic Medical Records and Genomics (eMERGE) network: past, present, and future. Genet Med 2013;15:761-71.

Haddow JE, Palomaki GE. ACCE: a model process for evaluating data on emerging genetic tests. In: Khoury M, Little J, Burke W, eds. Human Genome Epidemiology: A Scientific Foundation for Using Genetic Information to Improve Health and Prevent Disease. Oxford, UK: Oxford University Press, 2003:217-33.

Hetherington S, McGuirk S, Powell G, et al. Hypersensitivity reactions during therapy with the nucleoside reverse transcriptase inhibitor abacavir. Clin Ther 2001;23:1603-14.

Holmes DR Jr, Dehmer GJ, Kaul S, et al. ACCF/AHA clopidogrel clinical alert: approaches to the FDA "boxed warning": a report of the American College of Cardiology Foundation Task Force on Clinical Expert Consensus Documents and the American Heart Association. Circulation 2010;122:537-57.

Holmes MV, Perel P, Shah T, et al. CYP2C19 genotype, clopidogrel metabolism, platelet function, and cardiovascular events: a systematic review and meta-analysis. JAMA 2011;306:2704-14.

Ingelman-Sundberg M, Sim SC, Gomez A, et al. Influence of cytochrome P450 polymorphisms on drug therapies: pharmacogenetic, pharmacoepigenetic and clinical aspects. Pharmacol Ther 2007;116:496-526.

Ivanov M, Kacevska M, Ingelman-Sundberg M. Epigenomics and interindividual differences in drug response. Clin Pharmacol Ther 2012;92:727-36.

Johnson JA, Elsey AR, Clare-Salzler MJ, et al. Institutional profile: University of Florida and Shands Hospital Personalized Medicine Program: clinical implementation of pharmacogenetics. Pharmacogenomics 2013;14:723-6.

Kirchheiner J, Roots I, Goldammer M, et al. Effect of genetic polymorphisms in cytochrome p450 (CYP) 2C9 and CYP2C8 on the pharmacokinetics of oral antidiabetic drugs: clinical relevance. Clin Pharmacokinet 2005;44:1209-25.

Klein TE, Altman RB, Eriksson N, et al. Estimation of the warfarin dose with clinical and pharmacogenetic data. N Engl J Med 2009;360:753-64.

Lam YW. Scientific challenges and implementation barriers to translation of pharmacogenomics in clinical practice. ISRN Pharmacol 2013;2013:641089.

Lane HY, Tsai GE, Lin E. Assessing gene-gene interactions in pharmacogenomics. Mol Diagn Ther 2012;16:15-27.

Limdi NA, Veenstra DL. Expectations, validity, and reality in pharmacogenetics. J Clin Epidemiol 2010;63:960-9.

Malhotra AK, Goldman D, Buchanan RW, et al. The dopamine D3 receptor (DRD3) Ser9Gly polymorphism and schizophrenia: a haplotype relative risk study and association with clozapine response. Mol Psychiatry 1998;3:72-5.

Man CB, Kwan P, Baum L, et al. Association between HLA-B*1502 allele and antiepileptic drug-induced cutaneous reactions in Han Chinese. Epilepsia 2007;48:1015-8.

Martin MA, Hoffman JM, Freimuth RR, et al. Clinical pharmacogenetics implementation consortium guidelines for hla-B genotype and abacavir dosing: 2014 update. Clin Pharmacol Ther 2014;95:499-500.

Massie J, Castellani C, Grody WW. Carrier screening for cystic fibrosis in the new era of medications that restore CFTR function. Lancet 2014;383:923-5.

Mega JL, Simon T, Collet JP, et al. Reduced-function CYP2C19 genotype and risk of adverse clinical outcomes among patients treated with clopidogrel predominantly for PCI: a meta-analysis. JAMA 2010;304:1821-30.

Moorcraft SY, Smyth EC, Cunningham D. The role of personalized medicine in metastatic colorectal cancer: an evolving landscape. Ther Adv Gastroenterol 2013;6:381-95.

Morello JP, Valdes R Jr. Reporting and interpretation of pharmacogenetic test results. In: Valdes R Jr, Payne DA, Linder MW, eds. Laboratory Medicine Practice Guidelines: Guidelines and Recommendations for Laboratory Analysis and Application of Pharmacogenetics to Clinical Practice. Washington, DC: National Academy of Clinical Biochemistry (NACB), 2010:19-22.

National Institutes of Health (NIH) [homepage on the Internet]. 2008. Department of Health and Human Services. Realizing the Potential of Pharmacogenomics: Opportunities and Challenges. Available at http://osp.od.nih.gov/sites/default/files/SACGHS_PGx_report.pdf. Accessed March 12, 2014.

Newman WG, Payne K, Tricker K, et al. A pragmatic randomized controlled trial of thiopurine methyltransferase genotyping prior to azathioprine treatment: the TARGET study. Pharmacogenomics 2011;12:815-26.

Patrick AR, Avorn J, Choudhry NK. Cost-effectiveness of genotype-guided warfarin dosing for patients with atrial fibrillation. Circ Cardiovasc Qual Outcomes 2009;2:429-36.

Perlis RH, Ganz DA, Avorn J, et al. Pharmacogenetic testing in the clinical management of schizophrenia: a decision-analytic model. J Clin Psychopharmacol 2005;25:427-34.

PharmGKB.org [homepage on the Internet]. 2014. California: Translational Pharmacogenetics Project (TPP): Look Up Tables By Gene. Available at www.pharmgkb.org/page/tppTables. Accessed February 12, 2014.

Phillips KA, Van Bebber S, Issa AM. Diagnostics and biomarker development: priming the pipeline. Nat Rev Drug Discov 2006;5:463-9.

Powanda MC, Moyer ED. Some applications of pharmacogenomics and epigenetics in drug development and use in pursuit of personalized medicine. Inflammopharmacology 2012;20:245-50.

Pulley JM, Denny JC, Peterson JF, et al. Operational implementation of prospective genotyping for personalized medicine: the design of the Vanderbilt PREDICT project. Clin Pharmacol Ther 2012;92:87-95.

Ratner J, Mullins D, Buesching DP, et al. Pragmatic clinical trials: U.S. payers' views on their value. Am J Manag Care 2013;19:e158-65.

Relling MV, Klein TE. CPIC: Clinical Pharmacogenetics Implementation Consortium of the Pharmacogenomics Research Network. Clin Pharmacol Ther 2011;89:464-7.

Rieder MJ, Reiner AP, Gage BF, et al. Effect of VKORC1 haplotypes on transcriptional regulation and warfarin dose. N Engl J Med 2005;352:2285-93.

Roujeau JC, Kelly JP, Naldi L, et al. Medication use and the risk of Stevens-Johnson syndrome or toxic epidermal necrolysis. N Engl J Med 1995;333:1600-7.

Sackett DL, Rosenberg WM, Gray JA, et al. Evidence-based medicine: what it is and what it isn't. BMJ 1996;312:71-2.

Schildcrout JS, Denny JC, Bowton E, et al. Optimizing drug outcomes through pharmacogenetics: a case for preemptive genotyping. Clin Pharmacol Ther 2012;92:235-42.

Scott SA. Personalizing medicine with clinical pharmacogenetics. Genet Med 2011;13:987-95.

Scott SA, Jaremko M, Lubitz SA, et al. CYP2C9*8 is prevalent among African-Americans: implications for pharmacogenetic dosing. Pharmacogenomics 2009;10:1243-55.

Scott SA, Sangkuhl K, Stein CM, et al. Clinical Pharmacogenetics Implementation Consortium guidelines for CYP2C19 genotype and clopidogrel therapy: 2013 update. Clin Pharmacol Ther 2013;94:317-23.

Shon JH, Yoon YR, Kim KA, et al. Effects of CYP2C19 and CYP2C9 genetic polymorphisms on the disposition of and blood glucose lowering response to tolbutamide in humans. Pharmacogenetics 2002;12:111-9.

SNPedia [homepage on the Internet]. 2013. *CYP2C19*. Available at www.snpedia.com/index.php/CYP2C19. Accessed March 12, 2014.

Staa TP, Goldacre B, Gulliford M, et al. Pragmatic randomised trials using routine electronic health records: putting them to the test. BMJ 2012;344:e55.

Swen JJ, Huizinga TW, Gelderblom H, et al. Translating pharmacogenomics: challenges on the road to the clinic. PLoS Med 2007;4:e209.

Teutsch SM, Bradley LA, Palomaki GE, et al. The Evaluation of Genomic Applications in Practice and Prevention (EGAPP) initiative: methods of the EGAPP Working Group. Genet Med 2009;11:3-14.

Todd JN, Florez JC. An update on the pharmacogenomics of metformin: progress, problems and potential. Pharmacogenomics 2014;15:529-39.

Young B, Squires K, Patel P, et al. First large, multicenter, open-label study utilizing HLA-B*5701 screening for abacavir hypersensitivity in North America. AIDS 2008;22:1673-5.

Zhou K, Bellenguez C, Spencer CC, et al. Common variants near ATM are associated with glycemic response to metformin in type 2 diabetes. Nat Genet 2011;43:117-20.

Chapter 3

GENOMIC MEDICINE—CURRENT ADVANCES AND FUTURE PROJECTIONS

Lori A. Orlando, M.D., MHS; and Deepak Voora, M.D.

Learning Objectives
1. To show the scope of genomic medicine from the perspective of improving prognosis, diagnosis, monitoring, and therapy for disease.
2. To provide concrete examples of genomic medicine in current practice (at certain institutions) using family health history or DNA- or RNA-based technologies.

Keywords: Whole genome sequencing, whole exome sequencing, family health history, genome-wide testing, single-gene/single-syndrome testing, risk stratification, disease diagnosis, disease prognosis, disease monitoring, selecting optimal therapy

Abbreviations in This Chapter

CAD	Coronary artery disease
CF	Cystic fibrosis
EGFR	Epidermal growth factor receptor
FHH	Family health history
HBOC	Hereditary breast and ovarian cancer (syndrome)
HER2	Human epidermal growth factor receptor 2
JAK2	Janus tyrosine kinase 2 gene
SNP	Single nucleotide polymorphism
WES	Whole exome sequencing
WGS	Whole genome sequencing

Abstract

The ability to use the information in our genome to improve health is broadly used to describe genomic medicine. Recently, there has been an explosion in genomic technologies ranging from single nucleotide polymorphisms (SNPs) to the sequence of single genes to that of the entire genome to modifications to DNA that do not involve the sequence such as methylation (or epigenetics) to the diverse suite of molecules derived from our DNA such as RNA, proteins, and small molecules of metabolism. Either alone or in combination, these technologic platforms have slowly but definitively been integrated across the spectrum of disease: prevention, diagnosis, therapy, and monitoring. Pharmacogenomics is a specific application of genomic medicine that primarily pertains to the choice of pharmacologic therapy. The purpose of this chapter is to paint a wider picture of how genomics is applied to medicine using specific examples across a spectrum of diseases. Our discussion is not meant to be exhaustive, but instead, illustrative of examples of genomic medicine in current practice today. We anticipate that many new examples will develop over time, but we believe that the conceptual framework we have outlined will remain relevant.

Introduction

Genomic medicine is a broad term that encompasses DNA- and RNA-based technologies as well as family health history (FHH). The official National Human Genome Research Institute definition of genomic medicine is as follows: "an emerging medical discipline that involves using genomic information about an individual as part of their clinical care (e.g., for diagnostic or therapeutic decision-making) and the other implications of that clinical use." Although this definition does not specifically include FHH, it is widely accepted by the institute as part of genomic medicine. DNA-based technologies aim to identify allelic and structural differences in the genome and come in a variety of forms, including single-gene/single-syndrome tests such as *BRCA1* and *BRCA2* for breast cancer and *HFE* for hemochromatosis; copy number analysis for *CDKN2B* in pancreatic cancer; genome-wide scans looking for SNPs in common complex diseases such as diabetes and obesity; and sequencing. Sequencing can be performed on the whole genome (WGS), on the whole exome (WES) (the exome is the DNA that remains after transcription and splicing [the exons]), or on panels of targeted genes such as those in next-generation pharmacogenomic panels like the PGRN-Seq chip. RNA-based technologies can assess levels of gene expression for one or several genes and are used to evaluate the extent to which genes are "turned on" or "turned off" as well as the presence or absence of non-human gene expression (e.g., when viral load tests are measured in human immunodeficiency virus [HIV]). Family health history may seem as if it does not belong in the same discussion as DNA and RNA; however, FHH is considered by many the very first "genetic" test (Pyeritz 2012). Family health history provides information about inherited conditions as well as shared environmental factors and the interaction of the two. Although the relative contributions of genetics and environment can complicate the interpretation of a disease's etiology, FHH is a critically important tool in determining the heritability of a trait (Flossmann 2004). These constitute the major genomic medicine technologies; however, there are other emerging technologies that are not traditionally thought of in terms of genomic medicine because they do not rely on the measurement of DNA or RNA, but are directly related to the products of DNA and RNA. These other platforms include proteomics (measurement of protein levels), metabolomics (measurement of metabolites), and the microbiome (microbial organisms that reside either synergistically or pathologically within our bodies). In this chapter, examples will be limited to DNA- and RNA-based technologies because of their widespread use across a variety of conditions. However, the future of genomic medicine is anticipated to encompass several different technologies either alone or in combination.

All the above-listed DNA technologies can be used in clinical care across the spectrum of health and disease: disease risk stratification (predicting disease risk), disease diagnosis, disease prognosis, optimal therapy selection, and disease monitoring (see Figure 3.1). Currently, RNA technologies are approved for use in diagnosis, prognosis, therapy, and monitoring, but not for risk stratification, and FHH can only be used for risk stratification. To help explore each of these uses, each technology will be described in more detail below, together with examples that show how they are currently being used in clinical care. In addition, the sidebar provides examples of technologies applied at each stage for the following diseases: cardiovascular disease, breast cancer, and Lynch syndrome.

Disease Risk Stratification

Disease risk stratification is the process of using biological, clinical, and/or behavioral characteristics to identify an individual's future risk of developing disease. The exact combination of characteristics depends on the disease and the risk stratification algorithm. For example, FHH is the only characteristic used in risk stratification for hereditary cancer syndromes, hemochromatosis, and cystic fibrosis (CF), though genetic testing is available for diagnosis in all of these cases (ACOG 2011; Bacon 2011; Hampel 2004; Hunt 2003), whereas breast cancer risk prediction with the Gail model (Gail 1989) or the BRCAPRO score (Berry 2002) uses a combination of clinical and FHH variables (see Figure 3.2).

Family Health History

Specific examples of FHH for disease risk prediction are described in this section. In diabetes, several models for disease risk prediction have been developed, all of which incorporate some FHH information. The Hariri model (Hariri 2006) includes only different combinations of affected relatives, whereas the American Diabetes Association risk calculator (Bang 2009) includes FHH and clinical variables, and the James Meigs model includes FHH, clinical factors, and genotyping results (Meigs 2008). Melanoma is typically considered an acquired condition caused by frequent sunburns; however, dysplastic nevus syndrome, a condition that results in an individual having many moles with a high rate of conversion to melanoma, is hereditary. Even in individuals without a family history of this syndrome, a first-degree relative with melanoma can increase risk by 50%

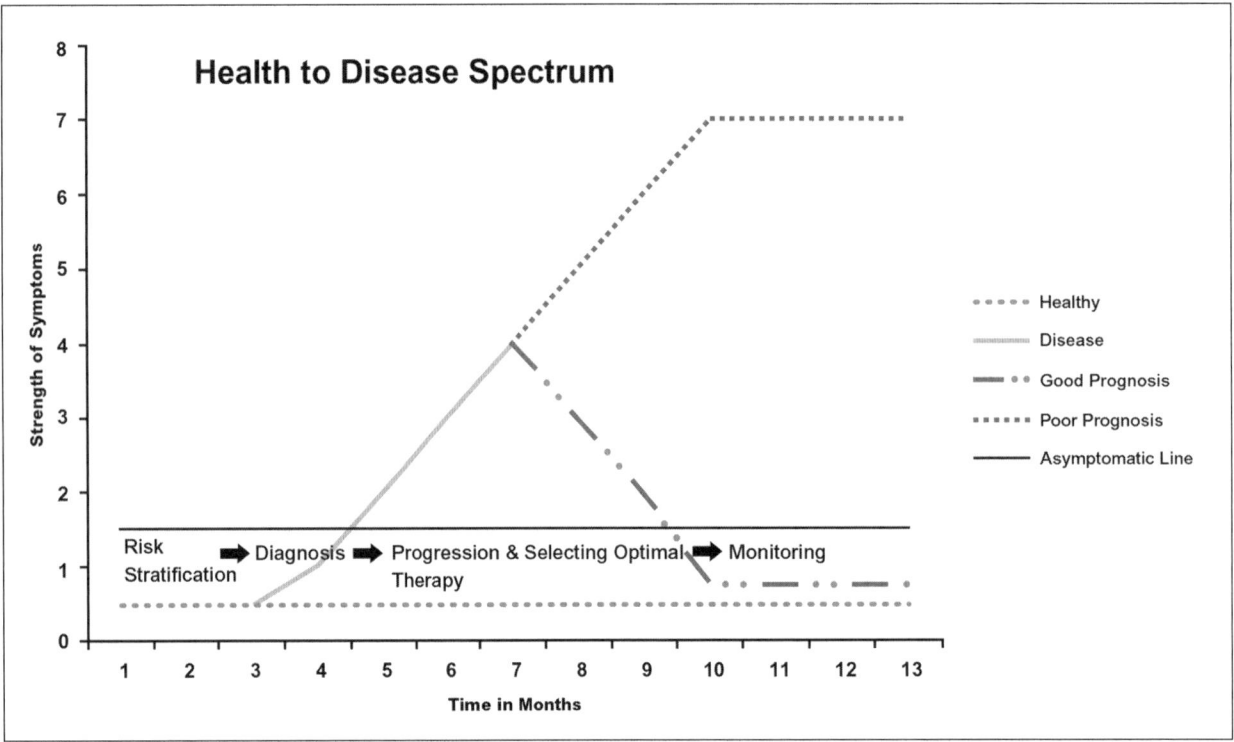

Figure 3.1. Health to disease spectrum.

In this figure, risk stratification occurs during the asymptomatic state (below the black line), and disease can be diagnosed before the patient becomes symptomatic or during the symptomatic state. At that point, disease can progress or regress, and optimal therapies are selected. Finally, disease is monitored.

(Holman 2014). In hereditary liver diseases such as α1-antitrypsin deficiency, hemochromatosis, and Wilson disease, having first-degree relatives with the condition can increase risk by 25%–50% and lead to recommendations for frequent screening and, in some cases, diagnostic genetic testing (EASL 2012; Bacon 2011; Hogarth 2008). In hereditary cancer syndromes such as hereditary breast and ovarian cancer (HBOC) syndrome, familial adenomatous polyposis, and Lynch syndrome (also known as hereditary nonpolyposis colon cancer), family history is critical to early risk assessment and diagnosis. The National Comprehensive Cancer Network (NCCN) has published complex but detailed algorithms for assessing patient risk using personal and family history of cancer (NCCN 2011). For the specific example of Lynch syndrome and breast cancer, see the sidebar.

DNA Single-Gene/Single-Syndrome Tests

One of the best-known and most widely discussed genetic tests ever, *BRCA*, falls into this category. On average, women have a 12% lifetime risk of developing breast cancer, and though only a minority of breast cancers occur because of a *BRCA* mutation (5%), women with a *BRCA1* mutation have a 65% risk of developing breast cancer, women with a *BRCA2* mutation have a 45% risk of developing breast cancer, and men with a *BRCA* mutation also have an increased risk of breast cancer (Teng 2008). In addition, women with a *BRCA* mutation have 3–4 times the risk of developing ovarian cancer as the average woman, and men have an increased risk of developing prostate cancer (Teng 2008). Because these risks are considerably higher than in the general population, women and men with *BRCA* mutations have options for risk management not recommended for those at low risk. These include mammography for men, breast MRI (magnetic resonance imaging) for women, chemoprevention with tamoxifen or raloxifene for women, and bilateral mastectomy with or without salpingo-oophorectomy for women (Teng 2008).

Another risk stratification test that has been widely incorporated into clinical practice is Lynch syndrome screening among patients with colon cancer. Colon cancer tumor specimens are examined with immunohistochemical staining (not a genomic test) and microsatellite instability (DNA testing of 7 loci), which predict the likelihood of having Lynch syndrome. Those with positive tests are recommended to undergo Lynch syndrome diagnostic DNA genetic testing.

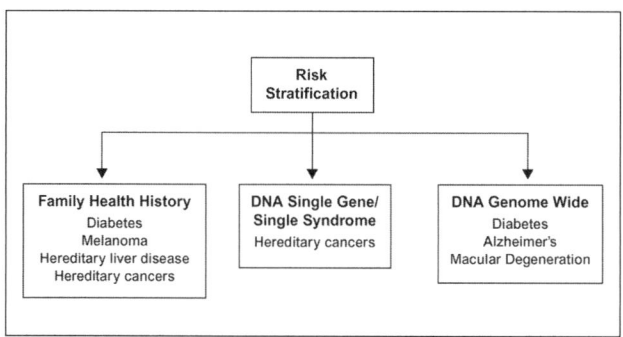

Figure 3.2. Disease risk stratification test examples.

DNA Genome-Wide Scans for SNPs

One consequence of the Human Genome Project has been the ability to assess an individual's entire genome at a cost that is acceptable to the patient. Initial experiences with this type of genome-wide testing were limited to high-profile consumers (e.g., Francis Collins [Francis Collins on Genetic Tests 2010], James Watson [Wheeler 2008]). The initial approach was aimed at genotyping most, if not all, of the validated genetic variants known to be associated with a variety of complex diseases. The advent of massively parallel sequencing (also called next-generation sequencing) heralded a precipitous decline in the cost of generating a WGS/WES. Although WGS/WES and analysis are now also available directly to consumers (i.e., outside a physician-patient interaction), limitations because of the lack of reproducibility and incomplete coverage (Dewey 2014) challenge the more widespread adoption of this technology.

In contrast to reading the sequence of the genome, as in WGS/WES, a more cost-effective approach is whole-genome genotyping in which a select number of SNPs are directly assayed across the genome. The number of genetic variants that can be detected with whole-genome genotyping (around 106) is smaller than with WGS/WES. However, whole-genome genotyping captures most common genetic variation associated with disease and is a robust and reproducible platform. Consequently, this technology has quickly become widely available in CLIA (Clinical Laboratory Improvement Amendments)-approved laboratories, thus ensuring high-quality genotype data to patients and providers as well as to the general public through direct-to-consumer advertising. Whole-genome genotyping tests can provide the relative risks of type 2 diabetes mellitus, Alzheimer dementia, and macular degeneration and include DNA analysis and interpretation. The absolute increase in risk of any given disease conferred by common genetic variants is small (less than 1%). However, anecdotal experience suggests that certain individuals can be motivated to make dramatic changes in lifestyle, diet, and behavior to mitigate their risk to a degree that is out of proportion to their individual risk. These experiences highlight a potentially powerful new paradigm in medicine in which patients are empowered not only to learn but also to use their genomic data. To date, no rigorous studies have shown that the information provided by these genome-wide scans can lead to improved health outcomes. However, the evidence to date suggests no significant psychological or behavioral harms in patients who receive such testing (Bloss 2011; Green 2009). Nevertheless, the full repertoire of potential responses to receiving genetic test information (both positive and negative) is not fully known. For example, although one person may be empowered to lose weight, diet, and exercise in response to an elevated genetic risk of type 2 diabetes mellitus, another may falsely conclude that the absence of such genetic risk justifies being overweight, eating poorly, and having a sedentary lifestyle. The balance of these positive and negative reactions to consumer-based genetic testing is not yet known. In the meantime, the U.S. Food and Drug Administration (FDA) has prohibited commercial entities from providing genetic risk data directly to consumers. Such testing, however, remains available to patients through their physicians.

Disease Diagnosis

Diagnostic testing is the use of a test to confirm the presence or absence of a disease. Before the era of genomic medicine, diagnosis was often a game of probabilities, with the probabilities being driven by a patient's age, sex, clinical features, and laboratory and/or imaging study results. An entire field of diagnostics was developed around the idea of pre- and posttest probabilities (Bayesian statistics) to guide whether a diagnostic test should be ordered. For example, if the pretest probability was so low that even a positive test would not raise the probability of disease enough to consider treatment, testing was discouraged. With the current knowledge about the genomic underpinnings for selected diseases, these diseases can now be diagnosed with almost complete certainty. Examples of diagnoses using DNA- and RNA-based technologies are provided in Figure 3.3.

DNA Single-Gene/Single-Syndrome Tests

Genetic testing in cancer diagnosis is common and has had a profound effect on the field. One common test is *JAK2*. The Janus tyrosine kinas 2 (*JAK2*) gene, located on the short arm of chromosome 9 (9p24), plays an important role in cell proliferation, especially for hematopoietic stem cells (the early progenitors for the red and white blood cells and platelets). Mutations in *JAK2* (primarily V617F) lead to overproduction and clonal proliferation of monoclonal cell lines of platelets (essential thrombocytosis) and red cells (polycythemia vera) and to bone marrow scarring (primary myelofibrosis) (ACOG 2011). Each of these conditions can be mimicked by other disorders, so the discovery of the *JAK2* gene and the ability to test for it was revolutionary in diagnosing such conditions. This gene is also a target for the therapeutics that are now in clinical trials.

Other widely used diagnostic DNA genetic tests are *MLH1*, *MSH2*, *MSH6*, *PMS2*, and *EPCAM* genes (all involved in mismatch repair) for Lynch syndrome. An important fact to remember is that a mutation in one of these genes is diagnostic of Lynch syndrome but is not diagnostic of a specific cancer—so they can be looked at as risk stratifiers for the cancers that make up the Lynch syndrome constellation (colon cancer, endometrial cancer, stomach cancer, and ovarian cancer). Fortunately, risk management strategies are specifically available for reducing cancer risk in patients with this syndrome (Woolf 1993).

DNA Single-Gene and Genome-Wide Testing Using Next-Generation Sequencing

Outside oncology, the field of medical genetics has recently undergone a paradigm shift using WGS/WES. Patients referred to medical genetics clinics are often children who typically have constellations of disorders affecting several organ systems but, most commonly, neurologic with a strong clinical suspicion for an underlying genetic disorder. However, typical approaches (i.e., chromosomal microarray or sequencing single genes for known mendelian disorders) can identify only a specific diagnosis in less than half of cases. Using WGS/WES, multidisciplinary teams of physicians, geneticists, statisticians, and bioinformaticians have begun the daunting task of identifying novel, rare, genetic mutations with the goal of identifying a causal diagnosis. Initial successes were anecdotal (Worthey 2011; Choi 2009) but showed the power of sequencing a patient and his or her unaffected parents in diagnosing rare diseases without the need to pool data from unrelated individuals. As medical-sequencing programs have developed across the United States and internationally, there is a growing body of evidence surrounding their potential benefits. Currently, WGS/WES can definitively diagnose about one-fourth of properly selected patients referred for sequencing and another one-fourth with mutations that are highly likely to be causal but that require additional evaluation (Yang 2013; de Ligt 2012; Gahl 2012; Need

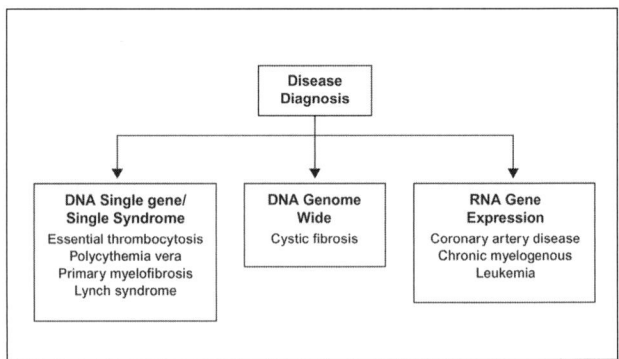

Figure 3.3. Disease diagnosis test examples.

2012). This exceeds the typical success rate with single-gene tests or gene panels currently in use, likely because, in part, children referred for WGS/WES have already undergone traditional testing that has failed; thus, the reported diagnostic yields of WGS/WES may be overestimated (Neveling 2013). Although current programs are focused on making diagnoses, this knowledge can sometimes lead to novel therapeutic approaches by applying the knowledge of other drugs and disorders that share the same genetic cause (Worthey 2011). One main limitation to the WGS/WES approach is our knowledge of how genetic variants lead to disease and our ability to filter through the thousands of potentially damaging genetic variants to those that are causal. Another limitation is the inability of WGS/WES to identify large copy number variations that underlie a significant proportion of cases and will continue to require chromosomal microarrays for diagnosis. With the cost of WGS/WES approaching $1000 U.S. dollars and with some insurers beginning to cover (Yang 2013) a portion of the sequencing costs, it is anticipated that WES will soon replace the current stepwise approach to genetic testing for undiagnosed medical conditions.

For other syndromes that can be categorized into more common, inherited diseases, next-generation sequencing has also contributed to advances. It has long been known that CF is caused by variants in the *CFTR* gene. However, many families carry "private" mutations that are not carried by other families with CF and that therefore would be missed by current genotyping panels. Although a few common variants account for two-thirds of diagnoses, the remaining cases are accounted for by almost 2000 rare genetic variants (Sosnay 2013). Therefore, *CFTR* sequencing is now the standard of care for clinical diagnoses when genotyping does not identify a mutation.

RNA Gene Expression Profiling

The information encoded by the human genome extends beyond the coding regions and includes epigenetic modifications and complex regulatory interactions between the genome and diet, medications, and the environment. Therefore, the genome also drives variation in the levels of gene expression, protein expression, and by-products of metabolic pathways. Together, these molecules are considered the "expressed genome," which is dynamic and represents the integrative effects of the genetic sequence as well as complex interactions with the environment. This integrative feature of the expressed genome makes it a potentially more powerful tool than individual genetic variants for identifying disease. For example, by studying the patterns of gene expression in peripheral blood RNA in patients with obstructive coronary artery disease (CAD) compared with those in patients without obstructive CAD, investigators were able to identify a select group of genes whose expression, in aggregate, could identify more individuals with CAD compared with traditional models (Thomas 2013; Rosenberg 2012; Wingrove 2008; Schirmer 2007). By narrowing the list of possible genes to a manageable number and using peripheral blood RNA that does not require any additional processing, these advances allow a novel, laboratory-based method to diagnose CAD. In contrast to traditional methods currently in use (e.g., exercise stress testing or imaging modalities) to diagnose CAD in patients with symptoms suggestive of CAD, an RNA-based test may prove to be a more cost-effective approach. RNA profiling has also changed the approach to diagnosis for some hematologic malignancies. The first cancer to be diagnosed by a genetic test—in this case, karyotyping for the Philadelphia chromosome—was chronic myeloid leukemia caused by a chromosomal translocation between *ABL1* on chromosome 9 and *BCR* on chromosome 22, resulting in a fusion gene, *BCR-ABL1*. Initial approaches to identifying the *BCR-ABL1* translocation relied on cytogenetics or fluorescence in situ hybridization. However, RNA profiling of peripheral blood for the RNA product of the fusion gene has allowed much greater sensitivity, which can now detect one cell positive for the translocation in 10^5–10^6 normal cells. This improvement in sensitivity ,improved detection, particularly when traditional approaches fail to identify the fusion gene because of cryptic translocations, for example. As a result, RNA profiling for *BCR-ABL1* is a "gold-standard" criterion for diagnosing chronic myeloid leukemia.

Disease Prognosis

When physicians provide a disease prognosis, they predict a patient's disease course and the likelihood they will improve. Many prognostic prediction tools are available for different diseases that consider biological and clinical factors. Currently, several DNA- and RNA-based technologies are useful for cancer prognosis. Examples are presented in Figure 3.4.

DNA Single-Gene/Single-Syndrome Tests

The *RAS* oncogene regulates cell growth and regulation and has three isoforms, of which *KRAS* is the most commonly mutated—86% of all *RAS* mutations are in *KRAS* (Arrington 2012). Although *KRAS* has been identified in lung cancer, pancreatic cancer, and thyroid cancers, it is best known for its prognostic implications in colon cancer. *KRAS* mutations occur in 30%–50% of colon cancer tumors and are associated with shorter survival and more aggressive tumors as well as a lower likelihood of responding to anti–epidermal growth factor receptor (EGFR) therapeutics (Arrington 2012). Of interest, mutations in the *BRAF* gene, a RAF kinase, in colon cancer tumors may also predict a lower likelihood of responding to anti-EGFR therapeutics (Arrington 2012), though the evidence for this is not yet sufficient to guide therapy.

Breast cancer is probably the best example of integrating testing for prognosis into clinical care pathways. For years, the estrogen receptor (ER) and progesterone receptor (PR) status of the tumor tissue (positive or negative) have been part of the standard testing used to assess prognosis as well as to select optimal therapy for invasive tumors requiring chemotherapy. The ER- and PR-positive tumors have lower relapse rates at 5 years and respond better to hormone therapeutics (e.g., tamoxifen and anastrozole). Receptor testing is not a DNA or RNA test but (like screening for Lynch syndrome) is done with immunohistochemical staining for proteins in the tissue sample. However, human epidermal growth factor receptor 2 (*HER2* or *ERBB2*), which is as prognostic as ER and PR and has now been incorporated into standard testing

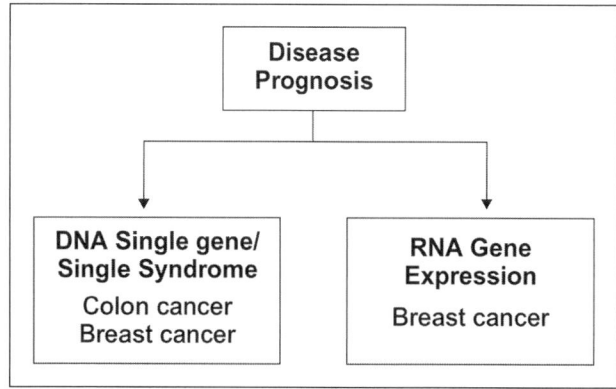

Figure 3.4. Disease prognosis test examples.

pathways on tumor tissue, can be done with either immunohistochemical staining or DNA testing (usually fluorescence in situ hybridization). Like other oncogenes, *HER2* plays an important role in cell growth and regulation, and tumors with high levels of *HER2* protein (*HER2* positive) tend to be more aggressive; however, they are much more likely respond to therapeutics that target *HER2* such as trastuzumab (which was specifically developed for this use) and lapatinib (Wolff 2013).

RNA Gene Expression Profiling
In breast cancer, 3′ gene expression profiling assays are available that can predict the likelihood of both relapse and benefit from selected therapeutics. All the tests are designed to be run on the breast cancer tumor tissue, so tumor specimens are required. The first, Oncotype DX, profiles 21 genes and is suitable for women with newly diagnosed stage 1 and 2 ER-positive/lymph node–negative tumors. The second, MammaPrint, profiles 70 genes and is suitable for women younger than 60 with invasive cancer that is lymph node negative. The third, the H/I ratio test, profiles two genes and is suitable for women with newly diagnosed breast cancer having ER-positive/lymph node–negative tumors. All three have been successfully used in clinical care, and the MammaPrint assay has received FDA clearance for this indication (Marchionni 2007). More recently, microarray profiling of invasive cancer has identified new subtypes of breast cancer that otherwise look similar, which may lead to the development of new tumor classification systems that rely on genomic characteristics rather than the current histologic system (Perou 2000).

Selecting Optimal Therapy
Clinical trials of drug efficacy are typically designed to assess the greatest benefit of a medication in the most optimal settings. The effect (both benefits and harms) is reported as a mean and standard deviation across the entire population, which does little to inform an individual's response to a given drug. In general, this means that identifying the most effective drug for a patient is empiric (i.e., trial and error). The advent of genomic tests to help identify optimal drug selection has vastly improved fields such as cancer and hepatitis C virus (HCV) therapies and shows great potential in other areas as well. In fact, in recent years, a proliferation of drug-diagnostic combinations have come onto the market.

Pharmacogenomics
The right dose of the right drug to the right individual is the often-stated goal of pharmacogenomics. The ability to tailor pharmacotherapy to an individual's genomic profile is one of the earliest and most tangible benefits of genomic medicine. To date, most pharmacogenomic indications are based on genetic variants affecting the pharmacokinetics and/or pharmacodynamics for a particular drug. However, variants outside pharmacology can also have profound effects on predicting drug response. For example, the FDA now requires prospective *HLA* (human leukocyte antigen) genotyping before prescribing abacavir because it eliminates hypersensitivity reactions to this drug (Mallal 2008). With recent advances in the ability to perform genome-wide scans of genetic variation, it is time to move beyond pharmacology to the variants underlying disease pathways that determine response to drug therapy. For example, genetic variants in *PEAR1* affect native platelet function (Johnson 2010), platelet function response to aspirin (Lewis 2013), and clinical outcomes in patients taking aspirin (Lewis 2013), yet this does not affect aspirin's pharmacology or ability to inhibit platelet COX-1 (cyclooxygenase 1). As clinicians move away from the one-gene, one-disease paradigm for understanding drug response, a shift toward a systems biology approach is now being embraced, as demonstrated by characterizing the expressed genome (i.e., RNA, protein, or metabolites) using unbiased screens of gene expression and metabolism, novel "signatures" of transcripts or metabolites that correlate with the response to antiplatelet therapy (Voora 2013; Yerges-Armstrong 2013; Seitz 2004) and statins (Kaddurah-Daouk 2011, 2010), for example. Although the remainder of this textbook will focus on well-described and clinically actionable examples of pharmacogenomics in clinical practice, this chapter highlights some additional uses of genomics in medicine to tailor drug therapy with the goal of improving health outcomes.

DNA Single-Gene/Single-Syndrome Testing
As our ability to understand disease mechanisms improves, so too does our ability to define distinct subgroups within a disease for targeted therapies. For example, an uncommon mutation (G551D, found in 4% of CF cases) in the *CFTR* gene leads to impaired chloride channel function. Ivacaftor is a novel small molecule that improves the function of the mutated *CFTR* channel as well as the laboratory and clinical outcomes in patients with CF having the G551D mutation (Ramsey 2011). Consequently, ivacaftor is an FDA-approved medication for patients with CF caused by the G551D mutation. Taking this concept further, novel pharmacologic approaches for overcoming the effects of the causal genetic variant, such as the viral delivery of the missing gene or antisense oligonucleotides to promote exon skipping, are

being tested for certain disorders such as Duchenne muscular dystrophy (Fairclough 2013). As previously mentioned, *HER2* testing in breast cancer can guide the use of anti-HER2 therapeutics. Those who are *HER2* positive have a worse prognosis but are more likely to respond to trastuzumab and lapatinib than to hormone therapies (Wolff 2013). Another example of single-gene testing for selecting optimal therapy is in adenocarcinoma lung carcinoma. More than 60% of these tumors express EGFR, and the 15%–20% of patients with DNA mutations of the receptor respond well to EGFR-targeting tyrosine kinase inhibitors (Siegelin 2014). This is an important opportunity for those with advanced-stage lung cancer because these carcinomas traditionally have poor outcomes.

RNA Gene Expression Profiling
As discussed earlier, the expressed genome is a powerful tool for understanding disease heterogeneity and improving disease diagnosis. Gene expression profiling can also be useful for guiding treatment decisions. For example, the decision to undergo adjuvant chemotherapy in women with invasive breast cancer is individualized depending on patient and tumor characteristics. As described earlier, the Oncotype DX gene expression test determines a recurrence score (RS) that helps identify women with a worse prognosis. In addition, the RS can be used to assist in the medical decision-making process between women and their oncologists. A high RS (greater than 31) is used as an additional piece of data that, combined with patient and tumor characteristics, can justify the use of adjuvant chemotherapy in certain women (Nguyen 2013).

Disease Monitoring
The same technologies that have revolutionized our ability to diagnose disease, predict disease risk, and tailor therapies using genomics have also improved our ability to monitor diseases. Much as the hemoglobin A1C value is used to monitor the quality of therapy in patients with diabetes, DNA- and RNA-based tests can be used to ensure therapeutic success (or failure). Consequently, patients can expect improved clinical outcomes by avoiding additional and potentially toxic therapeutic interventions or by identifying recurrent disease early before reaching an advanced stage.

DNA-Based Tests
Next-generation sequencing of patient tumors in parallel with adjacent noncancerous tissue is now a common research and emerging clinical tool in cancer genomics. This approach has revolutionized the ability to identify new somatic variations that underlie cancer biology.

At the same time, circulating tumor DNA detected in plasma from patients with solid tumor malignancies is a promising biomarker of tumor dynamics. Compared with currently available biomarkers, personalized assays to detect circulating tumor DNA are more sensitive and are strong predictors of survival (Dawson 2013). In this work by Dawson et al., many women with otherwise undetectable disease by traditional biomarkers (e.g., CA 15-3 or circulating tumor cells) are instead readily identified by circulating tumor DNA. In some women, serial monitoring reveals the emergence of new mutations not present in the primary tumor, thus enabling the ability to monitor for the emergence of resistant tumors without the need for repeated, invasive biopsies. Such a strategy may allow earlier detection of emerging resistance or tumor subclones that either were not present in the primary tumor or were not accessible by biopsy. However, use of this technology has not yet been shown to improve patient outcomes.

RNA Gene Expression Profiling
RNA-based diagnostics have long been used to monitor viral loads during treatment for HCV and/or HIV infection. Although these types of tests may not fall into the traditional definition of genomic medicine, they provide the foundation for using RNA-based tests to monitor diseases that are often detected through invasive methods or after irreversible damage has occurred. In the field of solid organ transplantation, for example, rejection of the transplanted organ is a chronic problem requiring long-term immunosuppression. The risk of rejection is highest in the early period immediately after transplantation and diminishes with time, but it never goes away. Rejection can also lead to permanent dysfunction in the transplanted organ. Therefore, patients are routinely monitored for rejection. For cardiac transplants, surveillance is performed by invasive biopsies of the endocardial lining of the heart. In renal transplant recipients, patients are monitored by blood/urine studies that show graft dysfunction, which may not reverse with treatment. In both settings, RNA profiling of peripheral blood (Horwitz 2004) (in cardiac transplants) or urine (Suthanthiran 2013) (in kidney transplants) has identified a key set of transcripts with sufficient sensitivity and specificity to detect rejection of the transplanted graft. For the urine RNA profile, the changes in urine preceded any clinical evidence of rejection by several weeks, opening the possibility for early treatment and preventing irreversible graft dysfunction. For cardiac transplantation, a randomized clinical trial showed that a strategy of monitoring for rejection by peripheral blood RNA was equivalent to one of routine biopsies, despite using many fewer invasive biopsies during follow-up (Pham 2010). In

contrast, the use of RNA profiling has not yet been tested to assess for a difference in transplant outcomes compared with usual care.

Near-Future Directions (1-3 years)

The explosion of genomic technologies will continue expanding the genomic medicine space. Each year, new uses for genomic testing quickly move from proof-of-principle research studies to commercially available tests. Pharmacogenomics is not anticipated to be an exception to this rapidly moving field. One area that holds major promise for human health, disease, and drug response is the microbiome. The community of organisms that lives on and within human beings has been finely tuned for thousands of generations to exist in a mutualistic relationship with us. The microbiome can rapidly adapt to changes in our diet (David 2014) and in response to a wide array of medications. For digoxin, which has long been known to undergo metabolism by the enteric microbiome, recent findings show that digoxin exposure can induce the proliferation of a specific bacterial species capable of inactivating this drug (Haiser 2013). With the ability to use PCR (polymerase chain reaction) to identify specific bacterial sequences in fecal samples, a novel fecal biomarker can be envisioned for an individual's ability to metabolize digoxin. Consequently, starting doses of this drug, which has a narrow therapeutic index could be individualized. Similarly, interactions between the enteric microbiome and statins have been suggested (Kaddurah-Daouk 2011). Therefore, researchers are only beginning to realize the powerful role of the microbiome and its impact on drug responses. Ultimately, however, showing that knowledge about the microbiome affects health outcomes will be required before widespread adoption can occur.

Our ability to generate genomic data has far outpaced our ability to fully understand the implications of the data. Just as radiologists had to grapple with incidental findings (i.e., abnormalities in areas of the body captured on the image but not related to the indication for testing) with advancing radiologic imaging, the modern genomic medicine physician will need to learn to accept incidental findings as an inevitable consequence of more advanced genomic scans. For example, a genome-wide screen for pharmacogenetic variants in cytochrome P450 (CYP) genes may inadvertently identify

an *APOE* variant that increases the risk of Alzheimer disease. A genome-wide analysis of peripheral blood gene expression for CAD may uncover a *BCR-ABL* fusion gene expression pattern, raising the question of undiagnosed chronic myeloid leukemia (Biernaux 1995). Are clinicians obligated to return such results to the patient? Does the patient want such results returned? If so, in what format? And under what circumstances? What type of follow-up is indicated? At whose expense? Thankfully, professional societies such as the American College of Medical Genetics and Genomics (ACME) have issued a document to guide the return of certain genetic findings known to cause disease back to patients (Green 2013). As evidenced by a recent update to this document ("Incidental Findings in Clinical Genomics: A Clarification"; ACME 2013), this is a rapidly changing landscape as more patients and their providers use genome-wide scans of DNA and/or RNA and demand more information from their genomes.

Finally, the ability to put genomic data into the hands of the providers caring for patients will ultimately be required for the full benefits of genomic medicine to be realized. Although medical education is increasingly including genomics in the educational pipeline (Demmer 2014), it is unlikely that an even a trained provider will be able to efficiently incorporate genomic data into clinical care without displacing other, equally important, elements of care. Consequently, we envision a medical environment with decision support tools embedded within electronic medical records that will not only house the patient's genomic data but also assist the provider in analyzing and interpreting the data (Hartzler 2013; Ury 2013). Of course, the provider will always be responsible for making the best decision, in cooperation with the patient. However, these tools will be specifically designed to support the provider in that decision-making process.

References

American College of Medical Genetics and Genomics (ACME). Incidental findings in clinical genomics: a clarification. Genet Med 2013;15:664-6.

American Congress of Obstetricians and Gynecologists (ACOG). ACOG Committee Opinion No. 486. Update on carrier screening for cystic fibrosis. Obstet Gynecol 2011;117:1028-31.

Arrington AK, Heinrich EL, Lee W, et al. Prognostic and predictive roles of KRAS mutation in colorectal cancer. Int J Mol Sci 2012;13:12153-68.

Bacon BR, Adams PC, Kowdley KV, et al.; American Association for the Study of Liver, Diseases. Diagnosis and management of hemochromatosis: 2011 practice guideline by the American Association for the Study of Liver Diseases. Hepatology 2011;54:328-43.

Coronary Artery Disease

Coronary artery disease is one of the most prevalent medical conditions in the United States and is the leading cause of death (24%), just slightly edging out all cancers combined (23.3%) (www.cdc.gov/nchs/nvss/mortality/lcwk2.htm). For risk stratification, CAD has well-characterized, strong environmental (e.g., diet and exercise) and hereditary (e.g., male first-degree relative with CAD at younger than 55 years of age or female at younger than 65) contributions; mildly contributory but very common DNA genome-wide risk markers (e.g., genetic variants on 9p21); and some rare but strongly contributory DNA single gene/single syndromes (e.g., familial hereditary hypercholesterolemia). For diagnosis, there is now novel RNA gene expression profiling as described in the text, and for therapy, there are several well-established and emerging pharmacogenomic tests. Examples include *CYP2C19*2* to guide $P2Y_{12}$ inhibitors (e.g., clopidogrel vs. prasugrel or ticagrelor) and *SLCO1B1*5* to guide the selection and dosing of statins.

Breast Cancer

Breast cancer is a complex disease, with multiple genetic and some environmental contributions, that beautifully exemplifies the roles of family history and DNA- and RNA-based testing across the spectrum of risk stratification to therapy. Risk stratification for breast cancer is divided into three levels: average or population-based risk (risk is the same as in the general population), familial risk (risk is definitely higher than in the general population, but there is no concern about a hereditary cancer syndrome), and hereditary risk (risk is caused by a hereditary cancer syndrome such as HBOC syndrome). Differentiating these risk levels is heavily driven by FHH, though familial risk levels may also partly account for environmental factors such as chest wall radiation in those 10–30 years of age, the number of years between the onset and end of menses, and hormone exposure. In addition, the combinations of factors that differentiate between familial and hereditary risk are very complex. For example, women with Ashkenazi Jewish heritage are recommended to undergo genetic testing as part of a hereditary risk assessment with a single first-degree relative with breast cancer, whereas women without Ashkenazi Jewish risk of hereditary cancer would not be considered unless the first-degree relative had developed breast cancer when younger than 40, developed bilateral breast cancer, or also developed ovarian cancer. The NCCN has published extensive guidelines on how to differentiate risk levels and how to manage risk in those given a diagnosis of a hereditary cancer syndrome (www.nccn.org). As discussed in the text, DNA single-gene/single-syndrome testing for *BRCA1* and *BRCA2* can identify the presence of a hereditary cancer syndrome, but that diagnosis only further characterizes the individual's risk level. Currently, diagnosis of breast cancer is still made using traditional pathology evaluations of tumor cell shape and structure; however, data are accumulating on how to molecularly distinguish breast cancer types so that, sometime in the future, diagnosis can be defined by molecular subtyping. Of note, the diagnosis of HBOC syndrome has implications for family members as well. They now are at risk of HBOC syndrome, and testing of the closest relatives is recommended to determine whether they carry the gene. If so, their closest relatives should be tested and so on. Once breast cancer has occurred, disease prognosis can be assessed with *HER2* DNA testing and RNA gene expression profiling with Oncotype DX, MammaPrint, or the H/I ratio, which also provide information about selecting optimal chemotherapies for those with invasive cancers (see text for details).

Lynch Syndrome

As a hereditary cancer syndrome, Lynch syndrome is a very important influencer of cancer risk; therefore, it is critical to identify when cancers are caused by this syndrome as opposed to non-hereditary causes. Risks of having the following cancers are significantly increased with Lynch syndrome: gastrointestinal tract (stomach, small bowel, colon, and rectum), especially for colon, uterine cancer, ovarian cancer, renal cell carcinoma, biliary ductal carcinomas, hepatocellular carcinoma, and skin cancers. The first step in risk stratification is family history—this is the initial tip that something more serious than the usual family cancers may be occurring in a family. The NCCN provides specific guidance on which combinations of FHHs are of concern, but in general, if several family members on the same side of the family have one or more of the listed cancers, red flags should be raised (www.nccn.org). In addition, patients with colon cancer are now routinely undergoing tumor testing to identify suspected cases of Lynch syndrome because FHH alone is able to detect only about half the cases (EGAPP 2009), as described in the text. Once suspicion is high enough to warrant genetic testing, single-gene/single-syndrome DNA testing (described in the text) is initiated to confirm the diagnosis. As with HBOC syndrome, diagnosis of Lynch syndrome is not a cancer diagnosis, but a diagnosis that significantly raises the individual's level of risk of the listed cancers. Management at this point focuses on prevention and early identification. For example, colon cancer screening with colonoscopies is performed every 1–2 years starting at age 20, endoscopy with small bowel imaging is performed every 2–3 years starting at age 30, and women may undergo prophylactic hysterectomy and bilateral salpingo-oophorectomy (Grover 2010). Given that both HBOC syndrome and Lynch syndrome are hereditary, diagnosis with these syndromes affects relatives, and testing is recommended, starting with the closest relatives, as with HBOC syndrome. In addition, as with breast cancer, cancer diagnosis is still based on the pathologic examination of tissue, but in the near future, it will be based more on molecular testing. And finally, after diagnosis is made, prognosis and therapeutic options are now guided by DNA testing of the *KRAS* gene, as discussed in the text.

Bang H, Edwards AM, Bomback AS, et al. Development and validation of a patient self-assessment score for diabetes risk. Ann Intern Med 2009;151:775-83.

Berry DA Jr, Iversen ES, Gudbjartsson DF, et al. BRCAPRO validation, sensitivity of genetic testing of BRCA1/BRCA2, and prevalence of other breast cancer susceptibility genes. J Clin Oncol 2002;20:2701-12.

Biernaux C, Loos M, Sels A, et al. Detection of major bcr-abl gene expression at a very low level in blood cells of some healthy individuals. Blood 1995;86:3118-22.

Bloss CS, Schork NJ, Topol EJ. Effect of direct-to-consumer genomewide profiling to assess disease risk. N Engl J Med 2011;364:524-34.

Choi M, Scholl UI, Ji W, et al. Genetic diagnosis by whole exome capture and massively parallel DNA sequencing. Proc Natl Acad Sci USA 2009;106:19096-101.

David LA, Maurice CF, Carmody RN, et al. Diet rapidly and reproducibly alters the human gut microbiome. Nature 2014;505:559-63.

Dawson SJ, Tsui DWY, Murtaza M, et al. Analysis of circulating tumor DNA to monitor metastatic breast cancer. N Engl J Med 2013;368:1199-209.

de Ligt J, Willemsen MH, van Bon BW, et al. Diagnostic exome sequencing in persons with severe intellectual disability. N Engl J Med 2012;367:1921-9.

Demmer LA, Waggoner DJ. Professional medical education and genomics. Annu Rev Genomics Hum Genet 2014 (doi: 10.1146/annurev-genom-090413-025522).

Dewey FE, Grove ME, Pan C, et al. Clinical interpretation and implications of whole-genome sequencing. JAMA 2014;311:1035-45.

European Association for Study of Liver (EASL). EASL clinical practice guidelines: Wilson's disease. J Hepatol 2012;56:671-85.

Evaluation of Genomic Applications in Practice and Prevention (EGAPP) Working Group. Recommendations from the EGAPP Working Group: genetic testing strategies in newly diagnosed individuals with colorectal cancer aimed at reducing morbidity and mortality from Lynch syndrome in relatives. Genet Med 2009;11:35-41.

Fairclough RJ, Wood MJ, Davies KE. Therapy for Duchenne muscular dystrophy: renewed optimism from genetic approaches. Nat Rev Genet 2013;14:373-8.

Flossmann E, Schulz UG, Rothwell PM. Systematic review of methods and results of studies of the genetic epidemiology of ischemic stroke. Stroke 2004;35:212-27.

Francis Collins on Genetic Tests. 2010. Available at www.genomeweb.com/blog/francis-collins-genetic-tests. Accessed June 19, 2014.

Gahl WA, Markello TC, Toro C, et al. The National Institutes of Health Undiagnosed Diseases Program: insights into rare diseases. Genet Med 2012;14:51-9.

Gail MH, Brinton LA, Byar DP, et al. Projecting individualized probabilities of developing breast cancer for white females who are being examined annually. Journal of the National Cancer Institute 1989;81:1879-86. Online risk calculator available at www.cancer.gov/bcrisktool/.

Green RC, Berg JS, Grody WW, et al. ACMG recommendations for reporting of incidental findings in clinical exome and genome sequencing. Genet Med 2013;15:565-74.

Green RC, Roberts JS, Cupples LA, et al. Disclosure of APOE genotype for risk of Alzheimer's disease. N Engl J Med 2009;361:245-54.

Grover S, Syngal S. Risk assessment, genetic testing, and management of Lynch syndrome. J Natl Compr Canc Netw 2010;8:98-105.

Haiser HJ, Gootenberg DB, Chatman K, et al. Predicting and manipulating cardiac drug inactivation by the human gut bacterium *Eggerthella lenta*. Science 2013;341:295-8.

Hampel H, Sweet K, Westman JA, et al. Referral for cancer genetics consultation: a review and compilation of risk assessment criteria. J Med Genet 2004;41:81-91.

Hariri S, Yoon PW, Qureshi N, et al. Family history of type 2 diabetes: a population-based screening tool for prevention? Genet Med 2006;8:102-8.

Hartzler A, McCarty CA, Rasmussen LV, et al. Stakeholder engagement: a key component of integrating genomic information into electronic health records. Genet Med 2013;15:792-801.

Hogarth DK, Rachelefsky G. Screening and familial testing of patients for alpha 1-antitrypsin deficiency. Chest 2008;133:981-8.

Holman DM, Berkowitz Z, Guy GP Jr, et al. The association between demographic and behavioral characteristics and sunburn among U.S. adults – National Health Interview Survey, 2010. Prev Med 2014 (doi: 10.1016/j.ypmed.2014.02.018).

Horwitz PA, Tsai EJ, Putt ME, et al. Detection of cardiac allograft rejection and response to immunosuppressive therapy with peripheral blood gene expression. Circulation 2004;110:3815-21.

Hunt SC, Gwinn M, Adams TD. Family history assessment: strategies for prevention of cardiovascular disease. Am J Prev Med 2003;24:136-42.

Johnson AD, Yanek LR, Chen MH, et al. Genome-wide meta-analyses identifies seven loci associated with platelet aggregation in response to agonists. Nat Genet 2010;42:608-13.

Kaddurah-Daouk R, Baillie RA, Zhu H, et al. Lipidomic analysis of variation in response to simvastatin in the Cholesterol and Pharmacogenetics Study. Metabolomics 2010;6:191-201.

Kaddurah-Daouk R, Baillie RA, Zhu H, et al. Enteric microbiome metabolites correlate with response to simvastatin treatment. PLoS One 2011;6:e25482.

Lewis JP, Ryan K, O'Connell JR, et al. Genetic variation in PEAR1 is associated with platelet aggregation and cardiovascular outcomes. Circ Cardiovasc Genet 2013 (doi: 10.1161/circgenetics.111.964627).

Mallal S, Phillips E, Carosi G, et al.; the PREDICT-1 Study Team. HLA-B*5701 screening for hypersensitivity to abacavir. N Engl J Med 2008;358:568-79.

Marchionni L, Wilson RF, Marinopoulos SS, et al. Impact of gene expression profiling tests on breast cancer outcomes. Evid Rep Technol Assess (Full Rep) 2007 Dec; (160):1-105.

Meigs JB, Shrader P, Sullivan LM, et al. Genotype score in addition to common risk factors for prediction of type 2 diabetes. N Engl J Med 2008;359:2208-19.

National Comprehensive Cancer Network (NCCN). 2011. Family History Risk Markers for Hereditary Cancer Syndrome. Available at www.nccn.org/index.asp. Accessed June 19, 2014.

Need AC, Shashi V, Hitomi Y, et al. Clinical application of exome sequencing in undiagnosed genetic conditions. J Med Genet 2012;49:353-61.

Neveling K, Feenstra I, Gilissen C, et al. A post-hoc comparison of the utility of sanger sequencing and exome sequencing for the diagnosis of heterogeneous diseases. Hum Mutat 2013;34:1721-6.

Nguyen MT, Stessin A, Nagar H, et al. Impact of oncotype DX recurrence score in the management of breast cancer cases. Clin Breast Cancer 2013 (doi: 10.1016/j.clbc.2013.12.002).

Perou CM, Sorlie T, Eisen MB, et al. Molecular portraits of human breast tumours. Nature 2000;406:747-52.

Pham MX, Teuteberg JJ, Kfoury AG, et al. Gene-expression profiling for rejection surveillance after cardiac transplantation. N Engl J Med 2010;362:1890-900.

Pyeritz RE. The family history: the first genetic test, and still useful after all those years? Genet Med 2012;14:3-9.

Ramsey BW, Davies J, McElvaney NG, et al. A CFTR potentiator in patients with cystic fibrosis and the G551D mutation. N Engl J Med 2011;365:1663-72.

Rosenberg S, Elashoff MR, Lieu HD, et al. Whole blood gene expression testing for coronary artery disease in nondiabetic patients: major adverse cardiovascular events and interventions in the PREDICT trial. J Cardiovasc Transl Res 2012;5:366-74.

Schirmer M, Rosenberger A, Klein K, et al. Sex-dependent genetic markers of CYP3A4 expression and activity in human liver microsomes. Pharmacogenomics 2007;8:443-53.

Seitz H, Royo H, Bortolin ML, et al. A large imprinted microRNA gene cluster at the mouse Dlk1-Gtl2 domain. Genome Res 2004;14:1741-8.

Siegelin MD, Borczuk AC. Epidermal growth factor receptor mutations in lung adenocarcinoma. Lab Invest 2014;94:129-37.

Sosnay PR, Siklosi KR, Van Goor F, et al. Defining the disease liability of variants in the cystic fibrosis transmembrane conductance regulator gene. Nat Genet 2013;45:1160-7.

Suthanthiran M, Schwartz JE, Ding R, et al. Urinary-cell mRNA profile and acute cellular rejection in kidney allografts. N Engl J Med 2013;369:20-31.

Teng LS, Zheng Y, Wang HH. BRCA1/2 associated hereditary breast cancer. J Zhejiang Univ Sci B 2008;9:85-9.

Thomas GS, Voros S, McPherson JA, et al. A blood based gene expression test for obstructive coronary artery disease tested in symptomatic non-diabetic patients referred for myocardial perfusion imaging: the COMPASS study. Circ Cardiovasc Genet 2013 (doi: 10.1161/circgenetics.112.964015).

Ury AG. Storing and interpreting genomic information in widely deployed electronic health record systems. Genet Med 2013;15:779-85.

Voora D, Cyr D, Lucas J, et al. Aspirin exposure reveals novel genes associated with platelet function and cardiovascular events. J Am Coll Cardiol 2013;62:1267-76.

Wheeler DA, Srinivasan M, Egholm M, et al. The complete genome of an individual by massively parallel DNA sequencing. Nature 2008;452:872-6.

Wingrove JA, Daniels SE, Sehnert AJ, et al. Correlation of peripheral-blood gene expression with the extent of coronary artery stenosis. Circ Cardiovasc Genet 2008;1:31-8.

Wolff AC, Hammond ME, Hicks DG, et al. Recommendations for human epidermal growth factor receptor 2 testing in breast cancer: American Society of Clinical Oncology/College of American Pathologists clinical practice guideline update. J Clin Oncol 2013;31:3997-4013.

Woolf SH. Practice guidelines: a new reality in medicine. III. Impact on patient care. Arch Intern Med 1993;153:2646-55.

Worthey EA, Mayer AN, Syverson GD, et al. Making a definitive diagnosis: successful clinical application of whole exome sequencing in a child with intractable inflammatory bowel disease. Genet Med 2011;13:255-62.

Yang Y, Muzny DM, Reid JG, et al. Clinical whole-exome sequencing for the diagnosis of mendelian disorders. N Engl J Med 2013;369:1502-11.

Yerges-Armstrong LM, Ellero-Simatos S, Georgiades A, et al. Purine pathway implicated in mechanism of resistance to aspirin therapy: pharmacometabolomics-informed pharmacogenomics. Clin Pharmacol Ther 2013 (doi: 10.1038/clpt.2013.119).

Chapter 4

IMPLEMENTING PHARMACOGENETICS IN THE CLINICAL SETTING AND COMPETENCIES FOR HEALTH CARE PROFESSIONALS

KRISTIN WEITZEL, PHARM.D.

Learning Objectives
1. Demonstrate the benefits of applying pharmacogenetic principles to patient care in clinical practice.
2. Develop practice-based strategies for implementing pharmacogenetics in the clinical setting.
3. Develop strategies to overcome real and perceived barriers to the clinical implementation of pharmacogenetics.
4. Evaluate existing pharmacogenetic competencies for health care professionals and their application to successful clinical implementation.
5. Design interdisciplinary educational strategies to support the clinical implementation of pharmacogenetics among diverse groups of health care professionals.

Keywords: Pharmacogenetics, pharmacogenomics, implementation, clinical, competencies, education

Abbreviations in This Chapter
CDS	Clinical decision support
CPIC	Clinical Pharmacogenetics Implementation Consortium
EHR	Electronic health record
TPMT	Thiopurine methyltransferase

Abstract
Recent scientific and research achievements increasingly support the clinical implementation of pharmacogenetic testing and genotype-guided drug therapy optimization. Clinical practice guidelines are available to support practice-based recommendations for many drug-gene pairs, and regulatory bodies increasingly recognize the importance of these data in prescribing information. However, the widespread adoption of clinical pharmacogenetics is not yet a reality because of the unique challenges of developing financially sustainable clinical models for pharmacogenetics implementation. Pharmacists are well poised to take on leadership roles and responsibilities in key areas supporting implementation, including stakeholder engagement, analysis, and interpretation of pharmacogenetic evidence to select optimal drug-gene pairs, pharmacogenetic test implementation and interpretation, clinical decision support development, reimbursement and cost-related issues, and health care professional education. Large-scale development of readily available educational and practice-based resources for health care professionals is needed to support the incorporation of pharmacogenetic scientific concepts into routine clinical practice.

Introduction
Significant professional and scientific advances have been made in recent years that support the incorporation of pharmacogenetics into routine clinical practice. Regulatory bodies are also

increasingly focusing on the clinical implications of these advances, with incorporation of pharmacogenetic data into the prescribing information for several medications. Clinical guidelines are available for many drug-gene pairs to advise clinicians on using pharmacogenetic test results to optimize patient care (Caudle 2014; Relling 2011). More health systems and pharmacies than ever before are exploring and developing innovative practice models and support for clinical pharmacogenetics services (Bielinksi 2014; Ferreri 2014; Goldspiel 2014; Hoffman 2014; O'Donnell 2014; Weitzel 2014; Gharani 2013; Pulley 2012; Crews 2011).

Despite these developments, however, the widespread adoption of clinical pharmacogenetics has lagged behind scientific advancement. This is partly because of the unique challenges that exist in creating financially sustainable clinical, administrative, and laboratory models for pharmacogenetics implementation. However, as medication experts who have historically led multidisciplinary initiatives that promote the optimal use of drug therapy to improve patient outcomes, pharmacists are well poised to champion these services within the health care system. The importance of pharmacist leadership and responsibility for clinical pharmacogenetics program development is increasingly being recognized (Haga 2014; Williams 2014; Crews 2011; McKinnon 2007). Educating pharmacists and other health care providers is an essential component and an area of need to support the adoption of clinical pharmacogenetics services.

This chapter presents a stepwise process and considerations for implementing common models of clinical pharmacogenetics services, with a focus on the health-systems pharmacy practice setting, and summarizes strategies for providing the necessary education and training for pharmacists and other health care providers.

Pharmacogenetics in the Clinical Setting

In many ways, clinical pharmacogenetics parallels the therapeutic drug monitoring services that use pharmacokinetic and pharmacodynamic data in the clinical setting (Crews 2011; McKinnon 2007). However, whereas therapeutic drug monitoring allows health care providers to optimize drug dosing according to the measurements of drug concentration, clinical pharmacogenetics enables clinicians to select the optimal medication and dose according to a patient's pharmacogenetic laboratory data. This process has the potential to improve patient outcomes, decrease risks of adverse events, and promote the use of targeted cost-effective therapy for patient care (Johnson 2013; Crews 2011; McKinnon 2007). The role of pharmacogenetics in drug safety, efficacy, and pharmacokinetics has been recognized by the U.S. Food and Drug Administration (FDA) through the inclusion of pharmacogenetic information in FDA-approved labels for more than 110 drugs. A listing of specific drugs can be found at the FDA Web site www.fda.gov/Drugs/ScienceResearch/ResearchAreas/Pharmacogenetics/ucm083378.htm and in the drug label sections of the Pharmacogenomics Knowledgebase (PharmGKB) Web site (www.pharmgkb.org).

Because of this potential to optimize patient care and the desire to address the unmet need for using these data to improve the safety and efficacy of targeted drug-gene pairs, several institutions have established best practices in implementing clinical pharmacogenetics services (Bielinksi 2014; Goldspiel 2014; Hoffman 2014; O'Donnell 2014; Weitzel 2014; Pulley 2012; Crews 2011; Gharani 2011). Although such programs vary in many aspects, including testing approach (e.g., preemptive or reactive), reimbursement strategies, and laboratory processing (e.g., internal or external commercial laboratory), many of these programs have similar elements (Box 4.1).

At the same time, significant and unique barriers exist to the widespread adoption of clinical pharmacogenetic testing (Box 4.2). Diverse strategies have been employed by successful programs to overcome these barriers, including proactive and early stakeholder engagement, provider and staff education, enlistment of multidisciplinary advocates for program development, and comprehensive planning to address logistic issues (Bielinksi 2014; Goldspiel 2014; Hoffman 2014; O'Donnell 2014; Weitzel 2014; Gharani 2013; Pulley 2012; Crews 2011). Representative strategies are described in the stepwise processes presented in this chapter.

Implementing Clinical Pharmacogenetics in the Institution and Developing a Pharmacogenetics Service

Although the scope and focus of clinical pharmacogenetics services vary among institutions, almost all such services require similar initial steps to establish the institutional framework and infrastructure for support. This process is complex and can present challenges related to working in an existing political or ethical framework, educating a diverse range of clinical and administrative groups, and developing the operational and physical infrastructure needed to support pharmacogenetic testing. However, once this infrastructure is in place, the way is paved for the future development of additional pharmacogenetic testing, and the long-term sustainability of a new program is promoted (McKinnon 2007).

Pre-implementation Planning

Planning, development, and outreach before launching a clinical pharmacogenetics implementation are essential. Pre-implementation planning may include developing clinical and administrative infrastructure, integrating evidence analysis into the medication-use system, and identifying the drug-gene pairs that will be targeted with each implementation.

Box 4.1. Representative Characteristics of Clinical Pharmacogenetics Services

Characteristics of Targeted Drug-Gene Pair(s)
- Clinical and economic consequences of inappropriate drug selection or dosing are significant and/or current strategies to optimize drug therapy or response are limited
- Evidence exists for a clinically relevant drug-gene effect
- An evidence-based alternative therapy or dosing strategy is available
- Pharmacogenetic information is provided in FDA-approved drug label
- Pharmacogenetic test exists that is likely to be reimbursed clinically according to evidence of clinical utility and FDA-approved labeling supportive of genotype-guided therapy

Institutional Oversight of Pharmacogenetic Evidence Analysis and Application to Patient Care
- Multidisciplinary evidence review body to oversee evidence interpretation and application to clinical implementation, with oversight by the Pharmacy and Therapeutics Committee in many cases
- Multidisciplinary body to oversee ethical, research, and patient privacy considerations

Provision of Clinical Support for Health Care Providers
- Integration of pharmacogenetic test ordering and documentation in the EHR
- Development of CDS to assist providers in pharmacogenetic test ordering and interpretation
- Assignment of appropriate clinical pharmacist support

Standardized Laboratory Ordering and Interpretation Procedures
- Standardized approach to laboratory processing and entering of pharmacogenetic tests in the EHR
- Adoption of evidence-based standardized clinical interpretation of pharmacogenetic data

Education of Pharmacists and Other Health Care Providers
- Development of formal and targeted educational initiatives to support specific implementations
- Development of easily accessible patient care educational materials

Quality Improvement and Economic Evaluations
- Use of continuous quality improvement methods to monitor implementation and uptake of services
- Evaluation of billing and reimbursement data to guide economic analyses

---

CDS = clinical decision support; EHR = electronic health record.

Adapted from the following: Weitzel KW, Elsey AR, Langaee TY, et al. Clinical pharmacogenetics implementation: approaches, successes, challenges. Am J Med Genet C Semin Med Genet 2014;166:56-67; Bielinksi SJ, Olson JE, Jyotishman P, et al. Preemptive genotyping for personalized medicine: design of the right drug, right dose, right time—using genomic data to individualize treatment protocol. Mayo Clin Proc 2014;89:25-33; Crews KR, Cross SJ, McCormick JN, et al. Development and implementation of a pharmacist-managed clinical pharmacogenetics service. Am J Health Syst Pharm 2011;68:143-50; Gharani N, Keller MA, Stack CB, et al. The Coriell personalized medicine collaborative pharmacogenomics appraisal, evidence scoring and interpretation system. Genome Med 2013;5:93; Pulley JM, Denny JC, Peterson JF, et al. Operational implementation of prospective genotyping for personalized medicine: the design of the Vanderbilt PREDICT project 2012;92:87-95; O'Donnell PH, Danahey K, Jacobs M, et al. Adoption of a clinical pharmacogenomics implementation program during outpatient care—initial results of the University of Chicago "1,200 Patients Project." Am J Med Genet C Semin Med Genet 2014;166:68-75; Goldspiel BR, Flegel WA, DiPatrizio G, et al. Integrating pharmacogenomics information and clinical decision support into the electronic health record. J Am Med Inform Assoc 2014;21:522-8; and Hoffman JM, Haidar CE, Wilkinson MR, et al. PG4KDS: a model for clinical implementation of pre-emptive pharmacogenetics. Am J Med Genet C Semin Med Genet 2014;166C:45-55.

Stakeholder Engagement

On a global level, clinical pharmacogenetic stakeholders include patients and their families, clinicians, pathology providers, insurers, health information technology groups, regulatory agencies, and ethicists (Patel 2014; Hartzler 2013). More applicable to clinical pharmacists, though, are stakeholders within the institution who have a significant influence on the success or failure of a new service. On this level, pharmacists should engage often and early with key groups in their organization that will be integral to launching a pharmacogenetics service (Table 4.1) (Hoffman 2014; Konstantinos 2014; Hartzler 2013; Pulley 2012; Crews 2011; McKinnon 2007). Many successful programs to date have consulted and engaged senior leadership early in the planning process to incorporate pharmacogenomic and genomic medicine initiatives into the broader strategic plan for the organization. On an organizational level, it is also important to consider areas of focus and expertise (e.g., oncology, pediatrics) for the organization as a whole.

Pharmacists should work to identify a "champion" in each group (e.g., pathology, patient care area) who can serve as an advocate and a point person for the new pharmacogenetics service. It is also important to be aware of and actively address stakeholders' concerns, which may vary among different groups. Differing approaches and educational strategies will be needed to engage and elicit the support of the many individuals needed to facilitate the adoption of a new service (McKinnon 2007).

Box 4.2. Barriers to the Clinical Implementation of Pharmacogenetics

Knowledge Barriers
- Limited awareness and/or understanding of pharmacogenetic data
- Uncertainty or lack of confidence in ability to interpret and apply pharmacogenetic test results to patient care

Logistic Barriers
- Concerns about lack of reimbursement for pharmacogenetic testing services
- Uncertainty about ordering and processing pharmacogenetic laboratory tests
- Turnaround time for pharmacogenetic testing
- Lack of research or clinical funding
- Lack of standardized EHR or available software to integrate data into the EHR
- Limited ability to develop and implement CDS in the existing EHR
- Provider and/or institutional resistance to practice change
- Ethical, legal, and social concerns surrounding pharmacogenetic testing, patient consent, and the return of results
- Concerns regarding processes for the return of results and patient follow-up
- Challenges in developing CDS

Evidence Barriers
- Clinical tests and applications evolve rapidly
- Varying thresholds for clinical utility and actionability
- Provider acceptance of pharmacogenetic data
- Conflicting interpretation of benefit/value

CDS = clinical decision support; EHR = electronic health record.

Adapted from the following: Bielinksi SJ, Olson JE, Jyotishman P, et al. Preemptive genotyping for personalized medicine: design of the right drug, right dose, right time—using genomic data to individualize treatment protocol. Mayo Clin Proc 2014;89:25-33; O'Donnell PH, Danahey K, Jacobs M, et al. Adoption of a clinical pharmacogenomics implementation program during outpatient care—initial results of the University of Chicago "1,200 Patients Project." Am J Med Genet C Semin Med Genet 2014;166:68-75; Goldspiel BR, Flegel WA, DiPatrizio G, et al. Integrating pharmacogenomics information and clinical decision support into the electronic health record. J Am Med Inform Assoc 2014;21:522-8; Johnson JA, Cavallari LH. Pharmacogenetics and cardiovascular disease—implications for personalized medicine. Pharmacol Rev 2013;65:987-1009; Manolio TA, Chisholm RL, Ozenberger B, et al. Implementing genomic medicine in the clinic: the future is here. Genet Med 2013;15:258-67; Lam YWF. Scientific challenges and implementation barriers to translation of pharmacogenomics in clinical practice. ISRN Pharmacol 2013;2013:641089; Reynolds KS. Achieving the promise of personalized medicine [editorial]. Clin Pharmacol Ther 2012;92:401-5; Crews KR, Hicks KJ, Pui C, et al. Pharmacogenomics and individualized medicine: translating science into practice. Clin Pharmacol Ther 2012;92:467-75; Perry CG, Shuldiner AR. Pharmacogenomics of anti-platelet therapy: how much evidence is enough for clinical implementation? J Hum Genet 2013;58:339-45; Scott SA. Personalizing medicine with clinical pharmacogenetics. Genet Med 2011;13:987-95; Overby CL, Erwin AL, Abul-Husn NS, et al. Physician attitudes toward adopting genome-guided prescribing through clinical decision support. J Pers Med 2014;4:35-49; Evaluation of Genomic Applications in Practice and Prevention (EGAPP) Working Group. The EGAPP initiative: lessons learned. Genet Med 2014;16:217-24; Shuldiner AR, Palmer K, Pakyz RE, et al. Implementation of pharmacogenetics: the University of Maryland Personalized Anti-Platelet Pharmacogenomics Program. Am J Med Genet C Semin Med Genet 2014;166C:76-84.

Table 4.1. Institutional Stakeholders and Roles to Support a Clinical Pharmacogenetics Implementation

Stakeholder	Example(s)	Potential Role Within a Clinical Pharmacogenetics Implementation	Comments
Health-system leaders and administrators	Financial and administrative providers and leaders within the institution Relevant pharmacy, medical, and/or nursing department heads	Establish or confirm buy-in and prioritizing of an implementation initiative within the health system Identify potential barriers that could affect health system–wide adoption (e.g., institutional costs) Demonstrate institutional support to help facilitate needed change within individual practices and/or departments	Engage early in pre-implementation planning to establish institutional buy-in If possible, identify an individual champion to lend support when challenges arise
Clinicians, administrators, and staff in targeted patient care areas	Physicians or other health care providers who will order pharmacogenetic test Clinic administrators who manage patient workflow, billing Nursing or phlebotomy staff who will collect sample for laboratory processing	Provide input on existing patient care workflow as well as clinician and staff needs and priorities during planning processes Provide support within the patient care area for adopting potential workflow changes Assist in patient education concerning specific pharmacogenetic test(s) Provide ongoing feedback on specific operational, billing, patient care, or other processes to help refine an existing implementation and support its sustainability	Buy-in is essential to program success If possible, may be helpful to insert a pharmacy clinician into the clinic for 6+ months to better understand the needs and patient care workflow Continue to maintain communication and elicit feedback for ongoing quality improvement
Clinicians and administrators engaged in medication-use and medication safety processes	Chair of Pharmacy and Therapeutics Committee Pharmacy liaison/ provider for medication safety or medication-use processes	Provide leadership, support, and guidance in developing new medication-use processes (e.g., CDS) within a health system Provide support for establishing a defined process for review and approval of clinical pharmacogenetics implementations Provide input on post-implementation analysis of changes to the medication-use process (i.e., determining whether a new best practice advisory alert promotes changes in prescribing practices)	Processes and/or stakeholders may differ in inpatient and outpatient settings, depending on organizational infrastructure
Pharmacy informatics, information technology, bioinformatics	Informatics pharmacist Information technology officer(s)	Interface with pathology and/or external laboratory provider to ensure pharmacogenetic test results are accessible in the EHR Build CDS tools Provide ongoing support for making changes to existing CDS tools	Engage early and identify a champion in this area, if possible, to minimize time delays after workload and competing priorities that influence changes to the EHR
Pathology providers, clinicians, and laboratory staff	Pathology leadership Laboratory staff	Prioritize pharmacogenetic tests within existing laboratory services Develop and validate testing processes Interface with informatics to ensure the appropriate documentation of pharmacogenetic tests in the EHR	If pharmacogenetic laboratory tests are developed, validated, and processed within the institution, pathology support and engagement are crucial to program success

Table 4.1. Institutional Stakeholders and Roles to Support a Clinical Pharmacogenetics Implementation *(continued)*

Stakeholder	Example(s)	Potential Role Within a Clinical Pharmacogenetics Implementation	Comments
Clinicians and administrators engaged in a review of ethics and/or the IRB	IRB chair Bioethicist with experience in genomic medicine Experienced clinical researcher	Provide guidance on research and ethical issues	Consider

CDS = clinical decision support; EHR = electronic health record; IRB = institutional review board.

Adapted from the following: Crews KR, Cross SJ, McCormick JN, et al. Development and implementation of a pharmacist-managed clinical pharmacogenetics service. Am J Health Syst Pharm 2011;68:143-50; Pulley JM, Denny JC, Peterson JF, et al. Operational implementation of prospective genotyping for personalized medicine: the design of the Vanderbilt PREDICT project 2012;92:87-95; Hoffman JM, Haidar CE, Wilkinson MR, et al. PG4KDS: a model for clinical implementation of pre-emptive pharmacogenetics. Am J Med Genet C Semin Med Genet 2014;166C:45-55; McKinnon RA, Ward MB, Sorich MJ. A critical analysis of barriers to the clinical implementation of pharmacogenomics. Ther Clin Risk Manag 2007;3:751-9; Hartzler A, McCarty CA, Rasmussen LV, et al. Stakeholder engagement: a key component of integrating genomic information into electronic health records. Genet Med 2013;15:792-801; and Konstantinos NL, McAllister TM, Babovic-Vuksanovic D, et al. Implementing individualized medicine into practice. Am J Med Genet C Semin Med Genet 2014;166C:15-23.

Drug-Gene Association(s)

When evaluating drug-gene associations for clinical implementation, pharmacists should consider the genetic variant(s) of interest (e.g., genotype or diplotype), how the genotype translates to a clinical phenotype, and the clinical action recommended according to a given phenotype. Although examples of successful implementations can often be found in genotype-guided therapy of cardiovascular or cancer medications, many factors influence the optimal scope and focus of a clinical pharmacogenetics service, including supporting evidence and/or existing clinical practice guidelines, likelihood of a clinically actionable result, patient outcomes associated with genotype-guided therapy, costs and logistics of pharmacogenetic testing, consequences of drug misuse or inappropriate dosing, national drug use characteristics, and institutional priorities (Bielinksi 2014; Konstantinos 2014; Weitzel 2014; Gharani 2013; McKinnon 2007). The Clinical Pharmacogenetics Implementation Consortium (CPIC) drug-gene association list, available on the PharmGKB Web site (www.pharmgkb.org/cpic/pairs), provides an updated list of drug-gene associations with level of evidence ratings and links to related clinical information sources.

Pharmacists are encouraged to use evidence-based decision-making strategies in identifying optimal drug-gene pair(s) for a new pharmacogenetics service, with consideration given to the factors supporting the value and long-term sustainability of the service within their institution. This process should also include an organizational analysis of medication-use patterns aligned with the drug-gene pairs being considered for implementation. Medication-use frequency and characteristics (e.g., inpatient vs. outpatient) and formulary considerations will influence the optimal choice of an initial drug-gene pair for clinical implementation.

Evidence Review Body and/or Process

A comprehensive evidence review process is an essential component of a clinical pharmacogenetics service, given the evolving nature of pharmacogenetic research (see also chapter 2). Programs should have sufficient resources to ensure an ongoing evaluation of the literature and its application to clinical translation (Pulley 2012).

Many clinical pharmacogenetics services use the model of a structured, multidisciplinary evidence review committee, work group, or task force with oversight by the institution's Pharmacy and Therapeutics Committee or other, similar regulatory body (Bielinksi 2014; Konstantinos 2014; Weitzel 2014; Gharani 2013; Pulley 2012; Crews 2011). Tasks of this evidence review body vary among institutions but may include reviewing evidence for candidate drug-gene pairs, stratifying implementations according to institutional priorities and existing evidence, identifying and monitoring emerging literature that may affect implementation, developing institutional guidance and nomenclature for interpreting pharmacogenetic test results, and developing clinical algorithms and clinical decision support (CDS) language for clinical implementations (Bielinksi 2014; Weitzel 2014; Gottesman 2013; Crews 2012, 2011).

Although the evidence review and application process is covered comprehensively in chapter 2 as mentioned previously, it is notable in the present chapter that the practical applications of pharmacogenetic data included in FDA-approved labels are somewhat limited in guiding clinical implementations. The quality and scope of included pharmacogenetic data vary widely, and the information provided is often insufficient to support explicit genotype-guided dosing recommendations in practice. In some cases, statements in the FDA-approved label have been questioned by experts because of the presence of conflicting published scientific data (e.g., cisplatin and the thiopurine methyltransferase gene [*TPMT*]) (Ratain 2013). In light of these limitations, pharmacists should consider the practical applications of FDA-approved labeling information when guiding clinical implementations and use additional evidence-based recommendations (e.g., CPIC guidelines) to support genotype-guided dosing in practice.

Incorporate Pharmacogenetic Testing into Patient Care Processes

Once the framework is in place to support clinical pharmacogenetics implementation, specific testing approaches, laboratory procedures, CDS, and educational strategies can be developed. Among the most important groups to engage in developing these operational processes are the clinicians and staff in the targeted patient care area(s). Early consideration of the existing patient care workflow and provider preferences will ultimately help support the uptake and use of pharmacogenetic information in clinical practice.

Pharmacogenetic Test Ordering and Laboratory Processing

Pharmacogenetic testing can be performed reactively (i.e., at the point of care to guide drug therapy) or preemptively (i.e., obtained for future use "when needed" in clinical care). Tests can be done to evaluate a single gene or multiple genes (e.g., a multiplex panel). Reactive or "point-of-care" testing is usually used to assess one or more genes that may influence drug therapy; it is often ordered at or near the time of prescribing or after the occurrence of an adverse drug reaction thought to be associated with a pharmacogenetic variant. In contrast, preemptive testing involves the analysis of one or more genetic variants to allow the future use of pharmacogenetic information, regardless of current drug therapy (Haga 2014). Preemptive data can be entered in a patient's electronic health record (EHR) at the time of the result (or later, in some cases) and are available for application when affected drugs are prescribed in the future (Bielinksi 2014; Haga 2014; Van Driest 2014; Johnson 2013, 2012; Schildcrout 2012).

Advantages of reactive pharmacogenetic testing include the ability to link the pharmacogenetic test(s) to a current diagnosis to establish medical necessity, increased likelihood of insurance reimbursement in the current payment environment, and the ability to target a defined patient and clinician population (Haga 2014). Reactive testing favors a staged implementation model in which institutions start small and subsequently expand to increase the scope of a service (McKinnon 2007). However, this approach has been criticized in clinical pharmacogenetics as being incompatible with the patient care process because of the time delay in ordering, processing, and resulting tests before a clinical decision can be made. In addition, repeated reactive testing quickly becomes relatively more costly than processing a single test within a broader preemptive array, particularly as the costs of preemptive testing continue to decrease (Bielinksi 2014; Haga 2014; Hoffman 2014; Van Driest 2014; Johnson 2012; Schildcrout 2012).

Conversely, preemptive testing can provide pharmacogenetic data when needed in the patient care process (because it is already available in the EHR), is associated with a lower cost-per-genotype, and leads to a decreased testing burden because data on several actionable pharmacogenetic variants are gathered at once (Van Driest 2014; Johnson 2012; Schildcrout 2012). This approach supports a comprehensive process in which relevant pharmacogenetic data can be incorporated cost-effectively as drugs are prescribed (Bielinksi 2014; O'Donnell 2014; Schildcrout 2012). Potential disadvantages of preemptive testing include its associated initial costs (and limited insurance reimbursement for array-based preemptive testing), concerns about clinician liability if test results are not used in drug selection and/or dosing, and the potential for generating incidental findings (that may or may not be reported to the patient at the time of testing) with future relevance, although this latter risk is minimal with pharmacogenomic (vs. genomic) data (Haga 2014).

Although preemptive testing is currently associated with limited clinical reimbursement and commercial availability, it is an important component of a comprehensive cost-effective pharmacogenetic testing model. It has been estimated that up to 65% of individuals would be exposed to at least one medication with an established pharmacogenetic association during a 5-year period, which could enable a preemptive approach to prevent hundreds of potential adverse events and lead to savings in a health care system (Schildcrout 2012). In a model that included almost 10,000 individuals, investigators found that more than 90% of patients genotyped on a preemptive array had one or more actionable variants, with almost 50% of these individuals exposed to a medication or drug class associated with an increased risk of adverse events with a variant genotype. Reactive genotyping done in this same patient group would have required performing almost 15,000 pharmacogenetic tests (Van Driest 2014).

Of note, the distinction between a reactive and a preemptive testing approach should not be thought of as an "either/or" proposition. Many institutions begin pharmacogenetics services with a limited, single-gene approach. As experience in implementation is gained, these services can more easily progress to a broader, multigene, preemptive approach. An institution may also use both approaches simultaneously in different patient care groups, depending on organizational and patient care needs and priorities. Reactive pharmacogenetic data also becomes somewhat preemptive over time, such as when a patient tested for cytochrome P450 2C19 *(CYP2C19)* variants post–percutaneous coronary intervention returns for a repeat procedure at a later date, at which time the patient's *CYP2C19* metabolism status has already been documented in the medical record during his or her previous encounter.

Whichever testing approach is taken, it is important that pharmacogenetic testing be integrated seamlessly into the existing clinical workflow to ensure provider and staff buy-in for new processes. Test information should be documented in the EHR in a meaningful and accessible manner, provided directly to patients whenever possible (e.g., through a patient portal), and made available for the patient's lifetime. Although results should ideally be available to other providers outside the institution, this is largely limited by the fragmentation of the health care system. However, this will be an important component in the widespread adoption of pharmacogenetic testing in the future to limit the costs of repeated testing when a previous lifetime result already exists for a patient.

Clinical Decision Support

Most established clinical pharmacogenetics implementations have used CDS integrated into the EHR to assist providers in quickly interpreting and acting on pharmacogenetic test results in a clinical environment (Figure 4.1). Integrating CDS into the clinical workflow has clear benefits and has been identified as a key factor associated with positive results of a personalized medicine implementation (Welch 2013). Clinical decision support facilitates the ongoing use of pharmacogenetic information over a patient's lifetime, given its continued availability to providers in the clinical environment. It also assists providers in quickly and accurately responding to pharmacogenetic data by providing the appropriate clinical actions for a given drug-gene pair. However, pharmacogenomic (and genomic) data can be challenging to incorporate into the EHR, given the need for lifetime access to these data in clinical care (i.e., in EHR systems that do not provide a seamless mechanism to store and access "lifetime" laboratory results), potential privacy and ethical concerns, and lack of existing standards for storage, interpretation, and communication of genomic data (Bell 2014; Peterson 2013; Ury 2013). In addition, these complex data must be accessible to clinicians in an appropriate, clear, and timely manner but communicated such that over-alerting busy providers (i.e., alert fatigue) is avoided.

Pharmacogenetic CDS alerts generally fall into one of two categories: (1) preemptive alerts that prompt providers to order a comprehensive pharmacogenetic panel when a certain risk threshold is met (e.g., a score that calculates the likelihood of a patient's needing a pharmacogenetic test result to guide drug therapy in the future) and (2) reactive alerts that prompt a specific clinical action because of predetermined criteria (Weitzel 2014; Peterson 2013). In the latter category, a CDS alert may be triggered by a specific drug order and prompt the clinician to obtain a pharmacogenetic test (e.g., an azathioprine prescription prompts the clinician to order a *TPMT* test), often referred to as a pretest CDS alert. A reactive CDS alert may also be triggered after a pharmacogenetic test has been documented in the EHR to guide clinical action with a new drug order (posttest CDS alert), such as in a patient post–percutaneous coronary intervention who is prescribed clopidogrel and has an existing variant *CYP2C19* test result documented in the EHR. Whichever type of CDS is needed, a decision algorithm should be developed to define specific rules and criteria governing when the alert will be triggered. Although alert language and content vary by institution, Box 4.3 lists questions that clinicians can use as a starting point to guide alert content development and rules.

Clinical decision support is an important piece of the implementation puzzle to help providers quickly process genotype-to-phenotype data translation and take the appropriate clinical action. Pharmacogenetic data that are reported from the laboratory in diplotype form (e.g., *CYP2C19*2/*2*), instead of as the resulting phenotypic translation (e.g., *CYP2C19* poor metabolizer), may be challenging for clinicians to interpret. Use of clinician-friendly language within the CDS (and related pharmacogenetic consult note or other communications) enables providers to focus on clinically actionable data that are relevant to patient care. In addition, CDS allows an automated mechanism to present genotype-guided drug therapy recommendations to clinicians. In the example of the *CYP2C19* poor metabolizer, the CDS may also include a recommendation for alternative therapies (i.e., ticagrelor or prasugrel). This component of CDS is especially important for the emerging science of pharmacogenetics, for which evidence evolves rapidly (Weitzel 2014; Crews 2012).

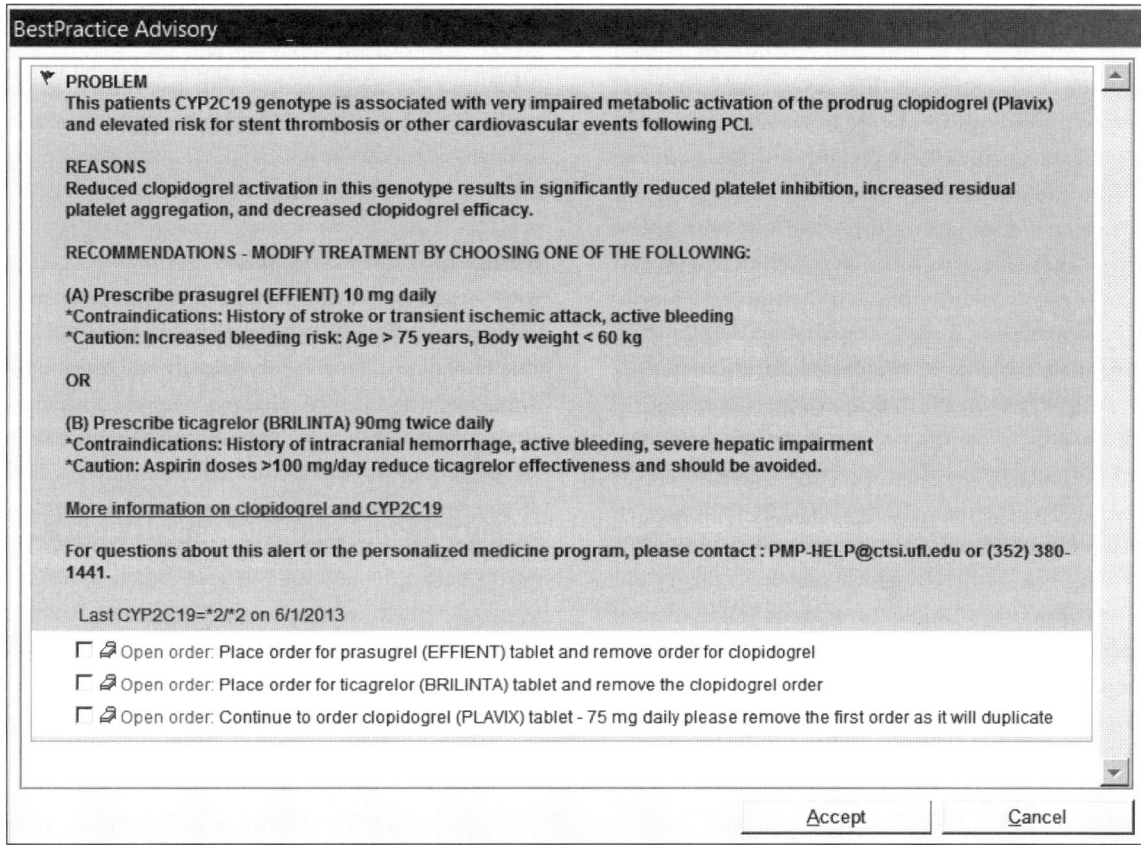

Figure 4.1. Sample clinical decision support alert for pharmacogenetics implementation.
PCI = percutaneous coronary intervention.

Source: Weitzel KW, Elsey AR, Langaee TY, et al. Clinical pharmacogenetics implementation: approaches, successes, challenges. AM J Med Genet C Semin Med Genet 2014;166:56-67.

Pharmacists are encouraged to collaborate with pharmacy informatics and information technology specialists within their organization or health system for support and resources to develop and modify existing CDS architecture. Detailed descriptions are available of strategies for developing and implementing CDS in pharmacogenetics services, including technical processes and sample genotype-to-phenotype translation tables (Bell 2014; Gottesman 2013; Peterson 2013; Ury 2013; Darcy 2011). In addition, CPIC created an informatics working group in 2013 to support the adoption of guideline recommendations in a clinical electronic environment (www.pharmgkb.org/page/cpicinformatics). With the establishment of this focused group, supplemental material for newly created or updated CPIC guidelines now includes CDS resources such as workflow diagrams showing the storage of test results in the EHR and CDS alert design information (Martin 2014).

Return of Results

Although pharmacogenetic data are associated with inherently fewer ethical, legal, and societal risks than genomic disease–based risk information, many unanswered questions remain regarding an optimal strategy and format for the return of results (Van Driest 2014; Gottesman 2013; Perry 2013). Similar to the communication of genomic findings, the communication of pharmacogenetic test results requires consideration of incidental findings, the need to interpret a lifetime test result outside the triggering clinical event, the relay of test results to other providers within and outside the originating health system, and the application of test results in the context of nongenomic clinical information and rapidly emerging science (Peterson 2013; Sturm 2013; van Schie 2011). In addition, preemptive pharmacogenetic testing may provide results that are not yet clinically actionable (and therefore should not be made immediately available in the patient's clinical record) (Weitzel 2014; Crews 2012).

Pharmacists can look to established clinical programs for guidance on best practices for communicating test results to health care providers and/or patients. Ideally, interpreted pharmacogenetic information should be made available in the patient's EHR as a lifetime result to the ordering provider and communicated to the patient and to other clinicians within and outside the originating health system. To accomplish this, programs have used CDS alerts to provide point-of-care interpretation, incorporated pharmacogenetic test results into the patient's problem list, and documented results and interpretation in a pharmacogenetic "tab" or other defined area in the EHR. Any of these mechanisms may also link to a formal pharmacogenetic consult note with clinician-friendly summary information, drug therapy recommendations, and/or printable patient or provider education materials (Weitzel 2014; Peterson 2013; Sturm 2013; Crews 2012, 2011). Patient education and test results have been transmitted to patients by a formal pharmacogenetic counseling session, in a mailed letter (which the patients can share with other health care providers), and/or electronically through the health system's patient information portal (Bielinksi 2014; Caudle 2014; Haga 2014; Weitzel 2014; Peterson 2013; Sturm 2013; Crews 2011).

Some pharmacy-led programs have incorporated or are exploring a multidisciplinary team-based approach to patient communication, together with a genetic counselor (Bielinksi 2014; Sturm 2013). A 2012 survey of genetic counselors showed that 52% anticipated they would play "some" role in delivering pharmacogenetic testing (Haga 2012). A pharmacist–genetic counselor partnership model has also been proposed that relies on the complementary knowledge and expertise of both disciplines (Mills 2013). Pharmacists are encouraged to develop and define procedures for the return of results to patients and health care providers that are consistent with these and other best practices and current guidance on the return of incidental findings (Green 2013).

Billing and Reimbursement

Although few data analyze the costs and benefits of pharmacogenetic testing, anticipated and actual test reimbursability play crucial roles in prioritizing, developing, and implementing clinical pharmacogenetics services (Lam 2013). Most current pharmacoeconomic literature is focused in therapeutic areas with the highest clinical use of pharmacogenetic testing (thromboembolic disorders [clopidogrel, warfarin] and cancer [e.g., *TPMT*]), with most payers identifying a link to conclusive evidence of improved health outcomes as the most important factor determining reimbursement (Phillips 2014; Cohen 2013; Lam 2013; Wong 2010). Cost-utility analyses of personalized medicine tests (including pharmacogenetic and genetic risk prediction tests) have shown that they can provide better health, albeit at higher costs, as measured by cost per quality-adjusted life-years (Phillips 2014). Some experts have argued that setting the "threshold" of acceptance for the clinical implementation of pharmacogenetics at proven cost-savings (i.e., pharmacogenetic tests

---

Box 4.3. Representative Questions for Clinicians When Developing Clinical Decision Support Language

- What criteria will cause the alert to fire?
- What message will displayed to the user?
- Will the alert be highly visible to providers?
- Will the alert fire at the point of clinical decision-making?
- What steps should be taken to prevent the alert from getting "lost" in the clinical record?
- Should the alert result in a hard stop (medication cannot be ordered without a documented genotype)?
- If an override is required, what are the allowable variables?
- Should ordering the pharmacogenetic test be facilitated by a preorder alert or by preselecting the laboratory order?
- Should ordering an alternative drug or dose be facilitated by medication order options within the alert?
- How will the alert language be updated as new evidence emerges?

Adapted from the following: Weitzel KW, Elsey AR, Langaee TY, et al. Clinical pharmacogenetics implementation: approaches, successes, challenges. Am J Med Genet C Semin Med Genet 2014;166:56-67; Crews KR, Cross SJ, McCormick JN, et al. Development and implementation of a pharmacist-managed clinical pharmacogenetics service. Am J Health Syst Pharm 2011;68:143-50; Goldspiel BR, Flegel WA, DiPatrizio G, et al. Integrating pharmacogenomics information and clinical decision support into the electronic health record. J Am Med Inform Assoc 2014;21:522-8; Ury AG. Storing and interpreting genomic information in widely deployed electronic health record systems. Genet Med 2013;15:779-85; Peterson JF, Bowton E, Field JR, et al. Electronic health record design and implementation for pharmacogenomics: a local perspective. Genet Med 2013;15:833-41; and Bell GC, Crews KR, Wilkinson MR, et al. Development and use of active clinical decision support for preemptive pharmacogenomics. J Am Med Inform Assoc 2014;21:e93-99.

should be adopted only if they have shown cost savings) is inconsistent with the standards for other preventive or medication safety initiatives (Phillips 2014; Cohen 2008). However, in a real-world implementation environment, the cost of testing and the potential cost-savings of implementation must be considered.

Few data are available regarding insurance coverage of preemptive pharmacogenetic testing, and reimbursement evidence for reactive pharmacogenetic testing is variable. It has been estimated that less than half of available pharmacogenetic tests are covered by insurers, with lack of evidence of clinical utility cited as a major determining factor in non-coverage (Hresko 2012). Conversely, the presence of pharmacogenetic information in the drug's labeling has been associated with an increased likelihood of test coverage (Hresko 2012). Some clinicians have cited early experience of reimbursement rates up to 85% of claims submitted in an outpatient environment with reactive CYP2C19 testing to guide antiplatelet therapy post–percutaneous coronary intervention (Weitzel 2014).

The Centers for Medicare & Medicaid Services (CMS) and other insurers define and publish the criteria for determining medical necessity and payment for laboratory testing, including targeted pharmacogenetic tests. If clinical reimbursement is a high priority for a given implementation, these criteria can guide the clinician in choosing pharmacogenetic tests, drugs, and patient populations. Insurance coverage for pharmacogenetic testing may be limited to a single test order, which should then be documented as a lifetime result in the patient's medical record. Coverage may also be influenced by concomitant drug therapy (e.g., *CYP2C19* in patients receiving clopidogrel), patient ancestry (e.g., the HLA-B*1502 allele in patients of Asian ancestry before beginning carbamazepine), diagnosis (e.g., the HLA-B*5701 allele in patients infected with HIV [human immunodeficiency virus] before beginning abacavir), and whether the pharmacogenetic test is FDA approved. Insurers have also designated specific tests as having research or investigational status, which may limit reimbursement. Such was the case with CMS's policy of restricting the coverage of pharmacogenetic testing to predict a patient's response to warfarin to a specific subset of patients enrolled in prospective randomized controlled clinical trials (Scott 2011; CMS 2009).

Reimbursement strategies and payment sources are also influenced by the patient care setting in which the test is ordered. As with other billable clinical services in the outpatient setting, some targeted pharmacogenetic tests (e.g., *CYP2C19, TPMT, CYP2D6*) are associated with existing Current Procedural Terminology codes for billing. Although coverage determination and reimbursement rates are based on many factors, pharmacogenetic tests that are associated with these codes may be reimbursed more consistently and at higher rates than those that are billed using nonspecific molecular testing codes. Conversely, although tests ordered in the outpatient setting are billed to the patient's major medical insurance provider, tests done in the inpatient setting are often absorbed into health-system diagnosis–related group codes or payments. Thus, reimbursement for pharmacogenetic tests done in an inpatient setting is essentially provided by the health system itself. In these cases, it is advisable to incorporate pharmacogenomic initiatives into strategic planning at the organizational level and proactively discuss with key stakeholders the potential costs and measures to contain such costs by monitoring and preventing inappropriate testing before implementation. Alternatively, some clinical programs have opted to develop a formal consultation process and seek reimbursement for the pharmacogenetic consult itself to partly offset unrecovered testing costs (Nutescu 2013; Crews 2011).

Whichever approach is adopted, pharmacists are encouraged to take a leadership role in proactively addressing reimbursement and cost-related concerns from key stakeholders, exploring payment mechanisms, and seeking reimbursement for pharmacogenetic testing. Cost concerns and trepidation about unrealized returns on investments in pharmacogenetic testing are important barriers to clinical implementation. A strategic approach to reimbursement can help allay stakeholder fears and increase the likelihood of sustainable clinical adoption. In addition, more frequent use of, and billing for, pharmacogenetic tests can increase awareness of the demand for these services among payers, ultimately contributing to long-term sustainability in this area.

Education
Clinical pharmacogenetics implementation requires participation and buy-in from a wide array of clinicians, administrators, and support staff, all of whom have varying educational needs. Developing appropriate educational programs to meet diverse needs creates unique challenges in this area (Weitzel 2014; Passamani 2013; Shuldiner 2013; Crews 2012; Stanek 2012; Scott 2011). Many established programs have addressed these challenges by developing targeted, ongoing, multidisciplinary educational programs in conjunction with the clinical service. Key educational needs have been identified as including both knowledge of pharmacogenomics and training for specific changes in workflow and/or patient care processes. Programs have reported the importance of (1) providing a spectrum of educational offerings that appropriately target the essential educational needs of clinical, support, and administrative staff through in-services,

grand rounds, patient care conferences, and other venues; (2) making such programs available and distributed in several formats (e.g., live or recorded training, development of brief competency documents, incorporation into health-system newsletter); (3) assessing the value of the educational intervention through postprogram surveys or assessments; (4) incorporating patient educational materials into the EHR, the electronic patient portal, and/or the institutional Web site to support clinicians in counseling patients; and (5) providing ongoing feedback to clinicians after implementation as a continuing education and development strategy (Bielinksi 2014; Goldspiel 2014; Hoffman 2014; Shuldiner 2014; Weitzel 2014; Gottesman 2013; Peterson 2013; Crews 2012, 2011; Scott 2011; McKinnon 2007).

An appreciation for the wide variance in educational needs among different target audiences is especially important in this process. Because of the depth and breadth of pharmacogenomics information, it is advisable to focus educational offerings precisely to targeted audience needs, such as offering an in-service to patient care staff on new laboratory processes compared with a comprehensive pharmacogenetic educational session with competency assessment for clinical pharmacists. When possible, it is helpful to engage a leader in the targeted area (e.g., attending physician) to provide guidance on educational needs, assist with developing the program focus, and participate in the delivery of the educational offering to further support participation and buy-in.

Evaluate Pharmacogenetics Implementations
Because of the emerging nature of pharmacogenetics implementation, evaluating and reporting data on clinical programs are helpful in individual and profession-wide program sustainability. Most programs to date have not been designed to capture patient outcomes data regarding the clinical effects of genotype-guided therapy. Instead, evaluation measures often encompass quantitative and qualitative metrics surrounding program implementation, adoption, perceived value, and/or sustainability (Box 4.4). Some programs have also reported instituting an ongoing review of logistic and quality measures data by a pharmacogenetic steering committee or other oversight body to refine existing initiatives and guide future development (Crews 2011).

Health Care Professionals and Competencies in Clinical Implementation

Competencies in genetics and genomics have been established for almost all health care professionals, and pharmacogenomics-specific competencies and outcomes exist for pharmacists (APHMG 2013; Genetics and Genomics Competency Center for Education, NHGRI 2012; Greco 2012; ANA 2008; NCHPEG 2007; Rackover 2007; Johnson 2002). Genetics and genomics competencies for non-pharmacy health care professionals focus primarily on genomic medicine and vary in their inclusion of specific pharmacogenomics competencies. Physician and physician assistant competencies include knowledge-based statements regarding practitioners' understanding and ability to define, discuss, or describe pharmacogenomics and its role in the range of genetic approaches to treating disease and/or predicting medication response and adverse drug reactions (APHMG 2013; Rackover 2007). Pharmacogenomics-related competency statements for nurses with graduate degrees are more extensive and include knowledge statements and clinical performance indicators stating that these practitioners should consider pharmacogenetic test results, if available, a component of safe medication administration (Greco 2012). Overall, however, pharmacogenomics competencies for non-pharmacist health care professionals overwhelmingly emphasize awareness and knowledge of pharmacogenomics, rather than clinical applications to optimize drug therapy regimens.

Within the pharmacy profession, competency and outcome statements related to pharmacogenomics were developed in 2002 and then updated in 2012 by representatives from 10 pharmacy and pharmacy-related organizations (Genetics and Genomics Competency Center for Education, NHGRI 2012; Johnson 2002). Through a consensus process, experts provided the pharmacist-specific competency statements for genetic concepts, genetics and disease, pharmacogenomics/pharmacogenetics, and ethical, legal, and social implications (Figure 4.2) that are needed for pharmacists and pharmacy graduates to achieve the following desired outcome:

Pharmacists and pharmacy graduates should be able to appreciate the contribution of genetic variability to inter-individual variations in drug response. With the knowledge of genetics/genomics pharmacists may recommend and interpret the results of pharmacogenetic/pharmacogenomics tests and make therapy recommendations based on test results. Pharmacists and pharmacy graduates should possess competent knowledge and skills to seek coordination and collaboration of care with an interdisciplinary team of health professionals when assessing genetic information (Genetics and Genomics Competency Center for Education, NHGRI 2012).

The American Society of Health-System Pharmacists has published a pharmacogenomics policy position and draft statement on the pharmacist's role in pharmacogenomics that further defines the role of pharmacists

> **Box 4.4.** Sample Evaluation Metrics for a Clinical Pharmacogenetics Implementation
>
> **Patient Care Process Metrics**
> - Number of patients eligible for pharmacogenetic test
> - Adoption rate of pharmacogenetic test order (test volume, order method, etc.)
> - Number and type of CDS rules fired
> - Rate of provider adoption and rejection of recommended drug therapy changes in patients with actionable variants
> - Reasons for rejection of drug therapy recommendations in applicable patients
> - Method of delivering and/or documenting pharmacogenetic therapy recommendations to provider
> - Method of delivering and/or communicating pharmacogenetic test results to patient
>
> **Pharmacogenetic Test Metrics**
> - Laboratory turnaround time and component breakdown (e.g., time spent in test processing vs. sample collection or transportation)
> - Genotype data by haplotype
> - Number/portion of patients with actionable variants
>
> **Economic/Health-System Metrics**
> - Costs and/or resourced needs for program development
> - Average time and/or resources spent on new clinical processes by providers
> - Burden of CDS implementation (e.g., calls to technical support)
> - Trends in test ordering, billing and reimbursement, and drug therapy changes in targeted patient populations
>
> **Qualitative Metrics**
> - Provider attitudes, knowledge, and beliefs surrounding the value and utility of pharmacogenomics
> - Perceived barriers and facilitators to the adoption of clinical pharmacogenetics
> - Effects of participation in clinical implementation on perceptions about clinical pharmacogenetics
>
> CDS = clinical decision support.
> Adapted from the following: Weitzel KW, Elsey AR, Langaee TY, et al. Clinical pharmacogenetics implementation: approaches, successes, challenges. Am J Med Genet C Semin Med Genet 2014;166:56–67; Crews KR, Cross SJ, McCormick JN, et al. Development and implementation of a pharmacist-managed clinical pharmacogenetics service. Am J Health Syst Pharm 2011;68:143-50; Shuldiner AR, Palmer K, Pakyz RE, et al. Implementation of pharmacogenetics: the University of Maryland Personalized Anti-Platelet Pharmacogenomics Program. Am J Med Genet C Semin Med Genet 2014;166C:76-84; Konstantinos NL, McAllister TM, Babovic-Vuksanovic D, et al. Implementing individualized medicine into practice. Am J Med Genet C Semin Med Genet 2014;166C:15-23; and Stanek EJ, Sanders CL, Taber KA, et al. Adoption of pharmacogenomics testing by U.S. physicians: results of a nationwide survey. Clin Pharmacol Ther 2012;91:450-8.

within a clinical implementation (ASHP 2014). Within this draft statement, the authors affirm that

Because of their distinct knowledge, skills, and abilities, pharmacists are uniquely positioned to lead interdisciplinary efforts to develop processes for ordering, interpreting, and reporting pharmacogenomic test results and for guiding optimal drug selection and drug dosing based on those results, as well as efforts to implement and improve those processes (ASHP 2014)

And in a 2011 white paper, the American Pharmacists Association addressed the pharmacist's role in pharmacogenomics as an extension of pharmacist responsibilities in existing medication therapy management services (APhA 2011).

Although clinical application–based competency statements in pharmacogenomics are lacking for most health care professionals, such statements and professional association support within the pharmacy profession can assist in standardizing pharmacists' roles and responsibilities in the clinical implementation of pharmacogenetics. To fulfill these roles and support clinical development in this area, pharmacists will have to be equipped to contribute throughout the medication-use system, including in formulary development, medication-use processes, patient safety, clinical pharmacy practice, pharmacy informatics and health information technology, research and ethics, education and training, evidence-based literature analysis, pathology, and quality assurance.

In addition to providing the roles and educational needs for pharmacists directly involved in developing an active pharmacogenetics implementation, the pharmacy profession should be aware of, and respond to, the educational and practice needs of frontline pharmacists in diverse practice environments. It is anticipated that pharmacogenetic and/or genetic information will be used on an increasingly larger scope to guide drug therapy decisions in the same way that other clinical data (e.g., drug-drug

---

Basic Genetic Concepts
- (B1) To demonstrate an understanding of the basic genetic/genomic concepts and nomenclature
- (B2) To recognize and appreciate the role of behavioral, social, and environmental factors (lifestyle, socioeconomic factors, pollutants, etc.) to modify or influence genetics in the manifestation of disease
- (B3) To identify drug and disease associated genetic variations that facilitate development of prevention, diagnostic and treatment strategies and appreciate there are differences in testing methodologies and are aware of the need to explore these differences in drug literature evaluation
- (B4) To use family history (minimum of three generations) in assessing predisposition to disease and selection of drug treatment

Genetics and Disease
- (G1) To understand the role of genetic factors in maintaining health and preventing disease
- (G2) To assess the difference between clinical diagnosis of disease and identification of genetic predisposition to disease (genetic variation is not strictly correlated with disease manifestation)
- (G3) To appreciate that pharmacogenomic testing may also reveal certain genetic disease predispositions (e.g. the Apo E4 polymorphism)

Pharmacogenetics/Pharmacogenomics
- (P1) To demonstrate an understanding of how genetic variation in a large number of proteins, including drug transporters, drug metabolizing enzymes, direct protein targets of drugs, and other proteins (e.g. signal transduction proteins) influence pharmacokinetics and pharmacodynamics related to pharmacologic effect and drug response
- (P2) To understand the influence (or lack thereof) of ethnicity in genetic polymorphisms and associations of polymorphisms with drug response
- (P3) Recognize the availability of evidence based guidelines that synthesize information relevant to genomic/pharmacogenomic tests and selection of drug therapy (e.g. Clinical Pharmacogenomics Implementation Consortium)

Ethical, Legal and Social Implications (ELSI)
- (E1) To understand the potential physical and/or psychosocial benefits, limitations and risk of genomic/pharmacogenomic information for individuals, family members and communities, especially with genomic/pharmacogenomic tests that may relate to predisposition to disease
- (E2) To understand the increased liability that accompanies access to detailed genomic patient information and maintain confidentiality and security
- (E3) To adopt a culturally sensitive and ethical approach to patient counseling regarding genomic/pharmacogenomic test results
- (E4) To appreciate the cost, cost-effectiveness, and reimbursement by insurers relevant to genomic or pharmacogenomic tests and test interpretation, for patients and populations
- (E5) To identify the need to refer a patient to a genetic specialist or genetic counselor

Figure 4.2. Pharmacist's pharmacogenomics competencies.

Reprinted with permission from The Genetics and Genomics Competency Center, funded by the National Human Genome Research Institute. Available at www.g-2-c-2.org.

interactions, weight-based dosing requirements) are currently applied. As such, all pharmacists will need to be knowledgeable about pharmacogenetic concepts and be able to apply these data to clinical patient care scenarios outside a specialized patient population.

Although most published descriptions of successful clinical pharmacogenetics implementations have focused on the health-system environment to date, the profession is beginning to witness the expansion of pharmacogenetics to community pharmacies and other practice settings (Ferreri 2014; Kisor 2014; Beier 2013; Fichter 2013). Researchers have reported the case of a 65-year-old man who underwent coronary stent placement after myocardial infarction who received pharmacogenetic testing as part of a community pharmacy–based medication therapy management service (Kisor 2014). In collaboration with the prescriber, the pharmacist initiated a drug therapy change from clopidogrel to prasugrel according to the patient's *CYP2C19* metabolizer status. Other researchers have described the development and implementation of a *CYP2C19* genotype–guided clopidogrel pilot program in a single location of a community pharmacy chain (Ferreri 2014). Pharmacists in this study identified genetic variants in 9 of 18 study patients; all pharmacist recommendations for genotype-guided drug therapy changes were accepted by prescribers. These and other examples show that, as with other intervention-based pharmacy services such as medication therapy management, the community pharmacist is well positioned to collaborate with other health care providers to identify patients that may benefit from genotype-guided therapy and to facilitate pharmacogenetic testing and the application of these data to optimize medication use and dosing. It is essential that the pharmacy profession support pharmacists in diverse roles and practice settings in the application of pharmacogenetic data to patient care by developing innovative practice models and educational and practice-based resources.

Educational Strategies to Support the Clinical Implementation of Pharmacogenetics

Comprehensive, multidisciplinary strategies will be needed to equip pharmacists, physicians, and other health care professionals with the knowledge and skills for the widespread adoption of clinical pharmacogenetics. Meeting these widespread educational needs comes with its own unique challenges. Surveys of practitioners show that pharmacists and physicians alike seem aware of the clinical implications of pharmacogenomics but believe they are inadequately prepared to apply this information to drug therapy selection, dosing, or monitoring (Stanek 2012; McCullough 2011). Pharmacogenomics topics are not consistently incorporated to a significant extent into medical or pharmacy school curricula, and drug-related pharmacogenetic information is poorly represented in commonly used drug information sources (Vaughan 2014; Daly 2013; Passamani 2013). The complexity of pharmacogenomics research and science creates further challenges. Formal continuing education programs have shown poor participant knowledge and retention after pharmacogenomics training (Formea 2013; Passamani 2013).

Fortunately, multidisciplinary educational initiatives are emerging that are designed to address these needs for practice-based education in clinical pharmacogenetics in innovative ways. An increasing number of pharmacy and medical schools are incorporating clinical genomic and pharmacogenomic concepts, with staff in some institutions using personal genotype evaluation by students to help them better understand the patient's perspective and apply genetic information to discussions and cases (Salari 2013). Education and training experiences are being developed in active clinical programs to help students, residents, and trainees gain experience practicing in an ongoing clinical pharmacogenetics program (Drozda 2013). At least two active PGY2 (postgraduate year two) pharmacogenetics specialty residency programs in the United States are designed to allow pharmacists to pursue advanced training within active clinical programs. In addition, the American Association of Colleges of Pharmacy has established a Pharmacogenomics Special Interest Group for faculty engaged and/or interested in teaching pharmacogenomics. National efforts and initiatives are also under way to provide additional resources for educational needs (Box 4.5).

These and other educational programs and resources are needed to support clinicians seeking to provide leadership and/or participate in clinical pharmacogenetics implementations within their institution and facilitate practice change within the health care system. High-quality education and training programs will continue to be needed that target multidisciplinary and interprofessional educational needs, emphasize clinical practice applications of pharmacogenetic research and science in a real-world environment, and are readily available to practitioners in many formats (e.g., live, online, and experiential).

Conclusion

Opportunities for achieving the promise of clinical pharmacogenetics to improve patient outcomes are becoming a reality with new research and scientific advances. As medication experts, clinical pharmacists are well poised to

> **Box 4.5. Resources to Support Health Care Professional Education in Pharmacogenetics**
>
> The Pharmacogenomics Knowledge Base: www.pharmgkb.org/
>
> Clinical Pharmacogenetics Implementation Consortium guidelines: www.pharmgkb.org/page/cpic
>
> American Society of Health-System Pharmacists Emerging Sciences Resource Center: www.ashp.org/EmergingSciences
>
> National Human Genome Research Institute: www.genome.gov/HealthProfessionals/
>
> The Centers for Disease Control and Prevention Office of Public Health Genomics: www.cdc.gov/genomics/default.htm
>
> UF Health Personalized Medicine Program Educational Resources: http://personalizedmedicine.ufhealth.org
>
> St. Jude Children's Research Hospital Healthcare Professional Resources for the Implementation of Pharmacogenomics: www.stjude.org/pg4kds
>
> Genetics/Genomics Competency Center for Education: www.g-2-c-2.org/index.php
>
> Global Genetics and Genomics Community: http://g-3-c.org/en

provide leadership in key implementation roles within the health care system. These efforts require the engagement of key administrative stakeholders, clinicians, and staff; the leadership of multidisciplinary internal and external educational initiatives; and the ability to apply knowledge of pharmacogenomics, evidence-based medicine, and medication-use processes to practice. Although steps are being taken to advance practitioners' readiness for clinical pharmacogenetics implementation, additional education and training will likely be needed for health care professionals interested in expanding their roles in this area.

References

American Nurses Association (ANA) [www.nursingworld.org/]. Essentials of Genetic and Genomic Nursing: Competencies, Curricula Guidelines, and Outcome Indicators, 2nd ed. Silver Spring, MD: ANA, 2008. Available at www.genome.gov/Pages/Careers/HealthProfessionalEducation/geneticscompetency.pdf. Accessed April 7, 2014.

American Pharmacists Association (APhA). Integrating pharmacogenomics into pharmacy practice via medication therapy management [white paper]. J Am Pharm Assoc 2011;51:e64-74.

American Society of Health-System Pharmacists (ASHP) [www.ashp.org]. ASHP Statement on the Pharmacist's Role in Clinical Pharmacogenomics. Bethesda, MD: ASHP, 2014. Available at www.ashp.org/DocLibrary/Policy/HOD/StPharmacogenomicsPrepress.aspx. Accessed April 6, 2014.

Association of Professor of Human and Medical Genetics (APHMG) Competencies Working Group [www.aphmg.org]. Medical School Core Curriculum in Genetics 2013. Bethesda, MD: APHMG, 2013. Available at www.aphmg.org/pdf/med-competencies.pdf. Accessed April 7, 2014.

Beier MT, Panchapagesan M, Carman LE. Pharmacogenetics: has the time come for pharmacists to embrace and implement the science? Consult Pharm 2013;28:696-711.

Bell GC, Crews KR, Wilkinson MR, et al. Development and use of active clinical decision support for preemptive pharmacogenomics. J Am Med Inform Assoc 2014;21:e93-99.

Bielinksi SJ, Olson JE, Jyotishman P, et al. Preemptive genotyping for personalized medicine: design of the right drug, right dose, right time—using genomic data to individualize treatment protocol. Mayo Clin Proc 2014;89:25-33.

Caudle KE, Klein TE, Hoffman JM, et al. Incorporation of pharmacogenetics into routine clinical practice: the Clinical Pharmacogenetics Implementation Consortium (CPIC) guideline development process. Curr Drug Metab 2014;15:209-17.

Centers for Medicare & Medicaid Services (CMS) [www.cms.gov]. National Coverage Determination (NCD) for Pharmacogenomic Testing for Warfarin Response (90.1). Washington, DC: CMS, 2009. Available at www.cms.gov/medicare-coverage-database/details/ncd-details.aspx?NCDId=333&ncdver=1&NCAId=224&NcaName=Pharmacogenomic+Testing+for+Warfarin+Response&IsPopup=y&bc=AAAAAAAACAAAAA%3d%3d&. Accessed March 25, 2014.

Cohen J, Wilson A, Manzolillo K. Clinical and economic challenges facing pharmacogenomics. Pharmacogenomics J 2013;13:378-88.

Cohen JT, Neumann PJ, Weinstein MC. Does preventive care save money? Health economics and the presidential candidates. N Engl J Med 2008;358:661-3.

Crews KR, Cross SJ, McCormick JN, et al. Development and implementation of a pharmacist-managed clinical pharmacogenetics service. Am J Health Syst Pharm 2011;68:143-50.

Crews KR, Hicks KJ, Pui C, et al. Pharmacogenomics and individualized medicine: translating science into practice. Clin Pharmacol Ther 2012;92:467-75.

Daly AK. Is there a need to teach pharmacogenetics [opinion]? Clin Pharmacol Ther 2013;95:245-7.

Darcy DC, Lewis ET, Ormond KE, et al. Practical considerations to guide development of access controls and decision support for genetic information in electronic medical records. BMC Health Serv Res 2011;11:294.

Drozda K, Labinov Y, Jiang R, et al. A pharmacogenetics service experience for pharmacy students, residents, and fellows. Am J Pharm Educ 2013;77:175.

Evaluation of Genomic Applications in Practice and Prevention (EGAPP) Working Group. The EGAPP initiative: lessons learned. Genet Med 2014;16:217-24.

Ferreri SP, Greco AJ, Michaels NM, et al. Implementation of a pharmacogenomics service in a community pharmacy. J Am Pharm Assoc 2014;54:172-80.

Fichter B, Heintzelman T, Hurst S, et al. Clinical utility of pharmacogenetic testing in a compounding pharmacy. Int J Pharm Compd 2013;17:452-7.

Formea CM, Nicholson WT, McCullough KB, et al. Development and evaluation of a pharmacogenomics educational program for pharmacists. Am J Pharm Educ 2013;77:10.

Genetics and Genomics Competency Center for Education, National Human Genome Research Institute (NHGRI) [www.g-2-c-2.org]. Bethesda, MD: Pharmacist Pharmacogenomic Competencies and Outcomes, 2012. Available at www.g-2-c-2.org/attachments/pdf/Pharmacist-Comp.pdf. Accessed April 6, 2014.

Gharani N, Keller MA, Stack CB, et al. The Coriell personalized medicine collaborative pharmacogenomics appraisal, evidence scoring and interpretation system. Genome Med 2013;5:93.

Goldspiel BR, Flegel WA, DiPatrizio G, et al. Integrating pharmacogenomics information and clinical decision support into the electronic health record. J Am Med Inform Assoc 2014;21:522-8.

Gottesman O, Scott SA, Ellis SB, et al. The CLIPMERGE PGx program: clinical implementation of personalized medicine through electronic health records and genomics-pharmacogenomics. Clin Pharmacol Ther 2013;94:214-7.

Greco KE, Tinley S, Seibert D. Essential Genetic and Genomic Competencies for Nurses with Graduate Degrees. Silver Spring, MD: American Nurses Association and International Society of Nurses in Genetics, 2012.

Green RC, Berg JS, Grody WW, et al. ACMG recommendations for reporting of incidental findings in clinical exome and genome sequencing. Genet Med 2013;5:565-74.

Haga SB, Moaddeb J. Comparison of delivery strategies for pharmacogenetic testing services. Pharmacogenet Genomics 2014;24:139-45.

Haga SB, O'Daniel JM, Tindall GM, et al. Survey of genetic counselors and clinical geneticists' use and attitudes toward pharmacogenetic testing. Clin Genet 2012;82:115-20.

Hartzler A, McCarty CA, Rasmussen LV, et al. Stakeholder engagement: a key component of integrating genomic information into electronic health records. Genet Med 2013;15:792-801.

Hoffman JM, Haidar CE, Wilkinson MR, et al. PG4KDS: a model for clinical implementation of pre-emptive pharmacogenetics. Am J Med Genet C Semin Med Genet 2014;166C:45-55.

Hresko A, Haga SB. Insurance coverage policies for personalized medicine. J Pers Med 2012;2:201-16.

Johnson JA, Bootman JL, Evans WE, et al. Pharmacogenomics: a scientific revolution in pharmaceutical sciences and pharmacy practice. Am J Pharm Educ 2002;66:12S-15S.

Johnson JA, Burkley BM, Langaee TY, et al. Implementing personalized medicine: development of a cost-effective customized pharmacogenetics genotyping array. Clin Pharmacol Ther 2012;92:437-9.

Johnson JA, Cavallari LH. Pharmacogenetics and cardiovascular disease—implications for personalized medicine. Pharmacol Rev 2013;65:987-1009.

Kisor DF, Bright DR, Conaway M, et al. Pharmacogenetics in the community pharmacy: thienopyridine selection post-coronary stent placement. J Pharm Pract 2014;27:416-9.

Konstantinos NL, McAllister TM, Babovic-Vuksanovic D, et al. Implementing individualized medicine into practice. Am J Med Genet C Semin Med Genet 2014;166C:15-23.

Lam YWF. Scientific challenges and implementation barriers to translation of pharmacogenomics in clinical practice. ISRN Pharmacol 2013;2013:641089.

Manolio TA, Chisholm RL, Ozenberger B, et al. Implementing genomic medicine in the clinic: the future is here. Genet Med 2013;15:258-67.

Martin MA, Hoffman JM, Freimuth RR, et al. Clinical Pharmacogenetics Implementation Consortium guideline for *HLA-B* genotype and abacavir dosing: 2014 update. Clin Pharmacol Ther 2014;95:499-500.

McCullough KB, Formea CM, Berg KD, et al. Assessment of pharmacogenomics educational needs of pharmacists. Am J Pharm Educ 2011;75:51.

McKinnon RA, Ward MB, Sorich MJ. A critical analysis of barriers to the clinical implementation of pharmacogenomics. Ther Clin Risk Manag 2007;3:751-9.

Mills R, Haga SB. Clinical delivery of pharmacogenetic testing services: a proposed partnership between genetic counselors and pharmacists. Pharmacogenomics 2013;14:957-68.

National Coalition for Health Professional Education in Genetics (NCHPEG) [www.nchpeg.org]. Core Competencies for All Health Professionals. Lutherville, MD: NCHPEG, 2007. Available at www.nchpeg.org/index.php?option=com_content&view=article&id=237&Itemid=84#. Accessed April 7, 2014.

Nutescu EA, Drozda K, Bress AP, et al. Feasibility of implementing a comprehensive warfarin pharmacogenetics service. Pharmacotherapy 2013;33:1156-64.

O'Donnell PH, Danahey K, Jacobs M, et al. Adoption of a clinical pharmacogenomics implementation program during outpatient care—initial results of the University of Chicago "1,200 Patients Project." Am J Med Genet C Semin Med Genet 2014;166:68-75.

Overby CL, Erwin AL, Abul-Husn NS, et al. Physician attitudes toward adopting genome-guided prescribing through clinical decision support. J Pers Med 2014;4:35-49.

Passamani E. Educational challenges in implementing genomic medicine. Clin Pharmacol Ther 2013;94:192-5.

Patel HN, Ursan ID, Zueger PM, et al. Stakeholder views on pharmacogenomic testing. Pharmacotherapy 2014;34:151-65.

Perry CG, Shuldiner AR. Pharmacogenomics of anti-platelet therapy: how much evidence is enough for clinical implementation? J Hum Genet 2013;58:339-45.

Peterson JF, Bowton E, Field JR, et al. Electronic health record design and implementation for pharmacogenomics: a local perspective. Genet Med 2013;15:833-41.

Phillips KA, Sakowski JA, Trosman J, et al. The economic value of personalized medicine tests: what we know and what we need to know. Genet Med 2014;16:251-7.

Pulley JM, Denny JC, Peterson JF, et al. Operational implementation of prospective genotyping for personalized medicine: the design of the Vanderbilt PREDICT project 2012;92:87-95.

Rackover M, Goldgar C, Wolpert C, et al. Establishing essential physician assistant clinical competencies guidelines for genetics and genomics. J Physician Assist Educ 2007;18:47-8.

Ratain MJ, Cox NJ, Henderson TO. Challenges in interpreting the evidence for genetic predictors of toxicity. Clin Pharmacol Ther 2013;94:631-5.

Relling MV, Klein TE. CPIC: Clinical Pharmacogenetics Implementation Consortium of the pharmacogenomics research network. Clin Pharmacol Ther 2011;89:464-7.

Reynolds KS. Achieving the promise of personalized medicine [editorial]. Clin Pharmacol Ther 2012;92:401-5.

Salari K, Karczewski KJ, Hudgins L, et al. Evidence that personal genome testing enhances student learning in a course on genomics and personalized medicine. PLoS ONE 2013;8:e68853.

Schildcrout JS, Denny JC, Bowton E, et al. Optimizing drug outcomes through pharmacogenetics: a case for preemptive genotyping. Clin Pharmacol Ther 2012;92:235-42.

Scott SA. Personalizing medicine with clinical pharmacogenetics. Genet Med 2011;13:987-95.

Shuldiner AR, Palmer K, Pakyz RE, et al. Implementation of pharmacogenetics: the University of Maryland Personalized Anti-Platelet Pharmacogenomics Program. Am J Med Genet C Semin Med Genet 2014;166C:76-84.

Shuldiner AR, Relling MV, Peterson JF, et al. The Pharmacogenomics Research Network Translational Pharmacogenomics Program: overcoming challenges of real-world implementation. Clin Pharmacol Ther 2013;94:207-10.

Stanek EJ, Sanders CL, Taber KA, et al. Adoption of pharmacogenomics testing by U.S. physicians: results of a nationwide survey. Clin Pharmacol Ther 2012;91:450-8.

Sturm AC, Sweet K, Manickam K. Implementation of a clinical research pharmacogenomics program at an academic medical center: role of the genetics healthcare professional. Pharmacogenomics 2013;14:703-6.

Ury AG. Storing and interpreting genomic information in widely deployed electronic health record systems. Genet Med 2013;15:779-85.

Van Driest SL, Shi Y, Bowton EA, et al. Clinically actionable genotypes among 10,000 patients with preemptive pharmacogenetic testing. Clin Pharmacol Ther 2014;95:423-31.

van Schie RMF, de Boer A, Maitland-van der Zee AH. Implementation of pharmacogenetics in clinical practice is challenging [editorial]. Pharmacogenomics 2011;12:1231-3.

Vaughan KTL, Scolaro KL, Anksorus HN, et al. An evaluation of pharmacogenomic information provided by five common drug information sources. J Med Libr Assoc 2014;102:47-51.

Weitzel KW, Elsey AR, Langaee TY, et al. Clinical pharmacogenetics implementation: approaches, successes, challenges. Am J Med Genet C Semin Med Genet 2014;166:56-67.

Welch BM, Kawamoto K. Clinical decision support for genetically guided personalized medicine: a systematic review. J Am Med Inform Assoc 2013;20:388-400.

Williams MS. Genomic medicine implementation: learning by example. Am J Med Genet C Semin Med Genet 2014;166C:8-14.

Wong WB, Carlson JJ, Thariani R, et al. Cost effectiveness of pharmacogenomics: a critical and systematic review. Pharmacoeconomics 2010;28:1001-13.

Section 2
Pharmacogenomics in Clinical Care

Chapter 5

PHARMACOGENETICS OF ANTIPLATELET DRUGS

Aniwaa Owusu Obeng, Pharm.D.; and Stuart A. Scott, Ph.D.

Learning Objectives
1. Develop competency with the indications for antiplatelet therapy and basic platelet biology.
2. Analyze and evaluate the effects of *CYP2C19* genotype on clopidogrel pharmacokinetics, pharmacodynamics, and clinical outcomes.
3. Identify clinically actionable genetic variants associated with clopidogrel response.
4. Assess the evidence necessary to evaluate the utility and feasibility of genotype-guided antiplatelet therapy.
5. Review the potential effects of genotype on other antiplatelet agents.
6. Assess the opportunities and challenges when implementing genotype-guided antiplatelet therapy.

Keywords: Antiplatelet agents, aspirin, clopidogrel, prasugrel, ticagrelor, acute coronary syndromes, percutaneous coronary intervention, *CYP2C19*

Abbreviations in This Chapter
ACS	Acute coronary syndromes
ADP	Adenosine diphosphate
COX-1	Cyclooxygenase-1
GWAS	Genome-wide association study
HTPR	High on-treatment platelet reactivity
MI	Myocardial infarction
PAD	Peripheral arterial disease
PCI	Percutaneous coronary intervention
RCT	Randomized controlled trial
TXA$_2$	Thromboxane A$_2$

Abstract
The currently available oral antiplatelet drugs include aspirin, clopidogrel, prasugrel, and ticagrelor, which typically are prescribed as dual antiplatelet therapy for the prevention of ischemic events in patients with acute coronary syndromes (ACS) and/or after percutaneous coronary intervention (PCI) and other indications. Potential genetic determinants of response variability for each of these agents have been widely studied, including associations with drug pharmacokinetics, pharmacodynamics, and clinical outcomes. However, the most robust pharmacogenetic effect among the antiplatelet drugs has been the role of cytochrome P450 (CYP) 2C19 (*CYP2C19*) loss-of-function alleles and clopidogrel response variability in ACS/PCI patients. This chapter reviews the antiplatelet pharmacogenetics field, with an emphasis on *CYP2C19* and clopidogrel response variability, and the feasibility of genotype-directed antiplatelet therapy.

Introduction
Antiplatelet agents are commonly prescribed to prevent ischemic events in patients with acute coronary syndromes (ACS) and other indications. Despite the continued development of new antiplatelet agents, variability in response is observed, which can result in adverse cardiovascular events. Pharmacogenetic determinants of response variability for each of the currently approved antiplatelet agents have been studied, but none more so than the antiplatelet prodrug clopidogrel. Before reviewing the antiplatelet pharmacogenetics literature, a brief overview of platelet biology and the available antiplatelet agents is presented.

Platelet Biology and Aggregation

Platelet activation and coagulation do not typically occur in an intact blood vessel. Rather, these processes are initiated in response to vascular injury or atherosclerotic plaque rupture, which also leads to vasoconstriction at the damaged site to attenuate blood loss. The endothelial cells of the compromised vessel subsequently secrete von Willebrand factor, collagen, and other tissue factors to the bloodstream that adhere to and activate circulating platelets (for a detailed review, see Sangkuhl 2011). Activated platelets change shape to facilitate further adhesion, initiate the arachidonic acid pathway to produce thromboxane A_2 (TXA_2), and excrete the contents of their granules, releasing adenosine diphosphate (ADP), serotonin, and other proteins. A positive feedback mechanism promotes further vasoconstriction, additional platelet localization, and ultimately aggregation to form a platelet plug. Activated platelets initiate the intrinsic and extrinsic coagulation pathways, which eventually leads to thrombin-mediated conversion of fibrinogen to fibrin, cross-linking and aggregation, strengthening of the platelet plug, and thrombus formation. Given this fundamental role of platelets in blood loss prevention and vasculature integrity, abnormal platelet counts and platelet dysfunction can lead to several different hematologic disorders. In addition, platelets are directly implicated in common cardiovascular diseases, including atherosclerosis, coronary artery disease, and myocardial infarction (MI), as well as cerebrovascular disease and stroke.

Antiplatelet Agents

Antiplatelet agents are commonly prescribed for primary and secondary prevention of ischemic events caused by ACS, ischemic stroke, and symptomatic peripheral arterial disease (PAD) (Faxon 2006; Tran 2004). Because platelet activation is influenced, in part, by TXA_2, ADP, serotonin, thrombin, epinephrine, and collagen (Jennings 2009), these factors have been widely studied as potential sites of action for antiplatelet therapies. The currently approved antiplatelet agents are detailed in the text that follows.

TXA2 Inhibitors

Acetylsalicylic acid, aspirin, is widely used for cardiovascular protection. In low doses, aspirin reduces the risk of MI, stroke, and cardiovascular death in patients with a history of atherosclerotic coronary, cerebral, or peripheral vascular disease and is standard of care for secondary prevention of coronary artery disease. It acts by irreversibly inhibiting the platelet cyclooxygenase-1 (COX-1) enzyme, which impedes the generation of TXA_2, a potent vasoconstrictor and activator of platelet aggregation (Faraday 2007a). Platelets exposed to aspirin are permanently disabled from generating TXA_2 and are inhibited throughout their life span. However, platelet activation can proceed along COX-1–independent pathways, prompting the need for supplementary blockage of ADP-$P2RY_{12}$–mediated activation with a thienopyridine derivative in patients at high risk of platelet-mediated thrombosis, such as in ACS and after PCI.

ADP Antagonists/$P2Y_{12}$ Receptor Inhibitors

Thienopyridines are antiplatelet agents that directly bind to $P2Y_{12}$ receptors and block ADP-induced platelet activation and aggregation. The first available thienopyridine for ACS and PCI was ticlopidine; however, it has largely been supplanted by newer antiplatelet agents because of the increased risks of neutropenia and thrombotic thrombocytopenia purpura (Quinn 1999).

Clopidogrel is a second-generation thienopyridine that is biotransformed in the liver to an active metabolite that binds specifically and irreversibly to the platelet $P2Y_{12}$ receptor, inhibiting ADP-mediated platelet activation and aggregation. Of note, most of the prodrug (around 85%) is hydrolyzed to inactive metabolites by esterases, leaving only 15% available for transformation to the active metabolite (Figure 5.1). Two sequential oxidative reactions are necessary to form the active metabolite, involving CYP1A2, CYP2B6, and CYP2C19, and CYP2B6, CYP2C9, CYP2C19, CYP3A4, and CYP3A5, respectively. (Figure 5.1) (Kazui 2010; Sangkuhl 2010). Dual antiplatelet therapy with clopidogrel and aspirin reduces cardiovascular death and ischemic complications in patients with ACS and those undergoing PCI (Sabatine 2005; Steinhubl 2002; Mehta 2001). However, wide interindividual variability in platelet aggregation is commonly observed during dual antiplatelet therapy, and some patients still experience thrombotic events (Gurbel 2003; Muller 2003; Jaremo 2002). Moreover, persistent high on-treatment platelet reactivity (HTPR) is a strong predictor of adverse cardiovascular events (Buonamici 2007). Other factors contributing to interindividual differences in clopidogrel response

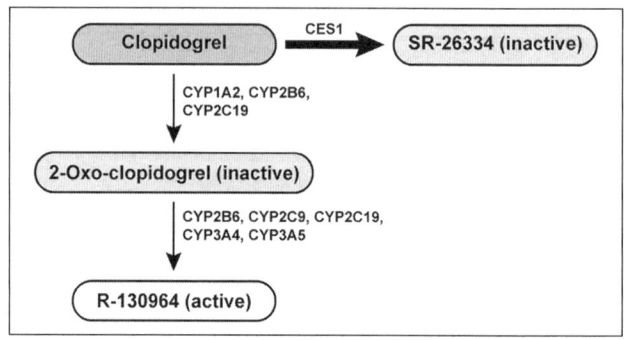

Figure 5.1. Schematic illustration of the hepatic metabolism of clopidogrel. The thickness of the arrows represents the relative contribution of the respective pathway.

include age, co-medications, diabetes, renal failure, cardiac failure, and reduced CYP2C19 enzyme activity as a result of germline *CYP2C19* loss-of-function alleles.

Prasugrel is a third-generation thienopyridine indicated with aspirin as dual antiplatelet therapy for the pharmaceutical management of patients with ACS undergoing PCI. Prasugrel is also a prodrug that undergoes hepatic bioactivation by CYP2B6, CYP2C9, CYP2C19, CYP2D6, and CYP3A4 to generate its active metabolite, which antagonizes the $P2Y_{12}$ receptor and impairs the ADP-mediated activation of the glycoprotein GPIIb/IIIa complex. Compared with clopidogrel, prasugrel is rapid acting and generates a higher level of active metabolite, resulting in a more potent and effective platelet inhibition (Mega 2009a; Wallentin 2008; Brandt 2007b; Payne 2007); however, these benefits are counterbalanced by the increased risk of major bleeding complications (Wiviott 2007). In addition, reduced dose requirements (i.e., 5 mg) have been recommended for patients weighing less than 60 kg and those 75 years and older because of increased production of active metabolites and bleeding risks (Roe 2013; Erlinge 2012; Wiviott 2007). Despite the advantages of prasugrel over clopidogrel, HTPR has also been reported among prasugrel-treated patients after PCI, which was associated with higher rates of thrombotic events (Bonello 2011).

Ticagrelor is a cyclopentyl-triazolo-pyrimidine agent that is an allosteric ADP antagonist, but unlike clopidogrel and prasugrel, it does not require hepatic bioactivation to generate an active metabolite (Htun 2013; Teng 2012). Ticagrelor has a faster onset and offset of action and achieves a more pronounced and consistent antiplatelet response than clopidogrel (Gurbel 2009), which has translated to ticagrelor's superior efficacy among patients with ACS, including reductions in stent thrombosis and all-cause mortality (Wallentin 2009). No significant difference in major bleeding rates was found between ticagrelor and clopidogrel, but ticagrelor was associated with a higher rate of major bleeding unrelated to coronary artery bypass grafting, dyspnea, and bradyarrhythmia (Yousuf 2011; Wallentin 2009). Contraindications include history of intracranial hemorrhage, active bleeding, severe hepatic impairment, and hypersensitivity.

Pharmacogenetic Determinants of Clopidogrel Response

Candidate Gene Studies
In an effort to identify genetic variants that influence interindividual variability in clopidogrel response, several candidate genes in the pharmacokinetic and pharmacodynamic clopidogrel pathways have been studied. Among them, the most robust association has been with the common *CYP2C19*2* loss-of-function allele (c.681G>A; rs4244285), which was initially reported in 2006 to be significantly associated with HTPR in healthy subjects (Hulot 2006). As detailed later in this chapter, since this initial observation, several studies have confirmed *CYP2C19* as the major genetic determinant of clopidogrel metabolite levels, on-treatment platelet reactivity, and adverse cardiovascular event risks. The *CYP2C19*2* allele frequencies are around 15% in whites and Africans and 29%–35% in Asians (Scott 2013b). Other *CYP2C19* alleles with established loss-of-function include *3–*8; however, their frequencies are typically below 1%, with the exception of *3 (c.636G>A; rs4986893) in Asians (2%–9%) (Scott 2013b). The *CYP2C19*17* allele (c.-806C>T; rs12248560) results in increased activity because of enhanced transcription and has multiethnic allele frequencies ranging from about 3% to 21% (Scott 2013b). These allele frequencies indicate that about 20%–30% of individuals from certain major racial and ethnic groups carry at least one *CYP2C19* loss-of-function allele, which can be as high as 50% among Asian individuals (Martis 2013). According to their *CYP2C19* genotype, individuals typically are categorized as ultrarapid (*1/*17, *17/*17), extensive (*1/*1), intermediate (e.g., *1/*2, *1/*3, *2/*17), or poor (e.g., *2/*2, *2/*3) metabolizers (Scott 2012).

Other studied candidate genes potentially implicated in clopidogrel response variability include *ABCB1* (Mega 2010a; Simon 2009), *CES1* (Lewis 2013a), *CYP2B6* (Mega 2009b), *CYP2C9* (Harmsze 2010; Gladding 2009; Brandt 2007a), *CYP3A4* (Lau 2004), *P2RY12* (Malek 2008; Fontana 2003), and *PON1* (Bouman 2011). However, these have not been adequately replicated, and their clinical validity remains uncertain. Of note, the reported association between *PON1* and clopidogrel pharmacokinetics, pharmacodynamics, and clinical outcomes has been refuted by several reports (Ancrenaz 2012; Gong 2012; Reny 2012; Hulot 2011; Lewis 2011; Sibbing 2011; Simon 2011b; Trenk 2011).

Genome-Wide Association Studies
To date, only one genome-wide association study (GWAS) of clopidogrel response has been reported (Shuldiner 2009). The studied population was a healthy Amish cohort treated with clopidogrel, and the response phenotype was ADP-induced platelet aggregation. Of importance, platelet response to clopidogrel was found to be highly heritable (h2 = 0.73), and the most significant associations with diminished clopidogrel response were variants in the *CYP2C18-CYP2C19-CYP2C9-CYP2C8* cluster

on chromosome 10q24. The most significant variant (rs12777823) was in strong linkage disequilibrium with *CYP2C19*2* in this population, which accounted for 12% of the variation in ADP-induced platelet aggregation (Shuldiner 2009). Although no other variants reached statistical significance for association with response, this agnostic GWAS validated previous *CYP2C19* candidate gene studies and confirmed its role as the major genetic determinant of interindividual clopidogrel response variability.

CYP2C19 and Evidence for Clinical Validity

Pharmacokinetics
Given the essential role that CYP2C19 plays in the two-step bioactivation of the active clopidogrel metabolite (R-130964) (Figure 5.1), defective *CYP2C19* activity caused by loss-of-function alleles (e.g., *2, *3) results in reduced formation of active metabolites in both healthy subjects (Gong 2012; Kelly 2012; Simon 2011a; Mega 2009b; Kim 2008; Umemura 2008; Brandt 2007a) and cardiac patients (Collet 2011; Varenhorst 2009) treated with clopidogrel.

Pharmacodynamics
Consistent with the association between *CYP2C19* and active clopidogrel metabolite levels, several studies have confirmed the major role of *CYP2C19* loss-of-function alleles with HTPR, typically measured by ex vivo ADP-induced platelet aggregometry, in both healthy subjects (Mega 2009b; Shuldiner 2009; Chen 2008; Kim 2008; Umemura 2008; Brandt 2007a; Fontana 2007; Hulot 2006) and cardiac patients (Zou 2013; Bonello 2012; Harmsze 2012; Price 2012; Cuisset 2011; Gurbel 2011; Hochholzer 2010; Collet 2009; Mega 2009b; Shuldiner 2009; Frere 2008; Trenk 2008; Giusti 2007). Some studies also have suggested that the common *CYP2C19*17* increased activity allele results in both enhanced platelet inhibition and clopidogrel response (Sibbing 2010a; Tiroch 2010; Frere 2009; Mega 2009b); however, other studies have not identified an independent effect of *CYP2C19*17* (Lewis 2013b; Sorich 2012; Shuldiner 2009; Simon 2009; Geisler 2008). This discrepancy suggests that altering antiplatelet therapy according to *17 genotype is not currently supported.

Clinical Outcomes
Given the reduced active clopidogrel metabolite formation and HTPR among *CYP2C19* loss-of-function allele carriers, substantial evidence exists linking *CYP2C19* genotype with clinical outcomes among clopidogrel-treated patients with ACS, particularly those undergoing PCI. For example, *CYP2C19*2*-mediated HTPR was initially reported in 2008 to be associated with poor clinical outcomes (1-year incidence of death and MI) after PCI (Trenk 2008). Direct association between *CYP2C19* loss-of-function alleles and adverse cardiovascular outcomes (e.g., cardiovascular death, MI, stroke, stent thrombosis) in clopidogrel-treated ACS/PCI patients has been replicated by several clinical studies (Cayla 2011; Harmsze 2010; Collet 2009; Giusti 2009; Mega 2009b; Shuldiner 2009; Sibbing 2009; Simon 2009). However, a significant effect of *CYP2C19* loss-of-function alleles on clopidogrel response has not generally been identified among lower-risk coronary patients (e.g., low frequencies of PCI) and other indications (e.g., atrial fibrillation) (Pare 2010; Wallentin 2010), underscoring the importance of clopidogrel indication when considering *CYP2C19* genetic testing for antiplatelet management (see text that follows) (Johnson 2012). As noted earlier, the *CYP2C19*17* allele has been reported to result in an enhanced clopidogrel response, which has translated to an increased risk of bleeding complications in some studies (Li 2012; Sibbing 2010b); however, this finding has not yet been adequately replicated to justify therapy modification according to *17 genotype.

CYP2C19 Meta-analyses and Clinical Outcomes
Large meta-analyses have shown that clopidogrel-treated ACS/PCI patients who are *CYP2C19*2* heterozygotes or homozygotes have an increased risk of major adverse cardiovascular events compared with *1 homozygotes (hazard ratio [HR] 1.55; 95% confidence interval [CI], 1.11–2.17 for heterozygotes; HR 1.76; 95% CI, 1.24–2.50 for homozygotes) and increased risks of stent thrombosis (HR 2.67; 95% CI, 1.69–4.22 for heterozygotes; HR 3.97; 95% CI, 1.75–9.02 for homozygotes) (Mega 2010b). Other meta-analyses have replicated the association between *CYP2C19* genotype and stent thrombosis, with reported odds ratios ranging from 1.75 to 3.82 among *2 heterozygotes and homozygotes (Yamaguchi 2013; Jang 2012; Singh 2012; Zabalza 2012; Bauer 2011; Jin 2011; Sofi 2011; Hulot 2010). Consistent with the studies that have not identified an effect of *CYP2C19* among lower-risk indications, meta-analyses that include studies with low frequencies of PCI, patients without coronary disease, follow-up periods beyond the duration of clopidogrel therapy, or non-cardiovascular outcomes have not supported a major role for *CYP2C19* in clopidogrel response variability in these patient populations (Holmes 2011).

Clopidogrel Dose Escalation Studies
Increased clopidogrel loading and maintenance dosing strategies in both healthy subjects and patients with ACS improve platelet inhibition among *CYP2C19*2* heterozygotes on the basis of ex vivo platelet aggregation, but only nominally among homozygotes after very high maintenance doses

(Collet 2011; Cuisset 2011; Mega 2011; Simon 2011). However, large clinical trials that evaluated higher-dose clopidogrel (e.g., 150-mg maintenance dose) in ACS/PCI patients with HTPR have concluded that doubling the clopidogrel dose on the basis of platelet function monitoring alone does not reduce the incidence of death from cardiovascular causes, nonfatal MI, or stent thrombosis (Collet 2012; Price 2011). Further clopidogrel dose escalation trials are still warranted given the significantly higher doses used in the pharmacodynamic studies compared with some of the clinical outcome trials. As such, implementing increased clopidogrel doses according to *CYP2C19* genotype is not currently supported by the available evidence.

CYP2C19 Genetic Testing Implementation

Antiplatelet Indication and *CYP2C19* Genetic Testing

Evidence from clinical trial genetic substudies and meta-analyses has reproducibly shown that the effect of *CYP2C19* loss-of-function alleles on clopidogrel efficacy is directly linked to the indication for antiplatelet therapy (Johnson 2012; Cayla 2011; Holmes 2011; Mega 2010b, 2009b; Pare 2010; Collet 2009; Giusti 2009; Shuldiner 2009; Sibbing 2009; Simon 2009). As previously noted, *CYP2C19* genotyping is not supported for all clopidogrel indications. Lower-risk indications (e.g., medical management of ACS, PAD) derive a lesser overall benefit from clopidogrel therapy than do higher-risk indications such as PCI (Holmes 2011; Pare 2010). As such, the influence of defective *CYP2C19* on clinical outcomes has been most evident among PCI patient cohorts, indicating that *CYP2C19* genotyping (if pursuing) should be restricted to ACS/PCI patients or elective PCI patients. Of note, electronic health records and other prescribing systems do not typically require that medication orders be linked to indication, which is important when implementing *CYP2C19* genotype clinical decision support for clopidogrel prescriptions. As such, best practice advisory content needs to underscore this caveat to avoid any inappropriate change in antiplatelet therapy for patients with low-risk indications carrying *CYP2C19* loss-of-function alleles. This should also be addressed by properly educating prescribing clinicians whenever possible.

CYP2C19-Directed Antiplatelet Therapy Practice Guidelines

The increasing evidence implicating defective *CYP2C19* activity and increased risks of adverse cardiovascular events among clopidogrel-treated ACS/PCI patients prompted the U.S. Food and Drug Administration (FDA) to add a boxed warning to the clopidogrel label in 2010 describing the importance of *CYP2C19* pharmacogenetics for clopidogrel response. In addition to this label revision, many professional societies and consortia have since published practice guidelines and recommendations for *CYP2C19*-genotype directed antiplatelet therapy (Table 5.1).

American College of Cardiology Foundation/ American Heart Association

In 2010, the American College of Cardiology Foundation (ACCF) and the American Heart Association (AHA), with endorsement from the Society for Cardiovascular Angiography and Interventions and the Society of Thoracic Surgeons, issued a clinical alert in response to the boxed warning added to the clopidogrel label (Holmes 2011). The expert committee urged clinicians to be aware of the association between *CYP2C19* loss-of-function alleles and impaired clopidogrel metabolism and reduced platelet inhibition; however, they did not recommend routine genetic testing before therapy initiation, citing insufficient evidence. Nevertheless, they did suggest that genetic testing for patients at moderate to high risk of poor outcomes (e.g., those undergoing elective high-risk PCI procedures) be considered and that identified poor metabolizers be prescribed an alternative antiplatelet regimen (Holmes 2011). Furthermore, although somewhat less clearly, the 2012 ACCF/AHA focused update guideline for management of unstable angina/non–ST-elevation MI noted that *CYP2C19* genetic testing "might be considered if results of testing may alter management" (Anderson 2013).

European Science Foundation— University of Barcelona

At the 2010 European Science Foundation—University of Barcelona Conference in Biomedicine on Pharmacogenetics and Pharmacogenomics, practical recommendations for 10 clinical pharmacogenetic interactions were developed (Becquemont 2011). For clopidogrel and *CYP2C19*, this expert panel recommended testing for the *2 and *3 loss-of-function alleles and the use of an alternative antiplatelet for patients with MI managed with PCI who are carriers (intermediate and poor metabolizers).

Royal Dutch Association for the Advancement of Pharmacy Pharmacogenetics Working Group

The Royal Dutch Association for the Advancement of Pharmacy Pharmacogenetics Working Group develops pharmacogenetic-driven dose guidelines for its nationwide computerized drug prescription and automated medication surveillance system (Swen 2008). Among the 163 drug-gene pairs this multidisciplinary working group evaluated in 2011 (involving 53 medications and

Table 5.1. Currently Available Practice Guidelines and Professional Recommendations on Pharmacogenetic Testing for Antiplatelet Agents

Organization	Drug	Gene	Recommendation	References
ACCF/AHA	Clopidogrel	CYP2C19	- Consider genetic testing for patients at moderate to high risk of poor outcomes (e.g., PCI) - Consider an alternative antiplatelet for poor metabolizers	Holmes 2010
ACCF/AHA	Clopidogrel	CYP2C19	- Genotyping CYP2C19 loss-of-function variants in patients with unstable angina/non–ST-elevation MI (or, after ACS and with PCI) on $P2Y_{12}$ receptor inhibitor therapy might be considered if results of testing may alter management	Anderson 2013
ESF-UB	Clopidogrel	CYP2C19	- Test patients for CYP2C19*2 and *3 loss-of-function alleles and use of an alternative antiplatelet agent for PCI-managed patients with MI who are either intermediate or poor metabolizers	Becquemont 2011
KNMP-PWG	Clopidogrel	CYP2C19	- Consider an alternative antiplatelet agent for intermediate and poor metabolizers (indication not explicitly stated)	Swen 2011
CPIC	Clopidogrel	CYP2C19	- Consider an alternative antiplatelet agent (in the absence of a contraindication) for PCI patients (ACS or elective) who are either intermediate or poor metabolizers	Scott et al. 2013b, 2011b

ACCF = American College of Cardiology Foundation; AHA = American Heart Association; CPIC = Clinical Pharmacogenetics Implementation Consortium; ESF-UB = European Science Foundation-University of Barcelona (Pharmacogenetics and Pharmacogenomics: Practical Applications in Routine Medical Practice Conference); KNMP-PWG = Royal Dutch Association for the Advancement of Pharmacy-Pharmacogenetics Working Group; MI = myocardial infarction; PCI = percutaneous coronary intervention.

11 genes), the group also concluded that CYP2C19 intermediate and poor metabolizers should avoid clopidogrel because of the increased risk of reduced efficacy (Swen 2011); however, the indication for antiplatelet therapy was not explicitly addressed.

Clinical Pharmacogenetics Implementation Consortium

The CPIC (Clinical Pharmacogenetics Implementation Consortium) guideline on CYP2C19 genotype-directed antiplatelet therapy and the 2013 update recommend alternative antiplatelet therapies (i.e., prasugrel or ticagrelor) for PCI patients (ACS or elective) who are CYP2C19 intermediate or poor metabolizers and standard clopidogrel dosing for those who are extensive and ultrarapid metabolizers (in the absence of a contraindication) (Figure 5.2) (Scott 2013b, 2011b). Of note, this recommendation is indication-specific for patients undergoing PCI because current evidence does not support CYP2C19-directed prescribing for patients with a lower risk of adverse cardiovascular outcomes (e.g., stroke, PAD) (Johnson 2012).

CYP2C19 Genotyping Platforms

Clinical genetic testing laboratories offer either FDA-approved tests or validated laboratory-developed tests. Table 5.2 summarizes the commercially available CYP2C19 genotyping tests that currently are FDA label approved for in vitro diagnostic testing. Of note, all of these tests interrogate the common *2 and *3 loss-of-function alleles, and the recently approved platforms also include *17. These approved tests do not include the CYP2C19 alleles with low frequencies in the general population (*4–*8), despite their established loss-of-function according to in vitro studies. However, all of these alleles are frequently included in CYP2C19 genotyping laboratory-developed tests, and some clinical laboratories interrogate CYP2C19 by Sanger or next-generation sequencing mutation scanning, which also identifies these rare alleles and other novel coding variants.

Genetic Testing Registry

The National Institutes of Health (NIH) Genetic Testing Registry (GTR; www.ncbi.nlm.nih.gov/gtr/) is a central location for voluntary submission of genetic test information by laboratory providers (Rubinstein 2013). The database provides details of each test (e.g., purpose, target populations, method, what it measures, analytic validity, clinical validity, clinical utility, ordering information) and each laboratory (e.g., location, contact information, certification, and licenses) used. However, because the GTR was only recently deployed and given the voluntary nature of the database, current genetic test availability, including *CYP2C19* genotyping, is only now emerging. As such, in addition to the GTR, testing menus from local CLIA-certified laboratories should be directly queried when seeking *CYP2C19* genetic testing availability.

Barriers to CYP2C19 Genetic Testing Implementation

Robust multiplexed genotyping techniques, rapidly advancing evidence linking medical conditions and therapeutic responses to germline genetic variation, and the ongoing publication of genome-directed practice guidelines are together facilitating the implementation of clinical pharmacogenetics, including *CYP2C19*-directed antiplatelet therapy. However, clinician adoption of *CYP2C19* testing has not been widespread because several barriers exist (Scott 2011a).

Clinical Utility and Randomized Controlled Trials

The most definitive evidence for changing medical practice is traditionally derived from randomized controlled trials (RCTs); however, these trials are challenging to execute for pharmacogenetic interventions because of ethical issues and the feasibility of conducting adequately powered studies for rare adverse events in heterogeneous patient populations. To date, no large prospective RCTs have been reported that tested the efficacy of genotype-guided antiplatelet therapy. However, a recent RCT assessing *CYP2C19*-directed clopidogrel therapy with 600 PCI patients concluded that personalized antiplatelet therapy according to *CYP2C19* genotype significantly decreases the incidence of major adverse cardiovascular events and

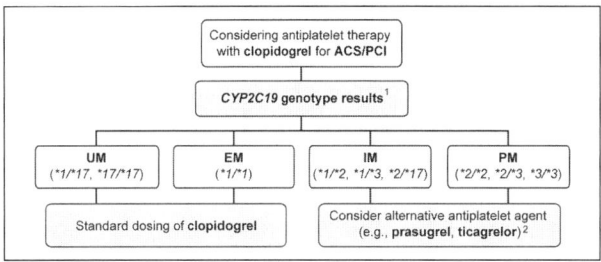

Figure 5.2. Clinical Pharmacogenetics Implementation Consortium (CPIC) algorithm for suggested clinical actions based on *CYP2C19* genotype when considering clopidogrel treatment for ACS/PCI patients.

[1] Other possible *CYP2C19* genotypes with rare loss-of-function alleles exist beyond those illustrated (see text and Scott 2013b).

[2] Note that prasugrel and ticagrelor are recommended only when not contraindicated clinically.

ACS = acute coronary syndromes; EM = extensive metabolizer; IM = intermediate metabolizer; PCI = percutaneous coronary intervention; PM = poor metabolizer. UM = ultrarapid metabolizer.

Image reproduced from Scott SA, Sangkuhl K, Stein CM, et al. Clinical Pharmacogenetics Implementation Consortium guidelines for *CYP2C19* genotype and clopidogrel therapy: 2013 update. Clin Pharmacol Ther 2013b;94:317-23.

Table 5.2. CYP2C19 Genotyping Tests Approved by the FDA for In Vitro Diagnostic Use[a]

Assay	Alleles Interrogated	Company	Date Approved
AmpliChip CYP450 Test	*2, *3	Hoffmann-La Roche	January 2005
INFINITI CYP2C19 Assay	*2, *3, *17	Autogenomics	October 2010
Verigene CYP2C19 Test	*2, *3, *17	Nanosphere	November 2012
Spartan RX CYP2C19 Assay	*2, *3, *17	Spartan Bioscience	August 2013
xTAG CYP2C19 Kit v3	*2, *3, *17	Luminex Molecular Diagnostics	September 2013

[a] As listed on the FDA In Vitro Diagnostic Product Database: www.fda.gov/MedicalDevices/ProductsandMedicalProcedures/InVitroDiagnostics/default.htm.

the risk of 180-day stent thrombosis in the Chinese population (Xie 2013). Moreover, preliminary results from the Genotyping Infarct Patients to Adjust and Normalize Thienopyridine Treatment (GIANT; ClinicalTrials.gov Identifier: NCT01134380) trial, presented at the 2013 TCT (Transcatheter Cardiovascular Therapeutics) annual meeting, suggest that *CYP2C19*-directed antiplatelet therapy post-PCI reduces ischemic events at 1 year (Chevalier 2013). Other ongoing RCTs evaluating genotype-directed clopidogrel therapy include the following studies: Reassessment of Anti-platelet Therapy Using an Individualized Strategy in Patients with ST-segment Elevation Myocardial Infarction (RAPID STEMI; NCT01452139), Tailored Antiplatelet Therapy Following PCI (TAILOR-PCI; NCT01742117), and Customized Choice of P2Y12 Oral Receptor Blocker Based on Phenotype Assessment via Point-of-Care Testing (PRU-MATRIX; NCT01477775).

Logistics for *CYP2C19* Genetic Testing

Point-of-Care CYP2C19 Genotyping
CYP2C19 genetic testing to direct antiplatelet therapy typically requires a rapid return of results for effective implementation. Consequently, commercial companies have developed genotyping platforms that offer rapid sample-to-result assays for possible use at the point of care (Dobson 2007). For example, a *CYP2C19*2* point-of-care genotyping assay was evaluated for PCI patients initiating clopidogrel in the Reassessment of Anti-platelet Therapy Using an Individualized Strategy Based on Genetic Evaluation (RAPID GENE) trial (Roberts 2012). Patients were randomized to either rapid *CYP2C19*2* point-of-care genotyping, where *2 carriers were treated with prasugrel and non-carriers were treated with clopidogrel, or no genotyping and standard treatment with clopidogrel. No carriers in the genotyping arm had HTPR at day 7 compared with 30% of patients undergoing standard treatment (Roberts 2012), indicating that *CYP2C19* point-of-care testing can be performed by nursing staff and that personalized antiplatelet therapy can reduce HTPR in this patient population. However, this trial was not designed to evaluate clinical outcomes, limiting definitive conclusions on the clinical utility of this treatment scheme. Although logistic and regulatory issues need to be considered when incorporating point-of-care *CYP2C19* genotyping into actual routine clinical care (Scott 2013a), the rapid turnaround times of these platforms is an advantage when implementing *CYP2C19* clinical testing.

Preemptive CYP2C19 Genotyping
Preemptive pharmacogenetic testing is another testing strategy that can circumvent the issue of rapid turnaround time genotyping. Although this model also has inherent challenges for effective clinical implementation, preemptive *CYP2C19* genetic testing has recently been deployed at several academic medical centers (Rasmussen-Torvik 2014; Gottesman 2013; Pulley 2012), in addition to other personalized medicine programs implementing real-time *CYP2C19* genetic testing to direct antiplatelet therapy (Shuldiner 2014; Johnson 2013). The preemptive approach deposits *CYP2C19* genotype data from patients not currently treated with clopidogrel into electronic health records through prospective or biobank patient recruitment and CLIA-certified genetic testing and alerts prescribers through electronic clinical decision support at the point of care if and when clopidogrel is ordered and the patient carries a *CYP2C19* loss-of-function allele. Despite the immediate delivery of pharmacogenetic information without a disruption in routine clinical care, the necessary investments in infrastructure, informatics, health care provider participation and education, and preemptive testing in a CLIA-certified environment currently limit the feasibility of this approach to large academic medical centers.

CYP2C19 Genetic Testing Reimbursement
The 2012 American Medical Association (AMA) Current Procedural Terminology (CPT) procedure code for *CYP2C19* testing is 81225. Insurance coverage for *CYP2C19* genetic testing is currently available from selected providers; however, this is a continually evolving issue, particularly for pharmacogenetic testing. A recent proposed draft for local coverage determination by the Centers for Medicare & Medicaid Services (CMS; Palmetto GBA, Virginia) concluded that *CYP2C19* (CPT 81225) genetic testing is medically necessary and covered for patients with ACS undergoing PCI initiating or reinitiating clopidogrel therapy, but not for the medical management of ACS without PCI, stroke, or PAD. How, or if, CMS will implement this draft more broadly is not yet determined. However, the University of Florida Health Personalized Medicine Program recently reported that around 85% of third-party payers (including Medicare) reimbursed *CYP2C19* genotyping for PCI patients (Weitzel 2014).

The reported cost-effectiveness studies on *CYP2C19* genotype-guided antiplatelet therapy have been inconsistent, but generally, they have concluded that this strategy is a more cost-effective approach (Kazi 2014; Lala 2013; Reese 2012). Cost-effectiveness evidence also exists supporting widespread use of ticagrelor regardless of *CYP2C19* status (Kazi 2014; Sorich 2013); however, real-world application of these approaches will also likely be significantly influenced by treatment indication, contraindications, clinician preference, and personal insurance coverage and policy.

Pharmacogenetic Determinants of Other Antiplatelet Agents

Aspirin Pharmacogenetics

Although aspirin is considered very effective at inhibiting COX-1–dependent platelet aggregation, variabilities in responses achieved, including aspirin "resistance," can occur. Of importance, residual platelet function variability after exposure to aspirin is heritable (Faraday 2007b), suggesting the potential for genetic determinants of aspirin response. Moreover, initial pharmacogenetic studies suggested a role for *COX-1* variant alleles in aspirin responsiveness (Lepantalo 2006; Halushka 2003); however, these findings have not been definitively substantiated. Additional candidate gene studies on aspirin response, often using healthy subjects, have assessed non–*COX-1* pathway genes, including platelet surface receptors (*F13A1*, *GP6*, *GPIBA*, *ITGA2*, *ITGB3*, *P2RY1*, *P2RY12*, *PEAR1*, *PLA2G7*, and *TBXA2R*), with inconsistent findings. (For recent reviews on aspirin pharmacogenetics, see Yasmina 2014; Tantry 2013; Wurtz 2013.) In addition, most aspirin pharmacogenetic studies have focused on pharmacodynamic measurements and have not included clinical end points. Moreover, the actual definition of aspirin "resistance" is not clearly established and can be described by treatment failure, or through lower-than-expected values of platelet function as measured by several different laboratory assays and agonists. Given the contradictory study results and the relatively sparse evidence for a clinically significant effect of the reported genetic variants on aspirin response, clinical genotyping to predict aspirin response is not currently supported.

Prasugrel Pharmacogenetics

Common variants in *CYP2C9*, *CYP2B6*, *CYP3A5*, *CYP1A2*, and *ABCB1* have been tested for their association with prasugrel pharmacokinetics and pharmacodynamics and for adverse cardiovascular event rates among patients with ACS; however, no significant attenuation of response or increased clinical risk was identified in variant carriers (Mega 2010a, 2009). Of note, these studies also did not detect any effect of *CYP2C19* variant alleles on prasugrel response (Braun 2013; Mega 2009a). In contrast, recent studies have reported a significant effect of *CYP2C19*2* and *17* on ex vivo platelet reactivity when measured by vasodilator-stimulated phosphoprotein but not by ADP-induced platelet aggregation among prasugrel-treated ACS/PCI patients (Grosdidier 2013; Cuisset 2012). In addition, a significantly increased bleeding risk was detected among *17 carriers compared with non-carriers (OR 2.5; 95% CI, 1.2–5.4; p=0.02) (Cuisset 2012). Given the inconsistencies in the literature regarding the potential influence of *CYP2C19* variant alleles on prasugrel response, it is currently premature to recommend for or against any modification of prasugrel therapy based on *CYP2C19* genotype.

Ticagrelor Pharmacogenetics

Although not studied as extensively as clopidogrel or prasugrel, the initial pharmacogenetic substudies of large ticagrelor clinical trials showed no difference in response among *CYP2C19* and *ABCB1* variant allele carriers (Wallentin 2010). However, a recent GWAS from the Platelet Inhibition and Patient Outcomes (PLATO) trial identified significant associations between *SLCO1B1*, *CYP3A4*, and *UGT2B7* variant alleles and plasma levels of ticagrelor and its major metabolite (AR-C124910XX) (Varenhorst 2014). These pharmacogenetic effects were limited to ticagrelor pharmacokinetics because the identified variants did not significantly affect the efficacy or safety of ticagrelor treatment (Varenhorst 2014). Of note, given the more prominent role of CYP3A4/5 in ticagrelor metabolism, the drug label currently advises against coadministration with potent CYP3A inhibitors or inducers.

Future Directions for Antiplatelet Pharmacogenetics

Although *CYP2C19* genotyping is increasingly being implemented by select medical centers to facilitate antiplatelet management, routine *CYP2C19* testing has not been widely adopted in the clinical cardiovascular community. The ongoing pharmacogenetic clinical trials designed to assess the clinical utility of this strategy will undoubtedly influence the future of *CYP2C19* genotype-directed antiplatelet therapy, including the position of third-party insurers. In addition, ongoing pharmacogenomic studies designed to identify additional genetic variants involved in clopidogrel response variability will potentially discover novel variants that may improve the positive predictive value of genotype-directed antiplatelet therapy.

Conflict of Interest/Disclosures

Dr. Scott receives support from the NIH for antiplatelet pharmacogenomics research, is a consultant to United States Diagnostic Standards (USDS) Inc., and is an associate director of a clinical laboratory that performs *CYP2C19* testing.

References

Ancrenaz V, Desmeules J, James R, et al. The paraoxonase-1 pathway is not a major bioactivation pathway of clopidogrel in vitro. Br J Pharmacol 2012;166:2362-70.

Anderson JL, Adams CD, Antman EM, et al. 2012 ACCF/AHA focused update incorporated into the ACCF/AHA 2007 guidelines for the management of patients with unstable angina/non-ST-elevation myocardial infarction: a report of the American College of Cardiology Foundation/American Heart Association Task Force on Practice Guidelines. J Am Coll Cardiol 2013;61:e179-347.

Bauer T, Bouman HJ, van Werkum JW, et al. Impact of *CYP2C19* variant genotypes on clinical efficacy of antiplatelet treatment with clopidogrel: systematic review and meta-analysis. BMJ 2011;343:d4588.

Becquemont L, Alfirevic A, Amstutz U, et al. Practical recommendations for pharmacogenomics-based prescription: 2010 ESF-UB Conference on Pharmacogenetics and Pharmacogenomics. Pharmacogenomics 2011;12:113-24.

Bonello L, Camoin-Jau L, Mancini J, et al. Factors associated with the failure of clopidogrel dose-adjustment according to platelet reactivity monitoring to optimize P2Y12-ADP receptor blockade. Thromb Res 2012;130:70-4.

Bonello L, Pansieri M, Mancini J, et al. High on-treatment platelet reactivity after prasugrel loading dose and cardiovascular events after percutaneous coronary intervention in acute coronary syndromes. J Am Coll Cardiol 2011;58:467-73.

Bouman HJ, Schomig E, van Werkum JW, et al. Paraoxonase-1 is a major determinant of clopidogrel efficacy. Nat Med 2011;17:110-6.

Brandt JT, Close SL, Iturria SJ, et al. Common polymorphisms of *CYP2C19* and *CYP2C9* affect the pharmacokinetic and pharmacodynamic response to clopidogrel but not prasugrel. J Thromb Haemost 2007a;5:2429-36.

Brandt JT, Payne CD, Wiviott SD, et al. A comparison of prasugrel and clopidogrel loading doses on platelet function: magnitude of platelet inhibition is related to active metabolite formation. Am Heart J 2007b;153:66.e9-16.

Braun OO, Angiolillo DJ, Ferreiro JL, et al. Enhanced active metabolite generation and platelet inhibition with prasugrel compared to clopidogrel regardless of genotype in thienopyridine metabolic pathways. Thromb Haemost 2013;110:1223-31.

Buonamici P, Marcucci R, Migliorini A, et al. Impact of platelet reactivity after clopidogrel administration on drug-eluting stent thrombosis. J Am Coll Cardiol 2007;49:2312-7.

Cayla G, Hulot JS, O'Connor SA, et al. Clinical, angiographic, and genetic factors associated with early coronary stent thrombosis. JAMA 2011;306:1765-74.

Chen BL, Zhang W, Li Q, et al. Inhibition of ADP-induced platelet aggregation by clopidogrel is related to *CYP2C19* genetic polymorphisms. Clin Exp Pharmacol Physiol 2008;35:904-8.

Chevalier B, ed. GIANT: evaluation of the genetic profile of *CYP2C19* in patients undergoing primary angioplasty. Twenty-Fifth Annual Transcatheter Cardiovascular Therapeutics Symposium. Moscone Center; 2013; San Francisco, CA.

Collet JP, Cuisset T, Range G, et al. Bedside monitoring to adjust antiplatelet therapy for coronary stenting. N Engl J Med 2012;367:2100-9.

Collet JP, Hulot JS, Anzaha G, et al. High doses of clopidogrel to overcome genetic resistance: the randomized crossover CLOVIS-2 (Clopidogrel and Response Variability Investigation Study 2). JACC Cardiovasc Interv 2011;4:392-402.

Collet JP, Hulot JS, Pena A, et al. Cytochrome P450 *2C19* polymorphism in young patients treated with clopidogrel after myocardial infarction: a cohort study. Lancet 2009;373:309-17.

Cuisset T, Loosveld M, Morange PE, et al. *CYP2C19*2* and **17* alleles have a significant impact on platelet response and bleeding risk in patients treated with prasugrel after acute coronary syndrome. JACC Cardiovasc Interv 2012;5:1280-7.

Cuisset T, Quilici J, Cohen W, et al. Usefulness of high clopidogrel maintenance dose according to *CYP2C19* genotypes in clopidogrel low responders undergoing coronary stenting for non ST elevation acute coronary syndrome. Am J Cardiol 2011;108:760-5.

Dobson MG, Galvin P, Barton DE. Emerging technologies for point-of-care genetic testing. Expert Rev Mol Diagn 2007;7:359-70.

Erlinge D, Ten Berg J, Foley D, et al. Reduction in platelet reactivity with prasugrel 5 mg in low-body-weight patients is noninferior to prasugrel 10 mg in higher-body-weight patients: results from the FEATHER trial. J Am Coll Cardiol 2012;60:2032-40.

Faraday N, Becker DM, Becker LC. Pharmacogenomics of platelet responsiveness to aspirin. Pharmacogenomics 2007a;8:1413-25.

Faraday N, Yanek LR, Mathias R, et al. Heritability of platelet responsiveness to aspirin in activation pathways directly and indirectly related to cyclooxygenase-1. Circulation 2007b;115:2490-6.

Faxon DP, Nesto RW. Antiplatelet therapy in populations at high risk of atherothrombosis. J Natl Med Assoc 2006;98:711-21.

Fontana P, Dupont A, Gandrille S, et al. Adenosine diphosphate-induced platelet aggregation is associated with *P2Y12* gene sequence variations in healthy subjects. Circulation 2003;108:989-95.

Fontana P, Hulot JS, De Moerloose P, et al. Influence of *CYP2C19* and *CYP3A4* gene polymorphisms on clopidogrel responsiveness in healthy subjects. J Thromb Haemost 2007;5:2153-5.

Frere C, Cuisset T, Gaborit B, et al. The *CYP2C19*17* allele is associated with better platelet response to clopidogrel in patients admitted for non-ST acute coronary syndrome. J Thromb Haemost 2009;7:1409-11.

Frere C, Cuisset T, Morange PE, et al. Effect of cytochrome p450 polymorphisms on platelet reactivity after treatment with clopidogrel in acute coronary syndrome. Am J Cardiol 2008;101:1088-93.

Geisler T, Grass D, Bigalke B, et al. The Residual Platelet Aggregation after Deployment of Intracoronary Stent (PREDICT) score. J Thromb Haemost 2008;6:54-61.

Giusti B, Gori AM, Marcucci R, et al. Relation of cytochrome P450 *2C19* loss-of-function polymorphism to occurrence of drug-eluting coronary stent thrombosis. Am J Cardiol 2009;103:806-11.

Giusti B, Gori AM, Marcucci R, et al. Cytochrome P450 2C19 loss-of-function polymorphism, but not *CYP3A4* IVS10 + 12G/A and *P2Y12* T744C polymorphisms, is associated with response variability to dual antiplatelet treatment in high-risk vascular patients. Pharmacogenet Genomics 2007;17:1057-64.

Gladding P, White H, Voss J, et al. Pharmacogenetic testing for clopidogrel using the rapid INFINITI analyzer: a dose-escalation study. JACC Cardiovasc Interv 2009;2:1095-101.

Gong IY, Crown N, Suen CM, et al. Clarifying the importance of *CYP2C19* and *PON1* in the mechanism of clopidogrel bioactivation and in vivo antiplatelet response. Eur Heart J 2012;33:2856-64.

Gottesman O, Scott SA, Ellis SB, et al. The CLIPMERGE PGx program: clinical implementation of personalized medicine through electronic health records and genomics-pharmacogenomics. Clin Pharmacol Ther 2013;94:214-7.

Grosdidier C, Quilici J, Loosveld M, et al. Effect of *CYP2C19*2* and **17* genetic variants on platelet response to clopidogrel and prasugrel maintenance dose and relation to bleeding complications. Am J Cardiol 2013;111:985-90.

Gurbel PA, Bliden KP, Butler K, et al. Randomized double-blind assessment of the ONSET and OFFSET of the antiplatelet effects of ticagrelor versus clopidogrel in patients with stable coronary artery disease: the ONSET/OFFSET study. Circulation 2009;120:2577-85.

Gurbel PA, Bliden KP, Hiatt BL, et al. Clopidogrel for coronary stenting: response variability, drug resistance, and the effect of pretreatment platelet reactivity. Circulation 2003;107:2908-13.

Gurbel PA, Shuldiner AR, Bliden KP, et al. The relation between *CYP2C19* genotype and phenotype in stented patients on maintenance dual antiplatelet therapy. Am Heart J 2011;161:598-604.

Halushka MK, Walker LP, Halushka PV. Genetic variation in cyclooxygenase 1: effects on response to aspirin. Clin Pharmacol Ther 2003;73:122-30.

Harmsze A, van Werkum JW, Bouman HJ, et al. Besides *CYP2C19*2*, the variant allele *CYP2C9*3* is associated with higher on-clopidogrel platelet reactivity in patients on dual antiplatelet therapy undergoing elective coronary stent implantation. Pharmacogenet Genomics 2010;20:18-25.

Harmsze AM, van Werkum JW, Hackeng CM, et al. The influence of *CYP2C19*2* and **17* on on-treatment platelet reactivity and bleeding events in patients undergoing elective coronary stenting. Pharmacogenet Genomics 2012;22:169-75.

Hochholzer W, Trenk D, Fromm MF, et al. Impact of cytochrome P450 *2C19* loss-of-function polymorphism and of major demographic characteristics on residual platelet function after loading and maintenance treatment with clopidogrel in patients undergoing elective coronary stent placement. J Am Coll Cardiol 2010;55:2427-34.

Holmes DR Jr, Dehmer GJ, Kaul S, et al. ACCF/AHA clopidogrel clinical alert: approaches to the FDA "boxed warning": a report of the American College of Cardiology Foundation Task Force on clinical expert consensus documents and the American Heart Association endorsed by the Society for Cardiovascular Angiography and Interventions and the Society of Thoracic Surgeons. J Am Coll Cardiol 2010;56:321-41.

Htun WW, Steinhubl SR. Ticagrelor: the first novel reversible P2Y(12) inhibitor. Expert Opin Pharmacother 2013;14:237-45.

Holmes MV, Perel P, Shah T, et al. *CYP2C19* genotype, clopidogrel metabolism, platelet function, and cardiovascular events: a systematic review and meta-analysis. JAMA 2011;306:2704-14.

Hulot JS, Bura A, Villard E, et al. Cytochrome P450 *2C19* loss-of-function polymorphism is a major determinant of clopidogrel responsiveness in healthy subjects. Blood 2006;108:2244-7.

Hulot JS, Collet JP, Cayla G, et al. *CYP2C19* but not *PON1* genetic variants influence clopidogrel pharmacokinetics, pharmacodynamics, and clinical efficacy in post-myocardial infarction patients. Circ Cardiovasc Interv 2011;4:422-8.

Hulot JS, Collet JP, Silvain J, et al. Cardiovascular risk in clopidogrel-treated patients according to cytochrome P450 *2C19*2* loss-of-function allele or proton pump inhibitor coadministration: a systematic meta-analysis. J Am Coll Cardiol 2010;56:134-43.

Jang JS, Cho KI, Jin HY, et al. Meta-analysis of cytochrome P450 *2C19* polymorphism and risk of adverse clinical outcomes among coronary artery disease patients of different ethnic groups treated with clopidogrel. Am J Cardiol 2012;110:502-8.

Jaremo P, Lindahl TL, Fransson SG, et al. Individual variations of platelet inhibition after loading doses of clopidogrel. J Intern Med 2002;252:233-8.

Jennings LK. Mechanisms of platelet activation: need for new strategies to protect against platelet-mediated atherothrombosis. Thromb Haemost 2009;102:248-57.

Jin B, Ni HC, Shen W, et al. Cytochrome P450 *2C19* polymorphism is associated with poor clinical outcomes in coronary artery disease patients treated with clopidogrel. Mol Biol Rep 2011;38:1697-702.

Johnson JA, Elsey AR, Clare-Salzler MJ, et al. Institutional profile: University of Florida and Shands Hospital Personalized Medicine Program: clinical implementation of pharmacogenetics. Pharmacogenomics 2013;14:723-6.

Johnson JA, Roden DM, Lesko LJ, et al. Clopidogrel: a case for indication-specific pharmacogenetics. Clin Pharmacol Ther 2012;91:774-6.

Kazi DS, Garber AM, Shah RU, et al. Cost-effectiveness of genotype-guided and dual antiplatelet therapies in acute coronary syndrome. Ann Intern Med 2014;160:221-32.

Kazui M, Nishiya Y, Ishizuka T, et al. Identification of the human cytochrome P450 enzymes involved in the two oxidative steps in the bioactivation of clopidogrel to its pharmacologically active metabolite. Drug Metab Dispos 2010;38:92-9.

Kelly RP, Close SL, Farid NA, et al. Pharmacokinetics and pharmacodynamics following maintenance doses of prasugrel and clopidogrel in Chinese carriers of CYP2C19 variants. Br J Clin Pharmacol 2012;73:93-105.

Kim KA, Park PW, Hong SJ, et al. The effect of CYP2C19 polymorphism on the pharmacokinetics and pharmacodynamics of clopidogrel: a possible mechanism for clopidogrel resistance. Clin Pharmacol Ther 2008;84:236-42.

Lala A, Berger JS, Sharma G, et al. Genetic testing in patients with acute coronary syndrome undergoing percutaneous coronary intervention: a cost-effectiveness analysis. J Thromb Haemost 2013;11:81-91.

Lau WC, Gurbel PA, Watkins PB, et al. Contribution of hepatic cytochrome P450 3A4 metabolic activity to the phenomenon of clopidogrel resistance. Circulation 2004;109:166-71.

Lepantalo A, Mikkelsson J, Resendiz JC, et al. Polymorphisms of COX-1 and GPVI associate with the antiplatelet effect of aspirin in coronary artery disease patients. Thromb Haemost 2006;95:253-9.

Lewis JP, Fisch AS, Ryan K, et al. Paraoxonase 1 (PON1) gene variants are not associated with clopidogrel response. Clin Pharmacol Ther 2011;90:568-74.

Lewis JP, Horenstein RB, Ryan K, et al. The functional G143E variant of carboxylesterase 1 is associated with increased clopidogrel active metabolite levels and greater clopidogrel response. Pharmacogenet Genomics 2013a;23:1-8.

Lewis JP, Stephens SH, Horenstein RB, et al. The CYP2C19*17 variant is not independently associated with clopidogrel response. J Thromb Haemost 2013b;11:1640-6.

Li Y, Tang HL, Hu YF, et al. The gain-of-function variant allele CYP2C19*17: a double-edged sword between thrombosis and bleeding in clopidogrel-treated patients. J Thromb Haemost 2012;10:199-206.

Malek LA, Kisiel B, Spiewak M, et al. Coexisting polymorphisms of P2Y12 and CYP2C19 genes as a risk factor for persistent platelet activation with clopidogrel. Circ J 2008;72:1165-9.

Martis S, Peter I, Hulot JS, et al. Multi-ethnic distribution of clinically relevant CYP2C genotypes and haplotypes. Pharmacogenomics J 2013;13:369-77.

Mega JL, Close SL, Wiviott SD, et al. Genetic variants in ABCB1 and CYP2C19 and cardiovascular outcomes after treatment with clopidogrel and prasugrel in the TRITON-TIMI 38 trial: a pharmacogenetic analysis. Lancet 2010a;376:1312-9.

Mega JL, Close SL, Wiviott SD, et al. Cytochrome P450 genetic polymorphisms and the response to prasugrel: relationship to pharmacokinetic, pharmacodynamic, and clinical outcomes. Circulation 2009a;119:2553-60.

Mega JL, Close SL, Wiviott SD, et al. Cytochrome p-450 polymorphisms and response to clopidogrel. N Engl J Med 2009b;360:354-62.

Mega JL, Hochholzer W, Frelinger AL III, et al. Dosing clopidogrel based on CYP2C19 genotype and the effect on platelet reactivity in patients with stable cardiovascular disease. JAMA 2011;306:2221-8.

Mega JL, Simon T, Collet JP, et al. Reduced-function CYP2C19 genotype and risk of adverse clinical outcomes among patients treated with clopidogrel predominantly for PCI: a meta-analysis. JAMA 2010b;304:1821-30.

Mehta SR, Yusuf S, Peters RJ, et al. Effects of pretreatment with clopidogrel and aspirin followed by long-term therapy in patients undergoing percutaneous coronary intervention: the PCI-CURE study. Lancet 2001;358:527-33.

Muller I, Besta F, Schulz C, et al. Prevalence of clopidogrel non-responders among patients with stable angina pectoris scheduled for elective coronary stent placement. Thromb Haemost 2003;89:783-7.

Pare G, Mehta SR, Yusuf S, et al. Effects of CYP2C19 genotype on outcomes of clopidogrel treatment. N Engl J Med 2010;363:1704-14.

Payne CD, Li YG, Small DS, et al. Increased active metabolite formation explains the greater platelet inhibition with prasugrel compared to high-dose clopidogrel. J Cardiovasc Pharmacol 2007;50:555-62.

Price MJ, Berger PB, Teirstein PS, et al. Standard- vs high-dose clopidogrel based on platelet function testing after percutaneous coronary intervention: the GRAVITAS randomized trial. JAMA 2011;305:1097-105.

Price MJ, Murray SS, Angiolillo DJ, et al. Influence of genetic polymorphisms on the effect of high- and standard-dose clopidogrel after percutaneous coronary intervention: the GIFT (Genotype Information and Functional Testing) study. J Am Coll Cardiol 2012;59:1928-37.

Pulley JM, Denny JC, Peterson JF, et al. Operational implementation of prospective genotyping for personalized medicine: the design of the Vanderbilt PREDICT project. Clin Pharmacol Ther 2012;92:87-95.

Quinn MJ, Fitzgerald DJ. Ticlopidine and clopidogrel. Circulation 1999;100:1667-72.

Rasmussen-Torvik LJ, Stallings SC, Gordon AS, et al. Design and anticipated outcomes of the eMERGE-PGx project: a multi-center pilot for pre-emptive pharmacogenomics in electronic health record systems. Clin Pharmacol Ther 2014 Jun 24. [Epub ahead of print]

Reese ES, Daniel Mullins C, Beitelshees AL, et al. Cost-effectiveness of cytochrome P450 2C19 genotype screening for selection of antiplatelet therapy with clopidogrel or prasugrel. Pharmacotherapy 2012;32:323-32.

Reny JL, Combescure C, Daali Y, et al. Influence of the paraoxonase-1 Q192R genetic variant on clopidogrel responsiveness and recurrent cardiovascular events: a systematic review and meta-analysis. J Thromb Haemost 2012;10:1242-51.

Roberts JD, Wells GA, Le May MR, et al. Point-of-care genetic testing for personalisation of antiplatelet treatment (RAPID GENE): a prospective, randomised, proof-of-concept trial. Lancet 2012;379:1705-11.

Roe MT, Goodman SG, Ohman EM, et al. Elderly patients with acute coronary syndromes managed without revascularization: insights into the safety of long-term antiplatelet therapy with reduced-dose prasugrel versus standard-dose clopidogrel. Circulation 2013;128:823-33.

Rubinstein WS, Maglott DR, Lee JM, et al. The NIH genetic testing registry: a new, centralized database of genetic tests to enable access to comprehensive information and improve transparency. Nucleic Acids Res 2013;41:D925-35.

Sabatine MS, Cannon CP, Gibson CM, et al. Effect of clopidogrel pretreatment before percutaneous coronary intervention in patients with ST-elevation myocardial infarction treated with fibrinolytics: the PCI-CLARITY study. JAMA 2005;294:1224-32.

Sangkuhl K, Klein TE, Altman RB. Clopidogrel pathway. Pharmacogenet Genomics 2010;20:463-5.

Sangkuhl K, Shuldiner AR, Klein TE, et al. Platelet aggregation pathway. Pharmacogenet Genomics 2011;21:516-21.

Scott SA. Clinical pharmacogenomics: opportunities and challenges at point of care. Clin Pharmacol Ther 2013a;93:33-5.

Scott SA. Personalizing medicine with clinical pharmacogenetics. Genet Med 2011a;13:987-95.

Scott SA, Sangkuhl K, Gardner EE, et al. Clinical Pharmacogenetics Implementation Consortium guidelines for cytochrome P450-*2C19* (*CYP2C19*) genotype and clopidogrel therapy. Clin Pharmacol Ther 2011b;90:328-32.

Scott SA, Sangkuhl K, Shuldiner AR, et al. PharmGKB summary: very important pharmacogene information for cytochrome P450, family 2, subfamily C, polypeptide 19. Pharmacogenet Genomics 2012;22:159-65.

Scott SA, Sangkuhl K, Stein CM, et al. Clinical Pharmacogenetics Implementation Consortium guidelines for *CYP2C19* genotype and clopidogrel therapy: 2013 update. Clin Pharmacol Ther 2013b;94:317-23.

Shuldiner AR, O'Connell JR, Bliden KP, et al. Association of cytochrome P450 *2C19* genotype with the antiplatelet effect and clinical efficacy of clopidogrel therapy. JAMA 2009;302:849-57.

Shuldiner AR, Palmer K, Pakyz RE, et al. Implementation of pharmacogenetics: the University of Maryland personalized anti-platelet pharmacogenetics program. Am J Med Genet C Semin Med Genet 2014;166:76-84.

Sibbing D, Gebhard D, Koch W, et al. Isolated and interactive impact of common *CYP2C19* genetic variants on the antiplatelet effect of chronic clopidogrel therapy. J Thromb Haemost 2010a;8:1685-93.

Sibbing D, Koch W, Gebhard D, et al. Cytochrome *2C19*17* allelic variant, platelet aggregation, bleeding events, and stent thrombosis in clopidogrel-treated patients with coronary stent placement. Circulation 2010b;121:512-8.

Sibbing D, Koch W, Massberg S, et al. No association of paraoxonase-1 Q192R genotypes with platelet response to clopidogrel and risk of stent thrombosis after coronary stenting. Eur Heart J 2011;32:1605-13.

Sibbing D, Stegherr J, Latz W, et al. Cytochrome P450 *2C19* loss-of-function polymorphism and stent thrombosis following percutaneous coronary intervention. Eur Heart J 2009;30:916-22.

Simon T, Bhatt DL, Bergougnan L, et al. Genetic polymorphisms and the impact of a higher clopidogrel dose regimen on active metabolite exposure and antiplatelet response in healthy subjects. Clin Pharmacol Ther 2011a;90:287-95.

Simon T, Steg PG, Becquemont L, et al. Effect of paraoxonase-1 polymorphism on clinical outcomes in patients treated with clopidogrel after an acute myocardial infarction. Clin Pharmacol Ther 2011b;90:561-7.

Simon T, Verstuyft C, Mary-Krause M, et al. Genetic determinants of response to clopidogrel and cardiovascular events. N Engl J Med 2009;360:363-75.

Singh M, Shah T, Adigopula S, et al. *CYP2C19*2/ABCB1*-C3435T polymorphism and risk of cardiovascular events in coronary artery disease patients on clopidogrel: is clinical testing helpful? Indian Heart J 2012;64:341-52.

Sofi F, Giusti B, Marcucci R, et al. Cytochrome P450 *2C19*2* polymorphism and cardiovascular recurrences in patients taking clopidogrel: a meta-analysis. Pharmacogenomics J 2011;11:199-206.

Sorich MJ, Horowitz JD, Sorich W, et al. Cost-effectiveness of using *CYP2C19* genotype to guide selection of clopidogrel or ticagrelor in Australia. Pharmacogenomics 2013;14:2013-21.

Sorich MJ, Polasek TM, Wiese MD. Systematic review and meta-analysis of the association between cytochrome P450 *2C19* genotype and bleeding. Thromb Haemost 2012;108:199-200.

Steinhubl SR, Berger PB, Mann JT III, et al. Early and sustained dual oral antiplatelet therapy following percutaneous coronary intervention: a randomized controlled trial. JAMA 2002;288:2411-20.

Swen JJ, Nijenhuis M, de Boer A, et al. Pharmacogenetics: from bench to byte—an update of guidelines. Clin Pharmacol Ther 2011;89:662-73.

Swen JJ, Wilting I, de Goede AL, et al. Pharmacogenetics: from bench to byte. Clin Pharmacol Ther 2008;83:781-7.

Tantry US, Jeong YH, Navarese EP, et al. Influence of genetic polymorphisms on platelet function, response to antiplatelet drugs and clinical outcomes in patients with coronary artery disease. Expert Rev Cardiovasc Ther 2013;11:447-62.

Teng R. Pharmacokinetic, pharmacodynamic and pharmacogenetic profile of the oral antiplatelet agent ticagrelor. Clin Pharmacokinet 2012;51:305-18.

Tiroch KA, Sibbing D, Koch W, et al. Protective effect of the *CYP2C19 *17* polymorphism with increased activation of clopidogrel on cardiovascular events. Am Heart J 2010;160:506-12.

Tran H, Anand SS. Oral antiplatelet therapy in cerebrovascular disease, coronary artery disease, and peripheral arterial disease. JAMA 2004;292:1867-74.

Trenk D, Hochholzer W, Fromm MF, et al. Paraoxonase-1 Q192R polymorphism and antiplatelet effects of clopidogrel in patients undergoing elective coronary stent placement. Circ Cardiovasc Genet 2011;4:429-36.

Trenk D, Hochholzer W, Fromm MF, et al. Cytochrome P450 *2C19* 681G>A polymorphism and high on-clopidogrel platelet reactivity associated with adverse 1-year clinical outcome of elective percutaneous coronary intervention with drug-eluting or bare-metal stents. J Am Coll Cardiol 2008;51:1925-34.

Umemura K, Furuta T, Kondo K. The common gene variants of *CYP2C19* affect pharmacokinetics and pharmacodynamics in an active metabolite of clopidogrel in healthy subjects. J Thromb Haemost 2008;6:1439-41.

Varenhorst C, Eriksson N, Johansson A, et al., eds. Ticagrelor plasma levels but not clinical outcomes are associated with transporter and metabolism enzyme genetic polymorphisms. American College of Cardiology (ACC) 63rd Annual Scientific Session and Expo; 2014; Washington, DC.

Varenhorst C, James S, Erlinge D, et al. Genetic variation of *CYP2C19* affects both pharmacokinetic and pharmacodynamic responses to clopidogrel but not prasugrel in aspirin-treated patients with coronary artery disease. Eur Heart J 2009;30:1744-52.

Wallentin L, Becker RC, Budaj A, et al. Ticagrelor versus clopidogrel in patients with acute coronary syndromes. N Engl J Med 2009;361:1045-57.

Wallentin L, James S, Storey RF, et al. Effect of *CYP2C19* and *ABCB1* single nucleotide polymorphisms on outcomes of treatment with ticagrelor versus clopidogrel for acute coronary syndromes: a genetic substudy of the PLATO trial. Lancet 2010;376:1320-8.

Wallentin L, Varenhorst C, James S, et al. Prasugrel achieves greater and faster P2Y12 receptor-mediated platelet inhibition than clopidogrel due to more efficient generation of its active metabolite in aspirin-treated patients with coronary artery disease. Eur Heart J 2008;29:21-30.

Weitzel KW, Elsey AR, Langaee TY, et al. Clinical pharmacogenetics implementation: approaches, successes, and challenges. Am J Med Genet C Semin Med Genet 2014;166C:56-67.

Wiviott SD, Trenk D, Frelinger AL, et al. Prasugrel compared with high loading- and maintenance-dose clopidogrel in patients with planned percutaneous coronary intervention: the Prasugrel in Comparison to Clopidogrel for Inhibition of Platelet Activation and Aggregation-Thrombolysis in Myocardial Infarction 44 trial. Circulation 2007;116:2923-32.

Wurtz M, Lordkipanidze M, Grove EL. Pharmacogenomics in cardiovascular disease: focus on aspirin and ADP receptor antagonists. J Thromb Haemost 2013;11:1627-39.

Xie X, Ma YT, Yang YN, et al. Personalized antiplatelet therapy according to *CYP2C19* genotype after percutaneous coronary intervention: a randomized control trial. Int J Cardiol 2013;168:3736-40.

Yamaguchi Y, Abe T, Sato Y, et al. Effects of VerifyNow P2Y12 test and *CYP2C19*2* testing on clinical outcomes of patients with cardiovascular disease: a systematic review and meta-analysis. Platelets 2013;24:352-61.

Yasmina A, de Boer A, Klungel OH, et al. Pharmacogenomics of oral antiplatelet drugs. Pharmacogenomics 2014;15:509-28.

Yousuf O, Bhatt DL. The evolution of antiplatelet therapy in cardiovascular disease. Nat Rev Cardiol 2011;8:547-59.

Zabalza M, Subirana I, Sala J, et al. Meta-analyses of the association between cytochrome *CYP2C19* loss- and gain-of-function polymorphisms and cardiovascular outcomes in patients with coronary artery disease treated with clopidogrel. Heart 2012;98:100-8.

Zou JJ, Xie HG, Chen SL, et al. Influence of *CYP2C19* loss-of-function variants on the antiplatelet effects and cardiovascular events in clopidogrel-treated Chinese patients undergoing percutaneous coronary intervention. Eur J Clin Pharmacol 2013;69:771-7.

Chapter 6

PHARMACOGENETICS OF ANTICOAGULANT DRUGS

Larisa H. Cavallari, Pharm.D., FCCP, BCPS; and
Jaekyu Shin, Pharm.D., M.S., BCPS

Learning Objectives

1. Estimate the effects of genotype on warfarin metabolism or pharmacodynamics.
2. Apply a warfarin pharmacogenetic dosing algorithm to estimate a warfarin maintenance dose for a patient.
3. Compose a warfarin dosing plan based on patient demographics, clinical factors, and genotype.
4. Resolve the data from clinical trials assessing genotype-guided warfarin dosing.
5. Justify provision of genotype-guided warfarin dosing based on existing literature.
6. Assess additional data necessary to evaluate the utility of genotype-guided warfarin dosing.
7. Infer candidate genes for response to novel oral anticoagulants based on their pharmacology.

Keywords: warfarin, genotype, *CYP2C9*, *VKORC1*, dabigatran

Abbreviations in This Chapter

CES1	carboxylesterase 1	IWPC	International Warfarin Pharmacogenetics Consortium
CPIC	Clinical Pharmacogenetics Implementation Consortium	PharmGKB	Pharmacogenomics Knowledge Base
COAG	Clarification of Optimal Anticoagulation through Genetics	NQO1	NAD(P)H quinone oxidioreductase 1
CYP	cytochrome P450	PTTR	percent time in therapeutic range
EU-PACT	European Pharmacogenetics of AntiCoagulant Therapy	QALY	quality adjusted life year
GWAS	genome-wide association study	SNP	single nucleotide polymorphism
ICER	incremental cost-effectiveness ratio	VKORC1	vitamin K epoxide reductase complex subunit 1
INR	international normalized ratio	VTE	venous thromboembolism

Abstract

Warfarin therapy is complicated by the drug's narrow therapeutic index and the significant inter-patient variability in dose requirements. Genetic polymorphisms are well recognized as important determinants of warfarin response, and guidelines are available to assist with interpretation and use of genotype data for warfarin dosing. Despite this, few institutions have implemented warfarin pharmacogenetics into clinical practice. In addition, variable results from recent clinical trials have called into question the efficacy of genotype-guided warfarin dosing. Further data are expected from on-going clinical trials, observational studies, and clinical experience with genotype guided dosing. In addition, pharmacogenetic data are emerging with regard to newer oral anticoagulant agents, which may eventually assist in predicting outcomes with these agents and tailoring anticoagulation therapy accordingly.

Introduction

Warfarin is commonly prescribed to treat and prevent thromboembolic events, and even with the approval of newer agents, it remains the mainstay therapy for oral anticoagulation (Kirley 2012). However, warfarin is a difficult drug to manage because of its narrow therapeutic index. Frequent monitoring is required to ensure safe and effective anticoagulation, guided by the international normalized ratio (INR) of 2 to 3 for most indications. The risks for bleeding and thrombosis during warfarin therapy increase when the INR rises above 4 or falls below 2, respectively (Hylek 2007; Hylek 2003). According to data from the National Electronic Injury Surveillance System, warfarin is the leading drug-related cause of hospitalization for adverse events among older adults in the United States (U.S.), accounting for 33% of such hospitalizations (Budnitz 2011). The risk for adverse events is greatest during the initial months of warfarin therapy (Hylek 2007). Therefore, achieving therapeutic anticoagulation efficiently is a major goal with warfarin initiation.

Further complicating warfarin use is the significant inter-patient variability in the dose necessary for optimal anticoagulation. Dose requirements may vary as much as 20-fold among warfarin-treated patients (Wadelius 2009). Current guidelines by the American College of Chest Physicians recommend a standard fixed-dose approach for warfarin initiation in the outpatient setting, with an initial dose of 5 mg to 10 mg for the first 2 days, followed by dosing based on INR response (Holbrook 2012). The problem with this approach is that it ignores patient-specific factors known to influence dose requirements. These include clinical factors such as age, body size, and medications that interfere with warfarin metabolism. In addition, genes are well recognized to influence warfarin pharmacokinetics and pharmacodynamics. This chapter will discuss the major genes contributing to the variability in warfarin dose requirements, data and guidelines addressing the incorporation of genotype in warfarin dosing decisions, and recent pharmacogenetic data with newer anticoagulant agents.

Clinical Translation of Candidate Genes

Genes Involved in Warfarin Pharmacokinetics and Pharmacodynamics

The primary genes contributing to the inter-patient variability in warfarin response are cytochrome P450 2C9 (*CYP2C9*), vitamin K epoxide reductase complex subunit 1 (*VKORC1*), and to a lesser extent, *CYP4F2*, as illustrated in Figure 6.1. The *CYP2C9* gene affects warfarin pharmacokinetics. Specifically, the CYP2C9 enzyme metabolizes the more potent *S*-enantiomer of warfarin to the inactive 7-hydroxy metabolite. The gene for CYP2C9 is located on chromosome 10q24.1, and deficiencies in CYP2C9 activity secondary to *CYP2C9* gene variation lead to reduced *S*-warfarin clearance.

Both the *VKORC1* and *CYP4F2* genes affect warfarin pharmacodynamics. Specifically, warfarin exerts its anticoagulant effects by inhibiting VKORC1, the enzyme responsible for reduction of vitamin K epoxide to vitamin K_1. Vitamin K_1 is further reduced to vitamin KH_2, a necessary co-factor for carboxylation and activation of clotting factors II, VII, IX, and X. The *VKORC1* gene is located on chromosome 16p11.2, and variation within the gene affects sensitivity to warfarin.

The CYP4F2 enzyme catalyzes metabolism of vitamin K_1 to an inactive form, thereby decreasing the concentration of vitamin K_1 available for reduction to vitamin KH_2. Variation in the *CYP4F2* gene reduces the concentration of the enzyme and its ability to metabolize vitamin K_1. As a result, more vitamin K_1 is available for conversation to vitamin KH_2 and subsequent clotting factor activation, leading to increased warfarin dose requirements.

Genetic Determinants of Warfarin Dose Requirements

Numerous candidate gene studies have demonstrated that the *CYP2C9*, *VKORC1*, and *CYP4F2* genotypes contribute to the inter-patient variability in warfarin dose requirements. This has been confirmed by several genome-wide association studies (GWAS) (Cooper 2008; Takeuchi 2009; Cha 2011; Perera 2013). The following sections highlight data with each gene and findings from GWAS in various ethnic groups.

CYP2C9 Genotype

Nearly 60 alleles have been identified in the *CYP2C9* gene (www.cypalleles.ki.se/cyp2c9.htm), and their location and frequency across ethnic groups are shown in Table 6.1. By far, the most commonly described are *CYP2C9*2* (p.Arg144Cys) and **3* (p.Ile359Leu), which reduce *S*-warfarin clearance by 40% and 75%, respectively (Takahashi 1998; Scordo 2002). The *CYP2C9*2* and **3* alleles are the most prevalent *CYP2C9* variants in Europeans, but occur less often in African Americans. The *CYP2C9*2* allele is virtually absent in some Asian populations. The *CYP2C9*5* (p.Asp360Glu), **6* (c.818delA), **8* (p.Arg150His), and **11* (p.Arg335Trp) alleles occur most often in persons of African descent and have also been associated with significant reductions in *S*-warfarin clearance (Dickmann 2001; Liu 2012; Redman 2004; Tai 2005). Other variants with potential effects on warfarin disposition have been described and include *CYP2C9*4, *14, *16,* and **35* (Niinuma 2013; Lee 2014).

Compared to individuals with the *CYP2C9 *1/*1* genotype, dose requirements are approximately 15% to 20% lower (per allele) in carriers of a **2* allele and approximately 30% to 40% lower with each **3* allele (Scordo 2002; Higashi 2002). Individuals with homozygous variant genotypes (e.g., **3/*3*) may require as much as 80% lower doses. For example, in a European population, the **1/*1, *1/*2, *2/*2, *1/*3,* and **3/*3* genotypes were associated with mean warfarin doses of 5.3, 4.6, 3.9, 2.9, and 1.6 mg/day, respectively (Higashi 2002). Data on the effects of the *CYP2C9*5, *6,* and **11* alleles on warfarin dosing are limited to case reports given their low frequencies (Dickmann 2001; Redman 2004; Tai 2005). However, at least two cohort studies have examined the effect of the *CYP2C9*8* allele, and both found that it conferred approximately 8 mg/week lower dose requirements compared to **1* allele homozygotes (Cavallari 2010; Mitchell 2011). Overall, the *CYP2C9*2* and **3* genotypes explain approximately 9% to 13% of the variability in dose requirements in Europeans, but less in African Americans (1% to 3%) and Asians (1% to 2%) due to the lower prevalence of the **2* and **3* alleles in these populations (Takeuchi 2009; Aquilante 2006; Limdi 2008; Cha 2010). When accounting for the **5, *6, *8,* and **11* alleles in addition to **2* and **3, CYP2C9* genotype explains approximately 7% of the variability in dose requirements in African Americans (Cavallari 2010).

VKORC1 Genotype

The *VKORC1* gene was first described in the context of warfarin resistance, whereby missense mutations in the gene coding region lead to exceptionally high warfarin dose requirements for effective anticoagulation (e.g., 20 mg/day or higher) (Rost 2004). The p.Asp36Tyr is the most commonly described warfarin-resistance mutation. It is rarely detected in most North American populations (frequency less than 0.1%) (Shahin 2013). An exception is in Ashkenzai Jewish individuals, in whom the allele frequency is approximately 4% and the allele accounts for a higher than usual occurrence of warfarin resistance (Scott 2008).

In addition to rare missense mutations, there are common polymorphisms in the non-coding, regulatory regions of the *VKORC1* gene that significantly contribute to variability in dose requirements observed in the general warfarin-treated population. The c.-1639G>A single nucleotide polymorphism (SNP) is located in the *VKORC1* 5' regulatory region and is in strong linkage disequilibrium with the c.1173C>T SNP, located in intron 1. The -1639A allele is associated with approximately 2-fold lower mRNA expression compared to the G allele (Wang 2008). As a result, the -1639 AA, AG, and GG genotypes are correlated with high, intermediate, and low sensitivity to warfarin which translates into low, intermediate, and high warfarin dose requirements, respectively. For example, daily warfarin dose requirements of approximately 3 mg, 5 mg, and 6 mg, have been reported with the AA, AG, and GG genotypes, respectively (Limdi 2010)..

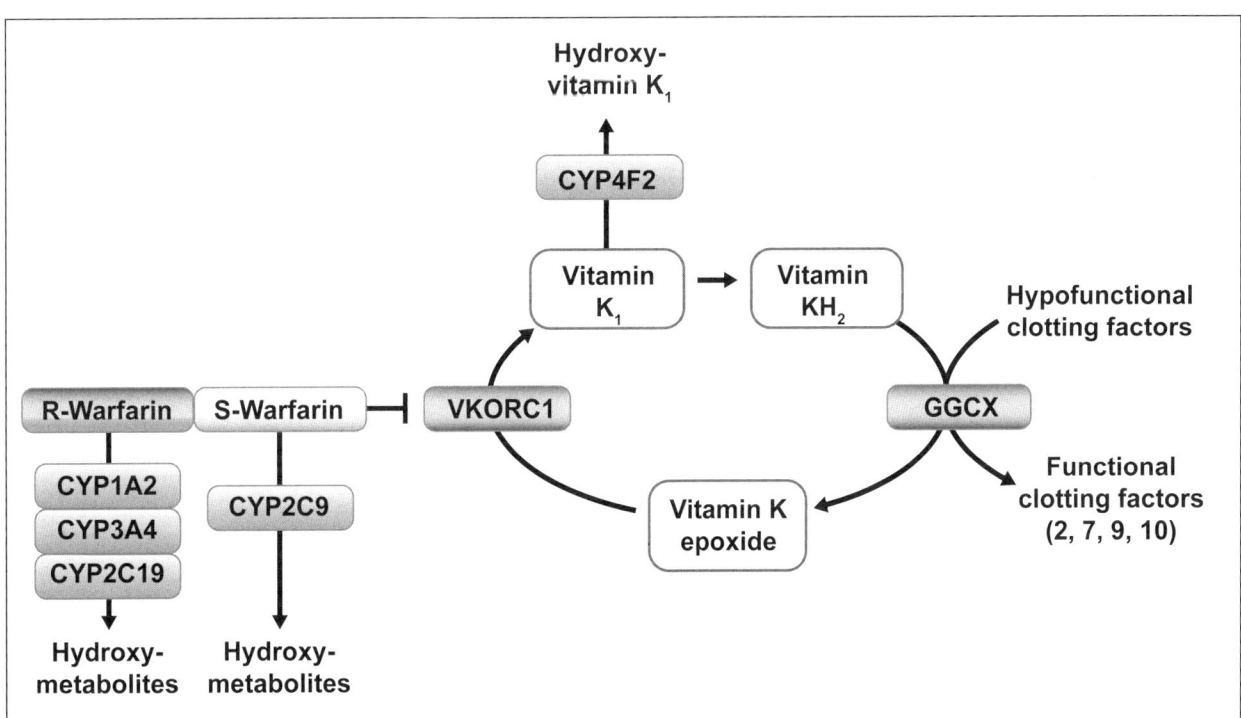

Figure 6.1. Genes influencing the pharmacokinetics and pharmacodynamics of warfarin.

The *VKORC1* -1639G>A genotype frequency distribution differs by ethnicity, with a higher prevalence of the AA genotype in Asians and a higher prevalence of the GG genotype in African Americans compared to Europeans (Table 6.1). This contributes to the comparatively low dose requirements observed in the Asian population, intermediate requirements in Europeans, and high requirements generally observed among African Americans. The *VKORC1* -1639G>A genotype explains 20% to 30% of the variability in dose requirements in Europeans and Asians (Wadelius 2009; Cha 2011). However, given the lower frequency of the -1639A allele in African Americans, -1639G>A genotype explains only about 5% to 7% of the variability in this ethnic group (Cavallari 2010; Limdi 2010).

CYP4F2 Genotype

The *CYP4F2* genotype provides additional contribution to warfarin dose requirements, explaining approximately 1% to 2% of variability in dose (Takeuchi 2009; Cha 2011). The primary variant described in the *CYP4F2* gene is the p.Val433Met SNP, located in exon 11. The Val433Met SNP leads to reduced CYP4F2 enzyme levels and a genotype-dependent decrease in vitamin K_1 metabolism (McDonald 2009). As a consequence, more vitamin K_1 is available for clotting factor activation, leading to increased warfarin dose requirements. In a study of 3 independent European cohorts, the Met/Met genotype was associated with 1 mg/day higher warfarin dose requirements than the homozygous Val/Val genotype (Caldwell 2008). Heterozygotes required intermediate doses. The 433Met allele is less common in African Americans than Europeans and Asians and thus appears to contribute minimally to dose requirements in this group (Cavallari 2010).

Genome-wide Association Studies of Warfarin Dose Requirements

Three GWAS have confirmed the *VKORC1* genotype as the single most important genetic determinant of warfarin dose requirements in European and Japanese populations (Cooper 2008; Takeuchi 2009; Cha 2011). The *CYP2C9* genotype provides additional, moderate contributions to dose requirements, with *CYP4F2* having minor effects (Cooper 2008; Takeuchi 2009). Together, *VKORC1* and *CYP2C9* genotypes plus clinical factors (e.g., age, sex, and amiodarone use) explained approximately 50% to 60% of the total variance in warfarin maintenance dose in European populations (Cooper 2008; Takeuchi 2009). The *VKORC1* -1639G>A, *CYP2C9*3*, and *CYP4F2* Val433Met genotypes in addition to clinical factors (age, body size, amiodarone use) explained 43% of dose variability in Asians (Cha 2011).

A more recent GWAS was conducted in African Americans (Perera 2013). Similar to GWAS in other populations, the *VKORC1* -1639G>A SNP was the most significant genetic predictor of warfarin dose requirements. After conditioning on the *VKORC1* SNP and *CYP2C9*2* and **3* alleles, a novel association emerged between dose requirements and the rs12777823 G>A SNP, located in chromosome 10 near the *CYP2C18* gene. Compared to individuals with the GG genotype, those with the AG and AA genotypes required 7 mg/week and 9 mg/week lower doses, respectively.

Table 6.1. Gene Location and Frequency of Polymorphisms Involved in Warfarin Pharmacokinetics and Pharmacodynamics (Cha 2011; Cavallari 2010; Limdi 2010; Caldwell 2008; Limdi 2008; Perera 2011)

Polymorphism	Gene Position	Minor allele frequency		
		Europeans	African Americans	Asians
*CYP2C9*2*	Exon 3	0.13	0.02	ND*
*CYP2C9*3*	Exon 7	0.06	0.01	0.02
*CYP2C9*5*	Exon 7	ND	0.01	ND
*CYP2C9*6*	Exon 5	ND	0.01	ND
*CYP2C9*8*	Exon 3	ND	0.06	ND
*CYP2C9*11*	Exon 7	ND	0.04	ND
VKORC1 -1639G>A	Promoter region	0.38	0.10	0.91
CYP4F2 Val433Met	Exon 11	0.29	0.07	0.29

*ND, not detected or occurring at very low frequency

A pharmacokinetic analysis revealed lower S-warfarin clearance with the minor allele at this position (Perera 2013). The rs12777823A allele is common, with a frequency of approximately 25% in African Americans, 14% in Europeans, and 30% in Asians. However, it was not associated with warfarin dose requirements in either of the latter two populations suggesting that its association with warfarin dose requirements is unique to persons of African descent. This may be because it is in linkage disequilibrium with a functional SNP or SNPs in Africans that underlies the association observed.

The CYP2C9*5, *6, *8, and *11 variants were not included on the GWAS platform used in the African American study, and only the *11 variant could be imputed (Perera 2013). However, *11 was associated with lower dose requirements confirming its importance in African Americans. A model containing the VKORC1 and CYP2C9*2 and *3 genotypes plus clinical factors explained 21% of the variability in dose in African Americans. This is less than that explained in Europeans but consistent with previous reports demonstrating lesser contribution of these genotypes to warfarin dose variability in African Americans, likely owing to their lower frequency in this population (Limdi 2008; Limdi 2010). The addition of the rs12777823 into the model increased the variability explained to 26%.

Genetic Associations with Anticoagulation Control

Several studies have assessed genetic associations with time to attain therapeutic anticoagulation and time spent within the therapeutic range upon warfarin initiation; however, the data are inconsistent in both regards. For example, variant CYP2C9 genotype was associated with a delay in time to reach therapeutic INR in a European population (Higashi 2002), but not in ethnically diverse cohorts (Limdi 2008; Schwarz 2008). Similarly, while some investigators reported early attainment of the first therapeutic INR with the VKORC1 -1639 AA genotype (Wadelius 2009; Schwarz 2008; Meckley 2008), others have not (Limdi 2008). Most studies found no association between genotype and time in range in the initial months of therapy (Wadelius 2009; Limdi 2008; Gong 2011).

Genetic Associations with Warfarin-Related Bleeding Risk

Warfarin is usually initiated using a fixed dose approach, with initiation at 5 to 10 mg and subsequent dosing based on INR response. Conceivably, this approach could lead to supra-therapeutic anticoagulation and increased bleeding risk in patients with variant VKORC1 or CYP2C9 genotypes. Thus, a number of investigators have examined the effect of genotype on bleeding risk with warfarin. Many of these studies evaluated the occurrence of INR values greater than or equal to 4 as a surrogate marker for bleeding risk, with a limited number of studies specifically examining hemorrhagic events.

Both the CYP2C9*2 and *3 alleles and VKORC1 -1639 AA genotype have been associated with supra-therapeutic anticoagulation during the initial weeks of warfarin therapy (Wadelius 2009; Schwarz 2008; Meckley 2008; Voora 2005; Lindh 2005). In one of the earliest studies to specifically examine bleeding events, investigators reported a significantly higher rate of serious and life-threatening bleeding with the CYP2C9*2 and *3 alleles compared to the *1/*1 genotype (Higashi 2002). However, this study was limited to Europeans. In a more recent study including both Europeans and African Americans, the variant CYP2C9 genotype conferred an increased risk of major hemorrhage after adjusting for VKORC1 genotype and clinical covariates (hazard ratio 3.7, 95% confidence interval 1.4 to 9.6 compared to the *1/*1 genotype) (Limdi 2008). The first major bleeding event also occurred earlier in therapy among variant allele carriers versus non-carriers. In addition to CYP2C9*2 and *3, African Americans were genotyped for the *5, *6, and *11 alleles, thus accounting for many of the variants important in this population. Interestingly, the same study found no association between the VKORC1 genotype and risk for major or minor bleeding.

Clinical Effectiveness of Genotype-Guided Warfarin Dosing

There are broadly two types of studies evaluating clinical benefit of genotype-guided warfarin dosing: clinical efficacy and clinical effectiveness studies. Clinical efficacy studies determine whether genotype-guided warfarin dosing "does more good than harm when delivered under optimum condition" (Flay 1986). Because a standardized implementation of genotype-guided warfarin dosing is delivered to a usually specific and homogenous population in this type of study, any difference in effect can be directly ascribed to the intervention (Glasgow 2003). Thus, the clinical efficacy of genotype-guided warfarin dosing is tested in a randomized, controlled trial. On the other hand, clinical effectiveness studies evaluate whether genotype-guided warfarin dosing "does more good than harm when delivered under real-world condition" (Flay 1986). The primary goal of these studies is to assess the applicability of genotype-guided warfarin dosing in a broad population. As a result, availability and access of genotype-guided warfarin dosing is standardized, whereas its implementation varies depending on the circumstance of the clinical setting (Glasgow 2003). Thus, the clinical effectiveness of genotype-guided warfarin dosing is typically evaluated through a non-randomized study design. As the clinical effectiveness of genotype-guided warfarin dosing was studied prior to its clinical efficacy, we will first discuss the MEDCO-MAYO Warfarin Effectiveness (MEDCO-MAYO WE) and COUMA-GEN II studies, the two largest clinical effectiveness studies (Epstein 2010; Anderson 2012). While these studies did not test the incremental value of genotyping when added to clinically-based dosing (i.e., use of a clinical algorithm), they provide insight into the potential benefit of genotyping compared to usual dosing done in clinical practice.

The MEDCO-MAYO WE study used administrative and pharmacy claims data from 23 large, national prescription benefit plans (Epstein 2010). Free genotyping was offered to 896 members of the plans who initiated warfarin as outpatients, and genotype results along with an interpretation of the results were provided to treating physicians. The study outcome was the hospitalization rate for the first 6 months of warfarin therapy, which was compared with age- and sex-matched historical controls who initiated warfarin in the previous year. The study found that the participants receiving free genotyping had a significantly lower hospitalization rate than the historical controls (14.0% vs. 20.5%; hazards ratio, 0.67; 95% confidence interval 0.55-0.81; p<0.001). In addition, the study participants had 43% fewer hospitalizations due to bleeding or thromboembolism (hazard ratio, 0.57: 95% confidence interval 0.39-0.83; p=0.003). It is interesting to note that, although the treating physicians were not obliged to change warfarin dosages based on the testing report, they responded to it by changing warfarin dosages during the 21 days after receiving it. In addition, the use of historical controls, who may have been differentially observed compared to the genotyped group, limit the interpretation of the results.

The COUMA-GEN II study had two parts (Anderson 2012). The first part was a randomized, single-blinded clinical trial that compared the percent of out-of-range INR values and percent of time in therapeutic range (PTTR) during the first 30 days of warfarin therapy between two different pharmacogenetic algorithms. The second part of the study compared outcomes between patients in part one receiving pharmacogenetic dosing and concurrent controls newly started on warfarin and dosed at the discretion of treating healthcare providers, although a fixed initial maintenance dose was generally used. Because the study did not find a significant difference in the study outcomes between pharmacogenetic algorithms in the first part, it combined the two groups with different algorithms for the second part. The study found that the genotype-guided warfarin dosing group (i.e., the combined group) had a significantly lower percent of out-of-range INR values (31.1% vs. 41.5%; p<0.001) and higher PTTR (68.9% vs. 58.4%; p<0.001) than the concurrent control group. In addition, the genotype-guided dosing group had a significantly lower incidence of serious adverse events defined as a composite of adjudicated death, myocardial infarction, stroke, thromboembolic event, and moderate or severe hemorrhage (4.5% in the genotype-guided dosing group vs. 9.4% in the concurrent control group; p=0.001).

Both the MEDCO-MAYO WE and COUMA-GEN II studies demonstrated better clinical effectiveness with pharmacogenetic-algorithm-guided warfarin dosing compared to traditional warfarin dosing. It is important to note that, in these studies, the control groups reflect the current practice in warfarin dosing because they were not obliged to follow a certain dosing method specified by the study protocol. As they did not strictly control implementation of the intervention, however, their results may have been influenced by unmeasured confounders and various biases including selection and misclassification biases. For example, retrospective medication record screening for adverse events may rely on diagnosis codes where misclassification of events may occur. In addition, the MEDCO-MAYO WE study had a median lag time of 32 days between a patient's warfarin initiation and the treating physician's receipt of the patient's testing report. Because of this long lag time, the observed difference in the study outcome may not reflect the effect of the genotype-guided warfarin dosing method.

Clinical Efficacy of Genotype-Guided Warfarin Dosing

Several randomized controlled trials have evaluated clinical efficacy of genotype-guided warfarin dosing (Table 6.2) (Caraco 2008; Anderson 2007; Huang 2009; Burmester 2011; Wang 2012; Jonas 2013; Kimmel 2013; Pirmohamed 2013). These trials have typically used time in therapeutic range or time to first therapeutic INR during the first 1–3 months of warfarin therapy as their outcome. In addition, most trials had a dosing algorithm without a genetic variable as a comparison group (i.e., clinical dosing algorithm). In this section, we will focus on the two largest and most rigorous trials—Clarification of Optimal Anticoagulation through Genetics (COAG) and the European Pharmacogenetics of AntiCoagulant Therapy (EU-PACT) trials (Kimmel 2013; Pirmohamed 2013).

The COAG trial is the only trial with double-blind methodology and a relatively diverse study population (27% African Americans and 5.4% Hispanics of 1015 total patients). Its primary outcome was PTTR from the completion of the dosing intervention (day 4 or 5) to day 28, and the intervention was the use of either pharmacogenetic algorithm-guided or clinical algorithm-guided warfarin dosing methods for the first 4 to 5 days of warfarin therapy. Warfarin dosages between days 4 and 28 were adjusted according to a standardized protocol. The www.warfarindosing.org algorithm was the pharmacogenetic algorithm, and the www.warfarindosing.org algorithm without a genetic variable was the clinical algorithm. The trial found that the primary outcome was not significantly different between groups (45.2% in the pharmacogenetic algorithm group vs. 45.4% in the clinical algorithm group; p= 0.91). Of note, a significant statistical interaction was found between race and dosing methods: African American patients in the pharmacogenetic algorithm-guided dosing group had an 8.3% lower primary outcome than those in the clinical algorithm-guided dosing group whereas non-African American patients in the pharmacogenetic algorithm-guided group had a 2.8% higher primary outcome (p for interaction = 0.003). The principal secondary outcome of time to any INR greater than or equal to 4, major bleeding, or thromboembolism was similar between pharmacogenetic and

Table 6.2. Summary of Prospective, Randomized Trials Evaluating Clinical Efficacy of Genotype-Guided Warfarin Dosing Method

Study	Study Population	Sample Size	Comparison Group	Primary Outcome	Results	Comment
Caraco et al. (2008)	Caucasians (Israelis)	191	Computerized warfarin dosing program	Time to reach first therapeutic INR range	Genotype-guided group: 14.1 ± 6.9 days Comparison group: 32.2 ± 21.1 days (p<0.001)	Only *CYP2C9* genotypes were used. Bleeding incidence[a]: Genotype-guided group: 3.2% (3/95) Comparison group: 12.5% (12/96)
COUMAGEN I (2007)	Caucasians	206	Published nomogram	Percent of out-of-range INRs in the first 90 days	Genotype-guided group: 30.7% Comparison group: 33.1% (p=0.47)	Serious clinical events[b]: Genotype-guided group: 4.0% (4/101) Comparison group: 5.0% (5/99)
Huang et al (2009)	Asians (Chinese) with heart valve replacement	121	Fixed starting dose (2.5 mg/day)	Time to a stable warfarin dose	Genotype-guided group: median 24 days Comparison group: median 35 days (p<0.001)	Warfarin dose for the first 3 days was ≤ 3.5 mg/day in the genotype-guided group. Bleeding incidence[c]: Genotype-guided group: 3.3% (2/61) Comparison group: 5% (3/60)
Burmester et al. (2011)	Caucasians	230	Clinical dosing algorithm	Absolute prediction error relative to therapeutic dose. PTTR during the first 14 days	Genotype-guided group: 0.80 mg/day Comparison group: 1.32 mg/day (p value not reported) PTTR: Genotype-guided group: 30.8 ± 28.4% Comparison group: 29.1 ± 15.5% (p=0.56)	Bleeding incidence[d]: Genotype-guided group: 2.6% (3/115) Comparison group: 3.5% (4/115)
Wang et al. (2012)	Asians (Chinese) with heart valve replacement	101	Fixed starting dose 2.5 mg/day)	Time to a stable warfarin dose	Genotype-guided group: median 24 days Comparison group: median 33 days (p<0.001)	Bleeding or INR > 3.5[d]: Genotype-guided group: (10% (5/50) Comparison group: 15.7% (8/51)

Table 6.2. Summary of Prospective, Randomized Trials Evaluating Clinical Efficacy of Genotype-Guided Warfarin Dosing Methods *(continued)*

Study	Study Population	Sample Size	Comparison Group	Primary Outcome	Results	Comment
Jonas et al. (2013)	80% Caucasians 20% AA	109	Clinical dosing algorithm	Number of anticoagulation visits for the first 90 days; PTTR for the first 90 days	Genotype-guided group: 6.96 Comparison group: 6.37 (p=0.51) PTTR: Genotype-guided group: 40% Comparison group: 43% (p = 0.59)	Bleeding incidence[a]; Genotype-guided group: 51% (28/55) Comparison group: 48% (26/54)
COAG (2013)	49% AA 27% Hispanics 5.4%	1015	Clinical dosing algorithm	PTTR from day 4-5 through day 28	Genotype-guided group: 45.2 ± 26.6% Comparison group: 45.4 ± 25.8% (p=0.91)	AA in the genotype-guided group had a significantly lower PTTR than those in the comparison group (35.2 ± 26.0% vs. 43.5 ± 26.5%; p=0.01) Bleeding incidence[e]: Genotype-guided group: 3.3% (17/514) Comparison group: 6.0% (30/501)
EU-PACT (2013)[f]	98.5% Caucasians	455	Clinical fixed dosing	PTTR for the first 12 weeks	Genotype-guided group: 67.4 ± 18.1% Comparison group: 60.3 ± 21.7% (p<0.001)	Bleeding incidence[g]: Genotype-guided group: 37% (78/211) Comparison group: 38% (82/216)

Abbreviations: INR, International normalized ratio; CYP, cytochrome P450; VKORC1, vitamin K epoxide reductase subunit 1; AA, African Americans; PTTR, percent of time in therapeutic range; COAG, Clarification of anticoagulation through genetics; EU-PACT, the European pharmacogenetics of anticoagulant therapy.
a: Both major and minor bleeding events were included.
b: Serious adverse events were defined as INR ≥ 4.0, use of vitamin K, major bleeding events, thromboembolic events, all-cause stroke, myocardial infarction, and death.
c: All bleeding events were minor bleeding.
d: Types of bleeding events were not reported.
e: Major and clinically relevant non-major bleeding were included.
f: Participants who remained in the study on day 13 or later were included in the analysis.
g: All events were minor bleeding because there was no major bleeding event in the study.

clinical algorithm dosing groups over the initial 4 weeks of therapy in the population overall (20% and 21%, respectively) and in African Americans (24% and 22%, respectively).

The EU-PACT trial studied a population with 98.6% European descent. Its primary outcome was PTTR for the first 12 weeks of therapy. As an intervention, the trial used a modified International Warfarin Pharmacogenetics Consortium (IWPC) algorithm capable of refining warfarin doses for the first 5 days, and, as a comparator, used a fixed warfarin dosing method (5 mg/day for the first 3 days except for 10 mg on the first day in patients less than or equal to 75 years old, followed by adjustments of warfarin doses on days 4 and 5 based on local practice). Warfarin dosages after day 5 were adjusted according to local practice in both groups. The trial found that the pharmacogenetic algorithm-guided warfarin dosing method significantly increased the primary outcome compared with the fixed dosing method (67.4 ± 18.1% vs. 60.3 ± 21.7%; p<0.001). In addition, the pharmacogenetic algorithm-guided warfarin dosing method produced a significantly higher PTTR for the first 4 weeks (54.6 ± 23.0% in the pharmacogenetic algorithm-guided dosing method vs. 45.7 ± 24.3% in the fixed dosing method; p<0.001). Compared to standard dosing, pharmacogenetic dosing was also associated with fewer instances of supra-therapeutic anticoagulation, defined as an INR greater than or equal to 4 (27% versus 37%, p=0.03), and reaching a therapeutic INR more quickly (median of 21 versus 29 days, p<0.001) than standard dosing.

Because of the inconsistent results from the COAG and EU-PACT trials, pharmacogenetic algorithm-guided warfarin dosing has not established clinical efficacy. The inconsistent results may be attributed, in part, to the difference in the study population, comparison group, and study duration. In addition, the COAG trial tested the incremental benefit of genotype-guided warfarin dosing whereas the EU-PACT trial did not. Moreover, the COAG trial masked both investigators and study subjects whereas the EU-PACT trial used an open-label design, which may have increased the likelihood of observing the significant difference between groups. Although the EU-PACT trial reported that the pharmacogenetic algorithm-guided dosing method resulted in a statistically significantly higher primary outcome than the clinical dosing method, the effect size (i.e., 7.1% difference) may not be large enough to justify routine use of warfarin pharmacogenetic testing in clinical practice (Furie 2013). In addition, despite better dosing prediction with pharmacogenetic dosing, factors such as diet and concomitant drug therapy may change over time and influence PTTR even after stable doses are reached.

There are several key issues with interpreting the results of trials evaluating clinical efficacy. First, instead of clinical outcomes, they measured surrogate outcomes and had insufficient power for clinical outcomes such as bleeding and thrombosis, although there tended to be a lower incidence of the clinical outcomes compared to previous clinical effectiveness studies and in the pharmacogenetic algorithm-guided dosing group. Second, because they determined clinical efficacy, they do not adequately represent current clinical practice of warfarin dosing. The mathematical clinical warfarin dosing algorithms adopted in the trials are not used routinely in clinical practice, although, the data show that they are superior at dosing warfarin and perhaps they should be routinely incorporated into practice (International Warfarin Pharmacogenetics Consortium 2009). In addition, Asian populations, who are more likely to benefit from genotype-guided warfarin dosing due to their increased sensitivity to warfarin, are underrepresented in these trials, although a trial in Asians is on-going. Moreover, these trials had frequent INR monitoring. For example, both the COAG and EU-PACT trials required INR monitoring 6 times over the first 4 weeks. In clinical practice, however, INR may not be as frequently monitored as in these trials. This frequent INR monitoring may have diluted the effect of pharmacogenetic algorithm-guided warfarin dosing (Zineh 2013). Finally, the trials generally enrolled patients from large tertiary teaching hospitals located in urban areas. Community hospitals and rural areas generally do not have sufficient resources in managing warfarin therapy. Therefore, it remains to be seen that community hospitals and rural areas may benefit from the use of warfarin pharmacogenetic testing.

Specific Implementation of Testing

Regulatory Stance on Warfarin Pharmacogenetics

Of the 3 revisions the warfarin label has undergone since 2007, two were directly related to pharmacogenetics. In 2007, general information on warfarin pharmacogenetics was added to the label. In 2011, "Three ranges of expected therapeutic warfarin doses based on the *CYP2C9* and *VKORC1* genotype," a pharmacogenetic dosing table, was incorporated into the label to help clinicians select an initial warfarin dose for a patient with known *CYP2C9* and *VKORC1* genotypes (Table 6.3). Based on the *CYP2C9* and *VKORC1* genotype, the table provides 3 expected therapeutic warfarin dose ranges: 0.5-2 mg/day, 3-4 mg/day, and 5-7 mg/day (Coumadin Package Insert 2010). According to the label, these dose ranges, which were constructed based on the data from multiple clinical studies, account for clinical factors influencing the warfarin dose variability (e.g., age, race, body weight, sex, co-morbidities, and concomitant medications). When implementing the dosing table in clinical practice, given a genotype, clinicians may choose a dose within the expected therapeutic warfarin dose range after determining the likely direction of warfarin dose requirement based on the patient's clinical factors. For example, if the requirement is likely increased based on the patient's clinical factors, then a dose near to the higher end of the range may be chosen.

It is important to note that, after the initial dose, the current labeling recommends daily INR monitoring until INR becomes stabilized in the therapeutic range. In addition, the current labeling leaves the decision to use warfarin pharmacogenetic testing to the treating physician because it does not require warfarin pharmacogenetic testing for warfarin initiation.

Clinical Pharmacogenetics Implementation Consortium Guideline

The Clinical Pharmacogenetics Implementation Consortium (CPIC) published guidelines to assist with incorporation of *CYP2C9* and *VKORC1* genotype data into warfarin dosing decisions in 2011 (Johnson 2011). As in the other CPIC guidelines, it focuses on how to interpret and apply warfarin pharmacogenetic testing results in clinical practice, not whether or not to perform genotyping. Key recommendations include 1) endorsing use of genetic information when available for dosing decisions; 2) using genetics-based algorithms over the pharmacogenetic dosing table in the warfarin label to assist with warfarin dosing (The pharmacogenetic dosing table is recommended if a genetics-based algorithm is not accessible);

Table 6.3. Three Ranges of Expected Therapeutic Warfarin Doses Based on *CYP2C9* and *VKORC1* Genotypes (mg/day) (Coumadin 2010)

VKORC1	*CYP2C9*					
-1639	*1/*1	*1/*2	*1/*3	*2/*2	*2/*3	*3/*3
GG	5-7	5-7	3-4	3-4	3-4	0.5-2
GA	5-7	3-4	3-4	3-4	0.5-2	0.5-2
AA	3-4	3-4	0.5-2	0.5-2	0.5-2	0.5-2

Abbreviations: *CYP2C9*, cytochrome P450 2C9; *VKORC1*, vitamin K epoxide reductase complex subunit 1; G, guanine; A, adenine.

and 3) using the www.warfarindosing.org algorithm as the genetics-based algorithm to estimate stable warfarin doses. The guideline also summarizes the important data on warfarin pharmacogenetics and provides practical information on the use of a warfarin pharmacogenetic test including a list of the warfarin pharmacogenetic tests that have been cleared by the FDA. It is an indispensable resource to assist clinicians in clinical application of warfarin pharmacogenetics and is freely available through the Pharmacogenomics Knowledge Base (PharmGKB) Web Site (www.pharmgkb.org). Because of the publication of the COAG and EU-PACT trials, it is expected that these new data will be incorporated into the guideline revision due out in the near future.

Tools for Warfarin Pharmacogenetic Dosing
The warfarin pharmacogenetic dosing table and warfarin pharmacogenetic dosing algorithms are two examples of dosing tools to help clinicians estimate stable therapeutic doses based on the patient's genetic and non-genetic factors that influence warfarin dose requirements. The warfarin pharmacogenetic dosing table (Table 6.3), which is currently included in the warfarin label, provides three expected dose ranges based on *CYP2C9* and *VKORC1* genotypes. Of note, the maximum daily warfarin dose in the table is 7 mg/day.

Warfarin pharmacogenetic dosing algorithms are linear regression models ($Y = aX + b$) where the dependent variable (i.e., the Y variable) is the stable warfarin dose, and the independent variables (i.e., the X variables) are genetic and non-genetic factors. For example:

Warfarin daily dose (mg/day) = exp[0.9751 − 0.3238 × *VKORC1* − 1639G>A + 0.4317 × body surface area − 0.4008 × *CYP2C9*3* − 0.00745 × age − 0.2066 × *CYP2C9*2* + 0.2029 × target INR − 0.2538 x amiodarone + 0.0922 × smokes − 0.0901 × African American race + 0.0664 × deep vein thrombosis/pulmonary embolism].

where the SNPs are coded 0 if absent, 1 if heterozygous, and 2 if homozygous; and race is coded as 1 if African American and 0 if otherwise (Gage 2008).

To estimate a therapeutic warfarin dose, known values of the independent variables are added into the model. For example, the model estimates a warfarin dose of 1.81 mg/day for a 70-year-old Caucasian woman who is 5'5" and 65 kg, carries *CYP2C9*1/*2* and *VKORC1* -1639A/A genotypes, does not smoke, is on amiodarone, and is treated for atrial fibrillation with a target INR of 2.5:

exp[0.9751 − 0.3238 × 2 + 0.4317 × 1.72 − 0.4008 × 0 − 0.00745 × 70 − 0.2066 × 1 + 0.2029 × 2.5 − 0.2538 × 1 + 0.0922 × 0 − 0.0901 × 0 + 0.0664 × 0] = 1.81 mg/day.

Of note, the algorithms provide an actual estimated dose instead of a dose range as provided by the table in the warfarin labeling. In addition, they can estimate a dose more than 7 mg/day.

Numerous warfarin pharmacogenetic dosing algorithms have been published. Some algorithms were derived from a racially diverse population whereas others were from a homogenous population (e.g., Slovenians, Chinese, Egyptians, Koreans, and Japanese, etc.) (International Warfarin Pharmacogenetics Consortium 2009; Gage 2008; Herman 2006; Kim 2009; Takahashi 2006; Lenzini 2010; Wei 2012; Ekladious 2013). Although an algorithm with a lower target INR of 1.5-2.5 is available, the vast majority of algorithms have a target INR of 2-3 (Liu 2012). Algorithms contain both genetic and non-genetic factors as independent variables. All of the algorithms have the *CYP2C9*3* and *VKORC1* -1639G>A (or 1173C>T) SNP and some includee additional SNPs. For example, an algorithm derived from African Americans has the *CYP2C9*5, *6, *8, *11*, and rs12777823 polymorphisms, and an algorithm derived from Israelis contains the *VKORC1* Asp36Tyr SNP (Hernandez 2014; Kurnik 2012). Algorithms derived from Asian populations do not have the *CYP2C9*2* SNP (Wei 2012). Many algorithms contain age, body size, amiodarone use and smoking status as clinical factors, and some additionally include the amount of vitamin K intake, heart failure, and the use of a CYP enzyme inducer. In addition, several algorithms can account for previous warfarin doses and INR responses (Horne 2012; Xu 2012). These algorithms, which may be called warfarin dose refinement algorithms, may be more practical because they can adjust warfarin doses based on genotype, clinical factors, previous warfarin doses, and INR responses as long as genotyping results are available within 6 to 11 days after warfarin initiation (Horne 2012).

Because warfarin pharmacogenetic dosing algorithms are linear regression models, the coefficient of determination (R^2) is used to measure how much they explain the variability in warfarin doses. Algorithms with high R^2 values explain a greater portion of the variability in warfarin doses than those with low R^2 values. In general, warfarin pharmacogenetic dosing algorithms have R^2 values ranging from 30%–60%. Algorithms with SNPs in addition to *CYP2C9*2*, **3* and *VKORC1*-1639G>A tend to have R^2 values in the upper part of this range (Wei 2012; Ekladious 2013). Although algorithms have low R^2 values (less than 30%) in African Americans, those containing SNPs common in African Americans such as *CYP2C9*5*, **6*, **8*, and **11* have R^2 values greater than 30% (International Warfarin Pharmacogenetics Consortium 2009; Gage 2008; Herman 2006; Kim 2009; Takahashi 2006; Lenzini 2010; Schelleman 2008). Because warfarin dose refinement algorithms account for previous warfarin doses and INR responses, they have R^2 values greater than 60%.

Accuracy of Prediction of Initial Warfarin Dosage by Various Warfarin Dosing Methods

The warfarin pharmacogenetic dosing table and algorithms have been compared with non-genetic warfarin dosing methods for accuracy of prediction of initial warfarin dosage. The non-genetic warfarin dosing methods include empiric initial (e.g., a fixed dose of 5 mg/day) and clinical dosing algorithm-guided dosing methods. Clinical algorithms are linear regression models without a genetic independent variable. As measures of the accuracy of the prediction of initial warfarin doses, studies have used percent of patients with estimated initial doses within 20% of actual stable doses and mean absolute error (i.e., absolute difference between actual and estimated doses).

When the pharmacogenetic dosing table was compared with two empiric fixed initial dosing methods (i.e., fixed 5 mg/day for all patients; 2.5 mg/day for Asians, 5 mg/day for Caucasians, and 7.5 mg/day for African Americans), it had a higher percent of patients with estimated initial doses within 20% of their actual stable doses and a lower mean absolute error than the empiric dosing methods (percent within 20%: 41.5% vs. 31.8%–32.7%; mean absolute error: 10.9 mg/week vs. 12.3–12.6 mg/week) (Shin 2011). In a direct comparison of various dosing methods, however, the dosing table was outperformed by a pharmacogenetic dosing algorithm (Finkelman 2011). The pharmacogenetic dosing algorithm had a higher percent of patients with estimated initial doses within 20% of their actual stable doses than the pharmacogenetic dosing table (52% vs. 43%). In addition, the pharmacogenetic dosing algorithm was more accurate than a clinical dosing algorithm (39%), and an empiric initial dosing method of 5 mg/day (37%) (Finkelman 2011). Several factors may have contributed to the superior performance of pharmacogenetic algorithms. First, pharmacogenetic dosing algorithms have 15%–40% higher R^2 values than the other methods (International Warfarin Pharmacogenetics Consortium 2009; Gage 2008). Second, pharmacogenetic algorithms outperform other dosing methods particularly in patients with actual stable doses less than or equal to 3 mg/day or greater than or equal to 7 mg/day because they better account for both genetic and non-genetic factors (International Warfarin Pharmacogenetics Consortium 2009). For example, the pharmacogenetic dosing table cannot predict warfarin doses less than 7 mg/day and may not accurately account for certain factors such as use of amiodarone or other interacting medications (Shin 2011). In addition, the table had only a 15% probability of its estimated initial dose being within 20% of the actual stable dose when the actual stable dose is less than or equal to 3 mg/day (Shin 2011). In summary, pharmacogenetic dosing algorithm-guided warfarin dosing is the most accurate method to predict initial warfarin doses.

Evaluation of Various Pharmacogenetic Dosing Algorithms

Which warfarin pharmacogenetic dosing algorithm is most accurate in estimating the initial warfarin dose relative to the stable actual dose? Studies have consistently shown that the www.warfarindosing.org and IWPC dosing algorithms are the two most accurate algorithms (Liu 2012; Shin 2011; Roper 2010; Langley 2009). Both were derived from racially diverse populations and have a high R^2 value (about 50%). Moreover, they are freely available on the internet. They have a 45%–54% probability of their estimated initial dose being within 20% of the actual stable dose and mean absolute errors ranging 0.7–1.4 mg/day for a wide range of target INR values (1.5–3) (Liu 2012; Shin 2011). Although the percent of patients with an estimated initial dose that is within 20% of the actual stable dose is similar across ethnic groups with each algorithm (47%–56% in Caucasians, 39%–52% in African Americans, and 37%–53% in Asians), the COAG trial data suggests that they do not perform reliably in African Americans (Kimmel 2013; Shin 2011). Their accuracy varies by the range of the actual stable dose: the percent of the estimated initial dose within 20% of the actual stable dose is 20%–45% for doses less than or equal to 3 mg/day, 54%–60% for dose greater than 3 and less than 7 mg/day, and 38%–48% for doses greater than or equal to 7 mg/day (Liu 2012; Shin 2011). In particular, for doses less than or equal to 1.5 mg/day, the range common in East Asian populations with a low intensity of anticoagulation, the percent of patients with the estimated initial dose within 20% of the actual dose was less than 5%, although no algorithm has a probability greater than 10% (Liu 2012). In addition, in a population with a high prevalence of *VKORC1* Asp36Tyr, the R^2 value is less than 30% (Kurnik 2012). The cause of the lower accuracy of the algorithms in the extreme warfarin dose ranges is multi-factorial. The majority of the population from whom algorithms were derived required greater than 3 mg/day and less than 7 mg/day. In addition, the SNPs included in the algorithms decrease warfarin dose requirements. Because they do not contain common genetic variables associated with warfarin dose variability in African Americans, they have a high mean absolute error and do not perform reliably in this population (Kimmel 2013; Shin 2011). Moreover, the algorithms do not account for all of the factors associated with warfarin dose variability. However, it is important to keep in mind that the www.warfarindosing.org and IWPC algorithms outperform non-genetic dosing methods in the warfarin dose ranges less than or equal to 3 mg/day or greater than or equal to 7 mg/day (International Warfarin Pharmacogenetics Consortium 2009). Finally, most algorithms have been evaluated for dosing accuracy instead of clinical outcomes.

Use of a Warfarin Pharmacogenetic Dosing Algorithm in Clinical Practice

At present, the www.warfarindosing.org algorithm is probably the most clinically applicable warfarin pharmacogenetic dosing algorithm because it is a warfarin dose refinement algorithm. The algorithm is easy to use and has an R^2 value of 70% on the 11th day of warfarin therapy in published studies (Horne 2012). However, the version freely available only estimates doses for up to 5 days after warfarin initiation. It provides mini loading doses for carriers of a *CYP2C9* variant allele who may have a delay in time to reach therapeutic anticoagulation if only maintenance doses are used. Based on previous warfarin doses and INR responses, it adjusts doses for patients who have warfarin pharmacogenetic test results available within the first 5 days of warfarin therapy.

When using a warfarin pharmacogenetic dosing algorithm, clinicians should be aware that its main role is to reduce the likelihood of dosing errors. Because even the most accurate algorithm has a probability of its estimated dose within 20% of the actual stable dose at about 50%, clinicians are likely to make dose adjustments according to INR responses until reaching a stable dose. However, clinicians are less likely to make a serious dosing error because it has a smaller mean absolute error than non-genetic warfarin dosing methods. Therefore, warfarin pharmacogenetic dosing algorithms are adjuncts to reduce the likelihood of dosing errors and do not replace regular INR monitoring and sound clinical judgment. In addition, until a reliable pharmacogenetic dosing algorithm is developed for African Americans, warfarin pharmacogenetic algorithms should be used cautiously in this population.

Structure and Feasibility of a Warfarin Pharmacogenetics Service

While there are limited examples of genotype-guided warfarin dosing in clinical practice, there are data showing the feasibility of such an approach. In particular, a multi-disciplinary warfarin pharmacogenetics consult service was established at the University of Illinois Hospital & Health Science System (UI-Health), which provided genotype-guided dosing recommendations for patients newly starting warfarin (Nutescu 2013). The pharmacogenetics consult team consisted of pharmacists and physicians specializing in anticoagulation management; experts in warfarin pharmacogenetics, information technology, and building clinical decision support rules; and personnel from the clinical pathology laboratory. Responsibilities of the consult team included the following:

- Obtaining institutional approval for genotype-guided warfarin dosing
- Coordinating provider education prior to pharmacogenetic implementation and periodically thereafter
- Selecting genotyping methodology in the clinical pathology laboratory
- Validating genotype tests
- Building the information technology infrastructure to support personalized warfarin dosing
- Providing consultative services, including daily warfarin dosing recommendations
- Developing resources for patient education
- Serving as an information resource to support clinical decision making
- Performing quality assurance assessment

The process for providing the service is outlined in Figure 6.2. To initiate the genotype-guided dosing approach, clinical decision support tools were built into the electronic medical record to automatically order genotyping and a consultation with the warfarin pharmacogenetics service for any hospitalized patient newly prescribed warfarin. New warfarin use was defined as no history of warfarin use in the previous 6 months per inpatient or outpatient medication orders or medication reconciliation on hospital admission. At the time of the warfarin order, the prescribing physician receives an electronic notification about the automatic genotype order and automatic consultation with the pharmacogenetics service to assist with genotype interpretation and provide dosing recommendations. An example of an alert is shown in Figure 6.3. In addition, an initial dose recommendation, calculated based on clinical factors utilizing the www.warfarindosing.org algorithm (Gage 2008), is provided.

A goal of the service was to obtain genotype results in time to inform the second warfarin dose. Blood for genotyping is drawn at the next scheduled phlebotomy draw, with genotyping for the *VKORC1* -1639G>A; *CYP2C9*2, *3, *5, *6, *11, *14, *15,* and *16;* and *CYP4F2* Val433Met polymorphisms conducted in the hospital's clinical pathology laboratory. Laboratory personnel performed genotyping at 10 am each day for any samples available. Once genotype results were available, the laboratory personnel notified the consult service pharmacist. The pharmacist and physician on the consult team then jointly provided a patient assessment, genotype interpretation, and warfarin dose recommendation. The pharmacist on the team continued to provide daily dose recommendation, refined based on INR responses to previous doses, until the patient reached a therapeutic INR or was discharged. The majority of patients were followed up at the UI-Health Antithrombosis Clinic, staffed by pharmacists knowledgeable in warfarin pharmacogenetics.

Evaluation of the service 6 months after its initiation showed that nearly 200 patients newly starting warfarin were genotyped during that period (Nutescu 2013). Nearly 80% of genotypes were available prior to the second warfarin dose; the median time from the initial warfarin order to the genotype result appearing in the medical record was 26 hours. These data demonstrate the procedural feasibility of providing routine genotype-guided warfarin dosing. There was also good adherence by the medical staff to dose recommendation provided by the pharmacogenetics consult team, with over 70% of doses ordered being within 0.5 mg of that recommended. Data comparing outcomes, including PTTR and percent time with an INR greater than 4, with genotype-guided dosing versus historical dosing, are forthcoming. In contrast to the COAG trial, which only genotyped for *VKORC1* -1639G>A and *CYP2C9*2* and *3*, data from the UI Health warfarin pharmacogenetics service will provide outcomes data with genotyping that captures many of the variants important for African American patients.

FDA Approved Genotyping Platforms

There are at least four in vitro diagnostic assays cleared by the FDA for clinical genotyping relevant to warfarin: Verigene System (Nanosphere, Inc., Northbrook, IL), INFINITI Warfarin Assay (AutoGenomics, Carlsbad, CA), eSensor platform (GenMark DX, Carlsbad, CA), and eQ-PCR LC Warfarin Genotyping Kit (TrimGe, Sparks, mDn). All four detect the *CYP2C9*2* and *3* alleles and either the *VKORC1* -1639G>A or 1173C>T genotype. Both AutoGenomics and GenMark also offer non-FDA cleared extended genotyping panels. The AutoGenomics panel includes the *CYP2C9*4, *5, *6,* and *11* SNPs and additional *VKORC1* polymorphisms. However, the benefit of including additional VKORC1 SNPs is questionable since many are in LD with the -1639G>A SNP, at least in Europeans, and genotyping for the -1639G>A SNP alone appears sufficient (Wang 2008; Limdi 2010). The GenMark panel is the only one to include the *CYP4F2* V433M SNP in addition to the *CYP2C9*5, *6, *11, *14, *15,* and *16* alleles on its extended panel. The platforms differ in the technology utilized and genotype turn-around times, which are summarized in Table 6.4.

Keys to Success (Specific to Warfarin)

Genotype Turnaround Time

A pharmacogenetic dosing algorithm has been shown to more accurately predict warfarin dose requirements than a clinical dosing algorithm, even up to 11 days after warfarin initiation (Lenzini 2010). However, the predictive ability of a pharmacogenetic algorithm over a clinical algorithm decreases over time. Thus, the earlier genotype information is available, the greater the likelihood that it will improve warfarin dosing over traditional methods.

In the EU-PACT trial, a point-of-care genotyping platform was utilized, which allowed genotype results to be available in time to influence the initial warfarin dose (Pirmohamed 2013). In the absence of point-of-care testing or preemptive genotyping, obtaining genotype results prior to the initial dose may be difficult, as demonstrated in the COAG trial, in which genotype was available before the first warfarin dose only 45% of the time, but available before the second dose 94% of the time (Kimmel 2013). However, data from clinical practice support the feasibility of obtaining genotype results prior to the second warfarin dose (Nutescu 2013). While awaiting genotype results, it would seem prudent to apply a clinical dosing algorithm, accounting for important factors such as age and body size, to assist in choosing warfarin dose. Indeed, there are data that clinical dosing algorithms for warfarin are superior to a fixed dose approach (e.g., 5 mg/day) (International Warfarin Pharmacogenetics Consortium 2009).

Personnel to Assist with Genotype Interpretation and Use of Algorithms

The success of genotype-guided warfarin dosing depends on accurate interpretation of genotype results and appropriate application of results to warfarin prescribing. Not only should personnel be able to interpret the genotype information available, but they should also be aware of the potential implications of genotypes not included on the testing panel. For example, if only the CYP2C9*2 and *3 alleles are detected for a patient of African descent, one must be cognizant that other genotypes not included on the testing panel (e.g., CYP2C9*5, *6, *8, or *11) could be present and have important effects on warfarin response.

Pharmacogenetic dosing algorithms are recommended over other approaches for genotype-guided dosing (Johnson 2011). Thus, clinicians should be familiar with recommended algorithms, but also recognize their limitations. Specifically, algorithms do not account for renal function, many medications with important effects on warfarin disposition, and a number of genotypes important for warfarin response. For example, some commercially available extended genotyping panels include the CYP2C9*11 allele, which is not included in the algorithm available through www.warfarindosing.org. Thus, clinicians should be able to modify algorithm-recommended doses for patients with significant renal impairment or concomitant medications or genotypes with important effects on warfarin that are not accounted for. In the absence of access to pharmacogenetic algorithms, then per CPIC guidelines, the table in the warfarin label should be used. Again, clinical judgment is needed to interpret genotype effects on warfarin response in the context of important clinical factors.

Barriers to Testing or Implementation

Questionable Clinical Utility

Both the COAG and EU-PACT trials limited genotyping to the CYP2C9*2 and *3 alleles and VKORC1 -1639G>A genotype. This was appropriate in EU-PACT, which was conducted in a relatively homogeneous European population in whom variants influencing warfarin dose variability are well defined. However, additional variants, namely the rs12777823 and CYP2C9*5, *6, *8, and *11 alleles, are common and impact warfarin dose requirements in African Americans, who comprised 27% of the COAG trial population. Therefore, it remains unknown whether genotype-guided dosing that takes into account variants important across ethnicities improves anticoagulation management. In addition, one cannot conclude from the COAG trial results that genotype-guided dosing is as good as clinically-guided dosing in Europeans, as the trial was not powered to detect differences in any one ethnic group but rather in the population as a whole.

Additional trials are on-going and expected to provide some clarification on the role of genotype-guided warfarin dosing. The Genetics Information Trial (GIFT) of warfarin to prevent deep vein thrombosis is examining genotype-guided warfarin dosing in an older (age greater than or equal to 65 years) population who receive warfarin for prophylaxis of venous thromboembolism after hip or knee arthroplasty (Do 2012). Investigators are targeting enrollment of approximately 1600 patients, which is more than the EU-PACT and COAG trials combined. The intervention is similar to that in the COAG trial, with participants randomized to warfarin dosing with a pharmacogenetic versus clinical algorithm. Patients are further randomized to an INR goal of 1.8 or 2.5. The targeted population of older patients undergoing major orthopedic surgery is at high risk for both venous thrombosis and warfarin-related bleeding, and thus, unlike earlier trials, GIFT will be powered to detect a difference in the outcomes of venous thromboembolism and major bleeding at 4 to 6 weeks between warfarin dosing strategies. The trial is expected to be completed in 2015, and while it will provide additional insight into the utility of genotype-guided warfarin dosing, it differs from previous trials in terms of patient characteristics, warfarin indication, and the intensity of anticoagulation. Thus, even if outcomes favor genotype-guided dosing, questions may remain on the utility of genotyping in patients requiring anticoagulation for the treatment (versus prophylaxis) of thromboembolism.

Another trial is underway in a Chinese population. Patients requiring warfarin for atrial fibrillation, venous thromboembolism, or heart valve replacement are being targeted and randomized to genotype-guided or standard dosing, with a primary endpoint of percent of out-of-range INRs.(ClinicalTrials.gov Identifier: NCT01610141) Each patient will participate for up to 3 months, and the study is estimated to be completed in 2015.

Further data will likely come from observational studies and clinical experience. As discussed above, UI-Health implemented genotype-guided warfarin dosing as the standard-of-care from August 2012 through January 2014 (Nutescu 2013). Genotypes detected included *CYP2C9* *5, *6, *11 and *CYP4F2* Val433Met, thus taking into account some of the variants important for African Americans. An assessment of outcomes between genotype-guided dosing and traditional dosing in a clinical practice setting such as this will provide further insight into the benefit of the former approach.

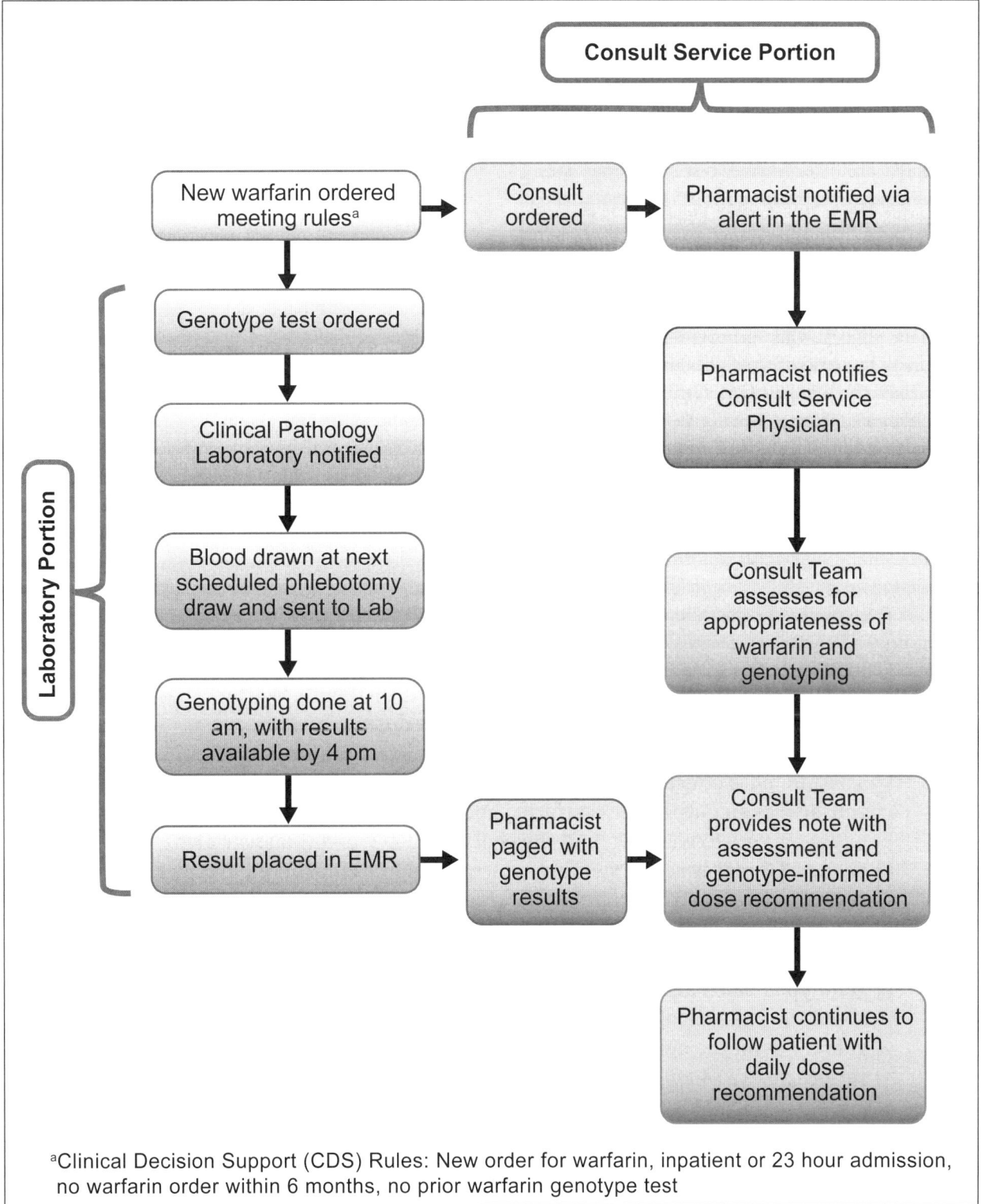

Figure 6.2. A process for providing personalized warfarin dosing.

Cost-Effectiveness of Genotype-Guided Warfarin Dosing

One of the important barriers to implementing genotype-guided warfarin dosing in clinical practice is its cost-effectiveness. Because new target-specific oral anticoagulants (e.g., dabigatran, rivaroxaban, apixaban etc.) are available and do not require regular laboratory monitoring, for wide clinical implementation, genotype-guided warfarin dosing should be cost-effective compared with traditional warfarin dosing as well as these new agents.

Studies comparing the cost-effectiveness of genotype-guided warfarin dosing with traditional warfarin dosing have produced mixed results. In one study, genotype-guided warfarin dosing was cost-effective 96.4% of the time (You 2014). Another study found that genotype-guided warfarin dosing has an incremental cost-effectiveness ratio (ICER) of £13,226 per quality adjusted life year (QALY) gained (Pink 2014). These data suggest that genotype-guided warfarin dosing may be cost-effective because this amount is lower than the $50,000 QALY gained, a threshold below which a society is willing to pay (Pink 2014). In contrast, other studies do not support its cost-effectiveness. In one study, genotype-guided warfarin dosing exceeds $170,000 QALY gained and has only a 10% chance of being less than $50,000 QALY gained (Eckman 2009). In addition, its incremental cost per unit outcome improved per QALY was greater than $50,000 for 54%–62% of the time (You 2009; Meckley 2010). When compared with the new target-specific oral anticoagulants, genotype-guided warfarin dosing appears to be less cost-effective. It has an incremental cost per unit outcome improved per QALY less than $50,000 for 46.2% of the time compared with dabigatran 150 mg twice daily (You 2012). In this study, it would be cost-effective if both genotype-guided warfarin dosing and dabigatran provide comparable quality of life, and genotype-guided warfarin dosing has time within therapeutic INR greater than 77% of the time. In another study, apixaban dominates genotype-guided warfarin dosing with an ICER of £13,226 per QALY gained (Pink 2014).

Pharmacoeconomic models rely on assumptions on input variables. However, many of these variables have not been reliably determined. For example, the incidence of major bleeding associated with the use of genotype-guided warfarin dosing is currently unknown and how it was assumed influenced the study results. For example, studies assuming a very low incidence of major bleeding with genotype-guided dosing tend to favor it over traditional warfarin dosing (You 2014; Pink 2014; McWilliam 2006). Another key variable is time spent within therapeutic INR range and studies with an

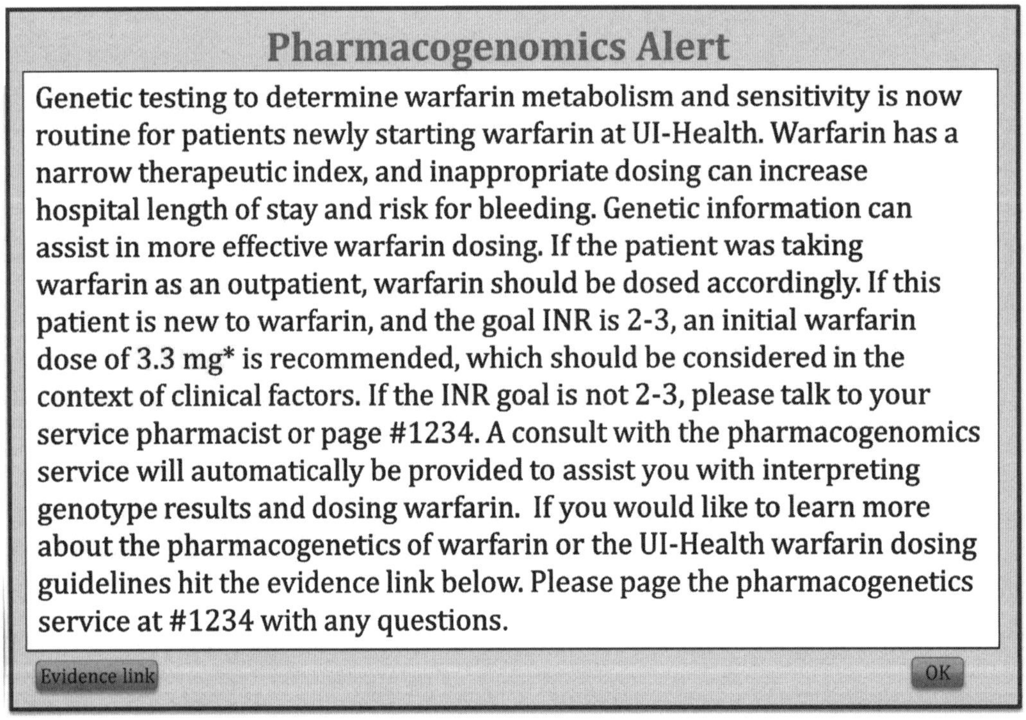

Pharmacogenomics Alert

Genetic testing to determine warfarin metabolism and sensitivity is now routine for patients newly starting warfarin at UI-Health. Warfarin has a narrow therapeutic index, and inappropriate dosing can increase hospital length of stay and risk for bleeding. Genetic information can assist in more effective warfarin dosing. If the patient was taking warfarin as an outpatient, warfarin should be dosed accordingly. If this patient is new to warfarin, and the goal INR is 2-3, an initial warfarin dose of 3.3 mg* is recommended, which should be considered in the context of clinical factors. If the INR goal is not 2-3, please talk to your service pharmacist or page #1234. A consult with the pharmacogenomics service will automatically be provided to assist you with interpreting genotype results and dosing warfarin. If you would like to learn more about the pharmacogenetics of warfarin or the UI-Health warfarin dosing guidelines hit the evidence link below. Please page the pharmacogenetics service at #1234 with any questions.

Evidence link   OK

Figure 6.3. Alert to the physician at the time of a new warfarin order for a patient followed by a warfarin pharmacogenetics consult service. *The dose recommendation provided is for a 65 year old, 5'4" Caucasian female, who weighs 80 kg and is taking amiodarone for atrial fibrillation.

Table 6.4. FDA-Cleared Warfarin Genotyping Platforms

Platform	Technology Used	Turn Around Time
Verigene (Nanosphere)	Hydridization to a gene-specific oligonucleotide capture strand with gold nanoparticle probe technology	2.5 hours
INFINITI Warfarin Assay (AutoGenomics)	Allele specific primer extension	10.6 hours
eSensor XT-8 Platform (GenMark DX)	Multi-plex PCR and a solid-phase electrochemical detection method	4 to 6 hours including PCR assay
eQ-PCR LC Warfarin Genotyping Kit (TrimGen Corp)	Real-time PCR	1.5 hours

assumption of more time within therapeutic INR range with genotype-guided warfarin dosing tend to favor it (You 2014; Pink 2014; Patrick 2009). Importantly, none of these studies used data from the COAG or EU-PACT trial. In addition, the cost of pharmacogenetic testing and turnaround time will likely be reduced with the rapid advance in genotyping technology and the wide availability of such technology.

Although the new target-specific oral anticoagulants may be more cost-effective than genotype-guided warfarin dosing in atrial fibrillation, it is unknown whether they are still more cost-effective in patients with an indication other than atrial fibrillation because the study population has been limited to patients with atrial fibrillation (Pink 2014; You 2012). Moreover, these agents are not recommended for certain populations such as those with reduced kidney function. Therefore, the current economic data on genotype-guided warfarin dosing should be viewed as exploratory instead of definitive. The ongoing Clinical and Economic Implications of Genetic Testing for Warfarin Management study (ClinicalTrials.gov ID NCT00964353) will provide more information on the cost effectiveness of genotype-guided warfarin dosing.

Third Party Reimbursement
An important barrier to clinical implementation of warfarin pharmacogenetics is the lack of coverage for genotyping by many third party payers. For example, the Center for Medicare & Medicaid Services cited insufficient evidence of improved health outcomes with genotype-guided dosing as their rationale for refusing genotype coverage outside of the clinical trial setting. However, genotyping is covered for patients enrolled in a clinical trial, such as GIFT, that assesses the effect of genotyping on the occurrence of important outcomes.

Special Populations

Minority Populations
African Americans and Hispanics are at especially high risk for poor outcomes as a result of non-therapeutic anticoagulation (White 2006; Shen 2007; Simpson 2010). For example, the risk for warfarin-related intracranial hemorrhage is higher in both African Americans and Hispanics compared to Europeans. Consequently, minority populations may have the most to gain from efforts to improve warfarin dosing. Despite this, minorities are underrepresented in warfarin pharmacogenetic studies. For example, only 9% of patients in the IWPC and 15% in the derivation cohort for the www.warfarindosing.org algorithm were African American (International Warfarin Pharmacogenetics Consortium 2009; Gage 2008). Fewer than 5% in either cohort were Hispanic.

African Americans and Hispanics are admixed populations with varying degrees of European, Amerindian, and West African ancestral components (Tian 2006; Gonzalez 2005). Because of differences in genotype frequencies and haplotype distribution by ancestry, genetic associations in Europeans may or may not replicate in African Americans and Hispanics. For example, the VKORC1 1542G>C SNP was significantly associated with warfarin dose requirements in Europeans, but not African Americans (Wang 2008). This is because the 1542G>C SNP is non-functional but in strong LD with the functional -1639G>A SNP in Europeans, but not in African Americans (Wang 2008). In addition, genome wide association studies in Europeans may fail to detect important variants for minorities. Specifically, without the recent GWAS in African Americans, the novel association between the rs12777823 SNP and warfarin dose requirements in this population would have gone undetected (Perera 2013). In addition, sequencing of the *VKORC1* loci in African Americans led to discovery of a novel association between the rs61162043 SNP, located approximately 8 Kb upstream of the *VKORC1* start site, and higher dose requirements (Perera 2011). This *VKORC1* SNP is not on conventional 1 million or 2.5 million SNP arrays nor is it in strong linkage disequilibrium with SNPs on these arrays, and thus it would not be detected with these arrays.

Both the IWPC and warfarindosing.org algorithms perform poorly in African Americans compared to Europeans, likely because they were derived from mostly European populations and do not include many variants important for African Americans (Limdi 2010; Schelleman 2008). Little is known about their performance in Hispanics. Recently, a novel pharmacogenetic algorithm was proposed for estimating warfarin dose requirements in African Americans (Hernandez 2013). In addition to the *CYP2C9 *2, *3*, and *VKORC1* -1639G>A variants, the algorithm contains the *CYP2C9*5, *8, *11*, rs12777823G>A, and rs61162043 variants. In an African American validation cohort, the dose predicted with the novel algorithm was better correlated with the observed (actual) dose (r=0.51 versus 0.38) and explained a higher portion of the variability in dose (R^2=0.27 versus 0.15) than the IWPC pharmacogenetic algorithm.

There are limited data on warfarin pharmacogenetics in U.S. Hispanics. However, preliminary data suggest that, similar to Europeans, the *CYP2C9*2, *3*, and *VKORC1* -1639G>A variants are major determinants of dose requirements (Bress 2012). There is evidence of unique genetic contributions in this population as well. Specifically, findings from a candidate gene study suggested that the NAD(P)H quinone oxidioreductase 1 (*NQO1*) and *CYP4F2* genes may have important implications for warfarin dose requirements in Hispanics (Bress 2012). Variants in both genes were associated with significantly higher warfarin dose requirements.

The data described above illustrate the importance of conducting warfarin pharmacogenetic studies in minorities. Without such, important variants would go undetected, and implementation of genotype-guided dosing based on data from Europeans may be of lesser benefit in minorities. Furthermore, failure to account for variants important in African Americans may lead to poorer outcomes with genotype-guided dosing versus other dosing approaches, as suggested by findings from the recent COAG trial.

Pediatrics
There is evidence from four studies, each including approximately 100 children (median age 4.3 to 11 years), that genetic associations with warfarin dose requirements extend to the pediatric population (Shaw 2014; Moreau 2012; Hamberg 2013; Biss 2012). Together, the *VKORC1*-1639G>A and *CYP2C9*2* and *3* genotypes explain approximately 20% of the overall variability in warfarin dose in children (Shaw 2014; Moreau 2012; Biss 2012). The combination of genotype and clinical factors, particularly body size and warfarin indication, explains approximately 70% of the variability in dose, more than explained in most adult populations (Moreau 2012; Biss 2012). In contrast, the *CYP4F2* genotype does not appear to be important for warfarin dosing in children (Hamberg 2013; Biss 2012). The accuracy of the IWPC algorithm was evaluated in children and shown to consistently over-predict warfarin dose, indicating the need for a dosing algorithm specifically developed for the pediatric population (Biss 2012).

Future Directions

Implications of Clinical Trial Data for Clinical Implementation of Warfarin Pharmacogenetics
The uncertainty of benefit with genotype-guided warfarin dosing and the lack of coverage for genotyping by many third party payers have had a negative impact on implementation of warfarin pharmacogenetics. For example, after the publication of the COAG trial results, showing no benefit with genotype-guided dosing and especially poor outcomes with this approach in African Americans, the automatic genotyping component of the warfarin pharmacogenetic service at UI-Health (described above) was discontinued. However, genotyping remains optional, and pharmacists continue to provide daily dosing recommendations based on clinical factors, emulating the clinical arm of the COAG trial. This practice will provide an opportunity to evaluate and compare outcomes with genotype-guided versus clinically-guided dosing in clinical practice.

Pharmacogenetics of Novel Oral Anticoagulants
Three new oral anticoagulant agents were approved in recent years. Dabigatran is a reversible direct thrombin inhibitor indicated for the prevention of stroke and thrombosis in patients with atrial fibrillation. Rivaroxaban and apixaban are direct factor Xa inhibitors. Similar to dabigatran, apixaban is approved for prevention of stroke and thromboembolism in atrial fibrillation. Rivaroxaban has the additional indications of treatment of venous thromboembolism (VTE) and VTE prophylaxis after hip or knee arthroplasty.

Candidate Genes for Influencing Response to Novel Oral Anticoagulants

Unlike either apixaban or rivaroxaban, dabigatran is administered as a prodrug, dabigatran etexilate, which must be hydrolyzed by esterases to its active form responsible for thrombin inhibition. There is significant inter-patient variability in plasma concentrations of dabigatran and dabigatran etexilate (Liesenfeld 2011). The *CES1* gene encodes for the liver carboxylesterase 1 enzyme responsible for conversion of dabigatran etexilate to its active form, and thus, is a candidate gene for influencing disposition of dabigatran.

The CYP450 enzymes do not appear to have a role in the metabolism of dabigatran etexilate or dabigatran, and thus, *CYP450* genotype is unlikely to affect dabigatran pharmacokinetics (Blech 2008). However, apixaban, and to a greater extent, rivaroxaban, are metabolized by CYP3A4 and CYP3A5, both of which are encoded by polymorphic genes with the potential to influence pharmacokinetic drug properties. All three agents are substrates for the polymorphic drug efflux transporter P-glycoprotein, which is encoded by the *ABCB1* gene (Stangier 2009). Thus, *ABCB1* represents an additional candidate gene for influencing disposition and response to any of the three novel agents.

Genetic Association with Dabigatran Pharmacokinetics

There are limited pharmacogenetic data with the new oral anticoagulant agents. However, a pharmacogenetic study was conducted using samples from participants in the Randomized Evaluation of Long-term Anticoagulation Therapy (RE-LY) trial, which compared dabigatran to warfarin for stroke prevention in non-valvular atrial fibrillation (Pare 2013). Investigators conducted a GWAS with samples from approximately 1500 participants of European descent who were randomized to dabigatran 110 or 150 mg twice daily to identify determinants of the inter-individual variability in concentrations of dabigatran. Of over 550,000 SNPs tested, the only ones to reach genome-wide significance for association with dabigatran concentrations were in *CES1* and *ABCB1*. The *CES1* rs2244613 SNP was present in 33% of participants and associated with lower dabigatran trough concentration. In a subsequent analysis, the rs2244613 SNP was associated with a lower risk for dabigatran-associated minor bleeding ($p = 4 \times 10^{-4}$), but not major bleeding (p=0.06). The *ABCB1* rs4148738 variant was associated with dabigatran peak concentrations, but not with bleeding risk. There was no genetic association identified with ischemic events. Given that the association was observed with dabigatran doses of 110 mg or 150 mg twice daily, and only the 150 mg dose is approved in the U.S., the clinical relevance of these findings is unclear.

Implications for Future Anticoagulant Management

Variable results from recent clinical trials of warfarin pharmacogenetics have led to uncertainly regarding the future of genotype-guided warfarin dosing. However, there are a number of questions that remain, such as:

- Would outcomes from the COAG trial have been different if the investigators accounted for variants important across ethnicities?
- Are their certain populations (e.g., children, Asians, minority populations) who may derive particular benefit from genotype-guided dosing, if appropriate variants are tested for?
- Would warfarin pharmacogenetics improve dosing and important outcomes for patients who are managed outside of anticoagulation clinics, with less frequent INR measurements?

Findings from the on-going GIFT and Asian trials and from clinical observations may address some of the uncertainly regarding genotype-guided warfarin dosing. As we await further data, it would certainly seem reasonable to dose warfarin based on clinical factors, utilizing either the IWPC or www.warfarindosing.org algorithms without genotypes included. As cost of genotyping decreases and preemptive testing becomes more commonplace, the question of whether or not to genotype at the time of warfarin initiation will become less important. Given the substantial and consistent evidence linking genotype to warfarin dose requirements, it would be difficult to argue against use of genotype results if available at the time of warfarin initiation, particularly if results reveal a genotype with major effects on warfarin metabolism *(CYP2C9*1/*3)* or sensitivity *(VKORC1-1639AA)*. The future for genotyping to predict response with alternative oral anticoagulant agents remains unclear.

References

Anderson JL, Horne BD, Stevens SM, et al. A randomized and clinical effectiveness trial comparing two pharmacogenetic algorithms and standard care for individualizing warfarin dosing (CoumaGen-II). Circulation 2012;125:1997-2005.

Anderson JL, Horne BD, Stevens SM, et al. Randomized trial of genotype-guided versus standard warfarin dosing in patients initiating oral anticoagulation. Circulation 2007;116:2563-70.

Aquilante CL, Langaee TY, Lopez LM, et al. Influence of coagulation factor, vitamin K epoxide reductase complex subunit 1, and cytochrome P450 2C9 gene polymorphisms on warfarin dose requirements. Clin Pharmacol Ther 2006;79:291-302.

Biss TT, Avery PJ, Brandao LR, et al. VKORC1 and CYP2C9 genotype and patient characteristics explain a large proportion of the variability in warfarin dose requirement among children. Blood 2012;119:868-73.

Blech S, Ebner T, Ludwig-Schwellinger E, et al. The metabolism and disposition of the oral direct thrombin inhibitor, dabigatran, in humans. Drug Metab Dispos 2008;36:386-99.

Bress A, Patel SR, Perera MA, et al. Effect of NQO1 and CYP4F2 genotypes on warfarin dose requirements in Hispanic-Americans and African-Americans. Pharmacogenomics 2012;13:1925-35.

Budnitz DS, Lovegrove MC, Shehab N, et al. Emergency hospitalizations for adverse drug events in older Americans. N Engl J Med 2011;365:2002-12.

Burmester JK, Berg RL, Yale SH, et al. A randomized controlled trial of genotype-based Coumadin initiation. Genet Med 2011;13:509-18.

Caldwell MD, Awad T, Johnson JA, et al. CYP4F2 genetic variant alters required warfarin dose. Blood 2008;111:4106-12.

Caraco Y, Blotnick S and Muszkat M. CYP2C9 genotype-guided warfarin prescribing enhances the efficacy and safety of anticoagulation: a prospective randomized controlled study. Clin Pharmacol Ther 2008;83:460-70.

Cavallari LH, Langaee TY, Momary KM, et al. Genetic and clinical predictors of warfarin dose requirements in African Americans. Clin Pharmacol Ther 2010;87:459-64.

Cha PC, Mushiroda T, Takahashi A, et al. Genome-wide association study identifies genetic determinants of warfarin responsiveness for Japanese. Hum Mol Genet 2011;19:4735-44.

Cooper GM, Johnson JA, Langaee TY, et al. A genome-wide scan for common genetic variants with a large influence on warfarin maintenance dose. Blood 2008;112:1022-7.

Coumadin (warfarin sodium) package insert. Princeton, NJ: Bristol-Myers Squibb; 2010 January.

Dickmann LJ, Rettie AE, Kneller MB, et al. Identification and functional characterization of a new CYP2C9 variant (CYP2C9*5) expressed among African Americans. Mol Pharmacol 2001;60:382-7.

Do EJ, Lenzini P, Eby CS, et al. Genetics informatics trial (GIFT) of warfarin to prevent deep vein thrombosis (DVT): rationale and study design. Pharmacogenomics J 2012;12:417-24.

Eckman MH, Rosand J, Greenberg SM, et al. Cost-effectiveness of using pharmacogenetic information in warfarin dosing for patients with nonvalvular atrial fibrillation. Ann Intern Med 2009;150:73-83.

Ekladious SM, Issac MS, El-Atty Sharaf SA, et al. Validation of a proposed warfarin dosing algorithm based on the genetic make-up of Egyptian patients. Mol Diagn Ther 2013;17:381-90.

Epstein RS, Moyer TP, Aubert RE, et al. Warfarin genotyping reduces hospitalization rates results from the MM-WES (Medco-Mayo Warfarin Effectiveness study). J Am Coll Cardiol 2010;55:2804-12.

Finkelman BS, Gage BF, Johnson JA, et al. Genetic warfarin dosing: tables versus algorithms. J Am Coll Cardiol 2011;57:612-8.

Flay BR. Efficacy and effectiveness trials (and other phases of research) in the development of health promotion programs. Prev Med 1986;15:451-74.

Furie B. Do pharmacogenetics have a role in the dosing of vitamin K antagonists? N Engl J Med 2013;369:2345-6.

Gage BF, Eby C, Johnson JA, et al. Use of pharmacogenetic and clinical factors to predict the therapeutic dose of warfarin. Clin Pharmacol Ther 2008;84:326-31.

Glasgow RE, Lichtenstein E and Marcus AC. Why don't we see more translation of health promotion research to practice? Rethinking the efficacy-to-effectiveness transition. Am J Public Health 2003;93:1261-7.

Gong IY, Tirona RG, Schwarz UI, et al. Prospective evaluation of a pharmacogenetics-guided warfarin loading and maintenance dose regimen for initiation of therapy. Blood 2011;118:3163-71.

Gonzalez Burchard E, Borrell LN, Choudhry S, et al. Latino populations: a unique opportunity for the study of race, genetics, and social environment in epidemiological research. Am J Public Health 2005;95:2161-8.

Hamberg AK, Wadelius M, Friberg LE, et al. Characterising variability in warfarin dose requirements in children using modelling and simulation. Br J Clin Pharmacol 2013.

Herman D, Peternel P, Stegnar M, et al. The influence of sequence variations in factor VII, gamma-glutamyl carboxylase and vitamin K epoxide reductase complex genes on warfarin dose requirement. Thromb Haemost 2006;95:782-7.

Hernandez W, Gamazon ER, Aquino-Michaels K, et al. Ethnicity-specific pharmacogenetics: the case of warfarin in African Americans. Pharmacogenomics J 2014;14:223-8.

Higashi MK, Veenstra DL, Kondo LM, et al. Association between CYP2C9 genetic variants and anticoagulation-related outcomes during warfarin therapy. JAMA 2002;287:1690-8.

Holbrook A, Schulman S, Witt DM, et al. Evidence-based management of anticoagulant therapy: Antithrombotic Therapy and Prevention of Thrombosis, 9th ed: American College of Chest Physicians Evidence-Based Clinical Practice Guidelines. Chest 2012;141:e152S-84S.

Horne BD, Lenzini PA, Wadelius M, et al. Pharmacogenetic warfarin dose refinements remain significantly influenced by genetic factors after one week of therapy. Thromb Haemost 2012;107:232-40.

Huang SW, Chen HS, Wang XQ, et al. Validation of VKORC1 and CYP2C9 genotypes on interindividual warfarin maintenance dose: a prospective study in Chinese patients. Pharmacogenet Genomics 2009;19:226-34.

Hylek EM, Evans-Molina C, Shea C, et al. Major hemorrhage and tolerability of warfarin in the first year of therapy among elderly patients with atrial fibrillation. Circulation 2007;115:2689-96.

Hylek EM, Go AS, Chang Y, et al. Effect of intensity of oral anticoagulation on stroke severity and mortality in atrial fibrillation. N Engl J Med 2003;349:1019-26.

International Warfarin Pharmacogenetics C, Klein TE, Altman RB, et al. Estimation of the warfarin dose with clinical and pharmacogenetic data. N Engl J Med 2009;360:753-64.

Jonas DE, Evans JP, McLeod HL, et al. Impact of genotype-guided dosing on anticoagulation visits for adults starting warfarin: a randomized controlled trial. Pharmacogenomics 2013;14:1593-603.

Johnson JA, Gong L, Whirl-Carrillo M, et al. Clinical Pharmacogenetics Implementation Consortium Guidelines for CYP2C9 and VKORC1 genotypes and warfarin dosing. Clin Pharmacol Ther 2011;90:625-9.

Kim HS, Lee SS, Oh M, et al. Effect of CYP2C9 and VKORC1 genotypes on early-phase and steady-state warfarin dosing in Korean patients with mechanical heart valve replacement. Pharmacogenet Genomics 2009;19:103-12.

Kimmel SE, French B, Kasner SE, et al. A pharmacogenetic versus a clinical algorithm for warfarin dosing. N Engl J Med 2013;369:2283-93.

Kirley K, Qato DM, Kornfield R, et al. National trends in oral anticoagulant use in the United States, 2007 to 2011. Circ Cardiovasc Qual Outcomes 2012;5:615-21.

Kurnik D, Qasim H, Sominsky S, et al. Effect of the VKORC1 D36Y variant on warfarin dose requirement and pharmacogenetic dose prediction. Thromb Haemost 2012;108:781-8.

Langley MR, Booker JK, Evans JP, et al. Validation of clinical testing for warfarin sensitivity: comparison of CYP2C9-VKORC1 genotyping assays and warfarin-dosing algorithms. J Mol Diagn 2009;11:216-25.

Lee YM, Eggen J, Soni V, et al. Warfarin dose requirements in a patient with the CYP2C9*14 allele. Pharmacogenomics 2014;15:909-14.

Lenzini P, Wadelius M, Kimmel S, et al. Integration of genetic, clinical, and INR data to refine warfarin dosing. Clin Pharmacol Ther 2010;87:572-8.

Liesenfeld KH, Lehr T, Dansirikul C, et al. Population pharmacokinetic analysis of the oral thrombin inhibitor dabigatran etexilate in patients with non-valvular atrial fibrillation from the RE-LY trial. J Thromb Haemost 2011;9:2168-75.

Limdi NA, Arnett DK, Goldstein JA, et al. Influence of CYP2C9 and VKORC1 on warfarin dose, anticoagulation attainment and maintenance among European-Americans and African-Americans. Pharmacogenomics 2008;9:511-26.

Limdi NA, McGwin G, Goldstein JA, et al. Influence of CYP2C9 and VKORC1 1173C/T genotype on the risk of hemorrhagic complications in African-American and European-American patients on warfarin. Clin Pharmacol Ther 2008;83:312-21.

Limdi NA, Wadelius M, Cavallari L, et al. Warfarin pharmacogenetics: a single VKORC1 polymorphism is predictive of dose across 3 racial groups. Blood 2010;115:3827-34.

Lindh JD, Lundgren S, Holm L, et al. Several-fold increase in risk of overanticoagulation by CYP2C9 mutations. Clin Pharmacol Ther 2005;78:540-50.

Liu Y, Jeong H, Takahashi H, et al. Decreased warfarin clearance associated with the CYP2C9 R150H (*8) polymorphism. Clin Pharmacol Ther 2012;91:660-5.

Liu Y, Yang J, Xu Q, et al. Comparative performance of warfarin pharmacogenetic algorithms in Chinese patients. Thromb Res 2012;130:435-40.

McDonald MG, Rieder MJ, Nakano M, et al. CYP4F2 is a vitamin K1 oxidase: an explanation for altered warfarin cose in carriers of the V433M variant. Mol Pharmacol 2009;75:1337-1346.

McWilliam A. Heath care savings from personalizing medicine using genetic testing: the case of warfarin. 2006;2013.

Meckley LM, Gudgeon JM, Anderson JL, et al. A policy model to evaluate the benefits, risks and costs of warfarin pharmacogenomic testing. Pharmacoeconomics 2010;28:61-74.

Meckley LM, Wittkowsky AK, Rieder MJ, et al. An analysis of the relative effects of VKORC1 and CYP2C9 variants on anticoagulation related outcomes in warfarin-treated patients. Thromb Haemost 2008;100:229-39.

Mitchell C, Gregersen N and Krause A. Novel CYP2C9 and VKORC1 gene variants associated with warfarin dosage variability in the South African black population. Pharmacogenomics 2011;12:953-63.

Moreau C, Bajolle F, Siguret V, et al. Vitamin K antagonists in children with heart disease: height and VKORC1 genotype are the main determinants of the warfarin dose requirement. Blood 2012;119:861-7.

Niinuma Y, Saito T, Takahashi M, et al. Functional characterization of 32 CYP2C9 allelic variants. Pharmacogenomics J 2013.

Nutescu EA, Drozda K, Bress AP, et al. Feasibility of implementing a comprehensive warfarin pharmacogenetics service. Pharmacotherapy 2013;33:1156-64.

Pare G, Eriksson N, Lehr T, et al. Genetic determinants of dabigatran plasma levels and their relation to bleeding. Circulation 2013;127:1404-12.

Patrick AR, Avorn J and Choudhry NK. Cost-effectiveness of genotype-guided warfarin dosing for patients with atrial fibrillation. Circ Cardiovasc Qual Outcomes 2009;2:429-36.

Perera MA, Cavallari LH, Limdi NA, et al. Genetic variants associated with warfarin dose in African-American individuals: a genome-wide association study. Lancet 2013;382:790-6.

Perera MA, Gamazon E, Cavallari LH, et al. The missing association: sequencing-based discovery of novel SNPs in VKORC1 and CYP2C9 that affect warfarin dose in African Americans. Clin Pharmacol Ther 2011;89:408-15.

Pink J, Pirmohamed M, Lane S, et al. Cost-effectiveness of pharmacogenetics-guided warfarin therapy vs. alternative anticoagulation in atrial fibrillation. Clin Pharmacol Ther 2014;95:199

Pirmohamed M, Burnside G, Eriksson N, et al. A randomized trial of genotype-guided dosing of warfarin. N Engl J Med 2013;369:2294-303.

Redman AR, Dickmann LJ, Kidd RS, et al. CYP2C9 genetic polymorphisms and warfarin. Clin Appl Thromb Hemost 2004;10:149-54.

Rieder MJ, Reiner AP, Gage BF, et al. Effect of VKORC1 haplotypes on transcriptional regulation and warfarin dose. N Engl J Med 2005;352:2285-93.

Roper N, Storer B, Bona R, et al. Validation and comparison of pharmacogenetics-based warfarin dosing algorithms for application of pharmacogenetic testing. J Mol Diagn 2010;12:283-91.

Rost S, Fregin A, Ivaskevicius V, et al. Mutations in VKORC1 cause warfarin resistance and multiple coagulation factor deficiency type 2. Nature 2004;427:537-41.

Schelleman H, Chen J, Chen Z, et al. Dosing algorithms to predict warfarin maintenance dose in Caucasians and African Americans. Clin Pharmacol Ther 2008;84:332-9.

Schwarz UI, Ritchie MD, Bradford Y, et al. Genetic determinants of response to warfarin during initial anticoagulation. N Engl J Med 2008;358:999-1008.

Scordo MG, Pengo V, Spina E, et al. Influence of CYP2C9 and CYP2C19 genetic polymorphisms on warfarin maintenance dose and metabolic clearance. Clin Pharmacol Ther 2002;72:702-10.

Scott SA, Edelmann L, Kornreich R, et al. Warfarin pharmacogenetics: CYP2C9 and VKORC1 genotypes predict different sensitivity and resistance frequencies in the Ashkenazi and Sephardi Jewish populations. Am J Hum Genet 2008;82:495-500.

Shahin MH, Cavallari LH, Perera MA, et al. VKORC1 Asp36Tyr geographic distribution and its impact on warfarin dose requirements in Egyptians. Thromb Haemost 2013;109:1045-50.

Shaw K, Amstutz U, Hildebrand C, et al. VKORC1 and CYP2C9 genotypes are predictors of warfarin-related outcomes in children. Pediatr Blood Cancer 2014.

Shen AY, Yao JF, Brar SS, et al. Racial/ethnic differences in the risk of intracranial hemorrhage among patients with atrial fibrillation. J Am Coll Cardiol 2007;50:309-15.

Shin J and Kayser SR. Accuracy of the pharmacogenetic dosing table in the warfarin label in predicting initial therapeutic warfarin doses in a large, racially diverse cohort. Pharmacotherapy 2011;31:863-70.

Simpson JR, Zahuranec DB, Lisabeth LD, et al. Mexican Americans with atrial fibrillation have more recurrent strokes than do non-Hispanic whites. Stroke 2010;41:2132-6.

Stangier J and Clemens A. Pharmacology, pharmacokinetics, and pharmacodynamics of dabigatran etexilate, an oral direct thrombin inhibitor. Clin Appl Thromb Hemost 2009;15 Suppl 1:9S-16S.

Tai G, Farin F, Rieder MJ, et al. In-vitro and in-vivo effects of the CYP2C9*11 polymorphism on warfarin metabolism and dose. Pharmacogenet Genomics 2005;15:475-81.

Takahashi H, Kashima T, Nomoto S, et al. Comparisons between in-vitro and in-vivo metabolism of (S)-warfarin: catalytic activities of cDNA-expressed CYP2C9, its Leu359 variant and their mixture versus unbound clearance in patients with the corresponding CYP2C9 genotypes. Pharmacogenetics 1998;8:365-73.

Takahashi H, Wilkinson GR, Nutescu EA, et al. Different contributions of polymorphisms in VKORC1 and CYP2C9 to intra- and inter-population differences in maintenance dose of warfarin in Japanese, Caucasians and African-Americans. Pharmacogenet Genomics 2006;16:101-10.

Takeuchi F, McGinnis R, Bourgeois S, et al. A genome-wide association study confirms VKORC1, CYP2C9, and CYP4F2 as principal genetic determinants of warfarin dose. PLoS Genet 2009;5:e1000433.

Tian C, Hinds DA, Shigeta R, et al. A genomewide single-nucleotide-polymorphism panel with high ancestry information for African American admixture mapping. Am J Hum Genet 2006;79:640-9.

Voora D, Eby C, Linder MW, et al. Prospective dosing of warfarin based on cytochrome P-450 2C9 genotype. Thromb Haemost 2005;93:700-5.

Wadelius M, Chen LY, Lindh JD, et al. The largest prospective warfarin-treated cohort supports genetic forecasting. Blood 2009;113:784-92.

Wang D, Chen H, Momary KM, et al. Regulatory polymorphism in vitamin K epoxide reductase complex subunit 1 (VKORC1) affects gene expression and warfarin dose requirement. Blood 2008;112:1013-21.

Wang M, Lang X, Cui S, et al. Clinical application of pharmacogenetic-based warfarin-dosing algorithm in patients of Han nationality after rheumatic valve replacement: a randomized and controlled trial. Int J Med Sci 2012;9:472-9.

Wei M, Ye F, Xie D, et al. A new algorithm to predict warfarin dose from polymorphisms of CYP4F2, CYP2C9 and VKORC1 and clinical variables: derivation in Han Chinese patients with non valvular atrial fibrillation. Thromb Haemost 2012;107:1083-91.

White RH, Dager WE, Zhou H, et al. Racial and gender differences in the incidence of recurrent venous thromboembolism. Thromb Haemost 2006;96:267-73.

Xu Q, Xu B, Zhang Y, Estimation of the warfarin dose with a pharmacogenetic refinement algorithm in Chinese patients mainly under low-intensity warfarin anticoagulation. Thromb Haemost 2012;108:1132-40.

You JH. Pharmacogenetic-guided selection of warfarin versus novel oral anticoagulants for stroke prevention in patients with atrial fibrillation: a cost-effectiveness analysis. Pharmacogenet Genomics 2014;24:6-14.

You JH, Tsui KK, Wong RS, et al. Potential clinical and economic outcomes of CYP2C9 and VKORC1 genotype-guided dosing in patients starting warfarin therapy. Clin Pharmacol Ther 2009;86:540-7.

You JH, Tsui KK, Wong RS, et al. Cost-effectiveness of dabigatran versus genotype-guided management of warfarin therapy for stroke prevention in patients with atrial fibrillation. PLoS One 2012;7:e39640.

Zineh I, Pacanowski M and Woodcock J. Pharmacogenetics and coumarin dosing--recalibrating expectations. N Engl J Med 2013;369:2273-5.

Chapter 7

PHARMACOGENETICS OF LIPID-LOWERING DRUGS WITH A FOCUS ON *SLCO1B1* AND STATIN THERAPY

Barry E. Bleske, Pharm.D., FCCP; Zachary L. Cox, Pharm.D., BCPS; and Orly Vardeny, Pharm.D., M.S., FCCP, BCACP

Learning Objectives

1. Determine in an individual patient modifiable and nonmodifiable risk factors for statin-induced myopathy.
2. Describe the mechanisms for increases in systemic exposure to a statin.
3. Determine an appropriate statin dose based upon a patient's *SLCO1B1* c.521T>C genotype.
4. Describe the challenges of establishing a genomic-based intervention for statin therapy in a patient care setting.
5. Develop a clinical implication plan incorporating *SLCO1B1* c.521T>C genotyping in a patient care setting.
6. Describe future genomic markers that may be associated with statin therapy.

Keywords: Myopathy; OATP1B1 (organic anion-transporting polypeptide 1B1); HMG-CoA Reductase Inhibitors; genotyping; electronic prescription; pharmacokinetics; education

Abbreviations in This Chapter

APOE	apolipoprotein	HPS	Heart Protection Study
ATP	adenosine triphosphate	LDL	low density lipoprotein
AUC	area under the curve	OATP	organic anion-transporting polypeptide
CDS	clinical decision support		
CPK	creatine phosphokinase	PCSK9	proprotein convertase subtilisin/kexin type 9
COQ10	co-enzyme Q10		
COQ2	co-enzyme Q2	SEARCH	Study of the Effectiveness of Additional Reductions in Cholesterol and Homocysteine
CPIC	Clinical Pharmacogenetics Implementation Consortium		
CPOE	computerized provider order entry	SNP	single-nucleotide polymorphisms
HMG-CoA	3-hydroxy 3-methylglutaryl coenzyme A	STRENGTH	Statin Response Examined by Genetic Haplotype Markers
HMGCR	3-hydroxy 3-methylglutaryl coenzyme A reductase	ULN	upper limit of normal

Abstract

HMG-CoA reductase inhibitors (statins) are one of the most common therapeutic drug class prescribed due to their benefit in treatment of atherosclerotic disease and for the most part an overall good safety profile. However, a limitation with statin therapy is development of myopathy, which is associated with an increased systemic exposure to statin therapy. An increase in systemic exposure to statin therapy may be related to modifiable and nonmodifiable risk factors including genomics. In particular, the *SLCO1B1* c.521T>C gene variant that encodes for the organic anion-transporting polypeptide 1B1 (OATP1B1 responsible for statin uptake) has been shown to influence statin pharmacokinetics. Specifically the CC genotype has been associated with increase in statin concentrations and secondarily to myopathy as compared to the TT and TC genotypes. Clinical guidelines have been proposed to incorporate the *SLCO1B1* c.521T>C gene variant into clinical setting. Best practices for translation into the clinical setting include use of electronic medical records and a clinical algorithm with structured and specific recommendations for the type of statin and starting dose.

Introduction

HMG-CoA reductase inhibitors (statins) are one of the most widely prescribed class of drugs in the world. Data from the Institute for Healthcare Informatics report that there were 94 million prescriptions for simvastatin in 2010 in the United States alone (IMS Institute for Healthcare Informatics 2010). Based on the recent updated treatment guidelines for managing low density lipoprotein (LDL) it has be extrapolated that over 1 billion patients worldwide may be considered for statin therapy based on these new guidelines (Ioannidis 2013). The wide spread use of statin therapy is not surprising given statin's overall safety profile and the global implications related to cardiovascular disease. Statins, through in part by decreasing LDL concentrations, can reduce cardiovascular risk by 25% (Mills 2011). Reduction in cardiovascular risk is seen for primary and secondary prevention and appears to be consistent across many subgroups (Mills 2011; Kearney 2008; Baigent 2010; Kostis 2012; Brugts 2009; Afilalo 2008). Statin therapy is also cost effective; statins are available as a generic drug and are a component of a "polypill" that may someday become widely available at minimal cost for developing countries with the purpose of treating cardiovascular disease on a global scale. Given statin's benefits in reducing cardiovascular risk, overall safety profile, and cost effectiveness, the use of statins since coming to the marketplace has been on a continuous growth curve making this class of drugs one of the most extensively used in the world.

Scope of the Problem—Myopathy

As mentioned, statins have an overall favorable to acceptable safety profile and are well tolerated, which contributes to their widespread use (Law 2006). However, one of the factors that can lead to nonadherence is the development of musculoskeletal symptoms (myopathy), which is one of the most common adverse effects associate with statin therapy. Statin associated myopathy include those associated with myalgia (pain), myositis (pain with evidence of muscle damage, perhaps manifest by modest elevations in creatine phosphokinase [CPK] from 3 to 10x ULN), more severe myopathy (myalgia, weakness and/or cramps plus elevations in CPK > 10x the ULN), and rhabdomyolysis (pain and severe muscle damage, CPK elevation typically > 10x ULN, and with elevated creatinine levels) (McKenney 2006). The incidence of these musculoskeletal symptoms varies widely and is dependent in part on the cohort evaluated, definition for a statin-induced muscle event, and the study design. Fortunately, in the case of rhabdomyolysis the incidence is low with one approximation at 1.6 patients per 100,000 person years (Law 2006). In the case of myopathy other than rhabdomyolysis, the incidence in randomized trials that often exclude higher risk patients is somewhere between 1%–5%. In larger studies, including analysis of large data bases and observational studies, that incidence is higher. One interesting large countrywide observational study is the PRIMO study (Prediction of Muscular Risk in Observational conditions), which reported a 10.5% incidence of muscular symptoms occurring with a median time of onset of one month (Bruckert 2005). This study also describes a number of other findings surrounding muscle pain including location, duration, triggers, and frequency of pain.

Risk Factors

From analysis of large randomized trials and systematic reviews, there have been a number of risk factors identified that may increase the chances for developing statin-induced myopathy. These risks factors can be classified as modifiable and nonmodifiable. Recognition and identification of these risk factors in an individual patient can promote the safe initiation of statin therapy through careful selection of the type of statin and appropriate starting and titration doses.

Modifiable Risk Factors

Strenuous exercise, extensive surgery involving high metabolic demand, and potentially extensive alcohol intake may increase the risk for developing myopathy (Rallidis 2012). Perhaps the most important modifiable risk factor is those factors that increase a statin systemic exposure (i.e., drug concentration). These factors include dose and drug interactions. There are a number of studies that have shown that the higher the dose of statin used the higher incidence of myopathy (Bruckert 2005; Harper 2007; McClure 2007; Pedersen 2005; de Lemos 2004). This dose-dependent effect appears not to be related to the degree of LDL reduction. Dose-related increase in muscle complaints was observed in a study evaluating statin use for secondary prevention after a myocardial infarction; the incidence of myalgia resulting in drug discontinuation was 2.2% for atorvastatin 80 mg versus

1.1% for simvastatin 20 mg (p<0.01) (Pedersen 2005). In another study in patients with acute coronary syndrome, 9 patients receiving simvastatin 80 mg developed myopathy (CK > 10 ULN associated with muscle symptoms) versus zero patients receiving 20mg simvastatin (de Lemos 2004). Dose related myopathy is associated regardless of the type of statin but may be even more sensitive with simvastatin (Bruckert 2005; Link 2008). Upon FDA review of simvastatin's safety profile in regard to dose-related myopathy a labeling change was made limiting the upper dosage limit to 40 mg in most instances.

Another important risk factor that needs to be avoided or managed is drug interactions that increase the systemic exposure to a statin. The importance of drug interactions is exemplified in the cases of cerivastatin-induced fatal myopathies (Staffa 2002). In majority of these cases patients were receiving gemfibrozil, which resulted in decrease metabolism and elevated concentrations of cerivastatin leading to rhabdomyolysis. The risk for drug interactions is dependent upon the individual statin. In general, lipophilic statins (e.g., atorvastatin, fluvastatin, lovastatin, and simvastatin) that undergo metabolism through mainly the P450 (CYP) pathway are at great risk for drug interactions due to the number of other drugs and natural products that utilize and inhibit this system (e.g., CYP3A4 inhibitors: amiodarone; azole antifungals; calcium channel blockers; etc.). In contrast, the more hydrophilic statins (e.g., pravastatin and rosuvastatin) that utilize other and sometimes multiple pathways for metabolism will be at less risk because there are fewer drugs that utilize and inhibit these pathways. A drug interaction concern for all statins, except perhaps fluvastatin, is with gemfibrozil. Statins also undergo glucoronidation (phase II conjugation), which may be inhibited by gemfibrozil.

Nonmodifiable Risk Factors
Patient-related risk factors include advanced age, female gender, low body mass index, comorbidites (hypothyroidism, diabetes, renal, and liver disease), race (patients of Asian and African descent), and rare hereditary metabolic muscle diseases. In addition, there are also genetic components that can increase the risk for the development of myopathy. The genetic factor that is perhaps best studied and is associated with an increase in risk for developing myopathy is the *SLCO1B1* gene polymorphism (rs4149056) (Link 2008). As discussed below the pathway for the increase risk associated with *SLCO1B1* polymorphism is similar to the pathway associated with the majority of other nongenetic risk factors, an increase in systemic exposure to a statin. The focus of this chapter is devoted to understanding and managing patients with this genetic trait.

Candidate Genes

The organic anion-transporting polypeptide 1B1 (OATP1B1) is responsible for uptake of statins, among other xenobiotics, into hepatocytes (Ho 2006). Activity of OATP1B1 can be modified in several ways, including drug-drug interactions as discussed previously, as well as through inherent variation in baseline function (Niemi 2011). The OATP1B1 enzyme is encoded by the *SLCO1B1* gene, located on chromosome 12. Over 40 nonsynonymous single nucleotide polymorphisms (SNP) have been identified in *SLCO1B1*, and a few confer reduced transporter activity. On exon 5, the c.521T>C SNP (rs4149056) results in impaired OATP1B1 activity and increased plasma exposure to various statins (Tirona 2001). This variant occurs in 5%–20% of most populations, and individuals homozygous for the minor C allele have been shown to exhibit significantly elevated plasma area under the curve (AUC) for simvastatin acid compared to those with the TT genotype (Pasanen 2006). Carriage of one variant allele has been correlated with intermediate enzyme activity. Two other noteworthy nonsynonymous SNPs within *SLCO1B1* include the c.388A>G (rs2306283) and c.463C>A (rs11045819) (Table 7.1). *SLCO1B1* alleles are defined by the combination of the three variants, which form several common haplotypes, although the c.521T>C SNP appears to drive the physiologic effect. As such, the *SLCO1B1**5 (c.521T>C alone) and *SLCO1B1**15 (c.521T>C and c.388A>G) haplotypes are associated with reduced transport activity and increased plasma concentrations of statins.

SLCO1B1 Variants and Pharmacokinetics
Several studies have examined associations with *SLCO1B1* variants and statin pharmacokinetics (Table 7.2). The effect of the *SLCO1B1* c.521T>C variant on pharmacokinetics of simvastatin was first shown in a single dose (40 mg) study of 32 healthy Caucasian individuals (Pasanen 2006). Participants were enrolled and assigned into one of three groups based on *SLCO1B1* haplotype. Evaluation of pharmacokinetic parameters revealed an association of mean simvastatin and simvastatin acid AUC with carriage of the c.521C allele; such that participants with the CC genotype exhibited a 221% higher concentration compared to TC and TT genotypes. Mean maximum concentration was also significantly higher among those with the CC genotype, and mean time to maximum concentration was shorter in this group. The c.521T>C variant was not associated with differences in simvastatin lactone (an inactive metabolite) kinetics, suggesting that simvastatin lactone is transported either by a different transporter, or by passive diffusion.

Pasanen and colleagues also examined effects of the c.521T>C variant on pharmacokinetics of atorvastatin and rosuvastatin in the same group of volunteers (Pasanen 2007). In this two-phase study, participants took a single dose of atorvastatin 20 mg, and following a one-week washout, took a single dose of rosuvastatin 10 mg. Similar to the previous study, individuals with the c.521CC genotype had a higher mean AUC of atorvastatin, but these findings were limited to the parent compound (atorvastatin acid) and only one of its metabolites, 2-hydroxyatorvastatin. For rosuvastatin, the CC genotype was also associated with higher mean AUC, but the effect was less pronounced.

For pravastatin, study findings of *SLCO1B1* variants and pharmacokinetics among Caucasian Europeans and Japanese cohorts have shown comparable results to the above described studies of simvastatin and rosuvastatin, with significant associations detected between pravastatin concentrations and the c.521CC genotype (Niemi 2004, 2006; Nishizato 2003; Mwinyi 2004). A few other notable similarities were observed among studies of simvastatin, atorvastatin, and rosuvastatin: the c.521T>C genotype was not associated with differences in elimination half-lives; and individuals heterozygous at c.521T>C (i.e., the TC genotype) and carrying different SLCO1B1

Table 7.1. *SLCO1B1* Variants

SNP rs Number and Genotype	Nucleotide Change	Amino Acid Change	Observed Phenotype and Examples of Diplotypes
rs4149056	c.521T>C	valine to alanine	
TT			Normal or high activity, reduced plasma concentration Diplotypes: *1a/*1a, *1a/*1b, *1b/*1b
TC			Intermediate activity Diplotypes: *1a/*5, *1a/*15, *1a/*17, *1b/*5, *1b/*15, *1b/*17
CC			Low activity, increased plasma concentration Diplotypes: *5/*5, *5/*15, *5/*17, *15/*15, *15/*17, *17/*17
rs2306283	c.388A>G	aspargine to aspartic acid	
AA			Low activity, increased plasma concentration
AG			Intermediate activity
GG			Normal or high activity, reduced plasma concentration
rs11045819	c.463C>A	proline to threonine	
CC			Low activity, increase plasma concentration
CA			Intermediate activity
AA			Normal or high activity, reduced plasma concentration
Haplotype			
*5	c.521T>C	Valine to alanine	Low activity, increased plasma concentration
*15	C521T>C & c.388A>G	Valine to alanine & aspargine to aspartic acid	Low activity, increased plasma concentration
1b	c.388A>G	Aspargine to aspartic acid	Normal or high activity, reduced plasma concentration
*14	c.388A>G & c.463C>A	Aspargine to aspartic acid & proline to threonine	Normal to high activity, reduced plasma concentration
*17	g.-11187G>A, & c.388A>G & c.521T>C	Aspargine to aspartic acid & valine to alanine	Low activity, increased plasma concentration

haplotypes (i.e., *15, *16, *17 haplotypes) did not demonstrate significant differences in statin pharmacokinetics, reinforcing that the c.521T>C variant portends the strongest effect on statin metabolism.

For fluvastatin, although the c.521CC genotype was associated with a 19% increased mean AUC, this finding was not statistically significant (Niemi 2006). In addition to hepatic transport by OATP1B1, rosuvastatin and fluvastatin are also substrates of OATP1B3 and OATP2B1 (Kopplow 2005). Additionally, rosuvastatin is transported by OATP1A2 and sodium-taurocholate co-transporting polypeptide uptake transporters (Ho 2006). Taken together, it is possible that the reduced effect of *SLCO1B1* variants on rosuvastatin and fluvastatin kinetics could result from alternative hepatic transport mechanisms.

Because the liver is the site of both the clearance and therapeutic effect of statins, it is logical to assume that genetic variation of the OATP1B1 would also result in reduced efficacy of statins, but studies of associations of *SLCO1B1* variants on LDL-lowering capabilities of statins have yielded mixed results. While single dose studies of pravastatin revealed higher cholesterol synthesis markers with the c.521CC genotype and *SLCO1B1**17 haplotype (Pasanen 2008; Niemi 2005), a three-week study of pravastatin did not demonstrate significant associations between *SLCO1B1* variants and LDL lowering (Igel 2006).

SLCO1B1 Variants and Myopathy

The increased plasma concentrations of statins potentiated by *SLCO1B1* variants may increase the risk for muscle-related symptoms, and a few analyses tested these associations (Table 7.3). The landmark Study of the Effectiveness of Additional Reductions in Cholesterol and Homocysteine (SEARCH) was a randomized study of simvastatin 20 mg daily versus 80 mg daily in 12,064 patients with a prior myocardial infarction (Link 2008). There were 49 cases of definite myopathy (defined as muscle symptoms and CPK levels > 10 times the upper limit of normal range) and 49 cases of incipient myopathy (CPK > 3 times the upper limit of normal range and > 5 times baseline values and alanine aminotransferase levels > 1.7 times the baseline value). A genome-wide association study was completed in participants taking simvastatin 80 mg daily, 85 individuals with myopathy; and in 90 matched controls without myopathy. As part of their analysis, investigators included a separate, genotyped cohort (n=16,664) from the Heart Protection Study (HPS). HPS assessed simvastatin 40 mg versus placebo in patients with known vascular disease (MRC/BHF Heart Protection Study 2002). This additional cohort was also included in order to explore the effect of genetic variants on LDL-lowering of simvastatin. Two *SLCO1B1* SNPs from the SEARCH cohort were associated with myopathy; rs4363657 (OR 4.3; 95% CI 2.5, 7.2 for heterozygotes), and c.521T>C, which was in strong linkage disequilibrium (r^2 > 95) with rs4363657 (OR for myopathy = 4.5; 95% CI 2.6, 7.7 for heterozygotes). In the replication cohort from the HPS, results were confirmed among 21 participants with myopathy who underwent genotyping for c.521T>C (RR 2.6; 95% CI 1.3, 5.0). *SLCO1B1* variants were not associated with LDL-lowering in the HPS.

In the Statin Response Examined by Genetic Haplotype Markers (STRENGTH) study, 509 participants were randomized to atorvastatin, simvastatin, or pravastatin (final doses of 80 mg/day for atorvastatin and simvastatin, and 40 mg/day for pravastatin) for 16 weeks (Voora 2009). Genetic associations were examined with a composite adverse event (study drug discontinuation, myalgia or muscle cramps, or CK elevations > 3 times the upper normal limit). Multivariable analyses revealed a significant association of the *SLCO1B1**5 haplotype in a gene-dose related fashion with composite adverse events (OR 2.2; 95% CI 1.4, 3.6). Effects were most pronounced among participants taking simvastatin, but a gene by

Table 7.2. Statins Affected by the *SLCO1B1* c.521T>C Variant

Statin	Affected by *SLCO1B1* c.521T>C Variant	Effects of *SLCO1B1* c.521T>C Variant
Simvastatin	Yes	Higher Cmax* & AUC of simvastatin acid, shorter time to Cmax, no effect on simvastatin lactone
Atorvastatin	Yes	Higher Cmax and AUC of atorvastatin and 2-hydroxyatorvastatin
Pravastatin	Yes	Higher Cmax & AUC
Rosuvastatin	Yes	Higher Cmax and AUC
Fluvastatin	No	N/A

*Cmax = Maximum concentrations

Table 7.3. Studies of *SLCO1B1* Variants and Associations with Myopathy

Study	Study Sample	Statin	Number of Myopathy-Related Adverse Events	Main Results
Link et al.	SEARCH cohort: 12,064 post-myocardial infarction; HPS cohort: 16,664 subjects with vascular disease or diabetes mellitus	Simvastatin 80mg or 20mg Simvastatin 40mg or placebo	85 21	From SEARCH: the *SLCO1B1* rs4363657, and c.521T>C were associated with myopathy (OR 4.3; 95% CI 2.5, 7.2 for heterozygotes; and OR 4.5; 95% CI 2.6, 7.7 for heterozygotes). Results confirmed in HPS cohort (RR 2.6; 95% CI 1.3, 5.0).
Voora et al.	STRENGTH cohort: outpatients with hypercholesterolemia	atorvastatin, simvastatin, or pravastatin: final doses 80mg/day for atorvastatin and simvastatin, and 40mg/day for pravastatin	99	Significant association of the *SLCO1B1*5* haplotype in a gene-dose related fashion with composite adverse events (OR 2.2; 95% CI 1.4, 3.6)
Brunham et al.	Retrospective chart review of ~9000 patients from two lipid clinics	Varied: simvastatin, atorvastatin, pravastatin, rosuvastatin	25	SLC01B1 c.521T>C genotype associated with myopathy only in patients taking simvastatin (OR 3.2; 95% CI 0.83, 11.96, χ^2 p=0.042)

treatment interaction was not significant. Interestingly, there was a significant association was detected between simvastatin acid concentrations and *SLCO1B1* c.521T>C genotype, in line with previous pharmacokinetic studies.

In another replication study, medical records of approximately 9000 patients from two lipid clinics in the Netherlands were examined for presence of biochemical myopathy (CK > 10 times upper normal limit) (Brunham 2012). For each of the 25 cases detected, 2-4 controls were identified, matched for age, gender, statin type, and dose. Non-significant trends were detected between *SLCO1B1* c.521T>C genotype and myopathy (TC or CC versus TT genotype, OR 1.5; 95% CI 0.58, 3.69). When patients taking simvastatin were examined, *SLCO1B1* c.521T>C genotype conferred a significantly increased risk for myopathy (OR 3.2; 95% CI 0.83, 11.96, χ^2 p=0.042).

In summary, the *SLCO1B1* c.521T>C polymorphism has been consistently associated with stain-induced myopathy in several large cohorts. A gene-dose effect suggested that individuals with the CC genotype were more likely to experience myopathy compared to those with the TC or TT genotypes. Results were most convincing for simvastatin, although findings were similar for pravastatin and atorvastatin. These data support the notion that genotyping for the *SLCO1B1* c.521T>C variant prior to statin initiation may identify patients at highest risk for myopathy, and thus help optimize lipid lowering therapy.

Clinical Translation and Implementation Barriers

Several critical issues need to be considered when incorporating patient genotype into the clinical decision-making processes providers employ when writing a statin prescription. Issues necessitating considerable thought and planning include: the presentation of genetic information to providers with varying levels of familiarity with genetic terminology, translation of genetic information into clinically useful assessments of OATP1B1 activity, recommendations for specific statins or doses classified by the strength of the supporting evidence, and the timing in the medication use process when genetic information is presented to providers (Relling 2011; Ramsey 2014). These issues can be reduced to the questions "what, where, and when will genetic information be presented," which are intrinsically interconnected.

What to Present

Successful implementation of *SLCO1B1* genotype (rs4149056, c.521T>C) into clinical practice hinges on presenting genetic information at a level understood by all health care providers. Providers cannot reasonably be expected to have an individual knowledge base that is comprehensive enough to enable interpretation and application of all genetic information in an era where pharmacogenomic information is rapidly expanding in scope and depth. The Clinical Pharmacogenetics Implementation Consortium (CPIC) acknowledges "the lack of sufficient awareness about genomics on the part of many clinicians" as a significant barrier to the adoption of pharmacogenetic tests in clinical practice (Relling 2011; Ramsey 2014). In a survey of family practitioners and specialists who commonly receive genetic results, the two major challenges to application reported were terminology unfamiliarity and the complexity of the results (Lubin 2009). Participants unanimously wanted a genetic test result to: (1) clearly state the result, (2) clearly state the significance of the result, and (3) provide guidance on the next steps (Lubin 2009). CPIC models report *SLCO1B1* results as diplotype (*1/*1), genotype (TT, TC, CC), and most important, the phenotype [high/normal enzymatic activity (TT), intermediate enzymatic activity (TC), or low enzymatic activity (CC)] when presenting genetic information to clinicians, while pioneer institutions present genomic results as diplotype and the corresponding phenotype (Relling 2011; Ramsey 2014; Pulley 2012; Bell 2014). While myopathy risk (high risk, intermediate risk, low risk) or percentage change in statin plasma areas under the curve could also be considered, it has been the experience of Vanderbilt University Medical Center that providers prefer a direct recommendation rather than a non-quantitative reference to risk (Pulley 2012; Peterson 2014). Additionally, caution is warranted when using "intermediate" as a risk descriptor. If unpaired with specific clinical recommendations, providers are often unclear on how to proceed with "intermediate risk" information. "The slow rate at which pharmacogenetic tests are being adopted in clinical practice is partly due to the lack of specific guidelines on how to adjust medications on the basis of the genetic test results" (Relling 2011). The CPIC provides a decision support algorithm template (Figure 7.1) to provide direct recommendations on alternative statins and serial CK monitoring when simvastatin is ordered (Relling 2011; Ramsey 2014). Of note, the template uses genotypes (CC, TC, TT) in the decision tree. Designers of decision support should use consistent *SCLO1B1* terminology in the lab results and decision support to avoid provider confusion from interchangeable descriptions of SLCO1B1 genotype. As in Figure 7.1, providers should be offered several statin options, with the preferred alternative statin noted, to individualize application to patient specific financial and medical considerations.

Where to Present

Equally as important as "what to present" is "where it is presented." Pharmacogenetic tests are unique in that their relevance does not diminish over time as with other laboratory tests. An *SLCO1B1* test result cannot be buried within the medical record because the odds of statin prescription increase over time in most patients. The best solution to this problem is the utilization of active clinical decision support (CDS) whenever possible. Active CDS examples include genotype alerts and treatment algorithms delivered though an electronic health record or computerized provider order entry (CPOE) at the time of statin prescription (Pulley 2012; Bell 2014; Goldspiel 2013). Active CDS is attractive because it can also amass other risk factors for statin myopathy that are dispersed throughout the medical record. Patient risk factors (age, gender), medication risk factors (drug interactions, past allergies/intolerances, statin dose), and laboratory risk factors (serum creatinine, creatinine phosphokinase) can also be summarized for the provider at the time *SLCO1B1* genotype is presented. Additionally, genotype information should be incorporated into a "patient summary profile" page within the medical record alongside diagnoses, allergies, and chronic medications (Relling 2011; Ramsey 2014; Pulley 2012). While active CDS as outlined above is ideal, this is not feasible for all institutions. In the absence of CDS, institutions should focus on incorporating genotype into a standard section on a "patient summary profile" and utilizing active pharmacy surveillance. Active pharmacy surveillance could include monitoring patients with new *SLCO1B1* results, patients with a history of statin myopathy to assess the need for *SLCO1B1* testing, and patients with simvastatin orders greater than 20 mg or concomitant interacting medications to assess the need for *SLCO1B1* testing.

When to Present

Lastly, "when genotype information is presented" must be considered. Broadly, two models of implementation exist. Genotyping "just in time" involves sending *SLCO1B1* genotype tests at the time when a provider wants to first prescribe a statin or in patients with prior statin intolerance. This model is labor intensive, requiring a provider to interpret test results after prescribing a statin and re-contact the patient to convey the results or alter a statin prescription, often taking a substantial amount of time. Furthermore, it could delay statin prescribing while awaiting test results in the setting of acute coronary syndrome, making it even more undesirable for *SLCO1B1* testing. As this model produces a barrier to the adoption of genetic testing in clinical practice (Relling 2011; Ramsey 2014), the preemptive genotyping or "just in case" model has been employed by pioneer institutions (Pulley 2012; Bell 2014; Goldspiel 2013).

In the preemptive model, *SLCO1B1* genotype would be performed on patients in anticipation of statin need in the future, so providers could be alerted of results at the time of prescription via CDS and CPOE. While preemptive *SLCO1B1* testing is ideal, this is not feasible for all institutions or reimbursed by all insurance providers. If *SLCO1B1* testing is performed "just in time," implementation strategies should consider: provider alerts to recommend *SLCO1B1* testing, the time to test results, protocol for informing patients and providers of the results and their meaning, and allowance of statin prescription before results are known (Brugts 2009).

Overall, in answering the questions "what, where, and when" regarding genetic information, the ideal scenario is to preemptively utilize *SLCO1B1* genotype testing, to record the results on a "patient summary profile" in the electronic medical record, and to display the genetic results with clear interpretation and direct clinical recommendations at the time of prescription. Unless *SLCO1B1* information can be easily seen and interpreted at the time of prescribing, providers will base therapeutic decisions on the standard parameters. Failure to do so will present artificial challenges as to whether preemptive *SLCO1B1* testing can improve statin safety in a cost-effective manner.

Front Line Practical Issues of Statin Genotyping (Table 7.4)

As outlined above, the ideal approach includes preemptive *SLCO1B1* genotyping, active clinical decision support integrated into CPOE and the electronic health record to alert providers of the *SLCO1B1* genotype at the time of statin prescribing, a consistent presentation of *SCLO1B1* genotype with clear descriptions of the phenotype, and direct recommendations for statin therapy and monitoring. However, with any implementation of clinical decision support not all scenarios are envisioned during planning. Challenges observed by institutions attempting to implement SLCO1B1 genotyping are outlined below.

Despite the clarity and place of genotype information presentation, simply presenting the information alone is likely not enough to totally avoid statin myopathy risk. Providers are often overloaded with computerized alerts, leading to alert fatigue and frequent overrides (van der Sijs 2006; McCoy 2012). At St. Jude Children's Research Hospital, all pharmacogenetic tests are accompanied by a written consultation, and pharmacists receive alerts for medications with pharmacogenetic interactions (Bell 2014). It has been our experience that pharmacist active surveillance is needed to alert and assist providers with

Table 7.4. *SLCO1B1* Genotype Implementation Clinical Pearls

SLCO1B1 Genotype Implementation	
Barriers	**Potential Solutions**
Provider understanding of *SLCO1B1* results	Display results as diplotype and phenotype
	Be consistent in terminology across laboratory reports and clinical decision support tools
Provider application of *SLCO1B1* results	Provide direct recommendations on statin therapy, doses, and laboratory monitoring
	Provide multiple, hierarchical options to allow individualization of therapy for diverse patient needs
	Avoid qualitative assessment of myopathy risks without direct statin therapy recommendations
SLCO1B1 results are not available at the time of prescribing	Employ preemptive *SLCO1B1* testing in patients likely to require future statin therapy
Preemptive *SLCO1B1* results are buried in past laboratory results	Utilize active clinical decision support to inform providers of the *SLCO1B1* genotype at the time of prescription with specific statin recommendations
	Active pharmacy surveillance with a report of patients who have an interaction between statin therapy and *SLCO1B1* genotype

pharmacogenomic interactions management to reduce statin myopathy risk.

Caution is also warranted when using "intermediate" as a risk descriptor in the phenotype description, as in "intermediate enzyme activity." Providers often view "intermediate enzyme activity" to be inconsequential. Additionally, since providers will see a much higher percentage of "intermediate" results than "low enzyme activity" results, they may consider *SLCO1B1* genotyping as futile for clinical decision making (Link 2008). Thus, "intermediate enzyme activity" should have clear statin dosing and monitoring guidance (Figure 7.1) to avoid incorrect interpretation and application.

SLCO1B1 genotyping can be useful in patients presenting with a new complaint of myalgia on statin therapy or a history of statin intolerance. While there are numerous other risk factors, providers should evaluate for low or intermediate *SLCO1B1* enzymatic activity as a potential cause before abandoning statin therapy. Institutions may find it useful to prompt providers to send a *SLCO1B1* panel on the order entry page where providers enter a statin "allergy" in the medical record for statins.

Lastly, a unique problem can arise under the issue of timing when a patient who is currently tolerating simvastatin 40mg daily has a preemptive genotype panel ordered for another indication that includes *SLCO1B1*, revealing a homozygous loss of function phenotype (diplotype *5/*5). Implementation planning should also consider how providers should be recommended to manage new genotype information in this situation [Recommended CDS alert contents are provided in Supplemental Table S11 and Supplemental Figure S3 of the 2014 update of CPIC guideline for *SLCO1B1* and statin-induced myopathy (Ramsey 2014)]. The SEARCH trial indicated that much of the increased myopathy risk with high-dose simvastatin was due to concomitant interacting medication use (Relling 2011; Ramsey 2014; Egan 2011). Providers should be reminded that despite historical tolerance of a statin, patients could reach the full potential of their high myopathy risk with the addition of an interacting medication in the future.

Future Directions

In considering the role of genomics in the management of statin therapy, there are a number of potential genomic targets beyond *SCLO1B1* polymorphism. These targets include those that may affect, similar to *SCLO1B1*, systemic exposure to statin therapy, but also targets that may contribute to statin's adverse effects and efficacy beyond elevation in drug concentrations. For each of these targets further studies are required to determine the clinical translational value.

Systemic Exposure

ABCG2

There are multiple pathways that may influence a patient's systemic exposure to a statin. One pathway involves the ATP-binding cassette transporter gene encoding for the ABCG2 protein (Gradhand 2008; Kerr 2011). This protein is expressed throughout the body including the liver, kidney, and intestinal wall. ABCG2 is an efflux transporter that limits the absorption and increases the excretion of a number of statins. To date the best described variation of the ABCG2 gene is 421 C>A variant (rs2231142) and is more commonly observed among individuals from East Asia (and may explain interethnic variation in statin response). This variant is associated with reduced function (exporter activity), which can increase absorption and decrease elimination resulting in higher systemic exposure to a statin. Studies have indicated for most statins (especially rosuvastatin) there is greater systemic exposure in carriers of 421 C>A variant (Keskitalo 2009; DeGorter 2013; Elsby 2012).

CYP Enzymes

The main metabolism pathway for lovastatin, simvastatin, and atorvastatin is the CYP3A4 pathway. There is wide variability in the expression and activity for this pathway. For example, the CYP3A4*22 (rs35599367) variant is associated with low hepatic expression of CYP3A4 and may be a future target for statin dosing (Elens 2013). In regard to CYP2D6, patients who are "poor metabolizers" may be at risk for higher systematic exposure to atorvastatin and simvastatin. In one study, patients with a non-functioning variant of CYP2D6 enzyme (CYP2D6*4 allele – rs3892097) had a higher incidence of myopathy associated with atorvastatin as compared to non-carriers (Mulder 2001; Frudakis 2007). However, not all studies have shown that CYP2D6 genotype influences statin exposure or response (Geisel 2002; Yin 2012). The clinical utility of this pathway still needs to be defined. The CYP2C9 pathway is an important route for fluvastatin metabolism. Reduced function genotypes (CYP2C9*3/*3, CYP2C9 *1/*3) have been associated with either higher drug concentrations of fluvastatin or greater efficacy (Kirchheiner 2003; Buzkova 2012).

Myopathy

A popular hypothesis for development of statin induced myopathy is a statin induced decrease in the synthesis of co-enzyme Q10 (COQ10) or ubiquinone. COQ10 is an

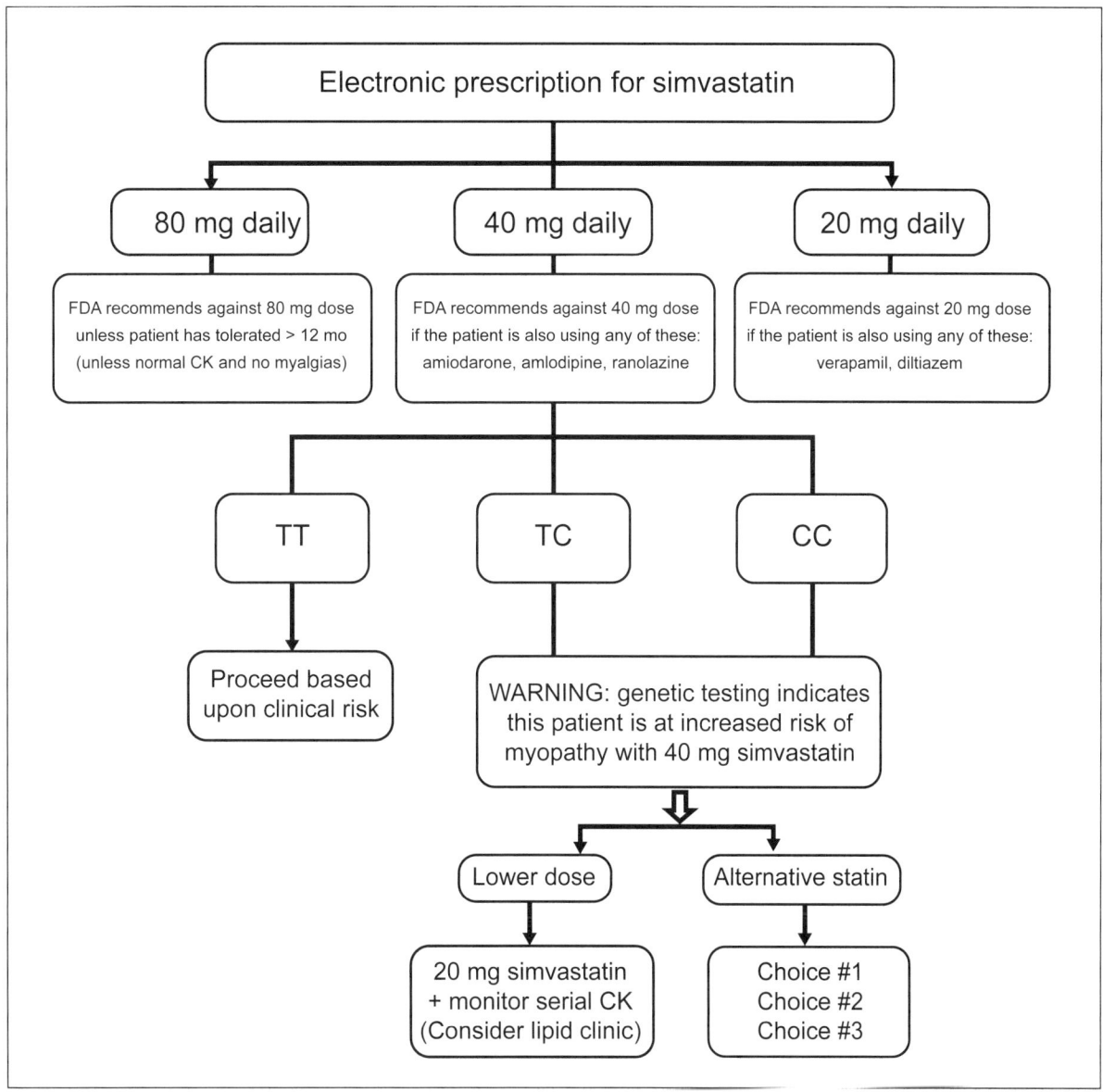

Figure 7.1. Clinical decision support algorithm for simvastatin by *SLCO1B1* genotype (c.521T>C).

essential cofactor in the mitochondrial electron transport system, which is essential for energy metabolism within the cell including mitochondrial adenosine triphosphate (ATP) production. A few studies have suggested that a COQ2 genetic variant (rs4693570) may be associated with statin induced myopathy (Oh 2007; Puccetti 2010; Ruano 2011). COQ2 is an essential enzyme in the biosynthesis of COQ10. Other factors that have been associated with statin-induced myopathy using a physiogenomic approach demonstrated not only COQ2 but additional candidate genes including one that is involved in calcium transport (ATP2B1, rs17381194) and another gene that encodes for a protein kinase that may be a factor for development of myotonic dystrophy (DMPK, rs672348) (Puccetti 2010).

Efficacy

Another area where gemomics may be used in the future is to optimize therapeutic efficacy. There are a number of candidate genes that may influence the effectiveness of statin therapy to lower LDL levels. These candidate genes include but not limited to polymorphisms associated with apolipoprotein E (APOE), proprotein convertase subtilisin/kexin type 9 (PCSK9), and 3-hydroxy 3-methylglutaryl coenzyme A (HMGCoA reductase–HMGCR). APOE plays a role in the hepatic clearance of lipoproteins. Major isoforms for APOE include E2, E3, and E4 (rs7412, rs429358 variant). Limited data suggests that carriers of the APOE E2 alleles have greater LDL response to statin therapy as compared to E4 carriers (Thompson 2009; Voora 2008; Mega 2009).

PCSK9 binds to the LDL receptor and is responsible for decreasing number of available LDL receptors at the cell surface (Lose 2013). The PCSK9 polymorphisms (rs11591147, rs17111584) have been associated with reduced response to statin therapy (Thompson 2009; Chasman 2012). Another interesting target is HMGCR, the pathway for statin's mechanism of action. There are several variants that may affect statin efficacy link to HMGCR (Thompson 2009; Medina 2008; Chung 2012; Yu 2014). There are two particular variants (rs3846662 and rs1920045) that have been shown in some trials to be associated with statin responsiveness (Medina 2008; Chung 2012; Yu 2014). Larger studies are required to validate these and other variants.

Conclusion

At this time the most practical and best evidenced approach to incorporating genomic testing for statin therapy is to determine a patient's *SLCO1B1* genotype. Best practices for the clinical translation for this genomic testing includes use of electronic medical records and a clinical algorithm resulting in structured recommendations for the practitioner. Future testing may include evaluating candidate genes for cellular risk for developing myopathy and to predict statin response.

References

Afilalo J, Duque G, Steele R, et al. Statins for secondary prevention in elderly patients: a hierarchical Bayesian meta-analysis. J Am Coll Cardiol 2008;51:37-45.

Baigent C, Blackwell L, Emberson J, et al. Cholesterol Treatment Trialists' (CTC) Collaboration. Efficacy and safety of more intensive lowering of LDL cholesterol: a meta-analysis of data from 170,000 participants in 26 randomised trials. Lancet 2010;376:1670-81.

Bell GC, Crews KR, Wilkinson MR, et al. Development and use of active clinical decision support for preemptive pharmacogenomics. J Am Med Inform Assoc. 2014;21(e1):e93-9.

Bruckert E, Hayem G, Dejager S, et al. Mild to moderate muscular symptoms with high-dosage statin therapy in hyperlipidemic patients—The PRIMO Study. Cardiovascular Drugs and Therapy 2005;19:403-14.

Brugts JJ, Yetgin T, Hoeks SE, et al. The benefits of statins in people without established cardiovascular disease but with cardiovascular risk factors: meta-analysis of randomized controlled trials. BMJ 2009;338:b2376.

Brunham LR, Lansberg PJ, Zhang L, et al. Differential effect of the rs4149056 variant in SLCO1B1 on myopathy associated with simvastatin and atorvastatin. Pharmacogenomics J 2012;12(3):233-7.

Buzkova H, Pechandova K, Danzig V, et al. Lipid-lowering effect of fluvastatin in relation to cytochrome P450 2C9 variant alleles frequency distributed in the Czech population. Med Sci Monit 2012;18:CR512-7.

Chasman DI, Guilianini F, MacFadyen J, et al. Genetic determinants of statin-induced low-density lipoprotein cholesterol reduction: the Justification for the Use of statins in Prevention: An Intervention Trial Evaluating Rosuvastatin (JUPITER) Trial. Circ Cardiovasc Genet 2012;5:257-64.

Chung JY, Cho SK, Oh ES, et al. Effect of HMGCR variant alleles on low-density lipoprotein cholesterol-lowering response to atorvastatin in healthy Korean subjects. J Clin Pharmacol 2012;52:339-46.

de Lemos JA, Blaxing MA, Wiviott SD, et al. Early intensive vs a delayed conservative simvastatin strategy in patients with acute coronary syndromes. JAMA 2004;292:1307-16.

DeGorter MK, Tiorna RG, Schwarz UI, et al. Clinical and pharmacogenetic predictors of circulating atorvastatin and rosuvastatin concentrations in routine clinical care. Circ Cardiovasc Genet 2013;6:400-8.

Egan A, Colman E. Weighing the benefits of high-dose simvastatin against the risk of myopathy. N Engl J Med 2011;365(4):285-7.

Elens L, van Gelder T, Hesselink DA, et al. CYP3A4*22: promising newly identified CYP3A4 variant allele for personalizing pharmacotherapy. Pharmacogenomics 2013;14:47-62.

Elsby R, Hilgendorf C, Fenner K. Understanding the critical disposition pathways of statins to assess drug-drug interaction risk during drug development: it's not just about OATP1B1. Clin Pharmacol Ther 2012;92:584-98.

Frudakis TN, Thomas M, Ginjupalli S, et al. CYP2D6*4 polymorphism is associated with statin-induced muscle effects. Pharmacogenet Genomics 2007;17:695-707.

Geisel J, Kivisto KT, Griese EU, et al. The efficacy of simvastatin is not influenced by CYP26D polymorphism. Clin Pharmacol Ther 2002;72:595-6.

Goldspiel BR, Flegel WA, Dipatrizio G, et al. Integrating pharmacogenetic information and clinical decision support into the electronic health record. J Am Med Inform Assoc. 2013.

Gradhand U, Kim RB. Pharmacogenomics of MRP transporters (ABCC1-5) and BCRP (ABCG2). Drug Metab Rev 2008;40:317-54.

Harper CR, Jacobson TA. The broad spectrum of statin myopathy: from myalgia to rhabdomyolysis. Curr Opin Lipidol 2007;18:401-8.

Ho RH, Tirona RG, Leake BF, et al. Drug and bile acid transporters in rosuvastatin hepatic uptake: function, expression, and pharmacogenetics. Gastroenterology 2006;130(6):1793-806.

Igel M, Arnold KA, Niemi M, et al. Impact of the SLCO1B1 polymorphism on the pharmacokinetics and lipid-lowering efficacy of multiple-dose pravastatin. Clin Pharmacol Ther 2006;79(5):419-26.

IMS Institute for Healthcare Informatics. The use of medicines in the United States: review of 2010. http://www.imshealth.com/cds/imshealth/Global/Content/Corporate/IMS%20Health%20Institute/Reports/Use_of_Meds_in_the_U.S._Review_of_2010.pdf. Accessed February 20, 2014.

Ioannidis JPA. More Than a Billion People Taking Statins? Potential Implications of the New Cardiovascular Guidelines. JAMA Published Online: December 2, 2013. doi:10.1001/jama.2013.284657.

Kearney PM, Blackwell L, Collins R, et al. Efficacy of cholesterol-lowering therapy in 18,686 people with diabetes in 14 randomized trials of statins: a meta-analysis. Lancet 2008;371:117-25.

Kerr ID, Haider AJ, Gelissen IC. The ABCG family of membrane-associated transporters: you don't have to be big to be mighty. Br J Pharmacol 2011; 164:1767-79.

Keskitalo JE, Zolk O, Fromm MF, et al. ABCG2 polymorphism markedly affects the pharmacokinetics of atorvastatin and rosuvastatin. Clin Pharmacol Ther 2009;86:197-203.

Kirchheiner J, Kudlicz D, Meisel C, et al. Influence of CYP2C9 polymorphisms on the pharmacokinetics and cholesterol-lowering activity of (-)-3S,5R fluvastatin and (+)-3S,5R fluvastatin in healthy volunteers. Clin Pharmacol Ther 2003;74:186-94.

Kopplow K, Letschert K, Konig J, et al. Human hepatobiliary transport of organic anions analyzed by quadruple-transfected cells. Mol Pharmacol 2005;68(4):1031-8.

Kostis WJ, Cheng JQ, Dobrzynski JM, et al. Meta-analysis of statin effects in women versus men. J Am Coll Cardiol 2012;59:572-82.

Law M, Rudnicka AR. Statin safety: a systematic review. Am J Cardiol 2006;97(8A):52C-60C.

Link E, et al. SEARCH Collaborative Group. SLCO1B1 variants and statin-induced myopathy—a genomewide study. N Engl J Med 2008;359:789-99.

Link E, Parish S, Armitage J, et al. SLCO1B1 variants and statin-induced myopathy--a genomewide study. N Engl J Med 2008;359(8):789-99.

Lose JM, Dorsch MP, Bleske BE. Evaluation of proprotein convertase subtilisin/kexin type 9: focus on potential clinical and therapeutic implications for low-density lipoprotein cholesterol lowering. Pharmacotherapy 2013;33:447-60.

Lubin IM, McGovern MM, Gibson Z, et al. Clinician perspectives about molecular genetic testing for heritable conditions and development of a clinician-friendly laboratory report. J Mol Diagn. 2009;11(2):162-71.

Mega JL, Morrow DA, Brown A, et al. Identification of genetic variants associated with response to statin therapy. Arterioscler Thromb Vasc Biol 2009;29:1310-15.

McClure DL, Valuck RJ, Glanz M, et al. Statin and statin-fibrate use was significantly associated with increased myositis risk in a managed care population. J Clin Epidemiol 2007;60:812-18.

McCoy AB, Waitman LR, Lewis JB, et al. A framework for evaluating the appropriateness of clinical decision support alerts and responses. J Am Med Inform Assoc. 2012;19(3):346-52.

McKenney JM, Davidson MH, Jacobson TA, et al. National Lipid Association Statin Safety Assessment Task Force. Am J Cardiol 2006;97:89C-94C.

Medina MW, Gao F, Ruan W, et al. Alternative splicing of 3-hydroxy-3-methylglutaryl coenzyme A reductase is associated with plasma low-density lipoprotein cholesterol response to simvastatin. Circulation 2008;118:355-62.

Mills EJ, O'Regan C, Eyawo O, et al. Intensive statin therapy compared with moderate dosing for prevention of cardiovascular events: a meta-analysis of >40000 patients. Eur Heart J 2011;32:1409-15.

MRC/BHF Heart Protection Study of cholesterol lowering with simvastatin in 20,536 high-risk individuals: a randomised placebo-controlled trial. Lancet 2002, 360(9326):7-22.

Mulder A, Van Lijf A, Bon M, et al. Association of polymorphism in the cytochrome CYP2D6 and the efficacy and tolerability of simvastatin. Clin Pharmacol Ther 2001;70:532-9.

Mwinyi J, Johne A, Bauer S, et al. Evidence for inverse effects of OATP-C (SLC21A6) 5 and 1b haplotypes on pravastatin kinetics. Clin Pharmacol Ther 2004;75(5):415-21.

Niemi M, Neuvonen PJ, Hofmann U, et al. Acute effects of pravastatin on cholesterol synthesis are associated with SLCO1B1 (encoding OATP1B1) haplotype *17. Pharmacogenet Genomics 2005;15(5):303-9.

Niemi M, Pasanen MK, Neuvonen PJ. Organic anion transporting polypeptide 1B1: a genetically polymorphic transporter of major importance for hepatic drug uptake. Pharmacological reviews 2011;63(1):157-81.

Niemi M, Pasanen MK, Neuvonen PJ. SLCO1B1 polymorphism and sex affect the pharmacokinetics of pravastatin but not fluvastatin. Clin Pharmacol Ther 2006;80(4):356-66.

Niemi M, Schaeffeler E, Lang T, et al. High plasma pravastatin concentrations are associated with single nucleotide polymorphisms and haplotypes of organic anion transporting polypeptide-C (OATP-C, SLCO1B1). Pharmacogenetics 2004;14(7):429-40.

Nishizato Y, Ieiri I, Suzuki H, et al. Polymorphisms of OATP-C (SLC21A6) and OAT3 (SLC22A8) genes: consequences for pravastatin pharmacokinetics. Clin Pharmacol Ther 2003;73(6):554-565.

Oh J, Ban MR, Miskie BA, et al. Genetic determinants of statin intolerance. Lipids Health Dis 2007;6:7.

Pasanen MK, Fredrikson H, Neuvonen PJ, et al. Different effects of SLCO1B1 polymorphism on the pharmacokinetics of atorvastatin and rosuvastatin. Clin Pharmacol Ther 2007;82(6):726-733.

Pasanen MK, Miettinen TA, Gylling H, et al. Polymorphism of the hepatic influx transporter organic anion transporting polypeptide 1B1 is associated with increased cholesterol synthesis rate. Pharmacogenet Genomics 2008;18(10):921-6.

Pasanen MK, Neuvonen M, Neuvonen PJ, et al. SLCO1B1 polymorphism markedly affects the pharmacokinetics of simvastatin acid. Pharmacogenet Genomics 2006;16(12):873-9.

Pedersen TR, Faergeman O, Kastelein JJP, et al. High-dose atorvastatin vs Usual-dose simvastatin for secondary prevention after myocardial infarction. JAMA 2005;294:2437-45.

Peterson JF. Personal email communication: Pharmacogenomic Implementation Issues. In: Cox ZL, ed; January 8, 2014.

Puccetti L, Ciani F, Auteri A. Genetic involvement in statins induced myopathy. Preliminary data from an observational case-control study. Atherosclerosis 2010;211:28-29.

Pulley JM, Denny JC, Peterson JF, et al. Operational implementation of prospective genotyping for personalized medicine: the design of the Vanderbilt PREDICT project. Clin Pharmacol Ther. 2012;92(1):87-95.

Rallidis LS, Fountoulaki K, Anastasiou-Nana M. Managing the underestimated risk of statin-associated myopathy. Int J Cardiol 2012;159:169-76.

Ramsey LB, Johnson SG, Caudle KE, et al. The Clinical Pharmacogenetics Implementation Consortium guideline for SLCO1B1 and simvastatin-induced myopathy: 2014 update. Clin Pharmacol Ther. Advance online publication 9 July 2014. doi:10.1038/clpt.2014.125.

Relling MV, Klein TE. CPIC: Clinical Pharmacogenetics Implementation Consortium of the Pharmacogenomics Research Network. Clin Pharmacol Ther 2011;89(3):464-7.

Ruano G, Windemuth A, Wu AHB, et al. Mechanisms of statin-induced myalgia assessed by physiogenomic associations. Atherosclerosis 2011;218:451-6.

Staffa JA, Chang J, Green L. Cerivastatin and reports of fatal rhabdomyolysis. N Engl J Med 2002;346:539-40.

Thompson JF, Hyde CL, Wood LS, et al. Comprehensive whole-genome and candidate gene analysis for response to statin therapy in the Treating to New Targets (TNT) cohort. Circ Cardiovasc Genet 2009;2:173-181.

Tirona RG, Leake BF, Merino G, et al. Polymorphisms in OATP-C: identification of multiple allelic variants associated with altered transport activity among European- and African-Americans. The Journal of Biological Chemistry 2001;276(38):35669-75.

van der Sijs H, Aarts J, Vulto A, et al. Overriding of drug safety alerts in computerized physician order entry. J Am Med Inform Assoc. 2006;13(2):138-47.

Voora D, Shah SH, Reed CR, et al. Pharmacogenetic predictors of statin-mediated low-density lipoprotein cholersterol reduction and dose response. Circ Cardiovasc Genet 2008;2:100-106.

Voora D, Shah SH, Spasojevic I, et al. The SLCO1B1*5 genetic variant is associated with statin-induced side effects. J Am Coll Cardiol 2009;54(17):1609-16.

Yu CY, Theusch E, Lo K, et al. HNRNPA1 regulated HMGCR alternative splicing and modulates cellular cholesterol metabolism. Hum Mol Genet 2014;23:319-32.

Yin OQ, Mak VW, Hu M, et al. Impact of CYP2D6 polymorphisms on the pharmacokinetics of lovastatin in Chinese subjects. Eur J Clin Pharmacol 2012;68:943-9.

Chapter 8

PHARMACOGENETICS IN MENTAL HEALTH

Jeffrey R. Bishop, Pharm.D., M.S., BCPP; James M. Stevenson, Pharm.D., M.S.; and Kyle J. Burghardt, Pharm.D.

Learning Objectives

1. Identify psychiatric medications with pharmacogenetic information listed in product labeling or testing guidelines.
2. Understand how genetic variability in drug metabolism, drug transport, or drug targets may influence treatment outcomes to psychiatric medications.
3. Understand how the relevance of pharmacogenetic testing may be influenced by clinical and demographic information.
4. Evaluate the state of the science surrounding pharmacogenetic testing in psychiatry and assess potential areas for future research.

Keywords: antidepressant, antipsychotic, mood stabilizer, stimulant, serotonin, dopamine, glutamate, schizophrenia, bipolar disorder, major depressive disorder, attention deficit hyperactivity disorder

Abbreviations in This Chapter

$5\text{-}HT_{1A}$	Serotonin-1A receptor	PM	Poor Metabolizer
$5\text{-}HT_{2A}$	Serotonin-2A receptor	SAM	S-Adenyl Methionine
5HTTLPR	Serotonin transporter gene-linked polymorphic region	SERT	Serotonin Transporter
		SGA	Second Generation Antipsychotic
ADHD	Attention-Deficit/Hyperactivity Disorder	SJS	Stevens-Johnson Syndrome
AUC	Area Under the Curve	SNP	Single Nucleotide Polymorphism
CL(int)	Intrinsic clearance	SNRI	Serotonin-Norepinephrine Reuptake Inhibitor
ConLiGen	The International Consortium on Lithium Genetics	SSRI	Selective Serotonin Reuptake Inhibitor
CPIC	Clinical Pharmacogenetics Implementation Consortium	STEP-BD	Systematic Treatment Enhancement Program for Bipolar Disorder
D_2	Dopamine-2 receptor	STin2	Second intron of the serotonin transporter affiliated with a variable number of tandem repeats polymorphism
DAT	Dopamine transporter		
DSM-5	Diagnostic and Statistical Manual of Mental Disorders 5th Edition	Taq1A	Restriction enzyme known to selectively cut at a single nucleotide polymorphism affiliated with the DRD2/ANKK1 gene region
EM	Extensive Metabolizer		
FGA	First Generation Antipsychotic		
GABA	gamma-Aminobutyric acid	TCA	Tricyclic Antidepressant
GWAS	Genome-Wide Association Study	TEN	Toxic Epidermal Necrolysis
HLA	Human Leukocyte Antigen	UGT	Uridine Diphosphate Glucuronosyltransferases
IM	Intermediate Metabolizer		
MAOI	Monoamine Oxidase Inhibitor	UM	Ultrarapid Metabolizer
MDMA	Methylenedioxymethamphetamine	UTR	Untranslated Region
NET	Norepinephrine Transporter	Vmax	Maximal reaction velocity (Michaelis-Menten kinetics)
PANSS	Positive and Negative Syndrome Scale		
P-gp	P-glycoprotein	VNTR	Variable Number of Tandem Repeats

Abstract

The treatment of major psychiatric illnesses is complex with numerous clinical and biological factors that influence disease risk and therapeutic strategies. Advances in our understanding of genetic associations with drug response and tolerability to psychotropic medications have introduced opportunities to integrate pharmacogenomics into patient care. Pharmacogenomic information currently included in product labeling as well as consensus guidelines largely focuses on drug metabolism or immune system markers. Ongoing work continues to evaluate the clinical implications of variants in pharmacodynamic genes. This chapter will review the pharmacogenomics of medications used to treat mental illnesses. Currently the bulk of potentially clinically relevant information is available for antidepressants, mood stabilizers, antipsychotics, and medications used to treat ADHD. Clinical applications as well as ongoing and future work will be discussed.

GENE LIST

GENE	DEFINITION
ABCB1	ATP-BINDING CASSETTE, SUBFAMILY B, MEMBER 1
ADRA2A	ALPHA-2A-ADRENERGIC RECEPTOR
ANKK1	ANKYRIN REPEAT AND KINASE DOMAIN CONTAINING 1
BDNF	BRAIN-DERIVED NEUROTROPHIC FACTOR
CES1	CARBOXYLESTERASE 1
COMT	CATECHOL-O-METHYLTRANSFERASE
CPS1	CARBAMOYL PHOSPHATE SYNTHETASE I
CYP1A2	CYTOCHROME P450, SUBFAMILY I, POLYPEPTIDE 2
CYP2C19	CYTOCHROME P450, SUBFAMILY IIC, POLYPEPTIDE 19
CYP2D6	CYTOCHROME P450, SUBFAMILY IID, POLYPEPTIDE 6
CYP3A4	CYTOCHROME P450, SUBFAMILY IIIA, POLYPEPTIDE 4
DRD2	DOPAMINE RECEPTOR D2
DRD3	DOPAMINE RECEPTOR D3
DRD4	DOPAMINE RECEPTOR D4
EPHX1	EPOXIDE HYDROLASE 1, MICROSOMAL
FKBP5	FK506-BINDING PROTEIN 5
GNB3	GUANINE NUCLEOTIDE-BINDING PROTEIN, BETA-3
HTR1A	5-HYDROXYTRYPTAMINE RECEPTOR 1A
HTR2A	5-HYDROXYTRYPTAMINE RECEPTOR 2A
HTR3A	5-HYDROXYTRYPTAMINE RECEPTOR 3A
HTR3B	5-HYDROXYTRYPTAMINE RECEPTOR 3B
MAOA	MONOAMINE OXIDASE A
MC4R	MELANOCORTIN 4 RECEPTOR
MTHFR	5,10-METHYLENETETRAHYDROFOLATE REDUCTASE
OTC	ORNITHINE CARBAMOYLTRANSFERASE
SLC6A3	SOLUTE CARRIER FAMILY 6 (NEUROTRANSMITTER TRANSPORTER, DOPAMINE), MEMBER 3
SLC6A4	SOLUTE CARRIER FAMILY 6 (NEUROTRANSMITTER TRANSPORTER, SEROTONIN), MEMBER 4
SLC6A2	SOLUTE CARRIER FAMILY 6 (NEUROTRANSMITTER TRANSPORTER, NORADRENALINE), MEMBER 2
SULT4A1	SULFOTRANSFERASE FAMILY 4A, MEMBER 1
TPH1	TRYPTOPHAN HYDROXYLASE 1
TPH2	TRYPTOPHAN HYDROXYLASE 2
UGT1A4	UDP-GLYCOSYLTRANSFERASE 1 FAMILY, POLYPEPTIDE A4
UGT2B7	URIDINE DIPHOSPHATE GLYCOSYLTRANSFERASE 2 FAMILY, MEMBER B7

Introduction

Mental illnesses are a significant source of disease morbidity world-wide. In 2010, mental illness and substance use disorders were estimated to be the leading cause of years lost to disability for non-fatal disease burden, and the fifth leading cause of global disability overall (Whiteford 2013). The onset of symptoms for mental illnesses ranges widely, from young children to the elderly, although the majority of patients present for care in the second to fourth decades of life (AP Association 2013). With an estimated 80% of suicides attributable to mental illness or substance use disorders (Harris 1997; Li 2011; Yoshimasu 2008), and only 32.7% of patients receiving "minimally adequate" treatment (Wang 2005), there is an urgent need to advance drug development efforts and optimize current treatment strategies for mental illness. Pharmacogenomic strategies may be helpful in these efforts.

Medications used to treat mental illness are often used long-term with a high degree of inter-individual variability in symptom improvement and tolerability. Undesirable side effects are reasons for medication switching and discontinuation. Early clinical observations that treatment response in depressed patients may be somewhat heritable or familial (Alexanderson 1969; Angst 1964; O'Reilly 1994; Pare 1971), followed by the recognition of subgroups of "metabolizer" types in pharmacokinetic studies, paved the way for genetic and molecular studies of psychotropic drug response. As a group, neuropsychiatric medications represent one of the top therapeutic drug categories with pharmacogenomic information contained in product information approved by the United States Food and Drug Administration (Drozda 2014). At this time, most potentially "clinically actionable" pharmacogenomic markers for neuropsychiatric medications involve drug metabolizing enzymes, with hypersensitivity reactions to some agents associated with rare variation in immune system genes (Drozda 2014). Genes hypothesized to be involved with the pharmacodynamic aspects of these medications have been investigated extensively but clinical applications are not well-established at this time.

Although the American Psychiatric Association recognizes numerous diagnostic categories for mental illnesses (AP Association 2013), pharmacological treatments overlap. Depression and anxiety disorders, bipolar disorders, schizophrenia and other psychotic disorders, as well as Attention-Deficit/Hyperactivity Disorder (ADHD) represent significant mental illness categories for which the bulk of currently available pharmacogenomic exists. To this end, this chapter will review pharmacogenomics in mental health with a focus on treatments for these disorders and information of potential clinical relevance.

Pharmacogenomics of Antidepressants

Antidepressants were originally used to treat unipolar depression. However, many of these agents are now used for a broad range of both psychiatric (e.g., mood and anxiety disorders) and non-psychiatric (e.g., neuropathic pain) indications. Resultantly, antidepressants are the third-most commonly used prescription medication class in the United States, with 8.7% of the population estimated to be receiving treatment with an antidepressant drug (NCfH Statistics 2013). Unfortunately, less than half of patients with major depressive disorder respond to first-line pharmacotherapy and only one-third of patients achieve symptomatic remission (Trivedi 2006A). Switching and augmentation strategies are viable alternative courses of action (Rush 2006; Trivedi 2006B) but predicting who may require which treatment strategy is difficult. Additionally, inter-individual variability in side effects and tolerability to treatments is a clinical challenge. Ongoing research has focused on identifying and determining the clinical utility of pharmacogenomic markers for patients requiring treatment with an antidepressant.

At present, the product labeling for all tricyclic antidepressants (TCAs), some selective serotonin reuptake inhibitors (SSRIs), and one serotonin and norepinephrine reuptake inhibitor (SNRI) contain considerations based upon genetic factors (see Table 8.1). However, these recommendations only relate to polymorphisms affecting drug metabolism. Antidepressant medications have differing pharmacodynamic profiles, but are generally understood to work through modulation of the metabolism, reuptake, or receptor activity of monoamine neurotransmitters such as serotonin, dopamine, and norepinephrine. Therefore, much additional work has also been completed to search for or characterize relationships between treatment response and pharmacodynamic genetic markers.

Antidepressant Pharmacokinetics

Antidepressants are largely metabolized through cytochrome P450 enzyme pathways. TCAs as a class share common metabolic pathways and have considerable toxicity at high doses; hence the potential for clinical use of genetically informed dosing is high for these agents. The Clinical Pharmacogenetics Implementation Consortium (CPIC) has recently published guidelines for the dosing of TCAs in patients with known *CYP2D6* or *CYP2C19* genotypes (Hicks 2013). TCAs (which can be tertiary or secondary amines) are metabolized by CYP2D6 to less active metabolites. Consequently, CYP2D6 poor metabolizer (PM) status may likely lead to higher concentrations of active

Table 8.1. Clinical Considerations of Pharmacogenomic Information for Psychiatric Medications

Drug	Clinical Recommendations or Considerations of Pharmacogenomic Information*
Amitriptyline	Recommend a 50% reduction in starting dose or consideration of alternative agents in CYP2D6 PM. Consider an alternative agent in CYP2D6 UM. Consider a 50% starting dose reduction in CYP2C19 PM.
Aripiprazole	Oral Dosage Form: Reduce to 50% of the usual dose in CYP2D6 PM. In patients with CYP2D6 PM status and concurrently receiving a strong CYP3A4 inhibitor (e.g., protease inhibitors, antifungals, etc), dose should be reduced to 25% of the usual dose. I.M. Dosage Form: Reduce to 300mg once monthly if they are a PM or to 200mg once monthly if they are also receiving a CYP3A4 inhibitor for 14 or more days.
Atomoxetine	Patients up to 70 kg body weight who are known to be CYP2D6 PM should be initiated at 0.5 mg/kg/day and only increased to the usual target dose of 1.2 mg/kg/day if symptoms fail to improve after 4 weeks and the initial dose is well tolerated. Patients over 70 kg body weight who are known CYP2D6 PM should be initiated at 40 mg/day and only increased to the usual target dose of 80 mg/day if symptoms fail to improve after 4 weeks and the initial dose is well tolerated.
Carbamazepine	Genetic testing is recommended in persons of Asian ancestry. Treatment with carbamazepine is not recommended in HLA-B*1502.
Citalopram	The maximum recommended daily dose in CYP2C19 PM is 20 mg/day.
Clomipramine	Recommend a 50% reduction in starting dose or consideration of alternative agents in CYP2D6 PM. Consider an alternative agent in CYP2D6 UM. Consider a 50% starting dose reduction in CYP2C19 PM.
Clozapine	It may be necessary to reduce the dose in CYP2D6 PM because higher levels can be seen at usual doses.
Desipramine	Recommend a 50% reduction in starting dose or consideration of alternative agents in CYP2D6 PM. Consider an alternative agent in CYP2D6 UM.
Doxepin	Recommend a 50% reduction in starting dose or consideration of alternative agents in CYP2D6 PM. Consider an alternative agent in CYP2D6 UM. Consider a 50% starting dose reduction in CYP2C19 PM.
Fluvoxamine	CYP2D6 PM are known to have altered pharmacokinetic properties (i.e. increased AUC, Cmax, and half-life) as compared to extensive metabolizers.
Haloperidol	The Dutch Pharmacogenetics Working Group recommends reducing the dose by 50% of finding an alternative drug in CYP2D6 PM.
Iloperidone	CYP2D6 PM should reduce their dose by 50%.
Imipramine	Recommend a 50% reduction in starting dose or consideration of alternative agents in CYP2D6 PM. Consider an alternative agent in CYP2D6 UM. Consider a 50% starting dose reduction in CYP2C19 PM.
Nortriptyline	Recommend a 50% reduction in starting dose or consideration of alternative agents in CYP2D6 PM. Consider an alternative agent in CYP2D6 UM.
Perphenazine	CYP2D6 PM have higher plasma levels of perphenazine at usual doses which may lead to a higher incidence of side effects.
Risperidone	Product labeling states that pharmacogenetic differences cause differing degrees of metabolizing of risperidone to 9-hyrdoxy-risperdone but that the pharmacokinetics are similar in poor and UM with no dosage recommendation. The Dutch Pharmacogenetics Working Group suggests choosing alternative drug or being extra alert for side effects in CYP2D6 poor, intermediate or UM.

Table 8.1. Clinical Considerations of Pharmacogenomic Information for Psychiatric Medications *(continued)*

Drug	Clinical Recommendations or Considerations of Pharmacogenomic Information*
Trimipramine	Recommend a 50% reduction in starting dose or consideration of alternative agents in *CYP2D6* PM. Consider an alternative agent in *CYP2D6* UM. Consider a 50% starting dose reduction in *CYP2C19* PM.
Valproic Acid	1) Avoid use in patients with the Urea Cycle Disorders as hyperammonemic encephalopathy can occur. Evaluation for genetic abnormalities (including ornithine transcarbamylase and carbamoyl-phosphate synthetase) should be considered in high-risk patients before initiation of therapy. 2) Genetic testing required for patients with suspected mitochondrial DNA polymerase or "POLG-related" disorder. The *POLG* A467T and W748S mutations are present in approximately two-thirds of patients with autosomal recessive POLG-related disorders
Vortioxetine	The maximum recommended daily dose is 10 mg/day in *CYPD26* PM.

*Defined as a specific, dose recommendation found in the drug's product labeling (http://www.accessdata.fda.gov/scripts/cder/drugsatfda/index.cfm) accessed on February 28, 2014 (FDA 2014), Dutch Pharmacogenetics Working Group (Swen 2011), or CPIC reference number..

drug and a higher incidence of side effects (Bertilsson 1981). Consideration of an alternate agent or a 50% reduction in starting dose is recommended for patients known to be PM (Hicks 2013). A 25% reduction in starting dose is recommended for intermediate metabolizers (IM). CYP2D6 ultrarapid metabolizers (UM) have been shown to require large doses of TCAs to achieve therapeutic concentrations (Bertilsson 1985), thus an alternate agent should be considered. If no acceptable alternative is available, a higher starting dose should be considered.

In addition to this CYP2D6 pathway, tertiary amines (amitriptyline, clomipramine, doxepin, imipramine, trimipramine) are metabolized by CYP2C19 to active secondary amines (desmethyl-metabolites of parent compounds, including the commercially marketed compounds desipramine and nortriptyline). Relative to secondary amines, tertiary amines have more pronounced effects on reuptake inhibition. Thus in patients treated with tertiary amines, variation in CYP2C19 activity will alter the parent drug to secondary amine metabolite ratio which may affect symptom response and tolerability. A 50% starting dose reduction is recommended for CYP2C19 PM patients beginning treatment with a tertiary amine. Because the magnitude of effect and clinical relevance of the *CYP2C19* *17 allele is not yet clear, prescribers may choose to consider alternative agents (such as a secondary amine) in patients with a UM diplotype (*1/*17 or *17/*17).

One important clinical consideration for the usefulness of genotype-guided therapy for TCAs is that therapeutic drug monitoring has long been an established practice for the optimization of TCA therapy (Muller 2003). Thus, it may be argued that the added benefit of *CYP2D6* and *CYP2C19* genotyping in these patients may be minimal except for proactively avoiding side effects in individuals just beginning therapy. It should be noted that the above recommended dose reductions may be less applicable to individuals being treated at lower doses for neuropathic pain. Because of the lower dose, PMs treated in this context are at less increased risk of side effects commonly observed in clinical practice for neuropathic pain.

SSRIs and SNRIs are also primarily metabolized by CYP enzymes. CYP2D6 is a major pathway for the metabolism of fluoxetine, paroxetine, fluvoxamine, sertraline, venlafaxine and duloxetine (Brandl 2014; Spina 2008). Some non-SSRI antidepressants such as mirtazapine and vortioxetine are also metabolized by CYP2D6 (Chen 2013; Crasmader 2004). CYP2C19 plays a role in the metabolism of citalopram, escitalopram, fluoxetine, and sertraline (Spina 2008). While some SSRIs and SNRIs are metabolized by other CYP enzymes, much of the existing pharmacogenetic literature focuses on *CYP2D6* and *CYP2C19* because of the high degree of genetic variation in these genes and the importance of the roles of their encoded enzymes in the metabolism of SSRIs. Studies of antidepressant substrates have associated loss-of-function variants with increased serum concentration (Huezo-Diaz 2012; Rudberg 2008A, 2008B; Suzuki 2011; Watanabe 2008) and UM diplotypes to 1-2 fold decreases in serum concentrations (Huezo-Diaz 2012; Rudberg 2008A, 2008B). However, the relationship between genotype and clinical outcomes is less clear as studies of these outcomes have produced mixed results (Gex-Fabry 2008; Mrazek 2011; Peters 2008; Tsai 2010; Van Nieuwerburgh 2009). The clinical implications for *CYP2D6* or *CYP2C19* genotype are still being investigated at this time. There is debate in this topic given that SSRI treatment response is not robustly associated with drug concentration, yet there are known dose-tolerability, and, dose-response relationships (Muller 2013; Rasmussen 2000).

Many antidepressants are substrates of P-glycoprotein (P-gp), which is encoded by *ABCB1*. Antidepressants known to be P-gp substrates include amitriptyline, nortriptyline, trimipramine, citalopram, sertraline, and venlafaxine (Porcelli 2011). P-gp is an efflux protein expressed in the intestines, liver, kidney, blood-brain barrier, placenta, and lymphocytes and facilitates the efflux of psychoactive substrates out of the central nervous system (Brinkmann 2001). Numerous pharmacogenetic studies of treatment outcomes and *ABCB1* have been conducted but have yielded inconsistent results in relation to clinical response (Porcelli 2011). Clarification of the functional nature of variants in *ABCB1* and larger studies of these variants on the pharmacokinetics and treatment outcomes of SSRIs are needed before the pharmacogenetics of *ABCB1* can be applied in clinical practice.

Antidepressant Pharmacodynamics

The mechanism of action varies widely between antidepressant classes (e.g., TCAs, SSRIs, SNRIs, MAOIs, and others). Most recent pharmacogenetic studies of antidepressants have focused on SSRIs because these are first-line agents for most indications in which antidepressants are used and have a relatively specific pharmacologic mechanism of action of blocking reuptake from the synapse via antagonism of transporters. Polymorphisms in the gene encoding the serotonin transporter (*SLC6A4*) have to date been the most extensively studied. Within *SLC6A4*, there is a well-described insertion/deletion polymorphism in the promoter region called 5HTTLPR. The long (L) allele is associated with increased expression of the *SLC6A4* relative to the short (S) allele (Heils 1996; Lesch 1996). SNPs identified within the insertion element may also influence the activity of this transporter.

A recent meta-analysis of 33 studies concerning the 5HTTLPR and antidepressant efficacy yielded interesting results (Porcelli 2012). The L/L genotype was associated with increased odds of remission (OR 1.37, p=0.007 for all antidepressants; OR 1.48, p=0.005 for SSRI studies only). This effect was more consistent in studies of Caucasian than in Asian populations. Asian ancestry may significantly confound the effect of 5HTTLPR, as a recent study of Korean patients showed both higher serotonin transporter activity and better antidepressant response in individuals with the S allele (Myung 2013). The mechanism(s) underlying this disparity is unclear and requires further study given the potential clinical implications of polymorphisms in this gene.

SLC6A4 has also been studied in regards to adverse events during antidepressant therapy, such as antidepressant-induced mania (Biernacka 2012), insomnia (Perlis 2003), and sexual dysfunction (Perlis 2003; Bishop 2009; Strohmaier 2011). A recent meta-analysis found significantly reduced odds of side effects for carriers of the L allele (OR 0.64, p=0.0005) (Kato 2010). However, heterogeneity between studies and poorly defined side effect phenotypes limit the clinical applicability of this association at this time.

Despite these associations, the small impact on odds of remission and possible differences in genetic effect based on ancestry indicate that further research is needed before 5HTTLPR genotyping is deployed in routine practice. Additional variants in *SLC6A4* may also impact transporter function. A polymorphism within intron 2 of *SLC6A4*, referred to as STin2, may affect expression of the *SLC6A4* and have an interacting effect with 5HTTLPR (Hranilovic 2004). This polymorphism may also influence antidepressant response but these findings have been very heterogeneous and may be difference across race or ancestry groups (Niitsu 2013). Although SSRIs are the most widely studied antidepressants in regards to *SLC6A4*, agents from other classes (SNRI and TCA as well as other agents) also block the serotonin transporter and may possess pharmacodynamics properties affected by variation in this gene, though studies focusing on these agents are less plentiful.

Polymorphisms in other candidate genes have also been associated with antidepressant treatment outcomes. *BDNF*, encoding a polypeptide precursor of brain-derived neurotrophic factor, contains a non-synonymous coding region SNP, Val66Met, that may be related to antidepressant efficacy (Niitsu 2013). The serotonin-2A receptor gene *HTR2A* may be linked to tolerability of SSRIs (Kato 2010), especially gastrointestinal side effects and sexual dysfunction (Stevenson 2013). The clinical usefulness of these markers is hampered by heterogeneous study results that may be due to differences in study population, treatment, genetic polymorphisms studied, or publication bias. Studies of polymorphisms in genes encoding the serotonin 1A, 3A, and 3B receptors (*HTR1A, HTR3A, HTR3B*), tryptophan hydroxylase isoforms 1 and 2 (*TPH1, TPH2*), catechol-O-methyltransferase (*COMT*), monoamine oxidase A (*MAOA*), G protein β3 subunit (*GNB3*), and a component of the glucocorticoid receptor heterocomplex (*FKBP5*) (Porcelli 2011; Kato 2010) have yielded compelling results that require further validation before clinical implementation.

In summary, investigations of antidepressant pharmacogenomics have been numerous (especially for SSRIs) and yielded some findings that are beginning to be used in practice. Guidelines now exist for genetic-informed dosing of TCAs (Hicks 2013) and the package labeling for many SSRIs now includes genetic considerations (see Table 8.1). Applying genetic information from *CYP2D6* and *CYP2C19* in patients treated with SSRIs may become common in the coming decade; however the clinical application of these genetic findings is complicated. Studies examining pharmacodynamic markers for efficacy and side effects are promising but will need additional evidence before clinical implementation. With commercially available genotyping platforms providing information on *SLC6A4, HTR2A, ABCB1, COMT , CYP1A2, CYP3A4* alongside more clinically-ready markers such as *CYP2D6* and *CYP2C19* (Drozda 2014), we may soon have data on many more patients that may help inform the clinical utility of these tests.

Antipsychotic Pharmacogenomics

Antipsychotics were originally developed for the treatment of schizophrenia and related psychotic disorders. However, their use now extends to other mental illnesses including bipolar disorders, autism spectrum disorder, as well as other off-label diagnoses. Antipsychotic medications are commonly used for long-term maintenance therapy, but tolerability and variable effectiveness are often limitations. Discontinuation rates as studied in patients with schizophrenia are high, ranging from 30% in first episode patients (Boter 2009; McEvoy 2007) to 74% in chronic patients (Lieberman 2005). The reasons for treatment discontinuation often include lack of symptom improvement and poor tolerability. Antipsychotic medications may cause movement disorders such as extrapyramidal side effects and tardive dyskinesia, metabolic side effects such as weight gain and associated morbidity, hormonal disturbances, anticholinergic side effects, cardiovascular effects, as well as others (Newcomer 2013). There is a high degree of patient variability with respect to symptom improvement and tolerability that has led to investigations searching for genetic contributors to this variance. Research in the field of antipsychotic pharmacogenomics has been ongoing since the 1990s and has focused both on symptom improvement and side effects, although largely on the later. This section will focus on the pharmacogenomics of antipsychotic pharmacokinetics and pharmacodynamics with an emphasis on pharmacogenomics in antipsychotic labeling information and findings from large and/or replicated trials.

Antipsychotic Pharmacokinetics

The antipsychotics are metabolized largely through the cytochrome P450 enzyme system with relatively unique CYP pathway combinations for most medications that includes, but is not limited to, CYP1A2, CYP2C19, CYP2D6, and CYP3A4. CYP2D6 is a primary metabolic pathway for several antipsychotics with language in the product labeling recommending dose adjustments or warning of the effects of PM status of six medications (see Table 8.1). Those with specific dosing adjustments listed include aripiprazole and iloperidone. Additionally, Dutch Pharmacogenetics Working Group guidelines mention dose adjustments for haloperidol and risperidone in persons who are CYP2D6 PM (Swen 2011). These labeling and international guidelines have not been evaluated to determine whether using genetic information to guide dose or drug selection improves outcomes over traditional care. To this end controversies exist regarding the utility of drug metabolism genetic information for antipsychotic agents (Grossman 2008). Like antidepressants, the therapeutic range for antipsychotics is thought to be broad, although clinical trials generally identify some dose-relationships with response and side effects (Bollini 1994; Davis 2004; Kinon 2004). Thus, if metabolizer status is available to clinicians, it may be helpful in assessing risk for dose-related adverse events from antipsychotics, including extrapyramidal side effects, orthostasis, and anticholinergic effects.

Antipsychotic Pharmacodynamics

The antipsychotics are often considered as two broad categories; the first-generation (FGA) or "typical" antipsychotics and the second generation (SGA) or newer, "atypical" antipsychotics. FGAs and SGAs are generally distinguished by their relative dopamine receptor 2 (D_2) antagonism and serotonin receptor subtype 2A ($5\text{-}HT_{2A}$) antagonism in the central nervous system. FGAs are generally thought to have higher D_2:$5\text{-}HT_{2A}$ binding affinities and levels of antagonism while SGAs generally have higher $5\text{-}HT_{2A}$:D_2 (Meltzer 2013). There are currently eight FGAs and 10 SGAs with varying indications for use in schizophrenia, bipolar disorder (acute and/or maintenance phase) and treatment resistant major depressive disorder. In addition to D_2 and $5\text{-}HT_{2A}$ binding affinities, each antipsychotic has a unique affinity profile for other dopamine, serotonin, histamine, muscarinic, and alpha receptors which may contribute to some differences in side effect and response profiles across medications. Numerous pharmacogenomic studies have been conducted to determine whether genetic contributions may also play a role in differential drug response and tolerability that is commonly seen in clinical environments. A larger body of pharmacogenomic antipsychotic research has focused on drug transporter, synaptic transmission, and specific pathway genes for antipsychotic efficacy and side effects. Targets for antipsychotic efficacy and side effects have consisted of both genome-wide studies as well as candidate gene work looking at pathways associated with the phenotype on interest. Candidate genes with replicated findings related to treatment outcomes involve a variety of serotonin, dopamine, as well as other genes. To date, pharmacodynamic findings have not been included in drug labels. This may change in the near future, as findings gain acceptance through replication and large cohorts.

Dopamine neurotransmission in the brain has historically been the central hypothesis regarding schizophrenia disease pathology, in part due to the effectiveness of D_2 blocking antipsychotics on improving symptoms. All antipsychotics have varying degrees of antagonistic action at dopamine D_2-like receptors in the brain. Pharmacogenomic investigations have studied variants of the dopamine receptor genes associated with antipsychotic treatment response and tolerability.

The dopamine-2 receptor gene (*DRD2*) has been examined as a pharmacogenetic candidate gene in many studies of antipsychotic response and side effects. While there are numerous functional variants in the *DRD2* region, the *Taq1A* (rs1800497) variant which exists in the neighboring ankyrin repeat and kinase domain containing 1 (*ANKK1*) gene has been the most studied variant from this region for antipsychotic pharmacogenomics as the C allele of this variant has been associated with higher striatal D2 receptor density (Thompson 1997). Two meta-analyses identified that patients carrying the C allele of *Taq1A* have an approximately 30% increased risk of tardive dyskinesia per allele (Bakker 2008; Zai 2007). Studies to examine the benefits of screening patients for this variant prior to treatment have not yet been done.

Dopamine-D3 receptors are similar in structure to D_2 receptors with some antipsychotics known to bind to both types (Girgis 2011; Graff-Guerrero 2009). Along with D_2 and dopamine-D4 receptors, the D3 receptors are considered to play an important mechanistic role in the movement disorders caused by antipsychotic binding (Meshul 1989). A relatively common missense variant resulting in a Serine to Glycine substitution at position 9 (Ser9Gly, rs6280) in the *DRD3* gene has also been studied for relationships with antipsychotic response. Initially, studies showed that patients carrying the Gly variant have an increased response to antipsychotic treatment likely due to the increased sensitivity for dopamine. However, alongside these findings, several meta-analyses have demonstrated an increased risk of tardive dyskinesia for patients with the Gly variant, which can be attributed to the increased dopamine sensitivity (Bakker 2006; Lerer 2002). An analysis by Lerer et al. identified an odds ratio of 1.33 (95% CI 1.04-1.70, p=0.02) and a pooled odds ratio of 1.52 (95% CI 1.08-1.68, p<0.001) for the glycine allele increasing the risk of tardive dyskinesia (Lerer 2002). A meta-analysis by Bakker et al. also identified an increased risk for tardive dyskinesia with the glycine allele (Odds Ratio = 1.17, 95% CI 1.01-1.37, p=0.04) (Bakker 2008). Despite the initial positive findings, many other studies have not identified associations between Ser9Gly and antipsychotic response or tardive dyskinesia (Tsai 2010A, B). The discordant findings across studies could be due to small sample sizes unable to detect modest effect sizes, lack of well-defined populations, complexity of the pathophysiology of tardive dyskinesia, and studying newer antipsychotic medications that may carry a lower risk of causing tardive dyskinesia. Thus, further work is needed in well defined, prospective samples to identify the true clinical utility of this association.

SGAs are believed to improve symptoms in schizophrenia in part by antagonizing several serotonin receptor subtypes with one primary contributor being 5-HT$_{2A}$. Variants within the *HTR2A* gene have been investigated for both antipsychotic response as well as side effects. As 5-HT2A receptors are tightly bound and antagonized by all SGAs and some FGAs, variants associated with expression (i.e., T102C rs6313, -1438G/A rs6311) (Falkenberg 2011), or altered protein conformation (i.e., His452Tyr rs6314) have been the most studied SNPs for antipsychotic response. Some studies have shown some promising results in associations with response, but conflicting or inconclusive data makes it difficult to draw conclusions about the clinical utility of these markers at this point (Serretti 2007). With respect to tolerability, associations for *HTR2A* variants and extrapyramidal symptoms have been inconsistent; however one genetic meta-analysis, pooling samples across several studies, found a strong statistical association between a 102-T/C and His452Tyr haplotype and tardive dyskinesia in patients treated with antipsychotics (Lerer 2005). Although the association was significant (p=0.0008), the pooled odds ratio of 1.64 was modest, signifying a small overall contribution of the haplotype to the risk for tardive dyskinesia.

Antagonism of 5-HT$_{2C}$ receptors is associated with weight gain pathways and has been the subject of several pharmacogenomic investigations of antipsychotic–induced metabolic side effects (Muller 2006). Repeated studies have found an association of a SNP in the promoter region of the *HTR2C* gene (-759-C/T, rs3813929) with antipsychotic-induced weight gain in both the pediatric and adult populations (Reynolds 2002). Patients carrying the minor T allele of -759-C/T appear to be protected from significant weight gain (greater than 7%) after initial antipsychotic treatment while those that carry a C allele are at a more than 2-fold risk for significant weight gain (Reynolds 2005). A meta-analysis of studies looking at *HTR2C* variants and antipsychotic-induced weight gain confirmed this effect of the -759-C/T variant (Sicard 2010). This meta-analysis identified the C allele as being associated significantly associated with weight gain with the strongest effect in the European population (p=0.006, 3.20 (95%CI 1.40-7.28).

The 1019C/G variant within the *HTR1A* gene (coding for 5-HT$_{1A}$) has been associated with negative symptom response in schizophrenia patients treated with various antipsychotic drugs (Mossner 2009; Reynolds 2006; Wang 2008). Carriers of the C allele were found to have larger responses as measured by negative symptom scales compared to subjects with the GG genotype. In the study by Mossner et al., carriers of the 1019C allele had an average of 2–3 points further reduction in the Positive and Negative Symptom Scale (PANSS) negative symptom scores, a finding that was confirmed in a second independent where the risk variant was responsible for approximately 7%–12% of the variance in improvement of negative observable after 4 weeks of antipsychotic treatment (Mossner 2009). Using this information in the clinical setting has not yet been evaluated. One important consideration with findings such as these is that treatment alternatives are not known for those who might be identified to have a less favorable response. Without actionable clinical alternatives this may limit the utility of this information as it related to drug selection and/or dosing at this time.

Glutamate Transmission

Glutamate is central to brain neurodevelopment and dysregulation of glutamate signaling in the central nervous system is thought to contribute to hyperactivity of the dopaminergic system (Moghaddam 2012). To date, several association studies have identified relationships between candidate variants in glutamate receptors with antipsychotic response (Bishop 2005; Bishop 2011; Fijal 2009; Lencer 2013). These findings may eventually prove useful in identifying patients at risk for efficacy and side effect risk and demonstrates the importance of the glutamate system in antipsychotic response. As variation in the glutamate system may also have implications for disease risk, it is important to clarify and understand these relationships if the findings are given consideration for patient care.

Other Variants Implicated in Metabolic Side Effects

Many studies have examined risk variants for antipsychotic-induced metabolic side effects. Of the metabolic side effects, antipsychotic-induced weight gain has been related to the aforementioned *HTR2C* -759-T/C variant as well as others. In particular, repeated studies have demonstrated significant associations with variants in the leptin–melanocortin pathway (Kao 2013; Lee 2011). Leptin is a protein produced by adipocytes during fasting in order to cause anorexigenic changes. The leptin protein is encoded by the *LEP* gene and the -2548-G/A (rs7799039) promoter variant has shown repeated positive findings for antipsychotic-induced weight gain (Lee 2011; Brandl 2012; Wu 2011). Interestingly, the direction of association for weight gain risk appears to be inconsistent across studies which limit the clinical usefulness of genotyping for this SNP until the reasons for and mechanisms underlying these discrepancies are identified.

The melanocortin four receptor (MC4R) is part of the leptin system, where, after a series of processes beginning with leptin, MC4R is activated and eventually modifies further downstream pathways leading to energy expenditures (Arch 2005). The *MC4R* gene was first identified as a potential candidate for antipsychotic-induced weight gain from a genome-wide association study of patients followed over the course of antipsychotic treatment. This relationship was subsequently replicated in three separate cohorts (Malhotra 2012) with associations findings from three other studies identifying relations ships with other *MCR4* SNPs and antipsychotic associated weight gain (Chowdhury 2013; Czerwensky 2013). These findings suggest that MCR4 likely plays an important role in predicting weight gain for patients treated with antipsychotics but additional studies are needed to establish how to clinically utilize variants in the *MCR4* gene.

The catechol-o-methyltransferase (*COMT*) rs4680 met allele polymorphism is associated with a 30%–50% decreased enzyme activity leading to increased dopamine levels in the prefrontal cortex and has been linked to cognitive function. Alternatively the COMT Val allele is associated with greater produciton of the homocysteine precursor S-adenosyl homocysteine from S-adenoyl methionine. Methylenetetrahydrofolate reductase (MTHFR) is closely linked to the *COMT* enzyme as it is responsible for a methyl group transfer reaction in the folate cycle which eventually acts as a substrate for *COMT*. The T allele of the *MTHFR* 677C/T variant confers up to a 70% reduction in enzyme activity. Both the *COMT* valine and *MTHFR* 677T variants (as well as combined effects of the variants) have been shown to be associated with weight gain and metabolic syndrome in three separate studies of patients on antipsychotics (Devlin 2012; Ellingrod 2012; Kuzman 2012). These findings require replication in larger cohorts however, the initial findings show promise as future candidates for antipsychotic-induced weight gain and metabolic syndrome.

Finally, recent investigations have found some evidence to support relationships between sulfotransferase-4A1

(*SULT4A1*) genetic variation and antipsychotic response (Ramsey 2011). This gene is involved in the metabolism of neurotransmitters and is particularly abundant in the cortex. Variants of the *SULT4A1* gene have been implicated in schizophrenia disease pathophysiology and presentation (Brennan 2005; Meltzer 2008) with pharmacogenetic relationships describing associations with olanzapine and risperidone as well as replication samples (Ramsey 2011).

Commentary and Future Work

Pharmacogenomic investigations of antipsychotic efficacy and side effects have advanced where we now see several variants being mentioned in FDA product labeling or included in commercially available testing panels (*CYP2D6, COMT, MTHFR*, etc.) (Drozda 2014). Despite these advances, certain factors have made it difficult for translation of antipsychotic pharmacogenomic research into clinical use. First, the main patient groups using antipsychotics are patients with severe mental illness, namely schizophrenia and bipolar disorder whose populations are relatively small. This makes it difficult to achieve high power in genome-wide association studies (GWAS). This limitation of the antipsychotic utilizing populations does not necessarily confine pharmacogenetic work to candidate-gene studies as is indicated by the important finding in the *MC4R* gene discussed above; however, it demonstrates the difficult barriers that must be overcome in order for GWAS to be effective. Furthermore, lack of replication and conflicting genetic findings coupled with smaller effect sizes, as well as low specificity and sensitivity have plagued translation into the clinical realm. This may be due to many factors including poorly defined populations and psychiatric polypharmacy. Research into predicting antipsychotic response is complicated by past antipsychotic trial histories, length of illness and medication adherence issues.

Investigations into antipsychotic side effects can be made more difficult if the side effect occurs in a relatively small number of patients, like clozapine-induced agranulocytosis. Clozapine is generally considered one of the most efficacious antipsychotics. However, the risk for agranulocytosis limits its use to treatment-resistance schizophrenia. Despite this disadvantage, there have been promising findings with various variants in the *HLA* gene of the Major Histocompatibility Complex Region for clozapine-induced agranulocytosis (Armar 1998; Dettling 2007; Dettling 2001; Lieberman 1990; Valevski 1998; Yunis 1995). Advances in our ability to predict risk for deadly blood cell abnormalities from clozapine have the potential for high impact on patient care. Due to the stringent monitoring requirements for clozapine, the utility of any resulting clinical tests is highly dependent on their ability alter the monitoring requirements for this medication.

Research into the pharmacogenomics of antipsychotic response and side effects show promising results. Much more work is needed to clarify how to best translate research findings into everyday clinical use. Although incorporation into the clinical realm has been slow, it is still considered a promising avenue to improve patient adherence since it is estimated that nearly 75% of patients will discontinue an antipsychotic largely due to lack of tolerability (Lieberman 2005). Along with discontinuation of treatment comes a higher likelihood of relapse and hospitalization which ultimately leads to increased healthcare costs. The ability to pre-determine which patients are most likely to have a given side effect to an antipsychotic may help to decrease the high rates of discontinuation, improve treatment outcomes and deliver the promise of personalized medicine in severe mental illness.

Pharmacogenomics of Mood Stabilizers

Bipolar disorders affect approximately 2% of the population (Merikangas 2011). Primary pharmacological treatment options include lithium, valproic acid and related formulations, carbamazepine, lamotrigine, and antipsychotic medications. The Systematic Treatment Enhancement Program for Bipolar Disorder (STEP-BD) study examined over 1500 patients in a prospective naturalistic study using standardized practice models and pharmacotherapy treatment guidelines (Perlis 2006). While nearly 60% of patients achieve symptomatic recovery from initial treatments in a naturalistic setting using standardized guidelines and treatment protocols (Perlis 2006), over a third of patients experience illness recurrence within one year (Gitlin 1995) and 50%–60% experience illness recurrence within two years (Perlis 2006; Gitlin 1995). Available treatment options have distinct mechanisms of action and side effect profiles. Side effect profiles can range from self-limiting to toxic. The consequences of inadequate treatment for bipolar disorder include functional disability and high rates of suicide, making the need for optimizing available treatment options and minimizing trial and error based drug and dose selection extremely important. This section will discuss pharmacogenomics aspects of traditional mood stabilizers (i.e., lithium, valproic acid, carbamazepine, and lamotrigine). Antipsychotics are increasingly used in bipolar disorder for acute and maintenance treatment and this class of medications is discussed earlier, with studies of tolerability, toxicity, and aspects related to drug metabolism generally not diagnosis-specific and thus applicable across patient populations. However, bipolar disorder does have unique characteristics in terms of genetics, lifestyle, and utilization of antipsychotics which justify continued research in this area. The bulk of the literature surrounding mood stabilizer pharmacogenomics focuses on examining treatment response to lithium and aspects of toxicity to carbamazepine. The primary clinical pharmacogenetic marker for this category is a black box warning which exists for carbamazepine and toxic dermatological events in persons of Asian ancestry who possess a risk allele of the human leukocyte antigen B (*HLAB*) gene.

Mood Stabilizer Pharmacokinetics
The pharmacokinetic properties and associated metabolic pathways of agents used as mood stabilizers are distinct from one another and largely do not involve cytochrome P-450 pathways with genetic variations described as clinically actionable at this time. Lithium is renally cleared and not hepatically metabolized. Valproic acid, carbamazepine, and lamotrigine are hepatically metabolized through some distinct and some overlapping metabolic pathways.

Carbamazepine is primarily metabolized hepatically through the CYP3A4 pathway to an active metabolite that is subsequently converted to an inactive compound by microsomal hydrolases (Bertilsson 1986). Association studies have been conducted to investigate relationships between genetic variants in *CYP3A4*, microsomal epoxide hydrolase (*EPHX1*), p-glycoprotein (*ABCB1*), *UGT2B7*, and carbamazepine pharmacokinetics as well as treatment outcomes (Hung 2012; Makmor-Bakry 2009; Meng 2011; Puranik 2013; Yun 2013). Most of the pharmacokinetic studies to date have been completed in epilepsy patients. While some evidence exists suggesting pharmacogenetic relationships with *EPHX1* (Hung 2012; Makmor-Bakry 2009), *ABCB1* (Meng 2011), *UGT2B7* (Hung 2012), *CYP3A4* (Puranik 2013; Yun 2013), these results are not uniformly consistent across all studies. Thus additional work is needed to clarify the pharmacogenomic aspects of carbamazepine metabolism to determine whether clinical applications may be appropriate.

Lamotrigine is metabolized both hepatically and renally with primary liver pathways involving glucuronic acid conjugation to inactive metabolites (Rambeck 1993). Not surprisingly, association studies involving genes encoding uridine 5'-diphosphate glucuronosyl-transferase (UGT) enzymes have been conducted (Gulcebi 2011; Zhou 2011). In vitro and in vivo studies of glucuronidation rates and two amino acid altering variants in the *UGT1A4* gene have identified a significant reduction in enzymatic activity on lamotrigine for both variants as compared to wild type with Vmax and CL(int) values reduced by 25% or greater (Zhou 2011). The L48V *UGT1A4* SNP was associated with 52% lower lamotrigine levels in carriers of the valine allele (Gulcebi 2011). One recent investigation additionally suggested a relationship between a promoter variant in the *UGT2B7* gene and lamotrigine (Singkham 2013). There may be clinical implications for variants in genes encoding UGTs, particularly *UGT1A4* given the demonstrated effects of amino acid variants on lamotrigine glucuronidation. This optimism is tempered by some of the inherent challenges in these types of studies that often involve co-treatment with other medications (i.e., vaproic acid) that may confound pharmacogenetic relationships with drug metabolism. Similarly, as they relate to treatments for bipolar disorder, lamotrigine is not often assessed with therapeutic drug monitoring as the effects may not be as dose-sensitive as in epilepsy. Thus, further investigation of UGT lamotrigine pharmacogenomics is needed in order to better understand any potential clinical utility.

Valproic acid is metabolized via glucuronide conjugation, beta-oxidation in mitochondria, as well as other oxidative pathways representing relatively minor contributions. Similar to lamotrigine, most pharmacogenomic investigations in valproic acid metabolism have examined the role of UGT gene families. There are many UGT genes that reside on different chromosomes with UGT1 and UGT2 families of genes identified as important in drug metabolism (Guillemette 2010). As it relates to valproic acid metabolism, at least five specific UGT1 or UGT2 enzymes are involved, with polymorphic loci throughout. A number of studies have examined variations in these genes on pharmacokinetic parameters of valproic acid, largely in epilepsy populations (Chatzistefanidis 2012). While relationships with altered metabolism variants have been described, the large number of genes involved and differing allele frequencies of functional variants across genes has likely contributed to heterogeneous findings. Nonetheless they represent an area of potential clinical importance given dose-related toxicities and some evidence for dose-related response in bipolar disorder.

Mood Stabilizer Pharmacodynamics
Agents covered in this mood stabilizer section comprise a variety of pharmacodynamic mechanisms thought to be related to symptom response as well as tolerability. Lithium is the most widely studied with respect to investigations of treatment response. Clinically relevant studies of the antiepileptic mood stabilizers have largely focused on risks for hypersensitivity dermatological reactions which are rare but deadly outcomes associated with carbamazepine and

lamotrigine. Clinical guidelines and a black box warning exist to guide carbamazepine use to minimize the risk of Stevens Johnson Syndrome (SJS) or toxic epidermal necrolysis (TEN) in some patient populations.

Despite being one of the oldest known psychotropic agents and in its purist form an elemental alkali metal, the therapeutic mechanisms of lithium are still being investigated. As it relates to the treatment of mood disorders, lithium may alter cation transport across cell membrane in nerve and muscle cells to influence and norepinephrine reuptake. It is also hypothesized to have a genetic mechanism of action by altering gene expression (Beech 2014). Lithium pharmacodynamics are also thought to include effects on second messenger systems involving the phosphatidylinositol cycle as well as glycogen synthase kinase 3 (GSKβ), which appear to be inhibited as a result of lithium exposure (Pettegrew 2001; Silverstone 2002). Not surprisingly a plethora of studies exist examining various associations between lithium response and a variety of pharmacodynamic genes in or related to these pathways (Rybakowski 2013). In line with many other areas of pharmacogenomics, heterogeneous results limit clinical interpretation or implementation at this time. The large scale Consortium on Lithium Genetics (ConLiGen) has recently been established to acquire the largest lithium-treated study sample to date to address pharmacogenomics in a genome-wide fashion (Schulze 2010). Pharmacogenomic studies of lithium toxicity are sparse as the primary contribution to toxicity appears to be dose or serum concentrations.

Valproic acid increases availability of gamma-aminobutyric acid (GABA) and may also mimic its action at postsynaptic receptor sites. It is also a known histone deactylase inhibitor, so it also affects DNA packaging (epigenetics) leading to downstream effects on gene expression (Machado-Vieira 2011). How these specific actions are related to treatment response in bipolar disorders is an area of active investigation. Pharmacogenetic studies of valproic acid response in bipolar disorder are sparse. It is unclear why, but clinically it is rare to identify persons on valproic acid monotherapy for the treatment of bipolar disorders, thus making it hard to parse out pharmacogenomic aspects of treatment response due to intended polypharmacy. Although not widely recognized, valproic acid does have pharmacogenomic information in its product labeling related to toxicity in persons who have known deficiencies in urea cycle genes encoding carbmyl phosphate synthetase 1 (*CPS1*) and ornithine carbamoyltransferase (*OTC*). Patients possessing these inborn errors in metabolism might have detrimental consequences if treated with valproic acid due to accumulation of urea cycle precursors. These may include cerebral edema, coma, and possibly death. "Testing" in the pharmacogenomic sense is not recommended at this time as these conditions are largely recognized by neonatal screening programs. Nonetheless, it is possible for milder forms to go unnoticed. Thus, while pharmacogenomic testing is not recommended, awareness of these potential contraindications to treatment is important.

The mechanism of action for carbamazepine is related to reducing the influx of sodium ions across cell membranes which is thought to reduce neuronal excitation. It is also thought to potentiate GABA receptor signaling which may be related to its effectiveness in the treatment of bipolar disorders (Bialer 2012). While pharmacogenomic studies of carbamazepine efficacy in bipolar disorders are sparse there is a plethora of evidence that has identified the importance of genetic testing to reduce the risk for toxic dermatological events in persons of Asian ancestry. In 2007 the product labeling for carbamazepine was updated to include an additional black box warning and recommendation for testing for the presence of the Human Leukocyte Antigen-B (*HLA-B*) *1502 allele in patients of Asian ancestry. The *HLA-B*1502* allele is linked to two polymorphisms (rs2844682, rs3909184) almost exclusively found in persons of Asian ancestry that are thought to result in an immune system reaction to carbamazepine and a substantial increase in the risk for SJS or TEN (Grover 2014). The specificity, and sensitivity of HLA-B*1502 testing was estimated to be 0.88 and 0.96 respectively with meta-analyses identifying that the odds of SJS/TEN were 113.4 times greater in *1502 carriers when compared to non-carriers. *HLA-B*1502* alleles are most frequently observed in persons of certain Asian descent and rarely in Caucasians or those of African ancestry. Thus the product labeling as well as published guidelines strongly support *HLA-B*1502* genotyping for patients of Asian ancestry who are being considered for the treatment with carbamazepine (FDA 2014). Carbamazepine may be used in non-*HLA-B *1502* carriers and in the case of those who carry one or two copies of *HLA-B*1502* alleles (*HLA-B*1502* positive), alternative agents should be considered.

Lamotrigine is another agent originally developed as an antiepileptic with demonstrated efficacy and an indication for maintenance therapy in bipolar disorder. Like carbamazepine, lamotrigine is associated with toxic dermatological events. Rash in the context of lamotrigine treatment is more likely to be observed early in treatment, and is correlated with rapid up-titration of the dose, as well as co-therapy with other agents such as valproic acid (Biton 2006). However, rash can develop after this initial period as well. Pharmacogenomic studies using candidate gene and genome-wide approaches have tested the hypotheses that variants in HLA genes are related to toxic rash from lamotrigine therapy but have yet to identify associations that are consistent and as robust as those found with carbamazepine. Therefore clinical pharmacogenomic testing for lamotrigine is not yet available at this time.

Commentary and Future Work

Pharmacogenomic studies of agents used for the treatment of bipolar disorders are complicated given the heterogeneous nature of disease presentation as well as mechanisms of action of medications with indications for treating these patients. Furthermore, psychotropic polypharmacy is common in bipolar disorder treatment which presents challenges in the conduct of pharmacogenomic association studies in well-defined samples difficult. Not surprisingly, association studies with clinical response have not yet identified pharmacogenomic markers to guide drug selection in this regard. Most of the agents discussed in this section are amenable to therapeutic drug monitoring, with some aspects of response and tolerability that are related to serum concentrations. Thus genetic factors that are related to drug metabolism and drug disposition may be important for carbamazepine, valproic acid, and lamotrigine. However, the complex nature of the metabolic pathways and related genes (e.g., glucuronidation) are such that reported associations are inconsistent and parsing the relevant contributions of multiple metabolism genes requires additional work. As with antipsychotics, mood stabilizing medications can have metabolic side effects. However, to date, a limited number of pharmacogenomic studies have looked at associations between mood stabilizing medications and metabolic side effects in bipolar disorder (Dols 2013). Genes encoding drug transporters (i.e., *ABCB1*) have also been investigated with mixed results in patients treated for epilepsy (Haerian 2010). Much of this work has been completed in patients treated with epilepsy, thus the importance of these factors in those with bipolar disorder needs clarification. Pharmacogenomic testing recommendations for variation in the *HLAB* gene is of clinically significant importance for carbamazepine. The identification of a rare yet highly influential variant related to toxic and deadly events lends hope to similar relationships with other medications such as lamotrigine. Elucidating potential genetic markers associated with these outcomes in other race/ethnicity groups is an important ongoing area of investigation.

Due to the episodic nature of bipolar illness and the increased risk for suicide or other risky behaviors during episodes, there may be benefit in studying the pharmacogenomic aspects of these outcomes in this patient population. Initial work in this area has focused on candidate gene studies, primarily *SLC6A4* and relationships to antidepressant-induced mania in bipolar disorders (Biernacka 2012), but more comprehensive studies are needed.

Pharmacogenomics of Stimulants and Related Medications for ADHD

ADHD has a prevalence of approximately 5% in children and 2.5% in adults (APA 2013). ADHD is characterized by a childhood onset of hyperactivity and/or inattention that affects functioning or development. Symptoms may extend from childhood through adolescence and into adulthood, but hyperactivity symptoms appear to lessen with age. Treatments for ADHD primarily include stimulants (e.g., methylphenidate, amphetamine, and related compounds) as well as the non-stimulant atomoxetine. While stimulants are generally effective in approximately 65%–75% (Dopheide 2009) of patients, there is variability in dose requirements needed for symptom relief as well as differential tolerability across patients. Atomoxetine is effective in approximately 75% of patients based on long-term studies evaluating discontinuation rates (Donnelly 2009). Differences in the pharmacodynamics, mechanisms, metabolic pathways, and response profiles for stimulants and non-stimulants has resulted in a growing body of research dedicated to identifying whether pharmacogenomics may be useful as a tool to guide drug selection, dosing, or monitoring. These studies have examined a variety of genetic markers hypothesized or known to influence pharmacodynamic and pharmacokinetic pathways associated with these medications. To date, only atomoxetine has dosing suggestions based on pharmacogenomic variables in the product labeling, although the clinical utility of this information is still being studied.

Pharmacokinetics of ADHD Medications

Stimulants and non-stimulants for the treatment of ADHD are largely hepatically metabolized. Methylphenidate formulations are metabolized to inactive ritalinic acid by deesterification. The carboxylesterase-1 enzyme hydrolyzes both the d- and l- isomers of methylphenidate (Sun 2004). The *CES1* gene encodes this enzyme with variants known to influence the carboxylase activity of this enzyme (Zhu 2008). Limited research investigating the relationships between these variants and methylphenidate outcomes has been completed. One initial study suggested a potential relationship between dose and one of these non-synonymous variants (Nemoda 2009), while a second found associations between other *CES1* variants and aspects of clinical response but not dose (Johnson 2013). These associations require replication, are not mentioned in product labeling for methylphenidate compounds, and thus possess limited clinical utility at this time.

Amphetamine-related compounds and the non-stimulant atomoxetine are hepatically metabolized with the CYP2D6 pathway accounting for the biotransformation of both of these compounds (de la Torre 2004; Sauer 2005) with additional contributions of the CYP2C family of enzymes in the deamination of amphetamines (de la Torre 2004). The pharmacogenomic influences of *CYP2D6* variation are better described and appear to have potential clinical implications for atomoxetine where PM have approximately a 10-fold higher area under the curve (AUC), higher peak plasma concentrations, and a half-life that is nearly five times longer in PM (24 hours versus approximately 5 hours in extensive metabolizers) (FDA 2014). The clinical implications of this have been described to increase the rates of side effects such as increased heart rate, increased blood pressure, decreased appetite, and increased incidence of tremor in CYP2D6 PM (Michelson 2007). However, these patients may also have increased rates of response. Collectively this information has resulted in dosing recommendations that differ based on CYP2D6 PM (see Table 8.1). Although clinical observations suggest possible utility, to date this has not been formally studied to determine the true benefit of pharmacogenomics testing, if any. This may arguably be due to the dichotomy between the increased risks for side effects balanced with perhaps an increased likelihood for response. Whether genotype-guided dosing results in better clinical outcomes than traditional dosing strategies has not yet been studied at the time of this review.

With respect to amphetamine and related compounds, CYP2D6 inhibition appears to influence d-amphetamine AUC in vitro (Tomkins 1997), indicating that genetically derived alterations in CYP2D6 metabolism may be important. There exists some evidence to suggest that CYP2D6 inhibition and genetic variation influence the metabolism of related compounds with potential implications for risk of overdose (i.e., methylenedioxymethamphetamine, MDMA) (Ramamoorthy 2001; Ramamoorthy 2002). Additional studies are needed to determine whether considering *CYP2D6* genetic information for clinical dosing of amphetamine-related compounds for ADHD is useful.

Pharmacodynamics of ADHD Medications
Central nervous system (CNS) stimulants and non-stimulants for the treatment of ADHD share a general common mechanism of action involving increased bioavailability of catecholamines such as dopamine and norepinephrine in the synapse and extraneuronal space (Dopheide 2009). Both methylphenidate and amphetamines inhibit the reuptake of dopamine and norepinephrine by blocking the transporters of each (NET and DAT, respectively). Amphetamines additionally promote the release of dopamine and norepinephrine. Atomoxetine is a selective norepinephrine reuptake inhibitor. While the onset of action and pharmacological effects of CNS stimulants are noticed almost immediately, the effects of atomoxetine take approximately two weeks to be noticed (Dopheide 2009).

Pharmacogenetic studies of these medications have evolved in a similar fashion as other areas of psychiatric pharmacogenomics beginning with candidate gene studies and a recent evolution to genome wide analyses. As anticipated from what is known about the pharmacodynamic actions of these medications, genes that influence catecholamine disposition or signaling have predominated in candidate gene studies. The most commonly studied gene in studies of stimulant response is the dopamine transporter gene (*SLC6A3/DAT1*). A VNTR element in the 3'UTR of the gene has been examined extensively (Contini 2013). The exact functional consequences of this VNTR are unclear, but it has been identified as a marker of risk for ADHD (Gizer 2009; Li 2006), brain function (Spencer 2013), and in some studies treatment response (Kambeitz 2014). When this data is looked at as a whole, there is distinct heterogeneity in the findings, with meta-analysis concluding that it is not a reliable predictor of treatment outcomes consistently across studies (Kambeitz 2014). While many studies have found relationships with response, and work in this area is ongoing, *SLC6A3* variation is not appropriate for guiding clinical decisions regarding drug selection or dosing at this time.

Other candidate genes that have been investigated in multiple studies include the dopamine-4 receptor gene (*DRD4*), the catechol-o-methyltransferase gene (*COMT*), norepinephrine transporter gene (*SLC6A2/NET1*), serotonin transporter gene (*SLC6A4/SERT*), alpha-2 adrenergeic receptor gene (*ADRA2A*). Of these genes, *ADRA2A* appears to have relatively consistent findings associating the G allele of the -1291C>G (rs1800544) SNP with better treatment response outcomes to some aspects of methylphenidate in three studies (Cheon 2009; da Silva 2008; Polanczyk 2007). All other candidate genes examined appear to have both positive and negative associations with response or tolerability to stimulants or atomoxetine. No pharmacodynamics genes are included in product labeling or have recommendations for clinical use as assessed by consensus guideline committees at this time.

Commentary and Future Work
Pharmacogenomic studies of CNS stimulants and non-stimulants for the treatment of ADHD follows a general

pattern consistent with other areas indicative of potential relationships yet with inconsistencies across studies that make extrapolating most findings to clinical use difficult at this time. Genetic variation in *CYP2D6* is included in the product labeling for atomoxetine as summarized earlier and in Table 8.1 with suggestions for dose adjustments in PM. The clinical uptake of this is slow due to the relatively lower utilization of this medication as compared to stimulants and also the duality of associations of PM with both increased risk for side effects and response. Work on pharmacodynamics genes requires additional information to describe the genetic influences on DAT function as well as other non-genetic factors that may influence response to stimulants. Continued investigations of other targets such as *SLC6A2/NET1* and *ADRA2A* along with larger scale genome-wide investigations are important.

Conclusions

There are currently over 30 neuropsychiatric medications with FDA-approved drug labeling or consensus treatment guidelines outlining the potential clinical utility of pharmacogenomic information (Drozda 2014). Many of these medications fall into the categories of antipsychotics, antidepressants, mood stabilizers, and stimulants or related medications to treat ADHD as outlined in this chapter. At this time, these pharmacogenetic markers almost exclusively represent variants related to drug metabolism. Notable exceptions include immune system gene variants as markers for increased risk for hypersensitivity reactions from carbamazepine. An abundance of work continues to characterize genetic variability in genes related to the pharmacodynamics of psychiatric medications, but the clinical implications of this work have not yet been realized. Treating psychiatric disorders is a challenging and complicated process with medications that may need to be used for extended periods of time. Using pharmacogenomics to better understand mechanisms underlying variability in response or tolerability has great potential to increase the quality of life for our patients as well as improve the clinician's ability to weigh the risks and benefits of potential treatments.

References

Alexanderson B, Evans DA, Sjoqvist F. Steady-state plasma levels of nortriptyline in twins: influence of genetic factors and drug therapy. Br Med J. 1969;4:764-8.

Amar A, Segman RH, Shtrussberg S, et al. An association between clozapine-induced agranulocytosis in schizophrenics and HLA-DQB1*0201. Int J Neuropsychopharmacol 1998;1:41-44.

Angst J. Effect of antidepressives and genetic factors. Arzneimittelforschung. 1964;14:SUPPL:496-500.

Arch JR. Central regulation of energy balance: inputs, outputs and leptin resistance. Proc Nutr Soc 2005;64:39-46.

Association AP. Diagnostic and statistical manual of mental health disorders: DSM-5. 5 ed. Washington, DC: American Psychiatric Publishing, 2013.

Bakker PR, van Harten PN, van Os J. Antipsychotic-induced tardive dyskinesia and polymorphic variations in COMT, DRD2, CYP1A2 and MnSOD genes: a meta-analysis of pharmacogenetic interactions. Mol Psychiatry 2008;13:544-56.

Bakker PR, van Harten PN, van Os J. Antipsychotic-induced tardive dyskinesia and the Ser9Gly polymorphism in the DRD3 gene: a meta analysis. Schizophr Res 2006;83:185-92.

Beech RD, Leffert JJ, Lin A, et al. Gene-expression differences in peripheral blood between lithium responders and non-responders in the Lithium Treatment-Moderate dose Use Study (LiTMUS). Pharmacogenomics J 2014;14:182-91.

Bertilsson L, Aberg-Wistedt A, Gustafsson LL, et al. Extremely rapid hydroxylation of debrisoquine: a case report with implication for treatment with nortriptyline and other tricyclic antidepressants. Ther Drug Monit 1985;7:478-80.

Bertilsson L, Mellstrom B, Sjokvist F, et al. Slow hydroxylation of nortriptyline and concomitant poor debrisoquine hydroxylation: clinical implications. Lancet 1981;1:560-1.

Bertilsson L, Tomson T. Clinical pharmacokinetics and pharmacological effects of carbamazepine and carbamazepine-10,11-epoxide. An update. Clin Pharmacokinet 1986;11:177-98.

Bialer M. Why are antiepileptic drugs used for nonepileptic conditions? Epilepsia 2012;53 Suppl 7:26-33.

Biernacka JM, McElroy SL, Crow S, et al. Pharmacogenomics of antidepressant induced mania: a review and meta-analysis of the serotonin transporter gene (5HTTLPR) association. J Affect Disord 2012;136:e21-9.

Bishop JR, Ellingrod VL, Akroush M, et al. The association of serotonin transporter genotypes and selective serotonin reuptake inhibitor (SSRI)-associated sexual side effects: possible relationship to oral contraceptives. Hum Psychopharmacol 2009;24:207-15.

Bishop JR, Ellingrod VL, Moline J, et al. Association between the polymorphic GRM3 gene and negative symptom improvement during olanzapine treatment. Schizophr Res 2005;77:253-60.

Bishop JR, Miller del D, Ellingrod VL, et al. Association between type-three metabotropic glutamate receptor gene (GRM3) variants and symptom presentation in treatment refractory schizophrenia. Hum Psychopharmacol 2011;26:28-34.

Biton V. Pharmacokinetics, toxicology and safety of lamotrigine in epilepsy. Expert Opin Drug Metab Toxicol 2006;2:1009-18.

Bollini P, Pampallona S, Orza MJ, et al. Antipsychotic drugs: is more worse? A meta-analysis of the published randomized control trials. Psychol Med 1994;24:307-16.

Boter H, Peuskens J, Libiger J, et al. Effectiveness of antipsychotics in first-episode schizophrenia and schizophreniform disorder on response and remission: an open randomized clinical trial (EUFEST). Schizophr Res 2009;115:97-103.

Brandl EJ, Frydrychowicz C, Tiwari AK, et al. Association study of polymorphisms in leptin and leptin receptor genes with antipsychotic-induced body weight gain. Prog Neuropsychopharmacol Bol Psychiatry 2012;38:134-41.

Brandl EJ, Tiwari AK, Zhou X, et al. Influence of CYP2D6 and CYP2C19 gene variants on antidepressant response in obsessive-compulsive disorder. Pharmacogenomics J 2014;14:176-81.

Brennan MD, Condra J. Transmission disequilibrium suggests a role for the sulfotransferase-4A1 gene in schizophrenia. Am J Med Genet B Neuropsychiatr Genet 2005;139B:69-72.

Brinkmann U, Eichelbaum M. Polymorphisms in the ABC drug transporter gene MDR1. Pharmacogenomics J 2001;1:59-64.

Chatzistefanidis D, Georgiou I, Kyritsis AP, et al. Functional impact and prevalence of polymorphisms involved in the hepatic glucuronidation of valproic acid. Pharmacogenomics 2012;13:1055-71.

Chen G, Lee R, Hojer AM, et al. pharmacokinetic drug interactions involving vortioxetine (Lu AA21004), a multimodal antidepressant. Clin Drug Investig 2013;33:727-36.

Cheon KA, Cho DY, Koo MS, et al. Association between homozygosity of a G allele of the alpha-2a-adrenergic receptor gene and methylphenidate response in Korean children and adolescents with attention-deficit/hyperactivity disorder. Biol Psychiatry 2009;65:564-70.

Chowdhury NI, Tiwari AK, Souza RP, et al. Genetic association study between antipsychotic-induced weight gain and the melanocortin-4 receptor gene. Pharmacogenomics J 2013;13:272-9.

Contini V, Rovaris DL, Victor MM, et al. Pharmacogenetics of response to methylphenidate in adult patients with Attention-Deficit/Hyperactivity Disorder (ADHD): a systematic review. Eur Neuropsychopharmacol 2013;23:555-60.

Czerwensky F, Leucht S, Steimer W. MC4R rs489693: a clinical risk factor for second generation antipsychotic-related weight gain? Int J Neuropsychopharmacol 2013;16:2103-9.

da Silva TL, Pianca TG, Roman T, et al. Adrenergic alpha2A receptor gene and response to methylphenidate in attention-deficit/hyperactivity disorder-predominantly inattentive type. J Neural Transm 2008;115:341-5.

Davis JM, Chen N. Dose response and dose equivalence of antipsychotics. J Clin Psychopharmacol 2004;24:192-208.

de la Torre R, Farre M, Navarro M, et al. Clinical pharmacokinetics of amfetamine and related substances: monitoring in conventional and non-conventional matrices. Clin Pharmacokinet 2004;43:157-85.

Dettling M, Cascorbi I, Opgen-Rhein C, et al. Clozapine-induced agranulocytosis in schizophrenic Caucasians: confirming clues for associations with human leukocyte class I and II antigens. Pharmacogenomics J 2007;7:325-32.

Dettling M, Cascorbi I, Roots I, et al. Genetic determinants of clozapine-induced agranulocytosis: recent results of HLA subtyping in a non-jewish caucasian sample. Arch Gen Psychiatry 2001;58:93-4.

Devlin AM, Ngai YF, Ronsley R, et al. Cardiometabolic risk and the MTHFR C677T variant in children treated with second-generation antipsychotics. Transl Psychiatry 2012;2:e71.

Dols A, Sienaert P, van Gerven H, et al. The prevalence and management of side effects of lithium and anticonvulsants as mood stabilizers in bipolar disorder from a clinical perspective: a review. Int Clin Psychopharmacol 2013;28:287-96.

Donnelly C, Bangs M, Trzepacz P, et al. Safety and tolerability of atomoxetine over 3 to 4 years in children and adolescents with ADHD. J Am Acad Child Adolesc Psychiatry 2009;48:176-85.

Dopheide JA, Pliszka SR. Attention-deficit-hyperactivity disorder: an update. Pharmacotherapy 2009;29:656-79.

Drozda K, Muller DJ, Bishop JR. Pharmacogenomic testing for neuropsychiatric drugs: current status of drug labeling, guidelines for using genetic information, and test options. Pharmacotherapy 2014;34:166-84.

Ellingrod VL, Taylor SF, Dalack G, et al. Risk factors associated with metabolic syndrome in bipolar and schizophrenia subjects treated with antipsychotics: the role of folate pharmacogenetics. J Clin Psychopharmacol 2012;32:261-5.

Falkenberg VR, Gurbaxani BM, Unger ER, et al. Functional genomics of serotonin receptor 2A (HTR2A): interaction of polymorphism, methylation, expression and disease association. Neuromolecular Med 2011;13:66-76.

FDA U. Drugs@FDA Database. Silver Spring, MD: United States Food and Drug Administration; 2014 [cited 2014 5/14/2014]; Available from: http://www.fda.gov/Drugs/InformationOnDrugs/default.htm.

Fijal BA, Kinon BJ, Kapur S, et al. Candidate-gene association analysis of response to risperidone in African-American and white patients with schizophrenia. Pharmacogenomics J 2009;9:311-8.

Gex-Fabry M, Eap CB, Oneda B, et al. CYP2D6 and ABCB1 genetic variability: influence on paroxetine plasma level and therapeutic response. Ther Drug Monit 2008;30:474-82.

Girgis RR, Xu X, Miyake N, et al. In vivo binding of antipsychotics to D3 and D2 receptors: a PET study in baboons with [11C]-(+)-PHNO. Neuropsychopharmacology2011;36:887-95.

Gitlin MJ, Swendsen J, Heller TL, et al. Relapse and impairment in bipolar disorder. Am J Psychiatry 1995;152:1635-40.

Gizer IR, Ficks C, Waldman ID. Candidate gene studies of ADHD: a meta-analytic review. Hum Genet 2009;126:51-90.

Graff-Guerrero A, Mizrahi R, Agid O, et al. The dopamine D2 receptors in high-affinity state and D3 receptors in schizophrenia: a clinical [11C]-(+)-PHNO PET study. Neuropsychopharmacology 2009;34:1078-86.

Grasmader K, Verwohlt PL, Kuhn KU, et al. Population pharmacokinetic analysis of mirtazapine. Eur J Clin Pharmacol 2004;60:473-80.

Grossman I, Sullivan PF, Walley N, et al. Genetic determinants of variable metabolism have little impact on the clinical use of leading antipsychotics in the CATIE study. Genet Med 2008;10:720-9.

Grover S, Kukreti R. HLA alleles and hypersensitivity to carbamazepine: an updated systematic review with meta-analysis. Pharmacogenet Genomics 2014;24:94-112.

Guillemette C, Levesque E, Harvey M, et al. UGT genomic diversity: beyond gene duplication. Drug Metab Rev 2010;42:24-44.

Gulcebi MI, Ozkaynakci A, Goren MZ, et al. The relationship between UGT1A4 polymorphism and serum concentration of lamotrigine in patients with epilepsy. Epilepsy Res 2011;95:1-8.

Haerian BS, Roslan H, Raymond AA, et al. ABCB1 C3435T polymorphism and the risk of resistance to antiepileptic drugs in epilepsy: a systematic review and meta-analysis. Seizure 2010;19:339-46.

Harris EC, Barraclough B. Suicide as an outcome for mental disorders. A meta-analysis. Br J Psychiatry 1997;170:205-28.

Heils A, Teufel A, Petri S, et al. Allelic variation of human serotonin transporter gene expression. J Neurochem 1996;66:2621-4.

Hicks JK, Swen JJ, Thorn CF, et al. Clinical Pharmacogenetics Implementation Consortium guideline for CYP2D6 and CYP2C19 genotypes and dosing of tricyclic antidepressants. Clin Pharmacol Ther 2013;93:402-8.

Hranilovic D, Stefulj J, Schwab S, et al. Serotonin transporter promoter and intron 2 polymorphisms: relationship between allelic variants and gene expression. Biol Psychiatry 2004;55:1090-4.

Huezo-Diaz P, Perroud N, Spencer EP, et al. CYP2C19 genotype predicts steady state escitalopram concentration in GENDEP. J Psychopharmacol 2012;26:398-407.

Hung CC, Chang WL, Ho JL, et al. Association of polymorphisms in EPHX1, UGT2B7, ABCB1, ABCC2, SCN1A and SCN2A genes with carbamazepine therapy optimization. Pharmacogenomics 2012;13:159-69.

Johnson KA, Barry E, Lambert D, et al. Methylphenidate side effect profile is influenced by genetic variation in the attention-deficit/hyperactivity disorder-associated CES1 gene. J Child Adolesc Psychopharmacol 2013;23:655-64.

Kambeitz J, Romanos M, Ettinger U. Meta-analysis of the association between dopamine transporter genotype and response to methylphenidate treatment in ADHD. Pharmacogenomics J 2014;14:77-84.

Kao AC, Muller DJ. Genetics of antipsychotic-induced weight gain: update and current perspectives. Pharmacogenomics 2013;14:2067-83.

Kato M, Serretti A. Review and meta-analysis of antidepressant pharmacogenetic findings in major depressive disorder. Mol Psychiatry 2010;15:473-500.

Kinon BJ, Ahl J, Stauffer VL, et al. Dose response and atypical antipsychotics in schizophrenia. CNS Drugs 2004;18:597-616.

Kuzman MR, Muller DJ. Association of the MTHFR gene with antipsychotic-induced metabolic abnormalities in patients with schizophrenia. Pharmacogenomics 2012;13:843-6.

Lee AK, Bishop JR. Pharmacogenetics of leptin in antipsychotic-associated weight gain and obesity-related complications. Pharmacogenomics 2011;12:999-1016.

Lencer R, Bishop JR, Harris MS, et al. Association of variants in DRD2 and GRM3 with motor and cognitive function in first-episode psychosis. Eur Arch Psychiatry Clin Neurosci 2013.

Lerer B, Segman RH, Fangerau H, et al. Pharmacogenetics of tardive dyskinesia: combined analysis of 780 patients supports association with dopamine D3 receptor gene Ser9Gly polymorphism. Neuropsychopharmacology 2002;27:105-19.

Lerer B, Segman RH, Tan EC, et al. Combined analysis of 635 patients confirms an age-related association of the serotonin 2A receptor gene with tardive dyskinesia and specificity for the non-orofacial subtype. Int J Neuropsychopharmacol 2005;8:411-25.

Lesch KP, Bengel D, Heils A, et al. Association of anxiety-related traits with a polymorphism in the serotonin transporter gene regulatory region. Science 1996;274:1527-31.

Li D, Sham PC, Owen MJ, et al. Meta-analysis shows significant association between dopamine system genes and attention deficit hyperactivity disorder (ADHD). Hum Mol Genet 2006;15:2276-84.

Li Z, Page A, Martin G, et al. Attributable risk of psychiatric and socio-economic factors for suicide from individual-level, population-based studies: a systematic review. Soc Sci Med. 2011;72:608-16.

Lieberman JA, Stroup TS, McEvoy JP, et al. Effectiveness of antipsychotic drugs in patients with chronic schizophrenia. N Engl J Med 2005;353:1209-23.

Lieberman JA, Yunis J, Egea E, et al. HLA-B38, DR4, DQw3 and clozapine-induced agranulocytosis in Jewish patients with schizophrenia. Arch Gen Psychiatry 1990;47:945-8.

Machado-Vieira R, Ibrahim L, Zarate CA, Jr. Histone deacetylases and mood disorders: epigenetic programming in gene-environment interactions. CNS Neurosci Ther 2011;17:699-704.

Makmor-Bakry M, Sills GJ, Hitiris N, et al. Genetic variants in microsomal epoxide hydrolase influence carbamazepine dosing. Clin Neuropharmacol 2009;32:205-12.

Malhotra AK, Correll CU, Chowdhury NI, et al. Association between common variants near the melanocortin 4 receptor gene and severe antipsychotic drug-induced weight gain. Arch Gen Psychiatry 2012;69:904-12.

McEvoy JP, Lieberman JA, Perkins DO, et al. Efficacy and tolerability of olanzapine, quetiapine, and risperidone in the treatment of early psychosis: a randomized, double-blind 52-week comparison. Am J Psychiatry 2007;164:1050-60.

Meltzer HY. Update on typical and atypical antipsychotic drugs. Annu Rev Med 2013;64:393-406.

Meltzer HY, Brennan MD, Woodward ND, et al. Association of Sult4A1 SNPs with psychopathology and cognition in patients with schizophrenia or schizoaffective disorder. Schizophr Res 2008;106:258-64.

Meng H, Guo G, Ren J, et al. Effects of ABCB1 polymorphisms on plasma carbamazepine concentrations and pharmacoresistance in Chinese patients with epilepsy. Epilepsy Behav 2011;21:27-30.

Merikangas KR, Jin R, He JP, et al. Prevalence and correlates of bipolar spectrum disorder in the world mental health survey initiative. Arch Gen Psychiatry 2011;68:241-51.

Meshul CK, Casey DE. Regional, reversible ultrastructural changes in rat brain with chronic neuroleptic treatment. Brain Res 1989;489:338-46.

Michelson D, Read HA, Ruff DD, et al. CYP2D6 and clinical response to atomoxetine in children and adolescents with ADHD. J Am Acad Child Adolesc Psychiatry 2007;46:242-51.

Moghaddam B, Javitt D. From revolution to evolution: the glutamate hypothesis of schizophrenia and its implication for treatment. Neuropsychopharmacology 2012;37:4-15.

Mossner R, Schuhmacher A, Kuhn KU, et al. Functional serotonin 1A receptor variant influences treatment response to atypical antipsychotics in schizophrenia. Pharmacogenet Genomics 2009;19:91-4.

Mrazek DA, Biernacka JM, O'Kane DJ, et al. CYP2C19 variation and citalopram response. Pharmacogenet Genomics 2011;21:1-9.

Muller DJ, Kekin I, Kao AC, et al. Towards the implementation of CYP2D6 and CYP2C19 genotypes in clinical practice: update and report from a pharmacogenetic service clinic. Int Rev Psychiatry 2013;25:554-71.

Muller DJ, Kennedy JL. Genetics of antipsychotic treatment emergent weight gain in schizophrenia. Pharmacogenomics 2006;7:863-87.

Muller MJ, Dragicevic A, Fric M, et al. Therapeutic drug monitoring of tricyclic antidepressants: how does it work under clinical conditions? Pharmacopsychiatry 2003;36:98-104.

Myung W, Lim SW, Kim S, et al. Serotonin transporter genotype and function in relation to antidepressant response in Koreans. Psychopharmacology (Berl) 2013;225:283-90.

Nemoda Z, Angyal N, Tarnok Z, et al. Carboxylesterase 1 gene polymorphism and methylphenidate response in ADHD. Neuropharmacology 2009;57:731-3.

Newcomer JW, Weiden PJ, Buchanan RW. Switching antipsychotic medications to reduce adverse event burden in schizophrenia: establishing evidence-based practice. J Clin Psychiatry 2013;74:1108-20.

Niitsu T, Fabbri C, Bentini F, et al. Pharmacogenetics in major depression: a comprehensive meta-analysis. Prog Neuropsychopharmacol Bol Psychiatry 2013;45:183-94.

O'Reilly RL, Bogue L, Singh SM. Pharmacogenetic response to antidepressants in a multicase family with affective disorder. Biol Psychiatry 1994;36:467-71.

Pare CM, Mack JW. Differentiation of two genetically specific types of depression by the response to antidepressant drugs. J Med Genet 1971;8:306-9.

Perlis RH, Mischoulon D, Smoller JW, et al. Serotonin transporter polymorphisms and adverse effects with fluoxetine treatment. Biol Psychiatry 2003;54:879-83.

Perlis RH, Ostacher MJ, Patel JK, et al. Predictors of recurrence in bipolar disorder: primary outcomes from the Systematic Treatment Enhancement Program for Bipolar Disorder (STEP-BD). Am J Psychiatry 2006;163:217-24.

Peters EJ, Slager SL, Kraft JB, et al. Pharmacokinetic genes do not influence response or tolerance to citalopram in the STAR*D sample. PLoS One 2008;3:e1872.

Pettegrew JW, Panchalingam K, McClure RJ, et al. Effects of chronic lithium administration on rat brain phosphatidylinositol cycle constituents, membrane phospholipids and amino acids. Bipolar Disord 2001;3:189-201.

Polanczyk G, Zeni C, Genro JP, et al. Association of the adrenergic alpha2A receptor gene with methylphenidate improvement of inattentive symptoms in children and adolescents with attention-deficit/hyperactivity disorder. Arch Gen Psychiatry 2007;64:218-24.

Porcelli S, Drago A, Fabbri C, et al. Pharmacogenetics of antidepressant response. J Psychiatry Neurosci 2011;36:87-113.

Puranik YG, Birnbaum AK, Marino SE, et al. Association of carbamazepine major metabolism and transport pathway gene polymorphisms and pharmacokinetics in patients with epilepsy. Pharmacogenomics 2013;14:35-45.

Ramamoorthy Y, Tyndale RF, Sellers EM. Cytochrome P450 2D6.1 and cytochrome P450 2D6.10 differ in catalytic activity for multiple substrates. Pharmacogenetics 2001;11:477-87.

Ramamoorthy Y, Yu AM, Suh N, et al. Reduced (+/-)-3,4-methylenedioxymethamphetamine ("Ecstasy") metabolism with cytochrome P450 2D6 inhibitors and pharmacogenetic variants in vitro. Biochem Pharmacol 2002;63:2111-9.

Rambeck B, Wolf P. Lamotrigine clinical pharmacokinetics. Clin Pharmacokinet 1993;25:433-43.

Ramsey TL, Meltzer HY, Brock GN, et al. Evidence for a SULT4A1 haplotype correlating with baseline psychopathology and atypical antipsychotic response. Pharmacogenomics 2011;12:471-80.

Rasmussen BB, Brosen K. Is therapeutic drug monitoring a case for optimizing clinical outcome and avoiding interactions of the selective serotonin reuptake inhibitors? Ther Drug Monit 2000;22:143-54.

Reynolds GP, Arranz B, Templeman LA, et al. Effect of 5-HT1A receptor gene polymorphism on negative and depressive symptom response to antipsychotic treatment of drug-naive psychotic patients. Am J Psychiatry 2006;163:1826-9.

Reynolds GP, Templeman LA, Zhang ZJ. The role of 5-HT2C receptor polymorphisms in the pharmacogenetics of antipsychotic drug treatment. Prog Neuropsychopharmacol Bol Psychiatry 2005;29:1021-8.

Reynolds GP, Zhang ZJ, Zhang XB. Association of antipsychotic drug-induced weight gain with a 5-HT2C receptor gene polymorphism. Lancet 2002;359:2086-7.

Rudberg I, Hermann M, Refsum H, et al. Serum concentrations of sertraline and N-desmethyl sertraline in relation to CYP2C19 genotype in psychiatric patients. Eur J Clin Pharmacol 2008A;64:1181-8.

Rudberg I, Mohebi B, Hermann M, et al. Impact of the ultrarapid CYP2C19*17 allele on serum concentration of escitalopram in psychiatric patients. Clin Pharmacol Ther 2008B;83:322-7.

Rush AJ, Trivedi MH, Wisniewski SR, et al. Bupropion-SR, sertraline, or venlafaxine-XR after failure of SSRIs for depression. N Engl J Med 2006;354:1231-42.

Rybakowski JK. Genetic influences on response to mood stabilizers in bipolar disorder: current status of knowledge. CNS Drugs 2013;27:165-73.

Sauer JM, Ring BJ, Witcher JW. Clinical pharmacokinetics of atomoxetine. Clin Pharmacokinet 2005;44:571-90.

Schulze TG, Alda M, Adli M, et al. The International Consortium on Lithium Genetics (ConLiGen): an initiative by the NIMH and IGSLI to study the genetic basis of response to lithium treatment. Neuropsychobiology 2010;62:72-8.

Serretti A, Drago A, De Ronchi D. HTR2A gene variants and psychiatric disorders: a review of current literature and selection of SNPs for future studies. Curr Med Chem 2007;14:2053-69.

Sicard MN, Zai CC, Tiwari AK, et al. Polymorphisms of the HTR2C gene and antipsychotic-induced weight gain: an update and meta-analysis. Pharmacogenomics 2010;11:1561-71.

Silverstone PH, Wu RH, O'Donnell T, et al. Chronic treatment with both lithium and sodium valproate may normalize phosphoinositol cycle activity in bipolar patients. Hum Psychopharmacol 2002;17:321-7.

Singkham N, Towanabut S, Lertkachatarn S, et al. Influence of the UGT2B7 -161C>T polymorphism on the population pharmacokinetics of lamotrigine in Thai patients. Eur J Clin Pharmacol 2013;69:1285-91.

Sokoloff P, Giros B, Martres MP, et al. Molecular cloning and characterization of a novel dopamine receptor (D3) as a target for neuroleptics. Nature 1990;347:146-51.

Spencer TJ, Biederman J, Faraone SV, et al. Functional genomics of attention-deficit/hyperactivity disorder (ADHD) risk alleles on dopamine transporter binding in ADHD and healthy control subjects. Biol Psychiatry 2013;74:84-9.

Spina E, Santoro V, D'Arrigo C. Clinically relevant pharmacokinetic drug interactions with second-generation antidepressants: an update. Clin Ther 2008;30:1206-27.

Statistics NCfH. Health, United States, 2012: With Special Feature on Emergency Care. National Center for Health Statistics [Internet]. 2013 May 14, 2014.

Stevenson JM, Bishop JR. Antidepressant tolerability and potential clinical implications of serotonin-2A genotypes. Clinical Pharmacology and Biopharmaceutics 2013;2.

Strohmaier J, Wust S, Uher R, et al. Sexual dysfunction during treatment with serotonergic and noradrenergic antidepressants: clinical description and the role of the 5-HTTLPR. World J Biol Psychiatry 2011;12:528-38.

Sun Z, Murry DJ, Sanghani SP, et al. Methylphenidate is stereoselectively hydrolyzed by human carboxylesterase CES1A1. J Pharmacol Exp Ther 2004;310:469-76.

Suzuki Y, Sugai T, Fukui N, et al. CYP2D6 genotype and smoking influence fluvoxamine steady-state concentration in Japanese psychiatric patients: lessons for genotype-phenotype association study design in translational pharmacogenetics. J Psychopharmacol 2011;25:908-14.

Swen JJ, Nijenhuis M, de Boer A, et al. Pharmacogenetics: from bench to byte--an update of guidelines. Clin Pharmacol Ther 2011;89:662-73.

Thompson J, Thomas N, Singleton A, et al. D2 dopamine receptor gene (DRD2) Taq1 A polymorphism: reduced dopamine D2 receptor binding in the human striatum associated with the A1 allele. Pharmacogenetics 1997;7:479-84.

Tomkins DM, Otton SV, Joharchi N, et al. Effect of CYP2D1 inhibition on the behavioural effects of d-amphetamine. Behav Pharmacol 1997;8:223-35.

Trivedi MH, Fava M, Wisniewski SR, et al. Medication augmentation after the failure of SSRIs for depression. N Engl J Med 2006B;354:1243-52.

Trivedi MH, Rush AJ, Wisniewski SR, et al. Evaluation of outcomes with citalopram for depression using measurement-based care in STAR*D: implications for clinical practice. Am J Psychiatry 2006A;163:28-40.

Tsai HT, Caroff SN, Miller DD, et al. A candidate gene study of Tardive dyskinesia in the CATIE schizophrenia trial. Am J Med Genet B Neuropsychiatr Genet 2010A;153B:336-40.

Tsai HT, North KE, West SL, et al. The DRD3 rs6280 polymorphism and prevalence of tardive dyskinesia: a meta-analysis. Am J Med Genet B Neuropsychiatr Genet 2010B;153B:57-66.

Tsai MH, Lin KM, Hsiao MC, et al. Genetic polymorphisms of cytochrome P450 enzymes influence metabolism of the antidepressant escitalopram and treatment response. Pharmacogenomics 2010;11:537-46.

Valevski A, Klein T, Gazit E, et al. HLA-B38 and clozapine-induced agranulocytosis in Israeli Jewish schizophrenic patients. Eur JImmunogenet 1998;25:11-3.

Van Nieuwerburgh FC, Denys DA, Westenberg HG, et al. Response to serotonin reuptake inhibitors in OCD is not influenced by common CYP2D6 polymorphisms. Int J Psychiatry Clin Pract 2009;13:345-48.

Wang L, Fang C, Zhang A, et al. The -1019 C/G polymorphism of the 5-HT(1)A receptor gene is associated with negative symptom response to risperidone treatment in schizophrenia patients. J Psychopharmacol 2008;22:904-9.

Wang PS, Lane M, Olfson M, et al. Twelve-month use of mental health services in the United States: results from the National Comorbidity Survey Replication. Arch Gen Psychiatry. 2005;62:629-40.

Watanabe J, Suzuki Y, Fukui N, et al. Dose-dependent effect of the CYP2D6 genotype on the steady-state fluvoxamine concentration. Ther Drug Monit 2008;30:705-8.

Whiteford HA, Degenhardt L, Rehm J, et al. Global burden of disease attributable to mental and substance use disorders: findings from the Global Burden of Disease Study 2010. Lancet 2013;382:1575-86.

Wu R, Zhao J, Shao P, et al. Genetic predictors of antipsychotic-induced weight gain: a case-matched multi-gene study. Zhong Nan Da Xue Xue Bao Yi Xue Ban 2011;36:720-3.

Yoshimasu K, Kiyohara C, Miyashita K, et al. Suicidal risk factors and completed suicide: meta-analyses based on psychological autopsy studies. Environ Health Prev Med. 2008;13:243-56.

Yun W, Zhang F, Hu C, et al. Effects of EPHX1, SCN1A and CYP3A4 genetic polymorphisms on plasma carbamazepine concentrations and pharmacoresistance in Chinese patients with epilepsy. Epilepsy Res 2013;107:231-7.

Yunis JJ, Corzo D, Salazar M, et al. HLA associations in clozapine-induced agranulocytosis. Blood 1995;86:1177-83.

Zai CC, Hwang RW, De Luca V, et al. Association study of tardive dyskinesia and twelve DRD2 polymorphisms in schizophrenia patients. Int J Neuropsychopharmacol 2007;10:639-51.

Zhou J, Argikar UA, Remmel RP. Functional analysis of UGT1A4(P24T) and UGT1A4(L48V) variant enzymes. Pharmacogenomics 2011;12:1671-9.

Zhu HJ, Patrick KS, Yuan HJ, et al. Two CES1 gene mutations lead to dysfunctional carboxylesterase 1 activity in man: clinical significance and molecular basis. Am J Hum Genet 2008;82:1241-8.

Chapter 9

GENETICS IN PAIN DISORDERS

Anita Gupta, D.O., Pharm.D.; and Lisa K. Lee, M.D.

Learning Objectives
1. Understand how genes are related to pain disorders and how sequence variations occur in a gene cause pain disorders.
2. Understand what lumbar disc degeneration is and learn about the genetic factors of the disorder.
3. Understand how migraine disorder occurs.
4. Learn about fibromyalgia and association between PTSD.
5. Understand the types of HSAN and the differences between the types of HSAN.
6. Learn different types of genes that are involved in pain and how sequence variations result in pain disorders.
7. Understand pharmacogenetic testing and ethical usage.

Keywords: Pharmacogenetics, pain disorders, genetics, pharmacogenetic testing, PGx guided therapy, LDD, migraine, HSAN, ethnicity

Abbreviations in This Chapter

CNS	Congenital Sensory Neuropathy	HSAN	Hereditary Sensory and Autonomic Neuropathy
CIPA	Congenital Insensitivity to Pain with Anhidrosis	IEM	Inherited Erythromelalgia
CIP	Congenital Indifference to Pain	LDD	Lumbar Disc Degeneration
CPIC	Clinical Pharmacogenetics Implementation Consortium	MEFV	Mediterranean Fever Gene
		PEPD	Paroxysmal Extreme Pain Disorder
COMT	Catechol-O-methyl tranferase		
CT	Computed Tomography	PGT	Pharmacogenetic Testing
EEG	Electroencephalogram	PGx	Pharmacogenomics
FD	Familial Dysautonomia	PTSD	Post Traumatic Stress Disorder
FHM	Familial Hemiplegic Migraine	SNP	Single Nucleotide Polymorphism
GI	Gastrointestinal Tract		

Abstract

Pain perception resulting from different chronic pain disorders may differ considerably between patients. Genotypic characteristics, along with the environmental factors and aging, contribute a significant portion of the cause of pain disorders. As a result, identification of genetic factors of pain and analgesic drug usage to treat pain disorders have been a concern within the past 15 years. To increase drug efficiency for the treatment of pain disorders, current research suggests obtainment of a patients' genetic make up to help in determining, the right drug and dosage for that patient. How a patient handles a drug is often based on genetic differences that affect two factors: pharmacokinetics, which involves with the metabolic activation or inactivation of drugs in the body and pharmacodynamics, which is the positive or negative response of the body to the given drug (Sadhasivam 2012).

Introduction

The occurrence and treatment of pain disorders have a correlation with patients' genetic characteristics. Genes that encode for certain enzymes provide information to the brain about the level of pain perception, which can vary by ethnic group. Previous research showed that some of the pain disorders potentially occur due to sequence variations occurring in genes that are pain related, resulting in chronic pain such as migraines, and/or increases or decreases in an individual's perception of sensitivity to pain. Many pain disorders can be treated with drugs. However, statistics show that each year billions of dollars are wasted on drugs that do not work in patients with certain genetic characteristics, as these patients do not receive full benefit from the drugs resulting in a loss of recovery from the illness. In addition, for some of the patients with specific genetic variants, the incidence of medication related adverse drug reactions (ADRs) negatively affects overall outcomes. This current research has sought to explore new solutions to the treatment of different pain disorders. Personalized medicine is a new paradigm for treating individuals who have had many failures with drugs and also provides a potential approach where medication efficacy is seen early on in treatment thus preventing poorer outcomes due to untreated pain syndromes. Specifically pharmacogenetic testing (PGT) is exceedingly important for the proper management of pain because finding the right drug and dose based on the genomic make up of the patient is critically important. The groundbreaking development of pharmacogenetic testing provides more individualized drug treatment while reducing the adverse events.

The human genome mapping research project paved the way to a wealth of information. Knowing the genetic code has helped scientists unlock many mysteries about diseases and treatments. And this research shows that of all the factors that alter a patient's response to drugs (such as age, sex, weight, general health, and liver function), genetic factors account for a substantial proportion. As part of the human body, it is your genes that provide your body specific instruction for making the enzymes necessary for drug metabolism as well as pain perception. Differences in the enzymes critical to drug metabolism can affect your body's ability to metabolize (break down) a drug as well as how long the drug stays your body, all culminating to affect how well drugs may work in an individual. In particular, common pain medications require activation by the enzyme CYP2D6 to become effective. Approximately half of patients have genes that alter the function of CYP2D6. Testing for these gene alterations allows for alteration of dosage regimens to compensate for altered metabolism and optimize the safety and efficacy of pain medications. Without knowing an individual's specific genetic code, physicians may often need to go through months of trial-and-error prescribing to find the right drug and dose. Physicians are often baffled when a drug will work for one person but not for another with the same diagnosis. The fact of the matter is that physicians really do not know how to predict drug effectiveness or toxicity because everyone is different. Therefore, pharmacogenetic testing helps assess drug responsiveness, using the individual's genes to serve as a map that guides physicians.

By utilizing a pharmacogenomics approach, it is hoped that physicians would be able to anticipate how one may respond to a drug instead of relying on a trial-and-error process. By knowing the specific way one may break-down drugs, a physician can tailor treatment according to an individual's unique metabolism and immediately find the right drug. Not only will this information help physicians predict which drug will best treat pain, a physician will also be able to predict the effective dose and potential for toxicity. In theory, this knowledge has the potential to save time, money, and lives.

With this in mind, the goal of this chapter is to describe some of the current research concerning the genetics of pain, as well as the role of pharmacogenomics in the treatment of pain.

Low Back Pain

Lumbar disc degeneration (LDD) is a common cause of low back pain, and has often been labeled as a product of old age. This commonly held assumption is not surprising given that the lumbar intervertebral discs have been shown to become worn down and dehydrated as the aging process occurs (Williams 2012). However, in recent years, research has begun to suggest that there are additional contributors to LDD besides age (Cheung 2010), which include nutritional, mechanical, and genetic factors. In looking at the genetic factors, we can see differences in both the structural function (responsible for maintaining the stability of the lumbar discs), and regulatory (responsible for the metabolism of the cells) affect LDD (Cheung 2010). Researchers have identified the methylation of the PARK2 (which is responsible for the synthesis of a protein called *parkin*) as an important genetic factor. Overall, parkin has many significant roles that include degradation of excess and damaged proteins via the ubiquitination pathway and the maintenance of mitochondria. It can also function as a tumor suppressor protein (Williams 2012).

In a more recent capacity, researchers have been using the association between low back pain and genetics and concentrating on the pain and its relation to genetics, rather than attempting to cure the disease that causes the pain. In a recent study conducted by Dai et al., genetic variation in the *PARK2* gene was used to identify a linkage between differences in protein expression and decreases in pain after LDD surgery. The results certainly provide new routes for alleviating low back pain (Cheung 2010) and may provide us with new therapeutic targets as we work to prevent pain related to LDD.

Migraine Disorder

In general, migraines are episodic neurovascular disorders and are characterized as common, chronic, and debilitating (Kors 2003; de Vries 2009). The occurrence or inheritance of a migraine disorder follows an autosomal dominant pattern, and because nearly 90% of patients afflicted by migraines have a family history of the condition, the genetic precedence is clear (Pestka 2013). However, identifying the genetic factors that eventually cause a migraine is quite a complex process (Kors 2003). Most of the genetic studies concerning migraines have been conducted using a rare strain, familial hemiplegic migraine (FHM) (de Vries 2009). The genes identified as part of this work appear to code for ion transporters such as *CACNA1A, ATP1A2,* and *SCN1A* and the mutations that occur in these genes appear to cause disruptions in neurotransmitter-ion gated channel levels, which have been hypothesized as a possible cause for common migraines (de Vries 2009).

While the work on FHM is promising, research on the genetic influences behind more common strains of migraines has not been promising, which is due to the broad phenotypes associated with migraines, including headaches, nausea, and visual disturbances (de Vries 2009; Kors 2003). Additionally, because the headaches produced by common strains of migraines are of neurovascular origin, the genetics behind both the powerful vasodilation and also the activation of the trigeminal nerve must be taken into account (de Vries 2009). A suggested solution for this better defines the phenotype to be examined by creating more subcategories and specifications for the headaches associated with each strain (de Vries 2009). However, within the past years, no progress has been made to update the phenotypes categories associated with migraines. Although the study of FHM has produced more insight into the genetic background of migraines the genetic results cannot be successfully translated to common migraine strains.

Fibromyalgia

Fibromyalgia is a common pain condition that is often attributed to a combination between genetic and environmental processes as determined by familial aggregation (Buskila 2005). The study of the genetics behind fibromyalgia has been conducted in two ways. One is by looking not at the disease itself, but diseases that accompany fibromyalgia and display similar phenotypes. For example, patients with post traumatic stress disorder (PTSD), which was thought to be a heritable disorder, are likely to also have fibromyalgia. Therefore, studying the genetic factors behind PTSD could lead to more conclusive genetic factors associated with fibromyalgia (Buskila 2005).

The second approach involves studying the genetic polymorphisms associated with fibromyalgia. Studies demonstrate that a missense mutation in the *MEFV* genes has an association with the cause of fibromyalgia (Feng 2009). Additionally work has centered on the genetic polymorphisms that affect the CNS serotonergic and catecholaminergic (particularly dopaminergic) processes, as the medication often used to treat fibromyalgia, pharmacology affect these systems (Guymer 2013; Buskila 2005). It does appear in looking at this research that a combination of the two approaches may prove to be successful in terms of studying the genetic factors behind fibromyalgia. In looking at this work related to PTSD, the ether dopamine transporter gene (*DAT1*) may be associated with the fact that dopaminergic neurotransmissions have a role in fibromyalgia (Buskila 2005). However, even with these developments, additional research must be conducted that considers other linking causes of fibromyalgia, such as environmental factors (Buskila 2005). This is exceedingly important as the statistics show that 3.4% of the patients with fibromyalgia are women, whereas 0.5% of the patients are men and interestingly, fibromyalgia occurs in nearly all countries and ethnic groups (Lempp 2009). Additionally, its ability to coexist with other diseases, such as PTSD, provides for not only difficult diagnosing, but for difficult determination of precise pathophysiology (Guymer 2013). Therefore, clearly more work is needed before the genetics of this illness can be identified.

Hereditary Disorders Insensitive to Pain

HSANs (hereditary sensory and autonomic neuropathies) are a diverse group of genetic disorders, which result in altered pain perception of varying degrees. At the time of this writing, there are six main groups with 11 genes having been identified as involved with this disease process (Takashima 2013). The frequency of HSAN differs depending on the type of the disorder.

HSAN Type I

Hereditary sensory radicular neuropathy is a rare autosomal dominant disorder that frequently presents in the second decade of life (Axelrod 2002). Sensorimotor axonal neuropathy is accompanied by painless injuries, chronic skin ulcers, and distal amputations. The genes involved code for the subunits of the enzyme serine palmitoyltransferase SPTLC1 (Garofalo 2011) and SPTLC2 (Rotthier 2010). Serine palmitoyltransferase catalyzes the reaction between L-serine and palmitoyl-CoA to form sphinganine, which eventually forms ceramides. Ceramides are important for the synthesis of cerebrosides, which are an important component of muscle and nerve cell membranes. Five missense mutations have been identified in the *SPTLC1* gene: C133W, C133Y, S331F, A352V, and V144D (Garofalo 2011). The C133W was the most common variation that was found in 18 out of 24 documented families with HSAN1 (Penno 2010). In the *SPTLC2* gene the following variations have been documented: G382V, V359M, and I504F. These allelic variants encode enzymes with decreased activity in vitro with respect to the use of L-serine as the substrate; these also have altered substrate selectivity toward L-alanine or L-glycine in the synthesis of sphinganine. This results in the formation and accumulation of abnormal sphingolipids that are thought to be the cause of HSAN I (Penno 2010). Not surprisingly, patients with HSAN Type I had a significantly decreased number of myelinated fibers in sural nerve biopsies. On the one hand, remaining fibers showed evidence of demyelinating damage, while unmyelinated fibers remained relatively normal in numbers (Houlden 2006).

HSAN Type II CNS or Congenital Sensory Neuropathy

Patients with this disorder have profound sensory loss and marked hypotonia. Onset of HSAN Type II occurs in infancy and the clinical course is non-progressive. This disease is inherited in an autosomal recessive pattern and has been documented in multiple ethnicities (Rotthier 2012). Sweating may be partially impaired but not as profound as in Congenital Insensitivity to Pain with Anhidrosis (CIPA) HSAN Type IV. Autonomic dysfunction is common, as is GERD and feeding problems. Corneal, gag, and deep tendon reflexes are diminished (Axelrod 2002).

HSAN Type II has been linked to sequence variations in WNK1, which is a serine/threonine kinase that plays an important role in the regulation of electrolyte homeostasis and FAM134B, which is a protein that may be necessary for the long-term survival of nociceptive and autonomic ganglion neurons, with additionally sequence variations in KIF1A (Rotthier 2012). Mutations in WNK1 are postulated to cause an increase in neuronal membrane excitability, which in excess can lead to cell cytotoxicity and neuronal death as a possible mechanism for HSAN Type II. The role of FAM134B and KIF1A in HSAN Type II is not well defined, but in mice silencing Fam134b resulted in apoptosis of small- and medium-sized neurons. KIF1A is a molecular motor protein associated with axonal transport of synaptic vesicle precursors and may be involved with the transport of WNK1, but its role in the development of HSAN Type II is unknown (Rotthier 2012).

HSAN Type III

Familial dysautonomia (FD) also known as Riley-Day is an autosomal recessive disorder that mainly affects one out of every 27 individuals of Ashkenazi Jewish descent. The incidence in American and Israeli Jews is about 1:3700 (Norcliffe-Kaufmann 2012). HSAN Type III is characterized by alacrima, absent fungiform papillae, depressed deep tendon reflexes, decreased pain perception that spares the neck, soles of the feet and genital area, and abnormal histamine test results, with a lack of axon flare after intradermal histamine injection of the patients with Ashkenazi Jewish ancestry.

Neuropathological studies in these individuals show markedly diminished unmyelinated neurons and sympathetic ganglia, with the size of these ganglia also being significantly smaller in diameter. Individuals with HSPA Type III can suffer from GI dysmotility. These attacks can be triggered by waking in the morning or by stress, sleep deprivation, fatigue, and visceral pain secondary to constipation or menses. Crises appear to be due to increased levels of plasma dopamine and norepinephrine (attributed to peripheral conversion of dopamine). There is also respiratory dysfunction secondary to chronic lung disease from frequent aspirations, restrictive lung disease secondary to scoliosis, and muscle weakness. There is also chemoreceptor dysfunction that blunts normal responses to hypoxia and hypercapnea. These patients also have blood pressure lability where they may experience postural hypotension, which can be can be worsened by infections and dehydration, or episodic hypertension triggered by emotional stress and may be a response to increased central release of catecholamines and denervation hypersensitivity. Hypertension can be treated with benzodiazepines and clonidine. Hypotension can be treated with fluid, fludrocortisone, and midodrine. As these patients age, there is also an increased loss of neurons in the dorsal root ganglia, loss of dorsal column myelinated axons, and atrophy of the brain leading to decrease in mental processing that ranges from poor concentration to phobias to dementia (Axelrod 2002).

The disease has been mapped to the *IKBAP* gene, which encodes the transcription elongation factor complex IKAP that possesses histone acetyltransferase activity and is predominantly expressed in neuronal tissues. Ninety-nine percent of patients diagnosed with HSAN Type III have a homozygous mutation in intron 20 (IVS20+6T>C) leading to altered splicing causing the deletion of exon 20. This causes a reduction in brain mRNA levels, which results in deprivation of the protein product, IKAP. The other 1% of mutations occur on exon 19 with the R696P (2390G>C) amino acid substitution leading to a disruption of the consensus serine/threonine kinase phosphorylation site and has been linked with HSANIII. Another amino acid substitution P914L (3051C>T) in exon 19 has also been associated with this phenotype (Lotsch 2008). Patients with HSAN Type III express equal amounts of mutant and normal forms of the IKAP protein in other tissues (Norcliffe-Kaufmann 2012). A recent study suggests that IKAP may cause precocious differentiation of precursor cells into neuronal cells and premature cell death (Hunnicutt 2012).

HSAN Type IV

Congenital insensitivity to pain with anhidrosis (CIPA) is a rare autosomal recessive disorder. Only several hundred cases have been reported since it was first described about 50 years ago. There are over 50 documented mutations in gene *NTRK1* (Gao 2013) of which the gene product is a tyrosine kinase receptor, which is one of the receptors for nerve growth factor (NGF). Individuals with this disorder have decreased sensation and autonomic dysfunction that has a congenital onset. The sensory abnormalities observed in this disorder are more profound than in FD (HSAN Type III) (Axelrod 2002). This disease is characterized by the lack of sweating, recurrent episodic fevers, decreased pain perception, self-mutilation, frequent accidents and injuries that are complicated by poor bone or joint healing, infections, and osteomyelitis (Indo 2002; Shorer 2013). Self-mutilation and orthopedic problems secondary to poorly healing bones and joints from repeated injury are common. There is widespread anhidrosis; however, lacrimation is normal. Developmental milestones in these patients are normal to mildly delayed, but hyperactivity and emotional lability are common (Axelrod 2002). EEG and CT scans are mostly normal. Routine motor and sensory conduction studies are also normal (Shorer 2013).

Neuronal survival requires trophic support by NGF (Nerve growth factor) and is needed to support growth and survival of neurons. Without this, immature sensory and sympathetic neurons die during development. This loss of function mutation in the NTRK1 gene was the first implicated in the NGF signal transduction pathway (Indo 2012).

HSAN Type V

Norrbottnian congenital insensitivity to pain is caused by a missense mutation in the *NGFB* gene, which results in the loss of trophic support of neurons that depend on NGF protein. Interestingly, these patients do not suffer from mental retardation or anhidrosis. This disorder is characterized by loss of pain perception, impaired temperature sensitivity, ulcers and self-mutilation and variable joint involvement. A severe reduction of unmyelinated nerve fibers and moderate loss of thin myelinated fibers is observed in histopathology studies of these patients (Minde 2004). Homozygotes are the most severely affected and some heterozygotes lack symptoms. Sural nerve biopsies show severe reductions in the numbers of unmyelinated C-fibers and moderate losses of thin myelinated A-delta fibers. Interestingly, heterozygotes, regardless of whether they show symptoms or not, also have a moderate loss of both A-delta and C-fibers. HSAN Type V was first described in a family from northern Sweden during 2004 (Einarsdottir 2004), with all three affected family members homozygous for 661C>T (R211W) (Lotsch 2008).

The *SCN9A* gene encodes voltage-gated sodium channel Nav1.7 that is expressed in the peripheral nervous system. Mutations in this gene have been linked to three different pain disorders: congenital indifference to pain (CIP), inherited erythromelalgia (IEM), and paroxysmal extreme pain disorder (PEPD). Mutations in the *SCN9A* gene have also been linked with a case of severe fibromyalgia (Klein 2013). There have been many reported mutations, although the incidence of these mutations is unknown (Klein 2013).

Congenital indifference to pain (CIP), also sometimes referred to as channelopathy-associated insensitivity to pain, is an extremely rare disorder that is the result of a loss-of-function mutation of the *SCN9A* gene that obliterates Nav1.7 sodium channel function (Cox 2006; Goldberg 2007). This disorder is inherited in an autosomal-recessive pattern and there have been nine families identified as having this disorder in Pakistan (Cox 2006), Argentina, Canada, United States, Italy, Switzerland, United Kingdom, and France with 10 unique non-sense mutations resulting in the truncation of the Nav1.7 protein (Goldberg 2007). Unlike the other hereditary sensory and autonomic neuropathies (HSANs), patients with this disorder have normal number of neurons, and histological studies and nerve conduction studies are usually normal (Cox 2006; Goldberg 2012). These patients are able to discriminate temperature, light touch, joint position, and vibration, and distinguish between sharp and dull (Goldberg 2012). However, they do not perceive to be unpleasant what others would consider painful. Consequently, these patients are unaware when they are injured themselves and frequently subject to fractures, dislocations, burns, and accidental self-mutilation.

Inherited Erythromelalgia (IEM) is a rare disorder that affects 2 out of 100 000 individuals per year, associated with a gain-of-function mutation of the *SCN9A* gene leading to increased activity of the Nav1.7 channel in nervous system tissue from a decreased threshold to activation. So far 20 different mutations have been reported in the literature (Eberhardt 2014). The disorder is inherited in an autosomal dominant fashion and is characterized by parosyxmal episodes of extremity pain associated with redness and warmth of the affected extremity (Klein 2013). Onset is usually during adolescence (Eberhardt 2014) but there have been reported cases of presenting as young as in infancy to as old as patients in their fifties and sixties (Klein 2013). These symptoms can be precipitated by exercise, prolonged standing, heat, and changes in humidity (Choi 2009; Goldberg 2012) and are relieved with immersion in ice-water (Eberhardt 2014) leading to possible complications such as tissue damage and wound infection. Patients with Nav1.7 V872G (Choi 2009) and A863P (Nathan 2005) amino acid substitutions have responded to mexiletine and a patient carrying V400M reportedly responded to treatment with carbamazepine (Fischer 2009). Further research is needed on the characterization of these variant channels to better understand the relationship between the presence of *SCN9A* sequence variations and drug response.

Paroxysmal extreme pain disorder (PEPD), formerly known as familial rectal pain syndrome is another rare disorder, approximately 80 affected patients were reported in the literature, associated with a gain-of-function mutation in the Nav1.7 sodium channel. This disorder is also inherited in an autosomal dominant fashion and is characterized by pain, flushing, and burning pain over the lower half of the body that can triggered, classically, by defecation, stimulation of the perianal area, or pain below the waist that is then followed by ocular or sub maxillary pain. The pain can also be triggered by emotional upset, temperature changes, and eating (Fischer 2009). The onset is usually in infancy (Fischer 2009; Choi 2011). The gain-of-function mutation in this disorder differs from IEM in that in PEPD, there is impaired inactivation of the Nav1.7 sodium channel, leading to persistent current, increasing DRG neuron excitability (Choi 2011). Most patients with PEPD respond to treatment with carbamazepine (Fischer 2009) or avoid triggers of the symptoms (Choi 2011).

Based upon this knowledge, it is not surprising that drugs known to have sodium channel-blocking properties, such as carbamazepine, lidocaine, and mexilitine, have had some utility in the treatment of a variety of nerve-mediated pain disorders such as trigeminal neuralgia, fibromyalgia and IEM. The non-specific nature of the aforementioned drugs limits the usage of these drugs at higher doses because of dizziness and sedation (Drenth 2007) and usually these agents do not work for patients with IEM (Choi 2009). As a result of research regarding SCN9A and its role in the perception of pain, the Nav1.7 sodium channel has been identified as a therapeutic target for small molecule compounds currently in development. At the time of this writing, one small molecule compound specifically targeted for Nav1.7, XEN402 (now TV-45070) (Goldberg 2012), has passed Phase II clinical trials and received orphan drug designation by the FDA for the treatment of inherited erythromelalgia, with the hopes that the drug may prove to be useful for other neuropathic pain indications that involve over-activation of the Nav1.7 channel.

Genes Involved in Metabolism of Pharmacotherapy for Treatment of Pain

Cytochrome P450 2D6

Cytochrome P450 2D6 is an important hepatic enzyme involved in the metabolism of many drugs, including a number of opioids: codeine, hydrocodone, tramadol, and oxycodone (Discipline of Chinese Med. RMIT University 2009). The CYP2D6 gene is highly polymorphic with over 660 allelic variants and subvariants being identified (Karolinska Institute, Department of Physiology and Pharm. 2013). In addition to numerous sequence variations including SNPs and insertions and deletions of single or few nucleotides, allelic variation also comprises gene deletions, duplications, multiplications and rearrangements with the *CYP2D7* pseudogene. Notably, duplications/multiplications have been reported for non-functional, reduced and functional gene copies. The complexity of the CYP2D6 gene locus and prediction of phenotype from genotype data has recently been reviewed (PMIDs 24151800, 24524666, and references therein).

Depending on a patient's diplotype (i.e., combination of allelic variants), his/her phenotype can be predicted and activity has been established for many variants in vivo or in vitro; however, the function for some remains unknown. To facilitate phenotype prediction alleles can be assigned an activity value (AS): 0 for non-functional, 0.5 for reduced function, and 1.0 for alleles encoding functional enzyme. The activity score is the sum of the values assigned to each allele. In subjects with more than two gene copies, the activity score is determined by the sum of the values given to each gene copy. CYP2D6 activity scores usually range from 0–3.0. Patients with two non-functional alleles (AS=0) are classified as poor metabolizers. Patients who have an AS of 0.5 are intermediate metabolizers and those with an AS of 1.0–2.0 are extensive metabolizers. Patients who have an AS greater than 2.0 are considered ultrarapid metabolizers. Allele frequencies differ widely among individuals as well as populations with about 7% of Caucasians being classified as poor metabolizers and 1%–2% classified as ultrarapid metabolizers. Additionally, poor metabolism occurs in about 3%–5% in Black Africans and their descendents, but can reach over 20% for ultrarapid metabolizers in East African populations. In contrast, both extreme phenotypes are rare in East Asians.

Codeine is a commonly prescribed opioid analgesic that is metabolized by CYP2D6 to morphine. Since morphine has a 200-fold higher affinity for mu opioid receptors than codeine (Volpe 2011; Thorn 2009), the majority of the analgesic effect of codeine is from the conversion of codeine to morphine. Patients who are poor metabolizers may not see clinical benefit from codeine as an analgesic, whereas patients who are ultrarapid metabolizers may suffer increased side effects such as nausea, vomiting, respiratory depression, and sedation as a result of high serum levels of morphine. There have been a number of case reports of patient deaths after receiving codeine associated with the ultrarapid metabolizer phenotype (Ciszkowski 2009; Koren 2006). The current recommendations from the Clinical Pharmacogenomics Implementation Consortium (CPIC) regarding dosing for codeine therapy is to avoid the use of codeine in patients who are known poor metabolizers (because of lack of efficacy) and in patients who are ultrarapid metabolizers (because of the potential for toxicity) while the recommended age or weight-based dosing should be followed for intermediate metabolizers and extensive metabolizers (Crews 2014). Because of a number of deaths in children with the ultrarapid metabolizer phenotype who received codeine after tonsillectomy and/or adenoidectomy (Voronov 2007; Ciszkowski 2009; Kelly 2012), the FDA has issued a black box warning on the use of codeine after tonsillectomy and/or adenoidectomy in children as of August 2013 (USFDA 2014).

Tramadol is another drug that is metabolized by CYP2D6 to O-desmethyltramadol. As with codeine, the parent drug shows significantly weaker affinity for mu-opioid receptors compared to the metabolite. Also similar to codeine, patients who are CYP2D6 poor metabolizers may not show an analgesic response to tramadol (Poulsen 1996; Stamer 2003), and therefore similar dosage adjustments may need to be made.

Oxycodone is another opioid analgesic that is metabolized by CYP2D6, however there are conflicting studies regarding whether CYP2D6 metabolizer phenotype affects efficacy of analgesia or if increased side effects are seen in ultrarapid metabolizers. Hydrocodone is also metabolized by CYP2D6 but due to the lack of pharmacokinetics data in ultrarapid metabolizers, no conclusions can be made regarding the connection between CYP2D6 metabolism and potential toxicity or lack of efficacy (Crews 2014).

COMT (Catechol-O-methyl transferase)

Gene: COMT

Allelic varaint 472G greater than A, amino acid substitution V158M

Catechol-o-methyl transferase (COMT) is an enzyme involved in the degradation of epinephrine, dopamine, and norepinephrine. Inhibition of COMT, which can be obtained with certain drugs such as entacapone and tolcapone and increased dopamine concentrations, has been suggested to decrease production of endogenous opioids leading to increased pain sensitivity while high COMT activity seems to decrease pain (Zubieta 2003). Within COMT is a common genetic variant (Val158Met) in which presence of the Met (M) allele results in reduced COMT activity due to the production of a more thermolabile protein. Carriers of the low activity variant 158M have been associated with increased sensitivity to pain (Tammimaki 2012). In one study of patients comparing homozygous wildtype with homozygous variant subjects, the latter reported higher pain scores after repeated pain stimuli (Jensen 2009). Interestingly, cancer patients homozygous for the V158M variant required less morphine than patients who were heterozygous for this allele or non-carriers (Rakvag 2005). One proposed mechanism as to why individuals with the variant required a lower dose of opioid is that native opioid receptor expression is upregulated in these patients, leading to a greater effect with exogenous opioid administration (Lotsch 2008).

Four SNPs within the COMT gene have been identified: rs6269 (G greater than A), rs4633 (C greater than T), rs4818 (G greater than C) and rs4680 (472G greater than A; V158M) and these SNPs determine the three COMT haplotypes associated with varying levels of pain tolerance. In looking at these the GCGG has been categorized as the low pain sensitivity haplotype, ATCA for average pain sensitivity, and ACCG is the high pain sensitivity haplotype. It is important to note that rs4680 appears to be the only nonsynonymous SNP and hence the only one responsible for functional consequences. In one study, carriers of the low pain sensitive GCGG haplotype also showed to have the least pain responsiveness to thermal, ischemic and mechanical stimuli. However an attempt to independently reproduce the low-pain phenotype associated with the particular COMT "low-pain" GCGG haplotype was unsuccessful (Kim 2006).

Receptors Involved in Pain

OPRM1

Mu-opioid receptor variant—associated with diminished opioid effects

Gene: OPRM1
Allelic variant 118A greater than G, amino acid substitution N40D

This variant has been associated with 0.8-fold decreased pressure pain intensity and 0.5-fold smaller amplitudes of pain related cortical potentials after stimulation of nasal nociceptors with CO_2. The frequency of the A118G variant varies between ethnicities. In European American populations its frequency is 30%–35% (Coller 2011; Cooper 2013) while it has been reported to be over 50% in Koreans (Kim 2009).

However, cancer patients carrying this SNP have been shown to require higher doses of morphine than those with the wildtype genotype. The variant decreases the effects of opioids mainly in those regions of the brain that are involved in the processing of pain (Lotsch 2008). Possibly, the effects seen may also be due to reduced mu-opioid receptor expression in the brain (Walter 2009). Other mechanisms that have been proposed are decreased beta-endorphin binding, protection from morphine-6-glucuronide toxicity, and susceptibility to drug addiction (Lotsch 2008).

In a study of women who had a cesarean section and were given hydrocodone for post-operative pain, pain relief was most closely associated with total hydrocodone dose and serum hydromorphone (the active metabolite of hydrocodone) levels in patients who were homozygous for the A allele. In patients who were homozygous or heterozygous for the G variant, these effects were not seen. Patients with one or two copies of the G allele also were more likely to experience adverse events than those homozygous for the A variant (Boswell 2013).

However, in a study involving 352 patients the OPRM1 118 A greater than G polymorphism was identified as the sole factor significantly influencing the 24-hour pain score. A gene dose-dependent increase in pain was observed and as a result opioid requirements decreased (Lotsch 2009). Also, in a study of 224 women in labor, those with at least one copy of the 118G allele required significantly less (ED50 17.7 micrograms) intrathecal fentanyl compared to patients homozygous for the 118A wild type allele (ED50 26.8 micrograms) (Landau 2008). Similar results were seen in pregnant women given epidural sufentanil for labor analgesia with the ED50 of patients with the 118A allele being 25.2 micrograms and patients with at least one copy of the 118G allele at 20.2 micrograms, for a relative potency of 1.25 in patients carrying at least one copy of the variant (Camorcia International Journal of Obstetric Anesthesia 2012). Thus it appears at least for opioids administered via the intrathecal or epidural route, carriers of the 118G variant require lower doses of opioids.

Melanocortin-1 receptor
Alpha melanocyte stimulating hormone receptor or Melanocortin-1 receptor—Greater analgesia by K-opioid agonist pentazocine, enhanced analgesic effects of morphine-6-glucuronide.

Gene: MC1R
Allelic variation 29insA,

451C greater than T, amino acid substitution R151C
478C greater than T, amino acid substitution R160W
880G greater than C, amino acid substitution D294H, all loss-of-function mutations

The MC1R gene is located on chromosome 16q24.3 and encodes a G protein-coupled receptor. It has been associated with anti-pyretic and anti-inflammatory effects

(Getting 2006). Subjects with loss-of-function mutations in the MC1 receptor, resulting in a redhead phenotype, had a 1.3 higher tolerance to electrical pain stimuli compared to subjects with functional MC1 receptors in one study. This was not reproduced in another study (Lotsch 2008). Also, patients with a loss-of-function mutation in the MC1 receptor had a lower tolerance to cold and heat pain. Mice studies though have consistently shown that mice with nonfunctional MC1 receptors have decreased sensitivity to pain. Women who have two copies of the nonfunctional gene had a greater analgesia with pentazocine than when compared to women carrying one or no copies, or when compared to men. This effect may be due to the inability of endogenous dynorphin to bind to non-functional MC1 receptors however there is no clear explanation for this difference. Activation of MC1 receptor by dynorphin exerts an anti-opioid effect. Without this, pentazocine has a greater effect in subjects without any functional MC1 receptors. This effect on kappa opioid receptors has only been observed in women. MC1R variants 29insA, 178G greater than T, 252C greater than A, and 274G greater than A c, or 488G greater than A also have been associated with functionally impaired MC1receptors. The analgesic effect of morphine-6-glucuronide was increased in both men and women by 1.3 times in subjects with non-functional MC1 receptors (Lotsch 2008).

MDR1

MDR1 or P-glycoprotein—Increased susceptibility to clinical fentanyl effects, and loperamide causes CNS effects due to decreased P-glycoprotein to excrete loperamide from the CNS.

Gene: ABCB1

Gene mutation: SNP 3435C greater than T (synonymous)
SNP1236C greater than T (synonymous)
2677G greater than T (Ala to Ser change),

The Multi-Drug Resistance (MDR1) gene is expressed mostly in the liver, kidney, GI tract and in the blood brain barrier, with the protein product of this gene, being thought to be an efflux transporter of the ATP-Binding Cassette transporter family (ABCB1) also known as P-glycoprotein. Patients with impaired p-glycoprotein-mediated transport would be expected to have an increased bioavailability of the drug and increased concentrations of the drug in the brain, due to the efflux function of this drug transporter. All three of the above listed sequence variations have been associated with increased efficacy of fentanyl. Loperamide, an antidiarrheal that pharmacologically exerts weak opioid effects, normally does not have CNS effects but has been shown to have increased central nervous opioid effects in patients with the 3435T/T genotype (Lotsch 2008). Linear regression analyses including numerous candidate genes (*OPRM1, COMT, MC1R, ABCB1,* and *CYP2D6*), identified the *ABCB1* synonymous 3435C>T SNP as the only variation that significantly modified opioid dosing with opioid doses decreasing in a gene dose-dependent manner (Lotsch 2009). Patients genotyped as 3435C/T or 3435T/T were also observed to have fewer adverse drug reactions with oxycodone use compared to those genotyped as 3435C/C. The same was observed with patients genotyped as 2677G/T and 2677T/T vs 2677G/G, although the presence of 2677T did not appear to confer

> **Case Scenario**
>
> A 51-year-old man is day one status post hernia repair. He experiences severe pain postoperatively. He receives hydromorphone IV PRN and a subcutaneous lidocaine pump and is discharged with equianalgesic dose of hydrocodone + acetaminophen every six hours for pain. The next day, his local analgesia wears off early the next day, and he is 10/10 severe pain. He calls his surgeon, takes another hydrocodone and goes to the ER.
>
> **Discussion:** Pharmacogenetic testing may help to determine if genetic polymorphisms may impact the metabolism of certain medications. Polymorphisms curing in the cytochrome P450 (CYP450) system, can result in metabolic changes which may lead to unexpected outcomes, such as lack of efficacy or increased or unexpected toxicity (Argoff 2010). In looking at this patient case, after pharmacogenetic testing is performed, it is determined that this patient is a CYP2D6 Poor Metabolizer. In looking at his drug regimen of hydrocodone, it is known that this medication is a prodrug with the parent compound having little activity, but after metabolism by 2D6, the metabolite formed (hydromorphone) is active against pain. Therefore a poor metabolizer would be expected to experience decreased efficacy, less toxicity, and higher doses may be necessary. In this case, the hydrocodone/acetaminophen drug combination he was given provided relatively no analgesia as a result of his CYP2D6 metabolizer status. These results may help explain why some patients may need higher doses, or why some specific medications don't work for some patients.

increased efficacy of oxycodone (Zwisler 2010). Fewer adverse events were also observed in carrying variant alleles (2677G/T, 2677T/T, 3435C/T or 3435T/T) and were treated with morphine (Fujita 2010). However, the question arises whether this evidence is sufficiently strong to clinically act upon ABCB1 genotype information.

Summary

The use of pharmacogenomics within the field of pain medicine may be a useful tool for identifying patients who may either require a higher analgesic dose or patients who may be susceptible to drug overdose at otherwise clinically acceptable dosing ranges. While current

Table 9.1. Relevant Literature on Pharmacogenomics and Pain

Title of the Study and Year	Authors	Type of Study	Study Population	Methods	Conclusion
Human Opioid Receptor A118G Polymorphism Affects Intravenous Patient-controlled Analgesia Morphine Consumption after Total Abdominal Hysterectomy/ Aug 2006	Wen-Ying Chou, M.D., Cheng-Haung Wang, M.D., Ping-Hsin Liu, M.D., Chien-Cheng Liu, M.D., Chia-Chih Tseng, M.D., Bruno Jawan, M.D.		80 female patients were enrolled	In this study, 80 female patients were enrolled and scheduled to undergo elective total hysterectomy surgery. All patients received general anesthesia and were screened for A118G by blood sample.	Genetic profile of the patients in terms of μ-opioid receptor may contribute to differences between the individuals in postoperative morphine consumption.
Effect of Catechol-o-Methyltransferase-gene (COMT) Variants on Experimental and Acute Postoperative Pain in 1000 Women Undergoing Surgery for Breast Cancer/Dec 2013	Kambur, Oleg M.Sc. (Pharm), Kaunisto, Mari A. Ph.D., Tikkanen, Emmi M.Sc., Leal, Suzanne M. Ph.D., Ripatti, Samuli Ph.D., Kalso, Eija A. M.D. Ph.D.	Article, Pain Medicine, Dec 13, Volume 119, Issue 6, p 1422-33	1,000 women patients were enrolled	Tolerance to cold pain and intensity of cold (+2-4 °C) and heat (+48 °C) pain were evaluated in 1,000 women undergoing breast cancer surgery. Patients' oxycodone requirements and acute postoperative pain measurements were recorded. 22 COMT SNPs were genotyped and their association with six pain phenotypes analyzed with linear regression.	Strongest effects on pain sensitivity were caused by SNPs rs887200 and rs165774 located in untranslated regions of the gene.
Sex-and Age-related Differences in Morphine Requirements for Postoperative Pain Relief/July 2005	Aubrun, Frédéric M.D., Salvi, Nadége M.D., Coriat, Pierre M.D., Riou, Bruno M.D., Ph.D.	Article, Pain and Regional Anesthesia, Volume 103, Issue 01, p156-60	Total of 4,317 patients; 54% male, 46% female were enrolled.	During the immediate postoperative period, intravenous morphine titration was dispensed as a bolus of 2 (body weight less than 60 kg) or 3mg (body weight greater than 60kg). The interval between each bolus was 5 min. The visual analog pain scale (VAS) threshold required to administer morphine was 30, and pain relief was defined as a VAS score of 30 or less.	In the immediate postoperative period, women experienced more severe postoperative pain and required more morphine than men. This sex-related difference disappeared in elderly patients.

work done in this area is fairly clear for some medications (i.e., codeine), further studies in pain genetics can also identify new targets for drug development.

The recognition of genetic factors modulating pain and analgesic drug effects has raised expectations that genotyping of pain patients not responding to analgesic treatment may provide: (i) explanations for the unintended responses to analgesics and (ii) advices for personalized analgesic therapy to achieve the intended responses. Genetic variants may influence pharmacotherapy by (i) altering the local availability of active molecules of the analgesic at its site of action, or (ii) affecting the interaction of analgesic molecules with their target structures and/or the consequences of this interaction. Ultimately, the influence of each of the individual genes on perception of pain remains a complex study and may be more of a scientific pursuit rather than one of a practical clinical use at this time (Walter 2009). Because of the complexity of analyzing individual genes as well as the interaction of these genes on top of a background of influencing non-genetic factors such as gender, medication compliance, underlying disease, age and concomitant medications, it remains very difficult to draw any hard and fast conclusions as to the clinical decisions that should be made on this data. Current data only allows us to make generalizations on how we expect patients to respond to opioid therapy or to pain but it is not stated that any one gene variant has so much influence on the patient's treatment plan.

At the patient level, genetic differences may explain an estimated 20%–95% of the variability in medication effects, such as poor response or serious adverse drug reactions which may be critical when utilizing opioid therapy for pain. Clinical validity and utility studies currently support the potential of PGT to help pain clinicians improve patient outcomes and thereby reduce associated healthcare costs for specific gene-drug pairs.

Possibilities for the future involve giving genetic testing only to those with high risk for adverse events associated with medications being utilized for pain. In the decade following the decoding of the human genome in 2001, substantial milestones have been reached on the path to genomic medicine specifically in improving the treatment of pain. Today, there are clear examples of clinically relevant and robustly supported gene-drugs pairs that merit implementation in the treatment of pain.

References

Argoff CE. Clinical Implications of Opioid Pharmacogenetics. Clin J Pain 2010;26(1):S16-20.

Axelrod F. Hereditary sensory and autonomic neuropathies: Familial dysautonomia and other HSANs. Clin Auton Res 2002;12(suppl 1):1/2-1/14.

Boswell M, Stauble E, Loyd G, et al. The role of hydromorphone and OPRM1 in postoperative pain relief with hydrocodone. Pain Physician 2013;16:e227-235.

Bunten H, Liang WJ, Pounder DJ, et al. OPRM1 and CYP2B6 gene variants as risk factors in methadone-related deaths. Clin Pharmacol Ther 2010;88(3):383-89.

Buskila, Dan, and Lily Neumann. Genetics of Fibromyalgia. Curr Pain Headache Rep 9.5 2005;313-15.

Camorcia M, Capogna G, Berritta C, et al. Effect of μ-opioid receptor A118G polymorphism on the ED50 of epidural sufentanil for labor analgesia. Int J Obstet Anesth 2012;21(1):40-4.

Cheung K. The relationship between disc degeneration, low back pain, and human pain genetics. Spine J 2010;958-960.

Choi JS, Boralevi F, Brissaud O, et al. Paroxysmal extreme pain disorder: a molecular lesion of peripheral neurons. Nat Rev Neurol 2011;7:51-5.

Choi JS, Zhang L, Dib-Hajj S, et al. Mexiletine-responsive erythromelalgia due to new Nav1.7 mutation showing use-dependent current fall-off. Exp Neurol 2009;216:383-89.

Ciszkowski C, Madadi P, Phillips M, et al. Codeine, Ultrarapid–Metabolism Genotype, and Postoperative Death. N Engl J Med 2009;361:827-28.

Coller JK, Cahill S, Edmonds C, et al. OPRM1 A118G genotype fails to predict the effectiveness of naltrexone treatment for alcohol dependence. Pharmacogenet Genomics 2011;21:902-5.

Cooper A, Rickels K, Lohoff FW. Association analysis between A118G polymorphism in the OPRM1 gene and treatment response to Venlafaxine XR in generalized anxiety disorder. Hum Psychopharmacol Clin Exp 2013;28:258-62.

Cox J, Reimann F, Nicholas A, et al. A SCN9A channelopathy causes congenital inability to experience pain. Nature 2006;444(14):894-98.

Dai, Feng, et al. Association of catechol-O-methyltransferase genetic variants with outcome in patients undergoing surgical treatment for lumbar degenerative disc disease. Spine J 2010;949-57.

de Vries, Boukje, et al. Molecular genetics of migraine. Hum Genet 126.1 2009;115-32.

Diatchenko L, Nackely AG, Slade GD, et al. Catechol-O-methyltransferase gene polymorphisms are associated with multiple pain-evoking stimuli. Pain 2006;125:216-24.

Drenth J, Waxman S. Mutations in sodium channel gene SCN9A cause a spectrum of human genetic pain disorders. J Clin Invest 2007;117(12):3603-9.

Eberhardt M, Nakajima J, Klinger A, et al. Sodium Channel Nav1.7 A1632T Mutation Causes Erythromelalgia Due to a Shift of Fast-Inactivation. J Biol Chem 2014;289(4):1971-80.

Einarsdottir E, Carlsson A, Minde J, et al. A mutation in the nerve growth factor beta gene (NGFB) causes loss of pain perception. Hum Mol Genet 2004;13(8):799-805.

Feng J, Zhang Z, and Sommer S. Missense Mutations in the MEFV genes are associated with fibromyalgia syndrome and correlate with elevated 1L-1β Plasma Levels. PlosOne2009;4(12);e8480.

Fischer TZ, Gilmore ES, Estacion M, et al. A novel Nav1.7 Mutation Producing Carbamazepine–Responsive Erythromelalgia. Ann Neurol 2009;65:733-41.

Fischer TZ, Waxman SG. Familial pain syndromes from mutations of Nav1.7 sodium channel. Ann N Y Acad Sci 2010;1184:196-207.

Fujita K, Ando Y, Yamamoto W, et al. Association of UGT2B7 and ABCB1 genotypes with morphine-induced adverse drug reactions in Japanese patients with cancer. Cancer Chemother Pharmacol 2010;65:251-58.

Gao L, Guo H, Ye N, et al. Oral and Craniofacial Manifestations and Two Novel Missense Mutations of the NTRK1 Gene Identified in the Patient with Congenital Insensitivity to Pain with Anhidrosis.

Garofalo K, Penno A, Schmidt BP, et al. Oral L-serine supplementation reduces production of neurotoxic deoxysphingolipids in mice and humans with hereditary sensory autonomic neuropathy type 1. J Clin Invest 2011;121(12):4735-45.

Getting SJ. Targeting melanocortin receptors as potential novel therapeutics. Pharmacol Ther 2006;111(1):1-15.

Goldberg YP, MacFarlane J, MacDonald ML, et al. Loss-of-function mutations in the Nav1.7 gene underlie congenital indifference to pain in multiple human populations. Clin Genet 2007;71:311–19.

Goldberg YP, Price N, Namdari R, et al. Treatment of Nav1.7-mediated pain in inherited erythromelalgiausing a novel sodium channel blocker. Pain 2012;153:80-5.

Grice GR, Seaton TL, Woodland AM, et al. Defining the opportunity forpharmacogenetic intervention in primary care. Pharmacogenomics 2006;7(1):61-5.

Guymer E, and Littlejohn G. Fibromyalgia. Aust Fam Physician 42.10 2013;690.

Houlden H, King R, Blake J, et al. Clinical, pathological and genetic characterization of hereditary sensory and autonomic neuropathy type I (HSAN I). Brain 2006; 129:411-25.

Hunnicutt BJ, Chaverra M, George L, et al. IKAP/Elp1 Is Required In Vivo for Neurogenesis and Neuronal Survival, but Not for Neural Crest Migration. PLoS One 2012;7(2):e32050.

Indo Y. Genetics of congenital insensitivity to pain with anhidrosis (CIPA) or hereditary sensory and autonomic neuropathy type IV. Clin Auton Res 2002;12(suppl 1):1/20-1/32.

Indo Y. Molecular basis of congenital insensitivity to pain with anhidrosis (CIPA): Mutations and polymorphisms in TRKA (NTRK1) gene encoding the receptor tyrosine kinase for nerve growth factor. Hum Mutat 2001;18:462–71.

Indo Y. Nerve growth factor and the physiology of pain: lessons from congenital insensitivity to pain with anhidrosis. Clin Genet 2012;82(4):341-50.

Ingelman-Sundberg M. Update on allele nomenclature for human cytochromes P450 and the Human Cytochrome P450 Allele (CYP- allele) Nomenclature Database. Section for Pharmacogenetics, Department of Physiology and Pharmacology, Karolinska Institute, Stockholm, Sweden. 2013.987:251-9.

Jensen K, Lonsdorf T, Schalling M, et al. Increased Sensitivity to Thermal Pain Following a Single Opiate Dose Is Influenced by the COMT val158met Polymorphism. PLoS One 2009; 4(6);e6016.

Kelly LE, Rieder M, van den Anker J, et al. More codeine fatalities after tonsillectomy in North American Children. Pediatr 2012;129(5):e1343-7.

Kim H, Mittal DP, Iadarola MJ, et al. Genetic predictors for acute experimental cold and heat pain sensitivity in humans. J Med Genet 2006;43:e40.

Kim SG, Kim CM, Choi SW, et al. A mu opioid receptor gene polymorphism (A118G) and naltrexone treatment response in adherent Korean alcohol-dependent patients. Psychopharmacol 2009;201:611-18.

Klein C, Wu Y, Kilfoyle D, et al. Infrequent SCN9A mutations in congenital insensitivity to pain and erythromelalgia. Ns. J Neurol Neurosurg Psychiatry 2013;84:386–91.

Koren G, Cairns J, Chitayat D, et al. Pharmacogenetics of morphine poisoning in a breastfed neonate of a codeine-prescribed mother. Lancet 2006;368:704.

Kors E, Haan J, and Ferrari M. Migraine genetics. Curr Pain Headache Rep 7.3 2003;212-17.

Landau R, Kern C, Columb MO, et al. Genetic variability of the mu-opioid receptor influences intrathecal fentanyl analgesia requirements in laboring women. Pain 2008. 139(1):5-14.

Oertel B, Lötsch J. Genetic mutations that prevent pain: implications for future pain medication. Pharmacogenomics 2008.9(2)179-94.

Lea DH, Williams J, and Donahue MP. Ethical issues in genetic testing. J Midwifery Womens Health 50.3 2005;234-40.

Lempp H, Hatch S, Carville S, et al. Patients' experience of living with and receiving treatment for fibromyalgia syndrome; a qualitative study.

BMC Musculoskeletal Disorders 2009. (doi: 10.1186/1471-2474-10-124).

Lotsch J, von Hentig N, Freynhagen R, et al. Cross-sectional analysis of the influence of currently known pharmacogenetic modulators on opioid therapy in outpatient pain centers. Pharmacogenet Genomics 2009;19:429-36.

Männistö PT, Tammimäki A. Catechol-O-methyltransferase gene polymorphism and chronic human pain: a systematic review and meta-analysis. Pharmacogenet Genomics 2012;22(9):673-91.

Minde J, Toolanen G, Andersson T, et al. Familial Insensitivity to Pain (HSAN V) and A Mutation in the NGFB Gene: A Neurophysiological and Pathological Study. Muscle Nerve 2004;30:752-60.

Norcliffe-Kaufmann L, Kaufmann H. Familial dysautonomia (Riley-Day Syndrome): When baroreceptor feedback fails. Autonomic Neuroscience 2012;172:26-30.

Penno A, Reilly MM, Houlden H, et al. Hereditary Sensory Neuropathy Type I is Caused by Accumulation of Two Neurotoxic Sphingolipids. J Biol Chem 2010;285(15):11178-87.

Pestka EL, and Nash VR. Genetic aspects of migraine headaches. Nurse Pract 38.5 2013;12-13.

Poulsen L, Arendt-Nielsen L, Brosen K, et al. The hypoalgesic effect of tramadol in relation to CYP2D6. Clin Pharmacol Ther1996;60:636-44.

Rakvag TT, Klepstad P, Baar C, et al. The Val158Met polymorphism of the human catechol-O-methyltransferase (COMT) gene may influence morphine requirements in cancer pain patients. Pain 2005;116:73–8.

Rotthier A, Auer-Grumbach M, Jannsens K, et al. Mutations in the SPTLC2 Subunit of Serine Palmitoyltransferase Cause Hereditary Sensory and Autonomic Neuropathy Type I. Am J Hum Genet 2010;87:513-22.

Rotthier A, Baets J, Timmerman V, et al. Mechanisms of disease in hereditary sensory and autonomic neuropathies. Nat Rev Neurol 2012;8:73-85.

Sadhasivam S, and Chidambaran V. Pharmacogenomics of Opioids and Perioperative Pain Management. Pharmacogenomics 2012;13(15):1719-40.

Shorer Z, Shaco-Levy R, Pinsk V, et al. Variation of Muscular Structure in Congenital Insensitivity to Pain. Pediatr Neurol 2013;48(4):311-13.

Stamer UM, Lehnen K, Hothker F, et al. Impact of CYP2D6 genotype on postoperative tramadol analgesia. Pain 2003;105(1-2):231-38.

Thorn CF, Klein TE, Altman RB. Codeine and morphine pathway. Pharmacogenet Genomics 2009;19:556-58.

Tong R. Ethical Concerns About Genetic Testing and Screening. N C Med J 74.6 2013;522-25.

U.S. Food and Drug Administration. Safety review update of codeine in children; new Boxed Warning and Contraindication on use after tonsillectomy and/or adenoidectomy. U.S. FDA Drug Safety Communication. Available at http://www.fda.gov/downloads/Drugs/DrugSafety/UCM339116.pdf. Accessed June 10, 2014..

Volpe DA. Uniform assessment and ranking of opioid mu receptor binding constants for selected opioid drugs. Regul Toxicol Pharmacol 2011;59:385-90.

Voronov P, Przbylo HJ, Jagannathan N. Apnea in a child after oral codeine: a genetic variant – an ultra-rapid metabolizer. Paediatr. Anaesth 2007;17:684-87.

Walter C, Lotsch J. Meta-analysis of the relevance of the OPRM1 118Agreater thanG genetic variant for pain treatment. Pain 2009;146:270-75.

Williams, Frances MK, et al. Novel genetic variants associated with lumbar disc degeneration in northern Europeans: a meta-analysis of 4600 subjects. Ann Rheum Dis 72.7 2013;1141-48.

Yuan J, Matsuura E, Higuchi Y, et al. Hereditary sensory and autonomic neuropathy type IID caused by an SCN9A mutation. Neurol 2013;80:1641.

Zubieta JK, Heitzeg MM, Smith YR, et al. COMT val158met genotype affects mu-opioid neurotransmitter responses to a pain stressor. Science 2003;299:1240–3.

Zwisler ST, Enggaard TP, Noehr-Jensen L, et al. The antinociceptive effect and adverse drug reactions of oxycodone in human experimental pain in relation to genetic variations in the OPRM1 and ABCB1 genes. Fundam Clin Pharmacol 2010;24:517-24.

Chapter 10
PHARMACOGENETICS OF SOMATIC MUTATIONS IN CANCER

JILL M. KOLESAR, PHARM.D., FCCP, BCPS

Learning Objectives

1. Understand the appropriate tissue for analysis when performing a test for a somatic mutation.
2. Differentiate between predictive and prognostic pharmacogenetic biomarkers.
3. Be able to assess the utility of a pharmacogenetic biomarker in clinical decision-making.
4. Integrate pharmacogenetic biomarkers into personalized care plans for oncology and hematology patients.
5. Assess the advantages and limitations of assay methodologies for the analysis of pharmacogenetic biomarkers.

Keywords: Biomarker, tyrosine kinase inhibitors, somatic mutations, solid tumors, hematologic malignancies

Abbreviations in This Chapter

ALK	Anaplastic lymphoma kinase (gene)	HER2	Human epidermal growth factor receptor 2
AML	Acute myelogenous leukemia	IHC	Immunohistochemistry
APL	Acute promyelocytic leukemia	KRAS	Kirsten RAS viral oncogene homolog
BRCA	Breast and ovarian cancer susceptibility gene	MDS	Myelodysplastic syndromes
BCR-ABL	Breakpoint cluster region–Abelson murine leukemia viral oncogene homolog 1	MEK1	Map kinase 1
		MEK2	Map kinase 2
		NRAS	Neuroblastoma RAS viral oncogene homolog
BRAF	V-raf murine sarcoma viral oncogene homolog B	NSCLC	Non–small cell lung cancer
CML	Chronic myelogenous leukemia	OS	Overall survival
EGFR	Epidermal growth factor receptor	PFS	Progression-free survival
		PML	Promyelocytic leukemia
EML4-ALK	Echinoderm microtubule-associated protein-like 4	PR	Progesterone receptor
		RARα	Retinoic acid receptor-alpha
ER	Estrogen receptor	RT-PCR	Reverse transcriptase polymerase chain reaction
FISH	Fluorescence in situ hybridization		

Abstract

What makes cancer unique is that it typically arises from genomic alterations, including from somatic mutations. These mutations can be either prognostic, providing information about the prognosis of the disease regardless of treatment, or predictive, providing information that predicts either response or toxicity to a medication. Mutations have been used in the diagnosis of cancer for decades, and more recently, several agents have been developed that are designed to inhibit the protein produced by a mutation. Currently, standard therapy for many cancers involves mutation assessment and the use of agents that target these mutations. Targeted therapies are typically more effective and have fewer adverse effects than standard cytotoxic chemotherapy. The diseases and therapies in which mutation assessment is routinely used to select therapy, including that for breast, colon, lung, melanoma, and acute and chronic leukemias, will be discussed.

Introduction

Cancer

Cancer is a name for more than 100 disorders that share the characteristic of genetic mutations resulting in uncontrolled cell growth and division. Cancers are typically named by either the cell type that is growing abnormally (e.g., acute myelogenous leukemia [AML], arising from blood cells of a myeloid origin) or the organ that is initially affected (e.g., breast cancer) (ACS 2014).

Somatic Mutations vs. Germline Mutations

Classically, genes are defined as a sequence of DNA that is transmitted as a unit of inheritance. When the sequence of DNA changes within a gene, it is defined as a mutation. Genetic variations in the germ cells (e.g., sperm or egg cells), called germline mutations, are present in almost every cell of the body and are then stably transmitted to offspring. In oncology, many, albeit rare, hereditary cancer syndromes are caused by germline mutations. Hereditary cancers are characterized by early onset and familial clustering. Some occur in childhood (retinoblastoma) (Jaradat 2012), whereas others, such as hereditary breast and ovarian cancer (Nelson 2013) as well as hereditary colon cancer (Balmaña 2013), typically present in adults.

For example, one of the most frequent hereditary cancer syndromes is breast cancer, the most commonly diagnosed cancer in women in the United States. The American Cancer Society (ACS) estimates that 235,000 new cases will be diagnosed and that 40,430 deaths will occur because of breast cancer in 2014 (ACS 2014). In addition, up to 20% of breast cancers are thought to be inherited because of family history and a smaller subset because of mutations in breast and ovarian cancer susceptibility genes 1 and 2 (*BRCA1* or *BRCA2*) (Nelson 2013). Risk assessment for hereditary breast cancer is limited to individuals from families known to be at high risk of breast cancer, and germline testing for the *BRCA1* and *BRCA2* mutations is sometimes performed as part of the risk assessment in these families (Friebel 2014). Individuals identified as high risk are then offered counseling and prevention strategies, which can include more frequent screening, prophylactic surgical procedures, or pharmacologic interventions to reduce the risk of developing cancer. This approach of risk assessment and prevention strategies is typical for families at high risk of other cancers as well.

A somatic mutation arises in a non–germ cell that cannot be transmitted to progeny; however, the mutation can be passed on through cell division. Moreover, cancer can sometimes develop when mutations affect cell division, DNA repair, proliferation, apoptosis, or essentially any growth control process. Somatic mutations in different cells and different genes are thought to cause most cancers, and therapeutic advances targeting specific mutations are occurring at a rapid pace.

Somatic Mutations as Biomarkers of Drug Response

A unique aspect of cancer is that the mutation both causes the disease and can be an actionable drug target. Somatic mutations, as well as other genetic changes, are frequently biomarkers, defined as "a characteristic that is objectively measured and evaluated as an indicator of normal biologic processes, pathogenic processes, or pharmacologic responses to a therapeutic intervention" (Dancey 2010).

A prognostic biomarker in oncology provides information about a disease outcome, regardless of the treatment. It can be used to diagnose disease or assess disease severity, which may help clinicians and patients select a more or less aggressive treatment plan. An example of this is Oncotype Dx, which is a 21-gene messenger RNA expression signature used to assess the risk of recurrence in women with early-stage breast cancer who are receiving hormonal therapy with tamoxifen (Paik 2004). According to the gene expression signature, a recurrence score is calculated. The recurrence score is a continuous score predicting the distant risk of breast cancer recurrence in 10 years; however, the risk is subsequently categorized as low, intermediate, and high. Low recurrence scores are defined as 0–17, with an average 10-year recurrence risk of 7%; intermediate risk ranges are 18–31, with an average 10-year risk of recurrence of 14%; and high-risk ranges are from 32 to 50, with an average risk of recurrence in this group of 31%. This assay is essentially a measure of disease severity and is used clinically, in combination with other characteristics, to assess patients most likely to benefit from additional chemotherapy. Guidelines recommend no chemotherapy for patients with a low recurrence score and chemotherapy for those with a high recurrence score (NCCN 2014a). For those with an intermediate score, there is currently no clinical consensus for appropriate management, and other factors may help determine the need for additional therapy. For example, an 85-year-old woman with an intermediate recurrence score has a 14% chance of her breast cancer recurring in the next 10 years and an almost 100% chance of experiencing adverse effects with a chemotherapy regimen; she may decide not to have additional chemotherapy. Alternatively, a 35-year-old woman with the same recurrence score may opt to have additional chemotherapy.

In a recent meta-analysis evaluating the impact of Oncotype Dx on chemotherapy decisions, 48.8% of women had a low recurrence score, 39% had an intermediate risk score, and 12.2% had a high score (Carlson 2013). The Oncotype Dx test results changed the chemotherapy recommendations in 33.4% of patients. After Oncotype Dx testing, 28.2% of women overall received chemotherapy—5.8% in the low-risk group, 37.4% in the intermediate-risk group, and 83.4% in the high-risk group. Patients with a high recurrence score were significantly more likely to follow the treatment suggested by the Oncotype Dx test and have chemotherapy compared with patients having a low recurrence score, who opted to have chemotherapy despite their low risk of recurrence with a risk ratio (RR) of 1.07 (1.01–1.14). Oncotype Dx testing is routinely recommended in women with early-stage breast cancer to assess the risk of recurrence and guide therapeutic decisions regarding the need for chemotherapy.

Predictive biomarkers are some of the most important developments in oncology and hematology during the previous 15 years. Predictive biomarkers provide evidence about the anticipated benefit or toxicity from a medication. The earliest example of a predictive biomarker in oncology and hematology is the story of the Philadelphia chromosome, chronic myelogenous leukemia (CML), and the tyrosine kinase inhibitor imatinib (Koretzky 2007).

The Philadelphia chromosome, a reciprocal translocation between chromosomes 9 and 22, was identified in cells obtained from a patient with CML in 1960 by Dr. Peter Nowell of the University of Pennsylvania. This was the first time a cancer had been linked with a specific genetic mutation, which led Nowell to hypothesize that mutations in cancer cells gave them a selective growth advantage. This fundamental understanding formed the basis of almost all subsequent and current work in cancer biology, though it took 30 years to prove that the Philadelphia chromosome caused CML, rather than being the consequence of CML. The Philadelphia chromosome results in a fusion protein, breakpoint cluster region–Abelson murine leukemia viral oncogene homolog 1 (*BCR-ABL*), which is an oncogenic tyrosine kinase. Imatinib was originally identified by high-throughput screening looking for tyrosine kinase inhibitors in the late 1980s (Sherbenou 2007). The first phase I clinical trial of imatinib started in 1998 (Druker 2001), and 53 of the 55 enrolled patients had a complete hematologic remission with little reported toxicity, forming the basis for imatinib's label approval by the U.S. Food and Drug Administration (FDA) in 2001.

Imatinib revolutionized the treatment of CML. Before its introduction, patients with CML were treated with interferon, an intervention that improved survival over no treatment by about 20 months, with substantial toxicity. The current treatment of CML is discussed extensively in the Chronic Myelogenous Leukemia section.

Imatinib also revolutionized drug development, going from discovery to approval in about 12 years, and from phase I testing to approval in less than 3 years, dramatically reducing the cost and duration of clinical trials required for a drug approval. Several targeted agents (discussed later in this chapter) have followed a similar developmental pathway.

Practical Implications

Tissue

A cancer tissue sample is essential for all analysis. Because somatic mutations do not occur in every cell of the body, mutations will not be present in typical germline DNA sources like buccal cells or peripheral blood mononuclear lymphocytes (Normanno 2011). Tissue is obtained by performing a biopsy of the cancerous region and may be prepared by flash freezing in liquid nitrogen or by formalin-fixed paraffin embedding. Although flash freezing is usually preferred for RNA-based applications because of the speed of freezing, paraffin embedding is certainly adequate for DNA-based sequencing applications and is preferred for analysis by immunohistochemistry (IHC) (Ross 2012).

Circulating tumor cells are thought to be shed from the primary tumor and detectable in the bloodstream. Many assays for assessing circulating tumor cells are available, including IHC, reverse transcriptase polymerase chain reaction (RT-PCR), and the most commonly used CellSearch system (Watanabe 2014). CellSearch is a commercially available system approved by the FDA for measuring circulating tumor cells that uses immunomagnetic purification from peripheral blood with antibodies against epithelial cell adhesion molecule. Currently, most circulating tumor cell detection methods give a count of the cells, which has been used as a prognostic indicator in several solid tumors, but experimental approaches are using isolated circulating tumor cells for downstream analysis, including mutation detection (Huang 2013).

Use of circulating cell-free DNA is an emerging strategy for assessing somatic mutations (Esposito 2014). Circulating cell-free DNA is readily accessible because it can be isolated from human plasma, serum, and other body fluids and does not require a repeat biopsy for analysis. Circulating cell-free DNA has been detectable in normal volunteers since the 1940s and, in patients with cancer, since the 1970s. In patients with cancer, plasma DNA is thought to originate primarily from tumor cells because these cells are expected to be undergoing more frequent necrosis and apoptosis and releasing DNA. Many studies have shown that somatic mutations can be identified in circulating cell-free DNA; however, further study of the clinical significance of mutations identified in circulating cell-free DNA is needed before routine clinical implementation.

Solid Tumors

Breast Cancer
Breast cancer is the most frequently diagnosed cancer in women in the United States. The treatment of breast cancer is quite complex, incorporating surgery, radiation, hormonal therapy, and chemotherapy (NCCN 2014a). Treatment is individualized in all patients, regardless of stage; all patients have their tumors assessed for hormone receptor status as well as human epidermal growth factor receptor 2 (HER2) expression. Both HER2-targeting therapy and hormonal therapy are used in the adjuvant setting (after complete surgical removal of the tumor) to prevent a local recurrence and for the treatment of metastatic cancer.

Estrogen and Progestin Receptors
The estrogen receptor (ER) is a central target in breast cancer treatment, and the prototypical antiestrogen targeting this pathway, tamoxifen, was approved for use in 1977 (Cadoo 2013). Inhibition of the ER through endocrine targeting (selective ER modulators) or by indirectly blocking the conversion of androgens to estrogen (aromatase inhibitors) remains a widely used therapy in breast cancer, and all women with a diagnosis of breast cancer are routinely assessed for ER and progesterone receptor (PR) positivity by IHC. Estrogen receptor positivity occurs in about 75% of all breast cancer cases. Estrogen-targeting therapies such as antiestrogens and aromatase inhibitors are indicated only in women whose tumor expresses ER (see Table 10.1). The PR is an estrogen-regulated gene, and its expression is therefore indicative of a functioning ER pathway. Progesterone receptor–positive tumors constitute between 55% and 65% of all breast cancer cases and have a better prognosis than PR-negative tumors. Estrogen receptor–targeting therapies can be used in PR-negative tumors if ER is expressed.

Although tumors are currently designated ER positive or negative, emerging data suggest that ER-positive tumors can be further subcategorized as weak (1+), moderate (2+), and strong (3+) (Bartlett 2011). Moreover, in postmenopausal women with ER-positive early breast cancer treated with either tamoxifen or exemestane, high PR or ER expression is associated with an improved patient prognosis, with a hazard ratio (HR) of high PR of 0.53 (95% confidence interval [CI], 0.43–0.65; $p<0.001$) and high ER HR of 0.66 (95% CI, 0.51–0.86; $p=0.002$), although high PR or ER versus low ER or PR do not predict response to exemestane or tamoxifen.

Everolimus
Everolimus is indicated for the treatment of postmenopausal women with advanced hormone receptor–positive, HER2-negative breast cancer in combination with exemestane after treatment with letrozole or anastrozole fails, according to the recently completed BOLERO-2 study (Baselga 2012). The trial compared exemestane and placebo with exemestane and everolimus in patients with HER2-negative metastatic breast cancer whose disease was resistant to letrozole or anastrazole. A recently identified mechanism of hormonal resistance is that a substrate of the mammalian target of rapamycin (mTOR) complex 1 called S6 kinase 1 phosphorylates the activation function domain 1 of the ER, which is responsible for ligand-independent receptor activation; therefore, the rationale of the BOLERO-2 trial was that inhibition of the mTOR pathway by everolimus could overcome hormonal resistance. The BOLERO-2 study met its primary end point, with a median progression-free survival (PFS) of 6.9 months for everolimus plus exemestane versus 2.8

Table 10.1. Routine Pharmacogenetic Testing in Solid Tumors

Disease	Target	Assay	Affected Medications	Clinical Implications
Breast cancer	ER	IHC	Antiestrogens, aromatase inhibitors, everolimus	Not indicated if ER negative
Breast cancer	HER2	IHC or FISH	Trastuzumab, pertuzumab, lapatinib	Not indicated if HER2 +1 or less
Colon cancer	*KRAS*	Sequencing	Cetuximab, panitumumab	Not indicated if *KRAS* mutant
Colon cancer	*NRAS*	Sequencing	Cetuximab, panitumumab	Not indicated if *NRAS* mutant
Lung cancer	*EGFR*	Sequencing	Erlotinib, gefitinib, afatinib	Indicated if activating mutation present
Lung cancer	*EGFR*	Sequencing	Afatinib	May be active if resistance mutation present
Lung cancer	*EML4-ALK*	FISH	Crizotinib, ceritinib[a]	Indicated if translocation present
Melanoma	*BRAF*	Sequencing	Dabrafenib, trametinib, vemurafenib	Indicated if mutation present

[a]Crizotinib is indicated in the first-line setting; ceritinib is indicated if crizotinib resistance develops.
FISH = fluorescence in situ hybridization; IHC = immunohistochemistry.

months for placebo plus exemestane (HR for progression or death 0.43; 95% CI, 0.35–0.54; p<0.001). Although ER status appears central to the mechanism of the synergy of exemestane and everolimus, HER2 status is likely unimportant. Individuals with HER2-negative disease were a convenient study population for the BOLERO-2 study, and the FDA indication is currently limited to HER2-negative disease; however, recent reports suggest that the combination of exemestane and everolimus is effective in patients with ER- and HER2-positive breast cancer as well (Yardley 2013).

HER2 Oncogene

The role of HER2 has been appreciated in breast cancer for decades (Gradishar 2013). HER2 is an oncogene that encodes the type I receptor tyrosine kinase HER2, which stimulates proliferation, migration, and invasion in breast cancer, with amplification in around 15%–20% of breast cancers.

Testing for HER2 is typically performed with IHC, with a confirmatory fluorescence in situ hybridization (FISH) test, and patients' disease is categorized as HER2 positive or negative for *HER2* gene amplification. Before the development of HER2-targeting therapies, HER2 negativity was used as a prognostic factor; however, the HER2-targeted agents trastuzumab, lapatinib, pertuzumab, and trastuzumab emtansine are now used exclusively in HER2-positive disease.

Trastuzumab, Ado-trastuzumab Emtansine, and Pertuzumab

Trastuzumab is a humanized monoclonal antibody that was originally approved in 1998 (Gradishar 2013). It targets HER2 by binding to the extracellular domain. By doing so, it possibly down-regulates HER2 and decreases cell signaling. In addition, it may possess general antibody-related cytotoxic effects. Trastuzumab is indicated for breast cancers overexpressing HER2, in both the metastatic and adjuvant setting. Pertuzumab is a more recent addition and is a recombinant monoclonal antibody that targets the extracellular dimerization domain of HER2. It is used in combination with trastuzumab to enhance HER2 inhibition. Ado-trastuzumab emtansine is also a recently approved HER2-targeted antibody but is conjugated to a microtubule inhibitor and used as a single agent in patients with HER2-positive breast cancer. Although each of the three antibodies has different indications, all require HER2 testing before use.

Lapatinib

Lapatinib is a small-molecule, reversible inhibitor of both epidermal growth factor receptor (*EGFR*) and HER2 tyrosine kinases that can be administered orally. Lapatinib is approved for combination therapy with capecitabine in the treatment of advanced HER2-positive breast cancer. On binding to the intracellular tyrosine kinase domain of *EGFR* and HER2, it inhibits autophosphorylation and blocks downstream signaling mechanisms, resulting in either apoptosis or growth arrest. Lapatinib causes tumor inhibition only in cases of *EGFR* and HER2 overexpression. Several ongoing and completed clinical studies also support lapatinib use as monotherapy or as an adjuvant with trastuzumab, though these uses have not received FDA label approval (Gradishar 2013).

Colon Cancer

Colorectal cancer remains the third most common cancer among men and women in the United States. The National Cancer Institute (NCI) estimates 96,830 new cases of colon cancer and 50,310 deaths caused by colon cancer in 2014 (ACS 2014). Standard first-line treatment of metastatic cancer is a fluorouracil-based regimen in combination with oxaliplatin (FOLFOX) or irinotecan (FOLFIRI) (NCCN 2014b). Cetuximab and panitumumab are monoclonal antibodies targeting the *EGFR* receptor that may be added to FOLFIRI regimens in patients who are wild type for the Kirsten RAS viral oncogene homolog (*KRAS*) or the neuroblastoma RAS viral oncogene homolog (*NRAS*), given that several studies and a meta-analysis have shown they are inactive in patients with colon cancer having a *KRAS* or *NRAS* mutation (Linardou 2008). Routine testing of all colon tumors for *KRAS* and *NRAS* is recommended (see Table 10.1).

KRAS and *NRAS*

The *KRAS* and *NRAS* are G proteins that are involved in downstream signaling in the activated *EGFR* pathway. Common point mutations in the *KRAS* gene lead to the replacement of an amino acid in codon 12, 13, 61, 117, or 146, which are critical to normal function. The most commonly mutated is codon 12, with G12D, G12C, and G12S accounting for around 35%, 20%, and 6% of all *KRAS* mutations, respectively. Codon 13 is the next most frequently mutated, with G13D accounting for about 20% of all *KRAS* mutations. Several additional mutations in codons 12, 13, 61, 117, and 146 constitute less than 1% of *KRAS* mutations individually. *KRAS* gene mutations result in increasing cell proliferation because of the decreased intrinsic GTPase activity of the protein, and it is assumed that they are involved early in tumorigenesis. About 40% of patients with colorectal cancer and 25% of patients with non–small cell lung cancer (NSCLC) are reported to have *KRAS* mutations (Russo 2014).

NRAS is also in the Ras family and is almost identical to *KRAS* with the exception of the 40-amino acid residues at the C-terminus. Like *KRAS*, it is mutated in codons 12, 13, and 61, with codon 12 the most frequently mutated, followed by codon 13. *NRAS* mutations are less frequent, occurring in around 2% of patients with colon cancer and 12% of patients with rectal cancer (Russo 2014).

Cetuximab and Panitumumab

Cetuximab and panitumumab are monoclonal antibodies that bind to the *EGFR* protein with high affinity and inhibit downstream signaling and *KRAS*. Cetuximab is approved for *KRAS* wild-type cancer either as monotherapy or in combination with irinotecan. Cetuximab is also FDA label approved in combination with FOLFIRI for first-line treatment of metastatic colon cancer. Panitumumab is approved in combination with FOLFOX for first-line treatment and as monotherapy following disease progression after prior treatment with fluoropyrimidine-, oxaliplatin-, and irinotecan-containing chemotherapy. Despite small differences in indications, the agents are considered interchangeable by clinical practice guidelines (NCCN 2014b).

Cetuximab's approval in the first-line setting was based on the CRYSTAL (Cetuximab Combined with Irinotecan in First-line Therapy for Metastatic Colorectal Cancer) trial, a phase III clinical trial that randomized 1217 patients with newly diagnosed *EGFR*-expressing metastatic colon cancer to FOLFIRI with or without cetuximab (Van Dutsem 2011). Of the 540 patients enrolled in the trial with tissue blocks for *KRAS* analysis, 192 (43%) had a *KRAS* mutation. The primary end point of this trial was PFS, with secondary end points of overall survival (OS), response rate, disease control rate, and safety. In patients with wild-type *KRAS*, PFS was 9.9 months in those treated with FOLFIRI plus cetuximab and 8.7 months in the control group (HR 0.68; p=0.017). For patients with mutant *KRAS*, PFS was 7.6 months in those treated with FOLFIRI plus cetuximab and 8.1 months in the control group (HR 1.07; p=0.47).

These results suggest that patients with a *KRAS* mutation do not benefit from adding cetuximab to FOLFIRI chemotherapy and may fare worse because of the increased toxicity of an ineffective treatment. However, this trial does not clearly establish the efficacy of FOLFIRI in combination with cetuximab because *KRAS* mutation was not a stratification criterion or primary end point of either trial and because the analysis was conducted retrospectively in a subset of patients. The difference between avoiding an *EGFR* inhibitor in a patient with a *KRAS* mutation, which is clearly supported by the available data, and using an *EGFR* inhibitor in all patients with wild-type *KRAS*, which is not supported by evidence, is a subtle yet critically important distinction.

In fact, adding cetuximab or panitumumab to oxaliplatin-based chemotherapy in first-line treatment of metastatic colon *KRAS* wild type did not improve efficacy, survival, or response rate. As shown in a recent meta-analysis (Zhou 2012) that included four clinical trials and 1270 patients, adding cetuximab or panitumumab does not result in significant improvement in OS compared with chemotherapy alone (HR 1.00, 95% CI, 0.88, 1.13; p=0.95) or in PFS (HR 0.86; 95% CI, 0.71, 1.04; p=0.13). Although guidelines recommend that all patients have their tumor tested for *KRAS*, cetuximab or panitumumab in combination with oxaliplatin is not recommended, and cetuximab- or panitumumab-based irinotecan combinations are not recommended over other regimens, even for patients with *KRAS* wild-type tumors (NCCN 2014b).

The v-raf murine sarcoma viral oncogene homolog B (*BRAF*), most commonly *BRAF* V600E, is mutated in about 6% of colon cancers. This mutation is clinically important in melanoma, with targeted agents clinically available. See the Melanoma section for a detailed description of *BRAF* and available agents. Clinical trials testing these agents in *BRAF*-mutated colon cancer are ongoing, and *BRAF* testing may be incorporated into routine clinical testing and therapeutic decision-making in the future if trial results are positive (NCCN 2014b). Microsatellite instability, thought to be related to mutations in genes responsible for DNA repair, occurs in about 10% of patients with colon cancer (NCCN 2014b). Microsatellite instability is typically assessed by sequencing five or more DNA regions in both the tumor and the adjacent normal tissue and comparing them. If novel alleles are identified in the tumor, it is taken as evidence of microsatellite instability. Although *BRAF* testing and assessment of microsatellite instability have prognostic value and are sometimes performed, neither is currently used for making treatment decisions (NCCN 2014b).

Lung Cancer

Non–Small Cell Lung Cancer

Lung cancer remains the leading cause of cancer death in the United States for both men and women. It has an estimated incidence of 224,210 and estimated deaths of 159,260 for 2014 (ACS 2014). Non–small cell lung cancer, which makes up 87% of lung tumor diagnoses, usually presents as an incurable locally advanced or metastatic disease. Despite major research efforts, OS remains dismally low because only 17% of all patients with lung cancer are expected to live 5 years after diagnosis (ACS 2014).

Historically, treatment of advanced NSCLC has been doublet chemotherapy with carboplatin and paclitaxel, cisplatin and paclitaxel, carboplatin and docetaxel, or cisplatin and docetaxel. All four regimens are considered equivalent to each other and more efficacious than placebo, but they have significant and sometimes fatal toxicity, including myelosuppression and neuropathy. The development of targeted therapies directed toward *EGFR* and echinoderm microtubule–associated protein-like 4 (*EML4-ALK*) mutations is one of the most significant advances in the treatment of NSCLC to date, and all patients with NSCLC will be tested for these mutations at diagnosis and sometimes if progression occurs (see Table 10.1). Only 15%–20% of patients have these mutations, and most patients will still receive standard cytotoxic chemotherapy (NCCN 2014c).

EGFR Mutations

Activating mutations in the *EGFR* gene are found in around 10%–20% of all patients with advanced NSCLC and in more than 50% of adenocarcinomas and tumors from East Asians, never-smokers, and women (Westwood 2014). Activating mutations primarily occur in exons 18–21 of *EGFR*, with a substitution at codon 858 (L858R) accounting for 40% of mutations and in-frame deletions at exon 19 accounting for another 40%–45%. The most common resistance mutation is located at T790M and is thought to arise from clonal selection during treatment with *EGFR* inhibitors. The mutation conveys resistance to erlotinib and gefitinib, though cells remain sensitive to afatinib, an irreversible inhibitor of *EGFR*.

Erlotinib, Gefitinib, and Afatinib

Erlotinib and gefitinib were both approved initially in unselected (*EGFR* mutation status unknown) populations; however, reanalysis after approval showed that almost all the benefit was observed in patients with an *EGFR*-activating mutation. After this observation, two landmark trials, one comparing gefitinib with standard chemotherapy and the other comparing erlotinib with standard chemotherapy for patients with NSCLC having an *EGFR* mutation, showed the superiority of targeted therapy with erlotinib or gefitinib over standard cytotoxic chemotherapy.

In a phase III trial (EURTAC) comparing erlotinib with standard doublet chemotherapy, the median PFS in the erlotinib group was 9.7 months (95% CI, 8.4–12.3) compared with 5.2 months (95% CI, 4.5–5.8) in the standard chemotherapy group (HR 0.37; 95% CI, 0.25–0.54; p<0.0001) (Rosell 2012). The most frequent grade 3 or 4 toxicities were rash, occurring in 13% of patients receiving erlotinib and none in the chemotherapy group, and neutropenia, occurring in 22% of the standard chemotherapy group and none in the erlotinib group. Five patients (6%) taking erlotinib had treatment-related severe adverse events compared with 16 patients (20%) receiving chemotherapy.

Similarly, in a phase III trial comparing gefitinib with standard chemotherapy with carboplatin and paclitaxel in patients with activating *EGFR* mutations, the gefitinib group had a significantly longer median PFS (10.8 months vs. 5.4 months in the chemotherapy group; HR 0.30; 95% CI, 0.22–0.41; p<0.001), as well as a higher response rate (Maemondo 2010). The incidence of grade 3 or higher adverse effects was significantly higher in the chemotherapy group than in the gefitinib group (71.7% vs. 41.2%, p<0.001). Of note, OS did not differ between the groups when the trial was initially reported or after long-term follow-up (Inoue 2013).

More recently, the irreversible *EGFR* inhibitor afatinib was studied in the LUX-Lung 3 study, a phase III trial comparing afatinib with cisplatin plus pemetrexed chemotherapy in patients having metastatic lung adenocarcinoma with *EGFR* mutations. Median PFS, the primary end point of the study, was significantly higher in the afatinib-treated patients (11.1 months) than in patients treated with cisplatin plus pemetrexed (6.9 months) (95% CI, 0.43–0.78; p=0.001), though OS did not differ between the groups (Sequist 2013). Afatinib also has activity in patients with a T790M resistance mutation (Katakami 2013), suggesting afatinib is a potential treatment option for patients developing resistance to other *EGFR* inhibitors.

Together, the results strongly suggest that that *EGFR* inhibitors, erlotinib, gefitinib, and afatinib are more effective with respect to PFS and response rate as well as less toxic than standard chemotherapy in patients with an activating *EGFR* mutation. In the United States, both erlotinib and afatinib are approved for first-line therapy in patients with an activating *EGFR* mutation (NCCN 2014c). Although they have not been compared with each other, afatinib's activity in erlotinib-resistant disease suggests it is more active; however, given erlotinib's expected generic debut, the anticipated place in therapy is first-line erlotinib, followed by afatinib at disease progression. Although generally more effective and less toxic than standard chemotherapy, *EGFR* inhibitors do not appear to change the overall disease course, with OS similar to that with standard therapy. A summary of each agent's place in therapy and required genetic testing can be found in Table 10.1.

EML4-ALK Translocations
The anaplastic lymphoma kinase (*ALK*) gene codes for a transmembrane tyrosine kinase receptor that, when activated, inhibits apoptosis and promotes cell growth. Several *ALK* fusion partners have been identified in cancer, though the most common fusion partner in *ALK*-positive NSCLC adenocarcinoma is *EML4-ALK* (Soda 2007). An inversion on the short arm of chromosome 2 generates a fusion between the N-terminus of the *EML4* and the C-terminus of the *ALK* gene, resulting in increased kinase activity. Truncations of *EML4* occurring at a variety of exons, including exons 2, 6, 13, 14, 15, 18, or 20, are most often fused to exon 20 (in the kinase domain) of the *ALK* gene. There are at least nine different *EML4-ALK* fusion combinations, called variants, with variant 1 (exon 13) and variant 3 (exon 6) being the most common, constituting around 80% of all *EML4-ALK* fusions (Pao 2011).

Crizotinib and Ceritinib
The activity of crizotinib was initially noted in a phase I clinical trial (Camidge 2012). Of the initial 37 patients enrolled in the trial, two patients with NSCLC and harboring *EML4-ALK* mutations had a response to therapy, which prompted the opening of an expansion cohort for patients with *ALK*-positive NSCLC. In this study of 81 patients, 71 (87%) had disease control after 8 weeks, with 46 (57%) having at least partial response and subsequent tumor shrinkage and an additional 27 (33%) having stable disease. One patient had a complete response. This trial led to the approval of crizotinib for patients with NSCLC and an *ALK* mutation.

This study was followed up with a phase III trial comparing crizotinib with standard chemotherapy, showing a median PFS of 7.7 months in the crizotinib group compared with 3.0 months in the chemotherapy group (HR for progression or death with crizotinib 0.49; 95% CI, 0.37–0.64; p<0.001). Response rates and quality of life were also improved in the crizotinib group, but no significant improvement occurred in OS. The incidence of grade 3 or greater adverse events was similar between the two groups (33% with crizotinib and 32% with chemotherapy) (Shaw 2013).

Acquired resistance mutations are implicated in about one-third of patients with *ALK*-rearranged NSCLC. Mutations typically occur within the *ALK* tyrosine kinase domain, or there may be amplification of the *ALK* fusion gene (Steuer 2014). A next-generation *ALK* inhibitor, ceritinib, is 20 times as potent as crizotinib against *ALK*, and in animal models, ceritinib had antitumor activity against both crizotinib-sensitive and crizotinib-resistant tumors (Marsilje 2013). Ceritinib was recently approved for the treatment of patients with *EML4-ALK*–positive, metastatic NSCLC who have disease progression or who are intolerant of crizotinib. The approval was based on a single phase I dose escalation and an expanded cohort study of single-agent ceritinib, where the maximal tolerated dose was determined to be 750 mg orally per day. All patients had *ALK* mutation–positive cancer and had previously received therapy, with 122 patients having NSCLC; of these patients, 83 had previously received crizotinib, suggesting that their disease was crizotinib resistant. Among 114 patients with NSCLC who received at least 400 mg of ceritinib per day, the overall response rate was 58% (95% CI, 48–67), and among 80 patients who had received crizotinib previously and received at least 400 mg of ceritinib, the response rate was 56% (95% CI, 45–67) (Shaw 2014).

Like *EGFR* inhibitors, *EML4-ALK* inhibitors are clearly more effective in terms of response rate and progression-free survival; and less toxic in patients with NSCLC having an *ALK* mutation. Given the FDA label-approved indications, crizotinib should be considered first-line therapy in patients with newly diagnosed cancers, and patients can be switched to ceritinib if they experience disease progression.

Melanoma

The ACS estimates the occurrence of 76,100 new cases of melanoma and 9710 deaths caused by melanoma in 2014 (ACS 2014). Before the approval of targeted agents, advanced melanoma was treated with single-agent dacarbazine, with reported median survival from historical phase III trials ranging from 5.6 to 9.7 months (NCCN 2014d). Three new agents are currently available targeting mutant *BRAF*, and all patients with melanoma are routinely tested (see Table 10.1).

BRAF Mutations

Mutations in the *BRAF* protein kinase are present in more than 50% of metastatic melanoma cases and in about 7%–8% of all cancers (Xia 2014). *BRAF* has more than 30 distinct mutations, with V600E being the most common variation in melanoma. V600K mutations are the next most common, and all other mutations are rare. *BRAF* is a serine/threonine protein kinase that activates the MEK/MAPK/ERK-signaling pathway, which mediates cellular responses to growth signals. A mutation in *BRAF* results in bypassing the RAS activation, which is a requirement for downstream signaling and regulation in the MEK/MAPK/ERK pathway; a *BRAF* mutation leads to cell proliferation, invasion, and survival (Safaee Ardekani 2012).

Dabrafenib, Trametinib, and Vemurafenib

Vemurafenib is an adenosine triphosphate (ATP)-competitive, reversible, and highly selective *BRAF* kinase inhibitor targeted at the mutant *BRAF* V600E. It was approved because of results from the BRIM-3 trial, which was a randomized phase III trial comparing vemurafenib with dacarbazine in patients with unresectable, previously untreated stage IIIc or IV melanoma that was positive for the *BRAF* V600 mutation (McArthur 2014). The co-primary end points were OS and PFS. Long-term follow-up was recently reported, with a median OS of 13.6 months (95% CI, 12.0–15.2) in the vemurafenib group compared with 9.7 months (7.9–12.8) in the dacarbazine group and an HR for death in the vemurafenib group of 0.70 (95% CI, 0.57–0.87; p=0.0008), showing the superiority of vemurafenib over standard chemotherapy.

Like vemurafenib, dabrafenib is a reversible, ATP-competitive selective inhibitor of *BRAF* V600E kinase. Dabrafenib was also compared with dacarbazine in patients with previously untreated metastatic melanoma that was positive for a *BRAF* V600 mutation in a randomized open-label phase III trial (Hauschild 2012). In this trial, investigator-assessed PFS was the primary end point, with an estimated median PFS for the dabrafenib group of 5.1 months compared with 2.7 months in the dacarbazine group, with an HR for progression in the dabrafenib group of 0.30 (95% CI, 0.18–0.51; p<0.0001). There was no difference in OS between the groups; however, the follow-up time was relatively short.

Trametinib is an orally available, small-molecule, selective inhibitor of map kinase 1 (MEK1) and map kinase 2 (MEK2), the downstream targets of *BRAF*, which inhibit tumors with *BRAF* V600E mutations in vitro. Trametinib was studied in a phase III trial that enrolled patients with unresectable stage IIIC or IV cutaneous melanoma with a V600E or V600K *BRAF* mutation (Flaherty 2012b). The primary end point was investigator-assessed PFS, which was 4.8 months in the trametinib group compared with 1.5 months in the chemotherapy group, with an HR for progression of 0.45 (95% CI, 0.33–0.63; p<0.001). Overall survival was also improved in the trametinib group, with a 6-month OS rate of 81% in the trametinib group and 67% in the chemotherapy group and an HR for death in the trametinib group of 0.54 (95% CI, 0.32–0.92; p=0.01). In addition to its approval as a single agent, trametinib is approved in combination with dabrafenib, a combination that inhibits the pathway at both *BRAF* V600E and MEK1 and MEK2 (NCCN 2014e).

Dabrafenib, trametinib, and vemurafenib are all recently approved agents for patients with unresectable metastatic melanoma that have a *BRAF* V600E mutation. All three agents are approved as initial therapy in this patient population and have been compared with dacarbazine but not with each other. Until comparative studies have been performed, choice of agent can be based on individual patient characteristics and drug toxicity profiles.

Hematologic Malignancies

Acute Myelogenous Leukemia

Acute myelogenous leukemia (AML) is the most common type of leukemia, with 18,860 new cases and 10,460 deaths estimated in the United States in 2014 (ACS 2014). The prognosis for AML is poor, with a 5-year survival of only 24%. Acute myelogenous leukemia is divided into several different types, one of which is acute promyelocytic leukemia (APL), which accounts for 5%–10% of AML cases (NCCN 2014e). Acute promyelocytic leukemia is caused by a translocation involving the retinoic acid receptor-alpha (RARα) locus on chromosome 17 and the promyelocytic leukemia (PML) gene on chromosome 15, which creates the fusion gene *PML/RAR*. This fusion gene then produces a chimeric protein that stops the maturation of myeloid cells at the promyelocytic stage, causing them not to differentiate (Saeed 2011). This translocation is evaluated in all patients with AML, either by FISH or RT-PCR, and is used to specifically identify the APL variant of AML as well as to monitor response to therapy (Tallman 2010) (see Table 10.2).

Arsenic Trioxide, Tretinoin

All-*trans* retinoic acid (tretinoin) is a retinoid that binds to the RAR, producing an initial maturation of the primitive promyelocytes. Tretinoin, which was initially approved in the United States in 1995, is generally regarded as one of the biggest success stories in hematology. When used as single-agent induction therapy in APL, response rates approach 70%, and when combined with standard cytotoxic chemotherapy, response rates exceed 90%, with an almost 80% cure rate (Wang 1999). This is in sharp contrast to the pre-tretinoin era, when APL was universally fatal within a few months, as well as the current prognosis for other variants of AML, which are still essentially untreatable.

Arsenic is thought to target the *PML* portion of the fusion protein, resulting in degradation (Huang 2012). Initially studied in patients with APL with relapsed disease, 85% achieved a clinical complete response, with 18-month overall and relapse-free survival estimates of 66% and 56%, respectively. Arsenic is indicated in patients who are refractory to, or have relapsed after, retinoid anthracycline induction therapy.

Because arsenic and tretinoin target different parts of the RAR/PML fusion protein, they have also been studied in combination, though trial results have been inconsistent (Wang 2011). A recently reported meta-analysis compared arsenic alone with the combination of arsenic and tretinoin, combining first-line and relapsed therapies. Results showed that combination therapy significantly increased the complete remission rate (RR 1.08; 95% CI, 1.00–1.17, p=0.04) and improved the 1-year disease-free survival rate (RR 1.22; 95% CI, 1.00–1.50, p=0.05), with no significant difference in early death rates or adverse effects. However, results were not significant for either first-line or relapsed populations when considered separately, and the combination should not be used routinely.

Chronic Myelogenous Leukemia

Chronic myelogenous leukemia is relatively uncommon, accounting for about 15% of all adult leukemias, with an anticipated 5980 new cases diagnosed and 810 deaths in 2014 (ACS 2014). Chronic myelogenous leukemia is characterized by a reciprocal translocation between chromosomes 9 and 22 t(9;22), known as the Philadelphia chromosome. This translocation generates a head-to-tail fusion gene, *BCR-ABL*, which in turn produces a constitutively active tyrosine kinase that increases proliferation, reduces differentiation, and inhibits apoptosis. All patients are assessed for the presence of the *BCR-ABL* fusion gene by FISH or RT-PCR, and transcript levels are monitored to assess drug response (see Table 10.2).

Imatinib

Imatinib, the first approved tyrosine kinase inhibitor for use in CML, inhibits the ATP binding site of *BCR-ABL*. Imatinib was compared with interferon and cytarabine, the previous standard therapy for CML, in the IRIS trial, a randomized phase III study of patients with newly diagnosed chronic-phase CML (O'Brien 2003). The trial was initially reported after a median follow-up of 19 months, and the primary end point was event-free survival, with an estimated rate of freedom from progression to accelerated-phase or blast-crisis CML of 96.7% in the imatinib group compared with 91.5% in the combination-therapy group (p<0.001). Major cytogenetic response was 87.1% (95% CI, 84.1–90.0) in the imatinib group compared with 34.7% (95% CI, 29.3–40.0) in the standard therapy group (p<0.001), and the estimated rates of complete cytogenetic response were 76.2% (95% CI, 72.5–79.9) compared with 14.5% (95% CI, 10.5–18.5), respectively (p<0.001), though there was no difference in survival. Grade III–IV adverse events were more common with combination therapy. Six-year follow-up was subsequently reported for this trial; however, because patients were allowed to cross over to imatinib for adverse effects or disease progression and most patients (65%) who were treated with combination therapy crossed over to imatinib, the survival follow-up was confined to patients initially randomized to imatinib. After 6 years of follow-up, the estimated freedom from progression was 93% (compared with 96.7% after 18 months of follow-up), and the OS rate of all patients randomized to imatinib was 88% (95% CI, 85%–92%). When deaths unrelated to CML were excluded and patients undergoing a transplant were censored at transplantation, the estimated OS rate was 95% at 6 years (95% CI, 92%–97%) (Hochhaus 2009).

Although most patients with CML treated with imatinib respond well to therapy for many years, a subset (16%–26%) of patients develop resistance. The first report of imatinib clinical resistance came in 2001, when investigators identified mutations in the imatinib binding site of *BCR-ABL* in 6 of the 11 patients studied with clinical resistance and in 3 of the 11 patients with genomic amplification of the *BCR-ABL* gene (Gorre 2001). Other reports soon followed, and currently, more than 90 imatinib resistance mutations with varying frequency and phenotypic effects have been reported. Moreover, several second-generation tyrosine kinase inhibitors were developed in an attempt to overcome imatinib resistance (Soverini 2011).

Imatinib, Dasatinib, and Nilotinib
Imatinib, dasatinib, and nilotinib are all approved for first-line treatment of CML, though dasatinib and nilotinib were originally approved for patients who developed imatinib resistance mutations; therefore, patients who develop resistance while on first-line imatinib therapy can be switched to dasatinib or nilotinib. However, if either dasatinib or nilotinib is used in the first-line setting, a change to imatinib is not recommended.

Evaluation of resistance mutations is recommended in all patients after 3 months of therapy if response is suboptimal, defined as more than a 10% increase in *BCR-ABL* transcripts or failure to achieve at least a partial cytogenetic response in the bone marrow. The resistance mutation identified can help direct subsequent therapy.

For example, mutations at positions F317L/V/I/C, T315A, and V299L are less sensitive to dasatinib, and nilotinib is preferred, whereas Y253H, E255K/V, and F359V/C/I mutations are less sensitive to nilotinib, and dasatinib is preferred (Cortes 2012a).

Bosutinib
Bosutinib is only approved for patients with CML who have either resistance to or intolerance of prior therapy. When was compared with imatinib in a trial of previously untreated patients with CML, the primary end point was not achieved; the complete cytogenetic response rate at 12 months was not significantly better for bosutinib (70%; 95% CI, 64%–76%) compared with imatinib (68%; 95% CI, 62%–74%; p=0.601) (Cortes 2013). However, in a patient with a resistance mutation at E255K/V, F317L/V/I/C, F359 V/C/I, T315A, or Y253H, bosutinib is an option.

Ponatinib and Omacetaxine
Ponatinib is a recently approved tyrosine kinase inhibitor that is active against unmutated and mutated *BCR-ABL*, including the threonine-to-isoleucine mutation at position 315 (T315I). This mutation is the most frequently identified resistance mutation, occurring in up to 20% of patients with resistant disease. The T315I mutation is resistant to all other approved *BCR-ABL* tyrosine kinase inhibitors. In a phase II clinical trial enrolling patients in all phases of CML who were resistant to or intolerant of nilotinib or dasatinib or had developed a T315I mutation, the primary end point of major cytogenetic response by 12 months was observed in 56% (95% CI, 50–62) of those in chronic phase (Cortes 2013). Unfortunately, ponatinib has a black box warning for both hepatoxicity and arterial thrombotic events, and use is restricted to patients with a T315I mutation.

Omacetaxine is a protein synthesis inhibitor that was recently approved for patients with CML for whom

Table 10.2. Routine Pharmacogenetic Testing in Hematologic Malignancies

Disease	Target	Assay	Affected Medications	Clinical Implications
AML	PML/RARα	FISH, RT-PCR	Arsenic, tretinoin	Indicated if translocation present
CML	*BCR-ABL*	FISH	Imatinib, dasatinib, nilotinib, bosutinib[a]	Indicated if translocation present
CML	T351I	Sequencing	Ponatinib, omacetaxine	Indicated if deletion present
MDS	5q deletion	FISH	Lenalidomide	Indicated if deletion present

[a]Imatinib, dasatinib, and nilotinib are indicated in the first-line setting; bosutinib is indicated for resistant disease.
FISH = fluorescence in situ hybridization; RT-PCR = reverse transcriptase polymerase chain reaction.

two prior tyrosine kinase inhibitors had failed (Cortes 2012b). Because its mechanism of action is independent of *BCR-ABL*, it has shown activity in patients with a T351I mutation. Although omacetaxine causes almost universal grade III–IV neutropenia, thrombocytopenia, and anemia, these are typically manageable, and no treatment-related deaths occurred in a clinical trial with 48 subjects. Given the adverse effects of ponatinib, most clinicians would recommend omacetaxine to patients with a T351I mutation.

Myelodysplastic Syndromes
Myelodysplastic syndromes (MDS) is a clonal myeloid disorder characterized by anemia, neutropenia, or thrombocytopenia or some combination of the three (NCCN 2014f). The biggest concern is the risk of progression to AML, which is usually untreatable. Myelodysplastic syndromes associated with a 5q deletion is one subtype of MDS in which patients typically present with macrocytic anemia and become transfusion-dependent.

Lenalidomide
Although the mechanism of lenalidomide is not completely defined, in vitro studies suggest it inhibits proliferation and induces apoptosis of 5q-MDS cells. Its clinical efficacy was evaluated in a randomized placebo-controlled phase III study comparing lenalidomide with placebo in transfusion-dependent patients with MDS with del5q31 (Fenaux 2011). The primary end point was transfusion independence for 26 weeks or more, which was achieved in 56.1% of patients taking lenalidomide 10 mg, 42.6% in patients taking lenalidomide 5 mg, and 5.9% in patients taking placebo, with both lenalidomide groups significantly better than placebo (p<0.001). Patients who were transfusion-independent for 8 weeks or more had 47% and 42% reductions in the relative risks of death and AML progression or death, respectively (p=0.021 and 0.048).

Future Directions
The current state of pharmacogenetically directed drug discovery in oncology uses a retrospective extreme phenotype-to-genotype approach. An investigational agent is tested in an unselected population, and exceptional responders are sequenced with whole genome or exome sequencing to identify the mutations potentially associated with response. Subsequent studies then focus on a patient population with the identified mutation. This approach has been successful for developing both crizotinib and erlotinib. Second-generation agents have typically focused on developing more potent agents or agents with a slightly altered binding site to overcome resistance. This approach has also proved very successful, leading to the approval of afatinib, ceritinib, bosutinib, dasatinib, nilotinib, and ponatinib. Depending on drug approvals for a given mutation and disease state, patients are tested clinically for a few actionable mutations.

Pharmacogenetic testing will soon move into prospective trials. The NCI MATCH (molecular analysis for therapy of choice) trial plans to use a prospective genotype-to-phenotype approach in which patient tumor samples will be analyzed for about 200 known mutations using a commercially available panel. From 20 to 25 FDA label-approved targeted agents will be tested outside their current indication in this trial. Subjects will be matched to an investigational agent according to their pharmacogenetic profile. For example, a patient with osteosarcoma having a *BRAF* mutation could receive vemurafenib. All treatment decisions are rule based and specified in the protocol, and if several mutations exist, a single prioritized agent will be used. This trial represents a significant change in disease categorization; cancer has historically been categorized by organ, and drug approvals have been disease-specific. If successful, targeted therapies may be approved for mutations, regardless of organ, in the future. The trial is anticipated to open in fall 2014 (Abrams 2014).

Whole genome or exome sequencing for the prospective selection of therapy in oncology has not yet been reported. This lack of reporting to date is likely related to several factors, including expense, time required to obtain results, and a lack of medications directed toward novel mutations. With advances in sequencing technology solving the expense and time problem, the remaining practical barrier is having a drug for newly discovered mutations. Therefore, the most likely near-term future direction is retrospective sequencing of existing tumors to identify additional mutations, followed by the development of new agents targeting these mutations. Whether whole genome or exome sequencing becomes a clinical reality or remains a research tool is an open question.

Conclusion

Current standard therapy for the treatment of many cancers involves mutation assessment and the use of agents that target these mutations. In all cases, targeted therapies are more effective and usually have fewer adverse effects than standard cytotoxic chemotherapy. In solid tumors, like lung cancer and melanoma, targeted therapies typically prolong the time to disease progression but do not change the disease course because OS remains unchanged. This is probably related to the heterogeneous nature of solid tumors; although a targeted therapy may eliminate the sensitive cells harboring the mutation, this environment allows for the selection of other cells with different mutations.

In contrast, hematologic malignancies like CML and APL have been treated successfully with targeted agents, with cures possible in APL and significant extensions of survival in CML. This may be because of the clonal nature of these hematologic malignancies. Resistance mutations sometimes develop in patients with CML treated with tyrosine kinase inhibitors; however, newer agents developed to treat patients with resistance mutations have been successful. Targeted therapies are clearly an important advance in the war on cancer.

References

Abrams J, Conley B, Mooney M, et al. National Cancer Institute's Precision Medicine Initiatives for the new National Clinical Trials Network. Am Soc Clin Oncol Educ Book 2014:71-6.

American Cancer Society (ACS). Cancer Facts & Figures 2014. Atlanta: ACS, 2014.

Balmaña J, Balaguer F, Cervantes A, et al. ESMO Guidelines Working Group. Familial risk-colorectal cancer: ESMO clinical practice guidelines. Ann Oncol 2013;24(suppl 6):73-80.

Bartlett JM, Brookes CL, Robson T, et al. Estrogen receptor and progesterone receptor as predictive biomarkers of response to endocrine therapy: a prospectively powered pathology study in the Tamoxifen and Exemestane Adjuvant Multinational trial. J Clin Oncol 2011;29:1531-8.

Baselga J, Campone M, Piccart M, et al. Everolimus in postmenopausal hormone-receptor-positive advanced breast cancer. N Engl J Med 2012;366:520-9.

Cadoo KA, Fornier MN, Morris PG. Biological subtypes of breast cancer: current concepts and implications for recurrence patterns. Q J Nucl Med Mol Imaging 2013;57:312-21.

Camidge DR, Bang YJ, Kwak EL, et al. Activity and safety of crizotinib in patients with *ALK*-positive non-small-cell lung cancer: updated results from a phase 1 study. Lancet Oncol 2012;13:1011-9.

Carlson JJ, Roth JA. The impact of the Oncotype Dx breast cancer assay in clinical practice: a systematic review and meta-analysis. Breast Cancer Res Treat 2013;141:13-22.

Cortes JE, Kim DW, Kantarjian HM, et al. Bosutinib versus imatinib in newly diagnosed chronic-phase chronic myeloid leukemia: results from the BELA trial. J Clin Oncol 2012b;30:3486-92.

Cortes JE, Kim DW, Pinilla-Ibarz J, et al. A phase 2 trial of ponatinib in Philadelphia chromosome-positive leukemias. N Engl J Med 2013;369:1783-96.

Cortes JE, Lipton JH, Rea D, et al. Phase 2 study of subcutaneous omacetaxine mepesuccinate after TKI failure in patients with chronic-phase CML with T315I mutation. Blood 2012a;120:2573-80.

Dancey JE, Dobbin KK, Groshen S, et al. Biomarkers Task Force of the NCI Investigational Drug Steering Committee. Guidelines for the development and incorporation of biomarker studies in early clinical trials of novel agents. Clin Cancer Res 2010;16:1745-55.

Druker BJ, Talpaz M, Resta DJ, et al. Efficacy and safety of a specific inhibitor of the *BCR-ABL* tyrosine kinase in chronic myeloid leukemia. N Engl J Med 2001;344:1031-7.

Esposito A, Bardelli A, Criscitiello C, et al. Monitoring tumor-derived cell-free DNA in patients with solid tumors: clinical perspectives and research opportunities. Cancer Treat Rev 2014;40:648-55.

Fenaux P, Giagounidis A, Selleslag D, et al. A randomized phase 3 study of lenalidomide versus placebo in RBC transfusion-dependent patients with low-/intermediate-1-risk myelodysplastic syndromes with del5q. Blood 2011;118:3765-76.

Flaherty KT, Robert C, Hersey P, et al. Improved survival with *MEK* inhibition in *BRAF*-mutated melanoma. N Engl J Med 2012;367:107-14.

Friebel TM, Domchek SM, Rebbeck TR. Modifiers of cancer risk in *BRCA1* and *BRCA2* mutation carriers: systematic review and meta-analysis. J Natl Cancer Inst 2014;106. First published online May 13, 2014.

Gorre ME, Mohammed M, Ellwood K, et al. Clinical resistance to STI-571 cancer therapy caused by *BCR-ABL* gene mutation or amplification. Science 2001;293:876-80.

Gradishar WJ. Emerging approaches for treating HER2-positive metastatic breast cancer beyond trastuzumab. Ann Oncol 2013;24:2492-500.

Hauschild A, Grob JJ, Demidov LV, et al. Dabrafenib in *BRAF*-mutated metastatic melanoma: a multicentre, open-label, phase 3 randomised controlled trial. Lancet 2012;380:358-65.

Hochhaus A, O'Brien SG, Guilhot F, et al. Six-year follow-up of patients receiving imatinib for the first-line treatment of chronic myeloid leukemia. Leukemia 2009;23:1054-61.

Huang BT, Zeng QC, Gurung A, et al. The early addition of arsenic trioxide versus high-dose arabinoside is more effective and safe as consolidation chemotherapy for risk-tailored patients with acute promyelocytic leukemia: multicenter experience. Med Oncol 2012;29:2088-94.

Huang J, Wang K, Xu J, Huang J, et al. Prognostic significance of circulating tumor cells in non-small-cell lung cancer patients: a meta-analysis. PLoS One 2013;8:e78070.

Inoue A, Kobayashi K, Maemondo M, et al. Updated overall survival results from a randomized phase III trial comparing gefitinib with carboplatin-paclitaxel for chemo-naïve non-small cell lung cancer with sensitive *EGFR* gene mutations (NEJ002). Ann Oncol 2013;24:54-9.

Jaradat I, Mubiden R, Salem A, et al. High-dose chemotherapy followed by stem cell transplantation in the management of retinoblastoma: a systematic review. Hematol Oncol Stem Cell Ther 2012;5:107-17.

Katakami N, Atagi S, Goto K, et al. LUX-Lung 4: a phase II trial of afatinib in patients with advanced non-small-cell lung cancer who progressed during prior treatment with erlotinib, gefitinib, or both. J Clin Oncol. 2013;31:3335-41.

Koretzky GA. The legacy of the Philadelphia chromosome. J Clin Invest 2007;117:2030-2.

Linardou H, Dahabreh IJ, Kanaloupiti D, et al. Assessment of somatic k-*RAS* mutations as a mechanism associated with resistance to *EGFR*-targeted agents: a systematic review and meta-analysis of studies in advanced non-small-cell lung cancer and metastatic colorectal cancer. Lancet Oncol 2008;9:962-72.

Maemondo M, Inoue A, Kobayashi K, et al. Gefitinib or chemotherapy for non-small-cell lung cancer with mutated *EGFR*. N Engl J Med 2010;24;362:2380-8.

Marsilje TH, Pei W, Chen B, et al. Synthesis, structure-activity relationships, and in vivo efficacy of the novel potent and selective anaplastic lymphoma kinase (*ALK*) inhibitor 5-chloro-N2-(2-isopropoxy-5-methyl-4-(piperidin-4-yl)phenyl)-N4-(2-(isopropylsulfonyl)phenyl)pyrimidine-2,4-diamine (LDK378) currently in phase 1 and phase 2 clinical trials. J Med Chem 2013;56:5675-90.

McArthur GA, Chapman PB, Robert C, et al. Safety and efficacy of vemurafenib in *BRAF*(V600E) and *BRAF*(V600K) mutation-positive melanoma (BRIM-3): extended follow-up of a phase 3, randomised, open-label study. Lancet Oncol 2014;15:323-32.

National Comprehensive Cancer Network (NCCN). NCCN Guidelines, 2014a. Breast Cancer v3.2014. Available at www.nccn.org/professionals/physician_gls/pdf/breast.pdf. Accessed June 4, 2014.

National Comprehensive Cancer Network (NCCN). NCCN Guidelines, 2014b. Colon Cancer v3.2014, Available at www.nccn.org/professionals/physician_gls/pdf/colon.pdf. Accessed June 4, 2014.

National Comprehensive Cancer Network (NCCN). NCCN Clinical Practice Guidelines, 2014c. Lung Cancer v3.2014. Available at www.nccn.org/professionals/physician_gls/pdf/nscl.pdf. Accessed June 4, 2014.

National Comprehensive Cancer Network (NCCN). NCCN Clinical Practice Guidelines, 2014d. Melanoma v4.2014. Available at www.nccn.org/professionals/physician_gls/pdf/melanoma.pdf. Accessed June 4, 2014.

National Comprehensive Cancer Network (NCCN). NCCN Clinical Practice Guidelines, 2014e. Acute Leukemia v2.2014. Available at www.nccn.org/professionals/physician_gls/pdf/aml.pdf. Accessed June 4, 2014.

National Comprehensive Cancer Network (NCCN). NCCN Clinical Practice Guideline, 2014f. Myelodysplastic Syndromes v2.2014. Available at www.nccn.org/professionals/physician_gls/pdf/mds.pdf. Accessed June 4, 2014.

Nelson HD, Fu R, Goddard K, et al. Risk Assessment, Genetic Counseling, and Genetic Testing for *BRCA*-Related Cancer: Systematic Review to Update the U.S. Preventive Services Task Force Recommendation [Internet]. Rockville, MD: Agency for Healthcare Research and Quality (US), 2013.

Normanno N, Pinto C, Castiglione F, et al. *KRAS* mutations testing in colorectal carcinoma patients in Italy: from guidelines to external quality assessment. PLoS One 2011;6:e29146.

O'Brien SG, Guilhot F, Larson RA, et al. Imatinib compared with interferon and low-dose cytarabine for newly diagnosed chronic-phase chronic myeloid leukemia. N Engl J Med 2003;348:994-1004.

Paik S, Shak S, Tang G, et al. A multigene assay to predict recurrence of tamoxifen-treated, node-negative breast cancer. N Engl J Med 2004;351:2817-26.

Pao W, Girard N. New driver mutations in non-small-cell lung cancer. Lancet Oncol 2011;12:175-80.

Rosell R, Carcereny E, Gervais R, et al. Erlotinib versus standard chemotherapy as first-line treatment for European patients with advanced *EGFR* mutation-positive non-small-cell lung cancer (EURTAC): a multicentre, open-label, randomised phase 3 trial. Lancet Oncol 2012;13:239-46.

Ross JS. Clinical implementation of *KRAS* testing in metastatic colorectal carcinoma: the pathologist's perspective. Arch Pathol Lab Med 2012;136:1298-307.

Russo AL, Borger DR, Szymonifka J, et al. Mutational analysis and clinical correlation of metastatic colorectal cancer. Cancer 2014;120:1482-90.

Saeed S, Logie C, Stunnenberg HG, et al. Genome-wide functions of PML-RARα in acute promyelocytic leukaemia. Br J Cancer 2011;104:554-8.

Safaee Ardekani G, Jafarnejad SM, Tan L, et al. The prognostic value of *BRAF* mutation in colorectal cancer and melanoma: a systematic review and meta-analysis. PLoS One 2012;7:e47054.

Sequist LV, Yang JC, Yamamoto N, et al. Phase III study of afatinib or cisplatin plus pemetrexed in patients with metastatic lung adenocarcinoma with *EGFR* mutations. J Clin Oncol 2013;31:3327-34.

Shaw AT, Kim DW, Mehra R, et al. Ceritinib in *ALK*-rearranged non-small-cell lung cancer. N Engl J Med 2014;370:1189-97.

Shaw AT, Kim DW, Nakagawa K, et al. Crizotinib versus chemotherapy in advanced *ALK*-positive lung cancer. N Engl J Med 2013;368:2385-94.

Sherbenou DW, Druker BJ. Applying the discovery of the Philadelphia chromosome. J Clin Invest 2007;117:2067-74.

Soda M, Choi YL, Enomoto M, et al. Identification of the transforming *EML4-ALK* fusion gene in non-small-cell lung cancer. Nature 2007;448:561-7.

Soverini S, Hochhaus A, Nicolini FE, et al. *BCR-ABL* kinase domain mutation analysis in chronic myeloid leukemia patients treated with tyrosine kinase inhibitors: recommendations from an expert panel on behalf of European LeukemiaNet. Blood 2011;118:1208-15.

Steuer CE, Ramalingam SS. *ALK*-positive non-small cell lung cancer: mechanisms of resistance and emerging treatment options. Cancer 2014;120:2392-402.

Tallman M, Douer D, Gore S, et al. Treatment of patients with acute promyelocytic leukemia: a consensus statement on risk-adapted approaches to therapy. Clin Lymphoma Myeloma Leuk 2010;10(suppl 3):S122-6.

Van Cutsem E, Köhne CH, Láng I, et al. Cetuximab plus irinotecan, fluorouracil, and leucovorin as first-line treatment for metastatic colorectal cancer: updated analysis of overall survival according to tumor *KRAS* and *BRAF* mutation status. J Clin Oncol 2011;29:2011-9.

Wang H, Chen XY, Wang BS, et al. The efficacy and safety of arsenic trioxide with or without all-*trans* retinoic acid for the treatment of acute promyelocytic leukemia: a meta-analysis. Leuk Res 2011;35:1170-7.

Wang Z, Sun G, Shen Z, et al. Differentiation therapy for acute promyelocytic leukemia with all-*trans* retinoic acid: 10-year experience of its clinical application. Chin Med J (Engl) 1999;112:963-7.

Watanabe M, Serizawa M, Sawada T, et al. A novel flow cytometry-based cell capture platform for the detection, capture and molecular characterization of rare tumor cells in blood. J Transl Med 2014;12:143.

Westwood M, Joore M, Whiting P, et al. Epidermal growth factor receptor tyrosine kinase (EGFR-TK) mutation testing in adults with locally advanced or metastatic non-small cell lung cancer: a systematic review and cost-effectiveness analysis. Health Technol Assess 2014;18:1-166.

Xia J, Jia P, Hutchinson KE, et al. A meta-analysis of somatic mutations from next generation sequencing of 241 melanomas: a road map for the study of genes with potential clinical relevance. Mol Cancer Ther 2014;13:1918-28.

Yardley DA, Noguchi S, Pritchard KI, et al. Everolimus plus exemestane in postmenopausal patients with HR(+) breast cancer: BOLERO-2 final progression-free survival analysis. Adv Ther 2013;30:870-84.

Zhou SW, Huang YY, Wei Y, et al. No survival benefit from adding cetuximab or panitumumab to oxaliplatin-based chemotherapy in the first-line treatment of metastatic colorectal cancer in *KRAS* wild type patients: a meta-analysis. PLoS One 2012;7:e50925.

Chapter 11

GERMLINE PHARMACOGENETICS IN ONCOLOGY

Daniel L. Hertz, Pharm.D., Ph.D.; and Anthony J. Perissinotti, Pharm.D., BCOP

Learning Objectives

1. Distinguish between the somatic (tumor) and germline (patient) genome and describe how the germline genome influences cancer treatment outcomes.
2. Assess the readiness of clinical implementation for each of the pharmacogenetic associations discussed.
3. Justify the lower threshold of evidence necessary for using existing genetic data compared with ordering a genetic test for a patient.
4. Design a prospective study for each of the pharmacogenetic associations discussed that would fill a critical gap in the current knowledge.
5. Argue that the clinical recommendations made for validated genetic variants should (or should not) be generalized to other variants that are known to have similar functional consequences.
6. Develop a hierarchy of factors that must be considered when evaluating whether to change a therapeutic recommendation because of a known genetic variant.

Keywords: Cancer, oncology, germline genome, *DPYD*, *TPMT*, *UGT1A1*, *CYP2D6*

Abbreviations in This Chapter

CPIC	Clinical Pharmacogenetics Implementation Consortium
DPYD	Dihydropyrimidine dehydrogenase (gene)
MAF	Minor allele frequency
NCCN	National Comprehensive Cancer Network
SLCO1B1	Solute carrier organic anion transporter family member 1B1 (gene)
SNP	Single nucleotide polymorphism
TA	Thymidine-adenosine
TGMP	Thioguanine monophosphate
TPMT	Thiopurine methyltransferase
UGT1A1	UDP glucuronosyltransferase 1A1

Abstract

The germline genome is relevant for predicting drug exposure and a patient's likelihood of experiencing efficacy and toxicity during cancer therapy. Several pharmacogenetic associations are either currently used in clinical practice or ready to be implemented, particularly in patients with existing genetic data (i.e., purine antimetabolites and thiopurine methyltransferase (TPMT, encoded by *TPMT*), pyrimidine antimetabolites, dihydropyrimidine dehydrogenase (gene) (*DPYD*), irinotecan, and UDP glucuronosyltransferase 1A1, encoded by *UGT1A1*). Some germline oncology pharmacogenetic associations are not at that point of maturity but may someday be clinically useful (i.e., methotrexate and solute carrier organic anion transporter family member 1B1 gene [*SLCO1B1*]), whereas others seem unlikely to ever be translated into clinical care (i.e., tyrosine kinase inhibitors and *UGT1A1*, tamoxifen, and cytochrome P450 2D6 [*CYP2D6*]). The number of patients with cancer for whom germline genetic information exists will continue to expand as the frequency of tumor genetic analyses to guide treatment selection increases. This uniquely positions oncology as an efficient translational opportunity for pharmacogenetics; however, clinicians are reluctant to make preemptive therapeutic modifications that may diminish cancer treatment efficacy. The ideal balance of treatment benefit and risk is a critical consideration in oncology and one in which pharmacists should have a substantial contribution to therapeutic decision-making.

Introduction

The accepted paradigm in cancer treatment is to treat patients at the threshold of tolerable toxicity in order to maximize treatment effectiveness. It is essential that oncology decision-making consider both effectiveness and toxicity, which are influenced by two genomes: that of the patient (germline) and that of the tumor (somatic). This chapter will cover germline pharmacogenetic factors; somatic pharmacogenetics is covered in chapter 10.

The influence of the germline genome on drug exposure has been investigated most extensively; thus, the pharmacogenetic associations that are closest to clinical implementation involve genetic variation in enzymes and transporters. Changes in the expression or activity of enzymes or transporters can influence systemic or cellular exposure of the active agent, leading to efficacy or toxicity (Figure 11.1). Understanding the role of the protein (enzyme/transporter) in drug disposition is necessary to predict the consequences of genetic variability and the appropriate therapeutic modification.

Pharmacists are ideally equipped to make pharmacogenetic-guided treatment decisions because of their expertise in pharmacotherapy and experience with integrating complex, often imperfect, information into therapeutic decision-making. Many pharmacogenetic decisions will require extrapolation from a highly structured clinical trial population to an individual patient with a different genetic variant, treatment regimen, or tumor type. When extrapolating to a different clinical situation, the pharmacist must integrate the genetic information with other factors, such as drug interactions, comorbidities, and patient preferences, to maintain the proper balance between treatment effectiveness and toxicity risk. This chapter covers several examples of germline genetic variation that may be useful for guiding cancer therapy to optimize treatment outcomes.

Purine Antimetabolites and TPMT

Background

Two purine antimetabolites, mercaptopurine and thioguanine, are used in oncology, where they remain key drugs in curative regimens for acute lymphoblastic leukemia and lymphomas. A third purine antimetabolite, azathioprine, is a direct mercaptopurine prodrug used in non-malignant conditions. Azathioprine will not be specifically discussed in this section, though some of the evidence contained in this section is extrapolated from studies of azathioprine in non-oncology patients. Mercaptopurine and thioguanine are metabolized through independent pathways to thioguanosine monophosphate (TGMP), which is then metabolized in several steps to active thioguanine nucleotides that are incorporated into DNA or RNA, interfere with DNA replication, and induce cytotoxicity. Thiopurine methyltransferase (TPMT, encoded by *TPMT*) catalyzes an alternative metabolic deactivation pathway for TGMP (Figure 11.2).

Genetic Variation

Several genetic variants with clinically relevant effects on TPMT activity have been identified (Table 11.1). The two best-studied non-synonymous variants (*2 and *3) (Tai 1996; Krynetski 1995) have decreased TPMT activity, partly because of faster protein degradation (Tai 1997). The *3A genotype is actually a haplotype containing two linked non-synonymous variants that individually are called *3B (rs1800460) and *3C (rs1142345). These three single nucleotide polymorphisms (SNPs) constitute greater than 90% of the inactivating mutations. The *TPMT*4 variant is a very rare splice site variant at exon 10 that creates a truncated, non-active protein (Otterness 1998). Other extremely rare, putatively low-activity variants have also been reported and curated by the TPMT Nomenclature Committee (Appell 2013).

Diminished TPMT activity leads to a shunting of TGMP metabolism toward greater production of active

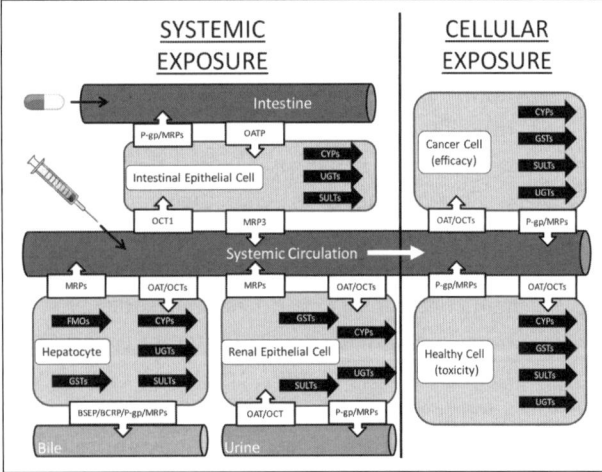

Figure 11.1. Germline genetics determines the expression and function of enzymes (black arrows) and transporters (white boxes with arrows), which dictate systemic and cellular drug exposure. Enzyme and transporter activity in organs responsible for drug absorption (intestine) or elimination (liver, kidney) dictate systemic drug exposure, which indirectly determines treatment efficacy and/or toxicity. Enzyme and transporter activity in cancer and healthy cells dictates cellular exposure, which is more directly relevant to efficacy and toxicity, respectively. Enzyme and transporter activity in cancer cells is also dictated by changes in the tumor (somatic) genome. The exact enzymes and transporters expressed in each cell type are still being catalogued, as are the functionally consequential polymorphisms in each gene.

Table 11.1. Validated *TPMT* Variants

rsID	* Designation	Change in TPMT Protein	Phenotype	Minor Allele Frequency (MAF)[a]			
				White	African	Mexican	Asian
rs1800462	*2	Ala80Pro	Low activity	0.00190	0.000873	0.00592	0
rs1800460/ rs1142345	*3A	Ala154Thr/ Tyr240Cys	Low activity	0.0354	0.00218	0.0533	0.000119
rs1800460	*3B	Ala154Thr	Low activity	0.000461	0	0.00690	0
rs1142345	*3C	Tyr240Cys	Low activity	0.004207	0.0480	0.00888	0.0157
rs1800584	*4[b]	Splice site variant	No activity	—	—	—	—

[a]MAF as reported in: Supplemental table S3 in Relling MV, Gardner EE, Sandborn WJ, et al. Clinical pharmacogenetics implementation consortium guidelines for thiopurine methyltransferase genotype and thiopurine dosing: 2013 update. Clin Pharmacol Ther 2013;93:324-5. See CPIC guidelines for additional variants and populations.

[b]*4 is an extremely rare non-functional allele found mainly in South American individuals (MAF = 0.000485).

thioguanine nucleotides (Lennard 2013), the accumulation of which enhances the toxicity (Relling 1999) and efficacy (Lennard 1990) of purine antimetabolite treatment. When treated with full-dose therapy, 35% of *TPMT* heterozygous patients and 100% of homozygous patients will require dose reductions, compared with only 7% of patients without mutations (Relling 1999). Coincident with this increased toxicity, however, is the enhanced treatment efficacy in patients carrying low-activity genotypes (Stanulla 2005), which complicates therapeutic decision-making. Some studies report that decreasing doses in patients with low-activity *TPMT* genotypes could limit toxicity without decreasing efficacy (Relling 2006), but recent prospective genotype-guided studies have reported a coincidental decrease in treatment efficacy (Levinsen 2014).

Clinical Use

In the United States, 1 in 300 patients has complete TPMT deficiency, and 11% carry a single defective allele (Weinshilboum 1980). With 6020 new cases of acute lymphoblastic leukemia each year, many *TPMT*-deficient patients are exposed to mercaptopurine and thioguanine (Siegel 2014). In clinical practice, TPMT deficiency is diagnosed according to either phenotypic testing of erythrocyte TPMT activity or genotyping for *TPMT* gene mutations. Several clinical considerations should be made before using phenotypic testing. First, patients who have received a red blood cell (RBC) transfusion within 60–90 days could have spurious results because of the donor's RBCs (Schwab 2001). Second, mercaptopurine by itself can increase erythrocyte TPMT activity by around 20%, whereas concomitant medications such as sulfasalazine and allopurinol can inhibit TPMT activity (McLeod

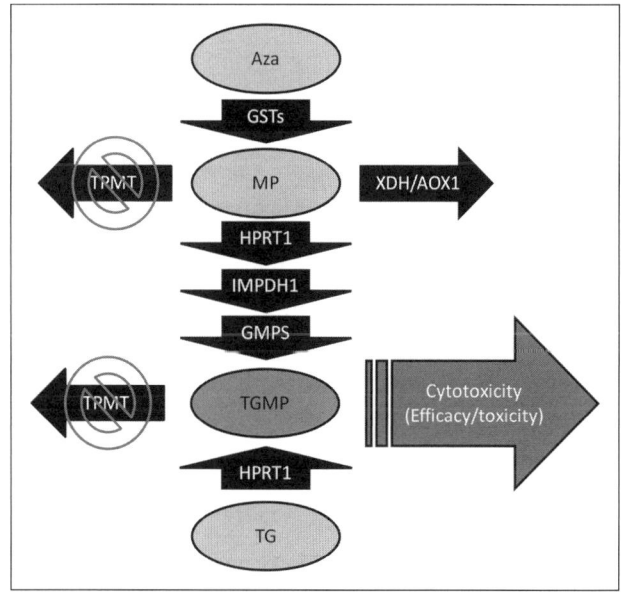

Figure 11.2. Azathioprine (Aza) is metabolized by glutathione S-transferases (GSTs) to mercaptopurine. Mercaptopurine and thioguanine are metabolized by independent pathways to TGMP, which is metabolized to active thioguanine metabolites that interfere with DNA replication and induce cytotoxicity. TPMT catalyzes an alternative metabolic pathway for deactivating TGMP and mercaptopurine. Genetic polymorphisms that diminish TPMT activity shunt metabolism toward TGMP and overproduction of active thioguanine nucleotides, increasing the toxicity as well as the efficacy of purine antimetabolite therapy. For a more comprehensive description of purine antimetabolite pharmacokinetics and pharmacodynamics, see the PharmGKB (Pharmacogenetics Knowledge Base) pathway diagram at https://www.pharmgkb.org/pathway/PA2040.

1995; Szumlanski 1995). In these situations, genotyping is preferred because it can assist in avoiding misclassification, though generally, results of TPMT genotyping and phenotyping are highly concordant (Hindorf 2012).

The U.S. Food and Drug Administration (FDA)-approved drug label includes a warning for patients with known TPMT deficiency, in whom treatment decisions should be individualized. The FDA has made no specific dosing recommendations, but guidelines that contain generally consistent recommendations have been published by the Clinical Pharmacogenetics Implementation Consortium (CPIC) and a similar Dutch consortium (Relling 2013; Swen 2011). The specific treatment recommendation depends on the patient's diplotype and the purine antimetabolite being administered (Table 11.2). Patients who carry a low-activity allele should initiate therapy at 30%–70% of the normal dose, with dose titration based on tolerability. Homozygous patients must be dosed cautiously and monitored closely, particularly with thioguanine treatment. Known homozygous patients should initiate therapy at 10% of the normal dose, with the dosing frequency changed from daily to 3 days per week, again with titration based on tolerability or possibly thioguanine metabolite levels.

Use of preemptive genotyping is becoming more common (Bell 2014; Teng 2014; Fusaro 2013), partly because of the support of consensus guidelines such as those of the National Comprehensive Cancer Network (NCCN). The NCCN Clinical Practice Guidelines in Oncology (NCCN Guidelines) for Acute Lymphoblastic Leukemia, V.1.2014, advocate for preemptive TPMT genotyping before initiating mercaptopurine or thioguanine as a routine safety measure (NCCN 2014a). In practice, patients identified as TPMT deficient receive either empiric dose reduction or full-dose therapy with enhanced myelotoxicity monitoring. The NCCN Guidelines further recommend TPMT testing in patients with unknown TPMT activity who experience disproportionate myelosuppression while undergoing treatment, particularly if it necessitates a delay in therapy of greater than 2 weeks. This can be particularly informative for patients who received combination therapy, such as with methotrexate. Diagnosis of TPMT deficiency in this patient would enable the reduction of purine antimetabolite dosing without the risk of thioguanine nucleotide concentrations decreasing below efficacious levels and maintenance of full-dose methotrexate (Relling 1999). Patients who experience severe myelosuppression on a purine antimetabolite and are negative for TPMT genotyping should be considered for phenotypic testing of erythrocyte TPMT activity or other potential causes of myelosuppression.

One of the main obstacles to the clinical use of prospective TPMT genotyping is the possibility that decreasing doses may limit treatment efficacy. Because more than 50% of heterozygous patients can tolerate full-dose therapy, it is critical that patients who are empirically dose reduced be monitored for myelosuppression to guide dose titration. This should assuage oncologists' apprehension regarding empiric dose reduction and further motivate the clinical uptake of preemptive TPMT genotyping.

Pyrimidine Antimetabolites and DPYD

Background

The pyrimidine antimetabolites are effective agents in the treatment of several solid tumors. Fluorouracil is an intravenously infused agent approved for use in metastatic colorectal cancer and head and neck cancer, whereas capecitabine is an oral fluorouracil prodrug used in colorectal, breast, and other solid tumors. Fluorouracil is not by itself efficacious; a small amount of the administered dose is intracellularly bioactivated to

Table 11.2. Recommended Starting Doses of Purine Antimetabolites in Patients Carrying Validated TPMT Variants

TPMT Genotype	Mercaptopurine Dosing	Thioguanine Dosing
Heterozygous	30%–70% of full dose	50%–70% of full dose
	2–4 weeks to reach steady state	2–4 weeks to reach steady state
Homozygous	10% of full dose given three times/week	10% of full dose given three times/week
	4–6 weeks to reach steady state	4–6 weeks to reach steady state

[a]Dosing recommendations adapted from: Table 2 in Relling MV, Gardner EE, Sandborn WJ, et al. Clinical pharmacogenetics implementation consortium guidelines for thiopurine methyltransferase genotype and thiopurine dosing: 2013 update. Clin Pharmacol Ther 2013;93:324-5. See CPIC guidelines for additional dosing information.

fluorodeoxyuridine monophosphate, which inhibits thymidylate synthase, an enzyme with a central role in folate metabolism that is necessary for cellular replication. Most administered fluorouracil is not bioactivated, but instead, it can cause severe, potentially life-threatening toxicities, including myelosuppression, gastrointestinal (GI) toxicity, mucositis/stomatitis, and hand-foot syndrome. The rate-limiting step for fluorouracil elimination is conversion to inactive metabolites, catalyzed by *DPYD* (Figure 11.3).

Genetic Variation

Patients who had low DPYD activity had higher fluorouracil exposure and consequently greater toxicity (Diasio 1998). Low DPYD activity is partly explained by genetic variability in the *DPYD* gene. Several rare SNPs have been verified to confer diminished DPYD activity in whites (Table 11.3). There is strong evidence linking these rare SNPs to decreased DPYD activity (Boisdron-Celle 2007), increased fluorouracil exposure (Morel 2006), and increased treatment toxicity (Terrazzino 2013; Schwab 2008). However, these variants do not account for the frequency of the DPYD deficiency phenotype (up to 5%), suggesting that additional non-functional genetic variants exist that are yet to be described. Several other variants that may account for the phenotypic diversity have been identified (Rosmarin 2014b; Offer 2013; Teh 2013), but their association with DPYD activity and toxicity risk have not been validated. Treatment recommendations for variants with demonstrated clinical validity should not be extrapolated to unvalidated variants until compelling evidence is gathered from in vivo, in vitro, or clinical association studies (Amstutz 2011).

Clinical Use

Fluoropyrimidines are a backbone chemotherapy class used in many malignancies involving the digestive system (e.g., colorectal, esophageal, pancreatic), breast cancer, and head and neck cancer. Consequently, 2 million patients worldwide receive fluoropyrimidine-based therapies each year (Scrip's Cancer Chemotherapy Report 2002). Although DPYD deficiency is rare, because of the high use of fluorouracil-based regimens, it is estimated that tens of thousands of DPYD-deficient patients will have fluoropyrimidine exposure. Despite this, prospective *DPYD* activity testing is not routinely performed in the clinical setting because of cost, complexity, and lack of availability. Demonstration of clinical utility has been hindered by the rarity of the causative SNPs and their inadequate sensitivity and specificity for predicting toxicity (Rosmarin 2014a). One report of preemptive genotyping identified four carriers of low-activity variants in 180 patients. The two variant carriers who received empiric dose decreases of 33% and 50% still experienced severe

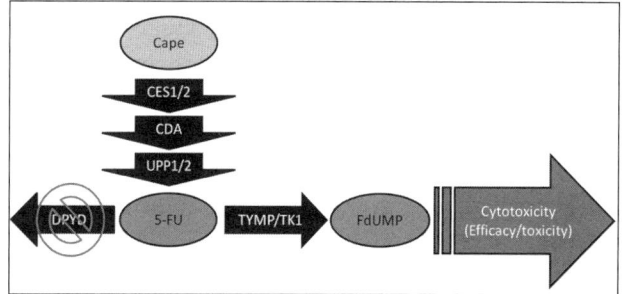

Figure 11.3. Capecitabine (cape) is an oral prodrug metabolized to fluorouracil (5-FU). 5-FU is intracellularly bioactivated through several steps to 5-fluorodeoxyuridine monophosphate (FdUMP), which inhibits thymidylate synthase, causing cellular cytotoxicity. The rate-limiting step for 5-FU elimination is an alternative metabolic pathway catalyzed by DPYD. Genetic variation in *DPYD* leads to diminished DPYD activity, resulting in increased 5-FU exposure and toxicity from pyrimidine antimetabolite therapy. For a more comprehensive description of pyrimidine antimetabolite pharmacokinetics and pharmacodynamics, see the PharmGKB (Pharmacogenetics Knowledge Base) pathway diagram at https://www.pharmgkb.org/pathway/PA150653776.

Table 11.3. Validated Nonfunctional *DPYD* Variants

| | | | | Minor Allele Frequency (MAF)[a] | | |
rsID	* Designation	Change in DPYD Protein	Activity	White	African	Asian
rs3918290	*2A	Splice site modification	None	0.00862	0	0.0015
rs55886062	*13	Ile560Ser	None	0.001	0.042	0.018
rs67376798	—	Asp949Val	None	0.0111	N/A	N/A

[a]MAF as reported in: Supplemental table S3 in Caudle KE, Thorn CF, Klein TE, et al. Clinical Pharmacogenetics Implementation Consortium guidelines for dihydropyrimidine dehydrogenase genotype and fluoropyrimidine dosing. Clin Pharmacol Ther 2013;94:640-5. See CPIC guidelines for additional variants with unknown function and allele frequencies in additional populations.

toxicity (Magnani 2013), suggesting that a further dose decrease is necessary in these high-risk patients.

The NCCN Guidelines have no specific recommendations for the treatment of DPYD-deficient patients or for *DPYD* testing (NCCN 2014c, 2014d). However, the FDA package insert includes a warning for patients with a known DPYD deficiency. The CPIC guidelines recommend treatment starting at less than 50% of the normal dose, followed by titration according to tolerability, for patients carrying a low-activity variant. Homozygous DPYD-deficient patients have complete DPYD deficiency, and treatment with a pyrimidine analog should not be attempted (Table 11.4) (Caudle 2013). Because prospective *DPYD* testing is not routinely performed, DPYD deficiency is usually diagnosed after a patient experiences severe toxicities during fluoropyrimidine treatment. *DPYD* testing is then performed while aggressive supportive care is initiated to manage toxicity. This genetic information is critical to the patient's care because of the large dose reduction required in DPYD-deficient patients. In addition, fluoropyrimidines are typically given in combination with other chemotherapies that cause similar toxicities, and diagnosing DPYD deficiency informs the appropriate selection of drug for reduction, preserving the dosing and efficacy of the others.

Several alternative methods of DPYD phenotypic activity assessment have been piloted (van Staveren 2013). A prospective study of 65 patients with head and neck cancers showed that in vivo DPYD activity–guided fluorouracil dosing diminished toxicity (9% vs. 22%) without decreasing efficacy compared with patients in a historical control group (Yang 2011). A prospective randomized study that confirms these findings could be practice changing, establishing the clinical utility of dosing pyrimidine antimetabolites by DPYD activity.

Table 11.4. Recommended Starting Doses of Pyrimidine Antimetabolites in Patients Carrying Validated *DPYD* Variants

DPYD Genotype	5-FU, Capecitabine, or Tegafur Dosing
Heterozygous	< 50% of full dose
Homozygous	Select alternative drug

^aDosing recommendations adapted from: Table 2 in Caudle KE, Thorn CF, Klein TE, et al. Clinical Pharmacogenetics Implementation Consortium guidelines for dihydropyrimidine dehydrogenase genotype and fluoropyrimidine dosing. Clin Pharmacol Ther 2013;94:640-5. See CPIC guidelines for additional dosing information.
5-FU = fluorouracil.

Irinotecan, Tyrosine Kinase Inhibitors, and UGT1A1

Background

Irinotecan is a chemotherapeutic agent used in combination therapy for several malignancies, including colon, rectal, esophageal, gastric, head and neck, lung, pancreatic, and other solid tumors. Treatment efficacy and toxicity, primarily neutropenia and diarrhea, are caused by exposure to the active metabolite of irinotecan, SN-38. Irinotecan is bioactivated to SN-38 by several carboxylesterases, and SN-38 is then deactivated through glucuronidation by UDP-glucuronosyltransferase isoform 1A1 (UGT1A1), encoded by *UGT1A1* (Figure 11.4).

Protein expression of UGT1A1 is affected by genetic variation in the promoter region, which contains a thymidine-adenosine (TA) repeat sequence. In wild-type patients, the TA is found six times, but some individuals have more than six TA repeats, which decreases protein expression and glucuronidation activity (Bosma 1995). The most common variant in whites has one additional repeat, $(TA)_7$, and is known as *UGT1A1*28* (rs8175347). This variant is relatively common in white (minor allele frequency [MAF] of about 0.30) and African (MAF of about 0.5) populations and somewhat less common in Asians (MAF of about 0.12).

Genetic Variation

The *UGT1A1*28* variant increases SN-38 exposure in vitro and in vivo (Iyer 1999; Ando 1998). The variant has also been associated, through this mechanism, with increased treatment toxicity (Innocenti 2004; Iyer 2002), particularly at higher doses (Hoskins 2007) and when combined with fluorouracil treatment (Liu 2014). Of interest, the association between the genotype and treatment efficacy, which might be expected to coincide with the toxicity association, has not been validated in meta-analyses of the reported studies (Liu 2013; Dias 2012), suggesting that genotype-guided dose adjustment minimizes toxicity without a corresponding decrease in treatment efficacy. Prospective dose-finding studies in preemptively genotyped patients confirm that *UGT1A1*28* homozygous patients can tolerate much lower doses (Marcuello 2011), whereas removal of these patients enables further escalation to doses in excess of those used clinically (Toffoli 2010). Similar studies have been carried out in Japanese patients, who cannot tolerate the doses used in whites (Takano 2013; Sunakawa 2012; Satoh 2011), perhaps because of the high frequency in Asian patients of another low-activity variant, *UGT1A1*6* (Cheung 2014).

Clinical Use

The association between *UGT1A1*28* and treatment toxicity, without a corresponding relationship with efficacy, supports the potential clinical utility of preemptive genotyping for irinotecan dose individualization. Doses have been identified for use in prospective phase II studies currently under way (NCT01523431, NCT01963182, NCT02138617), and future phase III studies showing improved efficacy, or decreased toxicity without a corresponding decrease in efficacy, would establish the clinical utility of *UGT1A1*-guided irinotecan dosing.

Currently, the evidence supporting preemptive *UGT1A1* genotyping is inadequate for clinical implementation, and the NCCN colon and rectal cancer panels recommend against routine *UGT1A1* testing, even in patients who experience irinotecan toxicity, because these patients will require a dose reduction regardless of the *UGT1A1* genotype result (NCCN 2014c, 2014d). Despite this recommendation, some clinicians may still choose to obtain *UGT1A1* genotype information in patients who are highly sensitive to treatment-related toxicity, primarily to determine which element of combination therapy is responsible for the toxicity being observed and should be dose reduced.

Alternatively, existing *UGT1A1* genetic data should be considered when making irinotecan treatment decisions. The FDA package insert specifically recommends a dose reduction in patients known to be homozygous for the *UGT1A1*28* allele, though it does not offer recommendations for the starting dose, and CPIC has not published guidelines for this gene-drug pair. The only treatment guidelines currently available are from the Dutch Pharmacogenomics Working Group, which recommends decreasing the dose by 30% in patients homozygous for *UGT1A1*28* who would otherwise receive high-dose (greater than 250 mg/m^2) therapy (Swen 2011). Of note, no recommendations exist for dosing changes for irinotecan administered at less than 250 mg/m^2, the typical doses used outside of a clinical trial.

Patients with a medical history of Crigler-Najjar or Gilbert syndrome or those presenting with an isolated unconjugated bilirubin elevation can be predisposed to irinotecan toxicity. These syndromes can be caused by genetic polymorphisms in *UGT1A1* and can be used as a surrogate marker for patients who potentially harbor a mutation. The NCCN recommends that irinotecan be used with caution and at a decreased dose in these patients; however, no dose reduction guidelines are given, and patients are typically treated with full doses and monitored more closely (NCCN 2014c, 2014d).

Tyrosine Kinase Inhibitors

Since the discovery that imatinib is a highly effective treatment for BCR-ABL translocated chronic myelogenous leukemia (O'Brien 2003), tyrosine kinase inhibitors have been developed for treating a variety of tumors driven by aberrant kinase signaling (Gschwind 2004). However, several of the small molecule tyrosine kinase inhibitors are known to cause hyperbilirubinemia in some patients, particularly those patients with high bilirubin levels before treatment. As described in the previous section, elevated bilirubin can be caused by low UGT1A1 expression secondary to the *UGT1A1*28* variant (Bosma 1995). Patients who are homozygous for *UGT1A1*28* are at an increased risk of hyperbilirubinemia when treated with tyrosine kinase inhibitors, including nilotinib (Shibata 2014; Singer 2007) and pazopanib (Motzer 2013; Xu 2010). Unlike the interaction with irinotecan, in which decreased enzyme expression increases drug exposure and toxicity, this interaction seems to be caused by direct UGT1A1 inhibition by this drug class (Fujita 2011). The FDA has included a warning about increased hyperbilirubinemia risk in patients carrying *UGT1A1*28* who receive nilotinib treatment; however, the benign clinical course of transient hyperbilirubinemia limits the clinical relevance of this pharmacogenetic association.

Other Notable Germline Oncology Pharmacogenetic Associations

Methotrexate and *SLCO1B1*

Methotrexate is used to treat several cancers, including both hematologic malignancies (leukemia and

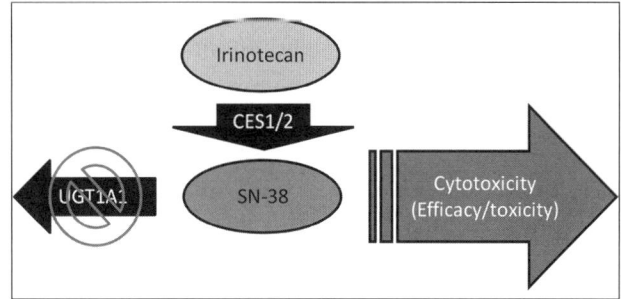

Figure 11.4. Irinotecan is an inactive prodrug that is metabolized by carboxylesterases 1 and 2 (CES1/2) to active SN-38. The rate-limiting step for SN-38 elimination is glucuronidation through UGT1A1. Genetic variation in *UGT1A1*, most notably the *UGT1A1*28* variant, leads to diminished UGT1A1 expression and results in increased SN-38 exposure. Increased SN-38 exposure causes an increase in neutropenia and possibly increased GI toxicity and/or treatment efficacy from irinotecan therapy. For a more comprehensive description of irinotecan pharmacokinetics, see the PharmGKB (Pharmacogenetics Knowledge Base) pathway diagram at https://www.pharmgkb.org/pathway/PA2001#.

lymphoma) and solid tumors (breast and bladder), in addition to autoimmune disorders. Methotrexate treatment is associated with various toxicities, including nephrotoxicity, hepatotoxicity, and GI toxicity. Across these indications, methotrexate is used in a variety of doses, regimens, and delivery forms. In cancers such as osteosarcoma, primary central nervous system lymphomas, and acute lymphoblastic leukemia, methotrexate is used in high dosages. In these regimens, drug exposure is monitored to guide the administration of leucovorin, which arrests the development of toxicity, particularly nephrotoxicity.

Targeting an optimal methotrexate level is a clinically accepted approach, as evidenced by algorithms that adapt methotrexate dosing in future cycles according to drug levels measured in prior cycles (Dombrowsky 2011; Barrett 2008). This suggests that pharmacogenetic predictors of drug clearance could be clinically useful. Methotrexate is eliminated from the systemic circulation primarily by renal elimination, but genetic variation in the solute carrier organic anion transporter family member 1B1 gene (*SLCO1B1*), which encodes the hepatic transporter OATP1B1, partly explains interpatient variability in drug exposure (Radtke 2013; Treviño 2009). A SNP that increases methotrexate exposure (rs4149056) is well known in the pharmacogenetics community because of its importance in causing statin-induced myopathy (covered in chapter 7). It is a non-synonymous variant (Val 174Ala) found across racial groups (MAF in whites is about 0.15, Asians, about 0.13, and Africans, about 0.01) that reduces transporter function and hepatic substrate uptake, resulting in increased systemic exposure and toxicity. Most other variants in *SLCO1B1* that have been associated with methotrexate pharmacokinetics can be explained by linkage disequilibrium with this SNP; however, there is evidence that rare SNPs (Ramsey 2012) and additional SNPs (Ramsey 2013) in this gene have an independent influence and may eventually be validated as causative variants. More supportive data are necessary to confirm that this SNP, which influences methotrexate exposure, is critical for predicting treatment toxicity (Fukushima 2013). Because of the paucity of prospective data, there are currently no treatment guidelines for these patients, though at some point, it is possible that this will be a clinically actionable association.

Tamoxifen and *CYP2D6*

Tamoxifen is used primarily to treat estrogen receptor–positive breast cancer, with some high-risk patients using it for cancer prevention as well. Tamoxifen works as a selective estrogen receptor modulator, antagonizing the estrogenic signaling that is driving cellular replication. The more active metabolite endoxifen is likely responsible for the efficacy of tamoxifen treatment, though this is an area of ongoing investigation.

Several CYPs are involved in the metabolic conversion of tamoxifen to endoxifen, but CYP2D6 is the predominant enzyme. *CYP2D6* is a highly polymorphic gene with many documented common SNPs in addition to relatively common gene duplications and deletions (Ingelman-Sundberg 2008). A phenotype classification system has been developed in which each allele is assigned an activity score of 0, 0.5, or 1. The patient's two alleles are then added to obtain an estimate of their CYP2D6 metabolic activity. Patients are then classified as poor metabolizers (no or little activity), intermediate metabolizers (diminished activity), extensive metabolizers (normal activity), or ultrarapid metabolizers (enhanced activity), as defined in the CPIC guidelines (Crews 2012) and described in chapter 1.

A patient's genotype is a modestly good predictor of their endoxifen exposure (Teft 2013); however, the reported association between endoxifen exposure and efficacy (Madlensky 2011) is not well established, and the endoxifen level may be completely irrelevant in some patients (Dowsett 2003). Many retrospective analyses investigating the association between *CYP2D6* genotype and tamoxifen treatment outcome have been published, with highly discordant results. A large retrospective study of more than 1300 patients with early-stage breast cancer receiving adjuvant tamoxifen found shorter disease-free survival in CYP2D6 poor metabolizers compared with extensive metabolizers and intermediate metabolizers (Schroth 2009). However, attempted replication in two large prospective, randomized, double-blind, phase III adjuvant breast cancer trials found no relationship between the *CYP2D6* genotype and clinical outcomes (Rae 2012; Regan 2012). This discordance in findings could be partly attributable to weaknesses in retrospective analyses, including lack of control for concomitant treatment or tamoxifen adherence (Province 2014; Hertz 2012) or possible issues introduced when genetic analyses are performed in preserved tumor tissue (Nakamura 2012).

Unless the relationship between treatment efficacy and CYP2D6 activity (or endoxifen exposure) is validated, prospective genotype-guided tamoxifen dose adjustment (Irvin 2011) or therapeutic substitution (Ruddy 2013) should not be used clinically. Consensus guidelines, such as those of the American Society of Clinical Oncology, specifically recommend against using *CYP2D6* genetic information to guide adjuvant hormonal therapy (Visvanathan 2009). Alternatively, several guidelines, including the NCCN Guidelines (2014b) and the FDA-approved tamoxifen package insert, recommend

Table 11.5. Available Genetic Tests Relevant to Oncology Drugs

Gene	Oncology Drugs	Type of Analysis	Laboratories Offering Genetic Testing[a]
TPMT	Mercaptopurine, thioguanine	Complete exome sequencing	Baylor Medical Genetics Laboratories
		Selected exome sequencing	Genetics Diagnostic Laboratory Boston Children's Hospital
		Targeted variant analysis	Molecular Genetics Laboratory Cincinnati Children's Hospital Medical Center
DPYD	Fluorouracil, capecitabine	Complete exome sequencing	Counsyl
			PreventionGenetics
		Targeted variant analysis	Quest Diagnostics Nichols Institute San Juan Capistrano
			Counsyl
			Molecular Genetics Laboratory ARUP Laboratories
UGT1A1	Irinotecan, tyrosine kinase inhibitors	Targeted variant analysis	Genetic Services Laboratory University of Chicago
			Quest Diagnostics Nichols Institute Chantilly
			Molecular Genetics Laboratory ARUP Laboratories
			Quest Diagnostics Nichols Institute San Juan Capistrano
			Molecular Genetics Diagnostic Laboratory Detroit Medical Center University Laboratories
			Molecular Diagnostics Laboratory Duke University Health System
		Deletion/duplication analysis	Genetic Services Laboratory University of Chicago
		Selected exome sequencing	Molecular Genetics Laboratory Children's Hospital of Philadelphia
		Complete exome sequencing	Genetic Services Laboratory University of Chicago
			Molecular Genetics Laboratory Mayo Clinic-Minnesota
SLCO1B1	Methotrexate	Targeted variant analysis	Molecular Genetics Laboratory ARUP Laboratories
CYP2D6	Tamoxifen	Targeted variant analysis	Molecular Genetics Laboratory ARUP Laboratories
			Baylor Medical Genetics Laboratories
			Molecular Genetics Laboratory Cincinnati Children's Hospital Medical Center
			Genelex Corporation
			Molecular Diagnostics and Toxicology Laboratory
			Molecular Diagnostics Laboratories PHD LLC
			Quest Diagnostics Nichols Institute San Juan Capistrano
			Quest Diagnostics Nichols Institute Chantilly

[a] Based on NIH Genetic Testing Registry (www.ncbi.nlm.nih.gov/gtr/) as of June 24, 2014.

avoiding concomitant administration of strong CYP2D6 inhibitors, including paroxetine or sertraline, in tamoxifen-treated patients. Clinical development of endoxifen, which would not require CYP2D6 activation, is currently under way (Endoxifen shows promise in breast cancer 2014; Ahmad 2010).

Concluding Remarks and Future Directions

This chapter highlights some of the most mature and promising germline pharmacogenetic associations in oncology. Several of these examples are beginning to be implemented, or they are ready for clinical use. Information regarding the availability of genetic testing for all of these genes can be found at the NIH Genetic Testing Registry (www.ncbi.nlm.nih.gov/gtr/). Few pharmacogenetic associations have surpassed the evidentiary threshold necessary to warrant preemptively ordering a genetic test; however, the threshold for using existing genetic data is substantially lower (Altman 2011). The proportion of patients with cancer having known genetic information continues to expand, driven primarily by somatic genetic analysis to guide treatment selection, providing an efficient framework for the clinical implementation of validated germline pharmacogenetic associations (Gillis 2014).

Oncology decision-making will continue to be driven primarily by treatment efficacy, with substantial risk of severe toxicity an accepted element of treatment. Health care practitioners who work with patients with cancer are accustomed to monitoring for and responding to severe toxicities. This adds some complexity to decision-making for oncology germline pharmacogenetic examples in which genetics similarly affects efficacy and toxicity. In this case, clinicians would be less inclined to decrease dosing to avoid a potential toxicity, fearing that efficacy would be diminished, but perhaps more willing to increase dosing to maximize efficacy, with the understanding that toxicity may be enhanced as well.

Continued work is needed in every facet of oncology pharmacogenetics: discovery of genetic markers in large patient cohorts, validation of promising markers in independent studies, and prospective translation to show the clinical utility of implementation. In addition to the translational work for individual germline pharmacogenetic associations, there is a critical need to establish efficient systems for integrating pharmacogenetics into clinical practice and determining the appropriate role for pharmacists in genotype-guided treatment decision-making (Crews 2012b, 2011). Finally, effort should be dedicated to establishing a paradigm for integrating germline analysis into tumor genetic assessment to increase the number of oncology patients with existing data that can be used to individualize their cancer treatment.

References

Amstutz U, Froehlich TK, Largiader CR. Dihydropyrimidine dehydrogenase gene as a major predictor of severe 5-fluorouracil toxicity. Pharmacogenomics 2011;12:1321-36.

Appell ML, Berg J, Duley J, et al. Nomenclature for alleles of the thiopurine methyltransferase gene. Pharmacogenet Genomics 2013;23:242-8.

Ahmad A, Shahabuddin S, Sheikh S, et al. Endoxifen, a new cornerstone of breast cancer therapy: demonstration of safety, tolerability, and systemic bioavailability in healthy human subjects. Clin Pharmacol Ther 2010;88:814-7.

Altman RB. Pharmacogenomics: "noninferiority" is sufficient for initial implementation. Clin Pharmacol Ther 2011;89:348-50.

Ando Y, Saka H, Asai G, et al. *UGT1A1* genotypes and glucuronidation of SN-38, the active metabolite of irinotecan. Ann Oncol 1998;9:845-7.

Barrett JS, Mondick JT, Narayan M, et al. Integration of modeling and simulation into hospital-based decision support systems guiding pediatric pharmacotherapy. BMC Med Inform Decis Mak 2008;8:6.

Bell GC, Crews KR, Wilkinson MR, et al. Development and use of active clinical decision support for preemptive pharmacogenomics. J Am Med Inform Assoc 2014;21:e93-9.

Boisdron-Celle M, Remaud G, Traore S, et al. 5-fluorouracil-related severe toxicity: a comparison of different methods for the pretherapeutic detection of dihydropyrimidine dehydrogenase deficiency. Cancer Lett 2007;249:271-82.

Bosma PJ, Chowdhury JR, Bakker C, et al. The genetic basis of the reduced expression of bilirubin UDP-glucuronosyltransferase 1 in Gilbert's syndrome. N Engl J Med 1995;333:1171-5.

Caudle KE, Thorn CF, Klein TE, et al. Clinical Pharmacogenetics Implementation Consortium guidelines for dihydropyrimidine dehydrogenase genotype and fluoropyrimidine dosing. Clin Pharmacol Ther 2013;94:640-5.

Cheng L, Li M, Hu J, et al. *UGT1A1*6* polymorphisms are correlated with irinotecan-induced toxicity: a system review and meta-analysis in Asians. Cancer Chemother Pharmacol 2014;73:551-60.

Crews KR, Cross SJ, McCormick JN, et al. Development and implementation of a pharmacist-managed clinical pharmacogenetics service. Am J Health Syst Pharm 2011;68:143-50.

Crews KR, Gaedigk A, Dunnenberger HM, et al. Clinical Pharmacogenetics Implementation Consortium (CPIC) guidelines for codeine therapy in the context of cytochrome P450 2D6 (CYP2D6) genotype. Clin Pharmacol Ther 2012a;91:321-6.

Crews KR, Hicks JK, Pui CH, et al. Pharmacogenomics and individualized medicine: translating science into practice. Clin Pharmacol Ther 2012b;92:467-75.

Dias MM, McKinnon RA, Sorich MJ. Impact of the *UGT1A1*28* allele on response to irinotecan: a systematic review and meta-analysis. Pharmacogenomics 2012;13:889-99.

Diasio RB, Beavers TL, Carpenter JT. Familial deficiency of dihydropyrimidine dehydrogenase. Biochemical basis for familial pyrimidinemia and severe 5-fluorouracil-induced toxicity. J Clin Invest 1988;81:47-51.

Dombrowsky E, Jayaraman B, Narayan M, Barrett JS. Evaluating performance of a decision support system to improve methotrexate pharmacotherapy in children and young adults with cancer. Ther Drug Monit 2011;33:99-107.

Dowsett M, Haynes BP. Hormonal effects of aromatase inhibitors: focus on premenopausal effects and interaction with tamoxifen. J Steroid Biochem Mol Biol 2003;86:255-63.

Endoxifen shows promise in breast cancer. Cancer Discov 2014;4:OF1.

Fujita K, Sugiyama M, Akiyama Y, et al. The small-molecule tyrosine kinase inhibitor nilotinib is a potent noncompetitive inhibitor of the SN-38 glucuronidation by human UGT1A1. Cancer Chemother Pharmacol 2011;67:237-41.

Fukushima H, Fukushima T, Sakai A, et al. Polymorphisms of MTHFR associated with higher relapse/death ratio and delayed weekly MTX administration in pediatric lymphoid malignancies. Leuk Res Treatment 2013;2013:238528.

Fusaro VA, Brownstein C, Wolf W, et al. Development of a scalable pharmacogenomic clinical decision support service. AMIA Summits Transl Sci Proc 2013;2013:60.

Gillis N, Patel J, Innocenti F. Clinical implementation of germ line cancer pharmacogenetic variants during the next-generation sequencing era. Clin Pharmacol Ther 2014;95:269-80.

Gschwind A, Fischer OM, Ullrich A. The discovery of receptor tyrosine kinases: targets for cancer therapy. Nat Rev Cancer 2004;4:361-70.

Hertz DL, McLeod HL, Irvin WJ. Tamoxifen and CYP2D6: a contradiction of data. Oncologist 2012;17:620-30.

Hindorf U, Appell ML. Genotyping should be considered the primary choice for pre-treatment evaluation of thiopurine methyltransferase function. J Crohns Colitis 2012;6:655-9.

Hoskins JM, Goldberg RM, Qu P, et al. *UGT1A1*28* genotype and irinotecan-induced neutropenia: dose matters. J Natl Cancer Inst 2007;99:1290-5.

Ingelman-Sundberg M, Daly AK, Nebert DW. Home page of the human cytochrome P450 (CYP) allele nomenclature database. Available at www.cypalleles.ki.se. Updated 2008. Accessed January 13, 2014.

Innocenti F, Undevia SD, Iyer L, et al. Genetic variants in the UDP-glucuronosyltransferase 1A1 gene predict the risk of severe neutropenia of irinotecan. J Clin Oncol 2004;22:1382-8.

Irvin WJ, Walko CM, Weck KE, et al. Genotype-guided tamoxifen dosing increases active metabolite exposure in women with reduced CYP2D6 metabolism: a multicenter study. J Clin Oncol 2011;29:3232-9.

Iyer L, Das S, Janisch L, et al. *UGT1A1*28* polymorphism as a determinant of irinotecan disposition and toxicity. Pharmacogenomics J 2002;2:43-7.

Iyer L, Hall D, Das S, et al. Phenotype-genotype correlation of in vitro SN-38 (active metabolite of irinotecan) and bilirubin glucuronidation in human liver tissue with *UGT1A1* promoter polymorphism. Clin Pharmacol Ther 1999;65:576-582.

Krynetski EY, Schuetz JD, Galpin AJ, et al. A single point mutation leading to loss of catalytic activity in human thiopurine S-methyltransferase. Proc Natl Acad Sci U S A 1995;92:949-53.

Lennard L, Cartwright CS, Wade R, et al. Thiopurine methyltransferase genotype-phenotype discordance and thiopurine active metabolite formation in childhood acute lymphoblastic leukaemia. Br J Clin Pharmacol 2013;76:125-36.

Lennard L, Lilleyman JS, Van Loon J, et al. Genetic variation in response to 6-mercaptopurine for childhood acute lymphoblastic leukaemia. Lancet 1990;336:225-9.

Levinsen M, Rotevatn EO, Rosthoj S, et al. Pharmacogenetically based dosing of thiopurines in childhood acute lymphoblastic leukemia: influence on cure rates and risk of second cancer. Pediatr Blood Cancer 2014;61:797-802.

Liu X, Cheng D, Kuang Q, et al. Association of *UGT1A1*28* polymorphisms with irinotecan-induced toxicities in colorectal cancer: a meta-analysis in Caucasians. Pharmacogenomics J 2014;14:120-9.

Liu X, Cheng D, Kuang Q, et al. Association between UGT1A1*28 polymorphisms and clinical outcomes of irinotecan-based chemotherapies in colorectal cancer: a meta-analysis in Caucasians. PLoS One 2013;8:e58489.

Madlensky L, Natarajan L, Tchu S, et al. Tamoxifen metabolite concentrations, CYP2D6 genotype, and breast cancer outcomes. Clin Pharmacol Ther 2011;89:718-25.

Magnani E, Farnetti E, Nicoli D, et al. Fluoropyrimidine toxicity in patients with dihydropyrimidine dehydrogenase splice site variant: the need for further revision of dose and schedule. Intern Emerg Med 2013;8:417-23.

Marcuello E, Paez D, Pare L, et al. A genotype-directed phase I-IV dose-finding study of irinotecan in combination with fluorouracil/leucovorin as first-line treatment in advanced colorectal cancer. Br J Cancer 2011;105:53-7.

McLeod HL, Relling MV, Liu Q, et al. Polymorphic thiopurine methyltransferase in erythrocytes is indicative of activity in leukemic blasts from children with acute lymphoblastic leukemia. Blood 1995;85:1897-902.

Morel A, Boisdron-Celle M, Fey L, et al. Clinical relevance of different dihydropyrimidine dehydrogenase gene single nucleotide polymorphisms on 5-fluorouracil tolerance. Mol Cancer Ther 2006;5:2895-904.

Motzer RJ, Johnson T, Choueiri TK, et al. Hyperbilirubinemia in pazopanib- or sunitinib-treated patients in COMPARZ is associated with *UGT1A1* polymorphisms. Ann Oncol 2013;24:2927-8.

Nakamura Y, Ratain MJ, Cox NJ, et al. Re: *CYP2D6* genotype and tamoxifen response in postmenopausal women with endocrine-responsive breast cancer: the Breast International Group 1-98 trial. J Natl Cancer Inst 2012;104:1264.

National Cancer Control Network (NCCN). 2014a. Clinical Practice Guidelines in Oncology (NCCN Guidelines®) for Acute Lymphoblastic Leukemia, V.1.2014. Available at www.nccn.org. Accessed August 22, 2014.

National Comprehensive Cancer Network (NCCN). 2014b. NCCN Clinical Practice Guidelines in Oncology (NCCN Guidelines®) for Breast Cancer, V.3.2014. Available at www.nccn.org. Accessed August 22, 2014.

National Comprehensive Cancer Network (NCCN). 2014c. Clinical practice guidelines in oncology (NCCN Guidelines®) for Colon Cancer, V.1.2015. Available at www.nccn.org. Accessed August 22, 2014.

National Comprehensive Cancer Network (NCCN). 2014d. Clinical Practice Guidelines in Oncology (NCCN Guidelines®) for Rectal Cancer, V.1.2015. Available at www.nccn.org. Accessed August 22, 2014.

O'Brien SG, Guilhot F, Larson RA, et al. Imatinib compared with interferon and low-dose cytarabine for newly diagnosed chronic-phase chronic myeloid leukemia. N Engl J Med 2003;348:994-1004.

Offer SM, Lee AM, Mattison LK, et al. A *DPYD* variant (Y186C) in individuals of African ancestry is associated with reduced DPD enzyme activity. Clin Pharmacol Ther 2013;94:158-66.

Otterness DM, Szumlanski CL, Wood TC, et al. Human thiopurine methyltransferase pharmacogenetics. Kindred with a terminal exon splice junction mutation that results in loss of activity. J Clin Invest 1998;101:1036-44.

Province MA, Goetz MP, Brauch H, et al. *CYP2D6* genotype and adjuvant tamoxifen: meta-analysis of heterogeneous study populations. Clin Pharmacol Ther 2014;95:216-27.

Rae JM, Drury S, Hayes DF, et al. *CYP2D6* and *UGT2B7* genotype and risk of recurrence in tamoxifen-treated breast cancer patients. J Natl Cancer Inst 2012;104:452-60.

Radtke S, Zolk O, Renner B, et al. Germline genetic variations in methotrexate candidate genes are associated with pharmacokinetics, toxicity, and outcome in childhood acute lymphoblastic leukemia. Blood 2013;121:5145-53.

Ramsey LB, Bruun GH, Yang W, et al. Rare versus common variants in pharmacogenetics: *SLCO1B1* variation and methotrexate disposition. Genome Res 2012;22:1-8.

Ramsey LB, Panetta JC, Smith C, et al. Genome-wide study of methotrexate clearance replicates *SLCO1B1*. Blood 2013;121:898-904.

Regan MM, Leyland-Jones B, Bouzyk M, et al. *CYP2D6* genotype and tamoxifen response in postmenopausal women with endocrine-responsive breast cancer: the Breast International Group 1-98 trial. J Natl Cancer Inst 2012;104:441-51.

Relling MV, Gardner EE, Sandborn WJ, et al. Clinical pharmacogenetics implementation consortium guidelines for thiopurine methyltransferase genotype and thiopurine dosing: 2013 update. Clin Pharmacol Ther 2013;93:324-5.

Relling MV, Hancock ML, Rivera GK, et al. Mercaptopurine therapy intolerance and heterozygosity at the thiopurine S-methyltransferase gene locus. J Natl Cancer Inst 1999;91:2001-8.

Relling MV, Pui C, Cheng C, et al. Thiopurine methyltransferase in acute lymphoblastic leukemia. Blood 2006;107:843-4.

Rosmarin D, Palles C, Church D, et al. Genetic markers of toxicity from capecitabine and other fluorouracil-based regimens: investigation in the QUASAR2 study, systematic review, and meta-analysis. J Clin Oncol 2014a;32:1031-9.

Rosmarin D, Palles C, Pagnamenta A, et al. A candidate gene study of capecitabine-related toxicity in colorectal cancer identifies new toxicity variants at *DPYD* and a putative role for ENOSF1 rather than TYMS. Gut 2014b Mar 19. [Epub ahead of print]

Ruddy K, Desantis S, Gelman R, et al. Personalized medicine in breast cancer: tamoxifen, endoxifen, and CYP2D6 in clinical practice. Breast Cancer Res Treat 2013;141:421-7.

Satoh T, Ura T, Yamada Y, et al. Genotype-directed, dose-finding study of irinotecan in cancer patients with *UGT1A1*28* and/or *UGT1A1*6* polymorphisms. Cancer Sci 2011;102:1868-73.

Schroth W, Goetz MP, Hamann U, et al. Association between CYP2D6 polymorphisms and outcomes among women with early stage breast cancer treated with tamoxifen. JAMA 2009;302:1429-36.

Schwab M, Zanger UM, Marx C, et al. Role of genetic and nongenetic factors for fluorouracil treatment-related severe toxicity: a prospective clinical trial by the German 5-FU Toxicity Study Group. J Clin Oncol 2008;26:2131-8.

Scrip's Cancer Chemotherapy Report. Scrip World Pharmaceutical News. London: PJB Publications, 2002.

Shibata T, Minami Y, Mitsuma A, et al. Association between severe toxicity of nilotinib and *UGT1A1* polymorphisms in Japanese patients with chronic myelogenous leukemia. Int J Clin Oncol 2014;19:391-6.

Siegel R, Ma J, Zou Z, et al. Cancer statistics, 2014. CA Cancer J Clin 2014;64:9-29.

Singer JB, Shou Y, Giles F, et al. *UGT1A1* promoter polymorphism increases risk of nilotinib-induced hyperbilirubinemia. Leukemia 2007;21:2311-5.

Stanulla M, Schaeffeler E, Flohr T, et al. Thiopurine methyltransferase (TPMT) genotype and early treatment response to mercaptopurine in childhood acute lymphoblastic leukemia. JAMA 2005;293:1485-9.

Sunakawa Y, Fujita K, Ichikawa W, et al. A phase I study of infusional 5-fluorouracil, leucovorin, oxaliplatin and irinotecan in Japanese patients with advanced colorectal cancer who harbor *UGT1A1*1/*1,*1/*6* or **1/*28*. Oncology 2012;82:242-8.

Schwab M, Schaeffeler E, Marx C, et al. Shortcoming in the diagnosis of TPMT deficiency in a patient with Crohn's disease using phenotyping only. Gastroenterology 2001;121:498-9.

Swen JJ, Nijenhuis M, de Boer A, et al. Pharmacogenetics: from bench to byte—an update of guidelines. Clin Pharmacol Ther 2011;89:662-73.

Szumlanski CL, Weinshilboum RM. Sulphasalazine inhibition of thiopurine methyltransferase: possible mechanism for interaction with 6-mercaptopurine and azathioprine. Br J Clin Pharmacol 1995;39:456-9.

Tai HL, Krynetski EY, Schuetz EG, et al. Enhanced proteolysis of thiopurine S-methyltransferase (TPMT) encoded by mutant alleles in humans (*TPMT*3A, TPMT*2*): mechanisms for the genetic polymorphism of TPMT activity. Proc Natl Acad Sci U S A 1997;94:6444-9.

Tai HL, Krynetski EY, Yates CR, et al. Thiopurine S-methyltransferase deficiency: two nucleotide transitions define the most prevalent mutant allele associated with loss of catalytic activity in Caucasians. Am J Hum Genet 1996;58:694-702.

Takano M, Goto T, Hirata J, et al. *UGT1A1* genotype-specific phase I and pharmacokinetic study for combination chemotherapy with irinotecan and cisplatin: a Saitama Tumor Board Study. Eur J Gynaecol Oncol 2013;34:120-3.

Teft WA, Gong IY, Dingle B, et al. CYP3A4 and seasonal variation in vitamin D status in addition to CYP2D6 contribute to therapeutic endoxifen level during tamoxifen therapy. Breast Cancer Res Treat 2013;139:95-105.

Teh LK, Hamzah S, Hashim H, et al. Potential of dihydropyrimidine dehydrogenase genotypes in personalizing 5-fluorouracil therapy among colorectal cancer patients. Ther Drug Monit 2013;35:624-30.

Teng K, DiPiero J, Meese T, et al. Cleveland Clinic's center for personalized healthcare: setting the stage for value-based care. Pharmacogenomics 2014;15:587-91.

Terrazzino S, Cargnin S, Del Re M, et al. DPYD IVS14+1G>A and 2846A>T genotyping for the prediction of severe fluoropyrimidine-related toxicity: a meta-analysis. Pharmacogenomics 2013;14:1255-72.

Toffoli G, Cecchin E, Gasparini G, et al. Genotype-driven phase I study of irinotecan administered in combination with fluorouracil/leucovorin in patients with metastatic colorectal cancer. J Clin Oncol 2010;28:866-71.

Treviño LR, Shimasaki N, Yang W, et al. Germline genetic variation in an organic anion transporter polypeptide associated with methotrexate pharmacokinetics and clinical effects. J Clin Oncol 2009;27:5972-8.

van Staveren MC, Jan Guchelaar H, van Kuilenburg AB, et al. Evaluation of predictive tests for screening for dihydropyrimidine dehydrogenase deficiency. Pharmacogenomics J 2013;13:389-95.

Visvanathan K, Chlebowski RT, Hurley P, et al. American Society of Clinical Oncology clinical practice guideline update on the use of pharmacologic interventions including tamoxifen, raloxifene, and aromatase inhibition for breast cancer risk reduction. J Clin Oncol 2009;27:3235-58.

Weinshilboum RM, Sladek SL. Mercaptopurine pharmacogenetics: monogenic inheritance of erythrocyte thiopurine methyltransferase activity. Am J Hum Genet 1980;32:651-62.

Xu CF, Reck BH, Xue Z, et al. Pazopanib-induced hyperbilirubinemia is associated with Gilbert's syndrome *UGT1A1* polymorphism. Br J Cancer 2010;102:1371-7.

Yang CG, Ciccolini J, Blesius A, et al. DPD-based adaptive dosing of 5-FU in patients with head and neck cancer: impact on treatment efficacy and toxicity. Cancer Chemother Pharmacol 2011;67:49-56.

Chapter 12

PHARMACOGENOMICS OF IMMUNOSUPPRESSANTS

Kinjal Sanghavi, M.Pharm.; and Pamala Jacobson, Pharm.D., FCCP

Learning Objectives

1. Assess the influence of genetic variants on the pharmacokinetics of tacrolimus, cyclosporine, azathioprine, and mycophenolate.
2. Design a tacrolimus dosage regimen using the *CYP3A5* genotype.
3. Design an azathioprine dosage regimen using the *TPMT* genotype.
4. Make inferences about pharmacogenetically influenced drug-drug interactions.
5. Demonstrate the importance of pharmacogenetics to physicians and patients and be able to guide them in drug therapy decisions using genetics.

Keywords: Tacrolimus, cyclosporine, CYP3A5, mycophenolate acid, azathioprine, TPMT, dosing, toxicity, drug interactions, *UGT*

Abbreviations in This Chapter

ABC	ATP (adenosine triphosphate) binding cassette
CL	Clearance
IMPDH	Inosine-5′-monophosphate dehydrogenase
MAF	Minor allele frequency
MRP2	Multidrug resistance protein 2
Pgp	P-glycoprotein
POR	Cytochrome P450 oxidoreductase
SLCO	Solute carrier organic anion transporter
TPMT	Thiopurine S-methyltransferase
UGT	Uridine diphosphate-glucuronyltransferase

Abstract

There is extensive literature on pharmacogenomic markers as tools to improve therapeutic drug monitoring and reduce immunosuppressive toxicity, acute rejection, and graft loss, though the clinical utility data are still emerging. In the future, pharmacogenetics will probably be most beneficial to the high-risk transplant patients who are at the highest risk of drug toxicity, graft injury, or loss; however, validation is required for clinical utility. Patients with high-risk pharmacogenetic markers may receive a different immunosuppressive drug or alternative dosage regimens. Although genetically guided therapy will not replace therapeutic drug monitoring, it will allow patients to achieve adequate immunosuppression faster, reduce the frequency of therapeutic drug monitoring, and possibly lower toxicity rates. This is important, especially because the focus for the past 15 years has been on designing immunosuppressive-sparing protocols that allow sufficient graft protection without over-immunosuppression. The chapter will review the pharmacogenomic findings that are most promising for clinical implementation. Most pharmacogenetic literature has focused on the genes involved in metabolism and transport (*CYP3A4, CYP3A5,* and *ABCB1*) of the calcineurin inhiitors. The *CYP3A5*1* genotype is highly significant toward tacrolimus pharmacokinetics and may soon have sufficient evidence for clinical implementation. Variants on the thiopurine S-methyltransferase (TPMT) gene are significantly associated with azathioprine hematologic toxicity. Clinical pharmacogenetic guidelines are available for azathioprine. Currently, evidence is limited for implementing pharmacogenetically guided cyclosporine and mycophenolic acid. Drug-drug interactions are common with tacrolimus and cyclosporine because of the broad substrate specificity of CYP enzymes and the pharmacogenetic variability in the genes encoding these enzymes. The combination of a genetic variant with a drug-drug interaction may result in an exaggerated pharmacokinetic effect or a loss in the interaction effect. Given that the immunosuppressants are narrow therapeutic index drugs with substantial toxicities, use of pharmacogenetics to guide dosing has significant promise in improving outcomes for transplant recipients.

Pharmacogenomics of Immunosuppressants

This chapter will summarize the important clinical studies and pharmacogenetic guidelines related to variants in genes involved in the pharmacokinetics and pharmacodynamics of the most commonly used immunosuppressants in transplantation. Because most pharmacogenetic literature is focused on tacrolimus, cyclosporine, azathioprine, and mycophenolic acid, the pharmacokinetics and pharmacodynamics of these agents will be reviewed in the following sections.

Tacrolimus

Tacrolimus is the primary immunosuppressant used to prevent rejection in patients undergoing organ transplantation, including kidney, liver, pancreas, lung, and heart. It is also used in autoimmune diseases and hematopoietic cell transplantation. In kidney transplants, tacrolimus is associated with better allograft survival than cyclosporine and is the preferred calcineurin inhibitor of choice in most transplant centers (Silva 2014; Ekberg 2009; Pirsch 1997). According to the most recent annual report of the U.S. Organ Procurement and Transplantation Network and the Scientific Registry of Transplant Recipients, 85% of all kidney and liver transplant recipients receive tacrolimus and mycophenolic acid as their initial immunosuppression (Matas 2014). Tacrolimus binds to immunophilins called FK-binding proteins in lymphocytes, forming a tacrolimus FK-binding protein complex. This complex inhibits calcineurin phosphatase, a serine-threonine phosphatase enzyme. Calcineurin inhibition blocks the dephosphorylation of NFATc (nuclear factor of activated T cells), thereby suppressing the transcription of interleukin-2 and other cytokines involved in the immune response and activation of T lymphocytes. Calcineurin inhibition also affects NFκβ (nuclear factor kappa beta light chain enhancer of activated B cell) activation of immunoregulatory genes and MAPK (mitogen-activated protein kinase) (Barbarino 2013).

Metabolism and Pharmacokinetics

Tacrolimus is extensively and primarily metabolized in the liver and, to a lesser extent, in the small intestine, where it is subject to high first-pass metabolism (Staatz 2004). It is metabolized by cytochrome P450 (CYP) 3A5 and CYP3A4 enzymes, and at least 15 metabolites have been identified (Staatz 2010a; Iwasaki 2007). CYP3A5 is a more efficient catalyst of tacrolimus metabolism relative to CYP3A4. The 13-O-desmethyl-tacrolimus is the major metabolite and is significantly less active than tacrolimus (Gonschior 1996; Vincent 1992; Christians 1991). The 31-O-desmethyl-tacrolimus is a minor metabolite but has activity similar to that of tacrolimus (Barbarino 2013). The metabolites are primarily eliminated in the bile. Only limited amounts of tacrolimus appear in the urine or feces. Tacrolimus and 13-O-desmethyl tacrolimus are inhibitors and/or substrates for the P-glycoprotein (Pgp) transporter. P-glycoprotein is an energy-dependent efflux transporter encoded by the ATP (adenosine triphosphate) binding cassette B1 (ABCB1) gene and is present in many tissues, including the gastrointestinal tract, hepatocytes, kidney, and lymphocytes (Saeki 1993; Christians 1991). Tacrolimus enters the enterocytes of the gut but is readily effluxed out of the enterocytes and back into the gastrointestinal tract by the Pgp transporter. Figure 12.1 shows the metabolic and transport pathway of the calcineurin inhibitors. Variable expression of Pgp and the resulting high first-pass metabolism has been associated with tacrolimus bioavailability (Vafadari 2013; Anglicheau 2003). The mean oral bioavailability of tacrolimus is about 20%–30%, and the interindividual variability is high (6%–89%) (Undre 1999; Venkataramanan 1995). Tacrolimus has a narrow therapeutic index and high interindividual variability in clearance (CL) (3–35 L/hour), which dramatically affects systemic exposure and the degree of immunosuppression (Staatz 2010b). High variability can negatively affect outcomes; therefore, therapeutic drug monitoring is routinely conducted. This variability is partly related to genetic variation in the CYP3A5 enzyme. Typical trough blood concentration targets in kidney transplants in the United States are 8–10 ng/mL in the first 3 months posttransplantation and 6–8 ng/mL after 3 months, depending on the indication and time posttransplantation (Ekberg 2007; Schiff 2007; Kershner 1996; Jusko 1995; McMaster 1995). Low blood concentrations increase the risk of immunologically mediated graft rejection, graft loss, and/or treatment failure. High concentrations lead to overimmunosuppression and are associated with a greater risk of toxicity, including nephrotoxicity, hyperglycemia, neurotoxicity, hypertension, infections, and malignancy (Hesselink 2010; Naesens 2009; Webster 2005; Kershner

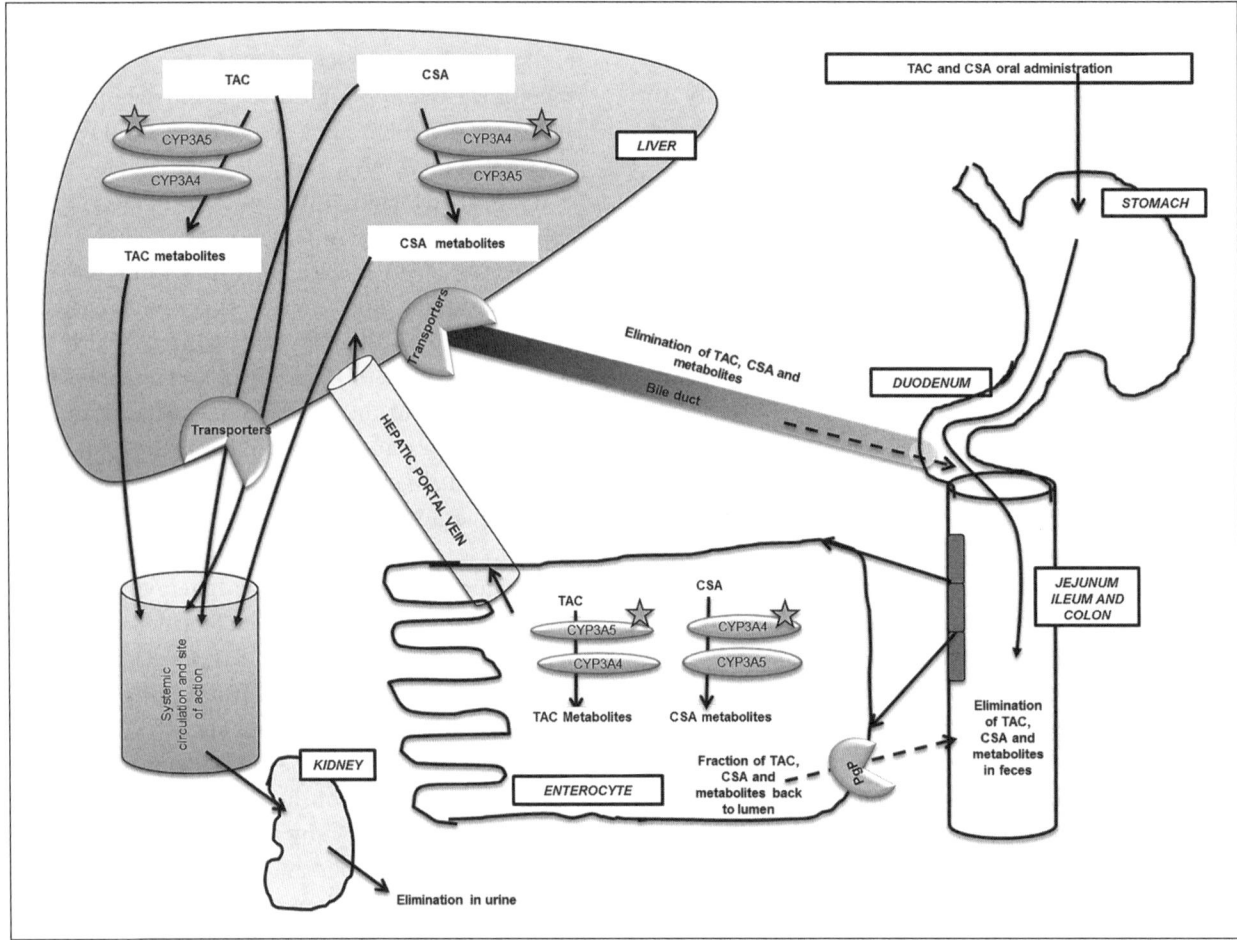

Figure 12.1. Major metabolic pathway of tacrolimus (TAC) and cyclosporine (CSA).

PgP = P-glycoprotein.

★Predominant enzyme

1996). Achieving a balance between immunosuppression and toxicity is of critical importance toward improving outcomes. Unfortunately, finding this balance currently occurs through a trial-and-error approach.

Tacrolimus Pharmacogenomics

Extensive reviews of tacrolimus pharmacogenomic data are available in recent publications and will not be reviewed in this chapter (Elens 2014a; Hesselink 2014; Barbarino 2013). Instead, the following section will discuss the data relevant to the clinical implementation of tacrolimus pharmacogenomics.

Cytochrome P450 3A5

Most tacrolimus pharmacogenomic data have been obtained from the kidney transplant population. A large and highly significant association between the *CYP3A5*3* variant and tacrolimus pharmacokinetics is well established and has been replicated in many studies (Birdwell 2012; Jacobson 2012, 2011; Staatz 2010b; Macphee 2005). *CYP3A5*3* (rs776746, 6986A>G) is a variant in intron 3 of the *CYP3A5* gene that generates splice variants containing stop codons leading to protein truncation. Individuals who carry two G alleles (*CYP3A5*3/*3*) have no *CYP3A5* messenger RNA expression, form no functional protein, and are considered non-expressers (Busi 2005; Hustert 2001). Individuals with one or more A alleles (*CYP3A5*1/*3* or **1/*1*) produce functional protein and are described as *CYP3A5* expressers. There is a highly significant difference in the *CYP3A5*1* allele frequencies by race. The *CYP3A5*1* (A allele) has a minor allele frequency (MAF) of 5%–15% in whites; 30% in Japanese, Chinese, and Koreans; and about 85% in African Americans (Abecasis 2012). This generates confusion when referring to the minor allele because A is the minor allele in whites but the major allele in African Americans. Tacrolimus concentrations are generally lower

Table 12.1. Factors Known to Influence Tacrolimus Pharmacokinetics and Their Level of Importance

Factor	Potential Degree of Importance
CYP3A5 genotype	High
Age	High
Time posttransplantation	High
Concomitant CYP3A inhibitors or inducers	High
Liver function	High, if significantly impaired
Comorbid conditions	High
Center practices	High
Food	Medium
Race	High, if genotype is not known Low, when CYP3A5 genotype is known
Hematocrit	Medium
Diarrhea	Medium
Liver allograft size	Medium
Weight	Medium
Albumin	Medium
Lipoproteins	Low
Circadian rhythm	Medium

Adapted from: Staatz CE, Tett SE. Clinical pharmacokinetics and pharmacodynamics of tacrolimus in solid organ transplantation. Clin Pharmacokinet 2004;43:623-53.

in African Americans (Figure 12.2), primarily because they are more likely to be *CYP3A5* expressers.

Patients expressing the CYP3A5 enzyme have a higher oral tacrolimus CL than do non-expressers (*CYP3A5*1/*1* about 1 L/hour/kg, *CYP3A5*1/*3* about 0.8 L/hour/kg, *CYP3A5*3/*3* about 0.5 L/hour/kg) (Passey 2011). Therefore, kidney transplant recipients who are *CYP3A5* expressers have about a 2-fold lower tacrolimus dose–normalized trough concentration compared with non-expressers (Figure 12.3). The dose requirements for expressers (*CYP3A5*1/*1* or *CYP3A5*1/*3*) are about 1.5- to 1.7-fold higher than those for non-expressers (*CYP3A5*3/*3*)

(Jacobson 2011). *CYP3A5*1* has been studied in heart, lung, and liver transplants for its association with tacrolimus pharmacokinetics, with similar conclusions (Diaz-Molina 2012; Gijsen 2011; Zheng 2004). The 13-*O*-desmethyl-tacrolimus metabolite is the major metabolite and is more likely to form in individuals who express *CYP3A5*; however, it is not yet understood whether this is clinically important (Dai 2004). Limited data are available for the once-daily tacrolimus product, but data suggest that genotype also has a significant effect on pharmacokinetics. Kidney transplant recipients who were *CYP3A5* expressers had 2-fold higher CL than non-expressers, and the *CYP3A5* genotype explained 25% of the interindividual variability in CL (Benkali 2010).

In liver transplants, both the donor and recipient *CYP3A5* genotype is important. Expression of *CYP3A5* in either the donor liver or the recipient intestine results in significantly lower tacrolimus concentrations than in non-expressers. In a meta-analysis, recipient genotype was most informative early posttransplantation, and donor genotype was important later (Buendia 2014). This is consistent with the clinical observation that the donor liver recovers over time and that there is a slow up-regulation of *CYP3A* activity in the liver in the weeks post-transplantation (Buendia 2014). Genetic-based dosage models for tacrolimus have not yet been developed for liver transplantation.

Many factors, in addition to genotype, influence tacrolimus blood concentrations (Table 12.1). For example, age greatly affects tacrolimus metabolism and pharmacokinetics (Table 12.2). Younger adults (18–34 years) who carry the *CYP3A5*3/*3* genotype have median tacrolimus dose requirements of 6 mg/day, whereas older adults (65–84 years) with the same genotype require only 4 mg/day. Therefore, considering genotypic and clinical factors in dose selection is important.

Clinical Implementation of *CYP3A5* Genotype for Tacrolimus Dose Determination

A randomized controlled trial using pretransplant genetic adaption compared individuals with tacrolimus dosing using the *CYP3A5* genotype together with body weight with individuals in a control group dosed on body weight alone (Thervet 2010). All patients were adult kidney transplant recipients at a low immunologic risk of rejection. Subjects receiving drugs known to interact with tacrolimus were excluded. Tacrolimus was initiated on day 7 posttransplantation. In the study arm, patients with the *CYP3A5*3/*3* genotype received tacrolimus 0.15 mg/kg/day, and those with one or more *CYP3A5*1* alleles received 0.30 mg/kg/day. Subjects in the control arm received 0.2 mg/kg/day. A greater proportion of patients in the

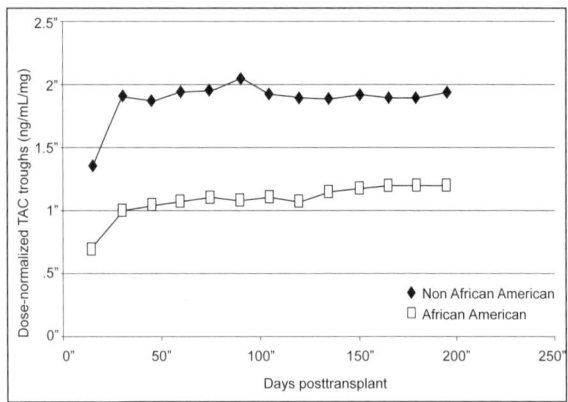

Figure 12.2. Effect of race on tacrolimus trough concentrations.

Adapted from: Jacobson PA, Oetting WS, Brearley AM, et al. Novel polymorphisms associated with tacrolimus trough concentrations: results from a multicenter kidney transplant consortium. Transplantation 2011;91:300-8.

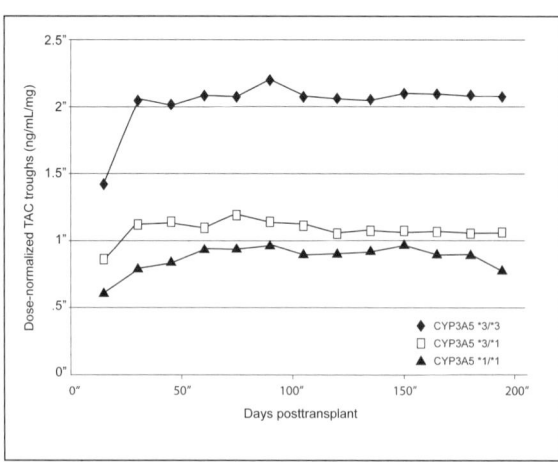

Figure 12.3. Effect of CYP3A5 genotype on tacrolimus trough concentrations.

genotype-dosed arm had trough concentrations in the targeted range (10–15 ng/mL) after six doses of tacrolimus compared with patients in the control arm (43.2% vs. 29.1%, p=0.03). The target trough was achieved by 75% of the patients in the genotype-dosed arm by day 8 and, in the control group, by day 25. The genotype-guided group required significantly fewer dose modifications (281 vs. 420, p=0.004). There were no differences in delayed graft function or acute rejection between the study arms. This may be because only subjects with a low immunologic risk were enrolled, and all received induction therapy, which minimized the need for aggressive tacrolimus exposure. There was no difference in nephrotoxicity between the groups. Compared with the dosing algorithm based on body weight alone, this study showed that genotype-guided dosing improved care by achieving the targeted tacrolimus concentrations, but for the other clinical outcomes, it appeared to have no positive or negative effects. The inclusion of other clinical factors into the genetic adaptation dosing algorithm may have substantially improved the number of patients within the target concentration. In the future, genotype-guided dosing algorithms for clinical implementation may improve precision when they also incorporate clinical factors, which are known modifiers of pharmacokinetics (Table 12.1).

Since the randomized trial, a tacrolimus dosage model was developed using 11,823 trough concentrations in the first 6 months posttransplantation from 681 racially diverse kidney transplant recipients enrolled through a multicenter consortium in the United States and Canada (Passey 2011). The model incorporated the *CYP3A5*1* genotype, days posttransplantation, age, transplantation at a steroid-sparing center, and use of a calcium channel blocker to estimate an individual's tacrolimus CL at any time in the first 6 months posttransplantation. From the CL estimate, a daily tacrolimus dose requirement was calculated. This dosage model was then validated in an independent cohort (Passey 2012). A British group found that the model predicted a lower CL than the observed CL at day 7 posttransplantation (Boughton 2013). This probably represents differences in clinical practice (e.g., all of their patients received steroids) and the need to include center- and population-specific factors in the model. Therefore, optimal dosage models will likely differ slightly from center to center. Defining and including the contributions of other important genetic variants (e.g., *CYP3A5*6, *7, CYP3A4*22*, cytochrome P450 oxidoreductase [*POR*] **28*) may also improve the predictability of dosage models (Elens 2014a). An example of a genotype dosage equation for adults that includes clinical factors is shown in Box 12.1 (Passey 2011).

Case Examples of Tacrolimus Pharmacogenetic Dosing

Case 1. An African American Patient

A.N. is a 56-year-old, 85-kg African American man with end-stage renal disease secondary to diabetes and hypertension. He has received hemodialysis three times weekly for the past 3 years. Other medical problems include hyperlipidemia, hyperphosphatemia, and anemia. Medications on admission are Renal Caps, sitagliptin, glipizide, simvastatin, carvedilol, diltiazem, losartan, daily aspirin, and calcium acetate. He is admitted for a deceased donor kidney transplant. The panel-reactive antibody is 10%, the cross match is negative, and HLA (human leukocyte antigen) typing shows a four-antigen

Table 12.2. Tacrolimus Dose and Troughs by Genotype and Age[a]

Tacrolimus	CYP3A5*1/*1 or *1/*3 Expresser (n=583 subjects)	CYP3A5*3/*3 Non-expresser (n=1384 subjects)
Daily dose (mg)		
18–34 years	10.0 (6.0–14.0)	6.0 (4.0–8.0)
35–64 years	8.0 (6.0–11.0)	5.0 (3.5–7.0)
65–84 years	7.0 (5.0–10.0)	4.0 (3.0–6.0)
Trough (ng/mL)		
18–34 years	6.7 (4.6–8.8)	8.5 (6.5–10.5)
35–64 years	6.9 (5.0–9.1)	8.4 (6.7–10.5)
65–84 years	7.3 (5.4–9.4)	8.4 (6.7–10.5)
Trough normalized for dose and weight (ng/mL per mg/kg)		
18–34 years	48.1 (34.9–68.4)	101.1 (69.6–161.9)
35–64 years	65.8 (45.4–100.8)	134.9 (89.4–207.3)
65–84 years	76.4 (51.3–108.9)	145.9 (102.0–227.3)

[a]Data are median (interquartile range).
Adapted from: Jacobson PA, Schladt D, Oetting WS, et al. Lower calcineurin inhibitor doses in older compared to younger kidney transplant recipients yield similar troughs. Am J Transplant 2012;12:3326-36.

match between the donor and the recipient. Recipient serology is negative for human immunodeficiency virus (HIV), hepatitis B surface antigen, hepatitis C, and cytomegalovirus. His *CYP3A5* (rs776746) genotype is *CYP3A5*1/*1*. A.N. receives induction therapy with intravenous rabbit antithymocyte globulin before surgery and on days 1 and 3, together with intravenous methylprednisolone 500 mg before surgery and 250- and 100-mg intravenous doses on days 1 and 2, respectively. His maintenance immunosuppression is mycophenolate mofetil 1.5 g orally twice daily and oral prednisone to be tapered to 5 mg daily within 1 month and continued indefinitely. Tacrolimus is initiated on day 2 when A.N. shows signs of good graft function, with a urine output of 1700 mL/day. A starting dose of oral tacrolimus could be determined using a (1) dosage equation that considers genotype status and clinical factors or (2) a simple algorithm that estimates the dose based on genotype status alone. Using these methods, the A.N.'s estimated doses would be calculated as follows.

Box 12.1. Example of a Tacrolimus Genotype Dosage Equation for Adults

Step 1: Estimate tacrolimus clearance.

$$\text{Tacrolimus CL/F (L/hour)} = 38.4 \times [(0.86, \text{if days 6–10 posttransplantation}) \text{ or}$$
$$(0.71, \text{if days 11–180})] \times [(1.69, \text{if CYP3A5*1/*3 genotype}) \text{ or}$$
$$(2.00, \text{if CYP3A5*1/*1 genotype})] \times (0.70, \text{if receiving a transplant at a steroid-sparing center})$$
$$\times [(\text{age in years}/50)^{-0.4}] \times (0.94, \text{if calcium channel blocker is present})$$

Step 2: Select the desired trough concentration. From the CL/F previously estimated and the desired trough concentrations, determine the TDD (total daily dose).

$$\text{TDD in mg} = [\text{CL/F (L/hour)} \times \text{tacrolimus trough goal (ng/mL)} \times 24 \text{ h}]/1000$$

Step 3: Check the trough concentration at steady state. Adjust the dose, if necessary, using the following equation:

$$\text{new tacrolimus dose (in mg)} = \frac{\text{current dose} \times \text{desired trough concentration (ng/mL)}}{\text{observed trough concentration (ng/mL)}}$$

Oral Dose Estimation for A. N. Using Genotype and Clinical Factors

There is one published equation for genotype-guided tacrolimus dosing that includes clinical factors. A.N.'s tacrolimus oral clearance (CL/F) would first be calculated using the following equations from Box 12.1 (Passey 2011).

tacrolimus CL/F (L/hour) = 38.4 × [(0.86, if days 6–10 posttransplantation) or

(0.71, if days 11–180)] × [(1.69, if *CYP3A5*1/*3* genotype)

or

(2.00, if *CYP3A5*1/*1* genotype)] × (0.70, if receiving a transplant at a steroid-sparing center) x [(age in years/50)$^{-0.4}$] × (0.94, if calcium channel blocker is present).

CL/F (L/hour) = [38.4 x 2.00 x (56/50)$^{-0.4}$ x 0.94].

CL/F (L/hour) = 68.99.

Once A.N.'s CL/F is estimated, the oral total daily dose (TDD) is calculated, assuming a desired trough concentration of 12 ng/mL, using the following.

TDD in mg = [CL/F (L/hour) × tacrolimus trough goal (ng/mL) × 24 hours]/1000.

TDD in mg/day = (68.99 L/hour × 12 ng/mL × 24 hours)/1000.

TDD in mg/day = 19.86 mg/day, rounded to 10 mg twice daily orally.

Oral Dose Estimation for A.N. Using Genotype Only

A second method for dosage determination is when expressers (*CYP3A5*1/*1* or *CYP3A5*1/*3*) receive 0.3 mg/kg/day and *CYP3A5*3/*3* non-expressers receive 0.15 mg/kg/day. A.N. carries the *CYP3A5*1/*1* high-expression genotype and would receive 0.3 mg/kg/day × 85 kg = 25.5 mg/day or 12–13 mg twice daily orally to achieve a trough of around 12 ng/mL.

The predicted doses from both methods are relatively close and represent the higher starting doses that individuals expressing *CYP3A5* typically require. In the first example, A.N.'s age and use of a calcium channel blocker are considered in the tacrolimus CL estimate. Increasing age and calcium channel blocker use are both associated with reductions in tacrolimus CL, which accounts for the lower estimated dose in example 1 relative to the second example. Note that the time posttransplantation and steroid use are not modifying factors for A.N. and that they thus drop out of the equation.

Case 2. A White Patient

M.A. is a 68-year-old, 76-kg white man with end-stage renal disease secondary to hypertension. He has received hemodialysis three times weekly for the past 1.5 years. Other medical problems include hyperlipidemia, anemia, and depression. His admission medications include Renal Caps, furosemide, calcium acetate, calcitriol, amlodipine, metoprolol, simvastatin, sodium bicarbonate, daily aspirin, and citalopram. He presents for a related living donor kidney transplant. His most recent panel-reactive antibody is 12%. Cross match is negative with a five-antigen match between the donor and the recipient. Serology is negative for HIV, hepatitis B surface antigen, and hepatitis C but positive for cytomegalovirus. Genotyping shows he carries the genotype *CYP3A5*3/*3* (non-expresser). He receives induction therapy with intravenous rabbit antithymocyte globulin before surgery and on days 1 and 3, with intravenous methylprednisolone 500 mg before surgery and 250- and 100-mg intravenous doses on days 1 and 2, respectively. He receives a steroid-sparing maintenance regimen of oral prednisone beginning on day 2, to be tapered off by day 7. M.A.'s maintenance immunosuppression is oral mycophenolate mofetil 1 g twice daily, starting 12 hours posttransplantation, and oral tacrolimus once he has good graft function. M.A. has good graft function at day 6 posttransplantation, producing 1500 mL of urine per day, and tacrolimus is initiated. Using the same equations as in the previous case, the following tacrolimus doses are estimated for M.A. to achieve a trough concentration of around 12 ng/mL.

Oral Dose Estimation for M.A. Using Genotype and Clinical Factors

CL/F (L/hour) = [38.4 × 0.86 × 0.70 × (68/50)$^{-0.4}$ × 0.94]

CL/F (L/hour) = 19.21.

TDD = (19.21 L/h × 12 ng/mL × 24 h)/1000.

TDD = 5.53 mg/day or 2.5–3 mg twice daily orally.

Oral Dose Estimation for M.A. Using Genotype Only

M.A. has a *CYP3A5*3/*3* genotype and would therefore receive 0.15 mg/kg/day × 76 kg = 11.4 mg/day or 5–6 mg twice daily orally.

The patient in case 2 has a substantially lower tacrolimus CL (19.21 vs. 68.99 L/hour) relative to the patient in case 1 and will require a lower tacrolimus dose. This is primarily because the patient in case 2 does not express *CYP3A5* and will metabolize tacrolimus slower than a *CYP3A5* expresser. The algorithm, which incorporates genotype and clinical factors, estimates a lower dose requirement than the genotype-only algorithm. This is because the clinical factors incorporate a reduction in CL for increasing time posttransplantation, transplantation at a steroid-sparing center (this center gives a short initial course of steroids, and these steroids only moderately influence CL; the CL would be lower if he were to receive standard daily steroid dosing), older age, and concomitant use of a calcium channel blocker.

Genotype-guided dosage algorithms for dosing tacrolimus may eventually become part of clinical practice. Because the trial-and-error approach to dosing tacrolimus is so widely accepted clinically, it will take time and additional studies for pharmacogenomic advances to become accepted methods for dose determination. Genetic-guided dosing will not eliminate the need for trough concentration monitoring but will reduce the frequency of monitoring in the early posttransplant period because target troughs will be reached faster. These types of dosage equations are flexible because any desired trough concentration can be targeted, allowing additional personalization. Genetic dosing may have a particularly important role in steroid-sparing protocols or induction-free regimens, which rely heavily on sufficient tacrolimus concentrations in the early days posttransplantation for sufficient immunosuppression. Algorithms for non-renal solid organ transplants and children have yet to be developed.

Cytochrome P450 3A4
Tacrolimus depends on *CYP3A4* and *CYP3A5* for metabolism; nevertheless, *CYP3A4* variants do not contribute substantially to interindividual variability in tacrolimus CL. Several *CYP3A4* variants have been evaluated for their effects on tacrolimus pharmacokinetics, which are small and/or have not consistently shown an effect. One of the first variants studied toward tacrolimus was *CYP3A4*1B* (rs2740574, -392A>G), which is a promoter variant associated with higher hepatic expression. Patients with this variant have lower dose-adjusted concentrations than do non-carriers (Barbarino 2013; Hesselink 2003). However, *CYP3A4*1B* is in linkage disequilibrium with the *CYP3A5*1* allele and does not contribute additional information if the *CYP3A5*1* genotype is known. The *CYP3A4*18* (rs28371759, 878T>C) variant has also been associated with increased tacrolimus CL but is also in linkage disequilibrium with the *CYP3A5*1* allele. *CYP3A4*22* (rs35599367, 15389 C>T) is a variant in intron 6 associated with lower CYP3A4 enzyme activity and may be important toward several CYP3A4 substrates used in transplantation, including statins and tacrolimus (Elens 2012). The *CYP3A4*22* allele is infrequent in whites (MAF 5%) and is not observed in African Americans (Abecasis 2012). Several studies have shown that patients with the *CYP3A4*22* variant have a lower tacrolimus CL, higher blood concentrations, and longer exposures to troughs greater than 15 ng/mL than do non-carriers (Elens 2013a, 2011; Gijsen 2013), whereas other data suggest a lack of association (Lunde 2014; Santoro 2013).

Cytochrome P450 Oxidoreductase
Cytochrome P450 oxidoreductase is a flavoprotein that transfers electrons from nicotine amine adenine dinucleotide to CYP enzymes and is essential in CYP activity (Pandey 2013; Agrawal 2008). The *POR*28* variant (rs1057868, 1508 C>T) is associated with increased *CYP3A* activity, thereby enhancing metabolism (Elens 2014b). This variant has a MAF of 30% in whites and 22% in African Americans (Abecasis 2012). The *POR*28* variant has been studied toward tacrolimus pharmacokinetics in kidney transplant recipients (Gijsen 2014; Lunde 2014; Saruwatari 2014; Elens 2013b; Woillard 2013; Zhang 2013; de Jonge 2011; Hesselink 2014). Individuals carrying *POR*28* have lower troughs and increased dose requirements, though the effect is modest.

ATP Binding Cassette B1
The *ABCB1* gene (also known as *MDR1* [multidrug resistance 1 gene]) encodes for the efflux transporter Pgp, which actively transports substrates out of cells. These proteins are expressed in the intestine, liver, and kidney. As efflux transporters, they reduce the intestinal absorption of tacrolimus and enhance biliary excretion from the liver and tubular excretion from the kidneys (Hawwa 2009). The large interindividual variability in the bioavailability of tacrolimus (Venkataramanan 1995) and in the variants within this gene has been proposed to be an important contributor. Some studies have shown that an *ABCB1* haplotype consisting of three variants, rs1045642 (3435 C>T) in exon 26, rs2032582 (2677 C>T) in exon 21, and rs1128503 (1236 C>T) in exon 21, affects tacrolimus transport (Hoffmeyer 2000). The influence of this haplotype on tacrolimus pharmacokinetics is conflicting. Some studies have shown a reduction in tacrolimus concentrations, whereas others have shown no effect. Because of this uncertainty, this haplotype is of limited clinical benefit in defining tacrolimus dose (Provenzani 2013; Staatz 2010b).

Tacrolimus Pharmacodynamics Genetic Variants
Most variants studied for associations with tacrolimus pharmacokinetics have also been evaluated for their effect on clinical end points such as delayed graft function, tacrolimus-related nephrotoxicity, new-onset diabetes, neurotoxicity, hypertension, acute rejection, and chronic kidney graft dysfunction (Hesselink 2014; Barbarino 2013; Tang 2011). Although studies have shown associations, the magnitude and direction of the effects have been

inconsistent. Toxicities have been particularly challenging to study because of the difficulty in accurately attributing toxicity to tacrolimus. For example, the cause of new-onset diabetes is also attributable to other immunosuppressive drugs, and nephrotoxicity is caused by many non–drug-related events. Studies also use varying toxicity definitions, further complicating the phenotypic-genotypic associations. In a meta-analysis of 23 studies of kidney and liver transplant recipients, *CY3A5* expressers were at a higher risk of acute rejection in the first month posttransplantation, but not after that (Tang 2011). However, the associations are not strong enough to recommend genotyping for the routine use as a predictor of acute rejection. The underlying biology and causes of acute rejection and chronic graft dysfunction is multifaceted, and it will be challenging to identify pretransplant genotypic predictors of rejection. Currently, no clear data support the use of genotyping in selecting immunosuppressive therapy to reduce rejection or toxicity.

Cyclosporine

Cyclosporine is a calcineurin inhibitor used primarily as maintenance immunosuppression after organ transplantation. It was first used in the 1980s, when it lowered the frequency of acute T cell–mediated rejection. During the past 15 years, it has been slowly replaced by tacrolimus. Cyclosporine has a mechanism of action similar to that of tacrolimus, where it binds to the immunophilin cyclophilin, forming a cyclosporine-cyclophilin complex, which inhibits calcineurin phosphatase. Cyclosporine then inhibits T-cell activation similar to tacrolimus.

Metabolism and Pharmacokinetics

Cyclosporine is mainly metabolized by intestinal and hepatic *CYP3A4* and *CYP3A5* (Dai 2004). Figure 12.1 shows the metabolic pathway. Cyclosporine is subject to high first-pass metabolism. At least 30 metabolites have been identified, of which AM1, AM9, and AM4N are primary, whose formation is mainly catalyzed by the CYP3A4 enzyme (Christians 1993). The CYP3A5 enzyme primarily forms the AM9 metabolite (Dai 2004). Cyclosporine metabolites are mainly eliminated in the bile. A limited amount of cyclosporine is found in the urine or feces (Kovarik 1999; Bleck 1989). Cyclosporine and its metabolites are substrates for the efflux transporter Pgp, encoded by the *ABCB1* gene (Saeki 1993). P-glycoprotein is present in many tissues, including the gastrointestinal tract, which may influence oral bioavailability. The mean oral bioavailability of the microemulsion formulation of cyclosporine is about 30%, with high interindividual variability (Kapturczak 2004; Akhlaghi 2002).

Like tacrolimus, cyclosporine has a narrow therapeutic index. Therapeutic drug monitoring of cyclosporine blood concentrations is standard of care because of the unpredictable nature of cyclosporine CL and the strong association between troughs and clinical outcomes (Jorga 2004; Thiel 1994).

Cyclosporine Pharmacogenomics

The pharmacogenomic data for cyclosporine have been reviewed in other publications and will not be repeated in this chapter (Barbarino 2013; Staatz 2010b). Instead, the following section will provide the status of clinical pharmacogenomic testing for cyclosporine.

Cytochrome P450 3A4

Although cyclosporine is highly dependent on *CYP3A4* metabolism, few genetic variants affect cyclosporine pharmacokinetics. *CYP3A4*1B* (rs2740574, -392A>G) is a genetic variant in the promotor region of the *CYP3A4* gene associated with higher *CYP3A4* expression. The frequency of the G allele in the African American population is about 80%, whereas its frequency is only around 2% in whites (Abecasis 2012). The G allele is in high linkage disequilibrium with the *CYP3A5*1* variant (Thervet 2005). Effects of this variant on cyclosporine metabolism are conflicting. Some studies have shown no effect of the variant (Bouamar 2011; von Ahsen 2001), whereas other data suggest lower dose-adjusted trough concentrations and higher cyclosporine dose requirements (Zochowska 2012; Crettol 2008). More recently, the *CYP3A4*22* variant has been studied, showing that carriers may have lower cyclosporine CL and higher troughs, though the effect was small (Lunde 2014; Moes 2014; Elens 2013c, 2011). Two hundred ninety-eight kidney transplant recipients (88% whites) were genotyped for the *CYP3A4*22* allele. *CYP3A4*22* carriers showed a 15% lower cyclosporine A CL compared with wild type (Moes 2014). Similar results were shown in a study of 174 kidney transplant recipients (Elens 2014a).

Cytochrome P450 3A5

The effects of the *CYP3A5* variants on tacrolimus are profound. This genotype has also been studied toward cyclosporine pharmacokinetics. Unfortunately, studies have shown mixed effects that are insufficient to justify clinical implementation for cyclosporine (Zochowska 2012; Bouamar 2011; Staatz 2010a; Haufroid 2004). A study of 100 white renal transplant recipients found that recipients who were *CYP3A5* non-expressers (*CYP3A5*3/*3*) had 1.6-fold higher cyclosporine trough concentrations than did the heterozygote *CYP3A5*1/*3* carriers (Haufroid 2004). Similarly, in another study of 100 renal transplant recipients, 7 patients carrying heterozygote *CYP3A5*1/*3* required a higher cyclosporine dose (400.65 mg vs. 263.52 mg) compared with 93 subjects who were *CYP3A5*3/*3* carriers (Zochowska 2012). However, in a population pharmacokinetic study of 298 kidney transplant patients, no association was observed between the *CYP3A5* variant and cyclosporine CL (Moes 2014). Cyclosporine is highly dependent on the *CYP3A4* enzyme for metabolism; therefore, it is not surprising that the *CYP3A5* variant has limited effect (Barbarino 2013; Dai 2004).

ABCB1 Variants

Three *ABCB1* variants—3435 C>T, 1236 C>T, and 2677 G>T/A (all in high linkage disequilibrium)—have been widely studied for their influence on immunosuppressant pharmacokinetics; however, few studies in the literature have observed a significant difference in cyclosporine pharmacokinetics between these genotype groups. A study of 44 liver transplant recipients found an association between *ABCB1* 3435 C>T and peaks in cyclosporine concentrations, in which recipients with the TT genotype had higher peak concentrations than the CC genotype recipients and required a 50% lower dose (Bonhomme-Faivre 2004). The T allele of the variant is associated with reduced *ABCB1* expression, and it can be speculated that individuals carrying the T allele efflux less cyclosporine, resulting in a higher peak concentration (Fung 2009). Another study found an association between the 2677G>T variant and cyclosporine dose-adjusted trough concentrations, in which transplant recipients with the GG genotype showed lower trough concentrations at 3, 6, and 12 months after transplant compared with recipients having the GT and TT genotypes. Thus, the first group required higher doses than the other two groups (Crettol 2008). Other studies, however, have found no association between the variants and cyclosporine pharmacokinetics (Bouamar 2011; Haufroid 2004; von Ahsen 2001).

Azathioprine

Azathioprine is a purine antimetabolite and is an imidazolyl derivative of mercaptopurine. Azathioprine is an older agent, and although it has excellent immunosuppressant activity, it has largely been replaced by mycophenolate in organ transplantation. It is still used in transplant recipients whose contemporary immunosuppressants have failed and in autoimmune diseases such as inflammatory bowel disease and dermatologic disorders. When mercaptopurine is used for leukemia indications, genetic testing to identify individuals at risk of severe myelosuppression has been adopted at many centers. The enthusiasm for the clinical implementation of genotyping for azathioprine to identify at-risk transplant recipients is still modest. The pharmacogenomics of azathioprine and mercaptopurine have been extensively reviewed elsewhere, and this section will focus on the clinical implementation of genotyping in transplantation when azathioprine is used as an immunosuppressant (Murray 2013; Chouchana 2012; Sahasranaman 2008). The CPIC (Clinical Pharmacogenomics Implementation Consortium) has published guidelines on genetic testing for thiopurines (mercaptopurine, azathioprine, and thioguanine) (Relling 2013,

Table 12.3. Dose Recommendations for Azathioprine According to *TPMT* Genotype

Likely Phenotypes	Genotypes	Azathioprine Dose Recommendation
Normal *TPMT* activity, homozygous wild type, (about 86%–97% of patients)	Two or more functional alleles (*1)	Start with a normal starting dose (e.g., 2–3 mg/kg/day), and adjust doses according to disease-specific guidelines. Allow 2 weeks to reach steady state after each dose adjustment
Intermediate *TPMT* activity, heterozygous (about 3%–14% of patients)	One functional allele (*1) plus one low-activity allele (*2, *3A, *3B, *3C, or *4)	If disease treatment normally starts at the full dose, consider starting at 30%–70% of the target dose (e.g., 1–1.5 mg/kg/day), and titrate according to tolerance. Allow 2–4 weeks to reach steady state after each dose adjustment
Deficient *TPMT* activity, homozygous for variants (about 1 in 178 to 1 in 3736 patients)	Two non-functional alleles (*2, *3A, *3B, *3C, or *4)	Consider alternative agents. If using azathioprine, start with drastically reduced doses (reduce daily dose by 10-fold and dose thrice weekly instead of daily), and adjust doses according to the degree of myelosuppression and disease-specific guidelines. Allow 2–4 weeks to reach steady state after each dose adjustment

Adapted from: Relling MV, Gardner EE, Sandborn WJ, et al. Clinical pharmacogenetics implementation consortium guidelines for thiopurine methyltransferase genotype and thiopurine dosing: 2013 update. Clin Pharmacol Ther 2013;93:324-5; and Relling MV, Gardner EE, Sandborn WJ, et al. Clinical Pharmacogenetics Implementation Consortium guidelines for thiopurine methyltransferase genotype and thiopurine dosing. Clin Pharmacol Ther 2011;89:387-91.

2011). These guidelines also apply to mercaptopurine and thioguanine, which are used in cancer therapies.

Metabolism

Azathioprine is a prodrug that is non-enzymatically reduced to mercaptopurine and then activated by HGPRT (hypoxanthine-guanine-phosphoribosyl transferase) to thioinosine monophosphate (TIMP). Thioinosine monophosphate is then converted to 6-thioguanine nucleotides (6-TGNs) and 6-methyl mercaptonucleotides (6-MeMPNs). The immunosuppressive activity of azathioprine is caused by the incorporation of 6-TGNs into the DNA and the inhibition of purine synthesis by 6-MeMPNs. Mercaptopurine is inactivated by two major pathways. One route is mediated by *TPMT*, where mercaptopurine is inactivated to 6-methyl mercaptopurine (6-MeMP), and the other route is mediated by xanthine oxidase, which converts mercaptopurine to 6-TU (6-thiouric acid). These reactions occur in erythrocytes and the liver. Figure 12.4 shows the azathioprine metabolic pathway. Genetic variants in the *TPMT* gene result in low-activity enzyme and reduce the conversion of mercaptopurine to 6-MeMP and 6-MePN. These variants can cause the accumulation of mercaptopurine and TIMP, thereby increasing the risk of hematologic toxicity.

Azathioprine Pharmacogenetics

Azathioprine is associated with several adverse effects, including bone marrow depression and gastrointestinal issues (nausea, vomiting, and hepatotoxicity). Leukopenia and thrombocytopenia are dose-dependent and often respond to dose reduction or temporary cessation of therapy. However, some individuals develop severe, life-threatening hematologic toxicity and require discontinuation of therapy. Genotyping will identify individuals at risk of the hematologic toxicities, though it does not eliminate the need for regular complete blood cell count monitoring. Variants in the gene encoding *TPMT* lead to varying functional activity of the enzyme. Several variants have been identified, most of which are responsible for reduced or severely deficient *TPMT* enzyme activity (Garat 2008). The variants *TPMT*2* (238G>C), *3A* (460G>A and 719A>G), *3B* (460G>A), *3C* (719A>G), and *4* (626-1G>A) have reduced activity. The *TPMT* variants 238C>G, 460G>A, and 719A>G account for 85%–90% of cases of reduced or severely deficient enzyme activity. They are commonly included in genetic testing, though some testing centers may include other variants (McLeod 2002; Krynetski 2000, 1996; Otterness 1997; Yates 1997). However, the star nomenclature that designates the genotypes is confusing. For example, the *TPMT*3A* genotype contains two variants (460G>A and 719A>G), whereas the *TPMT*3C* (719A>G) genotype contains only one variant. The official *TPMT* nomenclature is found at www.imh.liu.se/tpmtalleles.

Overall, about 10% of the population is heterozygous for one of the reduced-activity variant alleles, and 0.33% or less is homozygous (Krynetski 2000; Lennard 1989). The frequency of the variant alleles varies by race. The most common variant among whites and Southwest Asians is *TPMT*3A*, whereas the *TPMT*3C* variant predominates in African American and Chinese populations (Collie-Duguid 1999). Carriers of one or two variant alleles have reduced or severely deficient *TPMT* activity, respectively, which results in increased production of 6-MeMPNs and 6-TGNs. Cellular accumulation of 6-TGNs is primarily responsible for azathioprine-related hematologic toxicity (Vannprasaht 2009; Black 1998). Pharmacogenetic guidelines address dose management in individuals who are heterozygous or homozygous for *TPMT* variants (Relling 2013, 2011).

Clinical Implementation of the *TPMT* Genotype for Azathioprine Dosing

The evidence for an association between *TPMT* genotypes and toxicity is rated strong (the evidence is high quality, and the desirable effects clearly outweigh the undesirable effects), though most data for the association have been generated from cancer studies. Pharmacogenomic guidelines with recommended dosage guidelines for azathioprine have been developed (Table 12.3). Individuals who carry none of the tested variants should receive

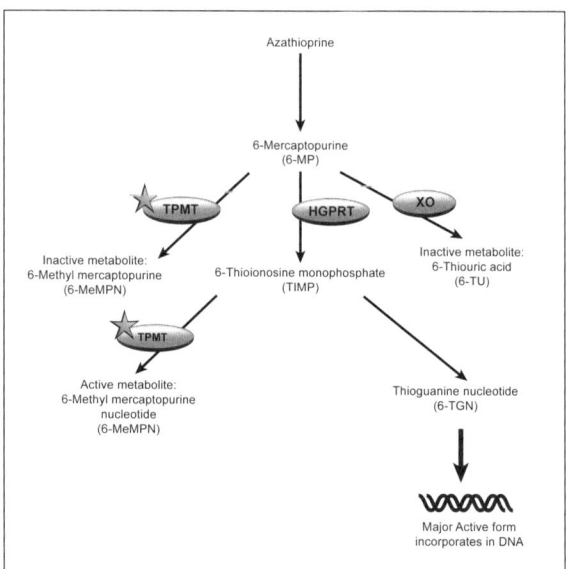

Figure 12.4. Metabolic pathway of azathioprine.

HGPRT = hypoxanthine-guanine-phosphoribosyl transferase; XO = xanthine oxidase; TPMT = thiopurine methyl transferase; .

★Predominant enzyme and pathway

the normal azathioprine starting dose. For individuals who are heterozygous and carry one variant allele, a starting dosage reduction of 30%–70% should be considered. Around 30%–60% of individuals who are heterozygous carriers will be unable to tolerate full doses. For managing individuals who are heterozygous, there is substantial debate. Patients who are homozygous carriers will have two variant alleles and a severe reduction in, or absent, *TPMT* activity. They are at a 100% risk of having a life-threatening hematologic toxicity. These individuals should receive a 10-fold reduction in planned dose and administration of the dose three times weekly instead of daily. For example, if the planned dose is 3 mg/kg/day, the new dose should be 3 mg/kg/day on Monday, Wednesday, and Friday. In these individuals, it is also reasonable to avoid the risk altogether and select another immunosuppressant. Genotyping is best used when performed before therapy is initiated. Some clinicians choose to genotype patients who develop hematologic toxicity to assist in attribution of the toxicity. Homozygous carriers are quite rare, and routine genotyping of all patients has not been considered cost-effective. However, as the cost of genotyping decreases and the availability of pharmacogenetic panels increases, genetic testing may become more common. Direct measurement of *TPMT* enzymatic activity is available in some laboratories, and some clinicians will choose to measure activity, instead of genotype, when making dosage determinations.

Mycophenolic Acid
Mycophenolic acid is a potent antiproliferative immunosuppressive agent that largely replaced azathioprine in organ transplantation after showing superiority in randomized trials (No authors listed 1996; Sollinger 1995). It is used in combination with a calcineurin inhibitor with or without prednisone for maintenance immunosuppressive therapy. Mycophenolic acid inhibits T- and B-lymphocyte proliferation by non-competitive and reversible inhibition of inosine monophosphate-5′-dehydrogenase (IMPDH) types 1 and 2. Type 2 *IMPDH* is more potently inhibited than is type 1 *IMPDH* (Allison 2000, 1977).

Metabolism and Pharmacokinetics
Mycophenolic acid areas under the curve (AUCs) and troughs have been associated with acute rejection and toxicity in organ transplant recipients in some studies, whereas in others, the association is weak or absent (Staatz 2007). Given this debate, not all centers perform therapeutic drug monitoring of mycophenolic acid. Thus, an evaluation of the genetic determinants of exposure would be a useful clinical tool to ensure that transplant recipients receive sufficient mycophenolic acid exposure and immunosuppression.

The two oral dosage forms of mycophenolic acid are an ester prodrug (mycophenolate mofetil) and an enteric-coated, delayed-release sodium salt (mycophenolic acid sodium). The mycophenolate mofetil prodrug is rapidly and completely hydrolyzed to mycophenolic acid by carboxylesterases (CES1 and CES2) (Fujiyama 2010; Rosso Felipe 2009). The mycophenolic acid metabolic pathway and transporters are the same for the two products. Figure 12.5 shows the mycophenolic acid metabolic pathway. The active component, mycophenolic acid, is metabolized by intestinal and hepatic uridine diphosphate-glucuronosyltransferases (*UGTs*) (*1A1, 1A7, 1A8, 1A9, 1A10,* and *2B7*) (Barraclough 2010; Staatz 2007). The liver is considered the primary site of metabolism, and mycophenolic acid-7-O-glucuronide (MPAG) is the major metabolite. Mycophenolic acid-7-O-glucuronide is an inactive metabolite formed mainly by *UGT1A9*. The acyl form of MPAG (Ac-MPAG) is a minor metabolite whose formation is mediated mainly by *UGT2B7*. The Ac-MPAG has immunosuppressive activity and may be associated with mycophenolic acid–related adverse reactions (Staatz 2007; Shipkova 1999). The DM-MPA (6-O-desmethyl-mycophenolic acid) is a minor metabolite that has not been well studied and may have limited clinical relevance (Picard 2005; Shipkova 1999). UGT1A8 and UGT1A10 are extrahepatic enzymes that contribute to mycophenolic acid metabolism in the gastrointestinal tract (Mojarrabi 1998). Mycophenolic acid-7-O-glucuronide, and possibly Ac-MPAG, are substrates for OATPs (organic anion transporter proteins), which are encoded by solute carrier organic anion transport protein 1B1 (SLCO1B1) and 1B3 (SLCO1B3) genes. These genes are involved in the uptake of mycophenolic acid and its metabolites into the liver (Michelon 2010; Miura 2007). Mycophenolic acid is primarily eliminated as MPAG through the urine, though MPAG and other metabolites are also eliminated in the bile. They are transported out of the liver and into the bile through the efflux transporter multidrug resistance protein 2 (MRP2) (Staatz 2007). Once in the bile, MPAG can undergo enterohepatic recirculation, where it is deconjugated by intestinal bacterial glucuronidase and is converted back to mycophenolic acid. Available mycophenolic acid is then reabsorbed into the systemic circulation. Enterohepatic recirculation may account for a substantial portion of the AUC, and it is a clinically relevant characteristic of mycophenolic acid (Bullingham 1998).

Mycophenolic Acid Pharmacogenomics
Mycophenolic acid pharmacogenomics has been primarily centered on the influence of variants in enzymes (*UGT1A9,*

UGT1A8, and *UGT2B7*) and transporters (*SLCO1B1*, *SLCO1B3*, and *MRP2*) on pharmacokinetics. Variants in the mechanistic pathway such as *IMPDH* have been evaluated toward rejection and toxicities, but to a lesser extent. An extensive review of mycophenolic acid pharmacokinetic and pharmacodynamic pharmacogenomic studies has been published and will not be reviewed in this chapter (Lamba 2014; Murray 2013; Barraclough 2010).

UDP Glucuronosyl Transferase 1A9 and 1A8

UGT1A9 -275T>A (rs6714486) and -2152C>T (rs17868320) are promoter region variants that have been most studied with respect to mycophenolic acid pharmacokinetics. They are in linkage disequilibrium. The MAF of -275T>A is 6% in whites and 18% in African Americans, and that of -2152C>T is 5% in whites and only 2% in African Americans (Abecasis 2012). In vitro, these variants increase glucuronidation (Girard 2004). Several clinical studies have reported lower mycophenolic acid exposure in carriers of these variants. *UGT1A9* -440C>T (rs2741045) and -331T>C (rs2741046) are other promoter variants in high linkage disequilibrium that have also been associated with lower mycophenolic acid exposure. Their MAF is 28% in whites and 2% in African Americans (Abecasis 2012). These data suggest that individuals with one of the promoter variants require higher starting doses of mycophenolate. *UGT1A9*3* (98T>C, rs72551330) is a low-frequency and low-activity variant associated with reduced mycophenolic acid metabolism and higher mycophenolic acid exposure (van Schaik

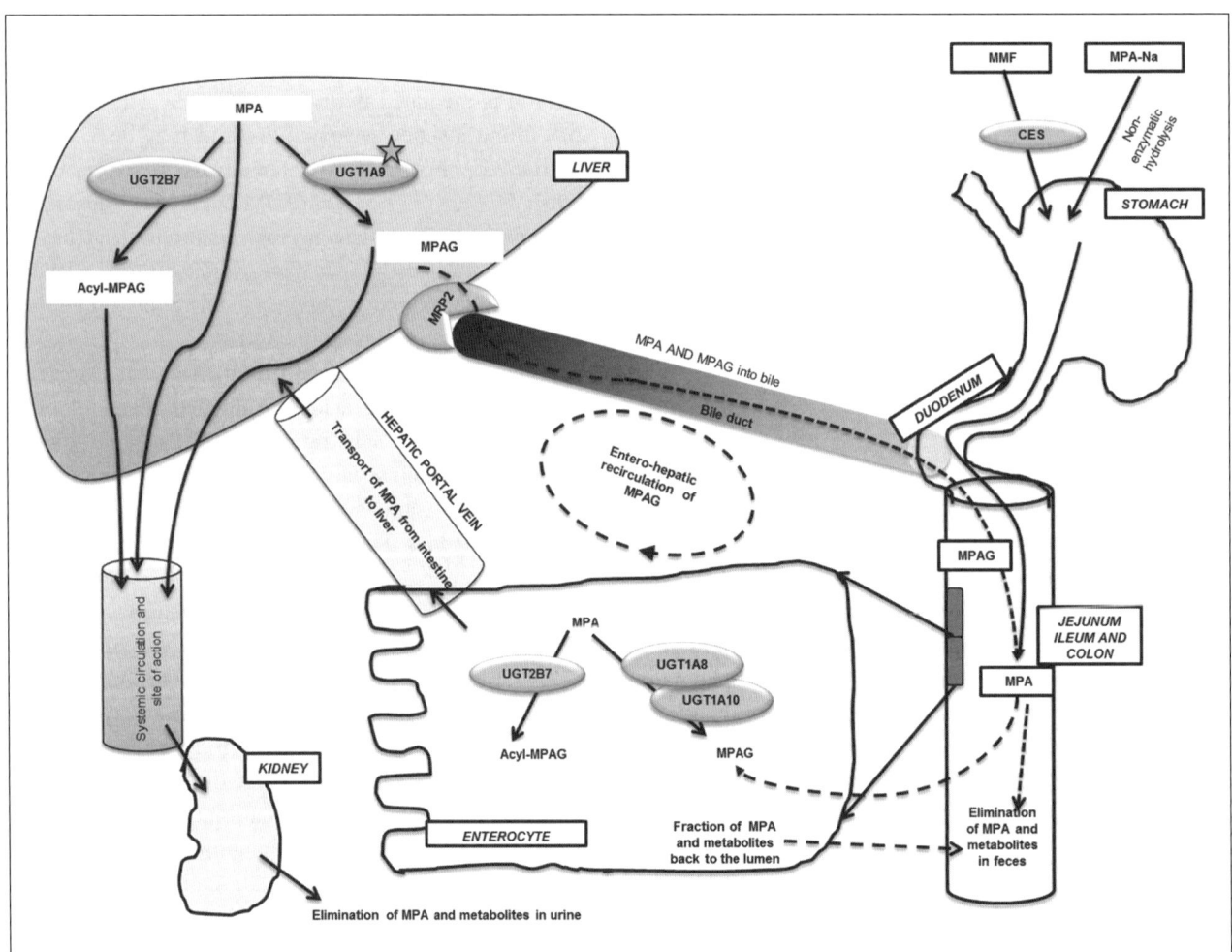

Figure 12.5. Major metabolic pathway of mycophenolic acid.

MMF = mycophenolate mofetil; MPA = mycophenolic acid; MPA-Na = mycophenolic acid sodium; CES = carboxylesterase; MPAG = MPA-7-O-methyl glucuronide; acyl MPAG = acyl-7-O-methyl glucuronide; UGT = uridine diphosphate-glucuronosyltransferase; MRP2 = multidrug resistance protein 2.

☆Predominant enzyme

2009). Although *UGT1A9* variants have been evaluated in several studies, their effects have not been consistently replicated.

UGT1A8*2 (rs1042597, 518C>T/G) has been studied for its effect on mycophenolic acid pharmacokinetics in transplant recipients. The MAF in whites is about 28%, and in African Americans, it is 2%. A 60% higher mycophenolic acid trough concentration occurred among the carriers of variant allele than among the non-carriers. This significance, however, occurred only among the patients also receiving tacrolimus and not among those receiving cyclosporine (Johnson 2008; Zucker 1999). Another study showed an 18% higher mycophenolic acid AUC_{0-12} in patients coadministered cyclosporine (van Schaik 2009).

UDP Glucuronosyltransferase 2B7
The frequently studied *UGT2B7*2* variant (rs7439366, 802C>T) is a common reduced-function variant with a high allele frequency (MAF 42% in whites and 69% in African Americans) (Abecasis 2012). UGT2B7 is the main enzyme responsible for Ac-MPAG formation. Some studies have shown an association between this variant and mycophenolic acid and Ac-MPAG pharmacokinetics, whereas others have not (van Schaik 2009; Bernard 2006). Some data suggest that this variant is associated with mycophenolic acid–related toxicity (Yang 2009).

SLCO Transporters 1B1 and 1B3
Several *SLCO* transporter variants have been evaluated toward MPAG and mycophenolic acid pharmacokinetics because mycophenolic acid glucuronide metabolites are substrates for this transporter. *SLCO1B1*5* (521T>C, rs4149056) is a low-function variant with a MAF of 15% in whites and 1% in African Americans (Abecasis 2012). Carriers of the *SLCO1B1*5* variant are at greater risk of developing statin-induced myopathy than are non-carriers because of reduced statin hepatic metabolism. In transplantation, data suggest that the *SLCO1B1*5* variant is protective from mycophenolic acid–related adverse drug reactions (Michelon 2010). The variant may limit the amount of glucuronide metabolite available for enterohepatic recirculation, thereby lowering mycophenolic acid concentrations and toxicity.

SLCO1B3 334T>G, rs4149117, and 669G>A, rs7311358, are variants within the *SLCO1B3* gene that are in complete linkage disequilibrium. They are common, with a frequency of 89% in whites, 37% in African Americans, and 66% in the Japanese population (Abecasis 2012). They are associated with increased hepatic mycophenolic acid uptake and its excretion in bile. This results in higher mycophenolic acid AUC_{6-12} values, indicative of higher mycophenolic acid enterohepatic circulation (Miura 2007). However, another study has suggested an opposite effect, in which the variants appear to decrease *SLCO* activity, thereby decreasing MPAG uptake. This resulted in decreased enterohepatic circulation and a decrease in mycophenolic acid dose-normalized exposure (Picard 2010; Miura 2008). Data for the *SLCO* variants are still limited and require confirmation before clinical use.

Multidrug Resistance Protein 2
The MRP2 transporter protein is encoded by the *ABCC2* gene and is expressed in many cells, including those in the liver and intestine. These transporters efflux MPAG and other metabolites into the biliary tract. Multidrug resistance protein 2 -24C>T (rs717620) is a promoter variant associated with increased MRP2 expression and has been evaluated in several pharmacokinetic studies. Its MAF is 17% in whites and 3% in African Americans (Abecasis 2012). This variant is thought to increase the amount of MPAG effluxed into the bile, where MPAG is then subject to enterohepatic recirculation. Associations have been inconsistent; some studies have shown no association, whereas others have found it to affect enterohepatic recirculation. The MRP2 variant 3972C>T (rs3740066) has also been evaluated. However, it is a synonymous variant in linkage disequilibrium with MRP2 -24C>T; therefore, its associations may be caused by the MPR2 -24C>T variant. Understanding the effects of MRP2 variants on mycophenolate is difficult. Cyclosporine is an inhibitor of MRP2 whereas tacrolimus is not an inhibitor. Inhibition of MRP2 and loss of enterohepatic recirculation is thought to be one of the primary reasons that mycophenolic acid concentrations are significantly lower in patients receiving cyclosporine relative to tacrolimus. Therefore, the inhibition may interfere with and/or confound the effect of the MRP2 variants. This may be why studies have found a differential effect of these variants, depending on the underlying calcineurin inhibitor.

Currently, the preemptive pretransplant determination of mycophenolate dose according to genetic variants to optimize mycophenolic acid pharmacokinetics cannot be recommended. The pharmacokinetics of mycophenolic acid is complex because of the many enzymes and transporters involved in the elimination pathway and enterohepatic recirculation. The *UGT* gene structure is complex, with many variants in linkage disequilibrium. In addition, nongenetic factors affect mycophenolic acid pharmacokinetics, including age, sex, and drug interactions such as cyclosporine inhibition of MRP2 and induction of *UGTs* by steroids.

Influence of Genetic Variants on Mycophenolic Acid Pharmacodynamics

Although mycophenolic acid has fewer adverse effects relative to other immunosuppressive agents, it has several clinically problematic adverse effects, mainly diarrhea, vomiting, leukopenia, and anemia, which, if severe, require dose reductions or discontinuation of therapy. Therefore, prospective markers to identify individuals at risk of toxic reactions would be clinically useful. Most of the variants investigated toward pharmacokinetics have also been evaluated for their association with mycophenolic acid–related toxicities and acute rejection, with mixed findings. Variants in *IMPDH1* and *IMPDH2* have also been evaluated. The *IMPDH2* variant 3735 T>C (rs11706052) was evaluated for its association with renal function and acute rejection. Patients carrying the minor allele had a 3-fold higher odds of experiencing biopsy-proven acute rejection than did those without the allele (Grinyo 2008). Another study also found that *IMPDH* enzymatic activity was higher in patients carrying the 3735 C allele (Sombogaard 2009). Within *IMPDH1*, two variants (rs2278293 and rs2278294) have been associated with a higher incidence of acute rejection and a higher risk of leukopenia (Gensburger 2010; Wang 2008). The clinical relevance of these variants is still emerging. Because the incidence of acute rejection with today's immunosuppression is low, large multicenter studies are required to clearly understand whether these variants will provide predictive information.

Modification of Drug Interactions by Genotypes

Many drug-drug pharmacokinetic interactions occur in transplantation. Data suggesting that genetic variants increase or decrease an individual's susceptibility to drug interactions are emerging. The presence of variants may partly explain the unpredictable nature of drug interactions. This section will outline examples of pharmacogenetic-modified drug interactions with immunosuppressants.

Tacrolimus and Azole Antifungals

The azole drugs are used posttransplantation for preventing or treating fungal infections. Azoles are potent inhibitors of the CYP3A enzyme and effectively reduce the metabolism of calcineurin inhibitors, resulting in significantly higher concentrations. Individuals taking tacrolimus who are *CYP3A5* non-expressers (*CYP3A5*3/*3*) and receive fluconazole have a significantly greater increase in tacrolimus concentrations and are more likely to have troughs greater than 15 ng/mL compared with *CYP3A5* expressers (*CYP3A5*1/*3*) (Kuypers 2008). Non-expressers and expressers require about 65% and 40% lower tacrolimus doses, respectively, when tacrolimus is given with fluconazole. Ketoconazole and itraconazole also have a greater inhibitory effect on tacrolimus in *CYP3A5* non-expressers (Nara 2013; Chandel 2009).

Tacrolimus and Calcium Channel Blockers

Calcium channel blockers are frequently used pre- and posttransplantation and are often coadministered with tacrolimus. One of the most commonly used calcium channel blockers, amlodipine, is an inhibitor of the CYP3A enzyme. In healthy volunteers, amlodipine and tacrolimus coadministration resulted in a decrease in the oral CL of tacrolimus by around 2-fold and an increase in tacrolimus concentrations (Zuo 2013). This interaction was observed only in *CYP3A5* expressers. Similarly, diltiazem and tacrolimus coadministration resulted in a greater rise in tacrolimus concentrations in *CYP3A5* expressers than in non-expressers (Li 2011). However, *CYP3A5* non-expressers are at a greater risk of high tacrolimus concentrations and are more likely to have tacrolimus doses held when continuous-infusion nicardipine is coadministered with tacrolimus (Hooper 2012). This may be because nicardipine is a more potent *CYP3A4* inhibitor than *CYP3A5*.

Tacrolimus and Dexamethasone

The effect of genotype on *CYP3A* induction has also been evaluated. Dexamethasone is a potent inducer of *CYP3A* basal activity. Dexamethasone induction of *CYP3A* activity is greater in *CYP3A5* non-expressers than in expressers (Roberts 2008). Therefore, steroid induction of CYP3A enzymes may result in enhanced tacrolimus metabolism and lower concentrations in *CYP3A5* non-expressers, whereas no change may occur in the tacrolimus concentrations of *CYP3A5* expressers.

Clinical Implementation of Genotype-Guided Drug Interactions

There are no clinical guidelines on how to manage drug interactions in patients with genetic variants. However, in patients with an exaggerated or complete lack of a drug-drug interaction, the possibility of a genetically modified interaction should be considered.

Conclusion

The influence of genetic variants within genes involved in the metabolism, transport, and activity of immunosuppressive agents has been investigated for the immunosuppressant for the past 15 years. Apart from variants within genes, clinical factors like race, age, and sex significantly affect pharmacokinetics and toxicity; therefore, any investigation of the genetic variants eventually moved into practice will have to consider these factors. Currently, pharmacogenetic dosage recommendations are available for azathioprine. Guidelines for tacrolimus are currently being written. As with most developing genetic data, findings for some variants are conflicting and must be confirmed in larger trials before clinical implementation. With the advent of new technologies like ADME (absorption, distribution, metabolism, and excretion) gene chips, all variants in known relevant drug genes will eventually be on one panel and will be obtained pretransplantation. Immunosuppression can then be personalized to improve efficacy and minimize toxicities.

References

Abecasis GR, Auton A, Brooks LD, et al. An integrated map of genetic variation from 1,092 human genomes. Nature 2012;491:56-65.

Agrawal V, Huang N, Miller WL. Pharmacogenetics of P450 oxidoreductase: effect of sequence variants on activities of *CYP1A2* and *CYP2C19*. Pharmacogenet Genomics 2008;18:569-76.

Akhlaghi R, Trull AK. Distribution of cyclosporin in organ transplant recipients. Clin Pharmacokinet 2002;41:615-37.

Allison AC, Eugui EM. Mycophenolate mofetil and its mechanisms of action. Immunopharmacology 2000;47:85-118.

Allison AC, Hovi T, Watts RW, et al. The role of de novo purine synthesis in lymphocyte transformation. Ciba Found Symp 1977:207-24.

Anglicheau D, Verstuyft C, Laurent-Puig P, et al. Association of the multidrug resistance-1 gene single-nucleotide polymorphisms with the tacrolimus dose requirements in renal transplant recipients. J Am Soc Nephrol 2003;14:1889-96.

Barbarino JM, Staatz CE, Venkataramanan R, et al. PharmGKB summary: cyclosporine and tacrolimus pathways. Pharmacogenet Genomics 2013;23:563-85.

Barraclough KA, Lee KJ, Staatz CE. Pharmacogenetic influences on mycophenolate therapy. Pharmacogenomics 2010;11:369-90.

Benkali K, Rostaing L, Premaud A, et al. Population pharmacokinetics and Bayesian estimation of tacrolimus exposure in renal transplant recipients on a new once-daily formulation. Clin Pharmacokinet 2010;49:683-92.

Bernard O, Tojcic J, Journault K, et al. Influence of non-synonymous polymorphisms of *UGT1A8* and UGT2B7 metabolizing enzymes on the formation of phenolic and acyl glucuronides of mycophenolic acid. Drug Metab Dispos 2006;34:1539-45.

Birdwell KA, Grady B, Choi L, et al. The use of a DNA biobank linked to electronic medical records to characterize pharmacogenomic predictors of tacrolimus dose requirement in kidney transplant recipients. Pharmacogenet Genomics 2012;22:32-42.

Black AJ, McLeod HJ, Capell HA, et al. Thiopurine methyltransferase genotype predicts therapy-limiting severe toxicity from azathioprine. Ann Intern Med 1998;129:716-8.

Bleck JS, Schlitt HJ, Christians U, et al. Urinary excretion of ciclosporin and 17 of its metabolites in renal allograft recipients. Pharmacology 1989;39:160-4.

Bonhomme-Faivre L, Devocelle A, Saliba F, et al. *MDR-1* C3435T polymorphism influences cyclosporine a dose requirement in liver-transplant recipients. Transplantation 2004;78:21-5.

Boughton O, Borgulya G, Cecconi M, et al. A published pharmacogenetic algorithm was poorly predictive of tacrolimus clearance in an independent cohort of renal transplant recipients. Br J Clin Pharmacol 2013;76:425-31.

Bouamar R, Hesselink DA, van Schaik RH, et al. Polymorphisms in *CYP3A5*, *CYP3A4*, and *ABCB1* are not associated with cyclosporine pharmacokinetics nor with cyclosporine clinical end points after renal transplantation. Ther Drug Monit 2011;33:178-84.

Buendia JA, Bramuglia G, Staatz CE. Effects of combinational *CYP3A5* 6986A>G polymorphism in graft liver and native intestine on the pharmacokinetics of tacrolimus in liver transplant patients: a meta-analysis. Ther Drug Monit 2014;36:442-7.

Bullingham RE, Nicholls AJ, Kamm BR. Clinical pharmacokinetics of mycophenolate mofetil. Clin Pharmacokinet 1998;34:429-55.

Busi F, Cresteil T. *CYP3A5* mRNA degradation by nonsense-mediated mRNA decay. Mol Pharmacol 2005;68:808-15.

Chandel N, Aggarwal PK, Minz M, et al. *CYP3A5*1/*3* genotype influences the blood concentration of tacrolimus in response to metabolic inhibition by ketoconazole. Pharmacogenet Genomics 2009;19:158-63.

Chouchana L, Narjoz C, Beaune P, et al. Review article: the benefits of pharmacogenetics for improving thiopurine therapy in inflammatory bowel disease. Aliment Pharmacol Ther 2012;35:15-36.

Christians U, Kruse C, Kownatzki R, et al. Measurement of FK 506 by HPLC and isolation and characterization of its metabolites. Transplant Proc 1991;23:940-1.

Christians U, Sewing KF. Cyclosporin metabolism in transplant patients. Pharmacol Ther 1993;57:291-345.

Collie-Duguid ES, Pritchard SC, Powrie RH, et al. The frequency and distribution of thiopurine methyltransferase alleles in Caucasian and Asian populations. Pharmacogenetics 1999;9:37-42.

Crettol S, Venetz JP, Fontana M, et al. *CYP3A7, CYP3A5, CYP3A4*, and *ABCB1* genetic polymorphisms, cyclosporine concentration, and dose requirement in transplant recipients. Ther Drug Monit 2008;30:689-99.

Dai Y, Iwanaga K, Lin YS, et al. In vitro metabolism of cyclosporine A by human kidney *CYP3A5*. Biochem Pharmacol 2004;68:1889-902.

de Jonge H, Metalidis C, Naesens M, et al. The P450 oxidoreductase *28 SNP is associated with low initial tacrolimus exposure and increased dose requirements in *CYP3A5*-expressing renal recipients. Pharmacogenomics 2011;12:1281-91.

Diaz-Molina B, Tavira B, Lambert JL, et al. Effect of *CYP3A5*, *CYP3A4*, and *ABCB1* genotypes as determinants of tacrolimus dose and clinical outcomes after heart transplantation. Transplant Proc 2012;44:2635-8.

Ekberg H, Bernasconi C, Tedesco-Silva H, et al. Calcineurin inhibitor minimization in the Symphony study: observational results 3 years after transplantation. Am J Transplant 2009;9:1876-85.

Ekberg H, Tedesco-Silva H, Demirbas A, et al. Reduced exposure to calcineurin inhibitors in renal transplantation. N Engl J Med 2007;357:2562-75.

Elens L, Bouamar R, Shuker N, et al. Clinical implementation of pharmacogenetics in kidney transplantation: CNIs in the starting blocks. Br J Clin Pharmacol 2014a;77:715-28.

Elens L, Capron A, van Schaik RH, et al. Impact of *CYP3A4*22* allele on tacrolimus pharmacokinetics in early period after renal transplantation: toward updated genotype-based dosage guidelines. Ther Drug Monit 2013a;35:608-16.

Elens L, Hesselink DA, Bouamar R, et al. Impact of *POR*28* on the pharmacokinetics of tacrolimus and cyclosporine A in renal transplant patients. Ther Drug Monit 2014b;36:71-9.

Elens L, Nieuweboer AJ, Clarke SJ, et al. Impact of *POR*28* on the clinical pharmacokinetics of *CYP3A* phenotyping probes midazolam and erythromycin. Pharmacogenet Genomics 2013b;23:148-55.

Elens L, van Gelder T, Hesselink DA, et al. *CYP3A4*22*: promising newly identified *CYP3A4* variant allele for personalizing pharmacotherapy. Pharmacogenomics 2013c;14:47-62.

Elens L, van Schaik RH, Panin N, et al. Effect of a new functional *CYP3A4* polymorphism on calcineurin inhibitors' dose requirements and trough blood levels in stable renal transplant patients. Pharmacogenomics 2011;12:1383-96.

Fujiyama N, Miura M, Kato S, et al. Involvement of carboxylesterase 1 and 2 in the hydrolysis of mycophenolate mofetil. Drug Metab Dispos 2010;38:2210-7.

Fung KL, Gottesman MM. A synonymous polymorphism in a common *MDR1* (*ABCB1*) haplotype shapes protein function. Biochim Biophys Acta 2009;1794:860-71.

Garat A, Cauffiez C, Renault N, et al. Characterisation of novel defective thiopurine S-methyltransferase allelic variants. Biochem Pharmacol 2008;76:404-15.

Gensburger O, Van Schaik RH, Picard N, et al. Polymorphisms in type I and II inosine monophosphate dehydrogenase genes and association with clinical outcome in patients on mycophenolate mofetil. Pharmacogenet Genomics 2010;20:537-43.

Gijsen V, Mital S, van Schaik RH, et al. Age and *CYP3A5* genotype affect tacrolimus dosing requirements after transplant in pediatric heart recipients. J Heart Lung Transplant 2011;30:1352-9.

Gijsen VM, van Schaik RH, Elens L, et al. *CYP3A4*22* and *CYP3A* combined genotypes both correlate with tacrolimus disposition in pediatric heart transplant recipients. Pharmacogenomics 2013;14:1027-36.

Gijsen VM, van Schaik RH, Soldin OP, et al. P450 oxidoreductase *28 (*POR*28*) and tacrolimus disposition in pediatric kidney transplant recipients—a pilot study. Ther Drug Monit 2014;36:152-8.

Girard H, Court MH, Bernard O, et al. Identification of common polymorphisms in the promoter of the *UGT1A9* gene: evidence that UGT1A9 protein and activity levels are strongly genetically controlled in the liver. Pharmacogenetics 2004;14:501-15.

Gonschior AK, Christians U, Winkler M, et al. Tacrolimus (FK506) metabolite patterns in blood from liver and kidney transplant patients. Clin Chem 1996;42:1426-32.

Grinyo J, Vanrenterghem Y, Nashan B, et al. Association of four DNA polymorphisms with acute rejection after kidney transplantation. Transpl Int 2008;21:879-91.

Haufroid V, Mourad M, Van Kerckhove V, et al. The effect of *CYP3A5* and *MDR1* (*ABCB1*) polymorphisms on cyclosporine and tacrolimus dose requirements and trough blood levels in stable renal transplant patients. Pharmacogenetics 2004;14:147-54.

Hawwa AF, McKiernan PJ, Shields M, et al. Influence of *ABCB1* polymorphisms and haplotypes on tacrolimus nephrotoxicity and dosage requirements in children with liver transplant. Br J Clin Pharmacol 2009;68:413-21.

Hesselink DA, Bouamar R, Elens L, et al. The role of pharmacogenetics in the disposition of and response to tacrolimus in solid organ transplantation. Clin Pharmacokinet 2014;53:123-39.

Hesselink DA, Bouamar R, van Gelder T. The pharmacogenetics of calcineurin inhibitor-related nephrotoxicity. Ther Drug Monit 2010;32:387-93.

Hesselink DA, van Schaik RH, van der Heiden IP, et al. Genetic polymorphisms of the *CYP3A4*, *CYP3A5*, and *MDR-1* genes and pharmacokinetics of the calcineurin inhibitors cyclosporine and tacrolimus. Clin Pharmacol Ther 2003;74:245-54.

Hoffmeyer S, Burk O, von Richter O, et al. Functional polymorphisms of the human multidrug-resistance gene: multiple sequence variations and correlation of one allele with P-glycoprotein expression and activity in vivo. Proc Natl Acad Sci U S A 2000;97:3473-8.

Hooper DK, Fukuda T, Gardiner R, et al. Risk of tacrolimus toxicity in *CYP3A5* nonexpressors treated with intravenous nicardipine after kidney transplantation. Transplantation 2012;93:806-12.

Hustert E, Haberl M, Burk O, et al. The genetic determinants of the *CYP3A5* polymorphism. Pharmacogenetics 2001;11:773-9.

Iwasaki K. Metabolism of tacrolimus (FK506) and recent topics in clinical pharmacokinetics. Drug Metab Pharmacokinet 2007;22:328-35.

Jacobson PA, Oetting WS, Brearley AM, et al. Novel polymorphisms associated with tacrolimus trough concentrations: results from a multicenter kidney transplant consortium. Transplantation 2011;91:300-8.

Jacobson PA, Schladt D, Oetting WS, et al. Lower calcineurin inhibitor doses in older compared to younger kidney transplant recipients yield similar troughs. Am J Transplant 2012;12:3326-36.

Johnson LA, Oetting WS, Basu S, et al. Pharmacogenetic effect of the *UGT* polymorphisms on mycophenolate is modified by calcineurin inhibitors. Eur J Clin Pharmacol 2008;64:1047-56.

Jorga A, Holt DW, Johnston A. Therapeutic drug monitoring of cyclosporine. Transplant Proc 2004;36:396s-403s.

Jusko WJ, Thomson AW, Fung F, et al. Consensus document: therapeutic monitoring of tacrolimus (FK-506). Ther Drug Monit 1995;17:606-14.

Kapturczak MH, Meier-Kriesche HU, Kaplan B. Pharmacology of calcineurin antagonists. Transplant Proc 2004;36:25s-32s.

Kershner RP, Fitzsimmons WE. Relationship of FK506 whole blood concentrations and efficacy and toxicity after liver and kidney transplantation. Transplantation 1996;62:920-6.

Kovarik JM, Koelle EU. Cyclosporin pharmacokinetics in the elderly. Drugs Aging 1999;15:197-205.

Krynetski EY, Evans WE. Genetic polymorphism of thiopurine S-methyltransferase: molecular mechanisms and clinical importance. Pharmacology 2000;61:136-46.

Krynetski EY, Tai HL, Yates CR, et al. Genetic polymorphism of thiopurine S-methyltransferase: clinical importance and molecular mechanisms. Pharmacogenetics 1996;6:279-90.

Kuypers DR, de Jonge H, Naesens M, et al. Effects of *CYP3A5* and *MDR1* single nucleotide polymorphisms on drug interactions between tacrolimus and fluconazole in renal allograft recipients. Pharmacogenet Genomics 2008;18:861-8.

Lamba V, Sangkuhl K, Sanghavi K, et al. PharmGKB summary: mycophenolic acid pathway. Pharmacogenet Genomics 2014;24:73-9.

Lennard L, Van Loon JA, Weinshilboum RM. Pharmacogenetics of acute azathioprine toxicity: relationship to thiopurine methyltransferase genetic polymorphism. Clin Pharmacol Ther 1989;46:149-54.

Li JL, Wang XD, Chen SY, et al. Effects of diltiazem on pharmacokinetics of tacrolimus in relation to *CYP3A5* genotype status in renal recipients: from retrospective to prospective. Pharmacogenomics J 2011;11:300-6.

Lunde I, Bremer S, Midtvedt K, et al. The influence of *CYP3A*, *PPARA*, and *POR* genetic variants on the pharmacokinetics of tacrolimus and cyclosporine in renal transplant recipients. Eur J Clin Pharmacol 2014;70:685-93.

Macphee IA, Fredericks S, Mohamed M, et al. Tacrolimus pharmacogenetics: the *CYP3A5*1* allele predicts low dose-normalized tacrolimus blood concentrations in whites and South Asians. Transplantation 2005;79:499-502.

Matas AJ, Smith JM, Skeans MA, et al. OPTN/SRTR 2012 annual data report: kidney. Am J Transplant 2014;14(suppl 1):11-44.

McLeod HL, Siva C. The thiopurine S-methyltransferase gene locus—implications for clinical pharmacogenomics. Pharmacogenomics 2002;3:89-98.

McMaster P, Mirza DF, Ismail T, et al. Therapeutic drug monitoring of tacrolimus in clinical transplantation. Ther Drug Monit 1995;17:602-5.

Michelon H, Konig J, Durrbach A, et al. *SLCO1B1* genetic polymorphism influences mycophenolic acid tolerance in renal transplant recipients. Pharmacogenomics 2010;11:1703-13.

Miura M, Kagaya H, Satoh S, et al. Influence of drug transporters and *UGT* polymorphisms on pharmacokinetics of phenolic glucuronide metabolite of mycophenolic acid in Japanese renal transplant recipients. Ther Drug Monit 2008;30:559-64.

Miura M, Satoh S, Inoue K, et al. Influence of *SLCO1B1*, *1B3*, *2B1* and *ABCC2* genetic polymorphisms on mycophenolic acid pharmacokinetics in Japanese renal transplant recipients. Eur J Clin Pharmacol 2007;63:1161-9.

Moes DJ, Swen JJ, den Hartigh J, et al. Effect of *CYP3A4*22*, *CYP3A5*3*, and *CYP3A* combined genotypes on cyclosporine, everolimus, and tacrolimus pharmacokinetics in renal transplantation. CPT Pharmacometrics Syst Pharmacol 2014;3:e100.

Mojarrabi B, Mackenzie PI. Characterization of two UDP glucuronosyltransferases that are predominantly expressed in human colon. Biochem Biophys Res Commun 1998;247:704-9.

Murray B, Hawes E, Lee RA, et al. Genes and beans: pharmacogenomics of renal transplant. Pharmacogenomics 2013;14:783-98.

Naesens M, Lerut E, de Jonge H, et al. Donor age and renal P-glycoprotein expression associate with chronic histological damage in renal allografts. J Am Soc Nephrol 2009;20:2468-80.

Nara M, Takahashi N, Miura M, et al. Effect of itraconazole on the concentrations of tacrolimus and cyclosporine in the blood of patients receiving allogeneic hematopoietic stem cell transplants. Eur J Clin Pharmacol 2013;69:1321-9.

[No authors listed]. A blinded, randomized clinical trial of mycophenolate mofetil for the prevention of acute rejection in cadaveric renal transplantation. The Tricontinental Mycophenolate Mofetil Renal Transplantation Study Group. Transplantation 1996;61:1029-37.

Otterness D, Szumlanski C, Lennard L, et al. Human thiopurine methyltransferase pharmacogenetics: gene sequence polymorphisms. Clin Pharmacol Ther 1997;62:60-73.

Pandey AV, Fluck CE. NADPH P450 oxidoreductase: structure, function, and pathology of diseases. Pharmacol Ther 2013;138:229-54.

Passey C, Birnbaum AK, Brundage RC, et al. Dosing equation for tacrolimus using genetic variants and clinical factors. Br J Clin Pharmacol 2011;72:948-57.

Passey C, Birnbaum AK, Brundage RC, et al. Validation of tacrolimus equation to predict troughs using genetic and clinical factors. Pharmacogenomics 2012;13:1141-7.

Picard N, Ratanasavanh D, Premaud A, et al. Identification of the UDP-glucuronosyltransferase isoforms involved in mycophenolic acid phase II metabolism. Drug Metab Dispos 2005;33:139-46.

Picard N, Yee SW, Woillard JW, et al. The role of organic anion-transporting polypeptides and their common genetic variants in mycophenolic acid pharmacokinetics. Clin Pharmacol Ther 2010;87:100-8.

Pirsch JD, Miller J, Deierhoi MH, et al. A comparison of tacrolimus (FK506) and cyclosporine for immunosuppression after cadaveric renal transplantation. FK506 Kidney Transplant Study Group. Transplantation 1997;63:977-83.

Provenzani A, Santeusanio A, Mathis E, et al. Pharmacogenetic considerations for optimizing tacrolimus dosing in liver and kidney transplant patients. World J Gastroenterol 2013;19:9156-73.

Relling MV, Gardner EE, Sandborn WJ, et al. Clinical pharmacogenetics implementation consortium guidelines for thiopurine methyltransferase genotype and thiopurine dosing: 2013 update. Clin Pharmacol Ther 2013;93:324-5.

Relling MV, Gardner EE, Sandborn WJ, et al. Clinical Pharmacogenetics Implementation Consortium guidelines for thiopurine methyltransferase genotype and thiopurine dosing. Clin Pharmacol Ther 2011;89:387-91.

Roberts PJ, Rollins KD, Kashuba AD, et al. The influence of *CYP3A5* genotype on dexamethasone induction of *CYP3A* activity in African Americans. Drug Metab Dispos 2008;36:1465-9.

Rosso Felipe C, de Sandes TV, Sampaio EL, et al. Clinical impact of polymorphisms of transport proteins and enzymes involved in the metabolism of immunosuppressive drugs. Transplant Proc 2009;41:1441-55.

Saeki T, Ueda K, Tanigawara Y, et al. Human P-glycoprotein transports cyclosporin A and FK506. J Biol Chem 1993;268:6077-80.

Sahasranaman S, Howard D, Roy S. Clinical pharmacology and pharmacogenetics of thiopurines. Eur J Clin Pharmacol 2008;64:753-67.

Santoro AB, Struchiner CJ, Felipe CR, et al. *CYP3A5* genotype, but not *CYP3A4*1b*, *CYP3A4*22*, or hematocrit, predicts tacrolimus dose requirements in Brazilian renal transplant patients. Clin Pharmacol Ther 2013;94:201-2.

Saruwatari J, Ogusu N, Shimomasuda M, et al. Effects of *CYP2C19* and P450 oxidoreductase polymorphisms on the population pharmacokinetics of clobazam and N-desmethylclobazam in Japanese patients with epilepsy. Ther Drug Monit 2014;36:309-9.

Schiff J, Cole E, Cantarovich M. Therapeutic monitoring of calcineurin inhibitors for the nephrologist. Clin J Am Soc Nephrol 2007;2:374-84.

Shipkova M, Armstrong VW, Wieland E, et al. Identification of glucoside and carboxyl-linked glucuronide conjugates of mycophenolic acid in plasma of transplant recipients treated with mycophenolate mofetil. Br J Pharmacol 1999;126:1075-82.

Silva HT Jr, Yang HC, Meier-Kriesche HU, et al. Long-term follow-up of a phase III clinical trial comparing tacrolimus extended-release/MMF, tacrolimus/MMF, and cyclosporine/MMF in de novo kidney transplant recipients. Transplantation 2014;97:636-41.

Sollinger HW. Mycophenolate mofetil for the prevention of acute rejection in primary cadaveric renal allograft recipients. U.S. Renal Transplant Mycophenolate Mofetil Study Group. Transplantation 1995;60:225-32.

Sombogaard F, van Schaik RH, Mathot RA, et al. Interpatient variability in *IMPDH* activity in MMF-treated renal transplant patients is correlated with *IMPDH* type II 3757T > C polymorphism. Pharmacogenet Genomics 2009;19:626-34.

Staatz CE, Goodman LK, Tett SE. Effect of *CYP3A* and *ABCB1* single nucleotide polymorphisms on the pharmacokinetics and pharmacodynamics of calcineurin inhibitors: part I. Clin Pharmacokinet 2010a;49:141-75.

Staatz CE, Goodman LK, Tett SE. Effect of *CYP3A* and *ABCB1* single nucleotide polymorphisms on the pharmacokinetics and pharmacodynamics of calcineurin inhibitors: part II. Clin Pharmacokinet 2010b;49:207-21.

Staatz CE, Tett SE. Clinical pharmacokinetics and pharmacodynamics of mycophenolate in solid organ transplant recipients. Clin Pharmacokinet 2007;46:13-58.

Staatz CE, Tett SE. Clinical pharmacokinetics and pharmacodynamics of tacrolimus in solid organ transplantation. Clin Pharmacokinet 2004;43:623-53.

Tang HL, Xie HG, Yao Y, et al. Lower tacrolimus daily dose requirements and acute rejection rates in the CYP3A5 nonexpressers than expressers. Pharmacogenet Genomics 2011;21:713-20.

Thervet E, Legendre C, Beaune P, et al. Cytochrome P450 3A polymorphisms and immunosuppressive drugs. Pharmacogenomics 2005;6:37-47.

Thervet E, Loriot MA, Barbier S, et al. Optimization of initial tacrolimus dose using pharmacogenetic testing. Clin Pharmacol Ther 2010;87:721-6.

Thiel G, Bock A, Spondlin M, et al. Long-term benefits and risks of cyclosporin A (sandimmun)—an analysis at 10 years. Transplant Proc 1994;26:2493-8.

Undre NA, Stevenson P, Schafer A. Pharmacokinetics of tacrolimus: clinically relevant aspects. Transplant Proc 1999;31:21s-4s.

Vafadari R, Bouamar R, Hesselink DA, et al. Genetic polymorphisms in *ABCB1* influence the pharmacodynamics of tacrolimus. Ther Drug Monit 2013;35:459-65.

Vannaprasaht S, Angsuthum S, Avihingsanon Y, et al. Impact of the heterozygous *TPMT*1/*3C* genotype on azathioprine-induced myelosuppression in kidney transplant recipients in Thailand. Clin Ther 2009;31:1524-33.

van Schaik RH, van Agteren M, de Fijter JW, et al. *UGT1A9* -275T>A/-2152C>T polymorphisms correlate with low MPA exposure and acute rejection in MMF/tacrolimus-treated kidney transplant patients. Clin Pharmacol Ther 2009;86:319-27.

Venkataramanan R, Swaminathan A, Prasad T, et al. Clinical pharmacokinetics of tacrolimus. Clin Pharmacokinet 1995;29:404-30.

Vincent SH, Karanam BV, Painter SK, et al. In vitro metabolism of FK-506 in rat, rabbit, and human liver microsomes: identification of a major metabolite and of cytochrome P450 3A as the major enzymes responsible for its metabolism. Arch Biochem Biophys 1992;294:454-60.

von Ahsen N, Richter M, Grupp C, et al. No influence of the *MDR-1* C3435T polymorphism or a *CYP3A4* promoter polymorphism (*CYP3A4-V* allele) on dose-adjusted cyclosporin A trough concentrations or rejection incidence in stable renal transplant recipients. Clin Chem 2001;47:1048-52.

Wang J, Yang JW, Zeevi A, et al. *IMPDH1* gene polymorphisms and association with acute rejection in renal transplant patients. Clin Pharmacol Ther 2008;83:711-7.

Webster AC, Woodroffe RC, Taylor RS, et al. Tacrolimus versus ciclosporin as primary immunosuppression for kidney transplant recipients: meta-analysis and meta-regression of randomised trial data. BMJ 2005;331:810.

Woillard JB, Kamar N, Coste S, et al. Effect of *CYP3A4*22, POR*28,* and *PPARA* rs4253728 on sirolimus in vitro metabolism and trough concentrations in kidney transplant recipients. Clin Chem 2013;59:1761-9.

Yang JW, Lee PW, Hutchinson IV, et al. Genetic polymorphisms of *MRP2* and *UGT2B7* and gastrointestinal symptoms in renal transplant recipients taking mycophenolic acid. Ther Drug Monit 2009;31:542-8.

Yates CR, Krynetski EY, Loennechen T, et al. Molecular diagnosis of thiopurine S-methyltransferase deficiency: genetic basis for azathioprine and mercaptopurine intolerance. Ann Intern Med 1997;126:608-14.

Zhang JJ, Zhang H, Ding XL, et al. Effect of the P450 oxidoreductase 28 polymorphism on the pharmacokinetics of tacrolimus in Chinese healthy male volunteers. Eur J Clin Pharmacol 2013;69:807-12.

Zheng H, Zeevi A, Schuetz E, et al. Tacrolimus dosing in adult lung transplant patients is related to cytochrome P4503A5 gene polymorphism. J Clin Pharmacol 2004;44:135-40.

Zochowska D, Wyzgal J, Paczek L. Impact of *CYP3A4*1B* and *CYP3A5*3* polymorphisms on the pharmacokinetics of cyclosporine and sirolimus in renal transplant recipients. Ann Transplant 2012;17:36-44.

Zucker K, Tsaroucha A, Olson L, et al. Evidence that tacrolimus augments the bioavailability of mycophenolate mofetil through the inhibition of mycophenolic acid glucuronidation. Ther Drug Monit 1999;21:35-43.

Zuo XC, Zhou YN, Zhang BK, et al. Effect of *CYP3A5*3* polymorphism on pharmacokinetic drug interaction between tacrolimus and amlodipine. Drug Metab Pharmacokinet 2013;28:398-405.

Chapter 13

HLA PHARMACOGENETICS AND SERIOUS CUTANEOUS ADVERSE DRUG REACTIONS

Ming Ta Michael Lee, Ph.D.; and Yanfei Zhang, M.D.

Learning Objectives

1. Assess the importance of human leukocyte antigen (HLA) pharmacogenetics.
2. Evaluate the relationship between HLA and serious cutaneous events.
3. Design genetic studies for cutaneous adverse drug reactions.
4. Apply *HLA* genotypes in clinical use.
5. Show the HLA function in the pathogenesis of serious cutaneous adverse drug reactions.

Keywords: SCAR, SJS/TEN, HSS, abacavir, carbamazepine, allopurinol, *HLA-B*5701, HLA-B*1502, HLA-B*5801, HLA-A*3101*

Abbreviations in This Chapter

ADR	Adverse drug reaction
AED	Antiepileptic drug
DIHS	Drug-induced hypersensitivity syndrome
DRESS	Drug reaction with eosinophilia and systemic symptoms
GWAS	Gnome-wide association study
HLA	Human leukocyte antigen
HSS	Hypersensitivity syndrome
LD	Linkage disequilibrium
MHC	Major histocompatibility complex
MPE	Maculopapular exanthema
NPV	Negative predictive value
PPV	Positive predictive value
SCAR	Severe cutaneous adverse drug reaction
SJS/TEN	Stevens-Johnson syndrome/toxic epidermal necrolysis
SNP	Single nucleotide polymorphism

Abstract

Severe cutaneous adverse drug reactions, including drug hypersensitivities and Stevens-Johnson syndrome/toxic epidermal necrosis (SJS/TEN), represent huge threats to patients' health and life. In recent decades, several associations between human leukocyte antigen (HLA) alleles and drug-induced hypersensitivities and SJS/TEN have been identified. Among these findings, abacavir-induced hypersensitivity is associated with *HLA-B*5701* across different populations, whereas carbamazepine-induced SJS/TEN is specifically associated with *HLA-B*1502* in Southeast Asian populations. Both identifications have been translated to clinical use, representing the two successful examples of implementing pharmacogenetics in clinical practice. Other findings include the associations between *HLA-B*5801* and allopurinol-induced SJS/TEN and between *HLA-A*3101* and carbamazepine-induced hypersensitivity, both of which have the potential to be used clinically. The study of pharmacogenetics is helpful in making clinical decisions with respect to avoiding severe adverse drug reactions.

Introduction

Adverse drug reactions (ADRs) represent a significant burden on hospital, health care, and societal perspective, constituting the fourth to sixth most common causes of death (Wester 2008; Pirmohamed 2004; Riedl 2003). Clinically, ADRs have been classified into two major types, type A and type B, according to the pharmacologic actions of the offending drug (Riedl 2003). Type A ADRs are dose-dependent and can be predicted, whereas type B ADRs are dose-independent and unpredictable from the known pharmacology of the drug.

Recent pharmacogenetic studies have evolved from a candidate gene approach to the genome-wide association study (GWAS), which has greatly advanced the discovery of genes associated with interindividual differences in drug response—genes that predispose individuals to ADRs. Type A ADRs are associated with genes involved in drug metabolism and drug transport, whereas many type B ADRs are associated with particular human leukocyte antigen (HLA) alleles.

These findings have furthered our understanding of the underlying mechanisms of ADRs. Because of these discoveries, the U.S. Food and Drug Administration (FDA) has relabeled more than 100 approved drugs to include genetic information (Wei 2012b). Clinical use of *HLA-B*5701* and *HLA-B*1502* has been implemented to reduce abacavir hypersensitivity syndrome (HSS) and carbamazepine-induced Stevens-Johnson syndrome/toxic epidermal necrolysis (SJS/TEN).

In this chapter, we focus on the pharmacogenetics of severe cutaneous adverse drug reactions (SCARs), as well as the potential use of these genetic biomarkers in clinical practice for preventing SCARs. We also discuss key challenges for the implementation of pharmacogenetics and future perspectives.

Severe Cutaneous Adverse Drug Reactions

Severe cutaneous adverse drug reactions constitute a significant proportion of type B serious ADRs (Pirmohamed 2011a). Severe cutaneous adverse drug reactions include SJS/TEN, HSS, and AGEP (acute generalized exanthematous pustulosis) (Pirmohamed 2011b). The Phenotype Standardization Project (PSP) has provided a detailed description of drug-induced skin injury (Pirmohamed 2011b); in this chapter, we only briefly introduce SJS/TEN and HSS. Many drugs can cause SJS/TEN and HSS; Table 13.1 lists the most frequently used causative drugs for SCARs (Borchers 2008; Mockenhaupt 2008).

Stevens-Johnson Syndrome/ Toxic Epidermal Necrolysis

Stevens-Johnson syndrome and TEN constitute the most severe SCARs. Both are characterized by the rapid development of blistering exanthema and mucosal involvement. Stevens-Johnson syndrome and TEN represent the same disease but with different extents of skin detachment, with TEN being the more severe form. In SJS, there is about 1%–10% body surface area detachment, whereas TEN involves more than 30% of the body surface area; SJS/TEN overlap is defined when 10%–30% of the body surface area is involved (Mockenhaupt 2008). The severity of epidermal detachment is also highly correlated with mortality, ranging from 1%–5% in SJS to 25%–35% in TEN (Roujeau 1994, 1990b; Guillaume 1987). In addition to skin manifestations, both SJS and TEN can cause serious complications, including mucous ulceration (can occur in trachea, bronchi, and gastrointestinal tract) and systemic involvement (e.g., hepatitis and lymphopenia), which also severely affect patients' health (Sugimoto 1998; McIvor 1996; Roujeau 1990a; Herman 1984; Kelin 1983).

Hypersensitivity Syndrome

Hypersensitivity syndrome may be the most complex phenotype in SCARs. Several types have been identified, including drug-induced hypersensitivity syndrome (DIHS) (Shiohara 2006), drug reaction with eosinophilia and systemic symptoms (DRESS) (Cacoub 2011; Chiou 2008; Kardaun 2007; Shiohara 2007), and drug-induced delayed

Table 13.1. Drugs That Frequently Cause SJS/TEN and HSS

Stevens-Johnson Syndrome/Toxic Epidermal Necrolysis	
Antigout	Allopurinol
Antiepileptic drugs	Carbamazepine Phenytoin Phenobarbital Lamotrigine
Anti-HIV	Nevirapine (NNRTI)
Nonsteroidal anti-inflammatory drugs (NSAIDs)	Ibuprofen Oxicams
Antibiotics	Sulfamethoxazole
Antiglaucoma	Methazolamide
Hypersensitivity Syndrome	
Antiepileptic drugs	Carbamazepine Phenytoin Phenobarbital
Anti-HIV	Abacavir (nucleoside reverse transcriptase inhibitor) Nevirapine (NNRTI)
Antigout	Allopurinol
Antibiotics	Aminopenicillin

HSS = hypersensitivity syndrome; NNRTI = nonnucleoside reverse transcriptase inhibitor; SJS/TEN = Stevens-Johnson syndrome/toxic epidermal necrolysis.

multiorgan hypersensitivity syndrome (Sontheimer 1998). The terms recommended to be used in articles and cases to avoid confusion by PSP are *HSS*, *DIHS*, and *DRESS* (Pirmohamed 2011b). Typically, HSS occurs within 1–8 weeks after drug exposure, with significant morbidity and mortality (Tennis 1997). This syndrome involves a combination of fever, morbilliform rash, and multiple internal organ involvement. Although several organs can be affected at the same time, the most frequently affected organ is the skin. Skin manifestations include exanthematous eruptions; urticarial plaques; exfoliative, pustular eruptions; and facial edema. The rash often generalizes into a severe exfoliative dermatitis or erythroderma, but usually without mucocutaneous involvement, which helps distinguish HSS from SJS/TEN (Pirmohamed 2011b; Shiohara 2006). The complexity of HSS increases the difficulty of making an accurate diagnosis; therefore, the RegiSCAR group has proposed criteria to be met or excluded to make a diagnosis of HSS (http://regiscar.uni-freiburg.de/downloads/inclusioncriteriaforscar.pdf), which were further replenished by PSP (Pirmohamed 2011b).

Human Leukocyte Antigen

The major histocompatibility complex (MHC), also called the "human leukocyte antigen" in humans, is located on the short arm of chromosome 6, the area in the human genome with high linkage disequilibrium (LD). This region contains genes for classical MHC class I genes *(HLA-A, HLA-B, and HLA-C)* and classical MHC class II genes *(HLA-DR, HLA-DQ, and HLA-DP)* (Horton 2004). *HLA* class I and class II genes are the most highly polymorphic genes within the human genome; of these, *HLA-B* is the most polymorphic gene, containing more than 3000 allelic variants (Figure 13.1).

The HLA molecules are of immense importance within the immune system. Both HLA class I and class II molecules can bind peptides through their peptide-binding grooves (Figure 13.2) and present them to T cells for inspection, resulting in the initiation of immune response against nonself antigens.

The HLA class I molecules are expressed in all nucleated cells and present peptides derived from intracellular proteins to CD8[+] T cells, which are mainly responsible for killing virally infected or abnormal cells such as tumor cells. The HLA class II molecules are expressed only in the professional antigen-presenting cells, mainly the dendritic cells, of the immune system. Antigen-presenting cells can take up the extracellular antigens through phagocytosis or pinocytosis and process them into antigen peptides, which bind to the HLA class II molecules. Processed peptides are then presented to CD4[+] helper T cells, which can assist B cells in making antibodies and enhance the cytotoxicity of CD8[+] T cells, thus playing an important role in both humoral immunity and cellular immunity.

The polymorphisms of *HLA* class I and class II genes result in significant differences in the peptide-binding grooves of HLA proteins. This ensures the successful

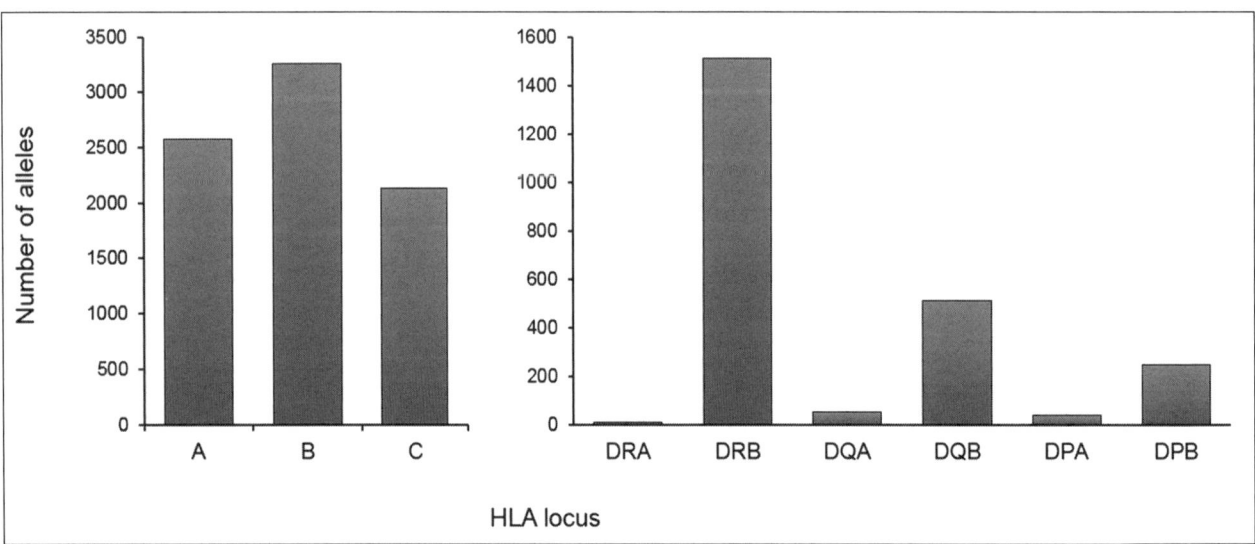

Figure 13.1. Polymorphisms within HLA class I and class II genes.

Number of distinct *HLA* class I (A, B, C) and class II (DRA, DRB, DQA, DQB, DPA, DPB) alleles at each locus (based on data released from IMGT/HLA Database, Release 3.15.0, 2014-01-17: www.ebi.ac.uk/ipd/imgt/hla/stats.html).

binding and presentation of a wide range of peptides and reduces the probability of failed presentations of peptides derived from pathogen proteins. However, in addition to their protective role during the immune responses against pathogens, HLA proteins can cause inappropriate immune responses leading to tissue injuries caused by, for instance, drugs. The mechanisms of this will be discussed later in this chapter.

Early Genetic Studies of SCARs

Idiosyncratic ADRs have been viewed as unpredictable until recently and have long been postulated to have an underlying genetic etiology. Case reports that describe the similar hypersensitivity reactions to the same drug among different family members, including monozygotic twins, have supplied evidence of genetic predisposition to these idiosyncratic ADRs (Johnson-Reagan 2003; Peyrieere 2001; Edwards 1999; Pellicano 1992; Gennis 1991).

The genetic determinants of TEN and SJS have been sought since the 1980s. Because drug-induced skin injuries were considered immune related, many of the early studies focused on *HLA* gene clusters, with several positive associations identified. *HLA-B12* was first found to be associated with TEN in white individuals of French ancestry (Roujeau 1986, 1987). When these patients were stratified according to the causative drugs, *HLA-A29, B12*, and *DR7* were linked to sulfonamide-related TEN (Roujeau 1986, 1987), whereas *HLA-A2* and *B12* were linked to oxicam-related TEN (Roujeau 1987). In a study of both white and black patients of SJS with ocular involvement, an association with *HLA-Bw44* was found (Mondino 1981). A subsequent study examined *HLA* class II typing using polymerase chain reaction (PCR)-based molecular techniques in white patients and suggested that *HLA-DQB1*0601* was involved (Power 1996). However, patients in these two studies were not stratified according to the drug used. In patients of Japanese and Korean decent, *HLA-B59* was found in three of four methazolamide-induced SJS cases (Shirato 1997). In the Italian population, patients with an aminopenicillin hypersensitivity had a higher prevalence of *HLA-A2* and *DRw52* but a lower frequency of *DR4* compared with the random population controls. In this study, the microlymphotoxicity standard test was used for *HLA* typing (Romano 1998).

Genes involved in drug metabolism (e.g., *CYP2C9*) were also investigated in drug-induced hypersensitivity. Although the *CYP2C9*2/*3* genotype and *CYP2C9*3* allele frequencies were 9- and 2.5-fold higher in the co-trimoxazole hypersensitive group than in the nonsensitive group, respectively, they were not statistically significant when corrected for multiple testing. This suggests that genetic polymorphisms in drug-metabolizing enzymes are unlikely to be major predisposing factors of co-trimoxazole hypersensitivity (Pirmohamed 2000).

Although there were several associations between HLA and SCARs, none could sufficiently predict these adverse events, nor could any be used clinically as prescreening tests before drug administration to prevent SCARs. Of note, most of these early findings were based on single studies without replication, and *HLA* typing was tested by low-resolution serologic methods. Meanwhile, the criteria for diagnosing SJS/TEN and hypersensitivities were not unified among studies. Many of the studies did not distinguish between the various types of SJS/TEN induced by different drugs, and the sample sizes were insufficient. Therefore, the weak *HLA* associations in the early studies might have occurred because imprecise diagnosis methods were used for SJS/TEN, because an insufficient number of patients were involved, and because a low resolution of *HLA* genotyping was used.

Pharmacogenetic Studies of SCARs

Recent Genetic Studies of SCARs

The International HapMap Project produced a genome-wide database of human genetic variation containing more than 10 million common single nucleotide polymorphisms (SNPs) and their LD patterns across all the major ethnic groups, laying the important groundwork for GWASs of common diseases and drug responses

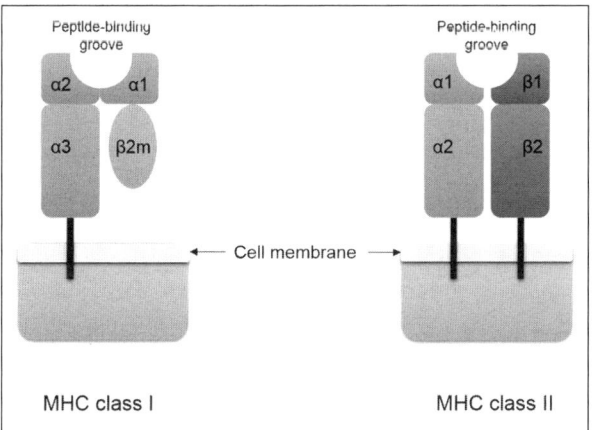

Figure 13.2. The structures of MHC class I and MHC class II proteins. MHC class I peptide complex is composed of an α chain and a β_2 microglobulin. The peptide binds to the peptide-binding grooves formed by the α_1 and α_2 domains. MHC class II peptide complex is composed of α and β chains. The peptide binds into the peptide-binding grooves formed by the α_1 and β_1 domains.

MHC = major histocompatibility complex.

(International HapMap C 2010, 2005). Genome-wide association studies have been defined as studies that identify genetic associations with observed traits by assessing common genetic variants (greater than 5% allelic frequency) across the entire human genome. Unlike knowledge-based candidate gene studies, GWASs offer a hypothesis-free approach that is unconstrained by the current understanding of genes and regulatory regions within the genome (Manolio 2009). In addition to GWASs, recent genetic studies of SCARs use the candidate gene approach, mainly *HLA* genes; however, this approach uses a higher-resolution *HLA* typing method, more precise diagnoses of SCARs, and larger sample sizes. Together, these two approaches have been applied to pharmacogenetic studies, with significant advances made in identifying genetic predictors of severe ADRs, drug dosing, and efficacy. Table 13.2 outlines the findings of *HLA* alleles associated with the serious type B ADRs and highlights the examples that have successfully been implemented in clinical application or have the potential to be translated (in bold).

Successful Clinical Implementation of *HLA* Pharmacogenetics

To date, studies of abacavir hypersensitivity and carbamazepine-induced SJS/TEN have provided two examples of how genetic investigations, combined with supportive clinical and basic science and laboratory systems, can lead to the successful translation of pharmacogenetics to the clinic.

*HLA-B*5701* and Abacavir-Induced HSS

Abacavir and Abacavir-Induced HSS

Abacavir is a nucleoside reverse transcriptase inhibitor used in conjunction with other antiretroviral agents in the treatment of human immunodeficiency virus (HIV) infection. Abacavir is generally well tolerated and has been used in most developed countries since 1998. However, it can cause delayed HSS characterized by fever, rash, systemic symptoms, gastrointestinal tract symptoms, and respiratory symptoms in about 5%–8% of patients during the first 6 weeks of treatment (Hetherington 2001). Symptoms worsen with continued abacavir use and can potentially be life threatening if the patient is rechallenged (Hetherington 2001; Shapiro 2001).

*HLA-B*5701 Is Highly Associated with Abacavir-Induced HSS*

An epidemiologic analysis revealed a different pattern of occurrence in abacavir-induced HSS according to ethnicity, with the lowest risk in African Americans (Symonds 2002). One case report also showed a familial predisposition of abacavir-induced HSS because a white HIV-infected patient and his 9-year-old daughter both developed abacavir-induced HSS after abacavir administration (Peyrieere 2001). These observations suggest that genetic factors are involved in abacavir-induced HSS.

In 2002, a strong association between abacavir-induced HSS and *HLA-B57* was reported independently by two research groups (Hethington 2002; Mallal 2002), the second group of which used a high-resolution *HLA* genotyping method and specifically mapped the risk allele to *HLA-B*5701* (Mallal 2008). Subsequent replication studies of different ethnic populations strengthened the association (Phillips 2005; Hughes 2004a; Martin 2004). Early studies showed a low sensitivity of *HLA-B*5701* for abacavir-induced HSS, especially in a black population with a lower prevalence of this allele, giving the misleading impression that the association between *HLA-B*5701* and abacavir-induced HSS was race restricted. Moreover, after adopting patch testing to improve the clinical diagnosis of true immunologically mediated abacavir-induced HSS, studies showed high *HLA-B*5701* sensitivity, but it was still restricted in mainly white population (Phillips 2005, 2002; Martin 2004). This race-restricted perception did not change until 2008, when the Study of Hypersensitivity to Abacavir and Pharmacogenetic Evaluation (SHAPE), a retrospective case-control study, first evaluated the sensitivity and specificity of the *HLA-B*5701* allele as a marker for abacavir-induced HSS in both the black and white populations in the United States. The SHAPE study showed that 100% of both black and white patch test–positive patients with abacavir-induced HSS were carriers of *HLA-B*5701* (i.e., the sensitivity was 100%) (Saag 2008). This finding supported the generalizability of *HLA-B*5701* screening for the prevention of abacavir-induced HSS in U.S. white and black populations.

*Implementation of HLA-B*5701 Screening in Clinical Use*

Cohort studies have shown that prospective genetic screening of *HLA-B*5701* can decrease the incidence of abacavir-induced HSS in a Western Australian cohort and an ethnically mixed French HIV population (Zucman 2007; Rauch 2006). However, none of these retrospective or prospective studies have garnered sufficient evidence for implementing *HLA-B*5701* screening because of their small sample sizes, lack of racial diversity, and nonuniform abacavir-induced HSS phenotype. To establish the effectiveness of prospective *HLA-B*5701* screening to prevent abacavir-induced HSS, the Prospective Randomized Evaluation of DNA Screening in a Clinical Trial (PREDICT-1) was conducted. This was a double-blind, prospective, randomized study of

1956 patients from 19 countries. This study showed that the prospective screening group had a lower incidence of hypersensitivity reaction than did the control group (3.4% vs. 7.8%, p<0.001), with a negative predictive value (NPV) of 100% and a positive predictive value (PPV) of 47.9%, suggesting that prescreening for *HLA-B*5701* can reduce the risk of hypersensitivity reaction to abacavir (Mallal 2008).

The results of PREDICT-1 and the huge volume of existing evidence prompted the FDA to amend the product information of abacavir, which included adding a black box warning about a high risk of *HLA-B*5701*-associated abacavir hypersensitivity reaction. Because of this change, it is recommended that all patients, irrespective of race, be screened for the presence of the *HLA-B*5701* allele before starting treatment with abacavir (Martin 2012).

Table 13.2. Association Between *HLA* Alleles and Severe ADRs[a]

Drug	Severe ADRs	*HLA* Alleles	References
Abacavir	**HSS**	***HLA-B*5701***	Saag 2008; Phillips 2005; Hughes 2004a; Martin 2004; Hetherington 2002; Mallal 2002
Allopurinol	**SJS/TEN/HSS**	***HLA-B*5801***	Cao 2012; Cristallo 2011; Jung 2011; Kang 2011; Tassaneeyakul 2009; Kaniwa 2008; Lonjou 2008; Hung 2005
Aminopenicillin	HSS	*HLA-A2; HLA-DRW52*	Romano 1998
Aspirin	Urticarial	*HLA-DRB1*1302-DQB1*0609-DPB1*0201*	Kim 2005
Carbamazepine	**SJS/TEN**	***HLA-B*1502***	Kulkantrakorn 2012; Chang 2011; Then 2011; Wang 2011; Zhang 2011; Tassaneeyakul 2010; Wu 2010; Mehta 2009; Locharernkul 2008; Man 2007; Hung 2006; Chung 2004
	HSS	***HLA-A*3101***	Genin 2014; Amstutz 2013; Kim 2011; McCormack 2011; Ozeki 2011; Hung 2006
Co-amoxiclav	DILI	*HLA-DRB1*1501*	Lucena 2011; O'Donohue 2000; Hautekeete 1999
Flucloxacillin	DILI	*HLA-B*5701*	Daly 2009
Lumiracoxib	DILI	*HLA-DRB1*1501-DQB1*0602-DRB5*0101*	Singer 2010
Methazolamide	SJS/TEN	*HLA-B*5901*	Kim 2010
Nevirapine	HSS	*HLA-Cw*04;HLA-B*3505* *HLA-Cw8-B14* *HLA-DRB1*0101*	Chantarangsu 2009; Likanonsakul 2009 Littera 2006 Vitezica 2008; Martin 2005a
Oxicams	SJS/TEN	*HLA-B*7301*	Lonjou 2008
Phenobarbital	SJS/TEN	*HLA-B*5101*	Kaniwa 2013
Phenytoin	SJS/TEN	*HLA-B*1502* *HLA-B*5101*	Locharernkul 2008; Man 2007 Kaniwa 2013
Sulfamethoxazole	SJS/TEN	*HLA-B*3802*	Lonjou 2008
Ximelagatran	DILI	*HLA-DRB1*0701*	Kindmark 2008
Zonisamide	SJS/TEN	*HLA-A*0207*	Kaniwa 2013

[a]*HLA* alleles in bold have been, or are strongly suggested to be, applied to clinical use.
ADR = adverse drug reaction; DILI = drug-induced liver injury; SCAR = severe cutaneous adverse drug reaction.

At the same time, a considerable amount of work was done to prepare for laboratory implementation. Several methods for *HLA-B* genotyping are now commercially available. These methods include direct sequence-based typing, in which the DNA coding for *HLA-B* is amplified and then fully sequenced; allele-specific PCR (Hammond 2007; Martin 2005b); flow cytometry (Kostenko 2011); real-time PCR (Dello Russo 2014); and the surrogate SNP detection method (Rodriguez-Novoa 2010; Colombo 2008).

A specific quality assurance program for *HLA-B*5701* screening has been validated and is currently actively administered by the Asia-Pacific Histocompatibility and Immunogenetics Association. Testing of *HLA-B*5701* has resulted in a reduced incidence of abacavir-induced HSS and has been shown to be cost-effective (Mallal 2008; Schackman 2008; Hughes 2004b). A cost-effectiveness analysis using a simulation model of HIV disease in the United States concluded that pharmacogenetic testing for *HLA-B*5701* is cost-effective if abacavir-based treatment is less expensive than tenofovir-based treatment but has comparable virologic efficacy (Schackman 2008). An analysis for a simulated cohort of 1000 HIV-infected patients in Spain based on the values used in the PREDICT-1 study also suggests that systematic *HLA-B*5701* testing to prevent abacavir-induced HSS in patients treated with abacavir is a cost-effective option for the Spanish National Health System, with $790 saved per abacavir-induced HSS. Moreover, the sensitivity analysis shows a savings of $127 or a cost-effectiveness ratio of $5300 (Nieves Calatrava 2010). Canada launched a pilot project in 2006 to offer *HLA-B*5701* prescreening as a standard of care for all patients with HIV infection. Concurrent preliminary data suggest that the *HLA-B*5701* test is cost-effective. Collaboration among clinicians, patients, laboratory workers, and the pharmaceutical industry has led to the successful implementation of a national *HLA-B*5701* genetic testing service in Canada (Lalonde 2010).

To date, *HLA-B*5701* testing for abacavir-induced HSS represents one of the best examples of pharmacogenetics being integrated into routine medical practice.

*HLA-B*1502* and Carbamazepine-Induced SJS/TEN

Carbamazepine and Carbamazepine-Induced SJS/TEN

Carbamazepine is an aromatic antiepileptic drug (AED) and mood stabilizer used to treat epilepsy, bipolar disorder, and trigeminal neuralgia. Although usually well tolerated, carbamazepine can cause cutaneous ADRs, including mild maculopapular exanthema (MPE), severe HSS, and SJS/TEN, in up to 10% of patients (Marson 2007). Carbamazepine-induced SJS/TEN occurs within 2 months after drug administration, with a median onset of 15 days (Hung 2006). Carbamazepine is one of the most common causative drugs of SJS/TEN (Roujeau 1994) and is estimated to contribute to 35% of drug-induced SJS/TEN cases in Han Chinese (Kamaliah 1998) but to only 6% of cases in whites (Rzany 1999).

*HLA-B*1502* Is Highly Associated with Carbamazepine-Induced SJS/TEN in Southeast Asians*

The population difference in the incidence of carbamazepine-induced SJS/TEN and case reports indicates an underlying genetic predisposition of this life-threatening adverse event (Edwards 1999). The first major breakthrough in identifying genetic factors associated with carbamazepine-induced SJS/TEN was reported in 2004 in Taiwan. The study showed a 100% association between *HLA-B*1502* and carbamazepine-induced SJS/TEN in all 44 Han Chinese patients residing in Taiwan, compared with the observed frequency of 3% (3 of 101) in carbamazepine-tolerant patients (Chung 2004). In the follow-up study, 59 of 60 patients (98.3%) with carbamazepine-induced SJS/TEN carried the *HLA-B*1502* allele, whereas only 6 of the 144 carbamazepine-tolerant controls (4.16%) tested *HLA-B*1502* positive (Hung 2006).

The association between *HLA-B*1502* and carbamazepine-induced SJS/TEN has been investigated in different populations by several independent groups. However, the results suggest that this association is ethnicity-specific because it is found only in Southeast Asian populations, with no associations in African, white, Japanese, or Korean populations (Table 13.3). The latest meta-analysis (Tangamornsuksan 2013), which included 227 SJS/TEN cases, 602 matched control subjects, and 2949 population controls, investigated the association between the *HLA-B*1502* allele and carbamazepine-induced SJS/TEN. The integrated odds ratio (OR) for the association between *HLA-B*1502* and carbamazepine-nduced SJS/TEN was 79.84 (28.45–224.06), and racial subgroup analyses yielded similar findings for Han Chinese (115.32; 18.17–732.13), Thai (54.43; 16.28–181.96), and Malaysian (221.00; 3.85–12,694.65) populations. However, among white or Japanese individuals, none of the patients with SJS/TEN were *HLA-B*1502* carriers. The ethnic difference is probably related to the difference in the *HLA-B*1502* allele frequencies of these populations. The allele frequency is much higher in Southeast Asian populations (2%–8%) than in in European, Japanese, and South Korean populations (less than 0.1%) (Lee 2010). Thus, the low *HLA-B*1502* allele frequency may explain the lower occurrence of carbamazepine-induced SJS/TEN in Japanese and European populations and therefore the lack of association for this allele.

Table 13.3. *HLA-B*1502* and CBZ-Induced SJS/TEN

Ethnicity	HLA-B*1502 Positive			P value	OR (95% CI)	Reference
	Case	Control	Population			
Han Chinese (Taiwan)	44/44 (100%)	3/101 (3%)	8/93 (8.6%)	3.13×10^{-27}	2504 (126–49,522)	Chung 2004
Han Chinese (Taiwan)	59/60 (98.3%)	6/144 (4.2%)	5.9%[a]	$< 1 \times 10^{-41}$	1357 (193.4–8838.3)	Hung 2006
Han Chinese (Hong Kong)	4/4 (100%)	7/48 (14.6%)	10.2%[a]	2.7×10^{-3}	46.9 (2.2–986.7)	Man 2007
Han Chinese (mainland)	8/8 (100%)	4/50 (8%)	6/71 (8.5%)	< 0.05	184 (33.2–1021)	Wu 2010
Han Chinese (mainland)	16/17 (94.1%)	2/21 (9.5%)	17/185 (9.2%)	2.6×10^{-4}	152 (12–1835)	Zhang 2011
Han Chinese (mainland)	9/9 (100%)	11/80 (13.8%)	11/62 (17.7%)	< 0.001	114.8 (6.3–2111)	Wang 2011
Thai	6/6 (100%)	8/42 (19%)	8.2%–8.5%[a]	6.1×10^{-4}	51 (2.6–1010.1)	Locharernkul 2008
Thai	37/42 (88.1%)	5/42 (11.9%)	8.2%–8.5%[a]	2.9×10^{-12}	54.8 (14.6–205.1)	Tassaneeyakul 2010
Thai	32/34 (94.1%)	7/40 (17.5%)	8.2%–8.5%[a]	< 0.001	75.4 (13.0–718.9)	Kulkantrakorn 2008
Malaysian	17/21 (81%)	N/A	47/300 (15.7%)	7.9×10^{-6}	16.2 (4.6–62.4)	Chang 2011
Malaysian	6/6 (100%)	0/8 (0%)	2%–16%[a]	3.0×10^{-4}	221 (3.9–12,694.7)	Then 2011
Hindu (India)	6/8 (75%)	N/A	0/10 (0%)	1.4×10^{-3}	71.4 (3–1698)	Mehta 2009
Japanese	0/7 (0%)	N/A	0.1%[a]	N/A	N/A	Kaniwa 2008
Whites (France)	4/12[b] (33.3%)	N/A	0[a]	N/A	N/A	Lonjou 2006
Whites (UK)	0/2 (0%)	0/43 (0%)	N/A	N/A	N/A	Alfirevic 2006
Canadians	3/9 (33.3%)	1/87 (1.1%)	N/A	0.0022	38.65 (2.68–2239.5)	Amstutz 2013
Han Chinese (Taiwan)	41/53 (77.4%)	4/72 (5.6%)	N/A	< 0.001	58.1 (17.6–192)	Genin 2014
Europeans	0/20 (0%)	0/43 (0%)[c]	N/A	NS	—	Genin 2014

[a]Population frequency is based on data from www.allelefrequencies.net/.
[b]The four patients carrying *HLA-B*1502* had Asian ancestry.
[c]Data are adapted from: Alfirevic A, Jorgensen AL, Williamson PR, et al. *HLA-B* locus in Caucasian patients with carbamazepine hypersensitivity. Pharmacogenomics 2006;7:813-8.
CBZ = carbamazepine; N/A = not applicable; NS = not significant.

Patients who develop drug-induced ADRs may also develop ADRs when they switch to other drugs with a similar chemical structure (Romano 2005). This cross-reactivity has been observed in patients who receive aromatic AEDs (Romano 2006). A 20%–30% probability of cross-reactivity was reported among carbamazepine, oxcarbazepine, phenytoin, and lamotrigine, all of which have similar chemical structures (Alvestad 2008; Hirsch 2008; Seitz 2006) (Figure 13.3). Therefore, it is necessary to investigate the association of *HLA-B*1502* with other structurally related aromatic AED-induced cases of SJS/TEN. Case-control studies involving subjects of Han Chinese and Thai origin have shown an association between *HLA-B*1502* and SJS/TEN induced by oxcarbazepine, phenytoin, and lamotrigine (Hung 2010; Locharernkul 2008; Man 2007).

In addition to causing SJS/TEN, carbamazepine can cause MPE and HSS. However, no association was found between *HLA-B*1502* and carbamazepine-induced MPE, which strongly suggests that this association is phenotype-specific (Genin 2014; Then 2011; Wang 2011; Wu 2010; Locharernkul 2008; Man 2007; Hung 2006). One meta-analysis that reviewed studies investigating *HLA-B*1502* and carbamazepine-induced cutaneous ADRs (SJS/TEN, MPE, and HSS) in Asian populations reinforced the association between *HLA-B*1502* and carbamazepine-induced SJS/TEN, but not MPE or HSS (Yip 2012). Moreover, another meta-analysis that investigated the *HLA* alleles and all AED-induced hypersensitivities arrived at a similar conclusion (Bloch 2014).

Implementation of HLA-B* 1502 Screening in Clinical Use

The association of *HLA-B*1502* with carbamazepine-induced SJS/TEN is one of the strongest genetic associations identified so far. The FDA updated the package labeling for carbamazepine, warned of a high risk of SJS/TEN occurring with carbamazepine administration in patients who carry the *HLA-B*1502* allele, and stated that "patients with ancestry in at-risk populations should be screened for the presence of *HLA-B*1502* allele before starting carbamazepine" (FDA 2007). Therapeutic recommendations state that *HLA-B*1502* carriers who are carbamazepine naive should not use carbamazepine. Patients who have been treated with carbamazepine for more than 3 months without the incidence of cutaneous ADRs, even though they are *HLA-B*1502* positive, are considered at low risk of SJS/TEN and can continue using carbamazepine (Leckband 2013; Ferrell 2008). Alternative AEDs should be used in *HLA-B*1502* carriers. However, patients with the *HLA-B*1502* allele have a possibility of experiencing cross-reactivity when using alternative AEDs (e.g., phenytoin, lamotrigine, and oxcarbazepine) (Hung 2010; Locharernkul 2008; Man 2007). Thus, "healthcare providers should consider avoiding phenytoin as alternatives for carbamazepine in patients who test positive for *HLA-B*1502*" (FDA 2008). In any case, all patients should be monitored for cutaneous adverse reactions.

Recently, a prospective cohort study in Taiwan showed that prescreening for the presence of the *HLA-B*1502* allele before initiating carbamazepine treatment and withholding carbamazepine from *HLA-B*1502*-positive patients can reduce the incidence of carbamazepine-induced SJS/TEN among Han Chinese (Chen 2011). *HLA-B*1502* screening would also provide benefit in other Southeast Asians. A systematic review of *HLA-B*1502* and carbamazepine-induced SJS/TEN calculated that the PPV and NPV for a screening test are 1.8% (7.7% in Han Chinese only) and 100%, respectively, according to an estimated incidence of carbamazepine-induced SJS/TEN

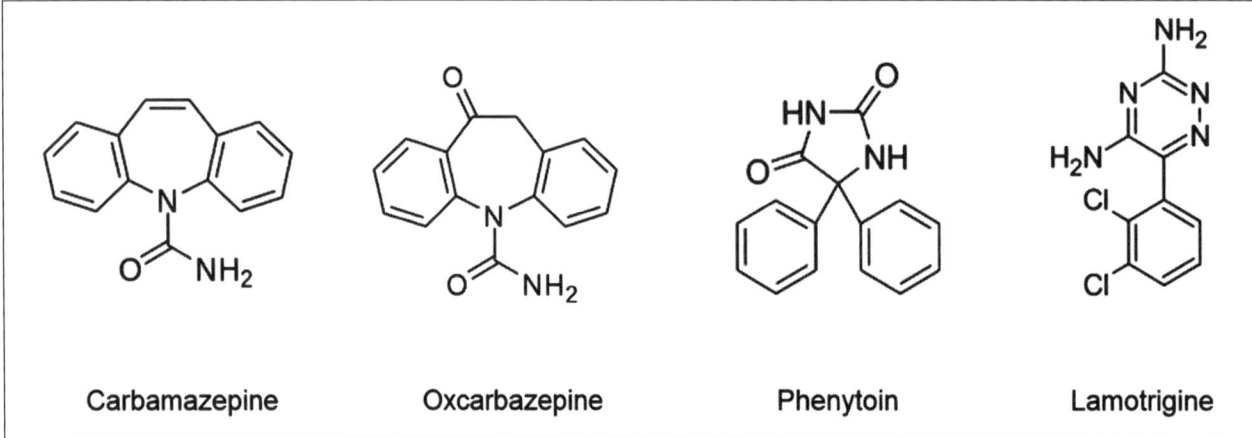

Figure 13.3. Chemical structure of carbamazepine, oxcarbazepine, phenytoin, and lamotrigine.

in a Southeast Asian population of 0.23%. Thus, to prevent one case of carbamazepine-induced SJS/TEN, 461 patients would need to be tested (Yip 2012). Considering the severity of SJS/TEN and the high NPV of the screening, a negative test result would represent very valuable information to avoid carbamazepine-induced SJS/TEN.

Cost-effectiveness is a prerequisite for adopting genetic screening into routine clinical practice. The prevalence of carbamazepine-induced SJS/TEN and the PPV are the main factors that influence cost-effectiveness. Thus, genetic screening would be less beneficial in populations with a low *HLA-B*1502* allele frequency. Recently, several cost-effective studies were organized to evaluate the economics of *HLA-B*1502* screening in Asian countries, including Thailand, Malaysia, and Singapore. Studies in Thailand found that tests specific to *HLA-B*1502* are more cost-effective compared with HLA-B genotyping or treating carbamazepine-induced SJS/TEN without screening (Locharernkul 2011, 2010). However, an active surveillance system was proposed to assess the prevalence of carbamazepine-induced SJS/TEN in the Thai population to enhance the generalizability of the results (Rattanavipapong 2013). A study in Malaysia reported that the in-house *HLA-B*1502* screening system set up in the UKM Medical Center is much more cost-effective than commercial kits, which will facilitate routine clinical typing (Then 2013). The cost-effectiveness of *HLA-B*1502* testing has also been shown in Singapore (Dong 2012). The Ministry of Health (MOH) in Singapore recommended *HLA-B*1502* genotyping before initiating carbamazepine therapy in new patients with Asian ancestry, and subsidized patients from the MOH-funded restructured hospitals and institutions will qualify for a flat-rate subsidy of 75% of the cost of the test (HSA 2013). In Taiwan, *HLA-B*1502* genotyping has become a routine clinical test before carbamazepine treatment, and the genotyping fee is covered by the government.

Other Potential Clinically Actionable Associations

HLA-B*5801 and Allopurinol-Induced SJS/TEN

Genetic Association of HLA-B*5801

Allopurinol, a xanthine oxidase inhibitor, has been used for the treatment of hyperuricemia and gout for decades. Although allopurinol is well tolerated by most patients, around 1%–5% of patients develop allopurinol hypersensitivity, a rare adverse reaction that is characterized by a combination of cutaneous manifestations (which may range in severity from relatively benign maculopapular rashes to more lethal SJS/TEN), fever, hepatic dysfunction, renal impairment, and eosinophilia (Ramasamy 2013). Allopurinol is the leading drug that induces SJS/TEN in Europe and Israel, overtaking other well-known drugs associated with such reactions, including carbamazepine, phenytoin, and the sulfonamides (Halevy 2008; Mockenhaupt 2008).

A case report showed familial predispositions to allopurinol-induced SJS/TEN, alluding to underlying genetic components (Melsom 1999). As early as 1989, a study of Southern Chinese patients with allopurinol-induced skin eruptions found an association with *HLA* haplotype *AW33-BW58* (Chan 1989). Until 2005, an extensive case-control study in Taiwan genotyped 823 SNPs in genes related to drug metabolism and immune response in the Han Chinese population and found a strong association between *HLA-B*5801* and allopurinol SCARs (including HSS and SJS/TEN) (Hung 2005). This association was subsequently confirmed in Japanese, European, Thai, and Korean populations, showing different strengths of association, with a strong association in the Han Chinese, Thai, and Korean populations and less strength in the Japanese and European populations (Table 13.4). In an Australian study, five of six patients with allopurinol-induced SJS/TEN were heterozygous for *HLA-B*5801*, four of which were of Southeast Asian origin (Lee 2012). In the mainland Han Chinese population, HLA-B*5801 is even associated with allopurinol-induced MPE, which is only a mild cutaneous ADR, with the highest OR among the four kinds of cutaneous ADRs (MPE/HSS/SJS/TEN) (Cao 2012). The lower association in whites may reflect differences in the prevalence of *HLA-B*5801*, which is about 15% in the Han Chinese population versus less than 6% in whites.

A GWAS of Europeans investigated genetic association at a genome-wide level in a large sample of patients with SJS/TEN. Six SNPs located in the HLA region showed significant evidence for association. However, the haplotype CACGAC, which is in incomplete LD with the *HLA-B*5801* allele, was more associated with the disease compared with any of the single SNPs and was even much stronger in patients exposed to allopurinol (Genin 2011). The GWAS of the Japanese population examined 14 patients with allopurinol-induced SJS/TEN and 991 ethnically matched healthy controls. Twenty-one SNPs on chromosome 6 were significantly associated with allopurinol-induced SJS/TEN. The strongest associations were rs2734583 in *BAT1*, rs3094011 in *HCP5*, and GA005234 in *MICC*. rs9263726 in *PSORS1C1*, also significantly associated with allopurinol-related SJS/TEN, is in complete LD with *HLA-B*5801* (Tohkin 2013). Of interest, both GWASs found associated SNPs in the *CCHCR1* and *PSORS1C1* gene known to be in LD with *HLA-C*. Both *CCHCR1* and *HLA-C* are reported to be associated with psoriasis (Gandhi 2011). The Japanese GWAS also showed associated SNPs in *BAT1* and *HCP5* (Tohkin 2013). *BAT1* has been associated with rheumatoid arthritis (Quinones-Lombrana 2008), and *HCP5* is in strong LD with *HLA-B*5701*, which is a predictor of abacavir-induced HSS (Columbo 2008). These findings suggest that the genes with potential function in keratinocytes or inflammatory diseases are other genetic factors of SJS/TEN.

A recent meta-analysis reviewed all studies that investigated an association between *HLA-B*5801* with allopurinol-induced SJS/TEN through a comprehensive search performed in 10 main databases from study inception until June 2011. Four studies with 55 SJS/TEN cases and 678 allopurinol-tolerant controls as well as five studies with 69 SJS/TEN cases and 3378 general population controls were identified. Cases of SJS/TEN were significantly associated with the *HLA-B*5801* allele in both groups, with an OR (95% confidence interval [CI]) of 96.60 (24.49–381.00) for tolerant controls and 79.28 (41.51–151.35) for population controls. The subgroup analysis for Asians and non-Asians yielded similar findings (Somkrua 2011). Therefore, *HLA-B*5801* allele screening should be considered for patients who will receive allopurinol treatment.

Clinical Application Efforts

Because of the strong association between *HLA-B*5801* and allopurinol-induced SJS/TEN in Han Chinese, the Taiwan Department of Health has updated the labeling for allopurinol to include information on *HLA-B*5801* and recommended *HLA-B*5801* testing before allopurinol treatment in allopurinol-naive patients. A prospective study of *HLA-B*5801* genotyping for the prevention of allopurinol-induced SCARs is under way in Taiwan; however, the results are not yet available (Hershfield 2013).

Table 13.4. Summary of Reports Regarding *HLA-B*5801* and Allopurinol-Induced cADRs

Ethnicity	Type of cADRs	HLA-B*5801 Positive		P value	OR	Reference
		Case	Control			
Han Chinese (Taiwan)	SJS/TEN/HSS	51/51 (100%)	20/135 15.0%	4.7×10^{-24}	580.3 (34.4–9780.9)	Hung 2005
Japanese	SJS/TEN	2/10 (20%)	6/986 (0.61%)[a]	$< 10^{-4}$	40.83 (10.5–158.9)	Kaniwa 2008
European	SJS/TEN	14/27 (51.9%)	28/1822 (1.5%)[a]	$< 10^{-6}$	80 (34–187)	Lonjou 2008
Thai	SJS/TEN	27/27 (100%)	7/54 (13.0%)	1.6×10^{-13}	348.3 (19.2–6336.9)	Tassaneeakul 2009
Korean	SJS/TEN/HSS	23/25 (92%)	6/57 (10.5%)	2.45×10^{-11}	97.8 (18.3–521.5)	Kang 2011
Korean[b]	SJS/TEN/HSS	9/9 (100%)	41/432 (9.5%)	1.6×10^{-9}	179.2 (10.2–3151.7)	Jung 2011
White[c]	SJS/TEN	3/7 (42.8%)	6/115 (5.2%)[a]	0.248	N/A	Cristallo 2011
Han Chinese (mainland)	SJS/TEN/HSS	16/16 (100%)	7/63 (11.1%)	7.40×10^{-12}	248.60 (13.48–4585.35)	Cao 2012
Han Chinese (mainland)	MPE	22/22 (100%)	7/63 (11.1%)	9.21×10^{-14}	339.00 (18.58–6186.39)	Cao 2012

[a]Population, but not allopurinol-tolerant, control.
[b]*HLA-B58* is thought identical with *HLA-B*5801* because a 100% coincidence of *HLA-B*5801* and serologic-type *HLA-B58* in the Korean population (Lee KW, Oh DH, Lee C, et al. Allelic and haplotypic diversity of *HLA-A, -B, -C, -DRB1*, and *-DQB1* genes in the Korean population. Tissue Antigens 2005;65:437-47).
[c]Although *HLA-B*5801* is not significantly associated with SJS/TEN, the *B*5801-DRB1*1302* haplotype is associated with allopurinol-induced SJS/TEN (p=0.00028).
cADR = cutaneous adverse drug reaction; MPE = maculopapular exanthema.

Table 13.5. Similarities of *HLA-B*5801* and *HLA-B*1502* Association with SJS/TEN

		*HLA-B*5801*	*HLA-B*1502*
Causative drug		Allopurinol	Carbamazepine
Incidence of SJS/TEN Prevalence		1–4/1000	1–6/1000 (0.23%)
	Chinese and Southeast Asians	6%–8%	2%–8%
	Whites	< 1%	< 0.1%
PPV		1.5%	1.8% (7.7% for Han Chinese)
NPV		100%	100%
Sensitivity		100%	100%
Specificity		85%	97%
Countries that recommend genetic testing		Taiwan	United States Taiwan Singapore

CBZ = carbamazepine; NPV = negative predictive value; PPV = positive predictive value.

Data are summarized from: Yip VL, Marson AG, Jorgensen AL, et al. *HLA* genotype and carbamazepine-induced cutaneous adverse drug reactions: a systematic review. Clin Pharmacol Ther 2012;92:757-65; Chen P, Lin JJ, Lu CS, et al. Carbamazepine-induced toxic effects and *HLA-B*1502* screening in Taiwan. N Engl J Med 2011;364:1126-33.

Table 13.6. *HLA-A*3101* and CBZ-Induced HSS

		*HLA-A*3101* Positive				
Ethnicity	Type of cADRs	Case	Control	P value	OR	Reference
Han Chinese (Taiwan)	MPE	6/18 (33.3%)	4/144 (2.8%)	2.2×10^{-3}	17.5 (4.6–66.5)	Hung 2006
Japanese[a]	MPE/HSS/SJS/TEN	45/77 (58.4%)	54/420 (12.9%)	1.09×10^{-16}	9.5 (5.6–16.3)	Ozeki 2011
European[a]	MPE/HSS/SJS/TEN	38/145 (26.2)	10/257 (3.9%)	1.0×10^{-7}	9.1 (4.0–20.7)	McCormack 2011
Korean	HSS/ SJS/TEN	13/24 (54.2%)	7/50 (14.0%)	0.001	7.3 (2.3–22.5)	Kim 2011
Canadian	HSS/MPE	9/32 (28.1%)	3/91 (3.3%)	2.6×10^{-4}	11.18 (2.53–69.27)	Amstutz 2013
Han Chinese (Taiwan)	HSS	7/10 (70%)	10/257 (3.9%)[b]	< 0.001	57.6 (11.0–341)	Genin 2014
European	HSS	5/10 (50%)	3/72 (4.2%)	< 0.001	23.0 (4.2–125)	Genin 2014

[a]Genome-wide association studies.

[b]Data are adapted from: McCormack M, Alfirevic A, Bourgeois S, et al. *HLA-A*3101* and carbamazepine-induced hypersensitivity reactions in Europeans. N Engl J Med 2011;364:1134-43.

CBZ = carbamazepine.

Given the high specificity for allopurinol-induced SCARs, the Clinical Pharmacogenetics Implementation Consortium (CPIC) guidelines for allopurinol dosing recommend that allopurinol not be prescribed for patients who have tested positive for *HLA-B*58:01*. For patients who have tested negative, allopurinol may be prescribed as usual. However, a negative test result does not entirely exclude the possibility of developing SCARs, especially in the European population (Hershfield 2013). The 2012 American College of Rheumatology guidelines for the management of gout also recommend considering *HLA-B*5801* testing before initiating allopurinol as a risk management component in subpopulations in which both the *HLA-B*5801* allele frequency is elevated and the *HLA-B*5801*-positive subjects are at high risk of experiencing severe allopurinol hypersensitivity reactions (Khanna 2012).

Although *HLA-B*5801* screening is recommended, especially in populations with a high prevalence (Hershfield 2013; Khanna 2012), investigators have commented that *HLA-B*5801* use alone as a population screening test is not effective. They argue that although the NPV and sensitivity of *HLA-B*5801* in allopurinol-induced SJS/TEN are very high, the PPV is low because of the very low incidence of allopurinol hypersensitivity itself and the relatively high prevalence of *HLA-B*5801* in the Southeast Asian population (Lee 2012).

The clinical translation of *HLA-B*1502* may serve as a referable example for *HLA-B*5801* because they share similarities in several terms (Table 13.5). Both alleles are more prevalent in Southeast Asians and less prevalent in whites; however, the association of *HLA-B*5801* with allopurinol-induced SJS/TEN is consistent across the populations (Somkrua 2011). A study aiming to determine the cost-effectiveness of genetic testing for *HLA-B*5801* compared with usual care before allopurinol administration in Thailand suggests that *HLA-B*5801* testing is a highly potential cost-effective intervention in Thailand. The findings are sensitive to several factors. They also state that other factors, including ethical, legal, and social implications, should be considered for informed policy-making (Saokaew 2014).

Whether *HLA-B*5801* should be tested before using allopurinol requires further study, including a prospective clinical study, the precise estimation of the incidence of allopurinol-induced SJS/TEN, and the evaluation of cost-effectiveness for genotyping. Regardless if genetic testing is performed or not, physicians should monitor patients closely.

*HLA-A*3101* and Carbamazepine-Induced HSS

Genetic Association of HLA-A*3101
As described earlier, carbamazepine can also cause mild MPE and more severe HSS in addition to the most severe SJS/TEN (Marson 2007). The distinct pathologic features of MPE, HSS, and SJS/TEN suggest that they use different genetic determinants.

Carbamazepine-induced SJS/TEN and MPE were first found to use different underlying genetic determinants in Han Chinese patients. *HLA-A*3101*, but not *HLA-B*1502*, was identified as a risk allele of MPE (Hung 2006). In 2011, two independent investigations using GWASs identified the *HLA-A*3101* association with carbamazepine-induced SCARs in Japanese and European populations (McCormack 2011; Ozeki 2011). Subsequent studies have confirmed this association in Korean and Canadian populations (Table 13.6). According to Table 13.6, the association with *HLA-A*3101* occurs in more diverse ethnic groups and predisposes to mild MPE as well as more severe HSS and even SJS/TEN. One meta-analysis suggested that *HLA-A*3101* had an extremely strong association with carbamazepine-induced DRESS ($p<0.001$, a pooled OR [95% CI] of 13.2 [8.4–20.8]) but a much weaker association with carbamazepine-induced SJS/TEN ($p=0.01$, OR [95% CI] 3.94 [1.4–11.5]) (Genin 2014). The other comprehensive meta-analysis suggested that *HLA-A*3101* is a universal risk marker, irrespective of cutaneous ADR type (SJS/TEN: $p=4.03 \times 10^{-6}$, OR [95% CI] 5.65 (2.70–11.78); HSS/MPE: $p=4.46 \times 10^{-22}$; OR [95% CI] 8.58 [5.55–13.28]) (Grover 2014).

Clinical Application Efforts
*HLA-A*3101* was only recently associated with carbamazepine-induced HSS. So far, only *HLA-B*1502* has been recommended for testing before carbamazepine administration in high-risk populations. Whether *HLA-A*3101* should be tested in all new users or whether *HLA-A*3101* combined with *HLA-B*1502* should be tested in high-risk populations requires further investigation. Japan registered a clinical trial program in December 2011 called the Genotype-Based Carbamazepine Therapy (GENCAT) study to assess whether *HLA-A*3101* genotyping can prevent carbamazepine-induced MPE and SCARs in the Japanese population (https://upload.umin.ac.jp/cgi-open-bin/ctr/ctr.cgi?function=brows&action=brows&type=summary&recptno=R000008044&language=E). It is recommended that patients stop taking carbamazepine if a cutaneous eruption occurs, even if it is mild reaction such as MPE, because it is not currently feasible to tell whether it will develop into SCARs.

Working Models for HLA-Linked ADRs
The symptoms of drug-induced cutaneous ADRs will be alleviated after drug withdrawal, but they may worsen and possibly become fatal after rechallenge with the

same drug after discontinuation. Thus, it has long been proposed that drug-induced cutaneous ADRs represent the delayed-type hypersensitivity (type IV hypersensitivity) that is mediated by T cells. As described earlier, HLA molecules play a pivotal role in T cell–mediated immune responses through the presenting antigen peptides. Are these *HLA* alleles merely genetic markers, or do they also participate in the development of SCARs? In support of the latter, drug-specific CD8[+] cytotoxic T cells have been found in the blister fluid of patients with drug-induced SJS/TEN (Nassif 2004).

To date, three immunologic working models/concepts have been proposed to explain the mechanism of drug-induced SCARs associated with *HLA* alleles. They are the hapten, p-i, and altered peptide repertoire concepts.

The Hapten Concept

T-cell activation by small molecules can be explained by the hapten concept. In this model, small molecules (less than 1000 Da, like drugs) are unlikely to be immunogenic; they need to covalently bind to a high-molecular-weight protein or peptide to become antigenic (Adam 2011). To be recognized by T cells as foreign antigens, haptenated proteins need to be processed by the antigen-presenting cells to produce haptenated peptides presented on MHC molecules. An example is the covalent binding of penicillin to the lysine residue of serum albumin and its presentation by HLA through the classical processing–required pathway to trigger T-cell activation, eliciting a penicillin allergy (Romano 1998; Padovan 1997).

The p-i Concept

Conversely, the p-i concept suggests a direct interaction between drugs and immune receptors such as the T-cell receptor and HLA, representing an unconventional presentation pathway that is independent of antigen presenting cell intracellular processing (Pichler 2006). For example, carbamazepine can interact directly with *HLA-B*1502* without drug-modified peptide formation and sufficiently trigger the cytotoxic activity of CD8[+] T cells (Wei 2012a; Yang 2007). Two other examples of this concept are T cells activated by sulfamethoxazole and lidocaine (Zanni 1998; Schnyder 1997).

The Altered Peptide Repertoire Concept

Recently, the altered peptide repertoire concept was proposed (Illing 2012), which is non-mutually exclusive with the two previous concepts. This concept suggests that drugs or their metabolites can bind non-covalently with the peptide-binding groove of particular HLA molecules, resulting in a modified structure and chemistry of the antigen-binding cleft, thereby altering the endogenous binding peptide repertoire. In this way, drugs and their metabolites guide the selection of new endogenous peptides, inducing a distinct alteration in the "immunological self." The resultant peptide-centric "altered self" activates drug-specific T cells, thereby driving ADRs. Accumulated findings regarding abacavir-induced HSS (Illing 2012; Norcross 2012; Ostrov 2012) and carbamazepine-induced SJS/TEN (Illing 2012; Wei 2012a) can be well explained by this working model.

Challenges and Future Perspectives

Genome-wide association studies have successfully identified several variants associated with severe ADRs, including SCARs (McCormack 2011; Ozeki 2011) and DILI (drug-induced liver injury) (Kim 2010; Singer 2010). However, although these studies identified variants that were highly penetrant and had a major genetic effect, they could not explain all cases of each specific severe ADR. Because current GWAS platforms can capture only the common genetic variants, some rare variants associated with drug-induced ADRs could be missed. With the rapid decrease in sequencing cost, next-generation sequencing is expected to be a solution in the future (di Iulio 2011).

The pharmacogenetic studies of SCARs are of great significance, not only in identifying genetic predictors but also in providing insights into the pathologic mechanism of these severe adverse events. Although *HLA-B*5701* and *HLA-B*1502* have successfully been used in the clinical setting to prevent abacavir-induced HSS and carbamazepine-induced SJS/TEN, these two examples are different; thus, it may be improper to simply apply their model to studies of other drug-induced severe ADRs. Translating pharmacogenetics into clinical practice is full of problems that must be solved to achieve a positive translation. We will describe this from two aspects, the study phase (preclinical/discovery) and the clinical implementation phase.

(1) Several factors must be considered for a pharmacogenetics study to successfully identify the genetic association with cutaneous ADRs: ensuring the accurate diagnosis of SCARs, establishing unified criteria for recruiting patients in multicenter studies, and ensuring adequate sample size. The most important of these is perhaps ensuring the accuracy of the phenotypes because a false-positive diagnosis can dilute and even mask true associations or provide false associations. For abacavir, the sensitivity of the *HLA-B*5701* association is enhanced after adopting patch testing to diagnose abacavir hypersensitivity (Mallal 2008; Saag 2008; Phillips 2005, 2002; Martin 2004). Moreover, sample size can become smaller to achieve the same statistical power with highly precise diagnoses (Zhou 2013).

(2) Although several *HLA* typing methods are used in clinical testing (Martin 2012), they are not widely accessible by clinical practitioners and are mainly available only in large research hospitals. A cost-effective, easy-to-use, and quality-assured genotyping method at the point of care will accelerate the implementation of pharmacogenetics in clinical practice.

(3) The practicing physicians, clinicians, and pharmacists must be educated and trained to understand pharmacogenetic tests and their proper use in clinical practice. One of the main roadblocks in the clinical implementation of pharmacogenetics lies in interpreting the genotypes. To address this issue, CPIC was established. To date, CPIC has published more than 10 guidelines (including those for *HLA-B*1502*, *HLA-B*5701*, and *HLA-B*5801*) that provide the recommended course of action when genotype information is available. These guidelines are all available online (https://www.pharmgkb.org/view/dosing-guidelines.do?source=CPIC#) (Relling 2011).

(4) Electronic health records (EHRs) are designed to digitally store health care information throughout an individual's lifetime so that continuity of care, education, and research will be supported. Electronic health records may include information such as observations, laboratory tests, medical images, treatments, therapies, drugs administered, patient-identifying information, and legal permissions (Ajami 2013). Genetic data obtained for clinical purposes, including pharmacogenetic testing, as for all clinical test results, are also expected to be included in EHRs (Shoenbill 2014). The full realization of EHRs will be of great value in improving quality of care, reducing medical errors, and avoiding potentially severe ADRs.

With the advances in genome technology and awareness of pharmacogenetics, more discoveries related to cutaneous ADRs will be made. Many researchers are giving their best efforts to remove the hurdles preventing the wide implementation of pharmacogenetics described in this chapter. As shown by the examples described in this chapter, clinical use of pharmacogenetics can not only significantly reduce the number of adverse events caused by these drugs, but can also save significant medical costs. In addition, the discoveries may help reveal the pathogenic mechanisms of these adverse events, which may assist in the design of safer drugs.

References

Adam J, Pichler WJ, Yerly D. Delayed drug hypersensitivity: models of T-cell stimulation. Br J Clin Pharmacol 2011;71:701-7.

Ajami S, Arab-Chadegani R. Barriers to implement electronic health records (EHRs). Mater Sociomed 2013;25:213-5.

Alfirevic A, Jorgensen AL, Williamson PR, et al. *HLA-B* locus in Caucasian patients with carbamazepine hypersensitivity. Pharmacogenomics 2006;7:813-8.

Alvestad S, Lydersen S, Brodtkorb E. Cross-reactivity pattern of rash from current aromatic antiepileptic drugs. Epilepsy Res 2008;80:194-200.

Amstutz U, Ross CJ, Castro-Pastrana LI, et al. *HLA-A 31:01* and *HLA-B 15:02* as genetic markers for carbamazepine hypersensitivity in children. Clin Pharmacol Ther 2013;94:142-9.

Bloch KM, Sills GJ, Pirmohamed M, et al. Pharmacogenetics of antiepileptic drug-induced hypersensitivity. Pharmacogenomics 2014;15:857-68.

Borchers AT, Lee JL, Naguwa SM, et al. Stevens-Johnson syndrome and toxic epidermal necrolysis. Autoimmun Rev 2008;7:598-605.

Cacoub P, Musette P, Descamps V, et al. The DRESS syndrome: a literature review. Am J Med 2011;124:588-97.

Cao ZH, Wei ZY, Zhu QY, et al. *HLA-B*58:01* allele is associated with augmented risk for both mild and severe cutaneous adverse reactions induced by allopurinol in Han Chinese. Pharmacogenomics 2012;13:1193-201.

Chan SH, Tan T. *HLA* and allopurinol drug eruption. Dermatologica 1989;179:32-3.

Chang CC, Too CL, Murad S, et al. Association of *HLA-B*1502* allele with carbamazepine-induced toxic epidermal necrolysis and Stevens-Johnson syndrome in the multi-ethnic Malaysian population. Int J Dermatol 2011;50:221-4.

Chantarangsu S, Mushiroda T, Mahasirimongkol S, et al. *HLA-B*3505* allele is a strong predictor for nevirapine-induced skin adverse drug reactions in HIV-infected Thai patients. Pharmacogenet Genomics 2009;19:139-46.

Chen P, Lin JJ, Lu CS, et al. Carbamazepine-induced toxic effects and *HLA-B*1502* screening in Taiwan. N Engl J Med 2011;364:1126-33.

Chiou CC, Yang LC, Hung SI, et al. Clinicopathological features and prognosis of drug rash with eosinophilia and systemic symptoms: a study of 30 cases in Taiwan. J Eur Acad Dermatol Venereol 2008;22:1044-9.

Chung WH, Hung SI, Hong HS, et al. Medical genetics: a marker for Stevens-Johnson syndrome. Nature 2004;428:486.

Colombo S, Rauch A, Rotger M, et al. The *HCP5* single-nucleotide polymorphism: a simple screening tool for prediction of hypersensitivity reaction to abacavir. J Infect Dis 2008;198:864-7.

Cristallo AF, Schroeder J, Citterio A, et al. A study of *HLA* class I and class II 4-digit allele level in Stevens-Johnson syndrome and toxic epidermal necrolysis. Int J Immunogenet 2011;38:303-9.

Daly AK, Donaldson PT, Bhatnagar P, et al. *HLA-B*5701* genotype is a major determinant of drug-induced liver injury due to flucloxacillin. Nat Genet 2009;41:816-9.

Dello Russo C, Lisi L, Fabbiani M, et al. Detection of *HLA-B*57:01* by real-time PCR: implementation into routine clinical practice and additional validation data. Pharmacogenomics 2014;15:319-27.

di Iulio J, Rotger M. Pharmacogenomics: what is next? Front Pharmacol 2011;2:86.

Dong D, Sung C, Finkelstein EA. Cost-effectiveness of *HLA-B*1502* genotyping in adult patients with newly diagnosed epilepsy in Singapore. Neurology 2012;79:1259-67.

Edwards SG, Hubbard V, Aylett S, et al. Concordance of primary generalised epilepsy and carbamazepine hypersensitivity in monozygotic twins. Postgrad Med J 1999;75:680-1.

Ferrell PB Jr, McLeod HL. Carbamazepine, *HLA-B*1502* and risk of Stevens-Johnson syndrome and toxic epidermal necrolysis: U.S. FDA recommendations. Pharmacogenomics 2008;9:1543-6.

FDA. 2007. Information for Healthcare Professionals: Dangerous or Even Fatal Skin Reactions—Carbamazepine (marketed as Carbatrol, Equetro, Tegretol, and generics). Available at www.fda.gov/Drugs/DrugSafety/PostmarketDrugSafetyInformationforPatientsandProviders/ucm124718.htm. Accessed November 4, 2014.

FDA. 2008. Phenytoin and Fosphenytoin Information. Available at www.fda.gov/Drugs/DrugSafety/PostmarketDrugSafetyInformationforPatientsandProviders/ucm110259.htm. Accessed November 4, 2014.

Gandhi G, Buttar BS, Albert L, et al. Psoriasis-associated genetic polymorphism in North Indian population in the *CCHCR1* gene and in a genomic segment flanking the HLA-C region. Dis Markers 2011;31:361-70.

Genin E, Chen DP, Hung SI, et al. *HLA-A*31:01* and different types of carbamazepine-induced severe cutaneous adverse reactions: an international study and meta-analysis. Pharmacogenomics J 2014;14:281-8.

Genin E, Schumacher M, Roujeau JC, et al. Genome-wide association study of Stevens-Johnson syndrome and toxic epidermal necrolysis in Europe. Orphanet J Rare Dis 2011;6:52.

Gennis MA, Vemuri R, Burns EA, et al. Familial occurrence of hypersensitivity to phenytoin. Am J Med 1991;91:631-4.

Grover S, Kukreti R. *HLA* alleles and hypersensitivity to carbamazepine: an updated systematic review with meta-analysis. Pharmacogenet Genomics 2014;24:94-112.

Guillaume JC, Roujeau JC, Revuz J, et al. The culprit drugs in 87 cases of toxic epidermal necrolysis (Lyell's syndrome). Arch Dermatol 1987;123:1166-70.

Halevy S, Ghislain PD, Mockenhaupt M, et al. Allopurinol is the most common cause of Stevens-Johnson syndrome and toxic epidermal necrolysis in Europe and Israel. J Am Acad Dermatol 2008;58:25-32.

Hammond E, Almeida CA, Mamotte C, et al. External quality assessment of *HLA-B*5701* reporting: an international multicentre survey. Antivir Ther 2007;12:1027-32.

Hautekeete ML, Horsmans Y, Van Waeyenberge C, et al. *HLA* association of amoxicillin-clavulanate–induced hepatitis. Gastroenterology 1999;117:1181-6.

Health Sciences Authority (HSA). 2013. Recommendations for *HLA-B*1502* Genotype Testing Prior to Initiation of Carbamazepine in New Patients. Available at www.hsa.gov.sg/publish/hsaportal/en/health_products_regulation/safety_information/product_safety_alerts/Safety_Alerts_2013/recommendations_for.html. Accessed October 22, 2014.

Herman TE, Kushner DC, Cleveland RH. Esophageal stricture secondary to drug-induced toxic epidermal necrolysis. Pediatr Radiol 1984;14:439-40.

Hershfield MS, Callaghan JT, Tassaneeyakul W, et al. Clinical Pharmacogenetics Implementation Consortium guidelines for human leukocyte antigen-B genotype and allopurinol dosing. Clin Pharmacol Ther 2013;93:153-8.

Hetherington S, Hughes AR, Mosteller M, et al. Genetic variations in *HLA-B* region and hypersensitivity reactions to abacavir. Lancet 2002;359:1121-2.

Hetherington S, McGuirk S, Powell G, et al. Hypersensitivity reactions during therapy with the nucleoside reverse transcriptase inhibitor abacavir. Clin Ther 2001;23:1603-14.

Hirsch LJ, Arif H, Nahm EA, et al. Cross-sensitivity of skin rashes with antiepileptic drug use. Neurology 2008;71:1527-34.

Horton R, Wilming L, Rand V, et al. Gene map of the extended human MHC. Nat Rev Genet 2004;5:889-99.

Hughes AR, Mosteller M, Bansal AT, et al. Association of genetic variations in *HLA-B* region with hypersensitivity to abacavir in some, but not all, populations. Pharmacogenomics 2004a;5:203-11.

Hughes DA, Vilar FJ, Ward CC, et al. Cost-effectiveness analysis of *HLA B*5701* genotyping in preventing abacavir hypersensitivity. Pharmacogenetics 2004b;14:335-42.

Hung SI, Chung WH, Jee SH, et al. Genetic susceptibility to carbamazepine-induced cutaneous adverse drug reactions. Pharmacogenet Genomics 2006;16:297-306.

Hung SI, Chung WH, Liou LB, et al. *HLA-B*5801* allele as a genetic marker for severe cutaneous adverse reactions caused by allopurinol. Proc Natl Acad Sci U S A 2005;102:4134-9.

Hung SI, Chung WH, Liu ZS, et al. Common risk allele in aromatic antiepileptic-drug induced Stevens-Johnson syndrome and toxic epidermal necrolysis in Han Chinese. Pharmacogenomics 2010;11:349-56.

Illing PT, Vivian JP, Dudek NL, et al. Immune self-reactivity triggered by drug-modified *HLA*-peptide repertoire. Nature 2012;486:554-8.

International HapMap C. A haplotype map of the human genome. Nature 2005;437:1299-320.

International HapMap C; Altshuler DM, Gibbs RA, et al. Integrating common and rare genetic variation in diverse human populations. Nature 2010;467:52-8.

Johnson-Reagan L, Bahna SL. Severe drug rashes in three siblings simultaneously. Allergy 2003;58:445-7.

Jung JW, Song WJ, Kim YS, et al. *HLA-B58* can help the clinical decision on starting allopurinol in patients with chronic renal insufficiency. Nephrol Dial Transplant 2011;26:3567-72.

Kamaliah MD, Zainal D, Mokhtar N, et al. Erythema multiforme, Stevens-Johnson syndrome and toxic epidermal necrolysis in northeastern Malaysia. Int J Dermatol 1998;37:520-3.

Kang HR, Jee YK, Kim YS, et al. Positive and negative associations of HLA class I alleles with allopurinol-induced SCARs in Koreans. Pharmacogenet Genomics 2011;21:303-7.

Kaniwa N, Saito Y, Aihara M, et al. HLA-B locus in Japanese patients with anti-epileptics and allopurinol-related Stevens-Johnson syndrome and toxic epidermal necrolysis. Pharmacogenomics 2008;9:1617-22.

Kaniwa N, Sugiyama E, Saito Y, et al. Specific HLA types are associated with antiepileptic drug-induced Stevens-Johnson syndrome and toxic epidermal necrolysis in Japanese subjects. Pharmacogenomics 2013;14:1821-31.

Kardaun SH, Sidoroff A, Valeyrie-Allanore L, et al. Variability in the clinical pattern of cutaneous side-effects of drugs with systemic symptoms: does a DRESS syndrome really exist? Br J Dermatol 2007;156:609-11.

Khanna D, Fitzgerald JD, Khanna PP, et al. 2012 American College of Rheumatology guidelines for management of gout. Part 1: systematic nonpharmacologic and pharmacologic therapeutic approaches to hyperuricemia. Arthritis Care Res 2012;64:1431-46.

Kim SH, Choi JH, Lee KW, et al. The human leucocyte antigen-DRB1*1302-DQB1*0609-DPB1*0201 haplotype may be a strong genetic marker for aspirin-induced urticaria. Clin Exp Allergy 2005;35:339-44.

Kim SH, Kim M, Lee KW, et al. HLA-B*5901 is strongly associated with methazolamide-induced Stevens-Johnson syndrome/toxic epidermal necrolysis. Pharmacogenomics 2010;11:879-84.

Kim SH, Lee KW, Song WJ, et al. Carbamazepine-induced severe cutaneous adverse reactions and HLA genotypes in Koreans. Epilepsy Res 2011;97:190-7.

Kindmark A, Jawaid A, Harbron CG, et al. Genome-wide pharmacogenetic investigation of a hepatic adverse event without clinical signs of immunopathology suggests an underlying immune pathogenesis. Pharmacogenomics J 2008;8:186-95.

Klein SM, Khan MA. Hepatitis, toxic epidermal necrolysis and pancreatitis in association with sulindac therapy. J Rheumatol 1983;10:512-3.

Kostenko L, Kjer-Nielsen L, Nicholson I, et al. Rapid screening for the detection of HLA-B57 and HLA-B58 in prevention of drug hypersensitivity. Tissue Antigens 2011;78:11-20.

Kulkantrakorn K, Tassaneeyakul W, Tiamkao S, et al. HLA-B*1502 strongly predicts carbamazepine-induced Stevens-Johnson syndrome and toxic epidermal necrolysis in Thai patients with neuropathic pain. Pain Pract 2012;12:202-8.

Lalonde RG, Thomas R, Rachlis A, et al. Successful implementation of a national HLA-B*5701 genetic testing service in Canada. Tissue Antigens 2010;75:12-8.

Leckband SG, Kelsoe JR, Dunnenberger HM, et al. Clinical Pharmacogenetics Implementation Consortium guidelines for HLA-B genotype and carbamazepine dosing. Clin Pharmacol Ther 2013;94:324-8.

Lee KW, Oh DH, Lee C, et al. Allelic and haplotypic diversity of HLA-A, -B, -C, -DRB1, and -DQB1 genes in the Korean population. Tissue Antigens 2005;65:437-47.

Lee MH, Stocker SL, Anderson J, et al. Initiating allopurinol therapy: do we need to know the patient's human leucocyte antigen status? Intern Med J 2012;42:411-6.

Lee MT, Hung SI, Wei CY, et al. Pharmacogenetics of toxic epidermal necrolysis. Exp Opin Pharmacother 2010;11:2153-62.

Likanonsakul S, Rattanatham T, Feangvad S, et al. HLA-Cw*04 allele associated with nevirapine-induced rash in HIV-infected Thai patients. AIDS Res Ther 2009;6:22.

Littera R, Carcassi C, Masala A, et al. HLA-dependent hypersensitivity to nevirapine in Sardinian HIV patients. AIDS 2006;20:1621-6.

Locharernkul C, Loplumlert J, Limotai C, et al. Carbamazepine and phenytoin induced Stevens-Johnson syndrome is associated with HLA-B*1502 allele in Thai population. Epilepsia 2008;49:2087-91.

Locharernkul C, Shotelersuk V, Hirankarn N. HLA-B* 1502 screening: time to clinical practice. Epilepsia 2010;51:936-8.

Locharernkul C, Shotelersuk V, Hirankarn N. Pharmacogenetic screening of carbamazepine-induced severe cutaneous allergic reactions. J Clin Neurosci 2011;18:1289-94.

Lonjou C, Borot N, Sekula P, et al. A European study of HLA-B in Stevens-Johnson syndrome and toxic epidermal necrolysis related to five high-risk drugs. Pharmacogenet Genomics 2008;18:99-107.

Lonjou C, Thomas L, Borot N, et al. A marker for Stevens-Johnson syndrome …: ethnicity matters. Pharmacogenomics J 2006;6:265-8.

Lucena MI, Molokhia M, Shen Y, et al. Susceptibility to amoxicillin-clavulanate–induced liver injury is influenced by multiple HLA class I and II alleles. Gastroenterology 2011;141:338-47.

Mallal S, Nolan D, Witt C, et al. Association between presence of HLA-B*5701, HLA-DR7, and HLA-DQ3 and hypersensitivity to HIV-1 reverse-transcriptase inhibitor abacavir. Lancet 2002;359:727-32.

Mallal S, Phillips E, Carosi G, et al. HLA-B*5701 screening for hypersensitivity to abacavir. N Engl J Med 2008;358:568-79.

Man CB, Kwan P, Baum L, et al. Association between HLA-B*1502 allele and antiepileptic drug-induced cutaneous reactions in Han Chinese. Epilepsia 2007;48:1015-8.

Manolio TA, Collins FS. The HapMap and genome-wide association studies in diagnosis and therapy. Annu Rev Med 2009;60:443-56.

Martin AM, Nolan D, Gaudieri S, et al. Predisposition to abacavir hypersensitivity conferred by HLA-B*5701 and a haplotypic Hsp70-Hom variant. Proc Natl Acad Sci U S A 2004;101:4180-5.

Martin AM, Nolan D, James I, et al. Predisposition to nevirapine hypersensitivity associated with *HLA-DRB1*0101* and abrogated by low CD4 T-cell counts. AIDS 2005a;19:97-9.

Martin AM, Nolan D, Mallal S. *HLA-B*5701* typing by sequence-specific amplification: validation and comparison with sequence-based typing. Tissue Antigens 2005b;65:571-4.

Martin MA, Klein TE, Dong BJ, et al. Clinical pharmacogenetics implementation consortium guidelines for HLA-B genotype and abacavir dosing. Clin Pharmacol Ther 2012;91:734-8.

Marson AG, Al-Kharusi AM, Alwaidh M, et al. The SANAD study of effectiveness of carbamazepine, gabapentin, lamotrigine, oxcarbazepine, or topiramate for treatment of partial epilepsy: an unblinded randomised controlled trial. Lancet 2007;369:1000-15.

McCormack M, Alfirevic A, Bourgeois S, et al. *HLA-A*3101* and carbamazepine-induced hypersensitivity reactions in Europeans. N Engl J Med 2011;364:1134-43.

McIvor RA, Zaidi J, Peters WJ, et al. Acute and chronic respiratory complications of toxic epidermal necrolysis. J Burn Care Rehabil 1996;17:237-40.

Mehta TY, Prajapati LM, Mittal B, et al. Association of *HLA-B*1502* allele and carbamazepine-induced Stevens-Johnson syndrome among Indians. Indian J Dermatol Venereol Leprol 2009;75:579-82.

Melsom RD. Familial hypersensitivity to allopurinol with subsequent desensitization. Rheumatology 1999;38:1301.

Mockenhaupt M, Viboud C, Dunant A, et al. Stevens-Johnson syndrome and toxic epidermal necrolysis: assessment of medication risks with emphasis on recently marketed drugs. The EuroSCAR-study. J Invest Dermatol 2008;128:35-44.

Mondino BJ, Brown SI, Biglan AW. *HLA* antigens in Stevens-Johnson syndrome with ocular involvement. Arch Ophthalmol 1982;100:1453-4.

Nassif A, Bensussan A, Boumsell L, et al. Toxic epidermal necrolysis: effector cells are drug-specific cytotoxic T cells. J Allergy Clin Immunol 2004;114:1209-15.

Nieves Calatrava D, Calle-Martin Ode L, Iribarren-Loyarte JA, et al. Cost-effectiveness analysis of *HLA-B*5701* typing in the prevention of hypersensitivity to abacavir in HIV+ patients in Spain. Enferm Infecc Microbiol Clin 2010;28:590-5.

Norcross MA, Luo S, Lu L, et al. Abacavir induces loading of novel self-peptides into *HLA-B*57: 01*: an autoimmune model for HLA-associated drug hypersensitivity. AIDS 2012;26:F21-9.

O'Donohue J, Oien KA, Donaldson P, et al. Co-amoxiclav jaundice: clinical and histological features and *HLA* class II association. Gut 2000;47:717-20.

Ostrov DA, Grant BJ, Pompeu YA, et al. Drug hypersensitivity caused by alteration of the MHC-presented self-peptide repertoire. Proc Natl Acad Sci U S A 2012;109:9959-64.

Ozeki T, Mushiroda T, Yowang A, et al. Genome-wide association study identifies *HLA-A*3101* allele as a genetic risk factor for carbamazepine-induced cutaneous adverse drug reactions in Japanese population. Hum Mol Genet 2011;20:1034-41.

Padovan E, Bauer T, Tongio MM, et al. Penicilloyl peptides are recognized as T cell antigenic determinants in penicillin allergy. Eur J Immunol 1997;27:1303-7.

Pellicano R, Silvestris A, Iannantuono M, et al. Familial occurrence of fixed drug eruptions. Acta Derm Venereol 1992;72:292-3.

Peyriere H, Nicolas J, Siffert M, et al. Hypersensitivity related to abacavir in two members of a family. Ann Pharmacother 2001;35:1291-2.

Phillips EJ, Sullivan JR, Knowles SR, et al. Utility of patch testing in patients with hypersensitivity syndromes associated with abacavir. AIDS 2002;16:2223-5.

Phillips EJ, Wong GA, Kaul R, et al. Clinical and immunogenetic correlates of abacavir hypersensitivity. AIDS 2005;19:979-81.

Pichler WJ, Beeler A, Keller M, et al. Pharmacological interaction of drugs with immune receptors: the p-i concept. Allergol Int 2006;55:17-25.

Pirmohamed M, Aithal GP, Behr E, et al. The phenotype standardization project: improving pharmacogenetic studies of serious adverse drug reactions. Clin Pharmacol Ther 2011a;89:784-5.

Pirmohamed M, Alfirevic A, Vilar J, et al. Association analysis of drug metabolizing enzyme gene polymorphisms in HIV-positive patients with co-trimoxazole hypersensitivity. Pharmacogenetics 2000;10:705-13.

Pirmohamed M, Friedmann PS, Molokhia M, et al. Phenotype standardization for immune-mediated drug-induced skin injury. Clin Pharmacol Ther 2011b;89:896-901.

Pirmohamed M, James S, Meakin S, et al. Adverse drug reactions as cause of admission to hospital: prospective analysis of 18,820 patients. BMJ 2004;329:15-9.

Power WJ, Saidman SL, Zhang DS, et al. *HLA* typing in patients with ocular manifestations of Stevens-Johnson syndrome. Ophthalmology 1996;103:1406-9.

Quinones-Lombrana A, Lopez-Soto A, Ballina-Garcia FJ, et al. BAT1 promoter polymorphism is associated with rheumatoid arthritis susceptibility. J Rheumatol 2008;35:741-4.

Ramasamy SN, Korb-Wells CS, Kannangara DR, et al. Allopurinol hypersensitivity: a systematic review of all published cases, 1950-2012. Drug Saf 2013;36:953-80.

Rattanavipapong W, Koopitakkajorn T, Praditsitthikorn N, et al. Economic evaluation of *HLA-B*15:02* screening for carbamazepine-induced severe adverse drug reactions in Thailand. Epilepsia 2013;54:1628-38.

Rauch A, Nolan D, Martin A, et al. Prospective genetic screening decreases the incidence of abacavir hypersensitivity reactions in the Western Australian HIV cohort study. Clin Infect Dis 2006;43:99-102.

Relling MV, Klein TE. CPIC: Clinical Pharmacogenetics Implementation Consortium of the Pharmacogenomics Research Network. Clin Pharmacol Ther 2011;89:464-7.

Riedl MA, Casillas AM. Adverse drug reactions: types and treatment options. Am Fam Physician 2003;68:1781-90.

Rodriguez-Novoa S, Cuenca L, Morello J, et al. Use of the HCP5 single nucleotide polymorphism to predict hypersensitivity reactions to abacavir: correlation with HLA-B*5701. J Antimicrob Chemother 2010;65:1567-9.

Romano A, De Santis A, Romito A, et al. Delayed hypersensitivity to aminopenicillins is related to major histocompatibility complex genes. Ann Allergy Asthma Immunol 1998;80:433-7.

Romano A, Gueant-Rodriguez RM, Viola M, et al. Cross-reactivity among drugs: clinical problems. Toxicology 2005;209:169-79.

Romano A, Pettinato R, Andriolo M, et al. Hypersensitivity to aromatic anticonvulsants: in vivo and in vitro cross-reactivity studies. Curr Pharm Des 2006;12:3373-81.

Roujeau JC, Bracq C, Huyn NT, et al. HLA phenotypes and bullous cutaneous reactions to drugs. Tissue Antigens 1986;28:251-4.

Roujeau JC, Chosidow O, Saiag P, et al. Toxic epidermal necrolysis (Lyell syndrome). J Am Acad Dermatol 1990a;23(6 pt 1):1039-58.

Roujeau JC, Guillaume JC, Fabre JP, et al. Toxic epidermal necrolysis (Lyell syndrome). Incidence and drug etiology in France, 1981-1985. Arch Dermatol 1990b;126:37-42.

Roujeau JC, Huynh TN, Bracq C, et al. Genetic susceptibility to toxic epidermal necrolysis. Arch Dermatol 1987;123:1171-3.

Roujeau JC, Stern RS. Severe adverse cutaneous reactions to drugs. N Engl J Med 1994;331:1272-85.

Rzany B, Correia O, Kelly JP, et al. Risk of Stevens-Johnson syndrome and toxic epidermal necrolysis during first weeks of antiepileptic therapy: a case-control study. Study Group of the International Case Control Study on Severe Cutaneous Adverse Reactions. Lancet 1999;353:2190-4.

Saag M, Balu R, Phillips E, et al. High sensitivity of human leukocyte antigen-b*5701 as a marker for immunologically confirmed abacavir hypersensitivity in white and black patients. Clin Infect Dis 2008;46:1111-8.

Saokaew S, Tassaneeyakul W, Maenthaisong R, et al. Cost-effectiveness analysis of HLA-B*5801 testing in preventing allopurinol-induced SJS/TEN in Thai population. PLoS One 2014;9:e94294.

Schackman BR, Scott CA, Walensky RP, et al. The cost-effectiveness of HLA-B*5701 genetic screening to guide initial antiretroviral therapy for HIV. AIDS 2008;22:2025-33.

Schnyder B, Mauri-Hellweg D, Zanni M, et al. Direct, MHC-dependent presentation of the drug sulfamethoxazole to human alphabeta T cell clones. J Clin Invest 1997;100:136-41.

Seitz CS, Pfeuffer P, Raith P, et al. Anticonvulsant hypersensitivity syndrome: cross-reactivity with tricyclic antidepressant agents. Ann Allergy Asthma Immunol 2006;97:698-702.

Shapiro M, Ward KM, Stern JJ. A near-fatal hypersensitivity reaction to abacavir: case report and literature review. AIDS Read 2001;11:222-6.

Shiohara T, Iijima M, Ikezawa Z, et al. The diagnosis of a DRESS syndrome has been sufficiently established on the basis of typical clinical features and viral reactivations. Br J Dermatol 2007;156:1083-4.

Shiohara T, Inaoka M, Kano Y. Drug-induced hypersensitivity syndrome (DIHS): a reaction induced by a complex interplay among herpesviruses and antiviral and antidrug immune responses. Allergol Int 2006;55:1-8.

Shirato S, Kagaya F, Suzuki Y, et al. Stevens-Johnson syndrome induced by methazolamide treatment. Arch Ophthalmol 1997;115:550-3.

Shoenbill K, Fost N, Tachinardi U, et al. Genetic data and electronic health records: a discussion of ethical, logistical and technological considerations. J Am Med Inform Assoc 2014;21:171-80.

Singer JB, Lewitzky S, Leroy E, et al. A genome-wide study identifies HLA alleles associated with lumiracoxib-related liver injury. Nat Genet 2010;42:711-4.

Somkrua R, Eickman EE, Saokaew S, et al. Association of HLA-B*5801 allele and allopurinol-induced Stevens Johnson syndrome and toxic epidermal necrolysis: a systematic review and meta-analysis. BMC Med Genet 2011;12:118.

Sontheimer RD, Houpt KR. DIDMOHS: a proposed consensus nomenclature for the drug-induced delayed multiorgan hypersensitivity syndrome. Arch Dermatol 1998;134:874-6.

Sugimoto Y, Mizutani H, Sato T, et al. Toxic epidermal necrolysis with severe gastrointestinal mucosal cell death: a patient who excreted long tubes of dead intestinal epithelium. J Dermatol 1998;25:533-8.

Symonds W, Cutrell A, Edwards M, et al. Risk factor analysis of hypersensitivity reactions to abacavir. Clin Ther 2002;24:565-73.

Tangamornsuksan W, Chaiyakunapruk N, Somkrua R, et al. Relationship between the HLA-B*1502 allele and carbamazepine-induced Stevens-Johnson syndrome and toxic epidermal necrolysis: a systematic review and meta-analysis. JAMA Dermatol 2013;149:1025-32.

Tassaneeyakul W, Jantararoungtong T, Chen P, et al. Strong association between HLA-B*5801 and allopurinol-induced Stevens-Johnson syndrome and toxic epidermal necrolysis in a Thai population. Pharmacogenet Genomics 2009;19:704-9.

Tassaneeyakul W, Tiamkao S, Jantararoungtong T, et al. Association between HLA-B*1502 and carbamazepine-induced severe cutaneous adverse drug reactions in a Thai population. Epilepsia 2010;51:926-30.

Tennis P, Stern RS. Risk of serious cutaneous disorders after initiation of use of phenytoin, carbamazepine, or sodium valproate: a record linkage study. Neurology 1997;49:542-6.

Then SM, Mohd Rani ZZ, Raymond AA, Jamal R. Pharmacogenomics screening of HLA-B*1502 in epilepsy patients: how we do it in the UKM Medical Centre, Malaysia. Neurol Asia 2013;18(suppl 1):27.

Then SM, Rani ZZ, Raymond AA, et al. Frequency of the HLA-B*1502 allele contributing to carbamazepine-induced hypersensitivity reactions in a cohort of Malaysian epilepsy patients. Asian Pac J Allergy Immunol 2011;29:290-3.

Tohkin M, Kaniwa N, Saito Y, et al. A whole-genome association study of major determinants for allopurinol-related Stevens-Johnson syndrome and toxic epidermal necrolysis in Japanese patients. Pharmacogenomics J 2013;13:60-9.

U.S. Food and Drug Administration (FDA). 2008. Drugs: Phenytoin and Fosphenytoin Information. Available at www.fda.gov/Drugs/DrugSafety/PostmarketDrugSafetyInformationforPatientsandProviders/ucm110259.htm. Accessed October 22, 2014.

Vitezica ZG, Milpied B, Lonjou C, et al. *HLA-DRB1*01* associated with cutaneous hypersensitivity induced by nevirapine and efavirenz. AIDS 2008;22:540-1.

Wang Q, Zhou JQ, Zhou LM, et al. Association between *HLA-B*1502* allele and carbamazepine-induced severe cutaneous adverse reactions in Han people of southern China mainland. Seizure 2011;20:446-8.

Wei CY, Chung WH, Huang HW, et al. Direct interaction between *HLA-B* and carbamazepine activates T cells in patients with Stevens-Johnson syndrome. J Allergy Clin Immunol 2012a;129:1562-9.e5.

Wei CY, Lee MT, Chen YT. Pharmacogenomics of adverse drug reactions: implementing personalized medicine. Hum Mol Genet 2012b;21(R1):R58-65.

Wester K, Jonsson AK, Spigset O, et al. Incidence of fatal adverse drug reactions: a population-based study. Br J Clin Pharmacol 2008;65:573-9.

Wu XT, Hu FY, An DM, et al. Association between carbamazepine-induced cutaneous adverse drug reactions and the *HLA-B*1502* allele among patients in central China. Epilepsy Behav 2010;19:405-8.

Yang CW, Hung SI, Juo CG, et al. *HLA-B*1502*-bound peptides: implications for the pathogenesis of carbamazepine-induced Stevens-Johnson syndrome. J Allergy Clin Immunol 2007;120:870-7.

Yip VL, Marson AG, Jorgensen AL, et al. *HLA* genotype and carbamazepine-induced cutaneous adverse drug reactions: a systematic review. Clin Pharmacol Ther 2012;92:757-65.

Zanni MP, von Greyerz S, Schnyder B, et al. *HLA*-restricted, processing- and metabolism-independent pathway of drug recognition by human alpha beta T lymphocytes. J Clin Invest 1998;102:1591-8.

Zhang Y, Wang J, Zhao LM, et al. Strong association between *HLA-B*1502* and carbamazepine-induced Stevens-Johnson syndrome and toxic epidermal necrolysis in mainland Han Chinese patients. Eur J Clin Pharmacol 2011;67:885-7.

Zhou K, Pearson ER. Insights from genome-wide association studies of drug response. Annu Rev Pharmacol Toxicol 2013;53:299-310.

Zucman D, Truchis P, Majerholc C, et al. Prospective screening for human leukocyte antigen-*B*5701* avoids abacavir hypersensitivity reaction in the ethnically mixed French HIV population. J Acquir Immune Defic Syndr 2007;45:1-3.

Chapter 14

PHARMACOGENETICS OF HEPATITIS C TREATMENT

MARINA KAWAGUCHI-SUZUKI, PHARM.D., BCPS, BCACP; DAVID R. NELSON, M.D.;
AND REGINALD F. FRYE, PHARM.D., PH.D., FCCP

Learning Objectives

1. Assess the clinical utility of polymorphisms shown to predict response to HCV therapy.
2. Evaluate the use of different HCV treatment options based on pharmacogenetic data.
3. Apply current pharmacogenetic evidence for the treatment of infection with different HCV genotypes.
4. Apply current available pharmacogenetic evidence for the care of special populations infected with HCV.
5. Justify the best clinical practice of HCV treatment by applying pharmacogenetic information.

Keywords: Hepatitis C virus, peginterferon alfa, ribavirin, sofosbuvir, simeprevir, boceprevir, telaprevir, IL-28B, IFNL3

Abbreviations in This Chapter

AASLD	American Association for the Study of Liver Disease	ISDR	interferon-sensitivity determining region
BOC	boceprevir	ISG	interferon-stimulated gene
CYP	cytochrome P450	ITPA	inosine triphosphatase
DAA	direct-acting antiviral	MAP	mitogen-activated protein
FDA	Food and Drug Administration	NS	nonstructural
HCV	hepatitis C virus	OR	odds ratio
HIV	human immunodeficiency virus	PEG	peginterferon alfa-2a or 2b
JAK-STAT	janus kinase-signal transducer and activator of transcription	RNA	ribonucleic acid
		RBV	ribavirin
IDSA	Infectious Disease Society of America	RVR	rapid virologic response
		SMV	simeprevir
IFNL3	Interferon lambda 3	SNP	single nucleotide polymorphism
IRRDR	interferon/ribavirin resistance determining region	SOF	sofosbuvir
		SVR	sustained virologic response
IAS-USA	International Antiviral Society-United States of America	TVR	telaprevir
		U.S.	United States

Abstract

The treatment strategy for hepatitis C virus (HCV) infection includes peginterferon alfa and ribavirin dual therapy, as well as direct-acting antivirals (DAAs). This chapter covers both first generation DAAs, boceprevir and telaprevir, as well as the second generation DAAs, simeprevir and sofosbuvir. Boceprevir, telaprevir, and simeprevir are nonstructural 3/4A serine protease inhibitors, whereas sofosbuvir is a nonstructural 5B polymerase inhibitor. The clinical guideline by the American Association for the Study of Liver Disease and the Infectious Disease Society of America recommends regimens including sofosbuvir as first-line therapy. The HCV is categorized into genotypes 1 to 7 and further into subtypes (e.g., 1a, 1b, etc.). In the United States, HCV genotype 1 is most commonly observed, followed by genotypes 2 and 3. The HCV genotype is essential for selecting the most appropriate treatment and duration of therapy. In addition, treatment response depends on the HCV genotype. With the emergence and use of DAAs, polymorphisms in HCV ribonucleic acid have been reported and associate with possible resistance to treatment, especially to the protease inhibitors. Polymorphisms in the Interferon lambda 3 (*IFNL3*) gene, formerly known as *IL28B*, have been shown to predict treatment response after peginterferon alfa and ribavirin dual therapy in HCV genotype 1 infection. This finding has been replicated in HCV non-genotype 1 infection, co-infections with hepatitis B virus or with human immunodeficiency virus, and re-infection of HCV after liver transplantation. Although the strength of the association between *IFNL3* genotypes and treatment response seems to lessen with the use of DAAs, the genotype is still believed to be informative to guide treatment decisions. A recently discovered polymorphism in Interferon lambda 4 (*IFNL4*) may also have implications for HCV treatment. It is important to consider both HCV and host genomes to apply pharmacogenetic information in HCV treatment, and polymorphisms in both genomes are discussed in this chapter.

Introduction: HCV Infection and Pharmacotherapy

According to the World Health Organization, 3% of the world's population or 170 million people are chronically infected with hepatitis C virus (HCV) (WHO 2002). The Centers for Disease Control and Prevention currently estimates that in the United States (U.S.), HCV chronically infects 3.2 million people and that the infection is most prevalent among those who were born during 1945–1965 (CDC 2013). The majority of HCV infection likely occurred during the 1970s and 1980s when the infection rates were highest (CDC 2013). The HCV infection is known to become chronic in 75%–85% of cases and is the most common blood-borne infection (CDC 2013).

The goals of treatment are to achieve eradication of the virus and to prevent liver-related complications and death. The complications of HCV infection include cirrhosis, hepatocellular carcinoma, and liver failure; HCV infection is the leading indication for liver transplant in the U.S. (CDC 2013; Davis 2003). Eradication of HCV or cure of HCV infection is indicated clinically as sustained virologic response (SVR), which was previously defined as undetectable HCV ribonucleic acid (RNA) 24 weeks after the discontinuation of treatment (Ghany 2009). More recently, other time frames are considered, such as the SVR at 12 weeks (or 48 weeks) (AASLD 2014). With current therapy which lasts for 12–24 weeks, SVR can be defined if HCV RNA is undetectable (or less than 15 IU/mL) at 12 or more weeks after completing treatment (AASLD 2014).

Dual therapy with peginterferon alfa-2a or 2b (PEG) and ribavirin (RBV) had been the standard of care for the past decade (Pacanowski 2012). The treatment duration was considerably longer at up to 72 weeks in HCV genotype-1 (HCV-1) infection and up to 24 weeks in HCV-2/3 infection (Ghany 2009). The rates of SVR with the conventional dual therapy were approximately 40% and 50% for HCV genotype-1 (HCV-1) infection in the U.S. and in Western Europe, respectively (Jacobson 2012). The SVR rates were about 80% with dual therapy for HCV-2/3 infection, while the SVR rates for HCV-4 infection were similar to or slightly higher than those for HCV-1 infection (Jacobson 2012). Patients who did not achieve SVR after previous treatment are categorized into relapsers or non-responders. Relapsers are patients who did not have any detectable HCV RNA during the previous treatment, but relapsed after the cessation of therapy (AASLD 2014). Non-responders are further classified into: (a) partial responders, who had a greater than or equal to $2\log_{10}$ IU/mL response in HCV RNA, but the virus remained detectable up to 24 weeks or at the end of treatment; and (b) null responders, who did not have the greater than or equal to $2\log_{10}$ IU/mL response in HCV RNA by week 12 during previous treatment (AASLD 2014).

In 2011, the first generation of direct-acting antivirals (DAAs), which includes boceprevir (BOC) and telaprevir (TVR), was approved in the U.S. (Pacanowski 2012). The American Association for the Study of Liver Disease (AASLD) practice guideline for treatment of HCV-1 infection was updated to include BOC or TVR as part of "triple therapy" in which the DAA was combined with the standard PEG/RBV dual therapy (Ghany 2011). Both BOC and TVR are "protease inhibitors" as they inhibit nonstructural (NS) 3/4A serine protease (Ghany 2011). The addition of BOC or TVR as triple therapy significantly improved the SVR rate from 40%–50% to approximately 80% (Ghany 2011). Although treatment outcomes were improved, BOC and TVR treatment regimens were complex, as both required multiple daily doses with food (Maasoumy 2013; Jesudian 2013), and associated with many adverse effects, including anemia and dysgeusia with BOC and rash, pruritus, and anorectal discomfort with TVR. The significant potential for drug-drug interactions added further to the major limitations of first generation DAA triple therapies (Maasoumy 2013; Jesudian 2013).

In 2013, the second-generation DAAs, simeprevir (SMV) and sofosbuvir (SOF), were approved (Scheel 2013; Vaidya 2013; Keating 2014). SMV is another serine protease inhibitor indicated for HCV-1 infection (Vaidya 2013). In treatment-naïve and prior relapser patients, simeprevir is given with PEG/RBV as triple therapy for 12 weeks, then treatment with PEG/RBV is continued for another 12 weeks for a total treatment duration of 24 weeks (Janssen Therapeutics 2013). Prior partial and null responders are treated with PEG/RBV therapy for 36 weeks after the initial 12-week SMV triple therapy for a total duration of 48 weeks (Janssen Therapeutics 2013). The adverse effects attributed to SMV are rash including occasional photosensitivity, pruritus, nausea, and dyspnea (Vaidya 2013). Because SMV is primarily metabolized by cytochrome P450 (CYP) 3A, drug interactions remain a concern (Vaidya 2013). However, SMV is dosed once daily, and SMV triple therapy was able to achieve SVR rates of about 80% in phase 3 trials (QUEST 1 and QUEST 2) (Vaidya 2013). The shorter therapy duration for treatment-naïve patients and relapsers and simpler dosing regimen are advantages that make SMV preferable to BOC and TVR when used with interferon-based regimens (Vaidya 2013).

Unlike SMV and first-generation DAAs, SOF is a nucleotide NS5B polymerase inhibitor (Keating 2014). Sofosbuvir is considered essential in current HCV treatment regimens for several reasons. First, SOF is believed to have HCV pan-genotypic activity and therefore can be used not only with HCV-1 infection, but also other HCV genotype infections (AASLD 2014; Asselah 2014). Second, PEG-free regimens officially became available with the approval of SOF for patients infected with HCV-2/3 (AASLD 2014; Gilead Sciences, Inc. 2013). It is important to note that SOF is still combined with PEG/RBV as triple therapy for HCV infections with genotypes 1 and 4-6 (AASLD 2014; Gilead Sciences, Inc. 2013). For interferon-ineligible patients, the clinical guideline recommends a combination of SOF and SMV with or without RBV in HCV-1 infection and combination of SOF and RBV in HCV-4 infection (AASLD 2014). Third, the treatment duration was shortened considerably to 12 weeks for all HCV genotypes except HCV-3, for which 24-week treatment is recommended (AASLD 2014; Gilead Sciences, Inc. 2013). Fourth, superior efficacy was seen with SOF (Keating 2014). In the phase 3 NEUTRINO trial, SOF triple therapy achieved an SVR12 rate of 91% in treatment-naïve patients infected with HCV-1/4/5/6 (Keating 2014; Lawitz 2013). The SVR rates with the oral-only SOF and RBV therapy were 93% in HCV-2 infection and 85% in HCV-3 infection (Zeuzem 2014). Finally, SOF is well-tolerated (i.e., no significant toxicity added to PEG/RBV or RBV therapy) and is dosed orally once daily without regard to food (Keating 2014; Asselah 2014). Sofosbuvir is considered a pro-drug and undergoes extensive hepatic metabolism to GS-461203, which is the pharmacologically active nucleoside analog triphosphate (Keating 2014; Gilead Sciences, Inc. 2013). Sofosbuvir is not metabolized by CYP enzymes and is not associated with clinically relevant CYP-mediated drug interactions. However, potential interactions with potent intestinal P-glycoprotein inducers are noted in the prescribing information (Keating 2014; Asselah 2014; Gilead Sciences, Inc. 2013).

In January 2014, after approval of the second-generation DAAs, the AASLD, Infectious Disease Society of America (IDSA), and International Antiviral Society-United States of America (IAS-USA) published an updated clinical guideline (AASLD 2014). According to the recommendations by the AASLD/IDSA/IAS-USA (Figure 14.1 and Figure 14.2), combination therapy including SOF is considered first-line in both treatment-naïve and treatment-experienced patients (AASLD 2014).

Hepatitis C Virus RNA

The hepatitis C virus is a 9,600-nucleotide positive-strand RNA virus that encodes polyprotein processed to structural (core, E1, and E2) and nonstructural (p7, NS2, NS3, NS4A, NS5A, and NS5B) proteins (Figure 14.3) (Scheel 2013). Hepatitis C virus isolates are classified into seven genotypes, 1–7 with ~70% sequence similarity, and further into a number of subtypes (e.g., 1a, 1b, etc., with ~80% sequence similarity) (Scheel 2013). The differences in HCV genotypes are observed based on geographic locations. In Americas, Europe, and Japan, 70%, 75%, and 50%–70% of cases are HCV-1 infection, respectively, while HCV-2/3 infection is also observed in these regions (Scheel 2013). Hepatitis C virus-4/5 infections are most commonly seen in Africa, but are also spreading to Europe, whereas HCV-3/6 infections are mostly found in South and Southeast Asia (Scheel 2013). Hepatitis C virus-7 was recently reported in

Central Africa, but it is not yet of major clinical importance (Scheel 2013). Hepatitis C virus genotyping is an important part of initial patient workup and an essential factor in selecting appropriate treatment and therapy duration, as well as in predicting the treatment response (AASLD 2014). In the U.S., the most common HCV genotype is genotype 1, which has also been considered the most difficult HCV genotype to treat.

The three regions of HCV associated with treatment outcome with PEG/RBV dual therapy are: (1) the core region, (2) the interferon-sensitivity determining region or ISDR, and (3) interferon/RBV resistance-determining region or IRRDR (Kawaguchi-Suzuki 2014). The ISDR and IRRDR are located in the NS5A region (Kawaguchi-Suzuki 2014). Amino acid substitutions at positions 70 and 91 of the core region have been associated with treatment resistance to PEG/RBV, whereas HCV with greater numbers of mutations found in the ISDR and IRRDR regions are generally considered to be more sensitive to PEG/RBV therapy (Kawaguchi-Suzuki 2014).

With the emergence of DAAs, increasing HCV resistance with viral mutations has been reported, especially with NS3/4A protease inhibitors. The mutations associated with HCV viral resistance to currently approved NS3/4A protease inhibitors are summarized in Table 14.1 (Wu 2013; Salvatierra 2013). As suggested in Table 14.1, cross-resistance is a major concern with first-generation DAAs. With SMV, mutations at NS3 positions 80, 122, 155, and/or 168 have been associated with viral breakthrough and relapse after therapy (Salvatierra 2013). Specifically, Q80K polymorphism is considered clinically important in HCV-1a infection when SMV+PEG+RBV combination is considered for treatment. Routine testing for other resistance-associated variants during or after therapy is not currently recommended for clinical practice (AASLD 2014). However, baseline testing for Q80K polymorphism is incorporated in the AASLD/IDSA/IAS-USA guideline recommendations and SMV prescribing information (AASLD 2014; Janssen Therapeutics 2013).

In the COSMOS phase 2 trial investigating SOF (400 mg daily) and SMV (150 mg daily) with or without RBV, viral relapse (detection of HCV after stopping therapy) was reported in patients infected with HCV-1a with the Q80K polymorphism (AASLD 2014). However, the SVR rate was high (~90%) in patients with HCV-1a and Q80K variant (AASLD 2014). Therefore, according to the AASLD/IDSA/IAS-USA, "the Q80K testing can be considered but not strongly recommended" for SOF+SMV±RBV combination (AASLD 2014). Virologic failure has not been observed yet in patients with HCV-1b and with HCV-1a in the absence of the Q80K polymorphism (AASLD 2014).

In the two phase 3 trials of SMV (150 mg daily) for the first 12 weeks with PEG/RBV for a total of 24 weeks, the overall SVR12 in the subgroup of treatment-naïve patients with the Q80K variant was 58%, compared to 84% in the

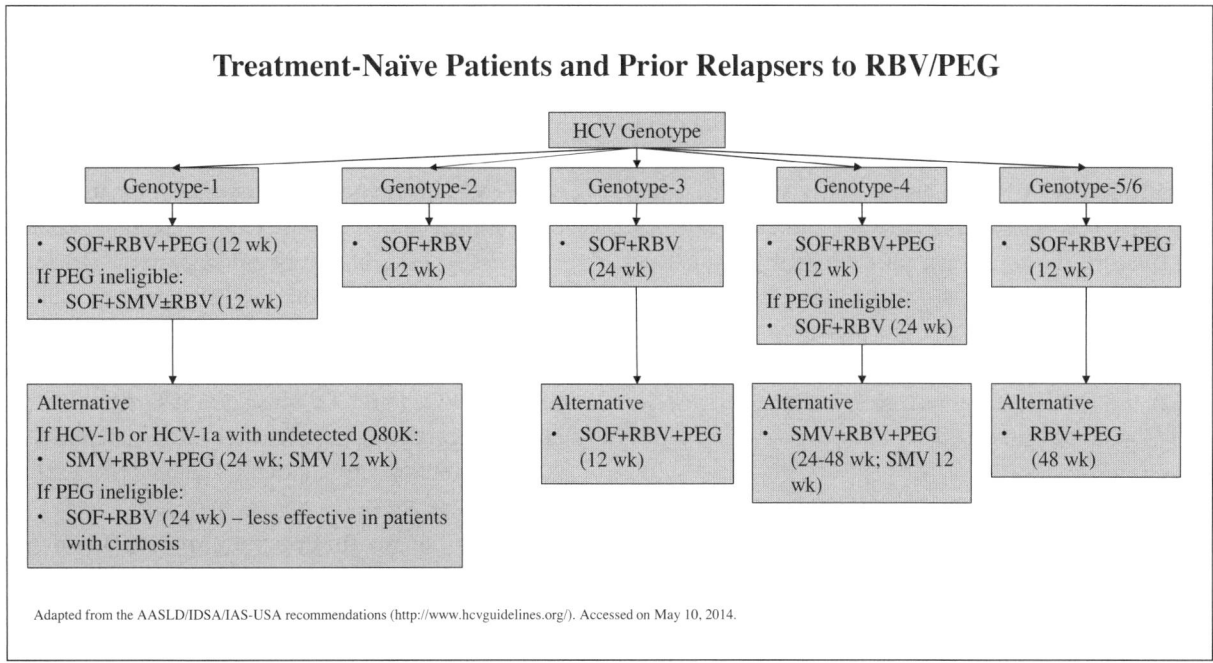

Figure 14.1. Recommended treatment by the AASLD/IDSA/IAS-USA clinical guideline for treatment-naïve patients and prior relapsers to PEG/RBV (AASLD 2014).

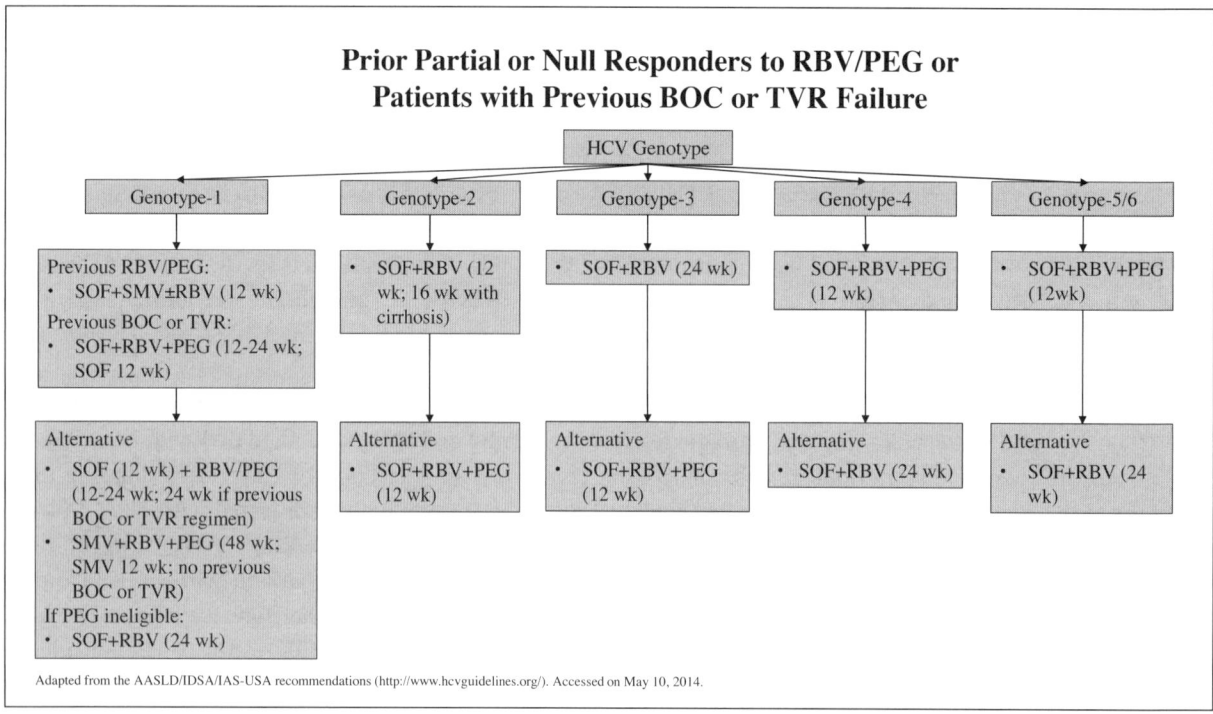

Figure 14.2. Recommended treatment by the AASLD/IDSA/IAS-USA clinical guideline for prior partial or null responders to PEG/RBV or patients with previous BOC or TVR failure (AASLD 2014).

patients without the Q80K polymorphism (Vaidya 2013; Jacobson 2014; Manns 2014). Furthermore, the SVR12 rate in these treatment-naïve patients with the Q80K variant was not better than the placebo plus PEG/RBV arm on which the SVR12 rate was 47% (Vaidya 2013; Jacobson 2014; Manns 2014). In the phase 3 PROMISE trial in which previous relapsers received SMV or placebo in addition to PEG/RBV, the rates of SVR12 with SMV were 78% without Q80K and 47% with Q80K in HCV-1a infection, and the SVR12 rate was 28% in the placebo arm with HCV-1a infection (Janssen Therapeutics 2013; Forns 2014). The AASLD/IDSA/IAS-USA guideline states that "baseline testing for Q80K is recommended for all patients before treatment with the simeprevir plus PEG/RBV regimen is initiated" (AASLD 2014). The observed prevalence of the Q80K variant during clinical trials of SMV was 35% (48% with HCV-1a and 0% with HCV-1b) (Janssen Therapeutics 2013). If the Q80K polymorphism is present, alternative therapy should be considered.

Host Genome: IFNL3 (IL28B) Gene

Background Information

Several genome-wide association studies have found and replicated that two single nucleotide polymorphisms (SNPs) near the *IFNL3* gene, previously known as the *IL28B* gene, predict SVR with PEG/RBV therapy (Table 14.2). Among genome-wide association studies and candidate gene studies, the favorable genotype is considered to be rs12979860 CC or rs8099917 TT, and the unfavorable response genotype is rs12979860 CT/TT or rs8099917 TG/GG (Kawaguchi-Suzuki 2014). The response to PEG/RBV in HCV-1 infection is predicted to be about 80% versus 20% on the basis of this genetic polymorphism at *IFNL3*. The *IFNL3* SNP is widely recognized as one of the strongest predictors of response to PEG/RBV in HCV treatment (Pacanowski 2012). The FDA acknowledged early that *IFNL3* genotype would be critical for the evaluation of treatment modalities under investigation (Pacanowski 2012). In the FDA 2010 draft guidance to the industry, routine genotyping of patients for *IFNL3* during clinical trials of HCV treatment was recommended (Pacanowski 2012). In 2011, the *IFNL3* genotype information was included in the PEG, BOC, and TVR package inserts, as well as the prescribing information of the second-generation DAAs in 2013 (Pacanowski 2012; Janssen Therapeutics 2013; Gilead Sciences, Inc. 2013).

In addition to findings in chronic HCV infection, both *IFNL3* rs12979860 and rs8099917 SNPs were demonstrated to predict spontaneous clearance in acute HCV infection (Matsuura 2014; Muir 2013). Among genome-wide association studies and candidate gene studies, spontaneous clearance is 2–4 times more likely with the

Table 14.1. Amino Acid Variations Implicated in HCV Viral Resistance to NS3/4A Protease Inhibitors (Janssen Therapeutics 2013; Wu 2013; Salvatierra 2013)

DAA	Amino Acid Variations Related to the DAA Resistance
First generation NS3/4A protease inhibitor	
Boceprevir	(NS3) V36M/A/L, T54S/A, V55A, R155K/T, A156S, V158I, V170A, I170T
Telaprevir	(NS3) V36M/A, T54A, R155K/T, A156V/T/S, V36M/A+R155K/T, V36M/A+A156V/T
Second generation NS3/4A protease inhibitor	
Simeprevir	(NS3) F43, Q80K/R, S122A/G/I/R/T, R155K/T/Q, A156S/T/V, D168A/E/F/H/T/V, I170T

favorable response genotype (Matsuura 2014). Therefore, *IFNL3* genotyping can be informative not only in chronic infection but also in acute infection.

The gene I*FNL3* is found on chromosome 19q.13.13, and SNPs rs12979860 and rs8099917 are located 3kb and 8kb upstream of this gene (Figure 14.7) (Arnaud 2014). *IFNL3* encodes interferon-λ 3, which is a member of the type 3 interferon-λ family known to possess antiviral, antiproliferative, and immune-modulatory properties (Kotenko 2011). Based on the location of the polymorphisms relative to the gene, the variants are likely to affect the expression of interferon-λ. Both interferon-λ and interferon-α induce the common janus kinase-signal transducer and activator of transcription (JAK STAT) and mitogen-activated protein (MAP) kinase pathways, leading to expression of interferon-stimulated genes (ISGs) and eventually to antiviral activity (Kotenko 2011; O'Brien 2009). The proposed mechanism by which the *IFNL3* polymorphism influences the response to HCV treatment involves the difference in baseline expression of ISGs between the *IFNL3* genotypes (Cariani 2011; Urban 2010; Honda 2010). The favorable *IFNL3* genotype is associated with low ISG expression at baseline, which makes greater ISG induction possible with HCV interferon-based treatment, resulting in better antiviral activity (Cariani 2011; Urban 2010; Honda 2010). In contrast, the unfavorable genotype is associated with high ISG expression at baseline, which makes the induction of ISG more difficult with the administration of HCV treatment (Cariani 2011; Urban 2010; Honda 2010). However, it remains controversial whether this mechanism fully explains the genetic effect because some studies report that *IFNL3* genotype may affect treatment outcome independent of ISG expression and may also relate to innate immune response (Muir 2013; Naggie 2012).

IFNL3 Polymorphism and Response to PEG/RBV

The association between *IFNL3* genotype and PEG/RBV treatment outcome has been demonstrated in several clinical studies. According to a meta-analysis, rs12979860 CC and rs8099917 TT genotypes independently predicted SVR in patients infected with HCV-1 and treated with PEG/RBV dual therapy (Luo 2013). The odds ratios (OR) from this meta-analysis were 4.47 with rs12979860 (95% confidence interval [95%CI] 3.81-5.25) and 5.17 with rs8099917 (95%CI 4.37-6.12) (Luo 2013). The majority of current evidence supports the *IFNL3* genetic effect to predict in HCV-1 treatment response with PEG/RBV.

Treatment outcomes are generally better in HCV-2/3 infection than in HCV-1 infection. Although the evidence is not entirely consistent, *IFNL3* genotyping is not

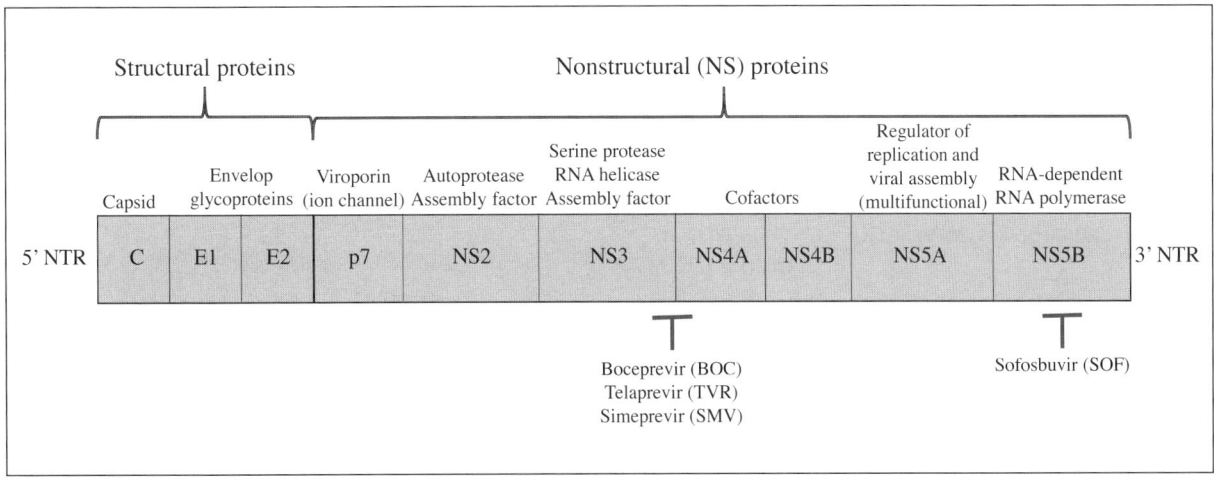

Figure 14.3. Hepatitis C virus genome structure (Scheel 2013; Asselah 2014; Salvatierra 2013).

recommended as a means to predict SVR in HCV-2/3 infection (Kawaguchi-Suzuki 2014; Muir 2013). According to a meta-analysis, the favorable response genotype was significantly associated with RVR in Caucasians (OR 1.82, 95%CI 1.12-2.96, P=0.02) and in Asians (OR 2.39, 95%CI 1.39-4.11, P=0.002) and with SVR in subgroup of Caucasians who did not show RVR (OR 3.29, 95%CI 1.67-6.51, P=0.001) (Rangnekar 2013).

Hepatitis C virus-4 infection has been considered more similar to HCV-1 than to HCV-2/3 in terms of treatment outcome with PEG/RBV. *IFNL3* genotyping can be considered to predict SVR with PEG/RBV therapy in HCV-4 infection (Kawaguchi-Suzuki 2014). However, sufficient data are not available in HCV-5/6 infection to make any definitive conclusions.

IFNL3 Polymorphism and Response to 1st-generation DAAs

Sustained virologic response rates based on *IFNL3* rs12979860 genotype from phase 3 clinical trials are shown in Figures 14.4 and 14.5 for the first-generation DAAs (BOC and TVR respectively).

SPRINT-2 is a phase 3 trial of BOC triple therapy in treatment-naïve patients (Poordad 2011A). In this study, patients with the *IFNL3* unfavorable response genotype benefited the most by the addition of BOC (Poordad 2011B; Clark 2012). However, the improvement in the SVR rate by BOC, compared to the dual therapy arm, was minimal in patients who had the *IFNL3* favorable response genotype (Poordad 2011B; Clark 2012). However, the *IFNL3* genotype was associated with shortened therapy; 90% of patients with the favorable response genotype were eligible for shortened 28-week treatment (Poordad 2011B; Clark 2012).

RESPOND-2 is a trial of BOC triple therapy in prior relapsers and partial responders, but null responders who had less than $2\log_{10}$ drop in HCV RNA at week 12 were excluded (Bacon 2011). In this study, major improvement in SVR rates were seen irrespective of *IFNL3* genotype (Poordad 2011B; Clark 2012). In addition, patients with the favorable response genotype were more likely to be eligible for the 36-week shortened therapy (Poordad 2011B; Clark 2012).

The ADVANCE trial evaluated TVR triple therapy in treatment-naïve patients (Jacobson 2011A). The addition of TVR improved SVR rates across the *IFNL3* genotype, and greater than a 2-fold increase in the SVR rate was seen in patients with unfavorable response genotype (Clark 2012; Jacobson 2011B). The rates of extended rapid virologic response, defined as undetectable HCV RNA at weeks 4 and 12, were 78%, 57%, and 45% for *IFNL3* rs12979860 CC, CT, and TT genotype, respectively, suggesting that eligibility for shortened therapy was also associated with the *IFNL3* genotype (Jacobson 2011B). A multivariate analysis of TVR triple therapy showed that *IFNL3* rs8099917 TT and rapid virological response (RVR), defined as undetectable HCV RNA at week 4, were independent predictors of SVR (Furusyo 2013).

The REALIZE trial examined TVR triple therapy in treatment-experienced patients, including prior relapsers, partial responders, and null responders (Zeuzem 2011). Improvements in the SVR rates by 40%-50% with TVR addition were observed irrespective of the *IFNL3* genotype (Pol 2011).

Overall, the utility of *IFNL3* genotyping in BOC or TVR triple therapy was convincing by evaluating the available data. A meta-analysis of PEG/RBV plus BOC or TVR therapy showed that patients with rs12979860 CC

Table 14.2. Results of Early Genome-Wide Association Studies (N = total number of patients)

Study	SNP	Population	N	HCV	OR	P-value
Ge (2009)	rs12979860 CC	White, African American, Hispanic	1,137	1	3.1	1.21×10^{-28}
McCarthy (2010)	rs12979860 CC	White, African American	231	1-3	5.8	9×10^{-6}
Thompson (2010)	rs12979860 CC	White, African American, Hispanic	1,671	1	5.2	$<1 \times 10^{-4}$
Suppiah (2009)	rs8099917	White	293	1	1.98	7.06×10^{-8}
Tanaka (2009)	rs8099917	Japanese	142	1	12.1	3.11×10^{-15}
Rauch (2010)	rs8099917	Swiss	465	1-4	5.2	5.47×10^{-8}

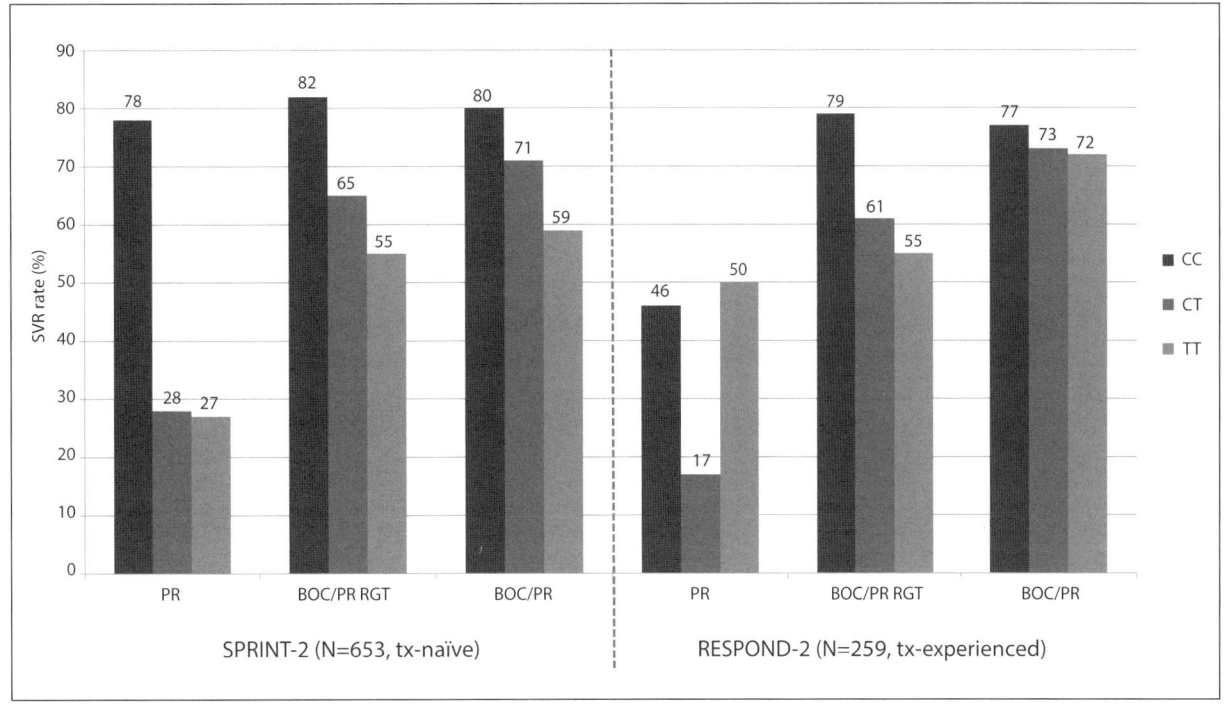

Figure 14.4. SVR rates (%) by IFNL3 genotype (rs12979860) with BOC therapy (Poordad 2011A, B; Clark 2012; Bacon 2011). (N=total number of patients. P=PEG, R=RBV, tx=treatment)

genotype had higher likelihood of achieving SVR irrespective of their prior treatment status (OR 3.91; 95%CI 2.11-7.28; p<0.0001) (Bota 2013).

IFNL3 Polymorphism and Response to 2nd-generation DAAs

Sustained virologic response rates based on *IFNL3* rs12979860 genotype from phase 3 clinical trials are shown in Figure 14.6 for SMV. The QUEST 1 and QUEST 2 phase 3 trials evaluated SMV triple therapy in treatment-naïve patients, and PROMISE is a study in prior relapsers (Janssen Therapeutics 2013). Improvement in SVR rates was observed irrespective of the prior treatment status and *IFNL3* genotype (Janssen Therapeutics 2013). In these studies of SMV triple therapy regimen, SVR rates were lower in *IFNL3* rs12979860 T minor allele carriers compared to patients who had the rs12979860 CC genotype (Janssen Therapeutics 2013). The pharmacogenetic data with SMV were consistent with findings of first-generation protease inhibitors by favoring the rs12979860 CC genotype in terms of efficacy.

In the NEUTRINO phase 3 trial of SOF triple therapy in treatment-naïve patients, sustained virologic response was achieved in 93 of 95 patients (98%) with *IFNL3* rs12979860 CC genotype whereas 202 of 232 patients (87%) had SVR with the unfavorable response genotype (Lawitz 2013). A multivariate logistic regression showed that rs12979860 T minor allele was predictive of reduced response to SOF triple therapy (OR 7.99, 95%CI 1.82-35.17, P=0.006) (Lawitz 2013). Although the SVR rates were high across the *IFNL3* genotype, compared to regimens with NS3/4A protease inhibitors, the difference in SVR rates between the genotype was still observed in the NEUTRINO study. In the phase 2 ATOMIC study, seven patients experienced relapse after completing regimens including SOF (Kowdley 2013). All but one of these relapsers were *IFNL3* rs12979860 T minor allele carriers (Kowdley 2013). *IFNL3* rs12979860 CC remains the favorable response genotype for treatment-naïve patients considering therapy with SOF+PEG+RBV combination.

The phase 3 trials POSITRON and FUSION investigated SOF/RBV dual therapy in PEG-ineligible and treatment-experienced patients infected with HCV-2/3 respectively (Jacobson 2013). In these two trials, the rates of SVR did not significantly differ between *IFNL3* rs12979860 CC and non-CC genotypes (Jacobson 2013). Similarly, in another study among patients infected with HCV-2/3, rapid viral suppression was achieved in all *IFNL3* rs12979860 genotypes after treatment regimens including SOF, regardless of previous treatment status and presence or absence of PEG (Gane 2013). The utility of *IFNL3* genotyping may be limited for patients with HCV-2/3 infection considering treatment with an interferon-free SOF regimen.

Pharmacogenetic data with the second-generation DAAs are scarce in HCV-4/5/6 infection. No definitive recommendation can be made with currently recommended regimens including SOF.

Role in the Treatment of Special Populations

Patients with HCV and Hepatitis B Co-infection

Hepatitis C virus and hepatitis B virus have the same mode of transmission, and co-infection is commonly seen (Guo 2013; Lee 2014). Interferon-λ coded by the *IFNL3* gene is also known to be effective against hepatitis B virus (Guo 2013; Lee 2014). Although the association of *IFNL3* genotype with spontaneous clearance of HCV was confirmed in acute infection, a meta-analysis found no association between the *IFNL3* genotype and spontaneous hepatitis B surface antigen (HBsAg) seroclearance (Lee 2014). In contrast, among patients with chronic co-infection, both *IFNL3* rs12979860 and rs8099917 was shown to be indicative of response to PEG/RBV therapy in this dual infection (Guo 2013; Coppola 2013). Although *IFNL3* genotyping may not help predicting the spontaneous resolution of hepatitis B virus acute infection, the genotype is believed to remain supportive in indicating HCV treatment outcome in the co-infection with hepatitis B virus.

Patients with HCV and Human Immunodeficiency Virus Co-infection

The co-infection of HCV and human immunodeficiency virus (HIV) is common since both viruses share the same mode of transmission (Barreiro 2012). This form of HCV infection is considered more aggressive because progression to cirrhosis occurs faster than HCV mono-infection (Barreiro 2012). Moreover, co-infection of HIV/HCV is considered more difficult to treat compared to HCV mono-infection because reported SVR rates after PEG/RBV were 30% in HIV/HCV-1/4 co-infection and 70% in HIV/HCV-2/3 infection (Barreiro 2012). Peginterferon alfa-based HCV treatment may not be appropriate for patients with CD4 counts less than 200 cells/mm^3 due to possible leukocyte depletion induced by PEG, and patients with CD4 counts greater than 350 cells/mm^3 demonstrated better eradication of HCV with treatment (Barreiro 2012). The findings of *IFNL3* genotype with PEG/RBV treatment outcome has thus far been replicated in HIV/HCV co-infection (Kawaguchi-Suzuki 2014; Muir 2013; Barreiro 2012).

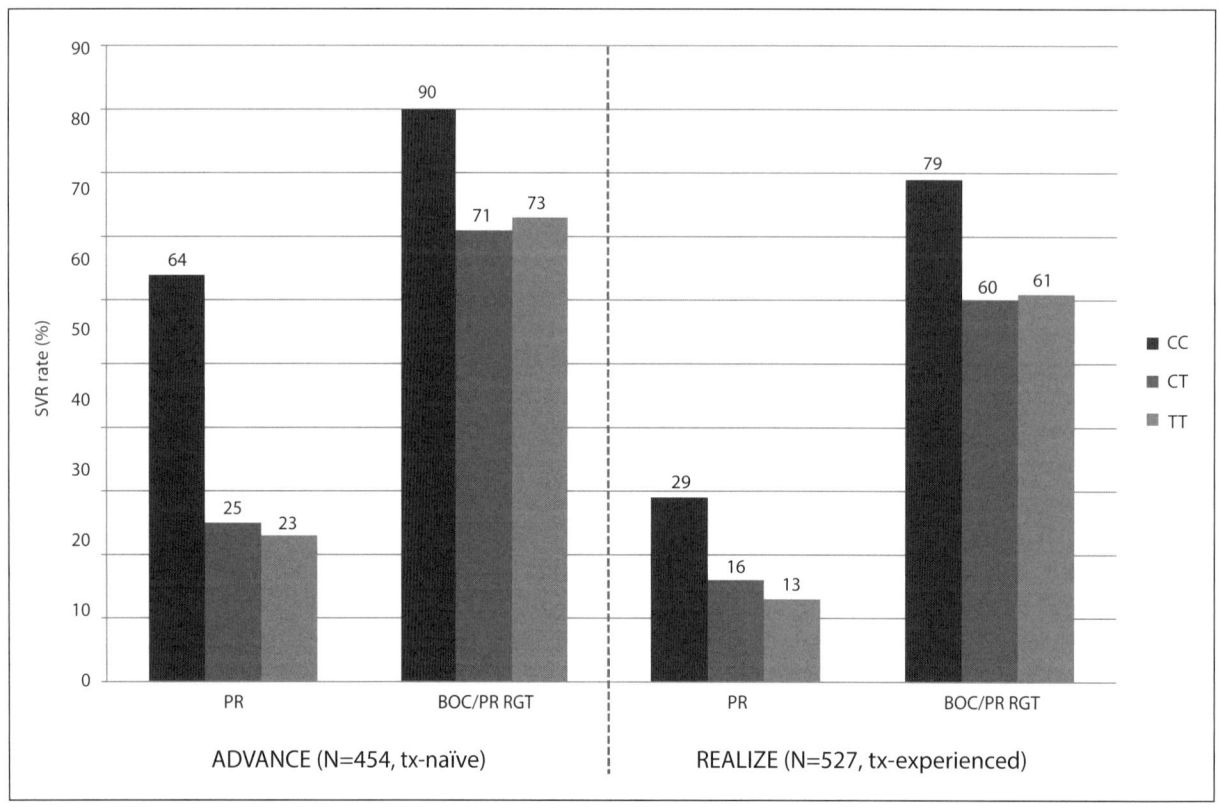

Figure 14.5. SVR rates (%) by IFNL3 genotype (rs12979860) with TVR therapy (Clark 2012; Jacobson 2011A, B; Zeuzem 2011; Pol 2011). (N=total number of patients. P=PEG, R=RBV, tx=treatment)

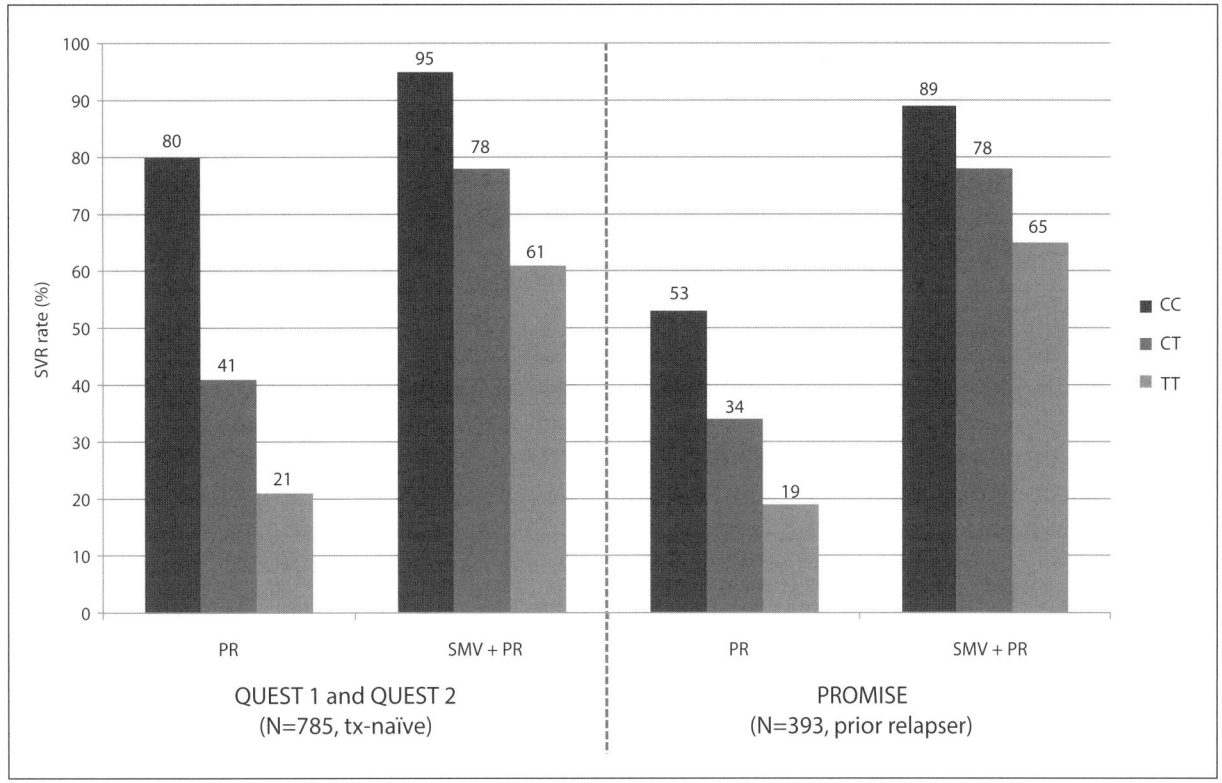

Figure 14.6. SVR rates (%) by IFNL3 genotype (rs12979860) with SMV therapy (Janssen Therapeutics 2013). (N=total number of patients. P=PEG, R=RBV, tx=treatment)

According to the AASLD/IDSA/IAS-USA guideline, treatment regimen including SOF is considered first-line in HCV/HIV co-infection (AASLD 2014). The study PHOTON-1 investigated SOF plus RBV in patients co-infected with HCV/HIV (Gilead Sciences, Inc. 2013). In this PHOTON-1, 24 of 30 patients (80%) achieved SVR with *IFNL3* favorable response genotype, and the SVR rate was slightly smaller (75%, 62 out of 83) in patients with the unfavorable response genotype (Gilead Sciences, Inc. 2013). At this time, the usefulness of *IFNL3* genotyping is undetermined for patients co-infected with HCV/HIV and considering treatment with SOF.

Patients with HCV Reinfection after Liver Transplantation

The chance of patient and graft survival is not as good for HCV-related liver transplantation as for liver transplantation undertaken due to other causes (Lange 2013). This poor prognosis after HCV-related liver transplantation is mostly due to recurrence of HCV reinfection, which happens almost universally in these patients (Lange 2013). Furthermore, PEG/RBV therapy was poorly tolerated and required dose adjustment in post-transplant patients with HCV reinfection, and virologic response rates were even lower than those in non-transplant patients (Lange 2013).

IFNL3 genotyping in a post-transplantation setting is unique because consideration has to be made for both donor and recipient genotypes. Previously, both donor and recipient *IFNL3* genotypes were shown to be independently associated with SVR rates after PEG/RBV therapy, while the evidence is strongest in a situation where both donor and recipient have the favorable response genotype (Kawaguchi-Suzuki 2014; Lange 2013). It remains controversial which genotype, donor versus recipient, is the stronger predictor of SVR. However, liver from a donor with the favorable response genotype may be preferable if liver transplantation is indicated for HCV infection, based on the fact that chance of recurring HCV eradication can be better post-transplantation (Lange 2013). This may justify an implementation of *IFNL3* genotyping in the setting of liver graft allocation for HCV-infected patients. Currently, SOF-based regimen is recommended by the AASLD/IDSA/IAS-USA: SOF plus SMV with or without RBV for HCV-1 infection and SOF plus RBV for HCV-2/3 infection (AASLD 2014). However, information is scarce regarding the usefulness of *IFNL3* genotyping with SOF in a post-transplantation setting.

Patients with Cirrhosis

Hepatitis C virus infected patients with cirrhosis have higher morbidity and mortality and require treatment to reduce and prevent cirrhosis-related complications, such as hepatocellular carcinoma and liver decompensation (Bourlière 2013; Boccaccio 2014). Unfortunately, these patients typically present lower SVR rates than patients without cirrhosis (Bourlière 2013). However, the treatment outcome is expected to improve with the use of DAAs (Bourlière 2013). It should be noted however that evidence is limited because patients with decompensated or end-stage liver disease were commonly excluded from clinical trials. According to the AASLD/IDSA/IAS-USA, the same treatment regimen including SOF as the first-line agent is recommended for patients with compensated cirrhosis (AASLD 2014).

Pharmacogenetic information with *IFNL3* has been reported among patients with cirrhosis treated with PEG/RBV combination. A multivariate analysis showed that *IFNL3* rs12979860 CC genotype was independently associated with SVR (OR 7.04, 95%CI 2.40-20.72, P<0.001) in patients with compensated cirrhosis (Di Marco 2013). Another study found that if cirrhotic patients have *IFNL3* rs12979860 CC genotype, the attainment of RVR can further identify patients more likely to achieve SVR with PEG/RBV (Aghemo 2013). When *IFNL3* genotype and ISDR mutant were considered together in cirrhotic and/or elderly patients treated with PEG/RBV, the positive predictive value of SVR was 70% with *IFNL3* rs8099917 TT genotype and ISDR mutant, and the negative predictive value of SVR was 89% with *IFNL3* rs8099917 G minor allele and wild-type ISDR (Tamai 2013). In this study, ISDR mutant was defined as amino acid substitution of glutamine or histidine for wild-type arginine at position 70 or substitution of methionine for wild-type leucine at position 91 (Tamai 2013). However, more evidence is necessary with response to DAAs.

Pediatric Patients

The studies of DAAs have not included pediatric patients, and the efficacy and safety of currently recommended regimen with SOF or SMV have not been established for patients under the age of 18 (Janssen Therapeutics 2013; Gilead Sciences, Inc. 2013). The standard of care in pediatrics remains PEG/RBV dual therapy for 24 or 48 weeks (El-Shabrawi 2013).

The association of *IFNL3* and PEG/RBV outcomes have been replicated in pediatric population. Spontaneous clearance of HCV in children with perinatal infection was more common in those with rs12979860 CC genotype (OR 2.7, 90%CI 1.3-5.8, P=0.02) when matched 1000 Genome Project data were compared (Indolfi 2014). Among pediatric patients with Caucasian ethnicity infected with HCV-1/4, higher SVR rates were observed with *IFNL3* rs12979860 CC (OR 6.81, 95%CI 1.98-23.42, P=0.001) and rs8099917 TT (OR 3.14, 95%CI 1.26-7.85, P=0.013) after PEG/RBV therapy (Domagalski 2013).

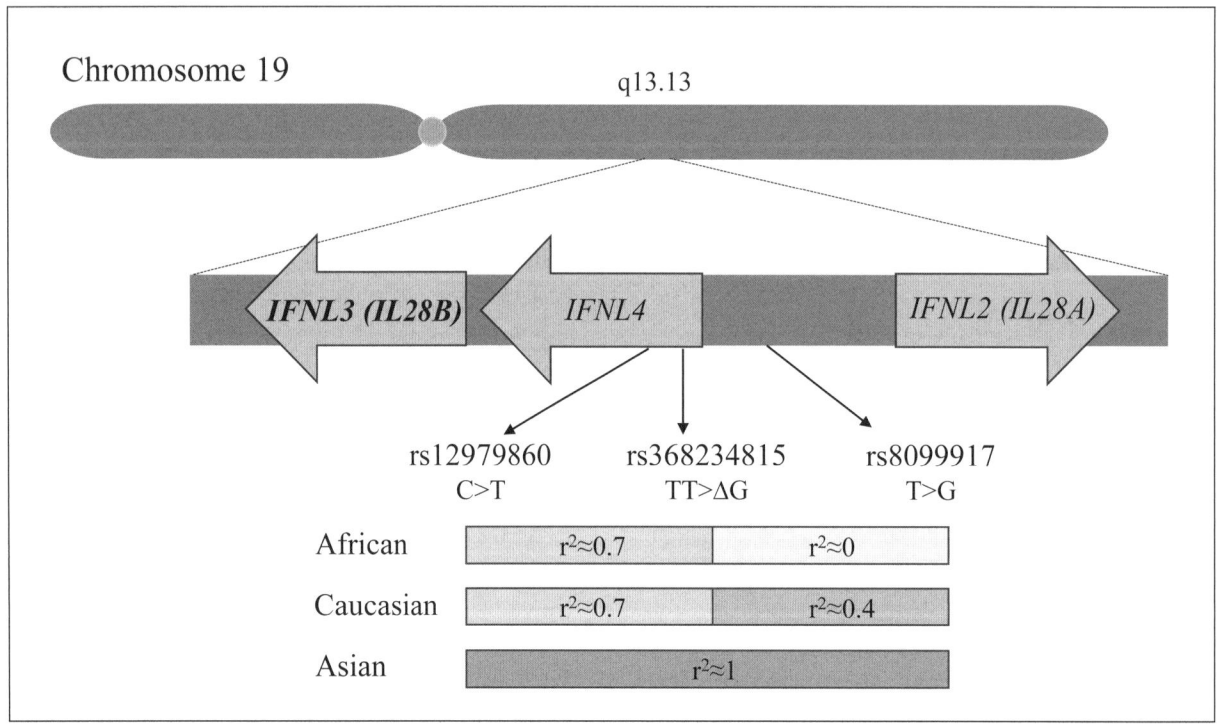

Figure 14.7. IFNL genes and the polymorphisms in the host genome (Hayes 2011; Booth 2013).

Sustained viral response rates after PEG/RBV treatment among Japanese pediatric patients infected with HCV-1/2 were 95.8% with rs8099917 TT genotype and 16.7% with rs8099917 TG/GG genotype (Tajiri 2013). Similarly, for Egyptian pediatric patients infected with HCV-4, SVR rates were 80% with rs12979860 CC genotype and 42% with rs12979860 CT/TT genotype after PEG/RBV therapy (Shaker 2013). The finding of *IFNL3* favorable response genotype was replicated in pediatric patients with different ethnicity.

Implementation of *IFNL3* (*IL28B*) Testing

Implementation guidelines regarding *IFNL3* genotyping are available from the Clinical Pharmacogenetics Implementation Consortium (Muir 2013). Commercially available testing for *IFNL3* genotype can be found at the PharmGKB Web Site (www.pharmgkb.org). Genetic testing is available for both rs12979860 and rs8099917 SNPs, and both SNPs were shown to predict treatment response independently. Although rs12979860 and rs8099917 are believed to be in strong linkage equilibrium, rs12979860 may be more reliable among African Americans and in BOC triple therapy if one SNP needs to be selected (Kawaguchi-Suzuki 2014; Muir 2013).

Future Directions

IFNL4 Gene

A dinucleotide variant, rs368234815, was found to be associated with HCV clearance (Prokunina-Olsson 2013). The rs368234815 SNP is located upstream of *IFNL3* and is in high linkage disequilibrium with rs12979860 ($r^2=0.71$ in African Americans, $r^2=0.92$ in Caucasians, and $r^2=1.00$ in Asians) (Figure 14.7) (Prokunina-Olsson 2013). The wild-type is TT, and ΔG allele is a frameshift variant that creates a novel gene, designated *IFNL4*, encoding the interferon-λ 4 protein, which is similar to interferon-λ 3 and can induce ISG expression (Prokunina-Olsson 2013). Interestingly, rs368234815 was a stronger indicator of HCV clearance than rs12979860 in patients with African ancestry; data were comparable for Europeans and Asians (Prokunina-Olsson 2013). Independent association of *IFNL4* TT/TT genotype with SVR was found in Caucasians treated with PEG/RBV for HCV-1 infection (OR 2.54, 95%CI 1.63-3.02, P<0.001) and HCV-4 infection (OR 12.57, 95%CI 3.43-46.13, P<0.001), but not for HCV-3 infection (Stättermayer 2014). In this study, consideration of *IFNL4* genotype in addition to *IFNL3* genotype rs8099917, but not rs12979860, improved the prediction of SVR (Stättermayer 2014). Similarly, ΔG allele was associated with treatment failure in Japanese patients infected with HCV-1 after PEG/RBV therapy (OR 4.73, 95%CI 2.43-9.20, P=0.019) (Nozawa 2014).

The *IFNL4* polymorphism was also investigated with TVR triple therapy. Among Japanese patients who were over 65 years of age and infected with HCV-1b, the SVR rate was significantly higher with the *IFNL4* TT/TT genotype than with TT/ΔG or ΔG/ΔG genotypes after TVR triple therapy (81.8% versus 42.9%, P=0.003) (Fujino 2013). Based on the multivariate analysis of this study, RVR (OR 36.60, P=0.002) and *IFNL4* TT/TT genotype (OR 19.50, P=0.009) were independent predictors for SVR in the Japanese elderly treated with TVR triple therapy (Fujino 2013).

Limited but consistent data were presented with SOF-based therapy for the *IFNL4* polymorphism. In a clinical study investigating 24-week SOF+RBV combination in treatment-naïve patients infected with HCV-1, *IFNL4* ΔG allele carriers had slower loss of free virus (P=0.039) and therefore slower early viral decay (Meissner 2014). In this study, SVR24 rates were 88% (95%CI 47%–100%) with TT/TT genotype and 67% (95%CI 52%–78%) with the ΔG allele (Meissner 2014). Although the study was not powered sufficiently to detect difference in the SVR24 rates, the modeled treatment efficacy of SOF+RBV was shown to be decreased with the presence of the ΔG allele (P=0.048), suggesting potentially functional relevance of the *IFNL4* polymorphism in PEG-free regimens (Meissner 2014).

IFNL4 polymorphism was examined also in HIV/HCV co-infection. When multivariate analysis was conducted in patients co-infected with HIV/HCV and treated with PEG/RBV, HCV load and *IFNL3* genotype were the independent and significant factors included in the final model to predict treatment response whereas *IFNL4* polymorphism did not remain significant in the model (Krämer 2013). In contrast, among the HCV mono-infected cohort of the same study, *IFNL4* rs368234815 was more strongly associated with treatment response than *IFNL3* rs12979860, suggesting possible difference in HCV clearance between HIV-(+) and HIV-(−) infection (Krämer 2013). The effect of *IFNL4* may provide further insight into the genetic regulation of HCV clearance and may have implications for clinical management in the future.

Other Polymorphisms

There is a significant toxicity burden with HCV therapy, and research has been conducted to investigate the contribution of genetics to the occurrence of adverse drug reactions. Polymorphisms (rs1127354 and rs7270101) in the *ITPA* gene have a protective effect against reduction in hemoglobin concentration or anemia, which is one of the major adverse drug reactions of PEG/RBV containing HCV therapy (Fellay 2010). The finding has been replicated in independent cohorts, but the clinical utility of prospective *ITPA* genotyping remains unclear (Kawaguchi-Suzuki 2014). Many other polymorphisms have been found in association with HCV treatment response or incidence of adverse drug reactions (Schlecker 2012), but data are lacking in the current setting where DAAs have become the key element of HCV therapy.

Treatment Regimens Under Investigation

Ledipasvir + SOF
Ledipasvir is NS5A inhibitor with antiviral activity against HCV-1a and -1b (Afdhal 2014A). In a phase 3 ION-1 trial, ledipasvir + SOF was studied in treatment-naïve patients infected with HCV-1 (Afdhal 2014A). This study randomized patients into four arms: ledipasvir + SOF for 12 weeks, ledipasvir + SOF + RBV for 12 weeks, ledipasvir + SOF for 24 weeks, or ledipasvir + SOF + RBV for 24 weeks (Afdhal 2014A). In this trial, the overall SVR12 rates were over 97%–99% in all arms (Afdhal 2014A). Similarly, the SVR12 rates were 97%–99% in patients with *IFNL3* rs12979860 non-CC genotype while 100% of patients with the CC genotype achieved SVR12 (Afdhal 2014A). The similar ION-2 phase 3 trial enrolled patients who were previously treated with PEG/RBV with or without NS3/4A protease inhibitor into the same four arms (Afdhal 2014B). The overall SVR12 rates were 94%–99% (Afdhal 2014B). The 84%–91% of patients were rs12979860 T allele carriers, and no significant difference in SVR12 rates between the CC and non-CC genotypes was found (Afdhal 2014B). In a phase 3 ION-3 trial, treatment length of 8 weeks was studied in treatment-naïve patients without cirrhosis (Kowdley 2014). This study had three arms: ledipasvir + SOF for 8 weeks, ledipasvir + SOF + RBV for 8 weeks, and ledipasvir + SOF for 12 weeks (Kowdley 2014). The overall SVR12 rates were 93%–95% (Kowdley 2014). In ION-3, patients with characteristics previously associated with a poor response to interferon-based therapy achieved similar SVR12 rates to patients without those characteristics (Kowdley 2014). The rates of SVR12 were 95.0%–96.4% with the CC genotype and 92.3%–95.0% across the three arms (Kowdley 2014).

During the ION-1 and ION-3 trials, the following viral polymorphisms were reported in patients with relapse in HCV-1a infection: M28A/T, Q30H/R/Y, L31M/P, S38F, Y93C/F/H/N located in the NS5A (Afdhal 2014A; Kowdley 2014). In HCV-1b, viral polymorphisms reported in patients with break-through or with relapse were: L31I and Y93H in the NS5A (Afdhal 2014A; Kowdley 2014). In ION-2, similar early viral kinetics were shown between patients with NS5A-resistant variants at baseline and those without NS5A-resistant variants at baseline, suggesting that development of resistant variants during treatment may be a concern (Afdhal 2014B). The NS5B S282T variant, known to reduce susceptibility to SOF, was not detected in these three studies (Afdhal 2014A, B; Kowdley 2014).

Daclatasvir + Asunaprevir
Daclatasvir is a NS5A replication complex inhibitor with pan-genotypic antiviral activity in vitro (Kumada 2014). Asunaprevir is a NS3 protease inhibitor with antiviral activity against HCV-1/4/5/6 in vitro (Kumada 2014). In a phase 3 study, daclatasvir and asunaprevir combination regimen was administered to interferon-ineligible/intolerant patients or previous non-responders with HCV-1b for 24 weeks (Kumada 2014). The rates of SVR24 were 87.4% in interferon-ineligible/intolerant patients and 80.5% in previous non-responders (Kumada 2014). Similar SVR24 rates were observed despite the rs12979860 genotype (CC 84.5%, non-CC 84.8%) (Kumada 2014). Out of 37 patients with NS5A L31M/V and/or Y93H variants at baseline, 12 interferon-ineligible/intolerant patients and 10 non-responders were not able to achieve SVR (Kumada 2014). Asunaprevir resistance-associated variant NS3 D168E was reported in a relapser in this study (Kumada 2014).

Similarly, in another phase 3 HALLMARK-DUAL study, SVR12 rates did not significantly differ based on the *IFNL3* genotype (Manns 2014). This study included patients infected with HCV-1b who were treatment-naïve, previous non-responders to PEG/RBV, or interferon-ineligible/intolerant (Manns 2014). The most common viral variants associated with treatment failure were NS5A L31, NS5A Y93, and NS3 D168 (Manns 2014). These three variants together accounted for 61 (77%) of 79 virological failure cases (Manns 2014). The resistance-associated variants reported in the two phase 3 studies were consistent.

Paritaprevir/ritonavir + Ombitasvir + Dasabuvir ± RBV
Paritaprevir is a NS3/4A serine protease inhibitor co-administered with ritonavir, a CYP3A4 inhibitor to facilitate once-daily dosing (Feld 2014). Ombitasvir is a NS5A inhibitor, while dasabuvir is a non-nucleoside NS5B polymerase inhibitor (Feld 2014).

In the phase 3 SAPPHIRE-I trial, the paritaprevir/ritonavir + ombitasvir + dasabuvir + RBV 12-week regimen was studied against matching placebos in treatment-naïve, non-cirrhotic patients with HCV-1 infection, and the overall SVR12 rate was 96.2% (Feld 2014). The rates of SVR12 were 96.5% in *IFNL3* rs12979860 CC genotype and 96.0% in the non-CC genotype (Feld 2014).

SAPPHIRE-II is another phase 3 trial, where the same regimen was investigated among HCV-1 infected non-cirrhotic patients who were previously treated with PEG/RBV (Zeuzem 2014). In this trial, the 263 out of 297 patients had rs12979860 non-CC genotype, and significant difference in the SVR12 rates were not detected between the CC and the non-CC genotypes (Zeuzem 2014).

PEARL-III and PEARL-IV are phase 3 trials comparing paritaprevir/ritonavir + ombitasvir + dasabuvir with RBV to those without RBV in treatment-naïve patients with HCV-1a and HCV-1b infections respectively (Ferenci 2014). The overall SVR12 rates were over 90% regardless of the HCV-1 subtype and RBV use (Ferenci 2014). The logistic-regression analyses of baseline demographic and clinical characteristics showed that only rs12979860 CC genotype was associated with increased rates of SVR in the HCV-1a infection (P=0.03) (Ferenci 2014). The rates of virologic failure were higher without RBV than those with RBV in HCV-1a infection, but not in HCV-1b infection (Ferenci 2014). The PEARL-II is phase 3 trial investigated the same combination with and without RBV in patients who previously treated with PEG/RBV and are infected with HCV-1b (Andreone 2014). In PEARL-II, SVR12 rates based on rs12979860 genotype were CC 100%, CT 96.4%, and TT 95.5% with RBV, while 100% of patients achieved SVR12 in the arm without RBV (Andreone 2014).

TURQUOISE-II is a phase 3 trial to compare 12-week or 24-week paritaprevir/ritonavir + ombitasvir + dasabuvir + RBV in patients with HCV-1 infection and with cirrhosis (Poordad 2014). SVR12 rates were 94.3% with rs12979860 CC genotype and 91.3% with the non-CC genotype after the 12-week regimen and 97.1% with the CC genotype and 95.7% with the non-CC genotype after the 24-week regimen (Poordad 2014).

In the SAPPHIRE-I, SAPPHIRE-II, PEARL-III, and TURQUOISE-II trials, virologic failure during treatment or relapse was reported with the following viral polymorphisms: D168V in NS3, M28T/V and Q30R in NS5A, and S556G in NS5B for HCV-1a and Y56H and D168A/V in NS3, L31M and Y93H in NS5A, and C316N/Y, M414I and S556G in NS5B for HCV-1b (Feld 2014; Zeuzem 2014; Ferenci 2014; Poordad 2014).

While current evidence supports the predictive utility of *IFNL3* (*IL28B*) genotyping, the future role of genetic testing as we move toward an era of all oral regimens is uncertain. The SVR rates with the new oral regimens are high (greater than 90%) for genotype 1 patients, suggesting a potential lack of utility for *IFNL3* (*IL28B*) as a pretreatment predictor of response. With all oral DAA regimens, polymorphism testing for the viral genome may play a more significant role to guide treatment decisions.

References

AASLD, IDSA, IAS–USA [homepage on the Internet]. Recommendations for testing, managing, and treating hepatitis C, 2014. Available at http://www.hcvguidelines.org/. Accessed May 10, 2014.

Afdhal N, Reddy KR, Nelson DR, et al. Ledipasvir and sofosbuvir for previously treated HCV genotype 1 infection. N Engl J Med 2014B;370:1483-93.

Afdhal N, Zeuzem S, Kwo P, et al. Ledipasvir and sofosbuvir for untreated HCV genotype 1 infection. N Engl J Med 2014A;370:1889-98.

Aghemo A, Degasperi E, Rumi MG, et al. Cirrhosis and rapid virological response to peginterferon plus ribavirin determine treatment outcome in HCV-1 IL28B rs12979860 CC patients. Biomed Res Int 2013;580796.

Andreone P, Colombo MG, Enejosa JV, et al. ABT-450, Ritonavir, Ombitasvir, and Dasabuvir Achieves 97% and 100% Sustained Virologic Response With or Without Ribavirin in Treatment-Experienced Patients With HCV Genotype 1b Infection. Gastroenterology 2014;147:359-65.e1.

Arnaud C, Trépo C, Petit MA. Predictors of the therapeutic response in hepatitis C. A 2013 update. Clin Res Hepatol Gastroenterol 2014;38:12-7.

Asselah T. Sofosbuvir for the treatment of hepatitis C virus. Expert Opin Pharmacother 2014;15:121-30.

Bacon BR, Gordon SC, Lawitz E, et al. Boceprevir for previously treated chronic HCV genotype 1 infection. N Engl J Med 2011;364:1207-17.

Barreiro P, Vispo E, Labarga P, et al. Management and treatment of chronic hepatitis C in HIV patients. Semin Liver Dis 2012;32:138-46.

Boccaccio V, Bruno S. Management of HCV patients with cirrhosis with direct acting antivirals. Liver Int 2014;34 Suppl 1:38-45.

Bota S, Sporea I, Şirli R, et al. Role of interleukin-28B polymorphism as a predictor of sustained virological response in patients with chronic hepatitis C treated with triple therapy: a systematic review and meta-analysis. Clin Drug Investig 2013;33:325-31.

Bourlière M, Wendt A, Fontaine H, et al. How to optimize HCV therapy in genotype 1 patients with cirrhosis. Liver Int 2013;33 Suppl 1:46-55.

Cariani E, Villa E, Rota C, et al. Translating pharmacogenetics into clinical practice: interleukin (IL)28B and inosine triphosphatase (ITPA) polymophisms in hepatitis C virus (HCV) infection. Clin Chem Lab Med 2011;49:1247-56.

Centers for Disease Control and Prevention [home page on the Internet]. Hepatitis C Information for Health Professionals, 2013. Available at http://www.cdc.gov/hepatitis/hcv/. Accessed January 11, 2014.

Clark PJ, Thompson AJ. Host genomics and HCV treatment response. J Gastroenterol Hepatol 2012;27:212-22.

Coppola N, Marrone A, Pisaturo M, et al. Role of interleukin 28-B in the spontaneous and treatment-related clearance of HCV infection in patients with chronic HBV/HCV dual infection. Eur J Clin Microbiol Infect Dis 2013.

Davis GL, Albright JE, Cook SF, et al. Projecting future complications of chronic hepatitis C in the United States. Liver Transpl 2003;9:331-8.

Di Marco V, Calvaruso V, Grimaudo S, et al. Role of IL-28B and inosine triphosphatase polymorphisms in efficacy and safety of Peg-Interferon and ribavirin in chronic hepatitis C compensated cirrhosis with and without oesophageal varices. J Viral Hepat 2013;20:113-21.

Domagalski K, Pawłowska M, Tretyn A, et al. Impact of IL-28B polymorphisms on pegylated interferon plus ribavirin treatment response in children and adolescents infected with HCV genotypes 1 and 4. Eur J Clin Microbiol Infect Dis 2013;32:745-54.

El-Shabrawi MH, Kamal NM. Burden of pediatric hepatitis C. World J Gastroenterol 2013;19:7880-8.

Feld JJ, Kowdley KV, Coakley E, et al. Treatment of HCV with ABT-450/r-ombitasvir and dasabuvir with ribavirin. N Engl J Med 2014;370:1594-603.

Fellay J, Thompson AJ, Ge D, et al. ITPA gene variants protect against anaemia in patients treated for chronic hepatitis C. Nature 2010;464:405-8.

Ferenci P, Bernstein D, Lalezari J, et al. ABT-450/r-ombitasvir and dasabuvir with or without ribavirin for HCV. N Engl J Med 2014;370:1983-92.

Forns X, Lawitz E, Zeuzem S, et al. Simeprevir with peginterferon and ribavirin leads to high rates of SVR in patients with HCV genotype 1 who relapsed after previous therapy: a phase 3 trial. Gastroenterology 2014;146:1669-79.e3.

Fujino H, Imamura M, Nagaoki Y, et al. Predictive value of the IFNL4 polymorphism on outcome of telaprevir, peginterferon, and ribavirin therapy for older patients with genotype 1b chronic hepatitis C. J Gastroenterol 2014;49:1548-56.

Furusyo N, Ogawa E, Nakamuta M, et al. Telaprevir can be successfully and safely used to treat older patients with genotype 1b chronic hepatitis C. J Hepatol 2013;59:205-12.

Gane EJ, Stedman CA, Hyland RH, et al. Nucleotide polymerase inhibitor sofosbuvir plus ribavirin for hepatitis C. N Engl J Med 2013;368:34-44.

Ge D, Fellay J, Thompson AJ, et al. Genetic variation in IL28B predicts hepatitis C treatment-induced viral clearance. Nature 2009;461:399-401.

Ghany MG, Nelson DR, Strader DB, et al. An update on treatment of genotype 1 chronic hepatitis C virus infection: 2011 practice guideline by the American Association for the Study of Liver Diseases. Hepatology 2011;54:1433-44.

Ghany MG, Strader DB, Thomas DL, et al. Diagnosis, management, and treatment of hepatitis C: an update. Hepatology 2009;49:1335-74.

Gilead Sciences, Inc. Sovaldi (sofosbuvir) package insert. Foster City, CA, 2013.

Guo X, Yang G, Yuan J, et al. Genetic variation in interleukin 28B and response to antiviral therapy in patients with dual chronic infection with hepatitis B and C viruses. PLoS One 2013;8:e77911.

Hayes CN, Kobayashi M, Akuta N, et al. HCV substitutions and IL28B polymorphisms on outcome of peg-interferon plus ribavirin combination therapy. Gut 2011;60:261-7.

Honda M, Sakai A, Yamashita T, et al. Hepatic ISG expression is associated with genetic variation in interleukin 28B and the outcome of IFN therapy for chronic hepatitis C. Gastroenterology 2010;139:499-509.

Indolfi G, Mangone G, Bartolini E, et al. Comparative Analysis of rs12979860 SNP of the IFNL3 Gene in Children with Hepatitis C and Ethnic Matched Controls Using 1000 Genomes Project Data. PLoS One 2014;9:e85899.

Jacobson IM, Pawlotsky JM, Afdhal NH, et al. A practical guide for the use of boceprevir and telaprevir for the treatment of hepatitis C. J Viral Hepat 2012;19 Suppl 2:1-26.

Keating GM, Vaidya A. Sofosbuvir: first global approval. Drugs 2014;74:273-82.

Kowdley KV, Lawitz E, Crespo I, et al. Sofosbuvir with pegylated interferon alfa-2a and ribavirin for treatment-naive patients with hepatitis C genotype-1 infection (ATOMIC): an open-label, randomised, multicentre phase 2 trial. Lancet 2013;381:2100-7.

Jacobson IM, Catlett I, Marcellin P, et al. Telaprevir Substantially Improved Svr Rates across All Il28b Genotypes in the Advance Trial. J Hepatol 2011B;54:S542-S3.

Jacobson IM, Dore GJ, Foster GR, et al. Simeprevir with pegylated interferon alfa 2a plus ribavirin in treatment-naive patients with chronic hepatitis C virus genotype 1 infection (QUEST-1): a phase 3, randomised, double-blind, placebo-controlled trial. Lancet 2014;384:403-13.

Jacobson IM, Gordon SC, Kowdley KV, et al. Sofosbuvir for hepatitis C genotype 2 or 3 in patients without treatment options. N Engl J Med 2013;368:1867-77.

Jacobson IM, McHutchison JG, Dusheiko G, et al. Telaprevir for previously untreated chronic hepatitis C virus infection. N Engl J Med 2011A;364:2405-16.

Janssen Therapeutics. Olysio (simeprevir) package insert. Titusville, NJ; 2013.

Jesudian AB, Jacobson IM. Optimal treatment with telaprevir for chronic HCV infection. Liver Int 2013;33Suppl1:3-13.

Kawaguchi-Suzuki M, Frye RF. The role of pharmacogenetics in the treatment of chronic hepatitis C infection. Pharmacotherapy 2014;34:185-201.

Kotenko SV. IFN-λs. Curr Opin Immunol 2011;23:583-90.

Kowdley KV, Gordon SC, Reddy KR, et al. Ledipasvir and sofosbuvir for 8 or 12 weeks for chronic HCV without cirrhosis. N Engl J Med 2014;370:1879-88.

Krämer B, Nischalke HD, Boesecke C, et al. Variation in IFNL4 genotype and response to interferon-based therapy of hepatitis C in HIV-positive patients with acute and chronic hepatitis C. AIDS 2013;27:2817-9.

Kumada H, Suzuki Y, Ikeda K, et al. Daclatasvir plus asunaprevir for chronic HCV genotype 1b infection. Hepatology 2014;59(6):2083-91.

Lange CM. The importance of IL28B genotype in hepatitis C virus-associated liver transplantation. Liver Int 2013;33:169-71.

Lawitz E, Gane EJ. Sofosbuvir for previously untreated chronic hepatitis C infection. N Engl J Med 2013;369:678-9.

Lee DH, Lee JH, Kim YJ, et al. Relationship between polymorphisms near the IL28B gene and spontaneous HBsAg seroclearance: a systematic review and meta-analysis. J Viral Hepat 2014;21:163-70.

Luo Y, Jin C, Ling Z, et al. Association study of IL28B: rs12979860 and rs8099917 polymorphisms with SVR in patients infected with chronic HCV genotype 1 to PEG-INF/RBV therapy using systematic meta-analysis. Gene 2013;513:292-6.

Maasoumy B, Manns MP. Optimal treatment with boceprevir for chronic HCV infection. Liver Int 2013;33Suppl1:14-22.

Manns M, Marcellin P, Poordad F, et al. Simeprevir with pegylated interferon alfa 2a or 2b plus ribavirin in treatment-naive patients with chronic hepatitis C virus genotype 1 infection (QUEST-2): a randomised, double-blind, placebo-controlled phase 3 trial. Lancet 2014;384:414-26.

Manns M, Pol S, Jacobson IM, et al. All-oral daclatasvir plus asunaprevir for hepatitis C virus genotype 1b: a multinational, phase 3, multicohort study. Lancet 2014.

Matsuura K, Watanabe T, Tanaka Y. Role of IL28B for chronic hepatitis C treatment toward personalized medicine. J Gastroenterol Hepatol 2014;29:241-9.

Meissner EG, Bon D, Prokunina-Olsson L, et al. IFNL4-ΔG genotype is associated with slower viral clearance in hepatitis C, genotype-1 patients treated with sofosbuvir and ribavirin. J Infect Dis 2014;209:1700-4.

McCarthy JJ, Li JH, Thompson A, et al. Replicated association between an IL28B gene variant and a sustained response to pegylated interferon and ribavirin. Gastroenterology 2010;138:2307-14.

Muir AJ, Gong L, Johnson SG, et al. Clinical Pharmacogenetics Implementation Consortium (CPIC) Guidelines for IFNL3 (IL28B) Genotype and PEG Interferon-α-Based Regimens. Clin Pharmacol Ther 2014;95:141-6.

Naggie S, Osinusi A, Katsounas A, et al. Dysregulation of innate immunity in hepatitis C virus genotype 1 IL28B-unfavorable genotype patients: impaired viral kinetics and therapeutic response. Hepatology 2012;56:444-54.

Nozawa Y, Umemura T, Katsuyama Y, et al. Genetic polymorphism in IFNL4 and response to pegylated interferon-α and ribavirin in Japanese chronic hepatitis C patients. Tissue Antigens 2014;83:45-8.

O'Brien TR. Interferon-alfa, interferon-lambda and hepatitis C. Nat Genet 2009;41:1048-50.

Pacanowski M, Amur S, Zineh I. New genetic discoveries and treatment for hepatitis C. JAMA 2012;307:1921-2.

Pol S, Aerssens J, Zeuzem S, et al. Similar Svr Rates in Il28b Cc, Ct or Tt Prior Relapser, Partial- or Null-Responder Patients Treated with Telaprevir/Peginterferon/Ribavirin: Retrospective Analysis of the Realize Study. J Hepatol 2011;54:S6-S7.

Poordad F, Bronowicki JP, Gordon SC, et al. Il28b Polymorphism Predicts Virologic Response in Patients with Hepatitis C Genotype 1 Treated with Boceprevir (Boc) Combination Therapy. J Hepatol 2011B;54:S6-S.

Poordad F, Hezode C, Trinh R, et al. ABT-450/r-ombitasvir and dasabuvir with ribavirin for hepatitis C with cirrhosis. N Engl J Med 2014;370:1973-82.

Poordad F, McCone J, Jr., Bacon BR, et al. Boceprevir for untreated chronic HCV genotype 1 infection. N Engl J Med 2011A;364:1195-206.

Prokunina-Olsson L, Muchmore B, Tang W, et al. A variant upstream of IFNL3 (IL28B) creating a new interferon gene IFNL4 is associated with impaired clearance of hepatitis C virus. Nat Genet 2013;45:164-71.

Rangnekar AS, Fontana RJ. IL-28B polymorphisms and the response to antiviral therapy in HCV genotype 2 and 3 varies by ethnicity: a meta-analysis. J Viral Hepat 2013;20:377-84.

Rauch A, Kutalik Z, Descombes P, et al. Genetic variation in IL28B is associated with chronic hepatitis C and treatment failure: a genome-wide association study. Gastroenterology 2010;138:1338-45, 45 e1-7.

Salvatierra K, Fareleski S, Forcada A, et al. Hepatitis C virus resistance to new specifically-targeted antiviral therapy: A public health perspective. World J Virol 2013;2:6-15.

Scheel TK, Rice CM. Understanding the hepatitis C virus life cycle paves the way for highly effective therapies. Nat Med 2013;19:837-49.

Schlecker C, Ultsch A, Geisslinger G, et al. The pharmacogenetic background of hepatitis C treatment. Mutat Res 2012;751:36-48.

Shaker OG, Nassar YH, Nour ZA, et al. Single-nucleotide polymorphisms of IL-10 and IL-28B as predictors of the response of IFN therapy in HCV genotype 4-infected children. J Pediatr Gastroenterol Nutr 2013;57:155-60.

Stättermayer AF, Strassl R, Maieron A, et al. Polymorphisms of interferon-λ4 and IL28B - effects on treatment response to interferon/ribavirin in patients with chronic hepatitis C. Aliment Pharmacol Ther 2014;39:104-11.

Suppiah V, Moldovan M, Ahlenstiel G, et al. IL28B is associated with response to chronic hepatitis C interferon-alpha and ribavirin therapy. Nat Genet 2009;41:1100-4.

Tajiri H, Tanaka Y, Takano T, et al. Association of IL28B polymorphisms with virological response to peginterferon and ribavirin therapy in children and adolescents with chronic hepatitis C. Hepatol Res 2014;44:E38-E44.

Tamai H, Mori Y, Shingaki N, et al. Low-dose pegylated interferon-α-2a plus ribavirin therapy for elderly and/or cirrhotic patients with hepatitis C virus genotype-1b and high viral load. Antivir Ther 2014;19:107-15.

Tanaka Y, Nishida N, Sugiyama M, et al. Genome-wide association of IL28B with response to pegylated interferon-alpha and ribavirin therapy for chronic hepatitis C. Nat Genet 2009;41:1105-9.

Thompson AJ, Muir AJ, Sulkowski MS, et al. Interleukin-28B polymorphism improves viral kinetics and is the strongest pretreatment predictor of sustained virologic response in genotype 1 hepatitis C virus. Gastroenterology 2010;139:120-9 e18.

Urban TJ, Thompson AJ, Bradrick SS, et al. IL28B genotype is associated with differential expression of intrahepatic interferon-stimulated genes in patients with chronic hepatitis C. Hepatology 2010;52:1888-96.

Vaidya A, Perry CM. Simeprevir: first global approval. Drugs 2013;73:2093-106.

World Health Organization [homepage on the Internet]. Hepatitis C. 2002. Available at http://www.who.int/csr/disease/hepatitis/whocdscsrlyo2003/en/. Accessed February 21, 2014.

Wu S, Kanda T, Nakamoto S, et al. Hepatitis C virus protease inhibitor-resistance mutations: Our experience and review. World J Gastroenterol 2013;19:8940-8.

Zeuzem S, Andreone P, Pol S, et al. Telaprevir for retirement of HCV infection. N Engl J Med 2011;364:2417-28.

Zeuzem S, Dusheiko GM, Salupere R, et al. Sofosbuvir and ribavirin in HCV genotypes 2 and 3. N Engl J Med 2014;370:1993-2001.

Zeuzem S, Jacobson IM, Baykal T, et al. Retreatment of HCV with ABT-450/r-ombitasvir and dasabuvir with ribavirin. N Engl J Med 2014;370:1604-1

Chapter 15

PHARMACOGENETICS OF HIV TREATMENT

Kimberly K. Scarsi, Pharm.D., M.S., BCPS; Sharon Seifert, Pharm.D.; Anthony T. Podany, Pharm.D.; Peter L. Anderson, Pharm.D.; and Courtney V. Fletcher, Pharm.D., FCCP

Learning Objectives

1. Justify the importance of performing viral resistance testing before initiating or changing antiretroviral therapy in HIV-infected patients.
2. Evaluate the chemokine CCR5 coreceptor tropism assay results to determine whether maraviroc is an appropriate treatment choice for an individual.
3. Assess an individual patient's genotypic test for HLA-B*5701 allele expression to recommend or avoid abacavir use in antiretroviral therapy.
4. Evaluate the role of *CYP2B6* polymorphisms in toxicities observed with efavirenz.
5. Discuss virus and host genomic tests that are currently recommended in national guidelines for the treatment of HIV infection.

Keywords: Antiretroviral therapy, drug resistance testing, HIV coreceptor tropism testing, abacavir hypersensitivity reaction, efavirenz, atazanavir, integrase strand transfer inhibitors

Abbreviations in This Chapter

ART	Combination antiretroviral therapy
CD4	CD4+ T lymphocyte
HIV	Human immunodeficiency virus
HLA	Human leukocyte antigen
HSR	Hypersensitivity reaction
UGT	Uridine 5'-diphospho-glucuronosyltransferase

Abstract

Combination antiretroviral therapy is highly effective against human immunodeficiency virus (HIV) infection, resulting in the evolution of HIV disease into a long-term illness. However, antiretroviral therapy is complicated by several factors, including the development of viral resistance that can render the antivirals ineffective, drug-associated toxicities, and complex drug distribution, transport, and metabolism that can result in significant drug-drug interactions. Using pharmacogenetics to assess viral genetic characteristics was incorporated into standard HIV care and treatment in the early 2000s, with HIV resistance testing to guide antiretroviral therapy selection. The role of pharmacogenetics in antiretroviral therapy has expanded significantly during the past decade through routine testing for HIV coreceptor tropism to determine the efficacy of maraviroc before initiating therapy and human leukocyte antigen testing to avoid a potentially life-threatening hypersensitivity reaction related to abacavir. Although not yet commonly used in clinical practice, individual polymorphisms in both cytochrome P450 (CYP) enzymes and uridine 5'-diphospho-glucuronosyltransferase metabolic pathways have emerged as a cause of wide interindividual variation in antiretroviral drug metabolism as well as antiretroviral-associated toxicities. Pharmacogenetics, which currently plays a substantial role in individualizing antiretroviral therapy, will continue to expand as more is discovered about optimizing the pharmacogenetics of drug transporters and metabolic pathways to minimize drug toxicities, manage potential drug-drug interactions, and optimize antiretroviral therapy. This chapter provides an overview of both viral and patient pharmacogenetic tests currently used in HIV therapy, together with pharmacogenetic considerations that are on the horizon.

Introduction

The impact of the human immunodeficiency virus (HIV) was first identified through a clinical report of *Pneumocystis carinii* pneumonia in five healthy young men in 1981 (CDC 1981), and it became a worldwide epidemic during the subsequent decade. Combination antiretroviral therapy (ART) has revolutionized the treatment of HIV infection by reducing the morbidity and mortality associated with HIV infection (Palella 1998). In fact, worldwide expanded access to ART decreased mortality from 2.3 million HIV-associated annual deaths in 2005 to 1.6 million annual deaths in 2012 (UNAIDS 2013). Therefore, current guidelines for the treatment of HIV in the United States recommend considering ART for all patients infected with HIV (Panel on Antiretroviral Guidelines for Adults and Adolescents 2014; Günthard 2014). This expanded use of ART, particularly in resource-rich settings, has resulted in HIV's transformation from a fatal disease to a lifelong, manageable, chronic condition (Samji 2013). In addition to improving the health of the HIV-infected individual, ART plays an essential role in HIV prevention. Effective treatment of an HIV-infected partner significantly reduces the risk of HIV transmission to uninfected partners (Cohen 2011), emphasizing the importance of access to HIV care services by HIV-infected individuals. Similarly, the preemptive use of antiretrovirals in HIV-negative persons at high risk of continued exposure to HIV can also reduce the risk of new HIV infection (preexposure prophylaxis), introducing a new population using antiretrovirals (WHO 2013).

HIV targets the host immune system, specifically the CD4+ T-lymphocyte (CD4) cells, and replicates inside the host cells. Antiretroviral agents target steps required for HIV replication, including viral entry into the CD4 cell (entry inhibitors), transcription of viral RNA into DNA (nucleoside/tide reverse transcriptase inhibitors and nonnucleoside reverse transcriptase inhibitors), incorporation of viral DNA into host DNA in the cell nucleus (integrase strand transfer inhibitors), and maturation of the virus (protease inhibitors). Combination antiretroviral therapy typically includes two nucleoside/tide reverse transcriptase inhibitors in combination with at least one medication from another antiretroviral class (Panel on Antiretroviral Guidelines for Adults and Adolescents 2014; Günthard 2014). The success of ART is measured by a decrease in the HIV-RNA (viral load) in plasma to less than that detectable by current HIV-RNA assays (less than 20–75 copies/mL of HIV-RNA, depending on the assay used). Consequently, this reduction in viral replication should result in sustained or improved immune function, measured by CD4 cell count, preventing HIV-associated morbidity and mortality (Panel on Antiretroviral Guidelines for Adults and Adolescents 2014).

Although highly effective, ART is not without clinical management challenges. Selection of the correct ART is influenced by both viral and patient factors, which can affect the efficacy and toxicity of the selected ART. Patient pharmacogenetic characteristics conferring a higher risk of drug-associated adverse events for three antiretrovirals—abacavir, atazanavir, and efavirenz—have been associated with premature treatment discontinuation (Lubomirov 2011). Similarly, viral genetic factors that influence the susceptibility of the virus to specific antiretrovirals may be detected before initiating therapy, allowing clinicians to choose regimens most likely to be effective against the patient's individual HIV virus (Panel on Antiretroviral Guidelines for Adults and Adolescents 2014). This chapter will focus on specific examples of both patient and viral genetic variation that can optimize HIV therapy outcomes.

Genetic Testing for Virus Characteristics

Drug Resistance Testing in HIV Therapy

HIV resistance mutations are known to occur frequently during the viral replication process because of spontaneous mutations, which are related to a lack of the 3′ to 5′ exonuclease proofreading mechanism during viral replication (Volberding 2012). Resistance mutations occur by amino acid base substitutions, deletions, and additions at the reverse transcriptase, protease, integrase, and envelope genes of the HIV DNA. Changes in viral coreceptor expression also cause resistance to some entry inhibitor antiretrovirals, which will be described in the Maraviroc: Coreceptor Tropism section. Individual mutations can cause complete resistance to several drugs in a class, such as the K103N mutation, which confers resistance to both efavirenz and nevirapine, or the M184V mutation, which confers resistance to both lamivudine and emtricitabine (Wensing 2014). Other antiretrovirals, including the protease inhibitors, dolutegravir, and etravirine, require multiple mutations to confer resistance to an individual agent. Examples of the reverse transcriptase mutations that can confer resistance to the nucleoside/tide and nonnucleoside reverse transcriptase inhibitors are described in Figure 15.1.

A virus that does not express any resistance mutations is commonly called wild-type virus. Each resistance mutation that a virus must maintain during the replication process slows the speed of viral replication; therefore, wild-type virus can replicate the most rapidly (Hinkley 2011; Martins 2010). Given this replication capacity, even if preexisting viral resistance exists, wild-type virus will become predominant in the viral population when ART is removed. Because of this predominance, wild-type virus is most likely to be transmitted during an exposure. However, in patients who are on a failing ART regimen (i.e., their HIV viral load

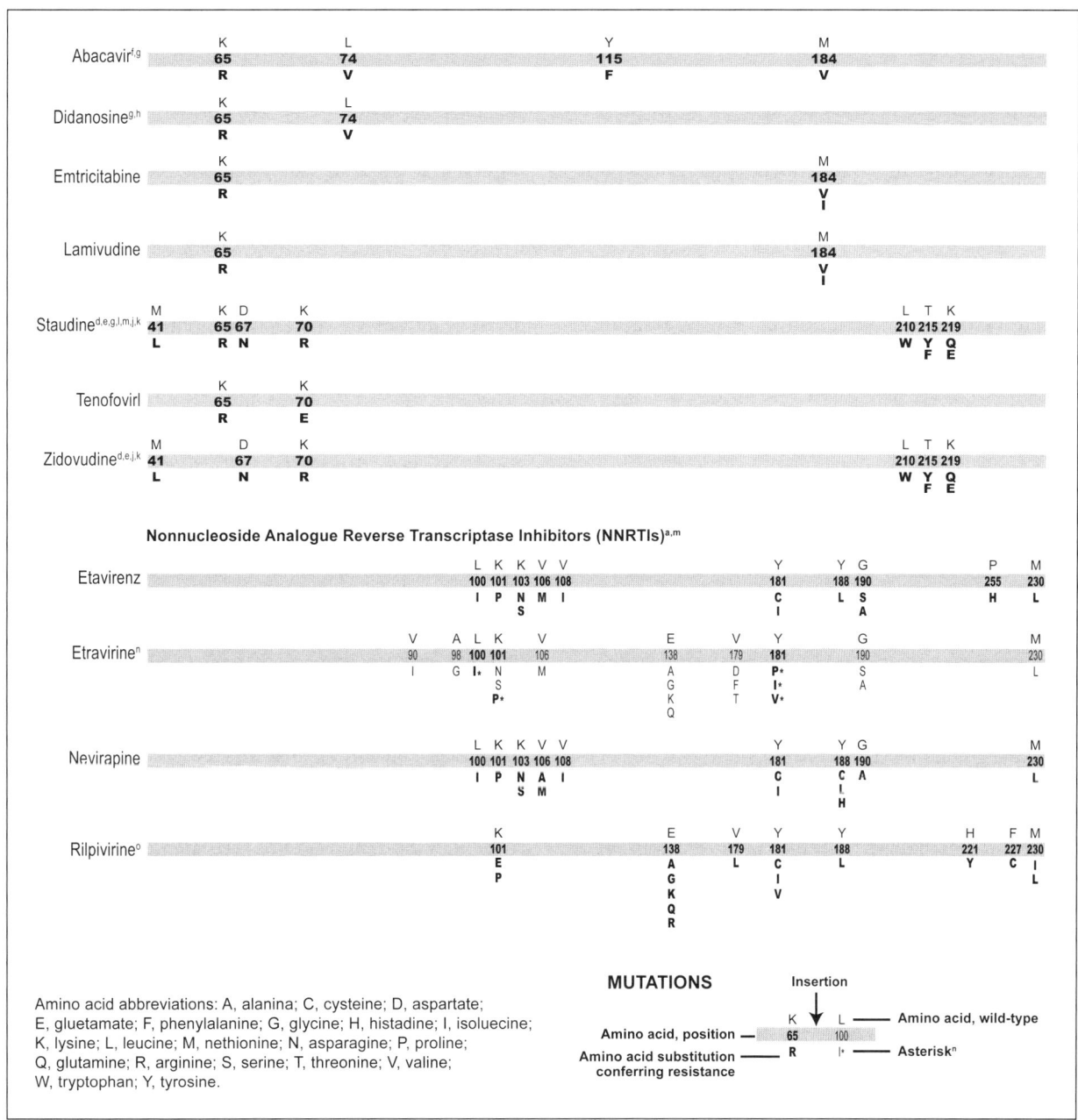

Figure 15.1. Nucleoside/tide and nonnucleoside reverse transcriptase inhibitor resistance mutations according to the International AIDS Society-USA.

The number represents the insertion site of the mutation. The letter above the bar represents the wild-type amino acid for HIV subtype B. The letter below the line indicates the amino acid substitution that creates the mutation. The mutation is expressed in script by the wild-type amino acid, followed by the insertion site number, followed by the mutation (e.g., K103N, M184V).
Reprinted with permission and updates from the International Antiviral Society—USA. Johnson VA, Calvez V, Günthard HF, et al. 2014 update of the drug resistance mutations in HIV-1. *Top Antivir Med* 2014;22(3):642-50. Updated information and User Notes are available at www.iasusa.org.

fails to suppress or resurges after initial suppression), HIV-expressing mutations that confer resistance to drugs contained in the current ART regimen may be the predominant virus. Whether acquired through transmitted drug resistance or caused by the development of resistance to specific components of ART, the virus may archive (integrate genetic material into long-lived cells) some mutations when selective pressure is removed (Volberding 2012). However, once reexposed to the antiretroviral for which it holds archived resistance, the virus can rapidly repopulate with that resistance mutation, resulting in virologic failure caused by drug resistance (Benson 2006).

Use of drug resistance testing in HIV therapy has been recommended by HIV guidelines since 2000 (Hirsch 2000). Before that, HIV drug resistance testing had not been shown to improve patient outcomes, and commercial laboratory tests were not readily available. However, in the late 1990s, commercial testing for drug resistance expanded, and clinical trials were designed to assess the impact of resistance testing on ART outcomes. Three prospective studies, conducted before the release of the guidelines, were all of patients whose existing ART strategy was currently failing (Cohen 2002; Baxter 2000; Durant 1999). Patients were randomized either to receive drug resistance testing to help guide subsequent ART selection, or to clinician selection of their next ART without the benefit of resistance testing. In each study, patients were more likely to achieve a virologic response to the new regimen when resistance testing was used to determine the ART. These studies, in addition to retrospective evaluations of drug resistance testing, established the foundation for the resistance testing guidelines (Hirsch 2000).

Although the original guidelines recommended consideration of testing ART-naive patients because of early cases of transmitted drug resistance in populations with a high HIV disease burden (Hirsch 2000), the current guidelines firmly recommend testing of all patients at entry into care, in addition to considering testing of patients with a significant delay between entering care and initiating ART (Panel on Antiretroviral Guidelines for Adults and Adolescents 2014; Panel on Antiretroviral Therapy and Medical Management of HIV-Infected Children 2014). Recent studies in the United States and Europe have shown that 6%–16% of patients may have transmitted drug resistance to at least one class of antiretroviral drugs and that transmitted resistance negatively affects ART outcomes. For patients initially infected with a resistant virus, the time required to revert to wild-type virus may be longer, allowing the detection of transmitted drug resistance mutations long after HIV transmission occurs; however, exactly how long it may take for these mutations to be detected has not been established (Hinkley 2011; Martins 2010).

Types of Resistance Testing

Genotype Resistance Test

Genotypic tests are the preferred method for HIV resistance testing in U.S. treatment guidelines (Panel on Antiretroviral Guidelines for Adults and Adolescents 2014; Panel on Antiretroviral Therapy and Medical Management of HIV-Infected Children 2014). All genotypic resistance tests sequence the protease and reverse transcriptase genes, and newer tests are available to test the integrase and envelope genes. The understanding of specific mutations and their relationship with antiretroviral efficacy is a quickly evolving specialty. The International AIDS Society-USA (Wensing 2014) and the Stanford University Drug Resistance Database (www.hivdb.stanford.edu) maintain a list of all known mutations as well as guidance for interpreting the mutations. Expert interpretation of the HIV resistance genotype is recommended for patients with extensive resistance because studies have found improved outcomes when HIV resistance specialists provided expert advice (Panel on Antiretroviral Guidelines for Adults and Adolescents 2014; Panel on Antiretroviral Therapy and Medical Management of HIV-Infected Children 2014). Genotypic testing is widely available, and test results are returned within around 1–2 weeks.

Phenotypic Resistance Testing

Phenotypic testing inserts the individual's virus genes of interest, including the protease, reverse transcriptase, integrase, and entry genes, into a pseudovirus. The pseudovirus is then replicated in vitro with various concentrations of antiretrovirals. The drug concentration that inhibits viral replication of the patient-specific virus by 50% (IC50) is identified, and the fold change in IC50 (patient virus to wild-type virus) is reported (Panel on Antiretroviral Guidelines for Adults and Adolescents 2014; Panel on Antiretroviral Therapy and Medical Management of HIV-Infected Children 2014). Although phenotypic testing is widely available, it is more expensive and takes longer to receive the results (2–3 weeks) compared with genotypic testing. Studies comparing ART regimen selection according to genotype with phenotypic resistance testing result in similar patient outcomes; therefore, phenotypic testing is currently not routinely recommended in treatment guidelines (Panel on Antiretroviral Guidelines for Adults and Adolescents 2014; Panel on Antiretroviral Therapy and Medical Management of HIV-Infected Children 2014). However, the phenotypic testing results can be useful to design the best regimen for a patient with resistance to many antiretroviral classes, particularly those with several protease inhibitor resistance mutations.

Practical Resistance Testing Considerations and Challenges

Genotypic and phenotypic drug resistance testing is widely available through commercial laboratories, and the incorporation of resistance testing into HIV treatment represents the successful implementation of a viral genetic test to optimize ART. If these tests are done to evaluate drug resistance as a case of ART failure, then testing is recommended either while the patient is still taking the failing ART or within 4–6 weeks of discontinuing the regimen. This timeframe is important to consider because after a failing regimen is discontinued, wild-type virus quickly reemerges (Miller 2000). Drug resistance testing may not detect the presence of resistant viral strains when they represent less than 10%–20% of the circulating virus population; therefore, delays between therapy discontinuation and resistance testing can result in suboptimal detection of all resistance mutations. Similarly, if a patient's antiretroviral regimen has previously failed because of the development of resistance, that mutation may be archived and not identified in resistance testing performed during ART that does not exert selective pressure for that mutation. Therefore, interpretation of resistance testing must always consider prior resistance results, prior ART exposure and outcomes, and patient adherence at the time of resistance testing (Panel on Antiretroviral Guidelines for Adults and Adolescents 2014; Panel on Antiretroviral Therapy and Medical Management of HIV-Infected Children 2014).

The patient's HIV viral load should ideally be 1000 copies/mL or greater and at least greater than 500 copies/mL at the time of testing. This is because commercial laboratories cannot reliably amplify low levels of virus, jeopardizing the reliability of the resistance test. Recently, some studies have found an association between low-level viremia (less than 1000 copies/mL) and ART failure because of the accumulation of viral resistance mutations (Panel on Antiretroviral Guidelines for Adults and Adolescents 2014; Swenson 2014). New methods for genotypic resistance testing in patients with low-level viremia may make the detection of emerging resistance possible; these tests are not yet commercially available, but they are on the horizon (Gonzalez-Serna 2014; Santoro 2014).

Maraviroc: Coreceptor Tropism

HIV entry into a cell is a multistep process: (1) attachment to the CD4 receptor; (2) binding of HIV to the chemokine receptor CCR5 or, alternatively, the chemokine receptor CXCR4 molecules on the cell surface (called the coreceptor); and (3) fusion of the viral and cell membranes (Panel on Antiretroviral Guidelines for Adults and Adolescents 2014). Viruses that use exclusively the CCR5 or CXCR4 coreceptor to bind to the cell are called R5-tropic or X4-tropic, respectively, whereas viruses that can use both CCR5 and CXCR4 are called dual- or mixed-tropic virus. Early in the course of HIV infection, R5-tropic virus is common, whereas over time and with accumulation of drug resistance, the virus is likely to shift exclusively to X4-tropic or dual/mixed-tropic virus (i.e., the virus population can bind to either CCR5 or CXCR4 coreceptors). Maraviroc is the only coreceptor antagonist available to date, and according to the package insert, it is specifically a CCR5 antagonist. Therefore, maraviroc is not effective against HIV that is X4-tropic or dual-tropic. The ability to change the coreceptor that the virus may use for cell entry is the mechanism by which the virus develops resistance to maraviroc.

The coreceptor used for cell entry by an individual's HIV is identified through phenotypic tropism testing, an in vitro process by which laboratory viruses are generated from the patient's virus envelope gene (surface and transmembrane proteins gp120 and gp41). These pseudoviruses infect target cells that express either CCR5 or CXCR4. The patient's virus is designated R5-tropic if only the cells expressing CCR5 chemokine receptors are infected, X4-tropic if the CXCR4 chemokine receptor cells are infected, or dual/mixed-tropic if both cell types are infected (Whitcomb 2007). Because of maraviroc's mechanism of action, a tropism assay was developed in conjunction with the drug and implemented in initial clinical trials (Trofile, Monogram Biosciences, South San Francisco, CA). The current test has 100% sensitivity when the X4 clones represent at least 0.3% of the overall virus population (Trinh 2008). Logistic challenges are common with the phenotypic tests, including a long turnaround time for test results (2–3 weeks) and high cost (greater than $1500). In addition, the patient's viral load must be 1000 copies/mL or greater at the time of testing (Panel on Antiretroviral Guidelines for Adults and Adolescents 2014; Whitcomb 2007).

More recently, genotypic assays that sequence the V3-coding region of the HIV envelope gene have become available (Trofile DNA). Genetic tests use cell-derived viral DNA from a 4-mL sample of whole blood (compared with the plasma viral RNA used in the traditional phenotypic tests) (Monogram Biosciences). After envelope gene sequencing, algorithms are used to predict the virus coreceptor tropism. To date, no prospective studies have been conducted with the genotypic tropism tests, but investigators have retrospectively compared the coreceptor genotypic results with the phenotypic results used in clinical trials of maraviroc. Although genotyping has a relatively high specificity (91%–93%), it has a much lower sensitivity (68%–81%) for detecting non-R5 virus (Swenson 2013; McGovern 2010). However, the clinical outcomes (virologic response) are consistent between phenotypic and genotypic selection of patients with R5-tropic virus.

Guidelines for the Use of Tropism Testing
Guidelines for the treatment of HIV infection in the United States recommend a phenotypic tropism test when maraviroc is being considered part of ART as well as in a patient whose maraviroc-based ART is currently failing to detect whether the patient's HIV is expressing R5-tropic virus alone (Panel on Antiretroviral Guidelines for Adults and Adolescents 2014). In the United States, the phenotypic test is recommended because of the predominance of prospective evidence with this test; however, genotypic testing is considered an alternative because of its faster return of results and lower cost compared with phenotypic tests. In addition, the genotypic method of coreceptor testing is the only method available for patients with a viral load of less than 1000 copies/mL, which may be useful in determining ART treatment options for patients who are intolerant of a regimen yet virologically suppressed on an existing regimen.

Although tropism testing is an essential tool in ensuring the efficacy of maraviroc, the necessity of this test is often cited as a reason for clinicians not choosing maraviroc as part of first-line ART. Maraviroc is not currently recommended as first-line therapy for ART-naive patients in the United States, but it may be used in special circumstances or as second-line therapy for subjects who are not naive to ART (Panel on Antiretroviral Guidelines for Adults and Adolescents 2014; Günthard 2014). Guidelines reference the high cost of tropism testing, in addition to the availability of several other efficacious ART options, as being the reason for maraviroc's disfavor in current guidelines.

Genetic Testing for Host Characteristics

Abacavir: HLA-B*5701
Abacavir is a nucleoside reverse transcriptase inhibitor that is used as part of ART for both antiretroviral-naive patients starting their first ART and treatment-experienced patients switching to a new ART regimen (Panel on Antiretroviral Guidelines for Adults and Adolescents 2014; Günthard 2014). In early clinical trials, an abacavir hypersensitivity reaction (HSR) was identified in 8% (range, 2%–9%) of subjects receiving abacavir-containing ART and was characterized by a combination of nonspecific, multiorgan clinical symptoms, according to the abacavir package insert. The HSR most often occurs within 6 weeks, with a median time of 9 days, after beginning abacavir therapy. Abacavir HSR is diagnosed by the occurrence of two or more of the following symptoms: fever, rash, and gastrointestinal, constitutional, or respiratory signs or symptoms. Fever, rash, and general malaise were reported in greater than 50% of patients reporting the HSR, but unlike other HSRs, a rash is not always present. Symptoms associated with abacavir HSR worsen if abacavir therapy is not discontinued, and they can result in rapid, life-threatening HSR on reinitiation of abacavir therapy (Hetherington 2001; Walensky 1999). For patients in whom an abacavir HSR is likely or cannot be ruled out, abacavir should be discontinued and not be reinitiated according to the package insert. Given the non-specific symptoms associated with abacavir HSR, misdiagnosis of a HSR is a critical issue in the treatment of HIV infection, resulting in fewer ART treatment options.

Pharmacogenetic Association with Abacavir HSR
In 2002, the first evidence emerged that linked the variation of human leukocyte antigen (HLA) alleles and the occurrence of abacavir HSR in a retrospective study of predominantly white male clinical trial participants (Hetherington 2002; Mallal 2002). More detail regarding the function of the HLA locus, located in the major histocompatibility complex, and its role in adverse drug reactions can be found in chapter 9, HLA Pharmacogenetics and Serious Cutaneous Adverse Drug Reactions. Briefly, *HLA* is part of the major histocompatibility complex class I genes, and its role is to form peptide complexes that present peptides on the cell surface. This presentation elicits a T-cell immune response if the peptide is recognized as non-self (Martin 2013). Given that specific *HLA* genes are known to predispose individuals to other cutaneous adverse drug reactions, early reports of lower risk of HSR in patients of African ancestry and a familial link for the occurrence of the HSR all led to the investigation of HLA alleles and their role in abacavir HSR (Symonds 2002; Peyrieere 2001).

Early studies identified the presence of HLA-B*5701 as predictive of abacavir HSR, and this allele was confirmed by follow-up work in patients who underwent HLA typing when starting or changing abacavir therapy between 2002 and 2005 (Rauch 2006). The frequency of the HLA-B*5701 allele is highly variable depending on geographic and racial backgrounds, with the highest rates reported in Southwest Asian and European populations, where the reported prevalence ranges from 3.8% to 19.6% and 1.4%

to 10.2%, respectively (Martin 2013). A lower prevalence of the allele is observed in patients of African ancestry (range, 0%–3.2%).

Although individual centers began implementing HLA-B*5701 screening after the retrospective associations were identified between the allele and the HSR, PREDICT-1 was the first prospective, randomized, double-blind study of any genetic test before drug therapy initiation (Mallal 2008). This multinational study randomly assigned 1956 subjects to either HLA-B*5701 screening, with subsequent exclusion of abacavir-containing ART in subjects with a positive HLA-B*5701 result (prospectively screened), or standard-of-care selection of abacavir-containing ART (control group). The PREDICT-1 study found no cases of immunologically confirmed HSR in the prospectively screened group compared with 2.7% in the control group (p<0.001). In addition, fewer patients in the prospectively screened group had clinically diagnosed HSR compared with the control group (3.4% vs. 7.8%, p<0.001). This study showed that the positive predictive value of HLA-B*5701 testing was 47.9%, suggesting that some patients who are positive for HLA-B*5701 do not experience HSR, although the negative predictive value is 100%. Therefore, this test is highly effective at identifying patients at risk of abacavir HSR.

Abacavir Hypersensitivity Screening

Results from the PREDICT-1 and other HLA-B*5701 studies prompted the U.S. Food and Drug Administration (FDA) to add a black box warning about the risk of HSR in subjects positive for HLA-B*5701 to all abacavir-containing products, according to the abacavir package insert. Guidelines from expert panels on HIV therapy and the Clinical Pharmacogenetics Implementation Consortium recommend routine testing for HLA-B*5701 before abacavir use (Panel on Antiretroviral Guidelines for Adults and Adolescents 2014; Martin 2012). Despite the variation in allele frequency across populations, the presence of the allele remains highly predictive of abacavir HSR; therefore, testing should be offered to all patients, irrespective of the likelihood of the presence of the HLA-B*5701 allele in the population. If patients test positive for HLA-B*5701, abacavir therapy should be avoided, and an abacavir allergy should be noted in their medical record (Panel on Antiretroviral Guidelines for Adults and Adolescents 2014). Rarely, patients with a negative HLA-B*5701 test may develop clinically diagnosed abacavir HSR, according to the package insert and the Panel on Antiretroviral Guidelines for Adults and Adolescents (2014). Of note, the Clinical Pharmacogenetics Implementation Consortium has published guidelines highlighting the data supporting the use of HLA-B*5701 testing and up-to-date commercial laboratories with HLA testing (Martin 2012).

Screening for abacavir HSR is one of the best examples of the widespread implementation of pharmacogenetic screening to prevent an adverse drug event, and it is widely available for commercial use. In addition, abacavir HSR screening has consistently been shown to be a cost-effective intervention. Specifically, one evaluation in 2008 found that HLA-B*5701 testing costs $36,700 per quality-adjusted life-year (Schackman 2008). The cost-effectiveness was maintained as long as abacavir-containing ART was less expensive than, and as effective as, alternative ART regimens. HLA-B*5701 testing is required only once per lifetime, and duplicating the test will negatively affect the cost-effectiveness (Panel on Antiretroviral Guidelines for Adults and Adolescents 2014). Therefore, to facilitate transitions in care from provider to provider, patients should be educated about the importance of this test result and be knowledgeable about the implications of their individual results.

Efavirenz and Nevirapine Metabolism and the Impact of *CYP2B6* Polymorphisms

Nonnucleoside reverse transcriptase inhibitor–containing antiretroviral regimens are first-line treatment options for both adults and children in both domestic and international HIV treatment guidelines (Panel on Antiretroviral Guidelines for Adults and Adolescents 2014; Panel on Antiretroviral Therapy and Medical Management of HIV-Infected Children 2014; WHO 2013; Günthard 2014). The nonnucleoside reverse transcriptase inhibitor efavirenz continues to be one of the most widely used antiretrovirals in the United States, and a second nonnucleoside reverse transcriptase inhibitor, nevirapine, has been widely used in international settings because it's efficacious as well as cost-effective. Both efavirenz and nevirapine are characterized by considerable pharmacokinetic variability; however, a large portion of this variability can be explained by patient-specific genetic factors (Saitoh 2007; Ward 2003). Nonnucleoside reverse transcriptase inhibitors are primarily oxidized by the cytochrome P450 (CYP) enzyme system. The *CYP2B6* pathway of efavirenz metabolism is thought to contribute to more than 90% of the overall metabolism of efavirenz, with *CYP2A6* as a secondary pathway (Di Iulio 2009; Ward 2003). Similarly, CYP3A and *CYP2B6* are the major CYP isoenzymes responsible for nevirapine metabolism (Erickson 1999; Riska 1999). Polymorphisms in *CYP2B6* significantly affect the plasma concentrations of both efavirenz and nevirapine (Ramachandran 2009; Penzak 2007).

The influence of T/T substitutions at the 516 position of the *CYP2B6* gene significantly affects nevirapine metabolism. Small studies of adult patients have shown a 1.5- to 1.7-fold increase in nevirapine 12-hour postdose concentrations in patients with *CYP2B6* 516 T/T alleles compared with 516 G/G alleles (p=0.011 and p<0.006, respectively) (Penzak 2007; Rotger 2005a). In a pediatric study of 126 children receiving nevirapine-based ART, the *CYP2B6* 516 T/T genotype was associated with a greater than 30% reduction in nevirapine clearance (Saitoh 2007). Even though the effects of patient-specific *CYP2B6* genetics on nevirapine metabolism have been consistently shown in various populations, pharmacogenetic screening and dosing strategies are not widely implemented in clinical practice. Therefore, the remainder of this section will focus on pharmacogenetic implications for efavirenz.

CYP2B6 Polymorphisms Associations with Efavirenz Efficacy and Toxicity

Efavirenz pharmacogenetics has been widely studied since its approval in 1998. The necessity of establishing relationships between patient-specific genetic characteristics and plasma concentrations of efavirenz is strengthened by the fact that efavirenz has a relatively narrow therapeutic range. Efavirenz plasma concentrations need to be maintained above a threshold associated with virologic efficacy, but below a threshold associated with an increased likelihood of adverse effects that may lead to therapy interruptions and discontinuations.

Mid-dosing–interval efavirenz plasma concentrations less than 1 mg/L have been associated with greater rates of virologic failure, whereas plasma concentrations greater than 4 mg/L have been associated with greater rates of central nervous system (CNS) adverse effects (Marzolini 2001).

Individuals with the *CYP2B6* 516 G/T or T/T polymorphism are often called "poor" or "slow" metabolizers of efavirenz (Di Iulio 2009). Patients with reduced efavirenz metabolism related to *CYP2B6* polymorphisms have higher plasma exposure to efavirenz. The *CYP2B6* 516 T/T genotype, which may occur in up to 20% of individuals with African ancestry, is associated with up to a 4.8-fold increase in 24-hour area under the curve (AUC) compared with the *CYP2B6 516* G/G genotype, which is more common in European-American populations (Haas 2004). Published data clearly depict the effect of *CYP2B6* polymorphisms on efavirenz clearance. In a study of 152 patients taking efavirenz, median elimination half-lives of 23, 27, and 48 hours (p<0.001) were associated with the 516 G/G, 516 G/T, and 516 T/T *CYP2B6* genotypes, respectively (Ribaudo 2006). Longer elimination half-lives associated with the *CYP2B6* 516 G/T and T/T genotypes increase the risk of developing resistance mutations during treatment discontinuations because lingering subtherapeutic concentrations of efavirenz allow residual HIV to select for efavirenz resistance mutations. This viral mutation and selection process is clinically important because efavirenz has a

Table 15.1. Investigational Efavirenz Dosing for Children 3 Months of Age to Younger Than 3 Years According to the *CYP2B6* Genotype

CYP2B6 GG and GT		*CYP2B6* TT	
Weight (kg)	EFV Dose (mg)	Weight (kg)	EFV Dose (mg)
3–4.99	200	3.3–6.99	50
5–6.99	300	7–13.99	100
7–13.99	400	14–16.99	150
14–16.99	500	≥ 17	150
≥ 17	600		

EFV = efavirenz.

Adapted from: Panel on Antiretroviral Therapy and Medical Management of HIV-Infected Children. Guidelines for the Use of Antiretroviral Agents in Pediatric HIV Infection. February 2014. Available at http://aidsinfo.nih.gov/contentfiles/lvguidelines/pediatric-guidelines.pdf. Accessed August 1, 2014.

low genetic barrier to resistance; even single nucleotide changes have the ability to render resistance to other nonnucleoside reverse transcriptase inhibitors.

Efavirenz use has been associated with high rates of CNS-related adverse effects. As many as 20%–40% of patients receiving the standard daily oral dose of 600 mg will experience some form of CNS adverse effects, including dizziness, somnolence, insomnia, abnormal dreams, and confusion (Gazzard 1998). Central nervous system toxicity is 3 times more frequent in patients with plasma efavirenz concentrations above 4 mg/L (Marzolini 2001). Although efavirenz concentrations are already known to be significantly elevated in subjects with the 516 T/T genotype (Haas 2004), the genotype has also been independently associated with higher rates of both sleep disorders and fatigue in European subjects receiving efavirenz (Rotger 2005a). These associations between efavirenz concentrations and the *CYP2B6* genotype have implications for the tolerability of efavirenz-based regimens, particularly among populations with a high prevalence of the 516 T/T genotype. This was shown in one recent study where subjects switched ART more frequently when receiving efavirenz-based ART compared with dolutegravir-based ART because of CNS-related adverse events (10% vs. 2%, respectively) (Walmsley 2013).

Clinical Application of CYP2B6 Genotyping
Current U.S. guidelines for use of antiretroviral agents in pediatric HIV infection list investigational efavirenz dosing recommendations for children from 3 months to 3 years of age according to their *CYP2B6* genotype (Panel on antiretroviral Therapy and Medical Management of HIV-Infected Children 2014). These recommendations are the result of a study of HIV- and HIV/tuberculosis-coinfected children younger than 3 years with variable efavirenz dosing by both weight and *CYP2B6* genotype. In children who received weight-based efavirenz dosing without respect to genotype, a high proportion with the *CYP2B6* 516 T/T genotype had excessive exposure to efavirenz. These results prompted the updated guidelines reflecting both weight- and genotype-based dosing regimens for pediatric patients younger than 3 years, with greatly reduced dosages for patients with the *CYP2B6* 516 T/T genotype, as shown in Table 15.1.

Although the U.S. pediatric antiretroviral guideline recommendation is a recent change, no accepted pharmacogenetic guidelines suggest the use of *CYP2B6* genotype for any drug or dose selection, and commercial access to *CYP2B6* genotyping is not widely available. In addition, although the *CYP2B6* 516 allele is most commonly cited, the occurrence of other *CYP2B6* and minor efavirenz metabolic pathway polymorphisms (e.g., *CYP2B6* 983 T/C and rs4803419) may contribute to the pharmacokinetic variability of efavirenz (Holzinger 2012). Clinically, slow metabolism of efavirenz because of pharmacogenetic variability will result in supratherapeutic concentrations, which may contribute to the severity and persistence of efavirenz CNS toxicities, as previously described. Therefore, clinicians may use therapeutic drug monitoring for efavirenz or clinical assessment of persistent or severe efavirenz-associated CNS toxicities as a surrogate measure of poor CYP2B6 metabolism.

Uridine 5′-diphospho-glucuronosyltransferase
Uridine 5-diphospho-glucuronosyltransferase (UGT) enzymes carry out glucuronidation reactions in the body by transferring glucuronic acid to one of several endogenous and drug compounds, allowing the compounds to be more readily eliminated from the body in bile and/or urine (Wolkoff 2012). UGT1A1 is the only known UGT enzyme involved in the glucuronidation of bilirubin, a non-polar compound produced by the degradation of hemoglobin. Variation in the *UGT1A1* gene leads to two clinical syndromes, Crigler-Najjar and Gilbert syndromes, resulting in unconjugated hyperbilirubinemia as a result of impaired bilirubin clearance (Wolkoff 2012; Sampietro 1999). Crigler-Najjar syndrome is an extremely rare disorder (1 in 1,000,000 births), resulting in complete loss of *UGT1A1* activity (Crigler-Najjar syndrome type I) or very low levels of *UGT1A1* activity (Crigler-Najjar syndrome type II) (Wolkoff 2012; Sampietro 1999). Gilbert syndrome is a far more common condition, occurring in around 10% of the population, and is caused by seven instead of six thymine adenine (TA) repeats in the promoter region of the *UGT1A1* gene (Zhang 2005; Bosma 1995). This variant allele, denoted *UGT1A1*28*, causes a reduction in the expression of the functional UGT1A1 enzyme, effectively decreasing the level of enzyme activity and subsequent clearance of bilirubin by around 40% for homozygous carriers (Wolkoff 2012; Zhang 2005). The frequency with which the *UGT1A1*28* allele occurs differs across populations; this allele occurs more commonly in people of African heritage (43%) than in those of European descent (36%–39%), and it is relatively uncommon in people of Asian ancestry (11%–19%) (Rotger 2005b).

Protease Inhibitors and UGT1A1
Protease inhibitors are used as part of ART in both treatment-experienced and treatment-naive patients (Panel on Antiretroviral Guidelines for Adults and Adolescents 2014; Günthard 2014). Two newer protease inhibitors, atazanavir and darunavir, in combination with ritonavir as a pharmacokinetic-enhancing agent, are most commonly

used because of their increased efficacy, reduced toxicities, and more convenient dosing requirements compared with older protease inhibitors. Several protease inhibitors have been shown to inhibit *UGT1A1* but have varying degrees of affinity for the enzyme (Zhang 2005). Clinically, atazanavir has most often been associated with unconjugated hyperbilirubinemia (Culley 2013).

This adverse effect is magnified as a result of increased atazanavir concentrations with the coadministration of ritonavir. When atazanavir 300 mg plus ritonavir 100 mg daily is used in patients with preexisting impairments in bilirubin conjugation caused by Gilbert syndrome, the risk and severity of hyperbilirubinemia is increased (Ribaudo 2013; Lubomirov 2011; Rotger 2005b). Patients treated with atazanavir/ritonavir who are homozygous for the *UGT1A1*28* allele tend to experience greater increases in unconjugated bilirubin levels compared with heterozygotes and non-carriers, with bilirubin levels that often exceed 5 times the upper limit of normal. This may lead to higher discontinuation rates because of the development of jaundice and/or scleral icterus (Lubomirov 20igure 11). The decision to discontinue treatment with atazanavir/ritonavir after the development of clinically significant hyperbilirubinemia, jaundice, and/or scleral icterus depends largely on the severity of the elevated bilirubin levels, the comfort level of the clinician, and the patient's willingness to tolerate the cosmetic effects (yellowing of the skin and eyes). The discontinuation of the boosting agent, ritonavir, and the subsequent increase in dose of atazanavir from 300 mg to 400 mg may lead to jaundice resolution in some patients (Ferraris 2012); however, it is not appropriate in patients with a history of protease inhibitor resistance or in those who are receiving atazanavir with concomitant agents that reduce atazanavir concentrations (e.g., efavirenz, tenofovir, gastric acid–reducing agents). Any other dose reduction of atazanavir is not recommended at this time because there are no long-term efficacy studies involving reduced doses, according to the atazanavir package insert and the Panel on Antiretroviral Guidelines for Adults and Adolescents (2014).

To date, there are no recommendations regarding routine genetic testing for the *UGT1A1*28* allele before initiating treatment with atazanavir according to the package insert. The cost-effectiveness of the preemptive use of the test remains uncertain, and there is a lack of prospective data showing the benefit of such genetic testing (Culley 2013; Schackman 2013). Before initiating treatment with atazanavir, clinicians should discuss the risk of developing hyperbilirubinemia and potential jaundice and/or scleral icterus with patients but should reassure them that these symptoms are not associated with liver damage and that they will reverse on discontinuing the medication. If possible, atazanavir use should be avoided in patients with previously diagnosed Gilbert syndrome, and a baseline chemistry panel that includes a serum bilirubin level should be done for all patients before initiating atazanavir to identify those with undiagnosed Gilbert syndrome or underlying liver disease (Panel on Antiretroviral Guidelines for Adults and Adolescents 2014).

Integrase Strand Transfer Inhibitors and UGT

Integrase strand transfer inhibitors (integrase inhibitors) are the newest class of antiretrovirals. This drug class includes raltegravir, which was the first integrase inhibitor approved in 2007; elvitegravir; and dolutegravir, the most recently approved antiretroviral. These agents are used as part of ART in treatment-naive and treatment-experienced individuals (Panel on Antiretroviral Guidelines for Adults and Adolescents 2014). Raltegravir and dolutegravir are primarily metabolized by *UGT1A1* and have increased concentrations in patients with Gilbert syndrome. Homozygous *UGT1A1*28* carriers have shown about 40% increases in AUC and maximum concentration values for both raltegravir and dolutegravir (Chen 2014; Siccardi 2012; Wenning 2009). However, both raltegravir and dolutegravir have wide therapeutic windows, and the increase in concentrations has not been associated with serious adverse effects.

A Role for Pretherapy Screening of UGT1A1?

Whole-blood testing for the *28 allele of the *UGT1A1* gene is commercially available (Invader UGT1A1 Molecular Assay, Third Wave Technologies, Madison, WI). *UGT1A1* testing has received the most attention for use in patients receiving irinotecan, where the *28 allele has been associated with severe neutropenia, as discussed in chapter 11, Germline Pharmacogenetics in Oncology. Currently, in ART therapy, no recommendations exist for preemptive genetic testing, but clinicians should be cognizant of the potential implications of *UGT* variations when monitoring response to therapy with these antiretroviral agents.

Future Directions of Pharmacogenetics in Clinical Practice of HIV Therapeutics

Pharmacogenetics of Drug Transporters

Drug transporters are known to affect the bioavailability, cellular accumulation, and penetration of antiretrovirals into several cells and sanctuary sites (e.g., brain and CNS; male and female genital tracts) throughout the body. The relationship between the drug concentration at various sites and in cells is related to a relationship between the influx and efflux transporters for which individual drugs are substrates (Michaud 2012). Understanding this complex relationship between various drugs and drug transporters is a rapidly expanding field of research. Although drug transporter pharmacogenetics is not yet used to guide clinical practice, it is helpful to understand how these transporters may affect both the beneficial and the adverse pharmacologic response to a medication.

One example of this relationship occurs with the nucleotide reverse transcriptase inhibitor tenofovir. Although infrequent, there is a dose-related nephrotoxicity associated with tenofovir use. The mechanism of nephrotoxicity is still unclear, but it may be related to the mitochondrial toxicity caused by the accumulation of tenofovir in renal proximal tubule cells. Tenofovir enters the renal cells through the organic anion transporters 1 and 3 (OAT-1 and OAT-3) and is then eliminated in the urine by MRP4-mediated active efflux from the cell (Ray 2007). Accumulating evidence suggests that polymorphisms in these drug transporter genes lead to higher rates of nephropathy as well as to disproportionate urine/plasma concentrations of tenofovir (Calcagno 2014; Michaud 2012). Although the presence of these transporter polymorphisms may help clinicians identify patients at risk of tenofovir-associated nephrotoxicity, the required tests are available only for research purposes.

Another important role that drug transporter pharmacogenetics may play in ART success pertains to emerging health concerns for patients living with long-term HIV. Continued viral replication in sanctuary sites, despite fully suppressed HIV-RNA in peripheral blood, contributes to persistent inflammation and immunologic activation, leading to cardiovascular complications and non–AIDS-related malignancies (Deeks 2013). Recent data suggest that suboptimal concentrations of antiretroviral drugs in the gastrointestinal lymphoid tissue correlate with continued viral replication in the same tissues (Fletcher 2014). Complementary to these findings, expressions of drug efflux transporter proteins P-glycoprotein and MRP2 were 1.9- and 1.5-fold higher, respectively, in the sigmoid colon of HIV-infected patients exposed to ART than in ART-naive individuals (De Rosa 2013). Up-regulation of these efflux pumps will significantly affect the tissue distribution of antiretrovirals proposed for rectally administered preexposure prophylaxis, as well as penetration of antiretrovirals into these sites for treatment. Understanding the additional variation in antiretroviral drug exposure related to drug transporter expression will be critical in fully characterizing antiretroviral exposure in sanctuary sites.

The Role of Pharmacogenetics in Drug-Drug Interactions

The role of pharmacogenetics in anticipating the pathways involved in drug-drug interactions is only beginning to be elucidated. An excellent illustration is the drug-drug interaction between efavirenz and tuberculosis therapy. Rifampin is a potent inducer of multiple CYP enzymes and drug transporters. Some studies, although not all, have identified that efavirenz concentrations are lower when coadministered with rifampin, leading the FDA to recommend dose escalations of efavirenz in patients receiving rifampin who weigh 50 kg or more (Panel on Antiretroviral Guidelines for Adults and Adolescents 2014). Recently, the STRIDE study evaluated efavirenz concentrations in a population that was coinfected with HIV and tuberculosis and found that the median minimum concentration of efavirenz was 1.96 mcg/mL while on tuberculosis therapy (including rifampin, isoniazid, pyrazinamide, and ethambutol), compared with 1.80 mcg/mL when off tuberculosis therapy (p=0.067) (Luetkemeyer 2013). Not only was the lack of a difference in overall efavirenz concentrations interesting, but also, more specifically, patients of African heritage had significantly higher efavirenz minimum concentrations while on tuberculosis therapy than when they were not

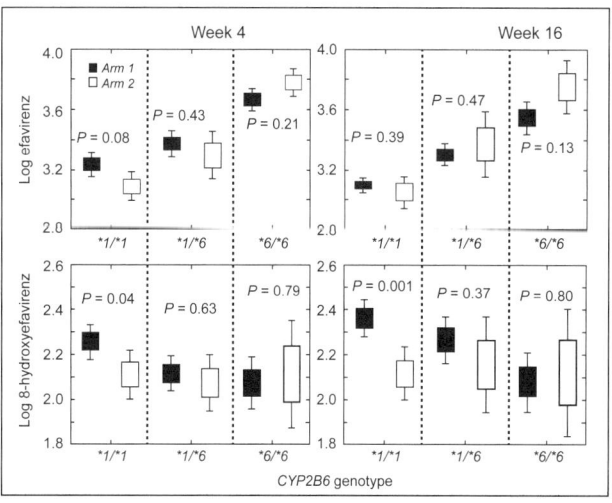

Figure 15.2. Comparison of the mean log efavirenz and 8-hydroxy efavirenz plasma concentrations stratified by the *CYP2B6* 516 genotype between patients with HIV receiving efavirenz-based ART alone (arm 1) and those receiving ART plus rifampin (arm 2) at week 4 and week 16.

*1/*1 = 516 G/G; *1/*6 = 516 G/T; *6/*6 = 516 T/T.

Reprinted by permission from Macmillan Publishers LTD. Ngaimisi E, Mugusi S, Minzi O, et al. Effect of rifampicin and *CYP2B6* genotype on long-term efavirenz autoinduction and plasma exposure in HIV patients with or without tuberculosis. Clin Pharmacol Ther 2011;90(3):406-13.

receiving tuberculosis therapy, which was contrary to the anticipated results (2.08 vs. 1.75, p=0.005).

Alternative routes of efavirenz metabolism may explain this finding. In patients with substantially reduced CYP2B6 metabolism of efavirenz, such as in those who carry the *CYP2B6* 516 T/T genotype, CYP2A6 may play a more dominant role in the disposition of the drug (Court 2014). Greater involvement of CYP2A6 in efavirenz metabolism may complicate tuberculosis therapy because isoniazid is a known CYP2A6 inhibitor. This genetic influence on the observed drug interaction is represented in Figure 15.2 (Ngaimisi 2011). Of note, for subjects with a homozygous wild-type *CYP2B6* genotype, the efavirenz and 8-hydroxy efavirenz concentrations are both lower in the first 4 weeks of efavirenz-rifampicin combination therapy, whereas only the 8-hydroxy efavirenz metabolite remains lower after 16 weeks of combination therapy. However, in subjects who are slow metabolizers (516 G/T or T/T),

Table 15.2. Summary of Pharmacogenetic Testing and Associated Clinical Characteristics of Antiretroviral Therapy

Antiretroviral Therapy	Pharmacogenetic Evaluation	Clinical or Laboratory Evaluation	Clinical Consequence
Testing for viral genetic characteristics			
HIV drug resistance testing	Genotypic or phenotypic tests	HIV-RNA: Failure to suppress HIV-RNA after initiating antiretroviral therapy, or HIV-RNA becomes detectable after suppression during antiretroviral therapy. Both indicate the potential development of resistance to the current antiretroviral therapy	Results from drug resistance testing will guide antiretroviral therapy selection. Without results from the resistance test, subjects may develop virologic failure because of suboptimal antiretroviral therapy
Maraviroc	CCR5 and CXCR4 coreceptor tropism test	HIV-RNA: Failure to suppress HIV-RNA after initiating maraviroc therapy, or HIV-RNA becomes detectable after suppression during maraviroc therapy. Both indicate the potential development of resistance to the current antiretroviral therapy	Use of maraviroc without coreceptor tropism testing may result in a suboptimal antiretroviral regimen if the patient has mixed- or CXCR4-tropic virus
Testing for host genetic characteristics			
Abacavir	HLA-B*5701	Clinical evaluation for signs/symptoms associated with abacavir hypersensitivity reaction, including two or more of the following: Fever, rash, and gastrointestinal, constitutional, or respiratory symptoms	Subjects with the HLA-B*5701 variant are most likely to develop abacavir hypersensitivity reaction, which can be severe and life threatening
Efavirenz	CYP2B6 516 position	Therapeutic drug monitoring of plasma efavirenz concentrations to detect supratherapeutic concentrations Monitor patients for CNS-associated toxicities	Supratherapeutic concentrations may result in severe or persistent efavirenz-related CNS adverse events
Atazanavir	UGT1A1*28	Bilirubin: Pretreatment and throughout atazanavir therapy	Presence of the UGT1A1*28 allele will result in a higher risk of symptomatic hyperbilirubinemia during atazanavir-based antiretroviral therapy

no differences in either efavirenz or 8-hydroxy efavirenz concentrations were observed at either time point. These data are particularly important for this therapeutic combination, given that the genotypes associated with slow efavirenz metabolism may be more common in regions of the world that are most affected by tuberculosis, and efavirenz-containing ART is the recommended first-line option for subjects who are coinfected with tuberculosis (WHO 2013; Haas 2004). This drug combination nicely illustrates the need for careful consideration of potential pharmacogenetic implications on concomitant medications when the literature related to drug-drug interaction studies is reviewed.

Summary

The ideal combination of antiretrovirals to be used for ART in an individual is significantly affected by pharmacogenetics. HIV drug resistance, coreceptor tropism, and HLA-B*5701 testing all represent successful implementation of pharmacogenetic tests into routine clinical practice. Current guidelines suggest all patients receive HIV drug resistance testing before starting therapy as well as when their ART is failing (Panel on Antiretroviral Guidelines for Adults and Adolescents 2014). In addition, clear guidance exists for coreceptor tropism and HLA-B*5701 testing before selecting either maraviroc or abacavir therapy, respectively, as part of ART therapy (Panel on Antiretroviral Guidelines for Adults and Adolescents 2014; Martin 2012). Additional clinical evidence is needed before the routine use of *CYP2B6* and *UGT1A1* pharmacogenetic testing can be supported. Table 15.2 summarizes the pharmacogenetic tests discussed in this chapter, the clinical and laboratory markers associated with these pharmacogenetic tests, and the potential clinical consequences of pharmacogenetic variation in an individual (Panel on Antiretroviral Guidelines for Adults and Adolescents 2014).

As scientists gain understanding of the pharmacogenetic impact of drug transporters and metabolizing enzymes, truly individualizing HIV therapy at each stage of the disease will become more feasible. In addition, new technologies such as dried blood spot and point-of-care tests will improve the cost and accessibility of pharmacogenetic and pharmacokinetic tests throughout the health care system. The role of clinical pharmacists in interpreting and applying pharmacogenetic information related to ART is clear, given the relationship among pharmacogenetic characteristics and drug toxicity and efficacy, dosing, and the potential for drug-drug interactions. Although there are clear associations between some pharmacogenetic characteristics and HIV treatment, new associations are continually emerging. Implementation of pharmacogenetic testing is not without limitations; usually, there are added costs related to testing, availability of testing, and delays in receiving test results for clinical use. Despite these limitations, clinical integration of pharmacogenetics into HIV care has been rapid, and the ability to update HIV guidelines frequently provides an evidence-based pathway for integrating pharmacogenetics into clinical care.

References

Baxter JD, Mayers DL, Wentworth DN, et al. A randomized study of antiretroviral management based on plasma genotypic antiretroviral resistance testing in patients failing therapy. CPCRA 046 Study Team for the Terry Beirn Community Programs for Clinical Research on AIDS. AIDS 2000;14:F83-93.

Benson CA, Vaida F, Havlir DV, et al. A randomized trial of treatment interruption before optimized antiretroviral therapy for persons with drug-resistant HIV: 48-week virologic results of ACTG A5086. J Infect Dis 2006;194:1309-18.

Bosma PJ, Chowdhury JR, Bakker C, et al. The genetic basis of the reduced expression of bilirubin UDP-glucuronosyltransferase 1 in Gilbert's syndrome. N Engl J Med 1995;333:1171-5.

Calcagno A, Cusato J, Marinaro L, et al. ABCC4 3348 T>C SNP affects tenofovir urinary output in HIV-positive patients. Conference on Retroviruses and Opportunistic Infections. John B. Hynes Veterans Memorial (Hynes) Convention Center; March 3–6, 2014; Boston, MA. Abstract 503.

Centers for Disease Control and Prevention (CDC). *Pneumocystis* pneumonia—Los Angeles. MMWR 1981;30:250-2.

Chen S, St. Jean P, Borland J, et al. Evaluation of the effect of *UGT1A1* polymorphisms on dolutegravir pharmacokinetics. Pharmacogenomics 2014;15:9-16.

Cohen CJ, Hunt S, Sension M, et al. A randomized trial assessing the impact of phenotypic resistance testing on antiretroviral therapy. AIDS 2002;16:579-88.

Cohen MS, Chen YQ, McCauley M, et al. Prevention of HIV-1 infection with early antiretroviral therapy. N Engl J Med 2011;365:493-505.

Court MH, Almutairi FE, Greenblatt DJ, et al. Isoniazid mediates the *CYP2B6*6* genotype-dependent interaction between efavirenz and antituberculosis drug therapy through mechanism-based inactivation of *CYP2A6*. Antimicrob Agents Chemother 2014;58:4145-52.

Culley CL, Kiang TK, Gilchrist SE, et al. Effect of the *UGT1A1*28* allele on unconjugated hyperbilirubinemia in HIV-positive patients receiving atazanavir: a systematic review. Ann Pharmacother 2013;47:561-72.

Deeks SG, Lewin SR, Havlir DV. The end of AIDS: HIV infection as a chronic disease. Lancet 2013;382:1525-33.

De Rosa MF, Robillard KR, Kim CJ, et al. Expression of membrane drug efflux transporters in the sigmoid colon of HIV-infected and uninfected men. J Clin Pharmacol 2013;53:934-45.

Di Iulio J, Fayet A, Arab-Alameddine M, et al. In vivo analysis of efavirenz metabolism in individuals with impaired CYP2A6 function. Pharmacogenet Genomics 2009;19:300-9.

Durant J, Clevenbergh P, Halfon P, et al. Drug-resistance genotyping in HIV-1 therapy: the VIRADAPT randomised controlled trial. Lancet 1999;353:2195-9.

Erickson DA, Mather G, Trager WF, et al. Characterization of the in vitro biotransformation of the HIV-1 reverse transcriptase inhibitor nevirapine by human hepatic cytochromes P-450. Drug Metab Dispos 1999;27:1488-95.

Ferraris L, Vigano O, Peri A, et al. Switching to unboosted atazanavir reduces bilirubin and triglycerides without compromising treatment efficacy in *UGT1A1*28* polymorphism carriers. J Antimicrob Chemother 2012;67:2236-42.

Fletcher CV, Staskus K, Wietgrefe SW, et al. Persistent HIV-1 replication is associated with lower antiretroviral drug concentrations in lymphatic tissues. Proc Natl Acad Sci U S A 2014;111:2307-12.

Gazzard B. Efavirenz in the management of HIV infection. Int J Clin Pract 1998;53:60-4.

Gonzalez-Serna A, Min JE, Woods C, et al. Performance of HIV-1 drug resistance testing at low-level viremia and its ability to predict future virologic outcomes and viral evolution in treatment-naive individuals. Clin Infect Dis 2014;58:1165-73.

Günthard HF, Aberg JA, Eron JJ, et al. Antiretroviral treatment of adult HIV infection: 2014 recommendations of the International Antiviral Society. JAMA 2014;312(4):410-25.

Haas DW, Ribaudo HJ, Kim RB, et al. Pharmacogenetics of efavirenz and central nervous system side effects: an Adult AIDS Clinical Trials Group study. AIDS 2004;18:2391-400.

Hetherington S, Hughes AR, Mosteller M, et al. Genetic variations in *HLA-B* region and hypersensitivity reactions to abacavir. Lancet 2002;359:1121-2.

Hetherington S, McGuirk S, Powell G, et al. Hypersensitivity reactions during therapy with the nucleoside reverse transcriptase inhibitor abacavir. Clin Ther 2001;23:1603-14.

Hinkley T, Martins J, Chappey C, et al. A systems analysis of mutational effects in HIV-1 protease and reverse transcriptase. Nat Genet 2011;43:487-9.

Hirsch MS, Brun-Vezinet F, D'Aquila RT, et al. Antiretroviral drug resistance testing in adult HIV-1 infection: recommendations of an International AIDS Society-USA Panel. JAMA 2000;283:2417-26.

Holzinger ER, Grady B, Ritchie MD, et al. Genome-wide association study of plasma efavirenz pharmacokinetics in AIDS Clinical Trials Group protocols implicates several *CYP2B6* variants. Pharmacogenet Genomics 2012;22:858-67.

Lubomirov R, Colombo S, di Iulio J, et al. Association of pharmacogenetic markers with premature discontinuation of first-line anti-HIV therapy: an observational cohort study. J Infect Dis 2011;203:246-57.

Luetkemeyer AF, Rosenkranz SL, Lu D, et al. Relationship between weight, efavirenz exposure, and virologic suppression in HIV-infected patients on rifampin-based tuberculosis treatment in the AIDS Clinical Trials Group A5221 STRIDE Study. Clin Infect Dis 2013;57:586-93.

Mallal S, Nolan D, Witt C, et al. Association between presence of HLA-B*5701, HLA-DR7, and HLA-DQ3 and hypersensitivity to HIV-1 reverse-transcriptase inhibitor abacavir. Lancet 2002;359:727-32.

Mallal S, Phillips E, Carosi G, et al. HLA-B*5701 screening for hypersensitivity to abacavir. N Engl J Med 2008;358:568-79.

Martin MA, Klein TE, Dong BJ, et al. Clinical pharmacogenetics implementation consortium guidelines for *HLA-B* genotype and abacavir dosing. Clin Pharmacol Ther 2012;91:734-8.

Martin MA, Kroetz DL. Abacavir pharmacogenetics—from initial reports to standard of care. Pharmacotherapy 2013;33:765-75.

Martins JZ, Chappey C, Haddad M, et al. Principal component analysis of general patterns of HIV-1 replicative fitness in different drug environments. Epidemics 2010;2:85-91.

Marzolini C, Telenti A, Decosterd LA, et al. Efavirenz plasma levels can predict treatment failure and central nervous system side effects in HIV-1-infected patients. AIDS 2001;15:71-5.

Michaud V, Bar-Magen T, Turgeon J, et al. The dual role of pharmacogenetics in HIV treatment: mutations and polymorphisms regulating antiretroviral drug resistance and disposition. Pharmacol Rev 2012;64:803-33.

Miller V, Sabin C, Hertogs K, et al. Virological and immunological effects of treatment interruptions in HIV-1 infected patients with treatment failure. AIDS 2000;14:2857-67.

Ngaimisi E, Mugusi S, Minzi O, et al. Effect of rifampicin and *CYP2B6* genotype on long-term efavirenz autoinduction and plasma exposure in HIV patients with or without tuberculosis. Clin Pharmacol Ther 2011;90:406-13.

Palella FJ Jr, Delaney KM, Moorman AC, et al. Declining morbidity and mortality among patients with advanced human immunodeficiency virus infection. HIV Outpatient Study Investigators. N Engl J Med 1998;338:853-60.

Panel on Antiretroviral Guidelines for Adults and Adolescents. Guidelines for the Use of Antiretroviral Agents in HIV-1-Infected Adults and Adolescents. Department of Health and Human Services. May 2014. Available at http://aidsinfo.nih.gov/ContentFiles/AdultandAdolescentGL.pdf. Accessed August 1, 2014.

Panel on Antiretroviral Therapy and Medical Management of HIV-Infected Children. Guidelines for the Use of Antiretroviral Agents in Pediatric HIV Infection. February 2014. Available at http://aidsinfo.nih.gov/contentfiles/lvguidelines/pediatricguidelines.pdf. Accessed August 1, 2014.

Penzak SR, Kabuye G, Mugyenyi P, et al. Cytochrome P450 2B6 (CYP2B6) G516T influences nevirapine plasma concentrations in HIV-infected patients in Uganda. HIV Med 2007;8:86-91.

Peyrieere H, Nicolas J, Siffert M, et al. Hypersensitivity related to abacavir in two members of a family. Ann Pharmacother 2001;35:1291-2.

Ramachandran G, Kumar AH, Rajasekaran S, et al. *CYP2B6* G516T polymorphism but not rifampin coadministration influences steady-state pharmacokinetics of efavirenz in human immunodeficiency virus-infected patients in South India. Antimicrob Agents Chemother 2009;53:863-8.

Rauch A, Nolan D, Martin A, et al. Prospective genetic screening decreases the incidence of abacavir hypersensitivity reactions in the Western Australian HIV cohort study. Clin Infect Dis 2006;43:99-102.

Ray AS, Cihlar T. Unlikely association of multidrug-resistance protein 2 single-nucleotide polymorphisms with tenofovir-induced renal adverse events. J Infect Dis 2007;195:1389-90; author reply 90-1.

Ribaudo HJ, Daar ES, Tierney C, et al. Impact of *UGT1A1* Gilbert variant on discontinuation of ritonavir-boosted atazanavir in AIDS Clinical Trials Group Study A5202. J Infect Dis 2013;207:420-5.

Ribaudo HJ, Haas DW, Tierney C, et al. Pharmacogenetics of plasma efavirenz exposure after treatment discontinuation: an Adult AIDS Clinical Trials Group Study. Clin Infect Dis 2006;42:401-7.

Riska P, Lamson M, MacGregor T, et al. Disposition and biotransformation of the antiretroviral drug nevirapine in humans. Drug Metab Dispos 1999;27:895-901.

Rotger M, Colombo S, Furrer H, et al. Influence of *CYP2B6* polymorphism on plasma and intracellular concentrations and toxicity of efavirenz and nevirapine in HIV-infected patients. Pharmacogenet Genomics 2005a;15:1-5.

Rotger M, Taffe P, Bleiber G, et al. Gilbert syndrome and the development of antiretroviral therapy-associated hyperbilirubinemia. J Infect Dis 2005b;192:1381-6.

Saitoh A, Sarles E, Capparelli E, et al. *CYP2B6* genetic variants are associated with nevirapine pharmacokinetics and clinical response in HIV-1-infected children. AIDS 2007;21:2191-9.

Samji H, Cescon A, Hogg RS, et al. Closing the gap: increases in life expectancy among treated HIV-positive individuals in the United States and Canada. PloS One 2013;8:e81355.

Sampietro M, Iolascon A. Molecular pathology of Crigler-Najjar type I and II and Gilbert's syndromes. Haematologica 1999;84:150-7.

Santoro MM, Fabeni L, Armenia D, et al. Reliability and clinical relevance of the HIV-1 drug resistance test in patients with low viremia levels. Clin Infect Dis 2014;58:1156-64.

Schackman BR, Haas DW, Becker JE, et al. Cost-effectiveness analysis of *UGT1A1* genetic testing to inform antiretroviral prescribing in HIV disease. Antivir Ther 2013;18:399-408.

Schackman BR, Scott CA, Walensky RP, et al. The cost-effectiveness of HLA-B*5701 genetic screening to guide initial antiretroviral therapy for HIV. AIDS 2008;22:2025-33.

Siccardi M, D'Avolio A, Rodriguez-Novoa S, et al. Intrapatient and interpatient pharmacokinetic variability of raltegravir in the clinical setting. Ther Drug Monit 2012;34:232-5.

Swenson LC, Dong WW, Mo T, et al. Use of cellular HIV DNA to predict virologic response to maraviroc: performance of population-based and deep sequencing. Clin Infect Dis 2013;56:1659-66.

Swenson LC, Min JE, Woods CK, et al. HIV drug resistance detected during low-level viraemia is associated with subsequent virologic failure. AIDS 2014;28:1125-34.

Symonds W, Cutrell A, Edwards M, et al. Risk factor analysis of hypersensitivity reactions to abacavir. Clin Ther 2002;24:565-73.

Wensing AM, Calvez C, Günthard HF, et al. 2014 update of the druf resistance mutations in HIV-1. Top Adivir Med 2014;22(3):642-650.

Trinh L, Han D, Huang W, et al. Technical validation of an enhanced sensitivity Trofile HIV coreceptor tropism assay for selecting patient for therapy with entry inhibitors targeting CCR5. Antivir Ther 2008;13:A128.

UNAIDS.org [homepage on the Internet]. Global Report: UNAIDS Report on the Global AIDS Epidemic. 2013. Available at www.unaids.org/en/media/unaids/contentassets/documents/epidemiology/2013/gr2013/UNAIDS_Global_Report_2013_en.pdf. Accessed August 1, 2014.

Volberding PA, Greene WC, Lange JMA, et al. Development and transmission of HIV drug resistance. In: Volberding PA, Greene WC, Lange JMA, eds. Sande's HIV/AIDS Medicine: Medical Management of AIDS 2013, 2nd ed. Beijing: Elsevier Saunders, 2012:155-67.

Walensky RP, Goldberg JH, Daily JP. Anaphylaxis after rechallenge with abacavir. AIDS 1999;13:999-1000.

Walmsley SL, Antela A, Clumeck N, et al. Dolutegravir plus abacavir-lamivudine for the treatment of HIV-1 infection. N Engl J Med 2013;369:1807-18.

Ward BA, Gorski JC, Jones DR, et al. The cytochrome P450 2B6 (CYP2B6) is the main catalyst of efavirenz primary and secondary metabolism: implication for HIV/AIDS therapy and utility of efavirenz as a substrate marker of CYP2B6 catalytic activity. J Pharmacol Exp Ther 2003;306:287-300.

Wenning LA, Petry AS, Kost JT, et al. Pharmacokinetics of raltegravir in individuals with *UGT1A1* polymorphisms. Clin Pharmacol Ther 2009;85:623-7.

Wensing AM, Calvez V, Günthard HF, et al. 2014 update of the drug resistance mutations in HIV-1. Top Antivir Med 2014; 22(3):642-50.

Whitcomb JM, Huang W, Fransen S, et al. Development and characterization of a novel single-cycle recombinant-virus assay to determine human immunodeficiency virus type 1 coreceptor tropism. Antimicrob Agents Chemother 2007;51:566-75.

Wolkoff AW. The hyperbilirubinemias. In: Longo DL, Fauci AS, Kasper DL, et al., eds. Harrison's Principles of Internal Medicine, 18e. New York: McGraw-Hill, 2012.

World Health Organization (WHO). Consolidated Guidelines on the Use of Antiretroviral Drugs for Treating and Preventing HIV Infection: Recommendations for a Public Health Approach. Geneva: WHO, 2013.

Zhang D, Chando TJ, Everett DW, et al. In vitro inhibition of UDP glucuronosyltransferases by atazanavir and other HIV protease inhibitors and the relationship of this property to in vivo bilirubin glucuronidation. Drug Metab Dispos 2005;33:1729-39.

Westby M, Lewis M, Whitcomb J, et al. Emergence of CXCR4-using human immunodeficiency virus type 1 (HIV-1) variants in a minority of HIV-1-infected patients following treatment with the CCR5 antagonist maraviroc is from a pretreatment CXCR4-using virus reservoir. J Virol 2006;80:4909-20.

Ghany M, Hoofnagle JH. Approach to the patient with liver disease. In: Longo DL, Fauci AS, Kasper DL, et al., eds. Harrison's Principles of Internal Medicine, 18e. New York: McGraw-Hill, 2012.

Section 3
Pharmacogenomics: Other Issues and Implications

Chapter 16

ETHICAL, LEGAL, AND SOCIAL CHALLENGES TO PHARMACOGENETICS

SUSANNE B. HAGA, PH.D.

Learning Objectives

1. Identify and understand the ethical, legal, and social challenges posed by the clinical integration of pharmacogenetics.
2. Understand how pharmacogenetics may introduce a new level of stratification in a targeted treatment population and how it may exacerbate existing stratifications associated with the disease.
3. Understand importance of patient-provider discussion prior to and after pharmacogenetic testing to promote informed decision-making and patient understanding of test results.
4. Understand barriers to the clinical use of pharmacogenetic testing, particularly lack of provider knowledge, disparate coverage policies.
5. Understand how pharmacogenetics may redefine the roles of health care providers.

Keywords: Pharmacogenetics, discrimination, stigmatization, familial implications, informed consent, genetic counseling

Abbreviations in This Chapter

HER2	Human epidermal growth factor receptor 2
HLA	Human leukocyte antigen
PPACA	Patient Protection and Affordable Care Act
TPMT	Thiopurine methyltransferase

Abstract

Pharmacogenetics presents an exciting opportunity to enhance the safe and effective use of drugs through the personalization of drug treatment based on an individual's genetic makeup. In addition to developing the clinical evidence to support the use of pharmacogenetic testing, several ethical, legal, and social must also be addressed to facilitate the successful implementation of clinical pharmacogenetic testing. This chapter aims to provide an overview of several of the ethical, legal, and social issues arising from the early use of pharmacogenetic testing. Although many of the issues are similar to traditional disease-based genetic testing, others are unique to pharmacogenetic testing and warrant careful attention to insure appropriate use and delivery of testing.

Introduction

Pharmacogenetic testing seeks to provide information about the safety and/or efficacy of drugs according to an individual's inherited factors (Wang 2011). Pharmacogenetic testing can be provided at different stages of treatment (pre-emptive or point-of-care testing when treatment is needed) and in different practice settings (e.g., inpatient hospital based, outpatient clinic based). Early experience has provided several insights regarding the risks, benefits, and public and professional perceived value of testing. Because pharmacogenetic tests do not provide information about disease risk, they are considered to have fewer ethical, legal, and social implications (Roses 2000). As the use of pharmacogenetic testing increases, enhanced public/

patient and professional awareness of these issues will help facilitate the safe and appropriate use of this testing. This chapter aims to briefly review many of these issues and the current literature as well as suggest some potential solutions for the appropriate use of pharmacogenetic testing in the clinical setting.

Public Interest and Awareness

Several studies have evaluated the attitudes and perceptions toward pharmacogenetics in the United States, the United Kingdom, and Europe. In general, public perceptions about pharmacogenetic testing have been favorable (Zhang 2014; Haga 2012b, 2012c; Payne 2011; Haddy 2010; Fargher 2007a). Among the reported concerns, however, are the development of race-targeted drugs (Bates 2004; Bevan 2003; Condit 2003), the affordability of targeted drugs (Almarsdottir 2005; Rothstein 2003), the cost of testing (Haddy 2010), and the privacy and confidentiality of pharmacogenetic test results (Haddy 2010). Participants of studies conducted in Germany and the United Kingdom were supportive of pharmacogenetic testing but had concerns regarding patient sovereignty, the unavailability of suitable drugs based on patient genetic makeup, and privacy (Rogausch 2006; Royal Society 2005).

Stigmatization

Societal norms and attitudes about various diseases are continuously shifting, influenced by several factors such as availability of treatment, outcome, burden, impact of lifestyle choices, and physical disability. Patients with a well-known and accepted common disease may be ostracized because of the unique characteristics of their condition or response to drugs. Patients with a test result that shows an abnormal drug disposition or disease subtype for which a targeted drug is unavailable may experience stress or anxiety, potentially leading to depression, anxiety, and/or failure to adhere to medication regimens. For patients who learn that a drug is available for their particular disease subtype, public perception may be positive if the condition is viewed as treatable (Foster 2003). However, those who have a disease subtype for which a targeted drug is not available may be labeled "untreatable," which may lead to the development of psychosocial sequelae (Paul 2006). Simply put, "[p]ersons who are affected by the physical symptoms of a disease but do not respond to a drug that alleviates them in others may stand out as different" (Foster 2003). For example, although pharmacogenetic testing to predict response to human immunodeficiency virus (HIV) drugs such as abacavir could minimize adverse responses, it could also further increase the stigmatization associated with HIV/AIDS (acquired immunodeficiency syndrome) for patients predicted not to respond (well) to available drugs. For patients with a psychiatric condition, a "negative" pharmacogenetic test result could compound their already compromised self-image (Morley 2004). "[O]ne could reasonably ask whether having a disease is more or less socially significant than being able to treat a disease" (Foster 2003).

Preemptive pharmacogenetic testing of healthy individuals could lead to the identification of "healthy ills" (see references within Paul 2006) and an increased psychological burden because of the uncertainty associated with an abnormal pharmacogenetic result for drug metabolism. The label of "difficult to treat" (Rothstein 2001), referring to a patient's ability (or lack thereof) to metabolize certain classes of drugs, would remain with the patient for his or her lifetime.

Genetic Discrimination

Some health care professionals believe that the use of pharmacogenetic information by insurers should be limited to developing formularies or justifying denial of coverage for targeted drugs and that this information should be prohibited from being used to set premiums or copays (Zachry 2002). Federal legislation prohibits discriminatory actions by health insurers and employers based on an individual's genetic information (Public Law 110-233); however, many providers are unaware of these protections (Laedtke 2012). Two major gaps in the protections offered by this legislation should be mentioned to patients considering pharmacogenetic testing. First, the law does not apply to disability, life, or long-term care insurance. One might imagine a life insurance company's interest in pharmacogenetic test results for an individual with a condition whose outcome could be influenced by his or her ability to metabolize drugs or the genetic characteristics of the disease itself relevant to treatment availability. Second, the law does not protect symptomatic individuals—the population most likely to undergo pharmacogenetic testing. For instance, pharmacogenetic testing can reveal not only an individual's genetic status but also his or her use of a targeted drug and, potentially, information about disease prognosis (Moldrup 2001). This is similar to situations involving the use of well-known drugs such as methadone or antidepressants, which can reveal an illness that would otherwise have been unknown or that an individual might choose not to disclose for fear of stigmatization or discrimination. These two gaps in the federal legislation should be disclosed to patients considering pharmacogenetic testing.

The recently enacted Patient Protection and Affordable Care Act (PPACA) overlaps with GINA (Genetic Information Nondiscrimination Act) with respect to group health insurance plans and prohibits the use of most types of health-related information (e.g., tobacco use, age) to adjust premiums. In addition, PPACA prohibits coverage exclusions based on preexisting health conditions.

Race-Based Drugs

Given the differences in the prevalence of genetic variance associated with disease risk or drug response, it is possible that a variant will be primarily associated with one population or have a particular phenotype in that population; hence, a drug may be developed that is primarily targeted to a population defined by race. Such a drug has in fact been developed, known as BiDil. BiDil is a combination of two old drugs, hydralazine and isosorbide dinitrate, approved by the U.S. Food and Drug Administration (FDA) to treat heart failure in 2005. Approval was based on a trial that enrolled only African Americans and compared the combination drug with standard therapy placebo, showing a significantly reduced mortality rate (Carmody 2007; Temple 2007; Taylor 2004). As a result, the approved drug label states that "BiDil is indicated for the treatment of heart failure as an adjunct to standard therapy in self-identified black patients to improve survival…."

A strong public and scholarly debate has ensued around the development and approval of BiDil. Physicians have expressed a range of opinions about race-based drugs (Frank 2010), though the public is generally skeptical (Butrick 2011; DeMarco 2010; Condit 2003). Although differences in drug response between racial groups have been noted in several drug labels (e.g., rosuvastatin [Crestor]), BiDil is the first drug approved for a patient population defined by race. Differences in drug response for heart disease have been reported for other drugs, though somewhat inconsistently (Exner 2001; Yancy 2001). The inconsistency in these results from comparison studies of heart failure drug response between black and white populations may partly may partly stem from the many factors that are subsumed under the variable of race. Without a further understanding of the biological mechanisms underlying the reportedly different drug responses between racial groups, the study and approval of future race-based drugs may raise further suspicion and reinforce beliefs of biological differences between races (Bibbins-Domingo 2007; Duster 2005; Kahn 2004). The significance of race and genetic variation has long been a subject of discussion in genetics and genomics (Cooper 2003; Burchard 2001), including pharmacogenetics. Genetic studies have reported differences in the prevalence of genetic variants associated with disease (Ni 2014) and drug response (Perera 2013). For future pharmacogenetic tests and drugs, these differences will remain important considerations for new clinical interventions, but further research is needed to understand the significance of these observations and their implications for patient care (Ortega 2014).

Informed Consent and Patient-Provider Communication

Because of the novelty of pharmacogenetic testing, patient-provider communication before test ordering will be important to promote informed decision-making as well as to promote comprehension of the test results, and impact of results on treatment decisions after testing is completed (immediate and future). Although patient familiarity with adverse responses or ineffective medications and general positive attitudes suggest overall patient interest (Haga 2012b), discussion about the benefits and limitations of testing is still warranted during this early stage of test use and limited public awareness.

As with any clinical test or procedure, the type of informed consent will vary depending on the risks and benefits of the intervention. It is currently unclear how pharmacogenetic testing will be viewed compared with other tests. For example, if pharmacogenetic testing is considered similar to other clinical (nongenetic) tests, the pretesting process will likely be similar, with minimal discussion and no need for written informed consent or genetic counseling (Wertz 2003; Roses 2000). In this case, a simple informed consent will suffice (Robertson 2002). However, some have questioned whether a simple consent process will truly constitute an informed consent (Netzer 2004). Potential risks of pharmacogenetic testing, including the familial implications of the information, the potential for incidental information, and the risk of stigmatization and discrimination, may warrant a longer discussion and a more formal consent process (Morley 2004). In addition, individual laboratories may have written consent requirements, and several states already require informed consent for genetic testing (Alaska, Arizona, Florida, Georgia, Massachusetts, Michigan, Nebraska, New Mexico, New York, South Carolina, South Dakota, and Vermont) (www.ncsl.org/research/health/genetic-privacy-laws.aspx).

Surveys of health professionals show that many do not routinely obtain informed consent for pharmacogenetic testing (Woelderink 2006). For example, informed consent was considered unnecessary by UK physicians ordering human epidermal growth factor receptor 2 (*HER2*) testing for patients with breast cancer because this test was viewed as part of a panel of tests routinely ordered (Hedgecoe 2006). Furthermore, informed consent was believed to be unnecessary because the test result did not reveal heritable information and because other types of clinical data provide prognostic information. Physicians did not consider the *HER2* test a traditional genetic test, but instead, another clinical test that does not involve the analysis of an inherited variant (Hedgecoe 2005). In particular, three factors influenced decisions about whether to seek informed consent for *HER2* testing: (1) clinical stage at diagnosis, (2) physician attitude toward testing, and (3) potential for false hopes based on test result (Hedgecoe 2005). If consent was sought, the test was not described as a "genetic" test for several reasons—it does not detect heritable changes; it does not look at DNA sequence but at protein expression through an immunohistochemistry assay; and patient concerns are associated with genetic tests. Regarding this last point, use of the term *genetic* alone may trigger negative public perceptions and an automatic association with traditional genetic tests familiar to the public (Wertz 2003).

The complexity of some pharmacogenetic tests and their impact on treatment decisions may be difficult for some patients to comprehend (Brewer 2014); use of informational sheets or brochures may be helpful (Nuffield Council on Bioethics 2003; Robertson 2001). Although some may choose to call pharmacogenetic tests "medicine response profiles" to distinguish them from disease risk genetic tests (Roses 2000), caution should be taken not to mislead patients. Some physicians purposely limit the information that they discuss with patients about testing to avoid possible psychosocial harms associated with the test results. For instance, women who learn that they do not overexpress *HER2* may be disappointed that they are not eligible for treatment with trastuzumab (Herceptin), despite the better prognosis associated with *HER2*-negative status (Hedgecoe 2005). In addition, physicians have withheld information about trastuzumab and *HER2* testing because of concerns about information overload (Hedgecoe 2005).

When testing is completed, providers should carefully review the results with patients and explain how the results will affect treatment decisions. For the testing of genes such as cytochrome P450 (CYP) 2D6 and 2C9 that affect several drugs, it will be critical for patients to understand that they have had testing and, if possible, to know the phenotype associated with their test result (e.g., poor metabolizer). Given that patients may seek care from many providers without access to a shared electronic medical record, patients will need to have a sufficient understanding of which test was performed, and optimally the outcome, to avoid duplicate testing and ensure the consideration of the test results for other prescription drugs; moreover, it is hoped that patients will share this information with other providers or the pharmacist. The portability of the test result may be improved with a results "card" or some other mechanism for the patient to store the results and promote sharing with other providers (Haga 2011). If patients cannot recall the result, particularly if they have been tested for a panel of genes, their awareness that the test has been done can enable them to initiate a request of the test results from the ordering provider. In addition, patient comprehension of the results and significance to care may reduce the risk of psychological harm or adverse behavior (Haga 2014).

Management of Incidental Information

Incidental information (also called collateral, ancillary, or secondary information) related to pharmacogenetic tests is defined as additional information pertaining to the predisposition to diseases or conditions, prognostic information, or information relevant to other classes of drugs for which the patient is not currently seeking treatment or manifesting symptoms (Netzer 2004; Buchanan 2002). All pharmacogenetic tests have the potential to reveal incidental information; one study reported that more than 50% of the 42 pharmacogenetic variants significantly associated with drug response had also been reported to be associated with common diseases other than that for which the patient was seeking treatment (Goldstein 2003). Although many pharmacogenetic tests analyze only one or a panel of genes per test related to given medication or commonly used medications used by a medical specialty (e.g., psychiatry), the growing use of comprehensive testing platforms like microarrays and sequencing will result in a higher potential for incidental information (Westbrook 2013; Kohane 2006).

Incidental findings are not unique to genetics; these findings have raised similar debates about clinical obligations to disclose the results and the need for follow-up testing (Grumbach 2003; Ravine 1993). In particular, clinical whole genome or whole exome sequencing has caused a huge debate in recent years as well as divisive recommendations regarding the management of incidental findings (Green 2013). The general criterion for the return of incidental findings is clinical utility (Brothers 2013; Berg 2011), but consensus on the definition of clinical utility and which conditions/traits meet the definition has been difficult to achieve. The obligation to disclose incidental information or inform patients of its existence will also be greatly influenced in the future by legal standards and case law (Clayton 2013). The potential burdens of a workup for a false-positive finding (including the possibility that further workup will not resolve the uncertainty) must be considered together with the potential risks of ignoring a finding that could be medically significant. In some cases, the potential for patient harms such as increased anxiety may be sufficient to argue against initial testing (Sonnenberg 2004). In addition, it will be critically important to consider patient preferences because additional factors may influence patients' interest in incidental findings (Daack-Hirsch 2013).

For pharmacogenetic testing, some recommend that patients be informed about the possibility that additional types of information will be revealed in addition to information about drug response/outcome (Netzer 2004). If sensitive incidental information is revealed by a pharmacogenetic test, the test may be considered high risk and require a detailed informed consent process at the outset of testing (Breckenridge 2004; Robertson 2001). A consent form that enables patients to indicate their preferences can help ensure that patients learn only the information they want (Netzer 2004). For example, testing for the apolipoprotein E gene could potentially help guide warfarin dosing (Sconce 2006; Visser 2005) or statin selection (Gerdes 2000; Ordovas 1995), but it would also provide information about the risk of Alzheimer disease (Corder 1993) or age-related macular degeneration (Thakkinstian 2006), diseases for which there are few preventive interventions or treatments. Incidental information should not necessarily be presumed to be one of the risks of pharmacogenetic testing. For example, variation in *CYP2D6* and *CYP2C19* has been associated with modest increases in the risk of lung, liver, and colorectal cancer (Rodriguez-Antona 2006; Agundez 2004). Even though the risks are slight, interventions are available for these conditions, and outcomes are often improved if the conditions are diagnosed early. If incidental results are to be returned, physicians then need to consider who should communicate results to the patient, the need for genetic counseling, and how the results should be documented in the medical record.

Storage and Portability of Pharmacogenetic Information

Given the lifelong relevance of a pharmacogenetic test result, the storage and portability of pharmacogenetic information is another critical aspect to consider. Issues of privacy and confidentiality are major concerns regarding access to test results, particularly as they relate to concerns about genetic discrimination. However, because this information is part of the patient's medical record, access to it is protected under medical privacy laws, including the Health Insurance and Portability and Accountability Act of 1996 (HIPAA) and other relevant state laws or regulations. Under HIPAA, medical information may be shared among health care providers for treatment purposes, although providers are bound by duty to maintain confidentiality. For example, when a patient with type 2 diabetes mellitus is being treated for vascular insufficiency, the internist, endocrinologist, and cardiologist are all likely to have access to the patient's record and information. However, given the wide range of information that may be present in a patient's medical record, a provider will have access to more information than may be needed for treatment, including genetic information.

In addition to the potential benefit to the patient when the physician who ordered testing considers the test results for each new medication prescribed, the patient may benefit if other health professionals such as urgent care physicians (Seefelder 2002), pharmacists, or other specialists have access to the test results for immediate or future care. Because the value of pharmacogenetic results is derived partly from their relevance to many drug classes that patients may need during their lifetime, nondisclosure of this type of information may result in redundant testing or adverse responses. In addition, access to a patient's pharmacogenetic test results by pharmacists could further improve safe and effective drug selection and dosing (Brushwood 2003), and patients have expressed an interest in pharmacist involvement (Haga 2011). Community pharmacists have expressed support for incorporating pharmacogenetic information into care (Tuteja 2013) but have noted some challenges, given their limited access and system constraints (Clemerson 2006).

Medical Liability

Currently, despite the more than 100 drugs with pharmacogenetic information included in the drug labels, many general practitioners have not ordered pharmacogenetic tests (Haga 2012a; Stanek 2012). With the anticipated improvements to clinical outcomes through tailored, genotype-informed treatments, a host of potential litigation accompanies this new field (Braff 2009; Evans 2007). This threat could potentially increase the uptake of pharmacogenetic tests (Rothstein 2001), even before testing has been shown to be clinically useful. The impact of this threat is confirmed by a survey of European physicians, which found that because of the life-threatening adverse events associated with genetic variants of the thiopurine methyltransferase (TPMT) gene, a higher proportion of *TPMT* than *HER2* test users for breast cancer treatment believe that testing reduces liability risk (Woelderink 2006). Lawsuits from patients alleging to have had adverse drug responses, including death, are familiar to the public, given the intense media coverage (e.g., COX-2 [cyclooxygenase-2] inhibitors such as rofecoxib [Vioxx]). The advent of pharmacogenetics adds a new claim that an adverse event was caused by a genetic variant that could have been detected through testing before the drug was prescribed (Marchant 2006). Several defendants may be possible in pharmacogenetic-related claims, including the drug manufacturer, insurance companies, physicians, and potentially pharmacists (Marchant 2006).

The first and only pharmacogenetic case known to be filed against a pharmaceutical company involved a vaccine for Lyme disease called LYMErix (Cassidy v SmithKline-Beecham 1999). LYMErix was the first FDA label-approved vaccine for tickborne Lyme disease, and more than 1 million doses were reportedly sold in its first year. The vaccine used a genetically engineered version of a surface protein, OspA, to induce an immune response against the bacteria. Some evidence suggested that the OspA protein triggers the development of autoimmune arthritis in individuals who carry a certain human leukocyte antigen (HLA) marker (*HLA-DR4+*) (Kalish 1993). In a lawsuit against the drug manufacturer, plaintiffs based their argument on the doctrine of negligence, alleging that the company failed to screen trial participants for the *HLA* variant and did not warn consumers of the increased risk caused by the variant. Even though the company maintained it had no evidence of an association with arthritis in the clinical trial or postmarket surveillance data, it reached an out-of-court settlement with the plaintiffs. The drug was later removed from the market.

In general, three categories of medical liability exist: medical malpractice, informed consent, and product liability (Palmer 2003). With respect to pharmacogenetics, the most likely cause of action against a pharmaceutical company is a "failure to warn" about genetic variability in drug response (Marchant 2006; Rothstein 2005). Specifically, a patient could claim that the genetic risk should have been disclosed in the drug label. In its defense, a pharmaceutical company could use many arguments. First, a company could claim that a warning could not be heeded even if disclosed because the patient could not have known that he or she carried a particular genotype that placed him or her at increased risk. Second, a company could argue that the association between the drug response and the genetic variant was invalid and therefore did not rise to the level requiring disclosure. Third, a company could respond that the physician, not the company, is in the best position to evaluate the patient and prescribe safe and effective treatment (known as "learned intermediary defense") (Kumorowski 2003). In effect, the company is arguing that it has already provided adequate information regarding the safe and effective use of the drug.

A second type of liability for a pharmaceutical company is called design-defect liability, whereon a plaintiff contends that a product is "defective" and therefore dangerous (Rothstein 2005). Some experts assert that test kit manufacturers as well as testing laboratories should not be exempt from defect liability (Kumorowski 2003; Ossorio 2003). Patient harms include discrimination, incorrect test interpretation, psychosocial risks, and harms because of subsequent actions based on test results (Ossorio 2003). Unlike drugs, however, genetic tests should not be considered "unavoidably unsafe" given the limited risks associated with testing (Kumorowski 2003; Ossorio 2003). Finally, a third type of liability for a pharmaceutical company is failure to test a drug adequately.

After pharmaceutical companies, physicians are the most vulnerable group to litigation related to pharmacogenetic testing. Specifically, physicians may be held liable for failure to order a pharmacogenetic test to screen patients if evidence suggests an association with a serious adverse response (Rothstein 2005). Several drug labels now include information about genetic variants that may affect drug responses or outcomes, although testing is not generally recommended. Physicians' limited knowledge of genetics increases the risk of test misinterpretation and liability (Ossorio 2003; Rothstein 2001; Caulfield 2000). Although physicians will likely depend on laboratory reports for test interpretation (Ossorio 2003), additional patient factors will probably have to be considered. In addition, potential liability exists for physicians' failure to provide genetic counseling or to warn at-risk family members (Rothstein 2005).

Moreover, private payers could be liable for refusing to authorize and cover a pharmacogenetic test. Although ERISA (Employee Retirement Income Security Act) provides extensive protection against liability for employer-sponsored group health plans, the individual insurance company may still be vulnerable to liability.

Finally, pharmacists may be at risk under the Omnibus Budget Reconciliation Act of 1990, whereby if a pharmacist is aware of or learns that a patient with a particular genetic variant may be at risk of an adverse event but does not inform the patient, the pharmacist is considered in breach of duty. Before 1990, it was highly unlikely that a pharmacist could be held liable if the prescription was accurately filled, regardless of how obvious the physician error might be. It was understood that the "the physician's responsibility superseded the pharmacist's responsibility" (Brushwood 2003). In 2000, however, an Institute of Medicine report highlighted the prescription-screening role of the pharmacist, raising the pharmacist's responsibility (and liability risk) in ensuring the safe and appropriate treatment of patients (IOM 2000). Although a pharmacist cannot prevent the incorrect drug from being given to a patient, given that the pharmacist is not privy to a patient's medical history and current diagnosis and is not trained to assess the medical appropriateness of a given drug, a pharmacist is still responsible for bringing obvious errors to the attention of the patient or physician (IOM 2000).

Testing in Children

About two-thirds of drugs prescribed for children have not been appropriately studied; therefore, they are not labeled for pediatric use (GAO 2007). The absence of data on the safety and effectiveness of drugs in children poses substantial challenges to health professionals in managing childhood illnesses. There has been a growing body of research and interest in pharmacogenetic testing in children to better understand the variability in drug response in children (Hawcutt 2013). Pharmacogenetic testing in children has been widely discussed (Vanakker 2013; Polanczyk 2010; Husain 2007), and parents have indicated interest in this testing for their children (Zhang 2014). Moreover, pharmacogenetic testing in children holds substantial promise to help predict serious adverse effects and guide drug therapy (Gasso 2014; Hamberg 2013; Kieling 2010; Madadi 2008), particularly given the increasing use of prescribed medications for children and the lack of data on drug safety and efficacy and resultant high off-label use. Most clinical guidelines discourage genetic testing in children unless it will result in a timely medical benefit, however, pharmacogenetic testing is considered acceptable given the immediate benefits (Ross 2013). With exploration into the use of whole genome sequencing during the newborn period, it may be possible to extract pharmacogenetic data that will potentially yield lifetime benefits. However, the use of sequencing in newborns is still being debated, and no consensus has yet been reached regarding the clinical and ethical issues involved (Knoppers 2014; Levenseller 2014).

Age-related drug sensitivities primarily attributed to the ontogeny of drug metabolic enzyme activity are further compounded by genetic variation in these enzymes (Rieder 2014; Blake 2005; Leeder 2004, 2003, 2001; de Wildt 1999). For example, low activity of drug-metabolizing enzymes such as CYP2D6 or CYP3A4 is typical in neonates, increasing to adult levels any time between early childhood and puberty (Leeder 2001). For CYP3A4, the levels spike before plateauing at adult levels such that children may fluctuate from low to very high levels of activity (Gow 2001; de Wildt 1999). The impact of genetic variability on expression of both of these genes can exacerbate normally fluctuating levels such that they may have little to no enzyme activity during the neonatal periods if they carry a variant that results in the poor metabolizer (PM) phenotype. Likewise, children could have ultra-high levels of enzyme activity once they reach the age of peak enzyme levels, which is further increased because of genetic variation.

Developmental changes in the regulation of gene expression of enzymes important to drug disposition and response make it difficult to extrapolate data from adults to children with respect to drug response. Therefore, for children to benefit from advances in pharmacogenetics, they must be included in clinical studies (Freund 2003). The federal regulations substantially limit research with more than minimal risk on healthy children, and affected children may participate in studies with more than minimal risk only if there is a likelihood of benefit or providing generalizable knowledge. One possibility to increase the available data about the impact of pharmacogenetic variants in children is to incorporate pharmacogenetic testing into protocols for therapeutic monitoring in children being treated with drugs lacking data on safety and effectiveness (Leeder 2004).

Familial Implications

Pharmacogenetics is similar to traditional genetic testing with respect to the potential familial implications of testing for germline mutations. The disclosure of genetic test results to at-risk family members has been debated for more than a decade, and many national and international guidelines have been developed in this regard (Forrest 2007; Godard 2006). In general, it is recommended that physicians encourage their patients to share the results of a genetic test with family members when preventable interventions can be taken (ASCO 2003; ASHG 1998). The American Society of Human Genetics (ASHG 1998) has identified two situations in which physicians should breach patient confidentiality to reduce harms to family members. (1) Patient confidentiality should be breached if attempts to encourage disclosure by the patient are unsuccessful; if harm is highly likely, serious, imminent, and foreseeable; if the at-risk relative(s) is identifiable; and if the disease is preventable or treatable. (2) Patient confidentiality should be breached if the harm from not disclosing is greater than the potential harms from disclosure. However, the ASHG recommendation was developed before the HIPAA Privacy Rule went into effect in 2003, and the disclosure of sensitive health information over the objection of the patient may well be found to violate the Privacy Rule. Furthermore, because these guidelines were developed for genetic disease risk information, it could be argued that the need to disclose or the desire to learn of pharmacogenetic test results differs than for disease-related information. However, family members with shared medical conditions may benefit from pharmacogenetic testing to inform their own treatment. In this regard, the American Medical Association (Taub 2004; Council on Ethical and Judicial Affairs of the American Medical Association 2003) urges physicians to discuss the familial implications of genetic testing and the circumstances under which the test result should be disclosed to family members. In particular, it states that "physicians should make themselves available to assist patients in communicating with relatives to discuss opportunities for counseling and testing, as appropriate."

Professional Readiness and Redefining Clinical Roles

In general, health professionals have a positive view of pharmacogenetics, believing that it will aid treatment decision-making and that it should be used to guide the development of treatment guidelines (Grant 2009; Shields 2005; Zachry 2002). However, despite the relatively early stage of clinical implementation, the uptake and use of a few tests have substantially increased during this short period by some groups. The decision to offer testing will likely rest on a combination of factors, including the availability of clinical guidelines, demonstrated benefits of testing, alternative drug monitoring options, drug labeling, physician knowledge and experience, and insurance coverage (Fargher 2006). Several surveys have been conducted in Europe, the United Kingdom, and the United States assessing physicians' knowledge and use of pharmacogenetic testing. Similar to the published literature of physician knowledge of genetics and the more traditional genetic testing for disease risk or diagnosis (Metcalfe 2002; Wilkins-Haug 2000), the results suggest room for improvement (Selkirk 2013; Haga 2012a; Stanek 2012; Dodson 2011). For example, one study reported that providers had very little knowledge of which pharmacogenetic tests were available and for which drugs, the purpose of testing, and test methods (Almarsdottir 2005). Poor knowledge of genetics, specifically pharmacogenetics, can contribute to test misinterpretation, potentially resulting in patient harms and medical liability. About 20% of physicians considered the interpretation of *HER2* test results "difficult," but less than 11% considered *TPMT* test interpretation difficult (Woelderink 2006).

With greater familiarity and recognition of the clinical benefits, the use of pharmacogenetic testing will likely increase. For example, in the United Kingdom in 1997, no dermatologists reported ordering *TPMT* tests before prescribing azathioprine (Tan 1997). Although *TPMT* clinical testing has been available since 1990, test orders did not begin to climb until 1998; more than 1000 tests were ordered in 2000, accounting for almost one-third of all pharmacogenetic tests ordered during a 10-year period in the United Kingdom (Holme 2002). In 2000, 54% of dermatologists reported ordering this test (Holme 2002). Other surveys have reported that more than 90% of dermatologists who prescribed azathioprine ordered *TPMT* testing pretreatment, even though the genotype-based test is not generally available in National Health Service laboratories (Fargher 2007b, 2006). The use of *TPMT* testing appears to vary between specialties, with the highest use reported by dermatologists (94%), followed by gastroenterologists (60%) and rheumatologists (47%) (Fargher 2007b). In Australia, TPMT and pseudocholinesterase were the most frequently ordered tests (400 and 250 tests ordered in 2003, respectively) (Gardiner 2005). A single hospital performed 4200 CYP2D6 assays for use in guiding perhexiline treatment, but the test was rarely used in other clinical centers, highlighting institutional differences (Gardiner 2005). In contrast, the use of *TPMT* testing by specialists in oncology/hematology and pediatrics in Ireland, the United Kingdom, Germany, and the Netherlands was lower—only 35% who prescribed thiopurine reported ordering *TPMT* testing for some patients (Woelderink 2006). Similarly, a survey of all registered Dutch gastroenterologists found that only a very small percentage of Dutch gastroenterologists (5%) determined *TPMT* status before initiating treatment but that 36% wished to use *TPMT* testing (de Boer 2006). Sixty-four percent of gastroenterologists believed the test provided no benefit and considered it superfluous to the mandatory blood monitoring.

All prescribing providers, nurses, genetic counselors, and pharmacists will need some knowledge of the field to provide pharmacogenetic testing safely and appropriately. Simply put, "a broad array of health care professionals, not just physicians, needs education about genetics" (Omenn 2003). Several professional organizations and professional schools, notably pharmacy (Drozda 2013; Krynetskiy 2013; Lee 2012; Murphy 2010), have recognized the importance of increasing pharmacogenetic knowledge. Published reports suggest, however, that medical education in the United States are lagging in revising their curricula to increase awareness of pharmacogenetics (Nickola 2012). In contrast, most respondents to a survey of British medical schools reported teaching pharmacogenetics, though for only a few hours total (Higgs 2008). Even for clinical geneticists and genetic counselors to provide adequate care to patients about pharmacogenetic testing, additional training related to pharmacotherapy (e.g., pharmacokinetics and pharmacodynamics) would be needed. Collaborations between pharmacy schools and genetic training programs could provide the necessary dual training requirements (Mills 2013a). With the widespread implementation of electronic health records, the development of clinical decision support for pharmacogenetic testing has been perceived favorably (Devine 2014) and a handful of hospital systems have begun to develop and implement such systems (Bell 2014; Overby 2012). However, some have argued that it would be easier to increase the number of clinical geneticists rather than try to educate physicians about genetics (Greendale 2001). Alternatively, others have suggested the creation of "genetic information specialists" to serve as liaisons between physicians and pharmacists (Morley 2004).

Professional Guidelines
Developing clinical guidelines will be essential to the appropriate use and application of pharmacogenetic test results for therapeutic decision-making. For example, analysis of the initial use of *HLA-B*5701* pharmacogenetic testing in the United States found that tests were often ordered for reasons other than prescreening for hypersensitivity (Faruki 2007). However, only a handful of professional guidelines have been developed. For example, several British medical organizations have issued guidelines with respect to *TPMT* testing—the British Society of Gastroenterology (Mowat 2011), the British Association of Dermatologists (Megitt 2011), the British Society for Paediatric and Adolescent Rheumatology (2010), and the British Society of Rheumatology (2008). For genotype-guided dosing for the anticoagulant drug Coumadin, professional organizations agree that because of insufficient evidence, testing is not recommended (Holbrook 2012; Flockhart 2008; Hirsh 2008). In contrast, a U.S. advisory group, the Panel on Antiretroviral Guidelines for Adults and Adolescents, recommends "screening for *HLA-B*5701* before starting patients on an abacavir-containing regimen, to reduce the risk of hypersensitivity reaction" (PAGAA 2012). Patients who test positive for *HLA-B*5701* should not be prescribed abacavir.

One of the most active groups to develop guidelines to inform how test results should be used in therapeutic decision-making is the Clinical Pharmacogenetics Implementation Consortium (CPIC). This group was established in 2009 as a partnership between the U.S.-funded Pharmacogenomics Research Network and PharmGKB. To date, CPIC has published 18 guidelines (see list at www.pharmgkb.org/views/listPublications.action?projectId=74).

Primary Care Providers
Although pharmacogenetic tests are relevant to all prescribing providers, primary care providers may benefit the most, given that they provide care for many of the common, complex disorders affected by pharmacogenetics as well as the many medications they prescribe. In 2006, 58% of office visits were made to primary care specialists, and 71% of physician visits involved drug therapy (Cherry 2008). It has been estimated that one-fourth of outpatients are taking medications that contain pharmacogenetic information in their labels, most of them prescribed by primary care practitioners (Frueh 2008). Another study estimated that 29% of primary care patients were taking at least one of 16 drugs that are metabolized by the polymorphic CYP enzymes (Grice 2006). A third study reported that more than half (59%) of 27 drugs with frequently reported adverse drug reactions were metabolized by at least one CYP enzyme linked to poor metabolism (Phillips 2001). However, primary care providers are less likely to have knowledge and/or experience with genetic testing than are other specialties such as pediatrics and obstetrics, and therefore, have a greater need for education and delivery support tools. The need for a solid understanding of genetics is underscored by the two roles envisioned for primary care physicians regarding genetic testing: (1) to assess and analyze genetic testing advertisements, particularly the "claims and recommendations of industry," and (2) to educate patients about the utility of genetic testing and potential social risks (Caulfield 2001). Both tasks require a firm understanding of the risks and benefits of testing, an awareness of the professional guidelines, and communication skills. If informed consent and counseling are deemed necessary for pharmacogenetic tests, primary care physicians without a firm understanding of the issues may need to refer patients to clinical genetic services (Morley 2004). Recognizing the time and knowledge limitations, investigators recently developed some talking points for discussion about pharmacogenetic testing before testing and when delivering test results (Mills 2013b).

Genetic Professionals
It is unclear whether pharmacogenetic testing always warrants consultation with genetics professionals or only when it is deemed particularly risky because of certain

test characteristics and incidental information. Some have suggested that genetic counseling would not be needed for most pharmacogenetic tests (Robertson 2001), and if routinely recommended or required, testing would likely not be feasible in many clinical settings, and an already limited workforce would be further strained (Netzer 2004).

Those making decisions about whether informal or formal genetic counseling is warranted should consider the type of incidental information generated by a given test. Furthermore, if the ordering physician finds the information in a laboratory report too complex with respect to interpretation or presence of incidental information, consultation by the ordering provider with a genetic professional might be helpful (Netzer 2004). For example, the test results for the first 100 *HLA-B*5701* abacavir hypersensitivity tests done by a major reference laboratory were reported to the ordering physician by a genetic counselor (Faruki 2007). The genetic counselor provided a "patient-specific interpretation" after considering the patient's race and noting the residual risk associated with a negative result. However, some physicians who ordered the test several times and understood how to interpret the results did not need to consult with the genetic counselor.

Pharmacists

Pharmacists play an important role in ensuring the safety of drug therapy by assessing potential adverse drug interactions and providing information about appropriate substitutions for patients with drug allergies and concomitant medications that should be avoided. In some clinical settings, the pharmacist scope of practice has expanded to incorporate identification of alternative therapies to reduce cost or increase safety (e.g., avoid drug-drug interactions), as well as other services such as case management for patients with complex drug regimens (Hinthorn 2002; Keely 2002). They have also taken on broader public health responsibilities, such as the provision of vaccinations (Hogue 2006), health screening (e.g., blood pressure, bone density), and, in some states, prescriptive authority to administer emergency contraception. Monitoring pharmacogenetic information to ensure appropriate drug dosing is a natural extension of the pharmacist's role (Morley 2004). Pharmacists will likely integrate pharmacogenetic test results to guide appropriate drug selection and dosage and play a broader role in treatment decision-making alongside the prescribing physician (Clemerson 2006). According to one expert, "pharmacogenomics will further expand the responsibility of pharmacists to screen prescriptions, because physicians will have more opportunities for error in prescribing, just as pharmacists will have more opportunities for error in dispensing" (Brushwood 2003). In addition, pharmacists may play a role in educating patients about pharmacogenetic testing and checking that testing has been offered (Sansgiry 2003; Robertson 2001). With respect to their role as educator, "the patient education role of pharmacists will be critical when dispensing and monitoring drug therapies based on pharmacogenomics, because of the unfamiliarity of patients with these therapies" (Brushwood 2003).

The expanded addition of pharmacogenetics to the pharmacist's scope of practice raises questions about the storage and access of pharmacogenetic information, pharmacist preparedness, and patient acceptance of this new service delivered by pharmacists. Despite success with the expanded scope of pharmacy practice at some institutions, largely in inpatient settings, the structure of an appropriate collaborative partnership between pharmacist and physician is not yet well defined, warranting further exploration, particularly as the use of pharmacogenetic testing increases. If any testing is conducted or required for a patient, the pharmacist can inform the treating physician about it (Woelderink 2006). These additional responsibilities would undoubtedly increase the amount of time spent with patients and could potentially increase overall costs (Clemerson 2006).

Although many pharmacists have expressed an interest in pharmacogenetics and have recognized a role for themselves in either providing testing or integrating the results into routine assessment, many have indicated that they currently lack adequate knowledge and skills to do so (de Denus 2013; Tuteja 2013). There is general agreement that more education for pharmacy students and practicing pharmacists is necessary to meet the expanding roles of pharmacists (Clemerson 2006; Johnson 2002; Vizirianakis 2002). The American Association of Colleges of Pharmacy and other professional organizations have called for the incorporation of pharmacogenomics into the pharmacy curriculum (Cavallari 2010; Murphy 2010). Specifically, the pharmacy school curriculum should be revised to include information on genomics, and pharmacists should be required to take continuing education in human genetics (Sansgiry 2003). Professional pharmacist organizations have recognized the need for increased pharmacogenomic education for students and practitioners alike to appropriately prepare pharmacists to use this information (Sansgiry 2003).

Equity and Access

Ensuring widespread access to and use of pharmacogenetic testing will hinge on several factors. In addition to provider awareness about pharmacogenetic tests, it is anticipated that coverage and reimbursement policies will substantially affect the use of these tests. Given the novelty of many pharmacogenetic tests and the ongoing assessment of their effectiveness compared with standard (non–genotype guided) care, insurers are divided in their coverage policies (Hresko 2012). It is likely that PPACA (Public Law 111-148) will positively influence the use of pharmacogenetics and other personalized medicine applications (Hays 2012). For example, PPACA will support the assessment of tests through the establishment of the Patient-Centered Outcomes Research Institute and its support of methods development and comparative effectiveness research to better inform coverage and appropriate use of diagnostic testing. In addition, PPACA may also benefit patients' use of pharmacogenetic and other genomic tests with the removal of coverage restrictions/caps (Hays 2012). If several tests are needed to inform treatment decisions or monitor response, patients can have the necessary testing without incurring additional expenses. Other policies regarding billing codes for diagnostic testing, however, may adversely affect the diagnostics market and availability of testing (PubMed Central 2010).

Related to the issue of reimbursement for testing is patient access to drugs indicated by a given test. For example, a molecular tumor analysis will provide insight into the characteristics of the cancer as well as inform potential drug options. Payers typically cover companion diagnostics (tests codeveloped and needed to determine patient suitability for a given drug), but issues may arise regarding which test is actually performed (the FDA-approved test or a similar laboratory-developed test) because reimbursement rates may differ. Another related issue is the possibility that a provider or patient could request treatment with a drug for an unapproved indication on the basis of a test result. For example, tumor sequencing may potentially identify genomic alterations for which treatments are available but for a different tumor type; in this case, use of the drug would be considered off-label. At this early stage, some large payers consider the use of sequencing to guide cancer therapies investigational, although coverage does occur on a case-by-case basis.

Conclusion

Pharmacogenetics presents an exciting opportunity to enhance the safe and effective use of drugs through the personalization of drug treatment based on an individual's genetic makeup. As with any new technology, several non-scientific issues must also be addressed to facilitate the successful translation of the science to the clinic and its safe and appropriate use. An overview of several of the ethical, legal, and social issues arising from the early uses of pharmacogenetics has been presented. With appropriate policies, research, and education, these issues can be addressed in a satisfactory manner; moreover, they should pose few challenges to the translation of pharmacogenetic testing. However, neglect of these issues could stall the use of these beneficial applications and/or cause harm to the user.

References

Agundez JA. Cytochrome P450 gene polymorphism and cancer. Curr Drug Metab 2004;5:211-24.

Almarsdottir AB, Bjornsdottir I, Traulsen JM. A lay prescription for tailor-made drugs—focus group reflections on pharmacogenomics. Health Pol 2005;71:233-41.

Healthcare reform law benefits children with genetic diseases: despite law's benefits, many questions about coverage remain. Am J Med Genet A 2012;158A:viii-ix.

American Society of Clinical Oncology (ASCO). American Society of Clinical Oncology policy statement update: genetic testing for cancer susceptibility. J Clin Oncol 2003;21:2397-406.

American Society of Human Genetics (ASHG). Professional disclosure of familial genetic information. The American Society of Human Genetics Social Issues Subcommittee on Familial Disclosure. Am J Hum Genet 1998;62:474-83.

Anstey AV, Wakelin S, Reynolds NJ, et al. Guidelines for prescribing azathioprine in dermatology. Br J Dermatol 2004;151:1123-32.

American (APhA). Report of the 2010 APhA House of Delegates: actions of the legislative body of the American Pharmacists Association. J Am Pharm Assoc 2010;50:471-2.

Bates BR, Poirot K, Harris TM, et al. Evaluating direct-to-consumer marketing of race-based pharmacogenomics: a focus group study of public understandings of applied genomic medication. J Health Commun 2004;9:541-59.

Bell GC, Crews KR, Wilkinson MR, et al. Development and use of active clinical decision support for preemptive pharmacogenomics. J Am Med Inform Assoc 2014;21:e93-9.

Berg JS, Khoury MJ, Evans JP. Deploying whole genome sequencing in clinical practice and public health: meeting the challenge one bin at a time. Genet Med 2011;13:499-504.

Bevan JL, Lynch JA, Dubriwny TN, et al. Informed lay preferences for delivery of racially varied pharmacogenomics. Genet Med 2003;5:393-9.

Bibbins-Domingo K, Fernandez A. BiDil for heart failure in black patients: implications of the U.S. Food and Drug Administration approval. Ann Intern Med. 2007 Jan 2;146:52-6.

Blake MJ, Castro L, Leeder JS, et al. Ontogeny of drug metabolizing enzymes in the neonate. Semin Fetal Neonatal Med 2005;10:123-38.

Braff JP, Chatterjee B, Hochman M, et al. Patient-tailored medicine, part two: personalized medicine and the legal landscape. J Health Life Sci Law 2009;2:1-3, 5-43.

Breckenridge A, Lindpaintner K, Lipton P, et al. Pharmacogenetics: ethical problems and solutions. Nat Rev Genet 2004;5:676-80.

Brewer NT, Defrank JT, Chiu WK, et al. Patients' understanding of how genotype variation affects benefits of tamoxifen therapy for breast cancer. Public Health Genomics 2014;17:43-7.

The British Society for Paediatric and Adolescent Rheumatology. Azathioprine Use in Paediatric Rheumatology. 2010. Available at https://www.bspar.org.uk/DocStore/FileLibrary/PDFs/BSPAR%20guidance%20for%20Azathioprine%202011.pdf. Accessed November 10, 2014.

Brock TP, Faulkner CM, Williams DM, et al. Continuing-education programs in pharmacogenomics for pharmacists. Am J Health Syst Pharm 2002;59:722-25.

Brock TP, Valgus JM, Smith SR, et al. Pharmacogenomics: implications and considerations for pharmacists. Pharmacogenomics 2003;4:321-30.

Brothers KB, Langanke M, Erdmann P. Implications of the incidentalome for clinical pharmacogenomics. Pharmacogenomics 2013;14:1353-62.

Brushwood DB. The challenges of pharmacogenomics for pharmacy education, practice, and regulation. In: Rothstein M, ed. Pharmacogenomics: Social, Ethical, and Clinical Dimensions. Hoboken, NJ: Wiley-Liss, 2003:207-25.

Buchanan A, Califano A, Kahn J, et al. Pharmacogenetics: ethical issues and policy options. Kennedy Inst Ethics 2002;12:1-15.

Burchard EG, Ziv E, Coyle N, et al. The importance of race and ethnic background in biomedical research and clinical practice. N Engl J Med 2003; 348:1170-75.

Butrick M, Roter D, Kaphingst K, et al. Patient reactions to personalized medicine vignettes: an experimental design. Genet Med 2011;13:421-8.

Carmody MS, Anderson JR. BiDil (isosorbide dinitrate and hydralazine): a new fixed-dose combination of two older medications for the treatment of heart failure in black patients. Cardiol Rev 2007;15:46-53.

Carter MJ, Lobo AJ, Travis SP, et al. Guidelines for the management of inflammatory bowel disease in adults. Gut 2004;53(suppl 5):V1-16.

Cassidy v SmithKline Beecham, 99-10423. (C.P. Chester County, Pa. 1999).

Caulfield TA. The informed gatekeeper?: a commentary on genetic tests, marketing pressure and the role of primary care physicians. Health Law Rev 2001;9:14-8.

Caulfield T. Genetic testing liability and regulatory policy: the Canadian situation. Jurimetrics 2000;41:7-21.

Cavallari LH, Overholser BR, Anderson D, et al. Recommended basic science foundation necessary to prepare pharmacists to manage personalized pharmacotherapy. Pharmacotherapy 2010;30:626-626.

Chakravarty K, McDonald H, Pullar T, et al. BSR & BHPR Guideline for Disease-Modifying Antirheumatic Drug Therapy (DMARD) in Consultation with the British Association of Dermatologists. 2008.

Cherry DK, Hing E, Woodwell DA, et al. National Ambulatory Medical Care Survey: 2006 summary. Natl Health Stat Rep 2008:1-39.

Clayton EW, Haga S, Kuszler P, et al. Managing incidental genomic findings: legal obligations of clinicians. Genet Med 2013;15:624-9.

Clemerson JP, Payne K, Bissell P, et al. Pharmacogenetics, the next challenge for pharmacy? Pharm World Sci 2006;28:126-30.

Condit C, Templeton A, Bates BR, et al. Attitudinal barriers to delivery of race-targeted pharmacogenomics among informed lay persons. Genet Med 2003;5:385-92.

Cooper RS, Kaufman JS, Ward R. Race and genomics. N Engl J Med 2003;348:1166-70.

Corder EH, Saunders AM, Strittmatter WJ, et al. Gene dose of apolipoprotein E type 4 allele and the risk of Alzheimer's disease in late onset families. Science 1993;261:921-3.

Council on Ethical and Judicial Affairs of the American Medical Association. AMA's Code of Medical Ethics, 2.131. Disclosures of Familial Risk in Genetic Testing. 2003. Available at www.ama-assn.org/ama/pub/physician-resources/medical-ethics/code-medical-ethics/opinion2131.shtml. Accessed November 6, 2014.

Daack-Hirsch S, Driessnack M, Hanish A, et al. "Information is information": a public perspective on incidental findings in clinical and research genome-based testing. Clin Genet 2013;84:11-8.

de Boer NK, Mulder CJ, van Bodegraven AA. Impracticalities of thiopurine S-methyltransferase determination in daily inflammatory bowel disease practice. Aliment Pharmacol Ther 2006;23:1278-9; author reply 79-80.

DeMarco M. Views on personalized medicine: do the attitudes of African American and white prescription drug consumers differ? Public Health Genomics 2010;13:276-83.

de Denus S, Letarte N, Hurlimann T, et al. An evaluation of pharmacists' expectations towards pharmacogenomics. Pharmacogenomics 2013;14:165-75.

Devine EB, Lee CJ, Overby CL, et al. Usability evaluation of pharmacogenomics clinical decision support aids and clinical knowledge resources in a computerized provider order entry system: a mixed methods approach. Int J Med Inform 2014;83:473-83.

de Wildt SN, Kearns GL, Leeder JS, et al. Cytochrome P450 3A: ontogeny and drug disposition. Clin Pharmacokinet 1999;37:485-505.

Dodson C. Knowledge and attitudes concerning pharmacogenomics among healthcare. Pers Med 2011;8:8.

Drozda K, Labinov Y, Jiang R, et al. A pharmacogenetics service experience for pharmacy students, residents, and fellows. Am J Pharm Educ 2013;77:175.

Duster, T. Race and reification in science. Science 2005;307:1050-51.

Evans BJ. Finding a liability-free space in which personalized medicine can bloom. Clin Pharmacol Ther 2007;82:461-5.

Exner DV, Dries DL, Domanski MJ, et al. Lesser response to angiotensin-converting enzyme inhibitor therapy in black as compared with white patients with left ventricular dysfunction. N Eng J Med 2001;344:303-10.

Fargher EA. Pharmacogenetic testing for azathioprine in the NHS: current uptake and implications for prescribing practice. Int J Pharm Pract 2006;14:B70-71.

Fargher EA, Eddy C, Newman W, et al. Patients' and healthcare professionals' views on pharmacogenetic testing and its future delivery in the NHS. Pharmacogenomics 2007a;8:1511-9.

Fargher EA, Tricker K, Newman W, et al. Current use of pharmacogenetic testing: a national survey of thiopurine methyltransferase testing prior to azathioprine prescription. J Clin Pharm Ther 2007b;32:187-95.

Faruki H, Heine U, Brown T, et al. *HLA-B*5701* clinical testing: early experience in the United States. Pharmacogenet Genomics 2007;17:857-60.

Flockhart DA, O'Kane D, Williams MS, et al. Pharmacogenetic testing of *CYP2C9* and *VKORC1* alleles for warfarin. Genet Med 2008;10:139-50.

Flordellis CS. The emergence of a new paradigm of pharmacogenomics. Pharmacogenomics 2005;6:515-26.

Forrest LE, Delatycki MB, Skene L, et al. Communicating genetic information in families—a review of guidelines and position papers. Eur J Hum Genet 2007;15:612-8.

Foster MW. Pharmacogenomics and the social construction of identity. In: Rothstein MA, ed. Pharmacogenomics: Social, Ethical and Clinical Dimensions. Hoboken, NJ: Wiley-Liss, 2003:251-65.

Frank D, Gallagher TH, Sellers SL, et al. Primary care physicians' attitudes regarding race-based therapies. J Gen Intern Med 2010;25:384-9.

Frueh FW, Amur S, Mummaneni P, et al. Pharmacogenomic biomarker information in drug labels approved by the United States Food and Drug Administration: prevalence of related drug use. Pharmacotherapy 2008;28:992-8.

Freund CL, Clayton EW. Pharmacogenomics and children: meeting the ethical challenges. Am J Pharmacogenomics 2003;3:399-404.

Gardiner SJ, Begg EJ. Pharmacogenetic testing for drug metabolizing enzymes: is it happening in practice? Pharmacogenet Genomics 2005;15:365-9.

Gasso P, Rodriguez N, Mas S, et al. Effect of *CYP2D6*, *CYP2C9* and *ABCB1* genotypes on fluoxetine plasma concentrations and clinical improvement in children and adolescent patients. Pharmacogenomics J 2014;14:457-62.

Gerdes LU, Gerdes C, Kervinen K, et al. The apolipoprotein epsilon4 allele determines prognosis and the effect on prognosis of simvastatin in survivors of myocardial infarction: a substudy of the Scandinavian simvastatin survival study. Circulation 2000;101:1366-71.

Godard B, Hurlimann T, Letendre M, et al. Guidelines for disclosing genetic information to family members: from development to use. Fam Cancer 2006;5:103-16.

Goldstein DB, Tate SK, Sisodiya SM. Pharmacogenetics goes genomic. Nat Rev Genet 2003;4:937-47.

Gow PJ, Ghabrial H, Smallwood RA, et al. Neonatal hepatic drug elimination. Pharmacol Toxicol 2001;88:3-15.

Grant RW, Hivert M, Pandiscio JC, et al. The clinical application of genetic testing in type 2 diabetes: a patient and physician survey. Diabetologia 2009;52:2299-305.

Green RC, Berg JS, Grody WW, et al. ACMG recommendations for reporting of incidental findings in clinical exome and genome sequencing. Genet Med 2013;15:565-74.

Greendale K, Pyeritz RE. Empowering primary care health professionals in medical genetics: how soon? How fast? How far? Am J Med Genet 2001;106:223-32.

Grice GR, Seaton TL, Woodland AM, et al. Defining the opportunity for pharmacogenetic intervention in primary care. Pharmacogenomics 2006;7:61-5.

Grumbach MM, Biller BM, Braunstein GD, et al. Management of the clinically inapparent adrenal mass ("incidentaloma"). Ann Intern Med 2003;138:424-9.

Haddy CA, Ward HM, Angley MT, et al. Consumers' views of pharmacogenetics—a qualitative study. Res Social Adm Pharm 2010;6:221-31.

Haga SB, Burke W, Ginsburg GS, et al. Primary care physicians' knowledge of and experience with pharmacogenetic testing. Clin Genet 2012a;82:388-94.

Haga SB, Kawamoto K, Agans R, et al. Consideration of patient preferences and challenges in storage and access of pharmacogenetic test results. Genet Med 2011;13:887-90.

Haga SB, Mills R, Bosworth HB. Striking a balance in communicating pharmacogenetic test results: promoting comprehension and minimizing adverse psychological and behavioral response. Patient Educ Couns 2014;97:10-5.

Haga SB, O'Daniel JM, Tindall GM, et al. Survey of U.S. public attitudes toward pharmacogenetic testing. Pharmacogenomics J 2012b;12:197-204.

Haga SB, Tindall G, O'Daniel JM. Public perspectives about pharmacogenetic testing and managing ancillary findings. Genet Test Mol Biomarkers 2012c;16:193-7.

Hamberg AK, Friberg LE, Hanseus K, et al. Warfarin dose prediction in children using pharmacometric bridging—comparison with published pharmacogenetic dosing algorithms. Eur J Clin Pharmacol 2013;69:1275-83.

Hawcutt DB, Thompson B, Smyth RL, et al. Paediatric pharmacogenomics: an overview. Arch Dis Child 2013;98:232-7.

Hays PV, Whence Social Determinants of Health?: Effective Personalized Medicine and the 2010 Patient Protection and Affordable Care Act. J Clinic Res Bioeth 2012;S:5.

Hedgecoe A. "At the point at which you can do something about it, then it becomes more relevant": informed consent in the pharmacogenetic clinic. Soc Sci Med 2005;61:1201-10.

Hedgecoe AM. Context, ethics and pharmacogenetics. Stud Hist Philos Biol Biomed Sci 2006;37:566-82.

Higgs JE, Andrews J, Gurwitz D, et al. Pharmacogenetics education in British medical schools. Genomic Med 2008;2:101-5.

Hinthorn DR, Generali JA, Godwin HN. Pharmacist scope of practice: response to position paper. Ann Pharmacother 2002;36:718-20.

Hirsh J, Guyatt G, Albers GW, et al. Antithrombotic and thrombolytic therapy: American College of Chest Physicians Evidence-Based Clinical Practice Guidelines (8th Edition). Chest 2008;133:110S-12S.

Hogue MD, Grabenstein JD, Foster SL, et al. Pharmacist involvement with immunizations: a decade of professional advancement. J Am Pharm Assoc (2003) 2006;46:168-79; quiz 79-82.

Holbrook A, Schulman S, Witt DM, et al. Evidence-based management of anticoagulant therapy: Antithrombotic Therapy and Prevention of Thrombosis, 9th ed: American College of Chest Physicians Evidence-Based Clinical Practice Guidelines. Chest 2012;141:e152S-84S.

Holme SA, Duley JA, Sanderson J, et al. Erythrocyte thiopurine methyl transferase assessment prior to azathioprine use in the UK. QJM 2002;95:439-44.

Hresko A, Haga S. Insurance coverage policies for personalized medicine. J Pers Med 2012;2:201-16.

Husain A, Loehle JA, Hein DW. Clinical pharmacogenetics in pediatric patients. Pharmacogenomics 2007;8:1403-11.

Institute of Medicine (IOM), Committee on Quality of Health Care in America. To Err Is Human. Washington, DC: IOM, 2000.

Johnson JA, Bootman JL, Evans WE, et al. Pharmacogenomics: a scientific revolution in pharmaceutical sciences and pharmacy practice. Report of the 2001-2002 Academic Affairs Committee. Am J Pharm Educ 2002;66:12S-15S.

Kahn J, Yale J. How a drug becomes "ethnic": law, commerce, and the production of racial categories in medicine. Health Policy Law Ethics 2004;4:1-46.

Kalish RA, Leong JM, Steere AC. Association of treatment-resistant chronic Lyme arthritis with *HLA-DR4* and antibody reactivity to OspA and OspB of *Borrelia burgdorferi*. Infect Immun 1993;61:2774-9.

Keely JL; American College of Physicians-American Society of Internal Medicine. Pharmacist scope of practice. Ann Intern Med 2002;136:79-85.

Kieling C, Genro JP, Hutz MH, et al. A current update on ADHD pharmacogenomics. Pharmacogenomics 2010;11:407-19.

Knoppers BM, Senecal K, Borry P, et al. Whole-genome sequencing in newborn screening programs. Sci Transl Med 2014;6:229cm2.

Kohane IS, Masys DR, Altman RB. The incidentalome: a threat to genomic medicine. JAMA 2006;296:212-5.

Krynetskiy E. Institutional Profile: Jayne Haines Center for Pharmacogenomics and Drug Safety: educating future generations of healthcare professionals. Pharmacogenomics 2013;14:465-8.

Kumorowski VM. Assessing legal liability in pharmacogenetic cases. Washburn Law J 2003;42:623.

Laedtke AL, O'Neill SM, Rubinstein WS, et al. Family physicians' awareness and knowledge of the Genetic Information Non-Discrimination Act (GINA). J Gen Couns 2012;21:345-52.

Lee KC, Ma JD, Hudmon KS, et al. A train-the-trainer approach to a shared pharmacogenomics curriculum for U.S. colleges and schools of pharmacy. Am J Pharm Educ 2012;76:193.

Leeder JS. Translating pharmacogenetics and pharmacogenomics into drug development for clinical pediatrics and beyond. Drug Discov Today 2004;9:567-73.

Leeder JS. Pharmacogenetics and pharmacogenomics. Pediatr Clin North Am 2001;48:765-81.

Leeder JS. Developmental and pediatric pharmacogenomics. Pharmacogenomics 2003;4:331-41.

Levenseller BL, Soucier DJ, Miller VA, et al. Stakeholders' opinions on the implementation of pediatric whole exome sequencing: implications for informed consent. J Genet Couns 2014;23:552-65.

Lunshof J. Teaching and practicing pharmacogenomics: a complex matter. Pharmacogenomics 2006;7:243-6.

Madadi P, Koren G. Pharmacogenetic insights into codeine analgesia: implications to pediatric codeine use. Pharmacogenomics 2008;9:1267-84.

Marchant GE, Milligan RJ, Wilhelmi B. Legal pressures and incentives for personalized medicine. Pers Med 2006;3:391-97.

Meggitt SJ, Anstey AV, Mohd Mustapa MF, et al. British Association of Dermatologists' guidelines for the safe and effective prescribing of azathioprine 2011. Br J Dermatol 2011;165:711-34.

Metcalfe S, Hurworth R, Newstead J, et al. Needs assessment study of genetics education for general practitioners in Australia. Genet Med 2002;4:71-7.

Mills R, Haga SB. Clinical delivery of pharmacogenetic testing services: a proposed partnership between genetic counselors and pharmacists. Pharmacogenomics 2013a;14:957-68.

Mills R, Voora D, Peyser B, et al. Delivering pharmacogenetic testing in a primary care setting. Pharmacogenomics Pers Med 2013b;6:105-12.

Moldrup C. Ethical, social and legal implications of pharmacogenomics: a critical review. Community Genet 2001;4:204-14.

Morley KI, Hall WD. Using pharmacogenetics and pharmacogenomics in the treatment of psychiatric disorders: some ethical and economic considerations. J Mol Med (Berl) 2004;82:21-30.

Mowat C, Cole A, Windsor A, et al. Guidelines for the management of inflammatory bowel disease in adults. Gut 2011;60:571-607.

Murphy JE, Green JS, Adams LA, et al. Pharmacogenomics in the curricula of colleges and schools of pharmacy in the United States. Am J Pharm Educ 2010;74:7.

Nawar R, Aron D. Adrenal incidentalomas—a continuing management dilemma. Endocr Relat Cancer 2005;12:585-98.

Netzer C, Biller-Andorno N. Pharmacogenetic testing, informed consent and the problem of secondary information. Bioethics 2004;18:344-60.

Ni X, Zhang J. Association between 9p21 genomic markers and ischemic stroke risk: evidence based on 21 studies. PLoS One 2014;9:e90255.

Nickola TJ, Green JS, Harralson AF, et al. The current and future state of pharmacogenomics medical education in the USA. Pharmacogenomics 2012;13:1419-25.

Nuffield Council on Bioethics. Pharmacogenetics: Ethical Issues. 2003. Available at www.nuffieldbioethics.org/sites/default/files/Pharmacogenetics%20-%20Summary%20and%20recommendations.pdf. Accessed December 2, 2014.

Omenn G, Motulsky AG. Integration of pharmacogenomics into medical practice. In: Rothstein M, ed. Pharmacogenomics: Social, Ethical, and Clinical Dimensions. Hoboken, NJ: Wiley-Liss, 2003:137-61.

Ordovas JM, Lopez-Miranda J, Perez-Jimenez F, et al. Effect of apolipoprotein E and A-IV phenotypes on the low density lipoprotein response to HMG CoA reductase inhibitor therapy. Atherosclerosis 1995;113:157-66.

Ortega VE, Meyers DA. Pharmacogenetics: implications of race and ethnicity on defining genetic profiles for personalized medicine. J Allergy Clin Immunol 2014;133:16-26.

Ossorio PN. Product liability for predictive genetic tests. Jurimetrics 2003;42:239-60.

Overby CL, Tarczy-Hornoch P, Kalet IJ, et al. Developing a prototype system for integrating pharmacogenomics findings into clinical practice. J Pers Med 2012;2:241-56.

Palmer LI. Medical liability for pharmacogenomics. In: Rothstein M, ed. Pharmacogenomics: Social, Ethical, and Clinical Dimensions. Hoboken, NJ: Wiley-Liss, 2003:187-206.

Panel on Antiretroviral Guidelines for Adults and Adolescents (PAGAA). Guidelines for the Use of Antiretroviral Agents in HIV-1 Infected Adults and Adolescents. 2012. Available at www.aidsinfo.nih.gov/ContentFiles/AdultandAdolescentGL.pdf. Accessed November 6, 2014.

Paul NW, Fangerau H. Why should we bother? Ethical and social issues in individualized medicine. Curr Drug Targets 2006;7:1721-7.

Payne K, Fargher EA, Roberts SA, et al. Valuing pharmacogenetic testing services: a comparison of patients' and health care professionals' preferences. Value Health 2011;14:121-34.

Perera MA, Cavallari LH, Limdi NA, et al. Genetic variants associated with warfarin dose in African-American individuals: a genome-wide association study. Lancet 2013;382:790-6.

Personalized Medicine Coalition. Issue Brief: The Adverse Impact of the U.S. Reimbursement System on the Development and Adoption of Personalized Medicine Diagnostics. Available at http://www.personalizedmedicinecoalition.org/Userfiles/PMC-Corporate/file/pmc_reimbursement_issue_brief.pdf.

Phillips KA, Veenstra DL, Oren E, et al. Potential role of pharmacogenomics in reducing adverse drug reactions: a systematic review. JAMA 2001;286:2270-9.

Polanczyk G, Bigarella MP, Hutz MH, et al. Pharmacogenetic approach for a better drug treatment in children. Curr Pharm Des 2010;16:2462-73.

Ravine D, Gibson RN, Donlan J, et al. An ultrasound renal cyst prevalence survey: specificity data for inherited renal cystic diseases. Am J Kidney Dis 1993;22:803-7.

Rieder MJ, Carleton B. Pharmacogenomics and adverse drug reactions in children. Front Genet 2014;5:78.

Robertson JA. Consent and privacy in pharmacogenetic testing. Nat Genet 2001;28:207-9.

Robertson JA, Brody B, Buchanan A, et al. Pharmacogenetic challenges for the health care system. Health Aff (Millwood) 2002;21:155-67.

Rodriguez-Antona C, Ingelman-Sundberg M. Cytochrome P450 pharmacogenetics and cancer. Oncogene 2006;25:1679-91.

Rogausch A, Prause D, Schallenberg A, et al. Patients' and physicians' perspectives on pharmacogenetic testing. Pharmacogenomics 2006;7:49-59.

Roses AD. Pharmacogenetics and the practice of medicine. Nature 2000;405:857-65.

Ross LF, Saal HM, David KL, et al. Technical report: ethical and policy issues in genetic testing and screening of children. Genet Med 2013;15:234-45.

Rothstein M. Liability issues in pharmacogenomics. LA Law Rev 2005;66:117-24.

Rothstein MA. Public attitudes about pharmacogenomics. In: Rothstein MA, ed. Pharmacogenomics: Social, Ethical, and Clinical Dimensions. Hoboken, NJ: Wiley-Liss, 2003:3-27.

Rothstein MA, Epps PG. Ethical and legal implications of pharmacogenomics. Nat Rev Genet 2001;2:228-31.

The Royal Society. Pharmacogenetics Dialogue: Findings from Public Workshops on Personalised Medicines Held by the Royal Society's Science in Society Programme. 2005. Available at https://royalsociety.org/~/media/Royal_Society_Content/policy/publications/2005/1111111400.pdf. Accessed December 2, 2014.

Sansgiry SS, Kulkarni AS. The human genome project: assessing confidence in knowledge and training requirements for community pharmacists. Am J Pharm Educ 2003;67:1-10.

Sansgiry SS, Kulkarni AS. Genetic testing: the community pharmacist's perspective. J Am Pharm Assoc (2003) 2004;44:399-402.

Sconce EA, Daly AK, Khan TI, et al. APOE genotype makes a small contribution to warfarin dose requirements. Pharmacogenet Genomics 2006;16:609-11.

Seefelder C, Leeder JS. Cytochrome P450 pharmacogenetics and anaesthesia. Paediatr Anaesth 2002;12:810-1.

Selkirk CG, Weissman SM, Anderson A, et al. Physicians' preparedness for integration of genomic and pharmacogenetic testing into practice within a major healthcare system. Genet Test Mol Biomarkers 2013;17:219-25.

Shields AE, Blumenthal D, Weiss KB, et al. Barriers to translating emerging genetic research on smoking into clinical practice. Perspectives of primary care physicians. J Gen Intern Med 2005;20:131-8.

Sonnenberg A. Personal view: "don 't ask, don't tell"—the undesirable consequences of incidental test results in gastroenterology. Aliment Pharmacol Ther 2004;20:381-7.

Stanek EJ, Sanders CL, Taber KA, et al. Adoption of pharmacogenomic testing by U.S. physicians: results of a nationwide survey. Clin Pharmacol Ther 2012;91:450-8.

Tan BB, Lear JT, Gawkrodger DJ, et al. Azathioprine in dermatology: a survey of current practice in the U.K. Br J Dermatol 1997;136:351-5.

Taub S, Morin K, Spillman MA, et al. Managing familial risk in genetic testing. Genet Test 2004;8:356-9.

Taylor AL, Ziesche S, Yancy C, et al. Combination of isosorbide dinitrate and hydralazine in blacks with heart failure. N Engl J Med 2004;351:2049-57.

Temple R, Stockbridge NL. BiDil for heart failure in black patients: the U.S. Food and Drug Administration perspective. Ann Intern Med 2007;146:57-62.

Thakkinstian A, Bowe S, McEvoy M, et al. Association between apolipoprotein E polymorphisms and age-related macular degeneration: a HuGE review and meta-analysis. Am J Epidemiol 2006;164:813-22.

Tuteja S, Haynes K, Zayac C, et al. Community pharmacists' attitudes towards clinical utility and ethical implications of pharmacogenetic testing. Per Med 2013;10(8).

U.S. Government Accountability Office (GAO). Pediatric Drug Research: The Study and Labeling of Drugs for Pediatric Use Under the Best Pharmaceuticals for Chilren Act. GAO Highlights. 2007. Available at www.gao.gov/highlights/d07898thigh.pdf. Accessed November 5, 2014.

Vanakker OM, De Paepe A. Pharmacogenomics in children: advantages and challenges of next generation sequencing applications. Int J Pediatr 2013;2013:136524.

Vizirianakis IS. Pharmaceutical education in the wake of genomic technologies for drug development and personalized medicine. Eur J Pharm Sci 2002;15:243-50.

Visser LE, Trienekens PH, De Smet PA, et al. Patients with an ApoE epsilon4 allele require lower doses of coumarin anticoagulants. Pharmacogenet Genomics 2005;15:69-74.

Wang L, McLeod HL, Weinshilboum RM. Genomics and drug response. N Engl J Med 2011;364:1144-53.

Wertz DC. Ethical, social and legal issues in pharmacogenomics. Pharmacogenomics J 2003;3:194-6.

Westbrook MJ, Wright MF, Van Driest SL, et al. Mapping the incidentalome: estimating incidental findings generated through clinical pharmacogenomics testing. Genet Med 2013;15:325-31.

Wilkins-Haug L, Hill LD, Power ML, et al. Gynecologists' training, knowledge, and experiences in genetics: a survey. Obstet Gynecol 2000;95:421-4.

Woelderink A, Ibarreta D, Hopkins MM, et al. The current clinical practice of pharmacogenetic testing in Europe: *TPMT* and *HER2* as case studies. Pharmacogenomics J 2006;6:3-7.

Yancy CW, Fowler MB, Colucci WS, et al. Race and the response to adrenergic blockade with carvedilol in patients with heart failure. N Engl J Med 2001;344:1358-65.

Zachry WM III, Armstrong EP. Health care professionals' perceptions of the role of pharmacogenomic data. J Manag Care Pharm 2002;8:278-84.

Zhang SC, Bruce C, Hayden M, et al. Public perceptions of pharmacogenetics. Pediatrics 2014;133:e1258-67.

Chapter 17

COST-EFFECTIVENESS, ECONOMIC INCENTIVES, AND REIMBURSEMENT ISSUES

WILLIAM J. CANESTARO, MSc.; JOSH J. CARLSON, PH.D., MPH;
LOUIS P. GARRISON, JR., PH.D.; AND DAVID L. VEENSTRA, PHARM.D., PH.D.

Learning Objectives

1. Distinguish exceptional aspects of evaluating the cost-effectiveness of PGx-based drug therapies.
2. Assess the key drivers of the clinical and economic value of a PGx application, and identify evidence gaps.
3. Evaluate the role of decision analysis in quantifying the benefits, harms, and costs associated with PGx tests and the uncertainty in these outcomes.
4. Judge the strengths and weaknesses of formal economic evaluation methods as applied to PGx.
5. Identify the opportunities and challenges in developing delivering, and paying for PGx tests in patient care and formulary decision making.

Keywords: Cost-effectiveness, cost-utility, decision analysis, patient quality of life

Abbreviations in This Chapter

AMCP	Academy of Managed Care Pharmacy
AMS	Academy of Medical Sciences
CYP P450	Cytochrome P450
CEA	Cost-effectiveness analysis
CUA	Cost-utility analysis
DDI	Drug-drug interaction
GDI	Gene-drug interaction
ICER	Incremental cost-effectiveness ratio
PGx	PGx
LDT	Laboratory-developed test
PI	Package insert
QALY	Quality-adjusted life-year
RCT	Randomized clinical trial
SNP	Single nucleotide polymorphism

Introduction

Previous chapters have described the scientific development of PGx technologies and their potential for clinical application. While the science of PGx may be developing rapidly, much of this work has yet to be translated to clinical care. Health economists raise questions that aim to address this gap: "How will the results of a test be used in real-world clinical practice?"; "When a test is used what will be the net impact on patient outcomes?"; "Does testing provide good value for the money spent?"; and "What incentives exist for stakeholders to adopt personalized medicine?"

General Background and Context

The FDA-approved drug package insert (PI) is extremely significant for the use of all pharmaceuticals. Not only is it an intensely negotiated contract between manufacturers and regulators on the scope of marketing claims for a drug but it is also the underlying evidence base for commonly used clinical decision aides such as Micromedex. While over 70% of the top 200 most prescribed drugs in U.S. have published PGx (PGx) associations in the scientific and clinical literature (Zineh 2006) less than 40% of FDA-approved drug PIs contain PGx information (Zineh 2004). Furthermore, despite the fact that the clinical effect of a drug-drug interaction (DDI) and slow metabolism genotype may be equivalent, among all prescriptions with drug-drug interactions mentioned in their PI greater than 25% do not describe the analogous gene-drug interaction (GDI) (Conrado 2013). Perhaps more shocking is that for those PI's that contain treatment recommendations for the DDI less than half of PI's will contain the analogous recommendation for the GDI (Conrado 2013). When the science is analogous, as you have learned in previous chapters, why are the clinical uses so different? What is exceptional about PGx information that causes its adoption to be so limited in clinical practice?

The answer may lie in the training of healthcare professionals and economic incentives for evidence generation. In a recent nationwide survey of U.S. physicians with more than 10,000 respondents, while greater than 97% agreed that genetic variation influences drug response, only 10% felt confident in using PGx testing (Stanek 2012). Furthermore even when testing and clinical guidance is offered, healthcare providers may act in ways that contradict the best available evidence (Desai 2013). Given how trained healthcare professionals can have such a variable response to this new clinical information it is no wonder that regulators are hesitant to recommend more testing.

With this lack of training and substantial uncertainty in real-world clinical effect, many have sounded the call for more and higher quality evidence for use of PGx in clinical practice (Shah 2012; Perry 2013; Chabner 2013). While this seems inherently reasonable that we should tackle uncertainty with more evidence, the incentives do not exist to generate it (Faulkner 2012; Towse 2013). Drugs are generally studied in large randomized clinical trials (RCTs), and the U.S. Food and Drug Administration (FDA) evaluates their benefits and risks before clinical use. This process provides an initial foundation of data and a framework for further expert evaluation of appropriate drug therapy. The generation of this evidence is sustained by the fact that pharmaceutical companies are granted periods of market exclusivity where they can act as a monopolist to recoup the large initial investment on their product. In fact, the World Health Organization has estimated that the pharmaceutical industry sales are between $300–400 billion U.S. dollars annually (World Health Organization 2014). This provides substantial opportunities to recoup large initial investments in research and development.

However, with PGx tests—as with most diagnostic or prognostic tests—the levels of evidence required by regulators, clinicians, and payers are more varied and often less rigorous. Only a small fraction of PGx tests are submitted to the FDA for formal review of their evidence and reliability while the rest are offered under the much less rigorous regulatory oversight of laboratory-developed test (LDTs or "home brews") (Weiss 2012). While laboratories must be licensed and certified to offer testing, there is no requirement for individual review of every LDT. Furthermore, unlike drugs, the same economic incentives do not exist to generate evidence for PGx tests. The entire molecular diagnostics market, of which PGx tests are a small fraction, has annual sales of only about $10 billion U.S. dollars (Price Waterhouse Coopers 2009). Even if a laboratory or test developer were to invest the resources to generate high-level evidence and go through the process of regulatory approval, there are no guarantees that they would not see competition from laboratories offering technologically-equivalent LDTs thereby undercutting their first-to-market advantage.

In the absence of rigorous clinical trials, a significant burden can thus be placed on medical experts trying to provide informed decisions for patients and health plans when faced with a PGx test. Clinician researchers and scientists working in the pharmaceutical and biotechnology industries also will be faced with substantial uncertainty regarding the influence of PGx on the drug development process. Health economists and outcomes researchers address this evidence gap by modelling the impact of new technology and resultant clinical information on patient care by synthesizing the best available evidence. In the practice of pharmacy, managed care pharmacists may perform their own analyses to aid in the process of developing a formulary. While not as rigorous as an RCT, modeling is faster, cheaper and more feasible given the current market structure for PGx tests. This work can have significant impact on patient outcomes as evidenced by a commercially marketed PGx test intended to help prevent aminoglycoside-induced hearing loss in patients with cystic fibrosis (Athena Diagnostics 2009).

Aminoglycoside antibiotics are commonly used to treat serious gram-negative bacterial infections such as Pseudomonas aeruginosa. Although this class of drugs is highly effective, it also causes adverse effects resulting in renal toxicity and ototoxicity (Munckhof 1996). It is estimated that 7% of all patients exposed to aminoglycosides experience some form of cochleotoxicity, although there is great uncertainty associated with the risk (Mulheran 2001). An association between the A1555G variant in the mitochondrial 12S ribosomal RNA gene and bilateral sensorineural hearing loss has been reported across several populations worldwide (Jaber 1992; Tang 2002; Nance 2003). Although there are relatively few studies, a recent review estimates the penetrance of this mutation to be 100% in individuals exposed to aminoglycosides (i.e., all people with this mutation who receive aminoglycosides will experience hearing loss) (Estivill 1998; Pandya 2004). The prevalence of the A1555G variant is low: A U.S. study estimated a population prevalence of 0.00086 (8.6 per 10,000) (Tang 2002). The test itself costs about $345 in the United States and the sensitivity and specificity are 99.9% and 87.0% (Athena Diagnostics 2009).

The health care intervention after a positive test result would be the use of an alternative regimen such as ciprofloxacin plus a β-lactam. Although no definitive studies have compared these two regimens, ciprofloxacin may have a higher risk of antibiotic resistance. A formal, quantitative cost-effectiveness analysis of this PGx test has been conducted (Veenstra 2007). While in the base-case analysis, the testing strategy led to a decrease in the incidence of severe aminoglycoside-induced hearing loss of 0.12% and an increase in quality of life, there was substantial uncertainty. In fact it is possible that changing antibiotic therapy according to the A1555G PGx test could result in harm to patients, on average. This would occur, in large part, because the A1555G variant is so rare that most test results would be false positive. At 0.50% prevalence and 87% specificity, 96% of children testing positive would be false positive and would inadvertently not receive first-line antibiotic therapy when they are indicated to.

While the test developer certainly had the best of intentions in offering the test, if users do not weigh the risks and benefits of the test's results they might unknowingly harm patients. It is important to note that even in the absence of cost information using new PGx tests may actually result in net harm if the information they provide is not reliable or if they led to overall lower-quality decision making by healthcare providers. By creating a decision model (using methods that will be described in this chapter) we can quantitatively weigh the risks and benefits of a new PGx test even when there is not a randomized controlled trial.

This chapter evaluates the use of analytic approaches used in economic evaluation, including decision modeling, cost-effectiveness evaluation, and outcomes research to address the myriad of questions that arise while the health care system incorporates PGx applications into clinical practice. Although these methods involve numerous factors ranging from health care costs to clinical effectiveness to patient quality of life, we will emphasize the importance of structuring the questions to be addressed, modeling clinical pathways, assessing data, and identifying key sources of uncertainty. Applications of decision modeling and cost-effectiveness evaluation for several PGx tests will be presented, and the incentives and challenges in the development of PGx explored.

Unique Challenges in Evaluating PGx Applications

Before exploring the application of economic evaluation to PGx, it is worthwhile to consider whether PGx is a fundamentally unique (i.e., exceptional) health care technology that might merit different methodological approaches. Economic evaluations and decision analyses have been conducted for decades in areas as diverse as highway construction (Sinha 2011), and education (Lee 2012), and for a wide variety of health care technologies and interventions, including testing strategies such as breast (Madan 2010) and colon (Gupta 2011) cancer screening, prostate-specific antigen testing and risk of prostate cancer (Garg 2013), HIV (human immunodeficiency virus) resistance testing (Sendi 2007), and radiology (Raymakers 2014). Test performance (i.e., sensitivity—the ability to detect those patients that truly have the condition; specificity—the ability to not falsely diagnose those who don't have the condition) readily can be incorporated into economic analyses, as well as the resulting changes in patient management and outcomes. The analytic approaches used in economic evaluation in health care are sufficiently adaptable that they can be used to evaluate PGx testing strategies.

There are certain aspects of PGx that do differ from other health care technologies—primarily with respect to the quantity and quality of data to inform cost-effectiveness analyses. Because genetic tests have fewer regulatory requirements than drugs or biologics, less evidence about their effectiveness or clinical utility is available when the test becomes available (Secretary's Advisory Committee 2006). Indeed, the time from the publication of a possible PGx association to marketing of that test may be as little as 1 or 2 weeks. Furthermore, the amount of data now available from genome-wide scans (up to 1 million SNPs) exceeds our understanding of the clinical implications of even a small fraction of these variants. It simply will not be feasible to generate RCT-quality data for all possible PGx tests, and traditional evidence-based guidelines will be challenging to develop. Yet the challenges for generating evidence should not mean that we recommend against the use of all tests without RCT-quality data. This creates a barrier to tests with likely benefits and low potential risks.

These evidence gaps may also arise disproportionately within the rubric of PGx. Certain types of tests (e.g., tests of inherited genetic variation versus those of acquired genetic variation) will have different levels of evidence of clinical utility, owing to differences in the types of tissue available, the health outcomes associated with their use, and the diseases involved. Tests of inherited genetic variation can be performed on most tissues of the body, so obtaining an appropriate tissue sample to test is not typically a barrier. This is not necessarily the case for tests of acquired genetic variation, which must be made on the specific tissues in which the variation/mutation has occurred (e.g., a cancerous tumor). If the tumor is in the pancreas or lung, for example, the difficulties in obtaining the appropriate tissue sample to test can become a barrier to evidence generation in both potential harms and costs. In addition, many tests of inherited variation are related to metabolizing enzymes and therefore tend to provide information on safety rather than efficacy. Because safety endpoints tend to be rare events, the number of patients needed to study to provide robust data of clinical utility can be prohibitively large. Finally, supporting evidence for PGx tests of acquired variation may be more prevalent because of the large number of clinical trials in oncology and the propensity to collect tumor tissue samples. Ultimately, the data available for the particular PGx testing situation will vary and, in turn, dictate how the evidence gaps are addressed.

Another distinctive aspect of genetic testing is the value that patients, in particular, place on knowing their genetic status, even if there are no health care or lifestyle changes that may be made based on the test result, sometimes referred to as "knowing for knowing's sake" (Grosse 2008). Because the inherent objective of PGx tests is to modify drug therapy, this issue tends to be less relevant, except when a PGx test also reveals information about disease risk (Henrikson 2008). Whether the disclosure of these so-called incidental findings is desirable and how they should be returned, if at all, is an area of considerable debate (Wolf 2013).

In summary, assessing the economic value of PGx tests will be particularly challenging because of the lack of data available on their impact in clinical use. Tests that have the potential, through subsequent treatment decisions, to *decrease* patient length or quality of life will require greater levels of evidence of clinical benefit. However, those with low risks—even if there is a relative paucity of effectiveness data—may warrant use (and evaluation) in clinical practice. Given the dramatically decreasing cost of obtaining genetic information, more and more data will be available to clinicians. The real cost, however, does not lie in the laboratory but instead in the clinic as clinicians and researchers attempt to understand the significance of the genotype information they receive. This means that assessing the potential clinical and economic outcomes of PGx tests will become an increasingly important undertaking because demand will increase with the removal of the cost barrier in obtaining genetic information—potentially increasing the negative impact of test adoption, if found harmful.

The Perspective of the Health Economist

Before diving too deeply into the methods of health economics it is useful to more fully appreciate the discipline's perspective. Decisions about resource allocation in healthcare can quickly become highly politicized as by its very nature it deals with rationing a highly valuable resource. This can be seen in the "death panel" rhetoric around the Affordable Care Act (Rocke 2014; Kliff 2013). Perhaps the most important concept to understand about the formal methods for economic evaluation in health care is that the primary objective of economic decision-making is not to save money. Rather:

> The goal of economic evaluations is to provide an assessment of the benefit to patients in life expectancy and quality of life in the context of real-world constraints on health care spending.

The objective is to assess the *value* of a technology or intervention. This concept should be separated from the actual cost of a technology. Just because a technology is inexpensive does not mean it has good value, and a technology with good value may be expensive. To assess the *value* of a novel drug, test, or procedure, the overall benefit to patients must be comprehensively assessed. Thus, not only are typical clinical endpoints assessed in clinical trials, but so are their longer-term implications—for example, the impact of reducing low-density lipoprotein cholesterol on a patient's risk of heart attacks during his/her remaining life span. The risks of unintended consequences such as adverse drug events are also weighed as well as the impacts on patient quality of life. Finally, costs are assigned to the various health care and/or societal resources used. The overall value of the intervention is then assessed by calculating the incremental cost per incremental benefit versus the next best alternative and comparing this value with those of other commonly used interventions across the health care spectrum. As will be seen later in this chapter, the choice of comparator is exceptionally important and one of the potential ways in which an evaluation might be misleading if done inappropriately.

Format of This Chapter

The methods described in this chapter can provide a flexible and transparent approach to quantitatively assessing the clinical benefits and risks, as well as the costs, of PGx tests—and in particular, providing a better understanding of the uncertainty and data gaps. These approaches can be applied or used by pharmacists at a variety of stages in the translational pathway for genetic tests (Khoury 2007) including early assessment of data requirements, research investment strategies, development of clinical guidelines, reimbursement decisions by health care payers, and clinical decision support. This chapter will describe five different methods for evaluating health care interventions and apply them to a well-known example of metabolism testing for clopidogrel. Each of these methods will build in terms of complexity on previously presented methods. It is important to note, however, that a more complex model does not necessarily make for a better analysis: instead, every analysis should aim to select a method that is "fit for purpose." For each method, there will be a description of the method's advantages and disadvantages, overall objective, in-depth methods, and final measure. While health economists and outcomes researchers can perform descriptive analyses that evaluate the cost or consequences of a new technology, we will only describe methods that employ the comparison of testing versus no testing scenarios.

Before delving into the methods, however, it is important to thoroughly consider the perspectives of the stakeholders for healthcare decisions so that we have a better appreciation of which method is most appropriate for our needs. Finally, we will conclude with a description of the current state of the science as well as reimbursement barriers and reforms.

Stakeholders in Healthcare and Their Perspectives

Healthcare is distinctly different from other economic markets in that the consumer ultimately using the product, the patient, is often not the one who makes the purchasing decisions or pays for the product. In healthcare the healthcare provider acts as the patient's agent recommending the best course of action. This is necessary given the breadth of knowledge that a physician or pharmacist requires but may create an information asymmetry between these two actors. Further complicating matters is the fact that health events that necessitate medical care are mostly unpredictable and can be more expensive than a single individual can afford necessitating the pooling of risk via insurance. All of these actors and their incentives are important in healthcare decision making and useful to consider when thinking about how to frame any analysis. Each of these groups is described more below.

Drug Manufacturers

The economic incentives to develop PGx tests vary depending on the stakeholder perspective. For the pharmaceutical industry, PGx could improve the clinical trials process by allowing smaller, shorter, faster, and cheaper trials with improved success rates. An understanding of the relationship between a genetic marker and drug efficacy could enable the design of a clinical trial enriched with likely responders, thus improving the chances of finding a clinically and statistically significant benefit. An additional benefit of PGx might be the exclusion of patients more likely to experience an adverse event, thus improving the risk-benefit ratio of the drug. However, the above benefits are counterbalanced by the difficulties identifying and validating a PGx marker and test to predict drug efficacy or toxicity. This is a process that may require a substantial amount of time and resources, including large clinical epidemiologic studies before targeted drug development. The most likely successes will be when the PGx marker is directly related to the mechanism of the drug—oncology being the most obvious area for application of this approach.

Currently, the pharmaceutical industry generally relies on a "blockbuster" model to finance most of its research and development: The 30% of drugs that earn more than the average cost of developing a new drug generate more than 50% of the revenues in a given year (Dimasi 2001). Let us first consider a drug that is on the market before the introduction of a test. Although a PGx test might reduce the number of patients who receive a drug, it is not necessarily true that the total revenues would decrease proportionally—or at all. The revenue that could be generated would depend on a variety of factors, including whether the drug is already on the market, pricing flexibility in the market, and other competitive factors (Garrison 2007; Vernon 2005). The effectiveness of such a drug in a targeted population would be higher, on average, than in a nontargeted population and would thus provide more value per patient. Assuming drug pricing is correlated to some extent with value, a portion of the "lost" revenues could be captured, although this effect might be limited by the ability to increase reimbursement levels for the drug. In a situation in which there are competing drugs, market share can also potentially be improved with PGx. For example, if a drug with a current 10% market share in a broad indication is highly efficacious in 20% of the patient population with a specific PGx marker—and most of that segmented market could be captured—revenues might be markedly increased. It is also possible that a new PGx test that targets a subgroup of responders could increase the sales of a drug that, because of its adverse drug events, physicians and patients are reluctant to use. However, for drugs "rescued" from safety problems, carefully developed risk management plans would require development and implementation to foster appropriate use.

Development of a novel drug in combination with a PGx test may present significant opportunities for drug manufacturers. If an accurate PGx test could be developed together with the drug, then the drug developer would have a better estimate of the market size and overall cost-effectiveness of the drug and could plan accordingly. If the drug developer also develops the test, additional value and revenue could be captured while screening (i.e., testing) the larger pool of potential patients (Garrison 2006, 2007A).

Test Manufacturers and Developers
For diagnostic manufacturers, the economic incentives for developing PGx technologies depend on a multitude of factors. These include diagnostic reimbursement policies, the timing of the launch of the diagnostic in relation to whether the therapeutic is on the market, and the level of regulation required to enter the market, gain FDA approval, or both (Garrison 2007B, 2006). The incentives for developing a PGx test by a diagnostic manufacturer are fundamentally related to the method used to establish test prices. Two primary approaches are administrative pricing (tied to the cost of performing the test) and one in which the supplier sets prices (tied to the value of, or willingness to pay for, the test). In the United States, administrative pricing is generally used for laboratory tests, and supplier pricing is used for medicines. With administrative pricing, the economic incentives are reduced because the payer or the patients as opposed to the test manufacturers would capture the benefit a PGx test could provide. This can present a major challenge for innovation in the diagnostic test field, a contention supported by an Institute of Medicine report suggesting that the current process for establishing reimbursement levels for new tests is not equipped to handle emerging diagnostic and genomic tests. The report states:

> Payments for some individual tests likely do not reflect the cost of providing services, and anticipated advances in laboratory technology will exacerbate the flaws in the current system. Problems with the outdated payment system could threaten beneficiary access to care and the use of enhanced testing methodologies in the future (Wolman 2000).

As the statement implies, the current reimbursement process, tied to the technical aspects of a test rather than value, may not be appropriate or sufficient to handle the next generation of laboratory tests of which PGx tests will be a part. A formal economic evaluation can help provide guidance on, and a quantitative evaluation of, the value created by PGx tests (Garrison 2007).

The timing of the launch of the diagnostic in relation to a drug has two important effects. First, if the drug is already on the market, a portion of the uncertainty in developing the PGx test associated with the given drug is removed. That is, there will likely be a market for the test, should it be developed. Furthermore, with respect to reimbursement, a co-marketed drug-test combination may have an easier hurdle to overcome than a test introduced several years after the drug because of data availability and quality of evidence from registration RCTs. The second factor is the amount of added benefit produced by the test and drug combination over the drug alone that the test could conceivably capture. If the drug is already on the market, its price has already been established; therefore, it will have decreased flexibility. Hence, a notable portion of the benefit (or added willingness to pay) could be captured with a higher test price (assuming the firm can set or negotiate the price or that the price is set with some relation to the value of the intervention).

For regulation, the higher the regulatory hurdle, the more evidence in support of the PGx test will have to be provided as well as the administrative costs and loss of time relative to the patent protection. The increased requirement for evidence in support of a test will increase the cost of research and development and therefore require a higher aggregate supply price to recoup the up-front costs. In addition, the loss of time on the market with patent protection will decrease the total profit that could be made during the lifetime of the patent. Therefore, tests developed alongside drugs during drug development are in a better economic position compared with tests developed after drugs have reached the market because the former have robust evidence from RCTs in support of regulatory approval.

Health Care Providers
Health care providers (HCPs) also have an interest in providing the best, most up-to-date diagnostic information to their patients, but this group has concerns about the evidence available for testing. In fact, only 1 in 10 U.S. physicians feels adequately informed about PGx testing (Stanek 2012). Presumably this is due to a confluence of a lack of extensive training in genetics and lack of clarity in testing reliability and utility (Haga 2012). Even in the pharmacy, where pharmacists are arguably better trained to handle the issues of personalized medicine, there is reluctance to fully embrace testing. In a survey of community pharmacists while greater than 75% expressed interest in offering personalized medicine services (counseling and testing) less 50% indicated they felt knowledgeable comfortable making therapy suggestions to physicians or confident counseling patients based on results of genetic screenings with their current level of training (Alexander 2014). Among the barriers that this group identified to offering the service, the two most challenging were cost of providing the service to their pharmacy and reimbursement issues (Alexander 2014). This highlights the importance of economics in providing these services to patients through a pharmacist.

Regulators

Regulators are charged with assessing the benefits versus risk of new technologies and deciding whether to allow them on the commercial market. The processes for evaluating new medicines are much more well developed than the processes for assessing new companion diagnostics (Olsen 2014). It has been suggested that the regulator's perspective is overwhelmingly risk averse as their incentives are to keep harmful technologies from entering the market rather than promote innovation (Shaw 2010).

Patients

Perhaps the most vulnerable stakeholder in PGx is the consumer or patient. Patients have several concerns, namely: that out-of-pocket costs are manageable, that they have access to most current available technologies, and that the test is accurate (Sun 2011). While patients are the ultimate consumer, they also may not have the training necessary to fully evaluate a test's performance characteristics. This creates an information asymmetry between them and their providers.

Framing the Question of Value: Qualitative Cost-Effectiveness Framework

Although the methods for assessing the value of PGx are not necessarily unique, testing introduces additional data requirements and added layers of complexity. The factors that can drive the cost-effectiveness of a PGx test can be organized into four categories: (1) those related to the gene(s) of interest, (2) those related to the test itself, (3) those associated with the disease of interest, and (4) those associated with the treatment that would arise because of testing (Flowers 2004). Table 17.1 lists these factors, together with the questions that clinical pharmacists should ask.

These criteria can be used to quickly identify any potentially serious limitations of the value of a PGx test. Not every factor must be researched in detail; rather, these questions should be used to identify areas warranting additional detailed investigation and evidence gaps. If a PGx test passes these criteria—that is, no red flags are raised—either more in-depth, quantitative analyses can be undertaken or cost-effectiveness may not be a

Table 17.1. Qualitative Framework for Evaluating the Potential Cost-Effectiveness of PGx Testing

Factors	Questions to Ask
Test	Accuracy—What are the specificity and sensitivity of the test for detecting the genetic variant of interest? Cost—What is the cost of the test and related services such as counseling? Timeliness—What is the time frame for obtaining test results?
Gene	*Prevalence* How common is the genetic variant? How many patients would have to be tested to identify a patient with a variant? What are the positive and negative predictive powers of the test in a patient population? *Penetrance* What is the relationship (association) between the genetic variant and drug response? What is the relative risk of an adverse event in patients with a variant genotype vs. those without? What is the probability of drug response in patients with a variant genotype vs. those without?
Clinical outcomes	*Prevalence and risk* How common is the drug-related adverse event that should be avoided? What is the difference in absolute risk for variant and nonvariant genotype patients? How common is drug nonresponse? What is the difference in likelihood of drug response in variant vs. nonvariant genotype patients? *Outcomes and economic impacts* How expensive is the adverse event or drug nonresponse? What is the impact of the adverse event or disease on quality of life?
Treatment	*Outcomes and economic impacts* Is there a clear intervention based on the result of the PGx test? How effective is the intervention? What risks are associated with the intervention? What is the cost of the intervention? What alternatives to individualized therapy are available other than PGx testing? What is the likelihood that treatment decisions suggested by test will be followed?

major concern and efforts can be directed toward other considerations such as implementation, access, and reimbursement. If major concerns arise when assessing these criteria, additional evidence may be sought, and if not available, the test may not have sufficient clinical and economic data supporting its routine use in clinical practice. In such situations, the test may be a candidate for a "coverage with evidence development" type of arrangement, in which the test is used and information on its clinical and economic impacts is collected. An example of this was the use of genetic testing to predict the therapeutic dose of the anticoagulant warfarin for patients covered by Medicare (Deverka 2010).

As Table 17.1 indicates, the data available to answer the relevant questions will likely come from multiple sources, especially if there are no relevant comparative clinical trials performed to date. To synthesize data from different sources, provide useful metrics on which decisions can be made, and evaluate the impact of uncertainty on such a decision, we should move beyond qualitative assessments of the evidence toward a quantitative framework—decision analysis. While a qualitative approach is a necessary first step and provides valuable insight, it is complemented by the addition of a rigorous approach to economic evaluation. Decision analysis has been commonly used to assist in the evaluation of a variety of health care technologies and will likely serve an important role in PGx.

Among the most important issues that can be addressed at this stage is the appropriate choice of comparator for the evaluation. If when evaluating a new treatment, the comparator chosen is a less effective and less commonly used drug, it may make the new drug appear to have greater value than if it were compared to a more effective therapy (Tsao 2012). This has been a criticism of previous economic analyses and may result in misleading the reader. To address this concern, the modeler must choose a comparator that most accurately reflects the decision facing health care providers (Gold 1996). As an example, if we were evaluating a new cholesterol lowering treatment the choice of branded fluvastatin would be a misleading choice of comparator as it is more expensive and less effective at cholesterol lowering than generic atorvastatin (Naci 2013).

Introduction to a Real-World Example

Throughout this chapter, the methods that are described will be applied to clopidogrel and *CYP2C19* testing. Numerous decision models have been published (Sorich 2013; Reese 2012; Panattoni 2012; Lala 2013; Kazi 2014) to investigate this question that we will draw from and generalize for our purposes (Table 17.2). First, we will describe the framing of the question of value but first it is necessary to have some clinical background. In patients with acute coronary syndrome, any one of a number of

Table 17.2. Published Decision Models for CYP2C19 Testing Prior to Clopidogrel Use

Study	Country	CYP2C19 Alleles	Base Case Strategy	Alternative Strategy	Result
Panattoni 2012	N.Z.	*2	All patients treated with clopidogrel	• All patients given prasugrel • *2 allele carriers (poor and intermediate metabolizers) given prasugrel all others on clopidogrel	ICER ranged from $6,706/QALY to $18,971/QALY depending on the underlying frequency of thromboses in the population
Lala 2013	U.S.	*2	All patients treated with clopidogrel	• All patients given prasugrel • *2 allele carriers (poor and intermediate metabolizers) given prasugrel all others on clopidogrel	78% of model simulations run reported an ICER below $100K/QALY
Reese 2012	U.S.	*2-*8, *17	All patients treated with clopidogrel	• All patients given prasugrel • *2-*8 allele carriers (poor and intermediate metabolizers) given prasugrel all others on clopidogrel	Genotype-guided therapy was cost saving compared with using only clopidogrel or prasugrel for all patients, ICERs of -$6760/QALY and -$11,710 respectively.
Kazi 2014	U.S.	*2-*8, *17	All patients treated with clopidogrel	• All patients given prasugrel • All patients given ticagrelor • *2-*8 allele carriers (poor and intermediate metabolizers) given prasugrel all others on clopidogrel • *2-*8 allele carriers (poor and intermediate metabolizers) given ticagrelor all others on clopidogrel	Compared to giving all patients generic clopidogrel, genotyping and giving *2-*8 carriers ticagrelor resulted in an ICER between $25K/QALY and $30K/QALY depending on the test's performance characteristics

conditions associated with artery obstruction, the standard therapy is aspirin plus clopidogrel as antithrombotic to prevent a subsequent thrombosis (O'Gara 2013). Clopidogrel is a pro-drug which has to be metabolized by the P450 isoenzyme *CYP2C19* in order to reach its active form. *CYP2C19*, the synonymous gene that encodes this enzyme, is polymorphic with 20% of patients being genetically defined poor metabolizers who ultimately receive reduced concentrations of the active drug (Zabalza 2012; Bauer 2011). Alternative approaches for poor metabolizers have been suggested and include tripling of the dose (Mega 2011) as well as treatment with prasugrel (Mega 2009), a branded agent that is not dependent on this enzyme for activation but also carries a higher bleeding risk. Another agent, ticagrelor, is also available for use in this indication but for sake of simplicity we will only use prasugrel as an alternative.

In a scenario where testing is not available, patients with acute coronary syndrome would all be treated with a standard aspirin plus clopidogrel therapy. In a scenario with testing, however, patients would be genotyped prior to treatment with poor metabolizers given prasugrel. This genotyping strategy could improve efficacy for patients who would have been poor metabolizers of clopidogrel while also reducing the number of bleeds of *CYP2C19* normal metabolizing patients on prasugrel who could effectively be treated with clopidogrel. Drug cost is a consideration in this comparison as clopidogrel is available as a generic drug while prasugrel is still under patent protection and carries a higher monthly price.

Decision Modeling and Risk-Benefit Analysis

Introduction
Assessing the many complex factors discussed above is challenging, particularly given the uncertainty that will exist in much of the data. Evaluating complex decisions by informal approaches is fraught with potential errors. For example, it has been shown that even highly educated individuals tend to overweight rare events (Burns 2004) and that responses to losses are greater than responses to gains (i.e., we tend to remember rare, bad events more than common, good ones). To address these and other related challenges, a formal approach termed *decision analysis* has been developed to provide a systematic, quantitative, and transparent process for evaluating and informing complex decision problems (Pettiti 2000; Weinstein 2003).

Objective
The essence of decision analysis is making explicit all relevant treatment strategies; their benefits, risks, and costs; and all sources of data. This is also the biggest challenge—primarily because it is rare that all necessary data are available, and the analysis must account for these limitations. Another advantage of decision analysis, in addition to making assumptions and data sources explicit, is the capability to incorporate data from disparate sources (e.g., efficacy estimates from clinical trials, likelihood of long-term disease progression from epidemiologic studies, patient quality of life, health care costs). Almost all cost-effectiveness evaluations in health care use some degree of decision-analytic modeling, even if most data are derived from a single source (e.g., a large RCT).

Method
The foundation of decision analysis is the development of a decision tree that depicts (1) the alternative treatment strategies to be considered, (2) the clinical pathways (events) that follow the selection of each strategy, and (3) the data that will be required to analyze the decision. Data required might include the prevalence of the genetic variant in the population, test performance characteristics, the strength of the genetic association, and the baseline probability of the clinical events of interest—the factors outlined in the qualitative framework described above.

The example in Figure 17.1 depicts the choice of antithrombotic agent based on a PGx test result. In this case, patients are normally treated with clopidogrel; those who carry one or more allele of *CYP2C19* associated with poor metabolism (*2-*8) may not receive the full benefit of therapy. With PGx testing, patients testing positive for any of these alleles are given prasugrel, whereas patients without any of these alleles are kept on clopidogrel. The test will not necessarily be 100% accurate, so true/false–positive/false-negative results are incorporated in the testing arm. The probabilities of having a good or poor response will depend on the "pathway" through the tree (i.e., the patient's genotype and the treatment he or she received). Finally, with the appropriate probabilities and data on the costs and outcomes patients experience on each pathway through the decision tree, it is possible to compare the two interventions (i.e., using a PGx test vs. standard care/no testing).

Another critical aspect of all decision analyses is the evaluation of uncertainty in the results. Because data are typically derived from a variety of sources and not available at the patient level, rather than using typical statistical techniques (e.g., calculating p values), uncertainty is assessed using sensitivity analysis, in which model inputs are varied over plausible ranges and the impact on results is evaluated.

Final Measure

In the most simplified version of a decision analysis we would use the inputs we have to estimate the total number of events (both "on-target" events prevented by the drug's mechanism of action as well as "off-target" adverse events) as well as total costs for each arm. The inherent limitation in this process is that while we may have counts of stent thrombosis and gastrointestinal bleeding under both scenarios for treatment, we have no means of quantitatively comparing them. Consider for example a situation in which 100 stent thromboses were prevented by switching poor metabolizers to prasugrel but this caused 50 extra bleeds. Assuming there is no mortality difference, is this still good value? What if introducing testing increases our costs fourfold? Our response will depend on how we weight the severity of these two events, an issue we will address further on in the chapter.

Application to Clopidogrel and C19 Testing

Below are some explicit assumptions for this worked example that are necessary for clarity despite lack of dataI

Table 17.3. Model Inputs[a]

Input	Value
Prevalence of poor metabolizer genotype	20%
Risk of stent thrombosis	
Wild type patients on clopidogrel	1.5%
Poor metabolizer on clopidogrel	3.0%
Any patient on prasugrel	2.0%
Risk of bleeding	
clopidogrel	1.8%
prasugrel	2.4%
Mortality	
stent thrombosis	50%
bleeding	10%
Costs	
drug costs for course of treatment	
clopidogrel	$100
prasugrel	$1000
clinical events	
stent thrombosis	$40,000
bleed	$20,000
genetic test	$300

[a]Inputs drawn from clinical data but highly generalized for sake of clarity in explanation

- The only clinical event for which drugs prevent is stent thrombosis; adverse events consist of bleeding only. These events are assumed to be mutually exclusive.
- The effect of carrying an allele associated with poor metabolism is on stent thrombosis risk only.
- C19 allele status is assumed to have no effect on response to prasugrel.
- Test sensitivity and specificity were assumed to be 100%.
- There is no background mortality, in other words those that do not have a stent thrombosis or bleed are not at risk due to death from other causes.

Given these inputs, a decision tree can be created to show the structure of the analysis (Figure 17.2). By simple algebra, we can determine the percent of the population that will travel down each path in the tree.

This model tells us that introducing testing decreases risk of stent thromboses and increases risk of bleeding as we might expect. We can see from the figure that the risk of bleeds is 1.8% and 1.85% in the standard of care and PGx strategy respectively. This slight increase in bleeding is caused by the 20% of patients who are allele carriers receiving prasugrel which has a slightly higher bleeding risk. Our risk of stent thrombosis is 1.8% and 1.6% in the standard of care and PGx strategy respectively, again driven by the switch of allele carriers to prasugrel. Most importantly, by using testing the mortality rate drops from 1.08% to 0.95%

While these results are certainly very interesting, there are numerous limitations for this overly simplistic approach. First, we do not capture the element of time. We assume clinical events happen over a set period and that the rate stays constant over this whole period. In practice we know this is rarely the case as many events happen with greater frequency earlier in treatment. Next, we have no estimate of quality of life for those that survive events. We know that after having an event patients do not immediately return to full health. While our model is able to capture mortality, it says nothing of morbidity or level of patient functioning.

Although overall health care costs may increase with testing, patient outcomes may be improved. Formal economic evaluation methods are required to provide a framework for interpreting results such as these. Although this illustrative example is rudimentary, it shows the potential magnitude of the benefit of PGx testing, given certain assumptions, and highlights specific data requirements. Finally, decision modeling approaches will not always be required when all key outcomes data are available directly from clinical trials or, in rare cases, database analyses. These situations are rare in health care, however, and some degree of decision modeling is usually used when evaluating the clinical and economic outcomes of health care technologies.

Budget Impact Analysis

Introduction
Although it is useful to have an estimate of the number of clinical events expected for health plan, they must also know the magnitude of expected expenditure on a new technology in order to set their premium for the upcoming year. To accomplish this task they must first estimate the number of individuals expected to be eligible for the technology as well as what the accounting cost is for various potential policies.

Objective
Budget impact analysis uses the known prevalence of a condition within a population as well as the average cost for treating patients with different courses of action to estimate the total accounting cost to the health plan for a new technology.

Method
The first step in a budget impact model is to have an estimate of the population or health plan size as well the prevalence of the condition. For our example, we will use a simplified health plan size of 1,000,000 members and 1% prevalence. This leaves us with 10,000 individuals who potentially could be treated with clopidogrel or prasugrel. We then multiply our affected population (10,000) by the percentages generated in our decision model to generate predicted number of individuals with each outcome. Next, we determine the costs that a hypothetical patient traveling down each path would accrue. For example, in our PGx strategy a wild type patient receiving clopidogrel who had a stent thrombosis would accrue the cost of the genetic test ($300), drug cost ($100), and clinical event ($40,000) for a total cost of $40,400. As a simplifying assumption we assume that patients generate costs even if they die in our simulation.

Once we do this for every path, we can multiply the percent that ended up in each final endpoint by the cost for getting to that endpoint and sum across the entire strategy to get our average cost per patient in that scenario. As is often the case, this benefit comes at a cost as the average cost of treating a patient with testing is $1,180 versus $1,590, an incremental cost of $410. This translates to a total cost of $11,800,000 in the standard care arm and $15,896,100 with testing. While all of the math is very simple, depending on how much nuance we want to capture, the number of calculations can expand exponentially.

Final Measure
For a health plan, these total costs must be translated to the cost of the increase in premium for the new technology. To do this, health plans convert the expense

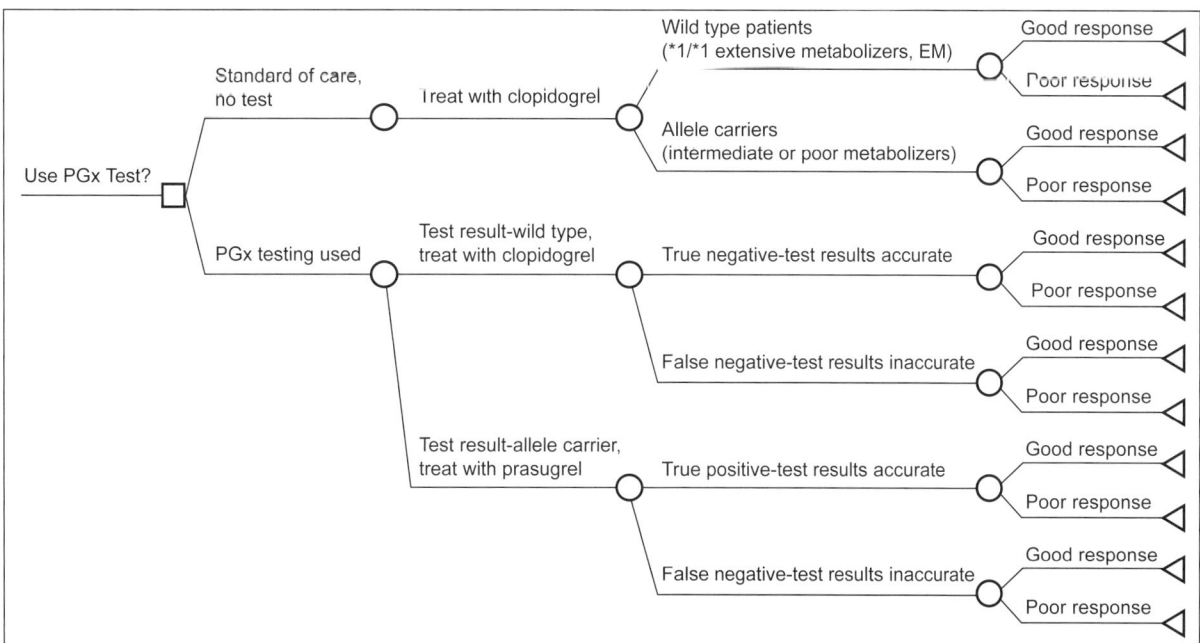

Figure 17.1. Decision tree depicting use of PGx test to select antithrombotic therapy.

.to "per-member per-month" or PMPM. If we assume that the events simulated in our model occurred over 3 months, then we can convert total costs to PMPM via algebra:

$11,800,000 ÷ 1,000,000 *Plan Members* ÷ 3 *months* = $3.93

$15,896,100 ÷ 1,000,000 *Plan Members* ÷ 3 *months* = $5.30

Incremental Cost = $5.30 - $3.93 = $1.37

This gives us an incremental cost of switching to a genotyping scenario of $1.37. This means that assuming all other spending stays the same, all million members of the health plan will have to contribute $1.37 more per month in order to accommodate PGx testing. Of course, this increased expense comes with added benefit (less mortality) but whether it is a good value will depend on what else that $1.37 per member per month could be spent on. As an example, perhaps that money could also be spent on a new and more effective cholesterol lowering medication that could be used in more individuals and ultimately have greater improvement on overall health.

Cost-Effectiveness Analysis

Introduction

Cost-effectiveness analysis and *cost-benefit analysis* are often used as general terms to refer to economic evaluation in health care. However, there are important differences in the specific methodological approaches used (Table 17.4) (Drummond 2005). Cost-minimization analysis, for example, considers only the difference in costs between strategies. The fundamental requirement for this approach is that the clinical outcomes (both effectiveness and adverse events) of the interventions are equal—which is rarely the case in health care. Thus, this approach is rarely used appropriately. In cost-benefit analysis, clinical outcomes are assigned a monetary value, and a total cost or return on investment can be calculated. Because of the substantial methodological challenges of assigning costs to clinical outcomes, as well as acceptability issues, this approach, like cost-minimization, is rarely used.

Objective

The objective of cost-effectiveness is to make a statement of relative value for an intervention in natural terms.

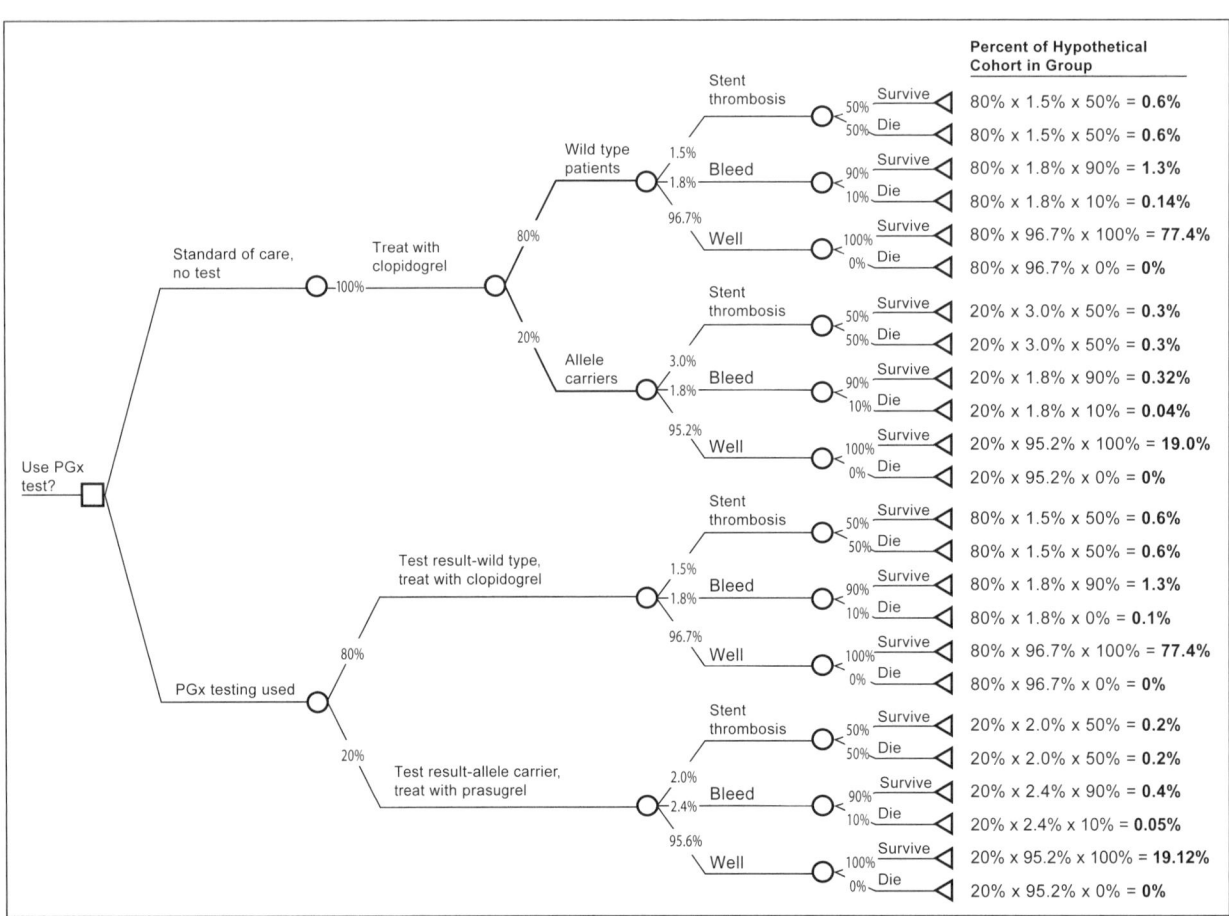

Figure 17.2. Decision tree with probabilities.

In other words, for a given scenario how much are we spending to get our estimated amount of clinical benefit.

Method

Cost-effectiveness analysis is similar to cost-benefit analysis because both costs and clinical outcomes are included; however, in this case, the clinical outcome such as a heart attack or life-years is included explicitly. A major limitation of using clinical events is that cost-effectiveness thresholds have not been assessed across a variety of disease areas—primarily because it is difficult to make such comparisons when the units of measure are different. For example, is $10,000 to prevent a heart attack more cost-effective than $10,000 to prevent breast cancer? The use of life-years as the clinical outcome prevents this limitation but does not capture the impact of interventions on patient quality of life. If one intervention is less costly and more effective, it is considered dominant because rational decision-makers will always opt for a better, cheaper intervention. Analogously, interventions that are less effective and more costly are considered dominated.

Cost-consequences analysis presents the difference in costs and the difference in *multiple* clinical outcomes and generally does not incorporate both into one measure such as cost per event prevented, as is done in a cost-effectiveness analysis. Although this approach does not provide a formal decision-making framework, clinicians, payers, and patients can more easily interpret the results. With limited health care resources, cost-effectiveness evaluations have become more common, with varying levels of implementation across health care systems globally. Such evaluations can provideEestimates of the long-term impact of health care interventions on patient life expectancy and quality of life, as well as costs, to assess the overall value of the intervention relative to other common technologies. These techniques have been applied to a wide variety of medical interventions, including pharmaceutical drugs, surgical techniques, health care services, and immunization programs. In theory, the cost-effectiveness of all interventions or technologies covered by a health care payer could be evaluated, and reimbursement could be approved for all that meet a certain cost-effective threshold or those that are most cost-effective until the budget is exhausted.

Final Measure

An incremental cost-effectiveness ratio (ICER) is calculated (e.g., the cost per heart attack prevented or the cost per life-year saved). The ICER provides a comparison between a new intervention and an alternative intervention, typically standard medical practice.

$$\text{ICER} = \frac{(\text{cost 2} - \text{cost 1})}{(\text{effectiveness 2} - \text{effectiveness 1})}$$

If the new intervention is both more costly and more effective, then the ICER can be compared with ICERs from other comparisons or with societal cost-effectiveness thresholds that represent a society's willingness to pay for health gains to make resource allocation decisions (e.g., $500 per heart attack prevented likely would be considered cost-effective).

Table 17.4. Types of Economic Evaluation in Health Care

Study Design	Costs Measured?	Outcomes Measured?	Strengths	Weaknesses
Cost-minimization	Yes	Not necessary	Easy to perform	Useful only if outcomes are the same for both interventions
Cost-benefit	Yes	Yes, in monetary terms	Good theoretical foudation; can be used within health care and across sectors of the economy	Less commonly accepted by health care decision-makers; evaluation of benefits methodologically challenging
Cost-effectiveness and cost-consequences	Yes	Yes, in clinical terms (e.g., events)	Relevant for clinicians; easily understandable	Cannot compare interventions across disease areas when using disease-specific end points; does not capture quality-of-life impacts
Cost-utility	Yes	Yes, in quality-adjusted life-years (QALY)	Incorporates quality of life; comparable across disease areas and interventions	Requires evaluation of patient preferences; can be difficult to interpret

Application to Clopidogrel and C19 Testing

In our example of C19 testing for choice of anti-thrombotic agent we have already calculated the number of clinical events (via our decision analysis) as well as our overall costs via budget impact. We could use any size population to calculate our ICER as the ratio would still hold but for sake of clarity we will draw the values directly from the budget impact model. Using the method described above, we can very easily calculate cost per death avoided.

Cost per Death Avoided (ICER) =
  ($15,896,100 − 11,800,000)
  (95 deaths − 108 deaths)
  = $315,084 per death avoided

Cost-Utility Analysis

Introduction

Cost-utility analysis (CUA) addresses many limitations of the above approaches and is the most often recommended method because it measures both length and quality of life, and it permits comparison among multiple interventions using a standard denominator—the quality-adjusted life-year (QALY).

Objective

The objective of a CUA is the same as that for a CEA but instead of determining value in terms of natural events (deaths, heart attacks, cancer recurrence, etc.) all of these clinical events are given relative meaning via utility weights. This frames the impact they have on quality of life on a 0 to 1 scale. When we give this a time component by considering this utility over time we arrive at the QALY. As opposed to natural events, using the QALY allows for easier comparison across different interventions that do not have comparable endpoints.

Method

Similar to how we calculate costs, for a CUA we must track the QALY's that our hypothetical cohort accrues as they travel down various paths in the decision tree. We can then sum QALYs within each strategy and compare via the ICER equation described above. Cost-utility thresholds (i.e., the willingness to pay for a QALY) have been identified, explicitly or implicitly, in several countries,

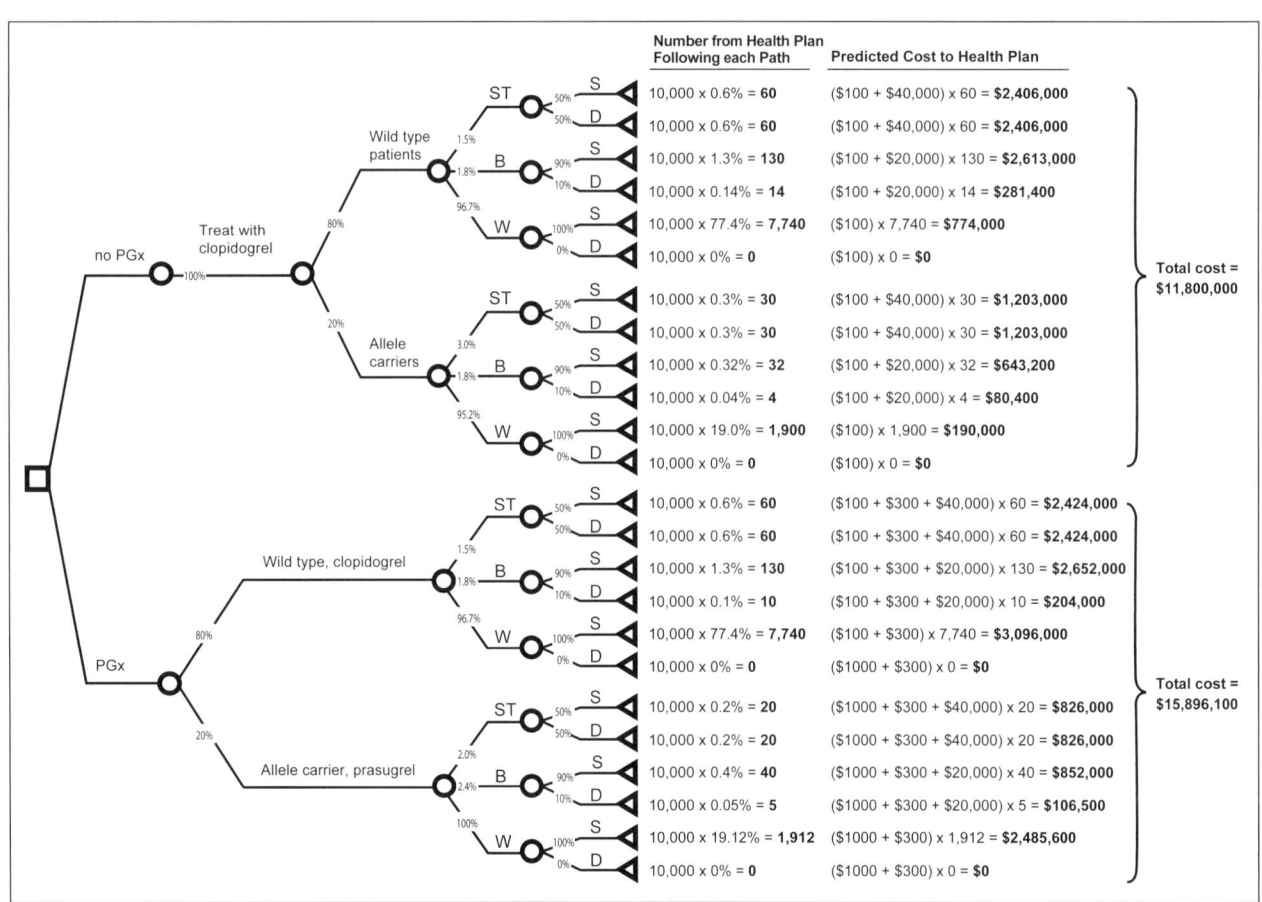

Figure 17.3. Decision tree with costs.

including Australia and the United Kingdom, because the government health care payers formally use such studies to inform reimbursement decisions. In the United States, however, because of the more fragmentary nature of the health care system, formal economic evaluations are not commonly used to inform reimbursement decisions, and there is no consensus on the appropriate threshold. Historically, $50,000 per QALY has been cited, supposedly based in part on Medicare coverage for dialysis, but the origins of this threshold are not exact (Grosse 2008). Recent analyses based on currently reimbursed technologies suggest the (implicit) threshold is closer to $100,000 per QALY (Braithwaite 2008). This threshold also likely varies across clinical indications or in the case of Medicare over $100,000 per QALY (Chambers 2010). For example, medical interventions for rare childhood cancers might be considered cost-effective at relatively higher thresholds than interventions in other clinical areas, because of the societal sensitivity to cancer in this age group and the lack of therapeutic options for such rare diseases.

Final Measure

The QALY, in simple terms, measures a year of life adjusted for the quality of life during that year. It is derived by applying health state preference scores, or utilities, which measure the strength of individuals' preferences for specific outcomes under conditions of uncertainty, to estimates of life-years. The utility score is measured on a scale from 0 to 1, with 1 being perfect health and 0 being death. These utilities are typically developed using community-based samples of unaffected individuals or through quality-of-life instruments applied during the course of clinical trials (Gold 1996). Although the use of utilities to estimate QALYs provides one of the most robust approaches to economic evaluation in health care, important challenges remain in the measurement and interpretation of preferences—for example, the method used to elicit preferences and the selection of individuals to survey.

This concept can be visualized in a two dimensional plane with time on the x-axis and utility on the y-axis. In a hypothetical patient who had originally high quality of life (utility = 1) for a year but then experienced significant morbidity from a health event for the second year (utility = 0.5), their total QALYs for the two years that they are followed is 1.5. While the use of utility weights and subsequently derived QALY's can be extremely useful for modeling, they have several limitations. Notably, the weighting of any given health state on a scale of 0–1 can be highly variable depending on numerous factors including whether the state is ranked by those with the condition or the general population, the instrument used to determine preference weight and the culture of population doing the weighting as well as the perception of quality of life for a given health state can vary widely from country to country (Craig 2009). While a thorough review of these issues is beyond the scope of this chapter, it is important to keep in mind that while the QALY appears very precise and quantitative, it is nonetheless subject to the complex milieu of social factors that define quality of life.

Application to Clopidogrel and C19 Testing

If we assume that patients accrue QALY's for 3 months in all paths whether they survive or not and that the utility weights for stent thrombosis and bleeds are 0.5 and 0.75 (again these are highly generalized utility assumptions for sake of clarity), respectively, then we can calculate the overall QALY estimates for each strategy as seen in Figure 17.5. As might be expected, the PGx-guided strategy results in a modest increase in QALYs

$$\begin{aligned}\text{Cost per QALY} &= (\$1{,}590 - \$1{,}180) / \\ &\quad (0.24677 \text{ QALY} - 0.24663 \text{ QALY}) \\ &= \$2{,}849{,}461 \text{ per QALY}\end{aligned}$$

It is important to note that this cost/QALY estimate is for illustrative purposes only as clinical inputs have been highly simplified to improve clarity. For a review of the cost/QALY estimates determined by several published CUA models please refer to Table 17.2.

This indicates that using our designated estimates and only tracking out 3 months, genotyping here does not appear like a good value at our standard willingness

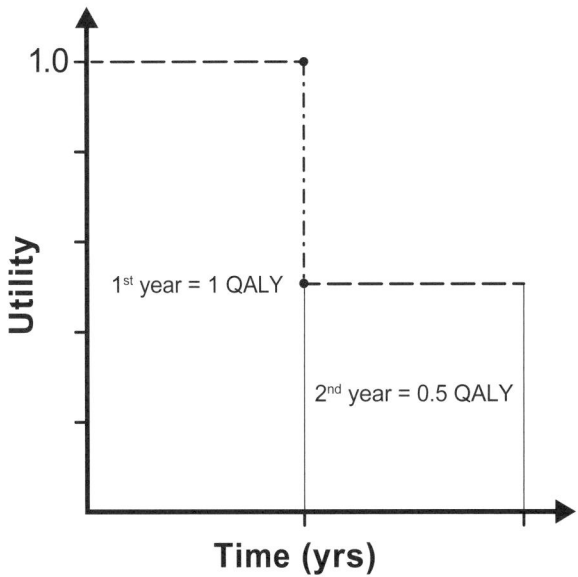

Figure 17.4. Hypothetical utility of a patient over time.

to pay thresholds of $50,000 and $100,000 per QALY. The estimate of value would improve if we believed thatuwe would continue to see reduced events due to genotyping after 3 months.

Economic Incentives and Challenges in Developing PGx Technologies

Current State of the Science

The current state of the science is somewhat different from that envisioned by early enthusiasts and assertions that PGx would soon transform the practice of medicine. Recent reports have suggested that PGx will unfold into patient care at a slower pace than previously predicted, with 10–15 new tests being used in routine clinical care in the next 10 years and primarily focused in oncology, as opposed to ubiquitous use and revolutionary impacts on drug development (Garrison 2007B; The Royal Society 2005). Why has PGx proceeded at an evolutionary rather than revolutionary pace? In large part, this is likely attributable to the complex interaction of genes, environment, and drugs, as well as the relatively small contribution of individual genes to drug response. However, it is also worth considering the roles that economic incentives, or the lack thereof, and cost-effectiveness have played in the process of bringing PGx to the bedside.

Reimbursement Barriers and Reforms

In the United States, however, economic evaluations are rarely, if ever, applied in such a fashion. Pragmatically, relying purely on such studies to establish reimbursement policies ignores important equity, access, and even simple budget impact issues. The most notable example of this issue is found in the treatment of rare diseases. While innovative drug therapies for these conditions can literally be life-saving for an individual patient, they come at an extremely high cost to the healthcare system (Coyle 2014). Furthermore, there is opportunity cost for devoting our research and development industry to conditions that may affect a handful of patients instead of more common conditions like influenza or diabetes where the gains might be smaller but many more

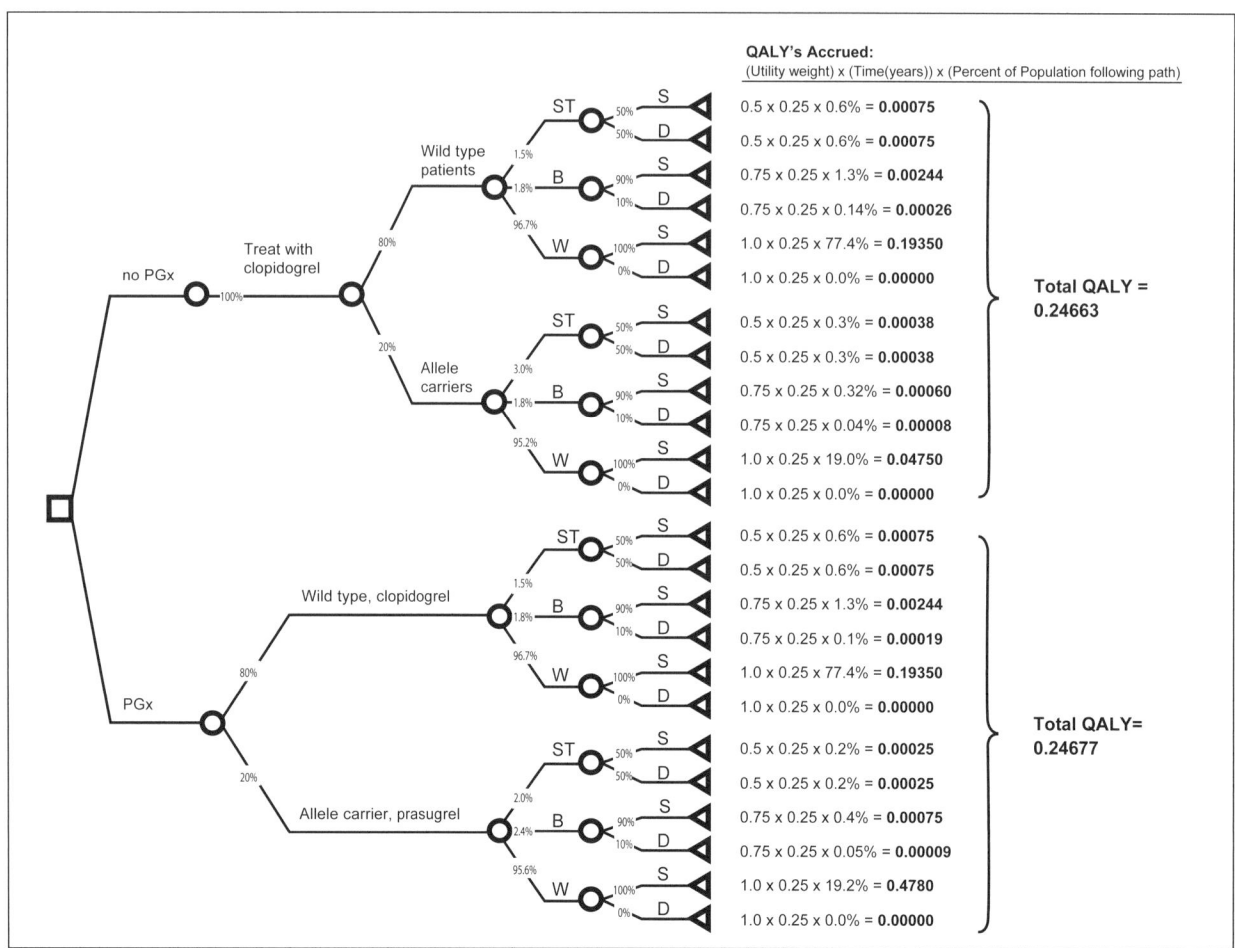

Figure 17.5. Calculation of QALY.

patients might be helped. Despite these barriers, we have decided as a society that these decisions must be made beyond a strict utilitarian framework and therefore have developed regulatory incentives to promote innovation for rare and orphan disease (Swinney 2014). Thus, cost-effectiveness analysis is reasonably viewed as an additional piece of information to inform decision-making. The Academy of Managed Care Pharmacy (AMCP) has developed an explicit process to enable the incorporation of value considerations into pharmaceutical reimbursement decision-making, rather than negotiating based on implicit arguments about drug price, effectiveness, and safety (Fry 2003). The AMCP Dossier Submission Format is a template that drug purchasers can use to request all information (published and unpublished) on the safety, efficacy, and value of a drug from the manufacturer. Cost-effectiveness modeling is also used by drug manufacturers in the early stages of drug development to project potential clinical and economic outcomes of their products and inform strategic decision-making (Clemens 1993). Specific to PGx, the Format requests information regarding the test's analytic validity, clinical validity, and clinical utility, as well as economic value. We presented a formal cost-effectiveness evaluation above to highlight the potential utility of such analyses in informing health care decision-making.

There are several additional stakeholder perspectives to consider regarding economic incentives related to the use of PGx tests, with payers, physicians, and patients being primary. Payers may be motivated to cover PGx testing if it limits the population of potential drug users—providing cost-savings—or PGx testing provides benefits to their policyholders—providing a competitive advantage in the market. Patients will likely be driven by the specific risks and benefits of a PGx application, but payer reimbursement levels and the potential for high copays and coinsurance policies could be a barrier to patient buy-in. Finally, physicians will primarily be driven by what is most likely to benefit their patients, but economic issues related to the requirement for and extent of additional training and ease of use (e.g., turnaround time, interpretability of test results) will likely play a role in their uptake or PGx testing. In particular, a better understanding is required of the potential role of clinical pharmacists in the process of delivering personalized medicine through PGx testing. Clearly, understanding the economic incentives for each stakeholder and adopting policies aimed at optimizing stakeholder buy-in will be integral to both the clinical and economic viability of PGx testing in the end.

The scientific, clinical development, and regulatory barriers to the translation of PGx discoveries into real-world interventions are, no doubt, substantial. Still, there is an open question as to whether the current reimbursement system is a significant barrier. However, it is clear that the current reimbursement system in most developed countries provides, at best, a limited mechanism for rewarding the value created by an innovative PGx test. As noted above, the reimbursement system in United States and Europe for companion diagnostic tests is not what we would call "value-based." Rather, reimbursement levels tend to be set on the perception of the costs of providing the test, reflecting laboratory process. The development of evidence to support a new companion diagnostic would require substantial clinical fixed costs for clinical trials. A reimbursement system based only on marginal cost will clearly not provide sufficient funds to support such trials.

In 2013 the Academy of Medical Sciences (AMS) in the United Kingdom reviewed the barriers to what they call "stratified medicine," which is more or less equivalent to personalized medicine. They issued a set of recommendations to address these reimbursement shortcomings as well as some of the other barriers (e.g., intellectual property). Not only did they recommend that companion diagnostics be reimbursed based on value, they also recognized the practical problem of attributing value to companion test versus the companion medicine that are combined in the stratified medicine product. From an economic perspective, the test and the drug are complementary economic goods, and the division of their combined value is arbitrary—at least, certainly from the buyer's perspective (Garrison 2007A). However, from the perspective of the manufacturers, the division has important implications for incentives to invest in product development and evidence generation. AMS made the following recommendation as a practical way forward for splitting the value, with an eye towards providing greater value for the diagnostics (AMS 2014).

> Recommendation 15. To incentivize stratification, at least in the short term, we recommend that health technology assessment bodies develop a model to separate the value between the drive and the companion diagnostic. The medicine should be considered as the primary source of the health gain in responders. The diagnostic should be valued in terms of the cost savings and improvements in quality and length of life from reduced adverse drug reactions in nonresponders, and in terms of increased certainty of response. Better patient adherence and greater overall appropriate use may also result, and this value could be divided similarly.

The "short term" mentioned in the recommendation is the recognition that this new approach would need some development and pilot testing. Also, the concept of a companion diagnostic test is evolving with the growth of whole gene sequencing technologies and their falling cost. The reimbursement model will most likely have to evolve as well. However, the key long-term issue remains providing sufficient incentives for efficient evidence generation for the right products over their life cycle, i.e., what economists refer to as dynamic efficiency (Garrison 2014).

Assessment of the value contribution of companion diagnostic or a drug will continue to require the use of the tools described above in terms of decision modeling, benefit-risk analysis, budget impact analysis, cost-effectiveness analysis, and cost-utility analysis. It is also important to remember that information has public good characteristics that require interventions, such as intellectual property or research subsidies, to achieve an appropriate level of investment from a societal perspective (Towse 2013). Finally, the evidence base regarding the effectiveness of a personalized medicine product changes over time as more real-world experience is gained. As this further information about the actual value delivered is gathered, it is important to develop reimbursement systems that are flexible as well as being valued-based. Prices, which serve as rewards for innovation, need to be able to move either up or down as more is learned about the real world cost-effectiveness of the test-drug combination.

In summary, it is clear that the role of economics in providing incentives or barriers to the development of PGx is complex. It is doubtful that PGx will bring about the widespread end of blockbuster drugs, dramatically speed the drug development process, or improve the cost-effectiveness of medicines overall. However, a key aim of the pharmaceutical industry is innovation, and as such, PGx will offer opportunities for incremental improvements in all of these areas. The challenge will be in providing appropriate economic incentives via reimbursements for drug treatments that improve over time with the introduction of PGx testing, and in discerning the allotment of rewards to competing drug and diagnostic manufacturers. Reimbursement for a PGx technology should be related to the value it provides to patients and society, and these reimbursement levels will have to be dynamic.

Summary

Is PGx cost-effective? This is an often-asked question, but as with most complex topics, the answer is, "it depends." Just as we do not categorize all drugs or technologies (e.g., biotechnology) as being cost-effective or not, we should not categorize PGx this way. Cost-effectiveness analysis provides a framework for addressing many of the issues discussed in this chapter. Such analyses do not necessarily have to be complex and some of the basic criteria presented can be used to evaluate, at least at an initial level, whether a PGx test offers benefits that are likely to outweigh risks and a reasonable value for the money.

Future challenges include creating a framework with categories of PGx tests with varying evidence requirements for clinical implementation and reimbursement, generating appropriate evidence for decision-making, establishing flexible, value-based reimbursement for PGx tests, and providing expertise in evidence- and value-based systematic decision-making. Clinical pharmacists and pharmaceutical scientists are in a unique position to meet these challenges given the appropriate analytic tools.

References

Academy of Medical Sciences. Realising the Potential of Stratified Medicine. 2013; Available at http://www.acmedsci.ac.uk/policy/policy-projects/Stratified-medicine/. Accessed July 16, 2014.

Alexander KM, Divine HS, Hanna CR, et al. Implementation of personalized medicine services in community pharmacies: Perceptions of independent community pharmacists. Journal of the American Pharmacists Association : JAPhA 2014;54(5):510-17.

Athena Diagnostics. Test Catalog. 2009; Available at http://www.athenadiagnostics.com/site/product_search/test_description_template.asp?id=206. Accessed April 17, 2009.

Bauer T, Bouman HJ, Werkum JWv, et al. Impact of CYP2C19 variant genotypes on clinical efficacy of antiplatelet treatment with clopidogrel: systematic review and meta-analysis. Bmj :48:31 2011;343.

Braithwaite RS, Meltzer DO, King JT, Jr., et al. What does the value of modern medicine say about the $50,000 per quality-adjusted life-year decision rule? Med Care 2008;46(4):349-56.

Burns BD, Corpus B. Randomness and inductions from streaks: "gambler's fallacy" versus "hot hand". Psychonomic bulletin & review 2004;11(1):179-84.

Chabner BA, Ellisen LW, Iafrate AJ. Personalized medicine: hype or reality. The oncologist 2013;18(6):640-43.

Chambers JD, Neumann PJ, Buxton MJ. Does Medicare have an implicit cost-effectiveness threshold? Med Decis Making 2010;30C(4):E14-E27.

Clemens K, Garrison LP, Jr., Jones A, et al. Strategic use of pharmacoeconomic research in early drug development and global pricing. PharmacoEconomics 1993;4(5):315-22.

Conrado DJ, Rogers HL, Zineh I, et al. Consistency of drug-drug and gene-drug interaction information in U.S. FDA-approved drug labels. Pharmacogenomics 2013;14(2):215-23.

Coyle D, Cheung MC, Evans GA. Opportunity Cost of Funding Drugs for Rare Diseases: The Cost-Effectiveness of Eculizumab in Paroxysmal Nocturnal Hemoglobinuria. Medical decision making : an international journal of the Society for Medical Decision Making. 2014.

Craig BM, Busschbach JJ, Salomon JA. Modeling ranking, time trade-off, and visual analog scale values for EQ-5D health states: a review and comparison of methods. Medical care 2009;47(6):634-41.

Desai NR, Canestaro WJ, Kyrychenko P, et al. Impact of CYP2C19 Genetic Testing on Provider Prescribing Patterns for Antiplatelet Therapy After Acute Coronary Syndromes and Percutaneous Coronary Intervention. Circulation: Cardiovascular Quality and Outcomes 2013;6(6):694-99.

Deverka PA, Vernon J, McLeod HL. Economic Opportunities and Challenges for Pharmacogenomics. Annual Review of Pharmacology and Toxicology 2010;50(1):423-37.

Dimasi JA. Risks in new drug development: approval success rates for investigational drugs. Clin Pharmacol Ther 2001;69(5):297-307.

Drummond F, Torrance GW. Methods for the Economic Evaluation of Health Care Programmes. Oxford University Press; 2005.

Estivill X, Govea N, Barcelo E, et al. Familial progressive sensorineural deafness is mainly due to the mtDNA A1555G mutation and is enhanced by treatment of aminoglycosides. Am J Hum Genet 1998;62(1):27-35.

Faulkner E, Annemans L, Garrison L, et al. Challenges in the development and reimbursement of personalized medicine-payer and manufacturer perspectives and implications for health economics and outcomes research: a report of the ISPOR personalized medicine special interest group. Value in health : the journal of the International Society for Pharmacoeconomics and Outcomes Research 2012;15(8):1162-71.

Flowers CR, Veenstra D. The role of cost-effectiveness analysis in the era of pharmacogenomics. Pharmacoeconomics 2004;22(8):481-93.

Fry RN, Avey SG, Sullivan SD. The Academy of Managed Care Pharmacy Format for Formulary Submissions: an evolving standard--a Foundation for Managed Care Pharmacy Task Force report. Value in health : the journal of the International Society for Pharmacoeconomics & outcomes research 2003;6(5):505-21.

Garg V, Gu NY, Borrego ME, et al. A literature review of cost-effectiveness analyses of prostate-specific antigen test in prostate cancer screening. Expert review of pharmacoeconomics & outcomes research 2013;13(3):327-42.

Garrison LP, Towse A. Personalized Medicine: Pricing and Reimbursement Policies as a Potential Barrier to Development and Adoption. In: AJ C, ed. Encyclopedia of Health Economics Vol 2: Elsevier, 2014:484-90.

Garrison LP, Jr., Austin MJ. Linking pharmacogenetics-based diagnostics and drugs for personalized medicine. Health affairs (Project Hope) 2006;25(5):1281-90.

Garrison LP, Austin MJ. The Economics of Personalized Medicine: A Model of Incentives for Value Creation and Capture. Drug Information Journal 2007A;41(4):501-9.

Garrison LP, Veenstra DL, Carlson RJ, et al. Backgrounder on Pharmacogenomics for the Pharmaceutical and Biotechnology Industries: Basic Science, Future Scenarios, Policy Directions. University of Washington, Pharmaceutical Outcomes Research and Policy Program;2007B.

Gold M, Seigel J, Russell L, et al. Cost-Effectiveness in Health and Medicine. Oxford: Oxford University Press, 1996.

Grosse SD, Wordsworth S, Payne K. Economic methods for valuing the outcomes of genetic testing: beyond cost-effectiveness analysis. Genetics in medicine : official journal of the American College of Medical Genetics 2008;10(9):648-54.

Gupta N, Bansal A, Wani SB, et al. Endoscopy for upper GI cancer screening in the general population: a cost-utility analysis. Gastrointestinal endoscopy 2011;74(3):610-24 e612.

Haga SB, Tindall G, O'Daniel JM. Professional perspectives about pharmacogenetic testing and managing ancillary findings. Genetic testing and molecular biomarkers 2012;16(1):21-4.

Henrikson NB, Burke W, Veenstra DL. Ancillary risk information and pharmacogenetic tests: social and policy implications. Pharmacogenomics J 2008;8(2):85-9.

Hockett RD, Close SL. Regulation of laboratory-developed tests: the case for utilizing professional associations. Clin Pharmacol Ther 2010;88(6):743-45.

Jaber J, Shohat M, Bu X, et al. Sensorineural deafness inherited at a tissue specific mitochondrial mutation. Journal of Medical Genetics 1992;29:86-90.

Joly Y, Koutrikas G, Tasse AM, et al. Regulatory approval for new pharmacogenomic tests: a comparative overview. Food and drug law journal 2011;66(1):1-24, i.

Kazi DS, Garber AM, Shah RU, et al. Cost-effectiveness of genotype-guided and dual antiplatelet therapies in acute coronary syndrome. Ann Intern Med 2014;160(4):221-32.

Khoury MJ, Gwinn M, Yoon PW, et al. The continuum of translation research in genomic medicine: how can we accelerate the appropriate integration of human genome discoveries into health care and disease prevention? Genetics in medicine: official journal of the American College of Medical Genetics 2007;9(10):665-74.

Kliff S. A quarter of head and neck surgeons think Obamacare has death panels. The Washington Post. December 19, 2013, 2013.

Lala A, Berger JS, Sharma G, et al. Genetic testing in patients with acute coronary syndrome undergoing percutaneous coronary intervention: a cost-effectiveness analysis. J Thromb Haemost 2013;11(1):81-91.

Lee S, Drake E, Pennucci A, et al. Economic evaluation of early childhood education in a policy context. Journal of Children's Services 2012;7(1):53-63.

Madan J, Rawdin A, Stevenson M, et al. A rapid-response economic evaluation of the UK NHS Cancer Reform Strategy breast cancer screening program extension via a plausible bounds approach. Value in health : the journal of the International Society for Pharmacoeconomics and Outcomes Research 2010;13(2):215-21.

Mega JL, Close SL, Wiviott SD, et al. Cytochrome P450 genetic polymorphisms and the response to prasugrel: relationship to pharmacokinetic, pharmacodynamic, and clinical outcomes. Circulation 2009;119(19):2553-60.

Mega JL, Hochholzer W, Frelinger AL, 3rd, et al. Dosing clopidogrel based on CYP2C19 genotype and the effect on platelet reactivity in patients with stable cardiovascular disease. JAMA 2011;306(20):2221-8.

Mulheran M, Degg C, Burr S, et al. Occurrence and Risk of Cochleotoxicity in Cystic Fibrosis Patients Receiving Repeated High-Dose Aminoglycoside therapy. Amtimicrob Agents Chemother 2001;45(9):2502-09.

Munckhof WJ, Grayson ML, Turnidge JD. A meta-analysis of studies on the safety and efficacy of aminoglycosides given either once daily or as divided doses. J Antimicrob Chemother 1996; 37(4):645-63.

Naci H, Brugts JJ, Fleurence R, et al. Dose-comparative effects of different statins on serum lipid levels: a network meta-analysis of 256,827 individuals in 181 randomized controlled trials. European journal of preventive cardiology 2013;20(4):658-70.

Nance W. The Genetics of Deafness. Mental Retardation and Developmental Disabilities Research Reviews 2003;9(2):102-09.

O'Gara PT, Kushner FG, Ascheim DD, et al. 2013 ACCF/AHA Guideline for the Management of ST-Elevation Myocardial InfarctionA Report of the American College of Cardiology Foundation/American Heart Association Task Force on Practice Guidelines. Journal of the American College of Cardiology 2013;61(4):e78-e140.

Olsen D, Jorgensen JT. Companion Diagnostics for Targeted Cancer Drugs - Clinical and Regulatory Aspects. Frontiers in oncology 2014;4:105.

Panattoni L, Brown PM, Te Ao B, et al. The cost effectiveness of genetic testing for CYP2C19 variants to guide thienopyridine treatment in patients with acute coronary syndromes: a New Zealand evaluation. PharmacoEconomics 2012;30(11):1067-84.

Pandya A. Nonsyndromic Hearing Loss. Mitochondrial2004. Available at http://www.genetests.com/servlet/access?id=8888891&key=kY45zXrm6ge11&gry=INSERTGRY&fcn=y&fw=hR-5&filename=/glossary/profiles/mt-deafness/index.html. Accessed 12/7/2005.

Perry CG, Shuldiner AR. Pharmacogenomics of anti-platelet therapy: how much evidence is enough for clinical implementation? Journal of human genetics 2013;58(6):339-45.

Pettiti D. Meta-Analysis, Decision Analysis, and Cost-Effectiveness Analysis: Methods for Quantitative Synthesis in Medicine.2nd ed. New York: Oxford University Press, 2000.

Price Waterhouse Coopers. The New Science of Personalized Medicine: Translating the promise into practice. 2009.

Raymakers AJ, Mayo J, Marra CA, et al. Diagnostic Strategies Incorporating Computed Tomography Angiography for Pulmonary Embolism: A Systematic Review of Cost-effectiveness Analyses. Journal of thoracic imaging 2014;29(4):209-16.

Reese ES, Daniel Mullins C, Beitelshees AL, et al. Cost-effectiveness of cytochrome P450 2C19 genotype screening for selection of antiplatelet therapy with clopidogrel or prasugrel. Pharmacotherapy 2012;32(4):323-32.

Rocke DJ, Thomas S, Puscas L, et al. Physician knowledge of and attitudes toward the Patient Protection and Affordable Care Act. Otalaryngology—head and neck surgery : official journal of American Academy of Otolaryngology-Head and Neck Surgery 2014;150(2):229-34.

The Royal Society. Personalised Medicine: Hopes and Realities. London: The Royal Society,2005.

Scott MG, Ashwood ER, Annesley TM, et al. FDA oversight of laboratory-developed tests: is it necessary, and how would it impact clinical laboratories? Clinical chemistry 2013;59(7):1017-22.

Secretary's Advisory Committee on Genetics Health and Society. Coverage and Reimbursement of

Genetic Tests and Services. 2006; Available at http://www4.od.nih.gov/oba/sacghs/reports/CR_report.pdf. Accessed December 1, 2007.

Sendi P, Gunthard HF, Simcock M, et al. Cost-effectiveness of genotypic antiretroviral resistance testing in HIV-infected patients with treatment failure. PLoS One 2007;2(1):e173.

Shah RR, Shah DR. Personalized medicine: is it a pharmacogenetic mirage? Br J Clin Pharmacol 2012;74(4):698-721.

Shaw PM, Zineh I. Generating and weighing evidence in drug development and regulatory decision making: 5th U.S. FDA-DIA workshop on pharmacogenomics. Pharmacogenomics 2010;11(12):1629-35.

Sinha KC, Labi S. Transportation decision making: Principles of project evaluation and programming. John Wiley & Sons, 2011.

Sorich MJ, Horowitz JD, Sorich W, et al. Cost-effectiveness of using CYP2C19 genotype to guide selection of clopidogrel or ticagrelor in Australia. Pharmacogenomics 2013;14(16):2013-21.

Stanek EJ, Sanders CL, Taber KA, et al. Adoption of pharmacogenomic testing by U.S. physicians: results of a nationwide survey. Clin Pharmacol Ther 2012;91(3):450-58.

Sun F, Bruening W, Erinoff E, et al. Addressing Challenges in Genetic Test Evaluation: Evaluation Frameworks and Assessment of Analytic Validity. Rockville, MD,2011.

Swinney DC, Xia S. The discovery of medicines for rare diseases. Future medicinal chemistry 2014;6(9):987-1002.

Tang HY, Hutcheson E, Neill S, et al. Genetic susceptibility to aminoglycoside ototoxicity: how many are at risk? Genetics in medicine: official journal of the American College of Medical Genetics 2002;4(5):336-45.

Towse A, Garrison LP, Jr. Economic incentives for evidence generation: promoting an efficient path to personalized medicine. Value in health: the journal of the International Society for Pharmacoeconomics and Outcomes Research 2013;16(6 Suppl):S39-43.

Tsao NW, Bansback NJ, Shojania K, et al. The issue of comparators in economic evaluations of biologic response modifiers in rheumatoid arthritis. Best practice & research. Clinical rheumatology 2012;26(5):659-76.

Veenstra DL, Harris J, Gibson RL, et al. Pharmacogenomic testing to prevent aminoglycoside-induced hearing loss in cystic fibrosis patients: potential impact on clinical, patient, and economic outcomes. Genetics in medicine: official journal of the American College of Medical Genetics 2007;9(10):695-704.

Vernon JA, Hughen WK. The Future of Drug Development: The Economics of Pharmacogenomics. National Bureau of Economic Research (NBER); 2005. NBER Working Paper No. 11875.

Weinstein MC, O'Brien B, Hornberger J, et al. Principles of good practice for decision analytic modeling in health-care evaluation: report of the ISPOR Task Force on Good Research Practices--Modeling Studies. Value in health: the journal of the International Society for Pharmacoeconomics and Outcomes Research 2003;6(1):9-17.

Weiss RL. The long and winding regulatory road for laboratory-developed tests. American journal of clinical pathology 2012;138(1):20-6.

Wolf SM, Annas GJ, Elias S. Patient Autonomy and Incidental Findings in Clinical Genomics. Science 2013 2013;340(6136):1049-50.

Wolman D, Kalfoglou A, LeRoy L. Medicare Laboratory Payment Policy: Now and in the Future. The National Academies Press,2000.

World Health Organization. Trade, Foreign Policy, Diplomacy and Health: Pharmaceutical Industry. 2014; Available at http://www.who.int/trade/glossary/story073/en/. Accessed July 7, 2014.

Zabalza M, Subirana I, Sala J, et al. Meta-analyses of the association between cytochrome CYP2C19 loss- and gain-of-function polymorphisms and cardiovascular outcomes in patients with coronary artery disease treated with clopidogrel. Heart 2012;98(2):100-08.

Zineh I, Gerhard T, Aquilante CL, et al. Availability of pharmacogenomics-based prescribing information in drug package inserts for currently approved drugs. Pharmacogenomics J 2004;4(6):354-58.

Zineh I, Pebanco GD, Aquilante CL, et al. Discordance between availability of pharmacogenetics studies and pharmacogenetics-based prescribing information for the top 200 drugs. The Annals of pharmacotherapy 2006;40(4):639-44.

Chapter 18

THE ROLE OF PHARMACOGENOMICS AND TARGETED THERAPEUTICS IN THE FDA DRUG APPROVAL PROCESS

HOBART ROGERS, PHARM.D., PH.D.; ELIMIKA PFUMA, PHARM.D., PH.D.; AND MICHAEL PACANOWSKI, PHARM.D., M.P.H.

Learning Objectives

1. Analyze the application of pharmacogenomics in the drug approval and regulation process.
2. Assess the regulatory challenges facing drug approval that have arisen in the age of pharmacogenomics and targeted therapeutics.
3. Analyze how the FDA incorporates an understanding of pharmacogenomics into regulatory decision making.
4. Demonstrate how the FDA utilizes genomic biomarkers to guide drug development for targeted therapy.
5. Produce examples of where the FDA has used pharmacogenomic biomarkers in prescription drug labeling.

Keywords: FDA, regulatory, pharmacogenomics, targeted, biomarker

Abbreviations in This Chapter

CYP	cytochrome P450
CYP1A2	cytochrome P450 1A2
CYP2D6	cytochrome P450 2D6
CYP3A4	cytochrome P450 3A4
EM	extensive metabolizer
IM	intermediate metabolizer
PM	poor metabolizer
UM	ultra-rapid metabolizer

Abstract

A critical mission of the FDA is to protect and promote the public health by ensuring the safety and efficacy of prescription and nonprescription drugs. Unfortunately, heterogeneity in drug response is a common theme in drug development. Pharmacogenomics and other personalized medicine approaches are valuable tools used by drug developers to further quantify sources of variability in individual patients. Both the FDA and drug developers are increasingly looking for pharmacogenomic biomarkers that can inform on the pharmacokinetics, pharmacodynamics, safety, and efficacy of drugs. The FDA works with drug developers to utilize these pharmacogenomic biomarkers to target drug development and further decrease variability in response. This chapter will highlight some key examples of how the FDA has implemented pharmacogenomic information in drug development and labeling. As the age of pharmacogenomics and targeted therapeutics moves forward, it will not only impact how drugs are utilized, but also how drugs are developed and regulated.

This publication reflects the views of the authors and should not be construed to represent the FDA's views or policies.

The Advent of Pharmacogenomics in the Regulation of Drug Development

The FDA is tasked with protecting the public health by ensuring the safety and efficacy of prescription and nonprescription drugs. Drug approvals are based on response in the specific indicated population, however interindividual variability exists in response to drug therapy leading to unwanted effects in the form of both toxicity and/or a lack of efficacy in some patients. Characterizing and addressing this variability is a problem that has constantly plagued clinicians, drug developers, and the FDA alike. As science and technology have progressed, new tools have become available to identify sources of this variability. One major advance is the advent of high throughput technologies to evaluate the human genome. Understanding how molecular factors influence variability in response can provide greater insight regarding sources of heterogeneity in drug response. The more heterogeneity that can be accounted for, the better the FDA can assess the risks and benefits of therapy.

Improvements in the technologies and our understanding of how the human genome relates both disease pathophysiology and drug response is changing how drugs are developed and regulated. Many drug developers have switched from the old paradigm of "blockbuster" to "targeted" approaches that incorporate molecular biomarkers to better select patients and doses that are safe and deffective. As such, clinical trials in narrower or more specific populations have become more common, and companion diagnostics (tests that are essential to the safe and effective use of the drug) are increasingly becoming an integral part of drug development and patient care.

Drug Development Is Becoming More Targeted

While the costs of drug development are likely prohibitive to develop drugs for certain individuals, drug developers are more routinely incorporating genetic information in the drug development process to personalize the drug before marketing. Personalized or targeted approaches are applied in several ways in drug development.

For the purposes of this chapter, "targeted" can refer to a drug that is specifically designed for a particular molecular target (e.g., CFTR potentiator), or a drug that is targeted to a certain subset of patients (e.g., those with a specific molecular marker), or a drug for which knowledge of patient "status" can inform and personalize treatment decisions (e.g., CYP2D6 poor metabolizers) (Zineh 2013). Drug developers pursue "targeted" approaches at different times during drug development and these approaches are sometimes prospective or retrospective. Prospective approaches typically occur when a developer has a thorough understanding of the underlying molecular pathophysiology of the disease and characteristics of the drug. Some examples of situations when prospective approaches have been used are as follows:

- The target of the drug is variably expressed.
- The disease has certain specific driving mutations.
- A subgroup of patients has a different prognosis.
- The metabolism of the drug is known to be via a polymorphic pathway.
- Diminished activity is observed in certain subjects in early phase studies.

Retrospective approaches often occur when the following become apparent during development or in the post marketing setting:

- Studies indicate differential efficacy in a subgroup that is mechanistically supported by the drug's mechanism of action.
- Substantial toxicity is observed in a certain subgroup.

As medicine is advancing from organ-based and histology-based disease classifications to more molecular-based classifications, subgroups of patients who share similar genetic characteristics are increasingly identified. These new genetic subgroups of patients with similar molecular characteristics represent new opportunities for drug developers. For example, biomarkers are routinely being used to identify subsets of responders and these biomarkers can be used to enrich future trials with patients likely to respond. Moreover, biomarkers are also being used to exclude subjects who are unlikely to respond to treatment, or modify doses of drug in those individuals who metabolize the drug to a different extent. The genetic biomarker is quickly becoming the driving force behind targeted development, and numerous examples of targeted therapies now exist. Some examples can be found in Table 18.1 below.

Pharmacogenomics in Drug Labeling

The first genomic-based information to appear in drug labeling was in 1949 regarding glucose-6-phosphate deficiency with the use of chloroquine (www.fda.com). Over the years, pharmacogenomic information has increasingly been included in prescription drug labeling (Huang 2008). Currently, over 120 drugs have pharmacogenomic information included in labeling across nearly all therapeutic areas (Figure 18.1). The therapeutic areas with the most labeling containing genomic information are oncology, infectious diseases, and psychiatric disorders. In oncology, the adoption of genomics has

Table 18.1. Examples of Targeted Therapies

Drug	Therapeutic Area	Biomarker	Label Timing
Cetuximab, Panitumumab	Oncology	EGFR; KRAS	Pre/Postapproval
Crizotinib	Oncology	ALK	Preapproval
Exemestane, Fulvestrant, Letrozole	Oncology	ER/PR	Preapproval
Imatinib	Oncology	C-Kit, PDGFR, FIP1L1	Preapproval
Ivacaftor	Pulmonary	CFTR	Preapproval
Lapatinib, Pertuzumab, Trastuzumab, Everolimus	Oncology	HER2	Preapproval
Vemurafenib	Oncology	BRAF	Preapproval
Lenalidomide	Hematology	Chromosome 5q	Preapproval
Maraviroc	Antivirals	CCR5	Preapproval
Nilotinib, Dasatanib, Imatanib	Oncology	Ph Chromosome	Preapproval
Arsenic, Trioxide, Tretinoin	Oncology	PML/RARα	Preapproval
Denileukin Diftitox	Oncology	CD25	Preapproval
Capecitabine, Fluorouracil	Oncology	DPD	Postapproval
Pimozide, Aripiprazole, Iloperidone, Tetrabenazine, Thioridazine	Psychiatry, Neurology	CYP2D6	Postapproval
Celecoxib	Analgesics	CYP2C19	Preapproval
Citalopram	Psychiatry	CYP2C19	Postapproval
Rasburicase	Oncology	G6PD	Preapproval
Valproic Acid	Psychiatry	UCD	Postapproval

likely been driven by reasons that include low response rates in larger populations, significant toxicities even in individuals that may not respond, increased scientific knowledge of biology of cancer, and a transition to more targeted therapies to limit toxicities. Some of the aforementioned reasons apply to the field of infectious diseases. It happens that in the field of psychiatric disorders, the majority of the examples of genomic information in labeling have been related to the drugs being metabolized by the polymorphic CYP2D6. A list of drugs that contain pharmacogenomic information in their labeling can be found here: www.fda.gov/Drugs/ScienceResearch/ResearchAreas/Pharmacogenetics/ucm083378.htm.

Pharmacogenomic information can appear throughout drug labeling depending on its implications for prescribing. As of 2014, the majority (~40%) of drug labeling with information on pharmacogenomic biomarkers is related to pharmacokinetics (PK) or drug metabolism, as this has been the primary context for pharmacogenomics. However, as pharmacogenomic information becomes more readily available, genetic differences in both biomarkers of disease and drug target pharmacodynamics (PD) are increasing incorporated in drug labeling.

In some instances, the biomarker information is not essential to the use of the drug, but rather serves to inform clinicians about the potential for extreme responders. For example, many drug labels acknowledge the potential for the CYP2D6 poor metabolizers (PMs) to increase plasma concentrations, though no explicit recommendations are provided with regard to dose. In this setting, drug-drug interactions (DDIs) in labeling can also inform about gene-drug interactions (GDIs). For instance, 73% of DDIs involving polymorphic enzymes in drug labeling describe the analogous GDIs, however the management recommendations are sometimes different (Conrado 2013).

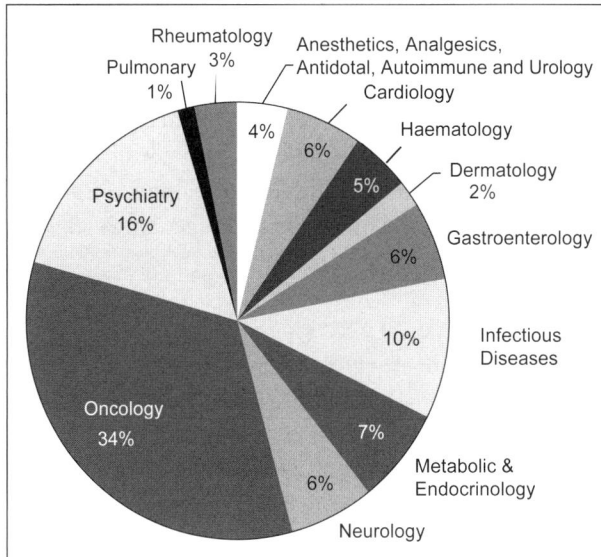

Figure 18.1. Pharmacogenomics in drug labeling by therapeutic area.

In addition to the inclusion of pharmacogenomic information in drug labeling at the time of initial approval, the FDA has also updated prescription drug labeling with pharmacogenomic information in the postmarketing setting. Notably, in some cases pharmacogenomic updates have been added more than 25 years after the initial drug approval (e.g., pimozide, carbamazepine, valproic acid). This has most often been the case for safety events, where risks and the genetic etiology are not well understood before marketing because of the low incidence of rare adverse events (see codeine example below). Hence, pharmacogenomic information is being used by the FDA to not only improve the drug development process, but to further refine therapy with approved drugs.

Regulatory Challenges of Pharmacogenomics and Targeted Therapeutics

The advent of the use of pharmacogenomic information in the drug development process is not without its unique challenges. We will highlight some regulatory considerations and opportunities that have been encountered in the premarket and postmarket settings.

Premarket Considerations

The FDA has recognized the importance of pharmacogenomics in the drug development process. In 2013, the FDA published guidance on pharmacogenomics: FDA's Guidance on Clinical Pharmacogenomics: Premarket Evaluation in Early-Phase Clinical Studies and Recommendations for Labeling (www.fda.gov/downloads/Drugs/GuidanceComplianceRegulatoryInformation/Guidances/UCM337169.pdf). This guidance calls for the collection of DNA, where possible, from all patients in clinical studies. Having collected this DNA at baseline in all subjects is a key first step in allowing both the FDA and drug developer to evaluate potential genetic sources of variability. However, this may not be possible for all programs particularly given the global nature of drug development and variability in local ethics board requirements and challenges in ascertaining IRB approval. Taking a more narrow view, collection of DNA samples is particularly relevant for drugs that display one or more of the following (Pacanowski 2014):

- High variability in PK or PD
- Multimodal distribution in PK or PD
- Polymorphic metabolism (e.g., via CYP2D6)
- Biological activation (i.e., prodrugs, drugs with active metabolites) by a polymorphic pathway
- Indication for a morbid/mortal or genetic disease (where no tools exist to monitor response)
- Poor tolerability
- Serious adverse events.

As developers collect more pharmacogenomic data and submit it to the FDA, some hurdles exist in the application and utilization of these data. Developing broadly applicable policies to address each of these emerging issues is essential for the FDA and for drug developers. Some of the challenges faced preapproval include the following:

- Assessment of benefit-risk in small subgroups/rare subsets
- Selection of appropriate biomarkers early in development and the validation and utilization of the biomarkers
- Selection of an appropriate trial design, marker, and test when moving from an overall population to one or more smaller subgroups
- Use and application of results from retrospective and/or unplanned subgroup analyses
- Lack of availability of appropriate test(s) for patient selection early in development
- The need for data in a biomarker-negative group
- Definition of biomarker positivity, including establishing a cut-off for a continuous variable

The following sections will focus on a discussion of the selection and utilization of pharmacogenomic biomarkers in drug development.

Selecting Appropriate Biomarkers in Early Phase Development

Utilizing pharmacogenomic information to best inform the drug development process is a challenge for both drug developers and the FDA. Biomarkers can be used for various purposes and can influence clinical trial design. Some of the roles for biomarkers are shown in Figure 18.2 below. Biomarkers can serve critical roles in both dose and patient selection as well as in the diagnosis and monitoring of disease states. However, identifying or selecting appropriate biomarkers early in the process can be formidable.

Genomic biomarkers can have varying utilities. Genomic biomarkers can be used from a prognostic standpoint to enrich the clinical trial for subjects more likely to experience the outcome of interest. Furthermore, genomic biomarkers can be predictive; that is, they are likely to indicate a response to the drug and as such the trial can be enriched for subjects who possess these specific biomarkers. Moreover, genomic biomarkers can also identify individuals that might be at risk for excessive drug exposure and may require a lower dose of the drug.

Often, drug developers conduct exploratory pharmacogenomic studies in early phase development. The focus of these exploratory studies can include genetic biomarkers related to the drug target pathway or those thought to have potential significance in the disease setting. This early exploration can aid in appropriate patient and dose selection in later phases of drug development. However, the limited sample sizes and limited data in early phase studies can limit the confidence about the medical relevance of the biomarker. Thus, developers are sometimes reluctant to pursue a targeted approach based upon these early phase data given a chance that a result is a false positive and the chance of unnecessarily limiting the treatment population. Effects of genetic variants on efficacy and safety that may be identified in these early studies often require confirmation in larger scale randomized trials. On the other hand, biomarkers that are more common, such as ADME gene variants that have effects on PK, can be more easily resolved in small, healthy subject studies.

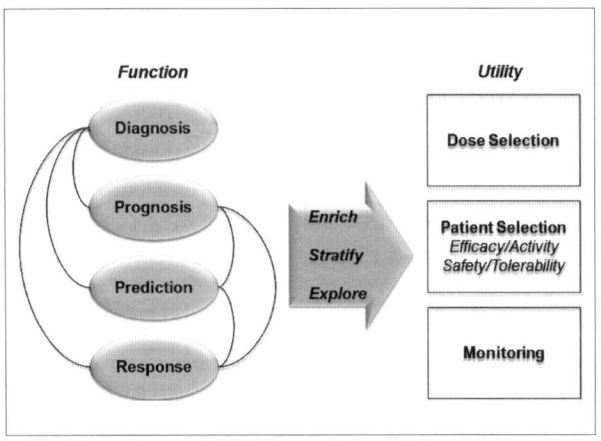

Figure 18.2. The role of biomarkers in clinical trial design.

Validation or Utilization of Selected Biomarkers

Once a pharmacogenomic biomarker is discovered, a new array of challenges arises. If the drug developer decides to utilize this biomarker to further target a population, then both the FDA and developer need to consider numerous issues. Some of the key development issues the FDA faces in the realm of pharmacogenomics and targeted therapeutics are as follows:

- How and when (at what phase of development) should one prospectively validate the biomarker(s) for use in the intended patient population, including how much and what type of data are required?
- Do patients with the biomarker need a dose modification or exclusion because they are at an increased risk of an adverse event?
- How much and what kind of data are required in the group of patients not possessing the biomarker?
- If the biomarker is continuous, how does one determine the appropriate cutpoint to select patients most likely to respond?
- Are FDA cleared or approved tests available or does one need to develop a test that can measure the selected biomarker?
- Do sufficient numbers of marker-positive patients exist to evaluate efficacy/safety in a clinical study?

Moreover, should a drug developer identify a genetic biomarker to select patients for—or exclude patients from—therapy, they will need to consider the potential need for a companion diagnostic that prescribers can use. The FDA regulates companion diagnostics through the premarket notification (510k) and premarket authorization (PMA). Also, much like new molecules are required to submit an investigational new drug application (IND) developers who are codeveloping a test may also need to submit an investigational device exemption (IDE).

When drug developers pursue a targeted approach for approval, if successful, the FDA must then decide how to label for the targeted population. Drugs are typically indicated for the populations studied in pivotal efficacy trials. In most cases, the trial populations are not selected on the basis of a molecular biomarker. In cases where a prognostic marker is used to select patients (e.g., a marker that informs the risk of poor outcome, which is sometimes used to increase event rates in clinical trials), it may still be possible to obtain broad, nontargeted indications because the drug activity may not depend on the presence or absence of the biomarker. However, when the biomarker is also predictive (i.e., it also informs differences in drug response), the indication may be limited to the subgroup of patients enrolled in clinical trials

(FDA Enrichment Guidance at www.fda.gov/downloads/ Drugs/GuidanceComplianceRegulatoryInformation/ Guidances/UCM332181.pdf) (FDA Draft Guidance for Industry 2012).

Characterizing risk/benefit in a targeted population or genetic subgroup can be difficult given that a paucity of data often exists. Often, the FDA must rely on additional sources of information including in vitro and animal models as well as PK/PD to further inform labeling. In addition, post hoc analysis of clinical trial data in genomic subgroups, although not ideal, can be used to further refine populations based on response. The FDA and developers sometimes have to make inferences about subgroups with little to no clinical data.

Postmarketing Considerations

As pharmacogenomic information has become more readily available, important discoveries have taken place that allow the FDA to further refine the safe and effective use of drugs in the postmarketing setting. Similar issues as observed in the premarketing setting exist. However, very rarely are prospective controlled clinical trial data available to support the postmarketing incorporation of pharmacogenomic information in the prescription drug label. Therefore, the FDA has to cautiously evaluate the sources of data, the robustness of the findings, and the availability of a test before updating the drug labeling. The FDA also relies on mechanistic PK/PD data to support the inclusion of pharmacogenomics to update the prescription drug label. Where possible, the FDA has asked developers to retain DNA collected in premarket trials so that it may be utilized in the event that a postmarketing safety/efficacy signal arises that may have a genomic association. In some cases, the FDA will require the developer to conduct additional studies to better understand a pharmacogenomic liability.

The following section contains illustrative examples that highlight the incorporation of pharmacogenomics information in drug development and in drug labeling. The examples also highlight the challenges discussed earlier.

Illustrative Examples

Premarket Examples

Ivacaftor: Developing Drugs for Rare Subsets

Ivacaftor was first approved by the FDA in January 2012 for the treatment of cystic fibrosis (CF) in individuals who have a G551D mutation in the Cystic Fibrosis Transmembrane Regulator (*CFTR*). In February 2014, ivacaftor was approved for use in eight additional *CFTR* mutations (G178R, S549N, S549R, G551S, G1244E, S1251N, S1255P, and G1349D). The highly targeted nature of ivacaftor development highlights many of the major considerations that arise in targeted development programs.

CF is a rare genetic disorder caused by mutations in the CFTR that reduce chloride transport. Over 1800 mutations have been reported to cause CF. The mutations present in a number of phenotypes that vary from severe disease to mild pancreatic insufficiency. Ivacaftor was developed to potentiate chloride transport via CFTR, potentially correcting the underlying pathophysiology of CF. In vitro models demonstrated that ivacaftor was able to increase chloride transport across the cell membrane for *CFTR* mutations that affect gating or conductance, where CFTR is expressed on the cell surface. Therefore, ivacaftor was initially studied in subjects who carried mutations that would likely be amenable to treatment with ivacaftor, specifically patients who carry at least one copy of the G551D allele, the most common of the gating mutation. Carriers of this allele make up about 4% of the CF population. Based upon the findings in two phase 3 clinical trials, ivacaftor was initially only indicated in patients with the G551D allele.

Subsequently, an additional Phase 3 trial in 39 subjects, the indication for ivacaftor was expanded to 8 additional CFTR mutations. Given the paucity of patients available for study, it was difficult to find substantial numbers of representative subjects of each of the genetic subgroups for which the developer was seeking an indication. Although in vitro findings suggested a response in each of the nine genetic mutations studied, consistent clinical and pharmacodynamic responses were only observed in eight of the nine mutations that were ultimately approved for labeling.

A large proportion of individuals with CF have mutations that cause CFTR to be broken down inside the cell and never expressed on the cell surface, such as F508Del. This mutation is carried by approximately 70% of CF patients (www.cftr2.org). While ivacaftor would not be expected to work on such mutations, based on its mechanism, F508del was a relevant "marker-negative" population to formally assess because it is present in most patients with CF. As such, a Phase 2 trial was conducted in F508del homozygotes, confirming the lack of benefit. Consequently, ivacaftor was not approved for treatment of patients with this particular mutation (Flume 2012).

The case of ivacaftor speaks to a couple of regulatory challenges when highly targeted, mechanistic approaches are used: (1) the need for data in the nontargeted group (those not expressing the genetic biomarker of interest, primarily subjects with the F508Del allele); (2) the ability to make informed regulatory decisions regarding safety and efficacy with small numbers of patients; and (3) the reliability of genomic subgroup analysis.

Had ivacaftor been developed prior to understanding the molecular genetics of the disease, it is likely that trials would not have demonstrated a clinical benefit, being that the responsive mutations are present in a minority of patients. A clear understanding of both the disease genetics and the drug activity across different mutations were critical to the success of the ivacaftor development program.

Crizotinib: Enrichment Strategies and Companion Diagnostic Codevelopment

Crizotinib is a kinase inhibitor approved in 2011 for patients with metastatic non-small cell lung cancer (NSCLC) whose tumors are anaplastic lymphoma kinase (ALK)-positive as detected by an FDA-approved test. This drug was approved along with a companion diagnostic to determine subjects ALK-status.

Nonclinical studies demonstrated that crizotinib inhibited several receptor tyrosine kinases. This included inhibition of ALK, ROS1, and c-Met phosphorylation in tumor cell-based assays and antitumor activity in tumor-bearing mouse models that expressed EML4-ALK or NPM-ALK fusion proteins or c-Met. Considering the nonclinical data, target biomarkers were analyzed in the Phase 1 trials. In the dose escalation portion of a Phase 1 trial, stable disease was observed in a small subset of patients that had ALK-positive NSCLC. This patient population (ALK+ NSCLC) was then further evaluated at the identified recommended Phase 2 dose in an expansion cohort of the same Phase 1 trial (n=119), and in a Phase 2 trial (n=136). Data from both trials were used to support an indication in this population and showed an investigator assessed overall response rate (ORR) of 61% in the Phase 1 trial cohort and 50% in the Phase 2 trial.

Local tests used at various study centers were used to identify ALK+ in the Phase 1 trial, but patients in the Phase 2 trial were selected using a central test. Data for the central test was submitted in support of a companion diagnostic approval that was approved concurrently with the drug. After the accelerated approval, the effect of the drug observed in these single arm trials was to be confirmed in a randomized Phase 3 trial comparing crizotinib to standard of care in metastatic ALK-positive NSCLC.

In this example, the selection of a certain subset of patients was supported by the crizotinib mechanism of action, nonclinical studies, and responses observed in patients in early studies. This provided rationale for trial enrichment. One could argue that since this drug is a multi-tyrosine kinase inhibitor it may have been appropriate to study it in a larger NSCLC population as it could have the potential to work in other patients. This is an important consideration because ALK + were estimated to only constitute 1%–7% of patients with NSCLC. However, such a large response rate may not have been observed if the trial was performed in an overall population. In this scenario, the developer enriched their trials for the subjects who early data determined were likely to respond to treatment. After the approval in this targeted population, the developer could explore other targeted populations that the drug is likely to benefit.

References: label and clinical pharmacology NDA review

Afatinib: Subset Analysis and Rare Mutations

Afatinib is a kinase inhibitor approved in 2013 in patients with metastatic NSCLC whose tumors have EGFR exon 19 deletions or exon 21 (L858R) substitution mutations as detected by an FDA-approved test.

In nonclinical studies, afatinib inhibited EGFR (ErbB1), HER2 (ErbB2), and HER4 (ErbB4). It inhibited cell lines expressing wild-type EGFR, EGFR exon 19 deletion mutations, EGFR exon 21 L858R mutations, and those overexpressing HER2. It also demonstrated activity in tumor bearing mouse models with wild type EGFR or HER2 or in an EGFR L858R/T790M double mutant model.

The primary trial used to support approval was a randomized trial comparing afatinib to cisplatin/pemetrexed chemotherapy doublet in first line metastatic or unresectable EGFR positive NSCLC. Patients were stratified by EGFR mutation status into 3 groups: exon 19 deletion (n=170), exon 21 L858R (n=138), or other (n=37). The other groups were composed of any EGFR mutations not in the first two groups and included several rare mutations. Some other known rare mutations were not detected in this trial.

Exploratory analyses showed varying treatment effects for different EGFR mutations, with the greatest effect in exon 19 deletions. The "other" group did not show overall benefit for afatinib and actually favored the chemotherapy group. Therefore, the indication was restricted to exon 19 deletions and exon 21 L858R substitution mutations. In the "other" group some of the patients appeared to have benefit, but it was difficult to assess the benefit-risk in each of the uncommon or rare mutations (10 different subtypes of mutations were observed in 37 patients; n=26 on the afatinib arm and n=11 on the chemotherapy arm). The mutations that were included in the "other" group were listed in the Clinical Studies section of the label with information regarding whether a response was observed or not in those patients.

In this example, the FDA made decisions regarding which mutation subtypes the drug would be approved for because of significant heterogeneity in response, which may have been driven by the presence of a known resistance mutation. Even though this example highlights the benefit of the use in pharmacogenomics to identify larger subgroups of patients most likely to respond to therapy, a great challenge still exists regarding how to deal with rare subgroups with the current trial design and statistical paradigms.

U.S. FDA Clinical Pharmacology Review 2012

Vemurafenib: Enrichment and Patient Selection Based Upon Molecular Biomarkers

Vemurafenib is a kinase inhibitor approved in 2011 for unresectable or metastatic melanoma with BRAF V600E mutation as detected by an FDA approved test. In nonclinical studies vemurafenib inhibited mutated forms of BRAF including V600E, and CRAF, ARAF, SRMS, ACK1, MAP4K5 and FGR and it demonstrated beneficial effects in BRAF V600E mutant melanoma animal models. Activation of MAP-kinase signaling and increased cell proliferation was observed in BRAF wild-type cells that were exposed to BRAF inhibitors. This suggested that the use of BRAF inhibitors in BRAF wild-type melanoma could result in tumor promotion. Activating BRAF mutations are observed in about 50% of patients with metastatic melanoma and the most common of the BRAF mutations are the V600 mutations, the most common being V600E (72%) and V600K (22%) (Eggermont 2011; Long 2011).

The first in human trial was in advanced solid tumors that contained an expansion cohort of BRAF V600E melanoma. Because of promising data in that subset, the FDA agreed to grant a special protocol agreement (SPA) for a trial in this patient population if overall survival (OS) was the primary endpoint. Although the developer chose to evaluate the drug in V600E, it was plausible that the drug would work in the V600K and other BRAF mutations. The submitted original New Drug Application (NDA) was supported by a randomized clinical trial in which vemurafenib showed a statistically significant advantage in overall survival compared to dacarbazine in patients with BRAF V600E mutation-positive unresectable or metastatic melanoma. At time of approval, the FDA issued a postmarketing commitment to the sponsor to develop an assay to reliably detect V600K mutations in unresectable or metastatic melanoma and perform an open label trial in patients selected using this test.

In this example, the developer chose to focus on the largest subset of patients with BRAF mutations by selecting V600E. The developer already had some early clinical data showing effect of their drug in this subgroup. By enriching for only the largest subgroup, analysis of the benefit risk ratio was simplified. However, the benefit risk ratio was not evaluated in patients with other BRAF mutations that could potentially benefit. Evaluation of effect in other subgroups can be performed in a post marketing setting. This example also highlights the use of nonclinical data in the selection of the patient population. Since the nonclinical data suggested that tumor promotion may occur in wild type melanoma, it was logical for the development to focus on BRAF mutations. *U.S. FDA Medical Review for Zelboraf 2011; U.S. Prescribing Information for Zelboraf 2014*

Postmarketing Examples

Pimozide: PK Variability with an Underlying Pharmacogenomic Cause

Pimozide was originally approved in 1984 for suppression of motor and phonic tics in patients with Tourette's Disorder. The labeling for pimozide was updated in 2011 to include pharmacogenomic information related to excessive exposures in a genetic subgroup of patients. Excessive exposure to pimozide potentially places patients at an increased risk for QT prolongation.

Pimozide is primarily metabolized by CYP3A4, with CYP1A2 and CYP2D6 previously thought to play minor roles in its disposition. Published drug-drug interaction studies alerted the FDA that CYP2D6 might play a significant role in the metabolism of pimozide. The FDA utilized modeling and simulation based on a small single-dose pharmacogenomic study to identify that CYP2D6 PMs had an approximate 2–2.5 fold increase in exposure compared to their intermediate metabolizer (IM) and EM counterparts in a multiple-dose setting. Based on published literature, concentration limits were set to not exceed the average exposure achieved with 10 mg in CYP2D6 EMs. As such, the FDA identified that at doses exceeding 4 mg (0.05 mg/kg in children) daily, CYP2D6 PMs would be exceeding the upper limit of exposure (10 mg daily) in EMs and IMs. Hence, the FDA incorporated this pharmacogenomic information into the drug labeling for pimozide. The labeling now recommends that at doses above 4 mg in adults or 0.05 mg/kg daily in children, CYP2D6 genotyping should be performed. Subjects who are identified as CYP2D6 PMs should not receive more than this dose, nor should doses be increased earlier than 14 days because of the prolonged half-life (Rogers 2012).

Pimozide represents a case where the FDA utilized postmarketing data and subsequent modeling and simulation to incorporate pharmacogenomic information into the prescription drug labeling. Because alternative (non CYP2D6 inhibitors) drugs exist, the use of pimozide is contraindicated in those receiving strong CYP2D6 inhibitors, however the FDA was cautious not to contraindicate for CYP2D6 PMs. In doing so, the FDA allowed this drug to still be dosed conservatively for patients in need by incorporating pharmacogenomics to improve safe use of this drug.

Codeine: Rare Adverse Event Supported by Mechanistic Pharmacogenomics

Codeine was first approved in 1950 for the treatment of mild to moderate pain. Codeine has since been widely used to treat various types of pain as well as cough. In

2013, the labeling for codeine was updated to restrict use in children post-adenotonsillectomy because of the increased risk of respiratory depression.

Codeine is a prodrug that is bioactivated to its active metabolite, morphine, by the hepatic enzyme CYP2D6. The FDA reviewed published case reports along with case reports that were submitted to the FDA's Adverse Event Reporting System (AERS) regarding adverse events surrounding respiratory depression and death subsequent to treatment with codeine. The review determined that the risk in children undergoing adenotonsillectomy was in part related to CYP2D6 ultra-rapid metabolism (UMs). Patients having two or more functional alleles of the CYP2D6 enzyme (UMs), which is present at a frequency of approximately 1% in Asians and up to 28% in some African and Middle Eastern populations (Crews 2012). Patients who are CYP2D6 UMs are at an increased risk of respiratory depression secondary to an increase in conversion of codeine to morphine.

Because such toxic reactions are rare and life-threatening CYP2D6 testing was not deemed adequate or efficient to decrease the potential for these events. Therefore, a boxed warning was added to the codeine labeling to warn of the increased risk of death in children undergoing adenotonsillectomy who are CYP2D6 UMs. In addition, the labeling now also bears a boxed warning indicating that subjects who already at an increased risk of respiratory depression such as those undergoing adenotonsillectomy should not receive codeine (Racoosin 2013).

Conclusion

This chapter highlights many of the regulatory challenges that the FDA faces in the age of pharmacogenomics and targeted therapeutics. In addition, this chapter demonstrates through examples across various therapeutic areas how the FDA utilizes pharmacogenomic information throughout the drug development process and has subsequently incorporated this information in prescription drug labeling. Pharmacogenomic and targeted therapies are undoubtedly changing the landscape of drug development and in some cases have significantly impacted the practice of medicine. As the era of pharmacogenomics and targeted therapies advances, the FDA will be faced with many novel issues in the evaluation of the safety and efficacy of these products. The FDA plans to incorporate the latest technology and regulatory flexibility to rise to this opportunity. The FDA continues to embrace pharmacogenomics and targeted approaches, while ensuring that the right patients receive the right drugs at the right dose.

References

Conrado DJ, Rogers HL, Zineh I, et al. Consistency of drug-drug and gene-drug interaction information in U.S. FDA-approved drug labels. Pharmacogenomics 2013;14(2):215-23.

Crews KR, Gaedigk A, Dunnenberger HM, et al. Clinical Pharmacogenetics Implementation Consortium (CPIC) guidelines for codeine therapy in the context of cytochrome P450 2D6 (CYP2D6) genotype. Clin Pharmacol Ther 2012;91(2):321-6.

Eggermont AM, Robert C. New drugs in melanoma: it's a whole new world. Eur J Cancer 2011;47(14):2150-7.

FDA Draft Guidance for Industry: Enrichment Strategies for Clinical Trials to Support Approval of Human Drugs and Biological Products, 2012. Available at www.fda.gov/downloads/drugs/guidancecomplianceregulatoryinformation/guidances/ucm332181.pdf%3Fsource%3Dgovdelivery.

Flume PA, Liou TG, Borowitz DS, et al. Ivacaftor in subjects with cystic fibrosis who are homozygous for the F508del-CFTR mutation. Chest 2012;142(3):718-24.

Huang SM, Temple R. Is this the drug or dose for you? Impact and consideration of ethnic factors in global drug

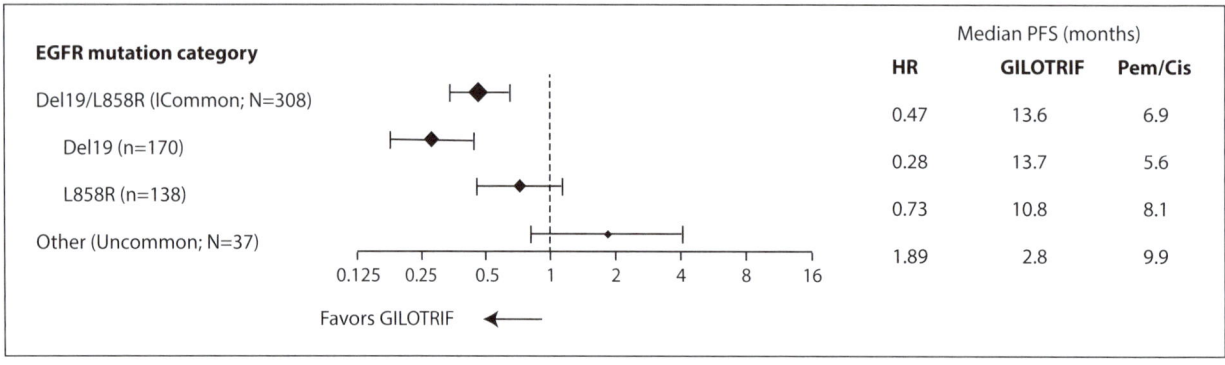

Figure 18.2. New evidence about an old drug risk with codeine after adenotonsillectomy.

Source: Figure 2 in afatinib labeling

development, regulatory review, and clinical practice. Clin Pharmacol Ther 2008;84(3):287-94.

Long GV, Menzies AM, Nagrial AM, et al. Prognostic and clinicopathologic associations of oncogenic BRAF in metastatic melanoma. J Clin Oncol 2011;29(10):1239-46.

Pacanowski MA, Leptak C, Zineh I. Next-generation medicines: past regulatory experience and considerations for the future. Clin Pharmacol Ther 2014;95(3):247-9.

Racoosin JA, Roberson DW, Pacanowski MA, et al. New evidence about an old drug—risk with codeine after adenotonsillectomy. N Engl J Med 2013;368(23):2155-7.

Rogers HL, Bhattaram A, Zineh I, et al. CYP2D6 genotype information to guide pimozide treatment in adult and pediatric patients: basis for the U.S. Food and Drug Administration's new dosing recommendations. J Clin Psychiatry 2012;73(9):1187-90.

U.S. FDA Medical Review for Zelboraf, 2011. Available at www.accessdata.fda.gov/drugsatfda_docs/nda/2011/202429Orig1s000MedR.pdf.

U.S. FDA Clinical Pharmacology Review for Gilotrif (Afatinib), 2012. Available at www.accessdata.fda.gov/drugsatfda_docs/nda/2013/201292Orig1s000ClinPharmR.pdf.

U.S. FDA Clinical Pharmacology Review for Xalkori (Crizotinib), 2011. Available at www.accessdata.fda.gov/drugsatfda_docs/nda/2011/202570Orig1s000ClinPharmR.pdf.

U.S. Prescribing Information for Zelboraf, 2014. Available at www.accessdata.fda.gov/drugsatfda_docs/label/2014/202429s004lbl.pdf.

Zineh I, Woodcock J. Clinical pharmacology and the catalysis of regulatory science: opportunities for the advancement of drug development and evaluation. Clin Pharmacol Ther 2013;93(6):515-25.

Chapter 19

PHARMACOGENOMICS IN DRUG DISCOVERY AND DRUG DEVELOPMENT

Alexander G. Vandell, Pharm.D., Ph.D.; and Joseph R. Walker, Pharm.D.

Learning Objectives

1. Argue for the importance of pharamcogenomic (PGx) research to scientists in a pharmaceutical company. Demonstrate an understanding of the current strengths and limitations of pharmacogenomic research.
2. Justify the value of toxicogenomic research in the overall development of a candidate drug in a pharmaceutical company.
3. Develop a clinical pharmacogenomic plan for any experimental compound at the preclinical stage.
4. Demonstrate the days in which pharmacokinetic PGx can streamline a clinical development plan and enrich an overall data package for a drug candidate.
5. Demonstrate the ways in which PGx and associated companion diagnostics can enhance oncology compound development in a pharmaceutical company.

Keywords: Pharmacogenomics, pharmaceutical company, pharmaceutical industry, drug discovery, drug development, toxicogenomics

Abbreviations in This Chapter

ABCB1	ATP-binding cassette subfamily B member 1	GWAS	Genome wide association study
		HR	Hazard Ratio
ALK	Anaplastic lymphoma kinase	HMG-CoA	3-hydroxy-3-methylglutarylcoenzyme A
ApoE2	Apolipoprotein E2		
ATP	Adenosine Triphosphate	IL28B	Interleukin 28B
BRAF	V-raf murine sarcoma viral oncogene homolog B1	I-PWG	International Pharmacogenomics Working Group
BRIM 3	BRAF in Melanoma 3	K-ras	Kirsten rat sarcoma 2 viral oncogene homologue
CI	Confidence interval		
CYP	Cytochrome P-450	LDL-C	Low-density lipoprotein cholesterol
CYP2C9	Cytochrome P-450 2C9	PCR	Polymerase chain reaction
CYP2C19	Cytochrome P-450 2C19	PD	Pharmacodynamic
EGFR	Epidermal growth factor receptor	PegINF	Pegylated interferon
EMA	European Medicines Agency	PGx	Pharmacogenomics
EML4	Echinoderm microtubule-associated protein-like 4	PK	Pharmacokinetic
		TGx	Toxicogenomics
FDA	Food and Drug Administration	VKORC1	Vitamin K epoxide reductase complex subunit 1
FXa	Factor Xa		

Abstract

Pharmacogenomics (PGx) is an innovative practice that is gaining in popularity and influence in drug development as the pharmaceutical industry seeks new ways to boost productivity. Currently, most pharmaceutical companies regularly collect PGx samples as a part of their clinical trials. As PGx research increasingly plays a part in drug development it is important that all industry scientists (current and aspiring) have at least a fundamental understanding of PGx. PGx research may be integrated into the drug development process at any time, from the earliest preclinical phases to late stage development or even post approval. PGx has already made several high profile contributions to drug development and is currently of great interest in oncology where greater personalization of therapies is urgently needed. This chapter aims to broadly describe how PGx fits into industry research, to describe the use of toxicogenomics in industry, to highlight specific examples where PGx research has made meaningful contributions to the pharmaceutical industry, and to briefly describe how some of the major PGx research techniques can be employed by industry scientists.

Introduction

Development and registration of safe and effective therapeutics is the primary goal of the pharmaceutical industry. However, while the need for novel innovative therapeutics remains, the industry is facing significant headwinds steadily working to decrease productivity in drug development. Pharmacogenomics (PGx) represents one innovative practice that may help boost productivity in drug development. PGx has made several high profile contributions to drug development and is currently of great interest in oncology where greater personalization of therapies is urgently needed. The aims of this chapter are to familiarize the reader with the strengths of PGx and how PGx fits into the broader scope of drug development.

For the last several decades, the pharmaceutical industry has followed a blockbuster model of drug development focused on creating single drugs for markets with large numbers of patients with chronic illnesses and an annual revenue potential of at least U.S. $1 billion. Unfortunately, despite breathtaking advances in research technology over the last few decades, the number of new drugs successfully brought to market per billion U.S. dollars spent has been steadily declining, with the number of new drugs approved by the FDA dropping by half every 9 years since 1950 (Scannell 2012). Predictably, as productivity declined, product portfolios have narrowed and the number of truly innovative compounds brought to market has decreased. Many reasons for the decline in productivity have been proposed. While a thorough discussion of them is beyond the scope of this chapter, we refer the reader to the following reference if they wish to learn more (Scannell 2012).

It is believed by many in industry that a greater focus on personalized medicine may represent one way to circumvent the productivity decline and reinvigorate drug development. Personalized medicine is a catch all term encompassing a wide variety of research avenues including biomarkers, pharmacogenomics, companion diagnostics, and other techniques for identifying subgroups of patients who are more likely to respond to a given therapy. PGx, like the other components of personalized medicine, has been gaining increased attention from the FDA and the pharmaceutical industry.

The Industry Pharmacogenomics Working Group

In response to the decline in industry productivity, a number of pharmaceutical companies came together to form the Industry Pharmacogenomics Working Group (I-PWG). The I-PWG was initially formed in 2000 as a response to regulator's requests for information from industry about PGx research. The I-PWG closely follows a variety of regulatory agencies and policy makers engaging and exchanging information with them pertaining to the use of PGx in pharmaceutical development. The I-PWG also provides information to the public at large via publications and informational programs regarding a variety of issues surrounding PGx.

To date, the I-PWG has produced a variety of white papers addressing issues pertinent to PGx research in industry. Some operational and scientific focused topics addressed by the I-PWG include a review of institutional review boards/ethics committee requirements for PGx sample collection across the globe (Ricci 2011), recommendations for best practices surrounding DNA sample collection and storage procedures (Franc 2011a; Warner 2011), suggestions for ensuring appropriate privacy protection for individuals who participate in PGx research (Franc 2011b), and suggestions for successfully conducting prospective-retrospective PGx analyses (Patterson 2011).

Preclinical PGx Research in the Pharmaceutical Industry

In the pharmaceutical industry, many scientists do important preclinical research across a wide range of scientific disciplines including toxicology, pharmacology, and pharmacokinetics (PK). Although a thorough review of their activities is beyond the scope of this chapter, these preclinical scientists use a wide variety of genetic techniques, such as small interfering RNA, genetically modified cell lines and transgenic animals (e.g., knockout/knockin mice), to clarify the contribution of a given gene to a disease phenotype or drug response. The knowledge obtained in these preclinical PGx experiments is invaluable for making critical go/no-go decisions and for optimizing the development strategies of drug candidates. The results of this important preclinical genomic research feed directly into the clinical PGx strategies.

Disease Research
Research into the underlying pathophysiology of diseases is the foundation of pharmaceutical industry research and development, and is the gateway to innovative therapies. Historically, when diseases were classified clinically, patients were grouped by presence of similar signs and symptoms. There was an assumption that most diseases were homogenous collections of symptoms with identical underlying pathogenesis. Now, it is becoming increasingly clear that many common diseases previously thought to be homogeneous are, in fact, a collection of unique pathogenic conditions with a similar phenotypic presentation (Roses 2007).

Disease subtypes are beginning to be distinguishable by using new techniques (e.g., genomics, gene expression signatures, and imaging) with or without necessarily providing a complete mechanistic explanation. The presence of different disease subtypes, each characterized by a different pathology, may help explain why some patients do not respond adequately to treatment. The nonresponders could have a different form of the disease. As researchers successfully identify molecular subpopulations of a specific disease, they increase the probability that the most efficacious therapy will be found for each subpopulation.

New information about disease subpopulations may benefit pharmaceutical industry scientists by allowing them to achieve the following:

- Recognize subdivisions of disease.
- Nominate predictive biomarkers of response.
- Gain a mechanistic understanding of a disease, which can lead to new drug targets.
- Identify a common link between diseases previously thought to be unrelated.

Genomics is one area in which clinical and translational researchers are able to make a substantial contribution to our understanding of the molecular underpinnings of disease. Clinical and translational scientists collect massive amounts of detailed information on the disease status of patients participating in clinical trials. Together with a DNA sample obtained with appropriate consent, these clinical trial data form the basis of a genomic study of disease. Throughout the pharmaceutical industry, translational researchers are engaged in this type of genomic study as part of ongoing clinical trials. The resulting associated genes serve to point clinical scientists toward candidate diagnostic or prognostic biomarkers, and point preclinical biologists toward new hypotheses around the pathogenesis of the disease. In this way, information and knowledge moves in both directions between bench and bedside.

An example of the use of genetic information to better understand underlying biology, identify new druggable targets, and discover drug candidates against those targets is illustrated in crizotinib. Researchers identified a chromosomal rearrangement present in about 4% of patients with non-small cell lung cancer that generated a fusion gene between *EML4* (echinoderm microtubule-associated protein-like 4) and *ALK* (anaplastic lymphoma kinase). The fusion product of this gene was shown to have constitutively active kinase activity leading to the activation of a variety of intracellular signaling pathways, resulting in promotion of tumor growth, migration, and invasion. Armed with this information, pharmaceutical scientists at Pfizer employed a combination of structure based drug design and further optimization based on kinase inhibitory activity. The resulting product, crizotinib, emerged as the lead compound for clinical development on the basis of its kinase specificity and its favorable PK and biopharmaceutical properties (Gandhi 2012).

Toxicogenomics
One specific preclinical research area in which PGx activities have expanded rapidly in the pharmaceutical industry is toxicology. Traditionally, toxicologists have used small animal studies both to understand the toxic potential of a drug candidate and to quantitatively describe the exposures required to observe these toxicities. The viability of a drug candidate is assessed by comparing the dose or exposure necessary to achieve efficacy with the dose or exposure that produces toxicity. The ratio of these doses or exposures is called the therapeutic index. A drug candidate with a narrow (or low) therapeutic index at this early stage will not be suitable for promotion to testing in humans.

In these preclinical animal studies, the presence of toxicity is assessed by assessing changes in gross physical examination, serum biochemistries, hematology, urinalysis, body weight, behaviors (e.g., feeding, sleeping, activity), and histopathology on necropsy after single or multiple doses of the drug candidate. These end points are nonspecific and rarely reveal information about the molecular mechanism behind any observed toxic changes.

Toxicogenomics (TGx) focuses on the use of expression microarray analysis in toxicology. It is rapidly being embraced by the pharmaceutical industry as a useful tool in preclinical toxicology to assess drug safety in a faster, more sensitive, more informative manner than traditional assessments. Almost all major pharmaceutical companies have devoted significant resources to developing and applying microarray analysis toward toxicology (Yang 2004). A typical TGx study might involve an experiment in animals or animal derived primary cell lines with three treatment groups: high-dose and low-dose treatment groups and a vehicle control group. The highest dose regimen is intended to produce an overtly toxic response that can be reliably detected. In a typical TGx experiment, expression microarrays generate lists of significantly differentially expressed genes for each biologic sample. Then, through literature mining, comparative analysis, and biologic modeling of gene expression data sets, it is possible to differentiate the adaptive biologic responses from the observed gene expression changes associated with adverse effects (Waters 2004).

Many public and proprietary database and pathway tools are available to facilitate interpretation of the results. These tools provide libraries of gene expression patterns from known toxins across a range of species and tissues. In addition, pathway tools help researchers physiologically link the constellation of genes whose expression patterns change in similar ways after exposure to the potential toxin. With this information, preclinical scientists are able to formulate hypotheses around the mechanism of toxicity and nominate and validate candidates for safety biomarkers before human testing. In summary, the strategic application of TGx by the pharmaceutical industry in the earliest stages of a drug discovery program offers a valuable opportunity to identify potential safety hurdles early and ultimately reduces the time required to uncover the optimal drug candidate (Ryan 2008).

Preclinical ADME Research
Preclinical ADME research involves the use of in vitro and in vivo models to obtain insights into the drug metabolism and pharmacokinetic properties of candidate drugs. Traditionally, these studies have focused on identifying drug metabolizing enzymes, metabolites, transporters, etc. Many drug metabolizing enzymes and transporters have well-known and well-characterized functional polymorphisms, the effects of which can be examined in preclinical work. If a drug candidate is found to be a substrate of a known polymorphic transporter or metabolizing enzyme, the effects of these polymorphisms on the drug can be investigated cell lines and animal models. Information gained from these preclinical studies may in turn prompt future PGx studies in humans to determine whether the preclinical findings have clinical significance. In addition, the lack of an observable PGx effect in preclinical studies is also valuable information as it may eliminate one potential source of variability in drug levels and response.

Clinical PGx Research in the Pharmaceutical Industry

Variability in Drug Response
There is a paradox in drug development. The clinical trial process provides evidence of the safety and efficacy of novel therapeutic candidates at usual doses in populations, whereas clinicians must treat individual patients, who can vary widely from population means in their responses to drug therapy. The critical questions in this paradox are, is it possible to better understand the true underlying causes of this variability, and if so, could better decisions in clinical medicine and drug development then also be made?

In fact, interindividual variability in drug response is a major barrier to successful drug development. As Sir William Osler said in 1892 about the practice of medicine, "If it were not for the great variability among individuals, medicine might as well be a science and not an art." At its core, the utility of PGx to pharmaceutical industry scientists lies in its ability to explain sources of observed variability in pharmacokinetics, pharmacodynamics, and clinical trial outcomes.

Clinical pharmacologists in the pharmaceutical industry have for a long time worked to identify and quantify all of the intrinsic (e.g., age, gender, body weight, renal function) and extrinsic (food, concomitant medications, etc.) factors that could have an effect on interindividual variability in the pharmacokinetics or pharmacodynamics of a candidate compound. Having a thorough understanding of both sets of factors (i.e., intrinsic and extrinsic) helps clinicians select doses most likely to provide positive outcomes and identify patients at greatest risk of negative outcomes. In this sense, the paradigm underlying the use of PGx in drug development has existed for a long time. In brief, the functional status of any gene is essentially just another patient-specific intrinsic factor like age, body weight, or renal function.

Comprehensive Clinical PGx Planning

For a compound progressing through the drug discovery and development process, detailed clinical PGx research plans are typically created and approved, together with other clinical research plans, the year before the compound enters the clinic. A comprehensive clinical PGx research plan includes details for DNA sample banking as well as the initial planned genomic analyses for all contemplated clinical studies. Table 19.1 provides an overview of a comprehensive clinical PGx plan, the early goals of which may include:

- to clarify the roles of variant genes already thought to affect human in vivo PK/PD based on completed preclinical pharmacology and PK research
- to identify novel variant genes not previously thought to be involved in human in vivo PK/PD

The later goals may include:
- to precisely quantify the contribution of any newly found variant genes to the overall observed variability in PK/PD
- to determine whether any variants affecting PK/PD exert an effect sufficient to affect clinical study outcomes
- to use SNP discovery techniques (such as genome wide association studies) to search for novel genomic markers that might be associated with the safety or efficacy outcomes of clinical trials

DNA Sample Banking in Industry Sponsored Trials

Collecting DNA samples with appropriate informed consent for PGx testing is a standard pharmaceutical industry practice for clinical studies. Although the goal is to collect a sample from every subject, the reality is that it varies from company to company when considering whether sample collection is a "required or optional" study activity. A recently published industry survey shows that many companies (~80%) routinely collect PGx samples in phase I, II, and III studies. However, only a few companies (~20%) have currently instituted PGx sample collection in phase IV studies. The companies cited a variety of reasons why they consider PGx sample collection worthwhile throughout the process of drug development. For phase I studies, PGx can aid in understanding PK variability and investigating adverse events. For phase II studies, PGx can be used to examine PD, efficacy, mechanism of action, and PK/PD effects. In phase III studies, PGx work is often focused on evaluating safety and efficacy biomarkers since the larger sample size allows for increased statistical power (Franc 2011a).

Providing a sample for possible PGx testing is typically an optional activity for individual subjects participating in large later-stage clinical trials. For large multicenter, multinational Phase III studies, the seemingly simple activity of PGx sample collection can be an

Table 19.1. Template for a Standard Clinical PGx Plan

A. Preclinical or "Phase 0"
   1. Preparation for human clinical studies
      i. SNP discovery in genes thought to be relevant (drug-metabolizing enzymes, drug target, etc.)
      ii. Additional SNP discovery in genes previously shown to influence the disease of interest
   2. Exploratory hypothesis-generating studies
      i. GWAS of patients with disease without drug treatment to identify novel biomarkers or molecular subtypes of disease
      ii. RNA microarray studies of the disease
B. Phase I
   1. PK/PD exploration in healthy volunteers
      i. Drug-metabolizing enzymes (CYPs, UGTs, esterases, etc.)
      ii. Drug transporters (P-glycoprotein, OCTs, etc.)
      iii. Drug targets (G protein–coupled receptors, ion channels, etc.)
C. Phase II/III
   1. Confirmation of healthy volunteer PK observations in patients
   2. PK/PD modeling and simulation supplemented with PGx data
   3. Drug response: early PD exploration and confirmation
      i. Candidate gene (candidate SNP and haplotypes) in drug target pathway
      ii. Genes/SNPs previously associated with disease of interest (e.g., SNPs from GWAS)
      iii. Genes/SNPs previously shown to modify response to similar drugs
   4. Exploratory hypothesis-generating approaches to identify novel biomarkers of response (GWAS, omics, etc.) in larger clinical studies (safety or efficacy outcomes)

CYP = cytochrome P450; GWAS = genome-wide association study; OCT = organic cation transporter; PD = pharmacodynamics; PGx = pharmacogenomics; PK = pharmacokinetics; SNP = single nucleotide polymorphism; UGT = UDP-glucuronosyltransferase; a family of drug-metabolizing enzymes.

onerous task. Difficulties include creating a PGx language suitable for all institutional review boards/ethics committees/national health authorities around the globe, following variable national regulations around the exportation of DNA samples, handling the differing cultural views on genetic testing, and facing fears that all of the above might negatively affect the overall project timelines and make the implementation of PGx sample collection in global Phase III clinical studies a much bigger feat than expected.

Using PGx Data to Augment Clinical PK/PD Research

The PK of a drug can be influenced by drug-metabolizing enzymes, transport proteins that shepherd drugs across cellular membranes, plasma proteins that bind drugs, and transcription factors that affect the expression of transporters and metabolic enzymes. Pharmaceutical industry scientists use PGx to augment PK/PD activities for many reasons, including the following:

- to recognize the involvement of previously underappreciated PK pathways
- to make decisions about the need for additional drug-drug interaction studies
- to enrich ongoing population (Pop) PK/PD activities
- to estimate the full range of variability in a population
- to better understand observed "outliers"
- to allow precise estimates of PK parameters in ethnic bridging studies
- to aid in dose selection for pivotal Phase II/III studies
- to write clarifying package insert statements that may help clinicians to achieve a marketing advantage over competition

The appropriate scientific tools exist to assess the contribution of each individual protein on the PK of a drug candidate. Unfortunately, the size and scope of the task of evaluating each of the scores of these proteins is daunting and expensive. Thus, for most drug candidates, pharmaceutical industry scientists typically are able to evaluate the contribution of only the most common enzymes and transporters before a drug is promoted to clinical development. Much of the important knowledge around the specific determinants of a candidate drug's PK is gained after a drug has been approved. Having this information available early in the drug development process could lead to much more informed decisions and much more efficiency in the process. There are hundreds of known or expected gene products that have a role in drug disposition, many of which contain functional polymorphisms (Li 2011). An important role for PGx in the pharmaceutical industry is to identify previously unsuspected proteins which could affect the human PK of a drug candidate.

PGx and Clinical ADME Studies

As an example, scientists at Abbott used PGx data from early Phase I studies to generate hypotheses about the specific drug transport proteins involved in the disposition of atrasentan, a selective endothelin A receptor antagonist being developed for prostate cancer. The investigators performed a PGx meta-analysis of PK data from individuals who participated in several clinical studies. After genotyping the DNA samples on a list of genes possibly involved in the PK of atrasentan, an analysis of covariance technique, which included genotype in addition to other patient-specific factors such as age, weight, gender, and ethnicity, was performed. Genetic variants in *OATP1B1* (organic anion transport protein isoform 1B1; also called SLCO1B1) significantly affected the total plasma exposure of atrasentan both after single doses and at steady state (Figure 19.1). In fact, in this study, genotype-predicted *OATP1B1* activity was the most influential covariate for atrasentan PK. The pharmacogenetic analysis suggested the testable hypothesis that atrasentan is an *OATP1B1* substrate. This analysis serves as a good example of how early PGx studies can be used by pharmaceutical industry scientists to gain insight into the PK of a new molecular entity (Katz 2006).

In addition to positive findings highlighting the importance of a particular drug-metabolizing enzyme or transport protein, negative results of a PK PGx study can help streamline a molecule's clinical development by demonstrating that certain pathways are not likely to significantly affect the PK of a drug. The FDA guidance on drug-drug interaction testing provides much more detail on the role that PGx data may play in determining the need for human drug-drug interaction studies (U.S. FDA 2012).

PGx and Late Development

Prasugrel and clopidogrel are thienopyridines that improve outcomes in patients with acute coronary syndrome because they inhibit platelet activation through an irreversible blockage of the platelet adenosine diphosphate G protein–coupled purinergic receptor found on the platelet surface receptor. Both are prodrugs that require biotransformation to active metabolites by CYP enzymes. Although the active metabolites of both drugs have similar affinity for the G protein–coupled purinergic receptor found on the platelet surface receptor in vitro, the metabolic pathways leading to their formation are different in vivo (Jakubowski 2007). The activation of clopidogrel from its prodrug requires two separate CYP-dependent oxidative steps, with *CYP2C19* involved in both steps, whereas prasugrel is oxidized to its active metabolite in a single step, which could be mediated by one of several CYP enzymes, with the greatest contributions from *CYP3A4* and *CYP2B6*, lesser contributions from *CYP2C9* and *CYP2C19*, and even less contributions from *CYP2D6* (Jakubowski 2007; Rehmel 2006; Kazui 2010). The PD response to clopidogrel displays substantial interpatient variability. It has been estimated that up to 30% of patients do not achieve an adequate antiplatelet effect from clopidogrel (Gurbel 2003; Angiolillo 2007), and patients with lesser degrees of platelet inhibition in response to clopidogrel appear to be at increased risk of cardiovascular events (Matetzky 2004; Gurbel 2005). Prasugrel displays greater degrees of platelet inhibition with much less interpatient variability (Brandt 2007a; Wiviott 2007).

To further understand this issue, researchers from Eli Lilly and Daiichi Sankyo (Brandt 2007b) tested the hypothesis that common genetic polymorphisms of the involved CYP enzymes contribute to decreased formation of the active metabolite of clopidogrel, with a corresponding effect on the PD response. Healthy volunteers participating in two clinical pharmacology studies involving clopidogrel and prasugrel were genotyped for commonly occurring functional variants in *CYP2B6, CYP2C19, CYP2C9, CYP3A4,* and *CYP3A5* in addition to the assessments of PK and PD (inhibition *CYP1A2* of platelet aggregation). The investigators found that subjects with variant genotypes of *CYP2C19* and *CYP2C9* had decreased exposure to the active metabolite of clopidogrel but not prasugrel. In addition, *CYP2C19*2* and the abnormal-functioning genotypes of *CYP2C9* were associated with a decreased PD response to clopidogrel but not prasugrel. These data suggest that differences in the metabolic pathways of these two molecules may account for this finding. Specifically, prasugrel requires esterases to form its thiolactone, followed by a single oxidative step that may be mediated by any one of several different CYPs (Figure 19.2) (Rehmel 2006). The lack of effect of the *CYP2C19* and *CYP2C9* variant genotypes on the response to prasugrel suggests that for the metabolism of prasugrel, the residual activity of other CYPs compensates for the reduced activity of one CYP (Brandt 2007B). Other studies have confirmed the association between *CYP2C19*2* and clopidogrel PK and PD (Fontana 2008; Giusti 2007; Hulot 2006; Kim 2008).

Of importance, these investigators were able to extend these PGx observations beyond the effects on PK and PD by genotyping a subset of patients participating in a global Phase III registrational study comparing the effects of clopidogrel and prasugrel on cardiovascular outcomes in patients with acute coronary syndrome with planned percutaneous

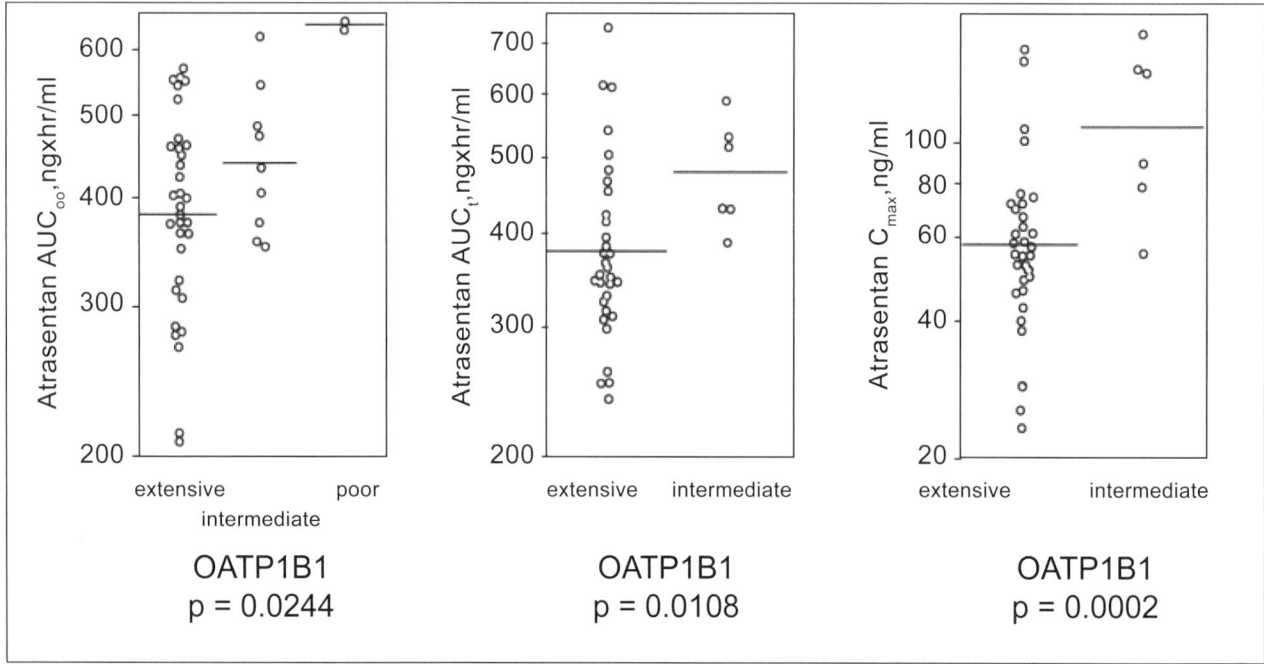

Figure 19.1. Individual atrasentan pharmacokinetic parameters by OATP1B1-predicted phenotype.

Horizontal lines represent the geometric mean of each group. AUC = area under the curve; Cmax = maximum concentration; OATP1B1 = organic anion transport protein isoform 1B1 (also called OATP-C or SLCO1B1). Adapted from Katz DA, Carr R, Grimm DR, et al. Organic anion transporting polypeptide 1B1 activity classified by SLCO1B1 genotype influences atrasentan pharmacokinetics. Clin Pharm Ther. 2006;79:186–96, © 2006. Reprinted by permission from Macmillan Publishers Ltd.

coronary intervention (Wiviott 2007). In this study, among clopidogrel-treated subjects, carriers of at least one *CYP2C19**2 allele had a relative increase of 53% in the composite primary efficacy outcome of the risk of death from cardiovascular causes, myocardial infarction, or stroke compared with noncarriers (12.1% vs. 8.0%). They also had an increase by a factor of 3 in the risk of stent thrombosis (2.6% vs. 0.8%). With prasugrel, there were no genetic variants that consistently influenced clinical outcomes (Mega 2009a, b). This study provided further compelling evidence for the chain linking genetics to PK, PK to PD response, and PD response to clinical study end points.

By providing compelling biologic linkage between *CYP2C19* genetic variants and PK, PD, and clinical study outcomes, these investigators were able to use PGx successfully to clarify the biology underlying the observation of compound-specific differences in interindividual variability. In this way, they were able to demonstrate the utility of PGx for providing the key differentiating factor between two competing drug products, clopidogrel and prasugrel.

In the past, interindividual variability in plasma drug concentrations and PK parameters was considered an inherent property of a drug. It was a fact to be accepted much like any other PK parameter. In some cases, this

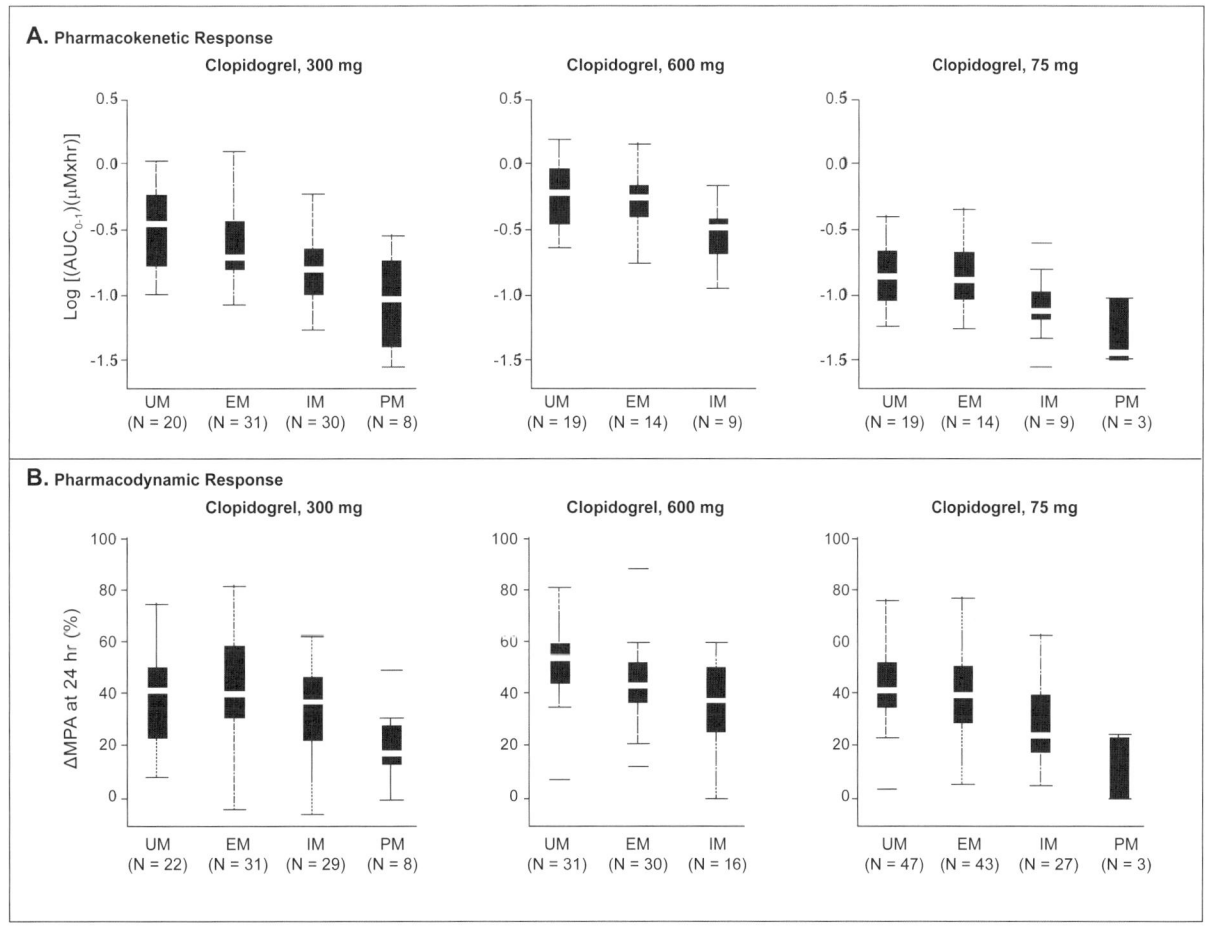

Figure 19.2. Relationship between *CYP2C19* genetic classification and PK and PD responses after the administration of loading and maintenance doses of clopidogrel in healthy subjects. Panel A shows box plots of the pharmacokinetic response of subjects after receiving a loading dose (either 300 mg or 600 mg) and during the administration of a 75-mg maintenance dose of clopidogrel, according to extended classification of metabolism genotypes into four subgroups: ultrarapid (UM), extensive (EM), intermediate (IM), and poor (PM). Panel B shows the pharmacodynamic response in the same group of healthy subjects, as assessed with the use of light transmission aggregometry in response to 20 μM of adenosine diphosphate, as the reduction in maximal platelet aggregation (Δ MPA) 24 hours after the administration of clopidogrel. The horizontal line within each box represents the median, and the lower and upper borders of each box represent the 25th and 75th percentiles, respectively. The single horizontal bars represent outliers that are more than 1.5 times the interquartile range from the border of each box, and the I bars represent the values farthest from the border of each box that are not outliers. PD = pharmacodynamic; PK = pharmacokinetic. Adapted from Mega JL, Close SL, Wiviott SD, et al. Cytochrome P-450 polymorphisms and response to clopidogrel. N Engl J Med 2009;360:354–62, © 2009 by Massachusetts Medical Society. All rights reserved.

made a drug candidate unsuitable for further development and registration. Today, the rapidly expanding knowledge base around the underlying factors that influence drug PK allows pharmaceutical industry scientists to develop drug candidates that were previously unsuitable. They are also able to do so more rapidly with lower costs and better outcomes. These benefits in drug PK have made widespread application of PGx an essential feature of almost all drug development programs.

Another example of PGx in late development is illustrated by recent work examining the influence of genetics on adverse events associated with two anticoagulants, edoxaban and warfarin. Warfarin is the most commonly used anticoagulant in the world; however it is one of the most clinically challenging drugs for both patients and prescribers. Warfarin doses must be individually tailored to each patient in order to keep their INR within therapeutic range. One major challenge is the high interindividual variability in warfarin response, which creates difficulty in reaching and maintaining therapeutic INRs. Over 40% of the variability in this response is associated with polymorphisms in two genes, *VKORC1* and *CYP2C9*. Many clinical trials have linked polymorphisms in these genes to the percent time spent in or out of therapeutic range. It has always been suspected that patients with genetic sensitivity to warfarin would spend more time with supra-therapeutic INRs and thus their likelihood of experiencing adverse bleeding events would be greater. However this increased bleeding risk had never demonstrated in a sufficiently powered analysis and remained only hypothetical.

Edoxaban is an oral anticoagulant under late stage development by Daiichi Sankyo, which works by directly inhibiting Factor Xa (FXa). Edoxaban is under investigation for reduction of stroke and systemic embolic events in subjects with atrial fibrillation, and for treatment of acute venous thromboembolism and prevention of recurrence in subjects with symptomatic deep vein thrombosis and/or pulmonary embolism. FXa inhibitors, such as edoxaban, have demonstrated a clear safety advantage over warfarin and are easier to administer due to their predictable PK and fast onset and offset of activity. Unlike warfarin, the mechanism of action and metabolism of edoxaban does not depend on *VKORC1* and *CYP2C9*, and thus are not affected by polymorphisms in these genes.

Industry scientists working at Daiichi Sankyo hypothesized that individuals with a greater genetic sensitivity to warfarin would be at increased risk for experiencing adverse bleeding events when treated with warfarin but not edoxaban. The Effective aNticoaGulation with factor xA next GEneration in Atrial Fibrillation (ENGAGE AF – TIMI 48) study was a registrational trial comparing edoxaban and warfarin for the prevention of stroke and systemic embolic events in subjects with atrial fibrillation. ENGAGE included 21,105 patients who were randomized to one of three treatment arms, edoxaban low exposure, edoxaban high exposure, or warfarin and followed for a median of 2.8 years (Giugliano 2013). As part of this trial, DNA samples were collected from over 14,000 individuals and genotyped for common polymorphisms in vitamin K epoxide reductase complex subunit 1 (*VKORC1*; -1639G greater than A) and *CYP2C9* (*2 and *3 haplotypes). Genotype information was then used to place patients into one of three genetic bins, normal responders (61.7% of patients), medium responders (35.4%), and highly warfarin sensitive responders (2.9%). This grouping is based on the genotype-guided warfarin starting dose recommendations endorsed by the Clinical Pharmacogenetics Implementation Consortium (CPIC) (Johnson 2011) and the warfarin prescribing information (Asher 2014).

A landmark analysis was conducted comparing the edoxaban high and low exposure groups to warfarin within each genetically defined warfarin sensitivity group during the first 90 days of the study. Among patients randomized to warfarin, medium and highly sensitive responders experienced significantly higher rates of bleeding compared with normal responders during the first 90 days (HR 1.31, 95% CI 1.05-1.64, P=0.018, and HR 2.66, 95% CI 1.69-4.19, P<0.001 respectively). Importantly, when compared to warfarin, treatment with edoxaban resulted in significantly less bleeding in sensitive and highly sensitive responders than in normal responders (Figure 19.3). Beyond 90 days, genotype was not associated with an increased risk of bleeding. Finally, when compared with normal responders, sensitive and highly sensitive responders spent more time over-anticoagulated (INR greater than 3) through the first 90 days (12.1%, 15.2%, and 20.9%, P<0.001). These data suggest a model where the presence of warfarin sensitivity risk alleles increase the time patients spend in an over-anticoagulated state, thus increasing their likelihood of experiencing adverse bleeding events (Mega in press).

PGx Research Techniques

PGx utilizes a variety of research techniques widely used in all areas of biomedical research for the study of complex traits, including disease etiology and response to an exogenous stimulus such as a drug. An association study looks for a statistical correlation between a specific genetic variant(s) and a specific phenotype (e.g., presence of a disease or response to a drug). A typical PGx association study uses a case-control design in which a population of subjects is dosed with a drug, and subjects with a specific response (cases) are compared with dosed subjects who do not display the response of interest (controls). For example, patients with a flushing reaction after a dose of niacin could be compared with those who tolerate niacin without incident. In this type of study, the primary comparison is of the frequency of a specific genotype or haplotype between the two groups. A significantly higher frequency of an allele or haplotype in the group with the phenotype of interest would be considered a positive association. Other types of study designs are also commonly used in genetic association studies including analyses that correlate continuous measures with the presence of specific genotypes or haplotypes.

Candidate Gene Association Studies

Pharmaceutical industry scientists often use candidate gene associations in their early PD and proof-of-concept studies. In these studies, groups of subjects with the disease or condition of interest are treated for the minimum duration to measure a drug effect on a biomarker of relevance. The goals are to get an early indication of whether the drug has any activity in humans and how that observed in vivo activity compares with the level of activity predicted from the preclinical animal models. In this type of study, scientists will select candidate genes, which include the drug target gene and other important genes in

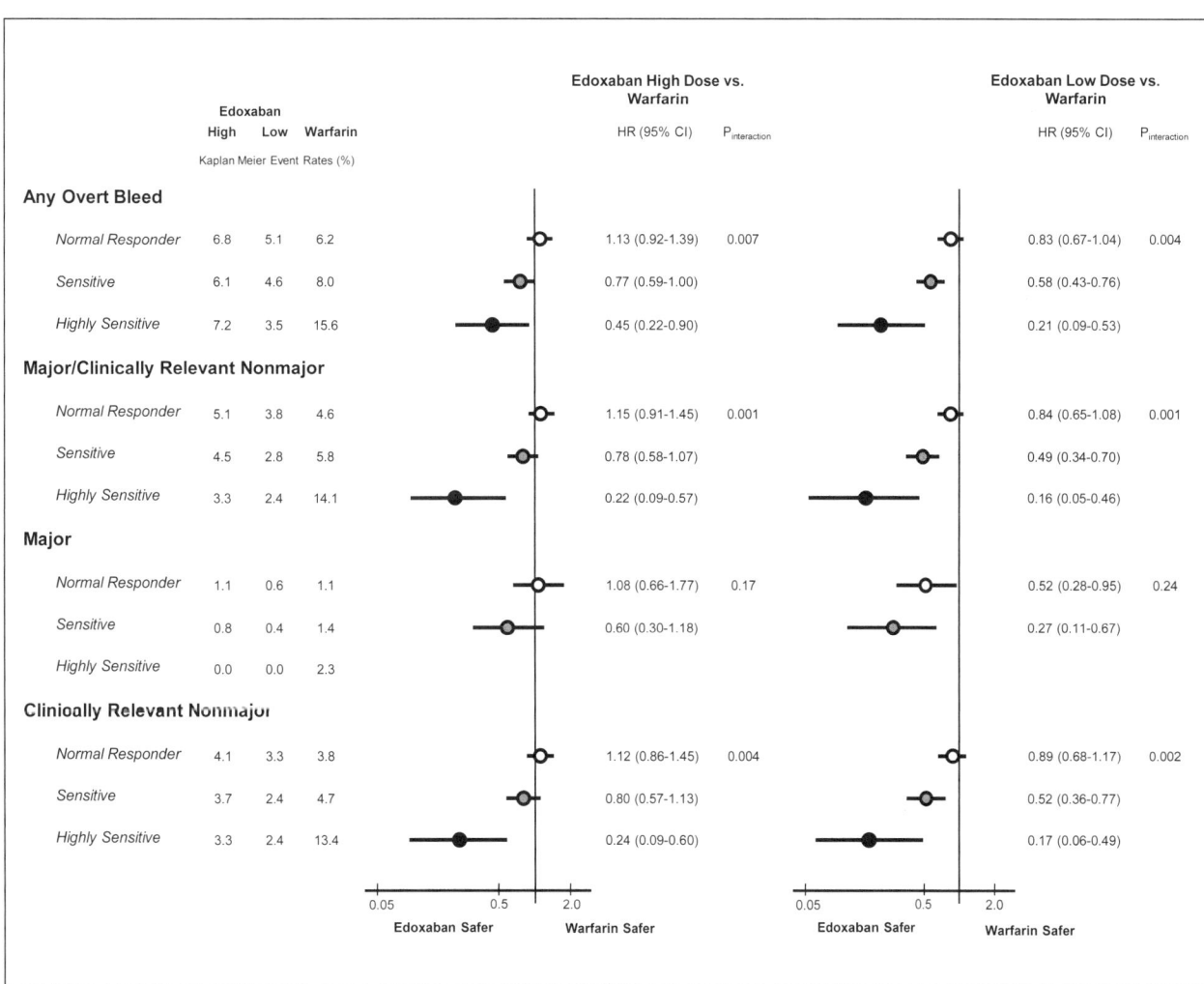

Figure 19.3. Bleeding outcomes among patients randomized to warfarin and edoxaban across genotype bins.

Event rates, HR (95% CI), and interaction terms are presented across genotype bins. The circles indicate the HR and the horizontal lines represent the 95% confidence bounds. During the first 90 days, the analyses included 2,982, 1,711, and 140 normal, sensitive, and highly sensitive responders among patients randomized to warfarin; 2,913, 1,627, and 186 among patients randomized to high-dose edoxaban; and 2,981, 1,632, and 176 among patients randomized to low-dose edoxaban. Beyond 90 days, the analyses included 2,803, 1,616, and 129 normal, sensitive, and highly sensitive responders among patients randomized to warfarin; 2,708, 1,517, and 172 among patients randomized to high-dose edoxaban; and 2,817, 1,541, and 166 among patients randomized to low-dose edoxaban. Adapted from Mega JL, Walker J, Ruff C, et al. Genetics and the Clinical Response to Warfarin and Edoxaban. Accepted for publication in Lancet.

the drug target pathway, genes that have been shown to be important to the PD of the other similar drugs in the class, and genes that have been associated with the disease. Because these early proof-of-concept studies are not powered to detect a subtle PGx effect, pharmaceutical industry scientists understand these investigations are not definitive. Replication is absolutely required (see below). Thus, these exploratory PGx investigations should be considered "hypothesis-generating" subgroup analyses. Nevertheless, these studies can be very useful because they allow drug development scientists to reliably observe or exclude very large, obvious genetic effects and to begin estimating the magnitude of any suspected genetic influences on the PD of their drug candidate.

Candidate gene association studies are often used by pharmaceutical industry scientists in both the early and late development phases as well as in drugs that have already gained marketing approval. Scientists at Pfizer, who conducted such research, studied 2,735 individuals, treated with HMG-CoA (3-hydroxy-3-methylglutarylcoenzyme A) reductase inhibitors, half on atorvastatin and the other half divided among fluvastatin, lovastatin, pravastatin, and simvastatin. The subjects were genotyped for 43 SNPs in 16 candidate genes thought to be involved in statin response or plasma lipid homeostasis. The 16 candidate genes were selected on the basis of previous positive associations published in the medical literature. Associations with low-density lipoprotein cholesterol (LDL-C) lowering, total cholesterol lowering, high-density lipoprotein cholesterol elevation, and triglyceride lowering were investigated. The only significant associations with LDL-C lowering were found with apoE2 (apolipoprotein E2) allele in which carriers of the rare allele who took atorvastatin lowered their LDL-C by 3.5% more compared with those homozygous for the common allele. In addition, a 3% difference in LDL-C lowering was observed between the two homozygous groups carrying the variant rs2032582 (S893A in *ABCB1* [ATP-binding cassette subfamily B member 1—also called multidrug resistance protein 1]) (Thompson 2005). This research was important because none of the other previously published genetic associations was positive in this study.

Genome Wide Association Studies
Genome Wide Association Studies (GWAS) are being used with increasing frequency in PGx research to identify novel genetic loci associated with drug response. Unlike candidate gene approaches, GWAS do not assume any a priori knowledge about candidate genes or polymorphisms. Instead, GWAS employ a SNP array to genotype hundreds of thousands, or even millions, of SNPs across the genome simultaneously. SNPs selected for inclusion in arrays are chosen agnostically in order to provide maximal coverage of the genome rather than because of any hypothesized genetic or PGx effect. Data from GWAS can be used to analyze a variety of endpoints such as drug response, PK endpoints, and adverse events. Because of its agnostic approach, GWAS is a powerful technique that can help identify genes not previously known or suspected of PGx involvement. However, GWAS studies do have important limitations that must be considered. Due to the small effect sizes usually observed, GWAS requires large numbers of patients, which may not be available to industry researchers until later in development. Also, like candidate gene studies, GWAS results require replication.

Genome wide association studies are increasingly being used by industry scientists, and have already contributed to clinical care. Scientists working with Schering-Plough studied 1,671 individuals, treated with either pegylated interferon-α-2b (PegINF-α-2b) or PegINF-α-2a combined with ribavirin for chronic hepatitis C virus infection. Using a SNP array, the patients were genotyped for over a half-million SNPs spread throughout the entire genome. Associations with treatment response and baseline viral load were tested. A significant association with sustained virological response found with rs12979860, a SNP near *IL28B*, the gene which encodes interferon-λ-3 was identified in patients of both European and African American ancestry. Patients with the CC genotype exhibited a 2-fold greater sustained virological response than patients with the TT genotype. This research was important because it identified a SNP as a novel major contributor to drug response. The PGx contribution to response is on par with conventional clinical factors such as baseline viral load and fibrosis. Furthermore, as the C allele is much more common in individuals of European ancestry than African Americans, this research helps explain the difference in response rate observed between these two patient populations (Ge 2009). Later studies have replicated the link between SNPs in the IL28B gene and drug response (Suppiah 2009; Tanaka 2009).

Emerging Technologies
In addition to candidate gene studies and GWAS, a variety of new "–omics" techniques are coming onto the scene. Some examples include whole exome sequencing, whole genome sequencing, epigenomics, and metabolomics. Each of these technologies has the potential to further increase our understanding of individual differences in drug response. However, like GWAS described above, these techniques produce an avalanche of data. The ability of researchers to generate additional "–omics" data as outpaced their ability to readily analyze it. For this reason, while there is great excitement in the pharmaceutical industry about the potential of employing novel "-omics" based approaches in drug development, most

companies are adopting a wait and see attitude until these technologies and their associated bioinformatics analyses become more established.

Replication in PGx Research

All PGx research tools have one major limitation that must be acknowledged by industry scientists: the failure to achieve consistent replication of findings. There are many reasons for the lack of reproducibility seen with PGx (or other genetic) studies. Discrepant findings are often owing to variations in study design. For example, studies might differ in the study population and in the definition of the phenotype. In addition, the nonreplication might result from real biologic differences or from the small magnitude of relative risks likely to be detected in many PGx studies. A criticism specific to candidate gene association studies is the challenge of selecting both the correct genes and the correct genetic variants within each tested gene. Some critics argue that current knowledge of the biology of human disease and human drug response is incomplete and insufficient to correctly select the influential genes.

PGx in Oncology Research

The prescription of a drug based on genomic biomarker profile is desirable because it limits drug exposure to patients most likely to benefit from the drug treatment. It also could minimize drug use in patients likely to experience harm by drug treatment, or it could enhance safe use by optimizing the dose. No therapeutic area has had more success in the hunt for useful biomarkers than oncology. With the success of the targeted therapies, there is enormous interest in finding more biomarkers that can help identify patients with the greatest likelihood of receiving benefit from a given therapy.

Panitumumab and cetuximab are monoclonal antibodies directed against the epidermal growth factor receptor (EGFR). Both are approved as single agents, and in combination, for the treatment of patients with metastatic colorectal cancer. At the beginning of preclinical and clinical development, it was expected that only EGFR-expressing tumors would respond to treatment. However, an increasing body of evidence emerged showing that the detection of EGFR on the tumor cells of patients was a poor marker for predicting efficacy (Lenz 2006), and that there were a significant number of nonresponders. After approval, additional research focused on detecting markers that were more reliably predictive for EGFR-targeted therapies.

Retrospective evaluations of banked, archived tumor samples from early registrational studies of panitumumab and cetuximab suggested that tumor K-ras (Kirsten rat sarcoma 2 viral oncogene homolog) mutations correlate with lack of response in colorectal cancer (De Roock 2008; Di Fiore 2007; Frattini 2007; Lievre 2008). K-ras, is a small G protein downstream of EGFR and an essential component of the EGFR signaling cascade. In 30% to 50% of colorectal tumors, K-ras has acquired mutations that may render the receptor constitutively active (Andreyev 1998) and confer resistance to anti-EGFR monoclonal antibody therapy.

Based on these early signals, scientists worked to expand the retrospective analysis of K-ras mutations from early studies to larger studies with archived tumor tissue. Treatment with anti-EGFR antibodies compared with supportive care alone was associated with almost a doubling in the median progression-free survival among patients with wild-type K-ras tumors. Of importance, there was no significant benefit among patients with tumors having K-ras mutation (Karapetis 2008; Messersmith 2008), suggesting that these therapies are ineffective in patients with mutant K-ras. These findings confirm that the mutation status of the K-ras gene is associated with overall survival among patients treated with cetuximab or panitumumab.

The compelling nature of these retrospective data convinced experts in this field to recommend that the standard of care change immediately rather than wait for prospective studies to be conducted and analyzed. The consistency of the results from more than 2,000 patients involving two different drugs clearly showed that the benefits of anti-EGFR antibodies are limited to the subgroup with wild-type K-ras tumors. These results led to the amendment of 10 planned or ongoing cetuximab studies to include K-ras testing (Messersmith 2008). These results also led the European Medicines Agency (EMA), the European regulatory agency, and the FDA to approve panitumumab and cetuximab for K-ras mutation negative colorectal cancer.

These drug development experiences highlight the value of banking biologic samples from clinical trials to allow the future analysis of patients who benefited, even if the putative biomarkers are unknown when the study is planned.

Vemurafenib is a small molecule targeted against the oncogene V-raf murine sarcoma viral oncogene homolog B1 (*BRAF*), and is approved for treatment of patients with unresectable or metastatic melanoma with the BRAF^{V600E} mutation as detected by an FDA-approved test. The BRAF^{V600E} mutation drives proliferation of cancer cells and is present in about half of melanomas and a variety of other cancer types. Vemurafenib was developed using a novel approach to identify potential kinase inhibitors that incorporated multiple iterations of chemistry, crystallography, computer stimulations, and biological assays directed to develop a *BRAF* specific inhibitor. Data from pre-clinical studies suggested that vemurafenib only modestly inhibited wild-type *BRAF*, however it exhibited pronounced inhibition of the BRAF^{V600E} mutant. Following additional optimization for target binding and PK properties, the decision was made to begin Phase 1 trials with concurrent development of a companion diagnostic to identify patients with BRAF mutations.

Industry scientists at Plexxikon and Roche collaborated to develop a companion diagnostic test designed to determine whether patients had vemurafenib sensitive tumors. Development of the PGx companion diagnostic test began during the initial phase 1 dose escalation study. A real time PCR based assay capable of determining BRAF^{V600E} status from formalin-fixed, paraffin-embedded tumor tissue was designed. This assay was then used to confirm the presence of the BRAF^{V600E} mutant form in patients recruited as part of the metastatic melanoma Phase 1 expansion cohort. Validation of the companion diagnostic suitable for approval by regulatory agencies would be based on its performance in subsequent Phase 2 and 3 trials.

Participation in the metastatic melanoma expansion cohort was limited to patients with the BRAF^{V600E} mutation. Results from the Phase 1 extension cohort were remarkable. Partial or complete response was observed in 26 of 32 (81%) of patients. Vemurafenib responders displayed a median progression-free survival of more than 7 months (Flaherty 2010). These results suggested the use of a genetic test would allow selection of patients most likely to respond to vemurafenib therapy.

Following the promising results from the Phase 1 study, Phase 2 and 3 trials were initiated. In the BRAF in Melanoma 3 (BRIM 3), Phase 3 trial, 675 patients with BRAF^{V600E} mutated metastatic or unresectable melanoma were enrolled. BRAF mutation status was determined using the companion diagnostic test developed in Phase 1. BRIM 3 compared 960 mg vemurafenib BID to 1000 mg/m^2 dacarbazine chemotherapy administered every three weeks. The response rate after 6 months of treatment was 48% for vemurafenib and 5% for dacarbazine. In response to these interim results, the data safety monitoring board recommended switching decarbazine treated patients to vemurafenib (Chapman 2011). Of the 338 patients randomized to decarbazine treatment, 83 (25%) crossed over to vemurafenib. Median overall survival and survival were significantly longer in vemurafenib treated patients compared to decabazine treated individuals (13.6 months [95% CI 12.0—15.2] vs 9.7 months [7.9—12.8]; hazard ratio [HR] 0.70 [95% CI 0.57—0.87]; p=0.0008) (McArthur 2014).

Conclusion

We are in the midst of a time of enormous change in the pharmaceutical industry. Pharmaceutical companies will no longer be able to support a blockbuster strategy. To survive, pharmaceutical portfolios will have to migrate from today's typically narrow product portfolio, composed of a few, albeit financially successful, drugs for large markets, to a larger and wider portfolio of treatments for targeted segments. Pharmacogenomics, as part of a broader personalized medicine approach, will be at the center of this critical transformation. Medication selection for patients based on obvious clinical characteristics must be replaced by a paradigm in which patients will be segmented into much smaller groups on the basis of biomarkers, genotypes, and other molecular measures. In this way, PGx has the potential to spearhead the transformation of the pharmaceutical industry, increasing the likelihood of its survival and bringing increased value to patients and payers.

References

Andreyev HJ, Norman AR, Cunningham D, et al. Kirsten ras mutations in patients with colorectal cancer: the multicenter "RASCAL" study. J Natl Cancer Inst 1998;90(9):675-84.

Angiolillo DJ, Fernandez-Ortiz A, Bernardo E, et al. Variability in individual responsiveness to clopidogrel: clinical implications, management, and future perspectives. J Am Coll Cardiol 2007 10;49(14):1505-16.

Asher E, Fefer P, Shechter M, et al. Increased mean platelet volume is associated with non-responsiveness to clopidogrel. Thromb Haemost 2014;112(1).

Brandt JT, Close SL, Iturria SJ, et al. Common polymorphisms of CYP2C19 and CYP2C9 affect the pharmacokinetic and pharmacodynamic response to clopidogrel but not prasugrel. J Thromb Haemost 2007b;5(12):2429-36.

Brandt JT, Payne CD, Wiviott SD, et al. A comparison of prasugrel and clopidogrel loading doses on platelet function: magnitude of platelet inhibition is related to active metabolite formation. Am Heart J 2007a;153(1):66e9-16.

Chapman PB, Hauschild A, Robert C, et al. Improved survival with vemurafenib in melanoma with BRAF V600E mutation. N Engl J Med 2011;364(26):2507-16.

De Roock W, Piessevaux H, De Schutter J, et al. KRAS wild-type state predicts survival and is associated to early radiological response in metastatic colorectal cancer treated with cetuximab. Ann Oncol 2008;19(3):508-15.

Di Fiore F, Blanchard F, Charbonnier F, et al. Clinical relevance of KRAS mutation detection in metastatic colorectal cancer treated by Cetuximab plus chemotherapy. Br J Cancer 2007;96(8):1166-9.

Flaherty KT, Puzanov I, Kim KB, et al. Inhibition of mutated, activated BRAF in metastatic melanoma. N Engl J Med 2010;363(9):809-19.

Franc MA, Cohen N, Warner AW, et al. Coding of DNA samples and data in the pharmaceutical industry: current practices and future directions—perspective of the I-PWG. Clin Pharmacol Ther 2011b;89(4):537-45.

Franc MA, Warner AW, Cohen N, et al. Current practices for DNA sample collection and storage in the pharmaceutical industry, and potential areas for harmonization: perspective of the I-PWG. Clin Pharmacol Ther 2011a;89(4):546-53.

Frattini M, Saletti P, Romagnani E, et al. PTEN loss of expression predicts cetuximab efficacy in metastatic colorectal cancer patients. Br J Cancer 2007;97(8):1139-45.

Fontana P, Senouf D, Mach F. Biological effect of increased maintenance dose of clopidogrel in cardiovascular outpatients and influence of the cytochrome P450 2C19*2 allele on clopidogrel responsiveness. Thromb Res 2008;121(4):463-8.

Gandhi L, Janne PA. Crizotinib for ALK-rearranged non-small cell lung cancer: a new targeted therapy for a new target. Clin Cancer Res 2012;18(14):3737-42.

Ge D, Fellay J, Thompson AJ, et al. Genetic variation in IL28B predicts hepatitis C treatment-induced viral clearance. Nature. 2009;461(7262):399-401.

Giugliano RP, Ruff CT, Braunwald E, et al. Edoxaban versus warfarin in patients with atrial fibrillation. N Engl J Med 2013;369(22):2093-104.

Giusti B, Gori AM, Marcucci R, et al. Cytochrome P450 2C19 loss-of-function polymorphism, but not CYP3A4 IVS10 + 12G/A and P2Y12 T744C polymorphisms, is associated with response variability to dual antiplatelet treatment in high-risk vascular patients. Pharmacogenetics and genomics 2007;17(12):1057-64.

Gurbel PA, Bliden KP, Hiatt BL, et al. Clopidogrel for coronary stenting: response variability, drug resistance, and the effect of pretreatment platelet reactivity. Circulation 2003;107(23):2908-13.

Gurbel PA, Bliden KP, Samara W, et al. Clopidogrel effect on platelet reactivity in patients with stent thrombosis: results of the CREST Study. J Am Coll Cardiol 2005;46(10):1827-32.

Hulot JS, Bura A, Villard E, et al. Cytochrome P450 2C19 loss-of-function polymorphism is a major determinant of clopidogrel responsiveness in healthy subjects. Blood 2006;108(7):2244-7.

Jakubowski JA, Winters KJ, Naganuma H, et al. Prasugrel: a novel thienopyridine antiplatelet agent. A review of preclinical and clinical studies and the mechanistic basis for its distinct antiplatelet profile. Cardiovasc Drug Rev 2007;25(4):357-74.

Johnson JA, Gong L, Whirl-Carrillo M, et al. Clinical Pharmacogenetics Implementation Consortium Guidelines for CYP2C9 and VKORC1 genotypes and warfarin dosing. Clin Pharmacol Ther 2011;90(4):625-9.

Karapetis CS, Khambata-Ford S, Jonker DJ, et al. K-ras mutations and benefit from cetuximab in advanced colorectal cancer. N Engl J Med 2008;359(17):1757-65.

Katz DA, Carr R, Grimm DR, et al. Organic anion transporting polypeptide 1B1 activity classified by SLCO1B1 genotype influences atrasentan pharmacokinetics. Clin Pharmacol Ther 2006;79(3):186-96.

Kazui M, Nishiya Y, Ishizuka T, et al. Identification of the human cytochrome P450 enzymes involved in the two oxidative steps in the bioactivation of clopidogrel to its pharmacologically active metabolite. Drug Metab Dispos 2010;38(1):92-9.

Kim KA, Park PW, Hong SJ, et al. The effect of CYP2C19 polymorphism on the pharmacokinetics and pharmacodynamics of clopidogrel: a possible mechanism for clopidogrel resistance. Clin Pharmacol Ther 2008;84(2):236-42.

Lenz HJ, Van Cutsem E, Khambata-Ford S, et al. Multicenter phase II and translational study of cetuximab in metastatic colorectal carcinoma refractory to irinotecan, oxaliplatin, and fluoropyrimidines. J Clin Oncol 2006;24(30):4914-21.

Li J, Bluth MH. Pharmacogenomics of drug metabolizing enzymes and transporters: implications for cancer therapy. Pharmgenomics Pers Med 2011;4:11-33.

Lievre A, Bachet JB, Boige V, et al. KRAS mutations as an independent prognostic factor in patients with advanced colorectal cancer treated with cetuximab. J Clin Oncol 2008;26(3):374-9.

Matetzky S, Shenkman B, Guetta V, et al. Clopidogrel resistance is associated with increased risk of recurrent atherothrombotic events in patients with acute myocardial infarction. Circulation 2004;109(25):3171-5.

McArthur GA, Chapman PB, Robert C, et al. Safety and efficacy of vemurafenib in BRAF(V600E) and BRAF(V600K) mutation-positive melanoma (BRIM-3): extended follow-up of a phase 3, randomised, open-label study. Lancet Oncol 2014;15(3):323-32.

Mega JL, Close SL, Wiviott SD, et al. Cytochrome p-450 polymorphisms and response to clopidogrel. N Engl J Med 2009a;360(4):354-62.

Mega JL, Close SL, Wiviott SD, et al. Cytochrome P450 genetic polymorphisms and the response to prasugrel: relationship to pharmacokinetic, pharmacodynamic, and clinical outcomes. Circulation 2009b;119(19):2553-60.

Mega JL, Walker JR, Ruff CT, et al. Genetics and the Clinical Response to Warfarin and Edoxaban. Accepted for publication in Lancet.

Messersmith WA, Ahnen DJ. Targeting EGFR in colorectal cancer. N Engl J Med 2008;359(17):1834-6.

Patterson SD, Cohen N, Karnoub M, et al. Prospective-retrospective biomarker analysis for regulatory consideration: white paper from the industry pharmacogenomics working group. Pharmacogenomics 2011;12(7):939-51.

Rehmel JL, Eckstein JA, Farid NA, et al. Interactions of two major metabolites of prasugrel, a thienopyridine antiplatelet agent, with the cytochromes P450. Drug Metab Dispos 2006;34(4):600-7.

Ricci DS, Broderick ED, Tchelet A, et al. Global requirements for DNA sample collections: results of a survey of 204 ethics committees in 40 countries. Clin Pharmacol Ther 2011;89(4):554-61.

Roses AD, Saunders AM, Huang Y, et al. Complex disease-associated pharmacogenetics: drug efficacy, drug safety, and confirmation of a pathogenetic hypothesis (Alzheimer's disease). Pharmacogenomics J 2007;7(1):10-28.

Ryan TP, Stevens JL, Thomas CE. Strategic applications of toxicogenomics in early drug discovery. Curr Opin Pharmacol 2008;8(5):654-60.

Scannell JW, Blanckley A, Boldon H, et al. Diagnosing the decline in pharmaceutical R&D efficiency. Nat Rev Drug Discov 2012;11(3):191-200.

Suppiah V, Moldovan M, Ahlenstiel G, et al. IL28B is associated with response to chronic hepatitis C interferon-alpha and ribavirin therapy. Nat Genet 2009;41(10):1100-4.

Tanaka Y, Nishida N, Sugiyama M, et al. Genome-wide association of IL28B with response to pegylated interferon-alpha and ribavirin therapy for chronic hepatitis C. Nat Genet 2009;41(10):1105-9.

Thompson JF, Man M, Johnson KJ, et al. An association study of 43 SNPs in 16 candidate genes with atorvastatin response. Pharmacogenomics J 2005;5(6):352-8.

U.S. FDA. Draft Guidance for Industry: Drug Interaction Studies — Study Design, Data Analysis, Implications for Dosing, and Labeling Recommendations, 2012. Available at http://wwwfdagov/downloads/Drugs/GuidanceComplianceRegulatoryInformation/Guidances/ucm292362pdf. Accessed May 17, 2014.

Warner AW, Bhathena A, Gilardi S, et al. Challenges in obtaining adequate genetic sample sets in clinical trials: the perspective of the industry pharmacogenomics working group. Clin Pharmacol Ther 2011;89(4):529-36.

Waters MD, Fostel JM. Toxicogenomics and systems toxicology: aims and prospects. Nat Rev Genet 2004;5(12):936-48.

Wiviott SD, Braunwald E, McCabe CH, et al. Prasugrel versus clopidogrel in patients with acute coronary syndromes. N Engl J Med 2007;357(20):2001-15.

Wiviott SD, Trenk D, Frelinger AL, et al. Prasugrel compared with high loading- and maintenance-dose clopidogrel in patients with planned percutaneous coronary intervention: the Prasugrel in Comparison to Clopidogrel for Inhibition of Platelet Activation and Aggregation-Thrombolysis in Myocardial Infarction 44 trial. Circulation 2007;116(25):2923-32.

Yang Y, Blomme EA, Waring JF. Toxicogenomics in drug discovery: from preclinical studies to clinical trials. Chem Biol Interact 2004;150(1):71-85.

Section 4
Fundamentals of Applied Human Genomics

Chapter 20

PRINCIPLES OF GENETICS AND GENETIC MEDICINE

CAITRIN W. MCDONOUGH, PH.D.

Learning Objectives
1. Assess the central assumptions made in genetics and genetic medicine.
2. Demonstrate how the Human Genome Project and the central dogma of molecular biology have informed genetic medicine.
3. Account for the consequences of genetic polymorphisms, the allele-genotype relationship, and the genotype-phenotype relationship.
4. Distinguish single-gene inheritance patterns from complex (multifactorial) inheritance patterns.
5. Apply the principles of the Hardy-Weinberg equilibrium and of linkage disequilibrium in genetic medicine.
6. Distinguish the technologies and databases used in genomic medicine.

Keywords: Genetics, genetic medicine, DNA, SNP, allele, genotype, phenotype, linkage disequilibrium, GWAS, next-generation sequencing

Abbreviations in This Chapter
3′ UTR	3′ untranslated region
5′ UTR	5′ untranslated region
A	Adenine
C	Cytosine
ENCODE	Encyclopedia of DNA Elements
G	Guanine
GTEx	Genotype-Tissue Expression (project)
GWAS	Genome-wide association study
HWE	Hardy-Weinberg equilibrium
Indel	Insertion/deletion
LD	Linkage disequilibrium
mRNA	Messenger RNA
PCR	Polymerase chain reaction
PharmGKB	Pharmacogenomics Knowledge Base
SNP	Single nucleotide polymorphism
T	Thymine
U	Uracil
VNTR	Variable number tandem repeat

Abstract

Understanding the principles of genetics and genetic medicine is essential in order to appreciate the vastly growing and changing field of pharmacogenomics. This chapter will review the key concepts of genetics and genetic medicine, including the Human Genome Project, the central dogma of molecular biology, and basic genetic nomenclature. This chapter will also examine the different types of genetic polymorphisms and their consequences, the basic inheritance patterns in single-gene genetic disorders, Hardy-Weinberg equilibrium, and the concept of linkage disequilibrium (LD). This chapter concludes by discussing and evaluating some of the current technologies and databases used in genetics and their applications to genomic medicine.

Introduction

As the modern health care paradigm shifts further toward personalized medicine, it is essential that practitioners be familiar with the principles of genetics and genetic medicine, the concepts and vocabulary used to discuss these principles, and the databases and technologies in genomic medicine. This introductory chapter provides a broad overview of genetics and genetic medicine; however, this is an extensive topic with continued ongoing study. The references and list of relevant Web sites should provide additional information and access to new data as this information is presented, developed, and applied to therapeutics. All health care professionals must strive to be lifelong learners. With the advances in pharmacogenomics, pharmacists and pharmacologists are positioned at the forefront of this new era of incorporating genetic medicine into personalized medicine and patient care.

This chapter reviews the central assumptions in genetics and genetic medicine, the Human Genome Project, and the central dogma of molecular biology. In addition, the chapter discusses common terms and concepts used in genetic medicine and different inheritance patterns. Finally, this chapter ends with an examination of some of the technologies that are central to genomic medicine and some of the databases that provide scientists with the tools necessary to conduct clinically meaningful research.

Central Assumptions in Genetics and Genetic Medicine

What Is Genetics?
Genetics is the study of biologic variation. Many nongeneticists may think of variation as something deleterious. However, in the living world, variation is the rule rather than the exception. Without it, species would neither evolve nor survive. In *Homo sapiens*, as in other species, some variations are advantageous and contribute to our survival in certain environments. In addition, some variations are disadvantageous in certain environments and can even influence disease and drug response. Finally, some variations are neutral and neither advantageous nor disadvantageous to our survival but may contribute to the features and traits that make each human being unique.

What Is Genetic Medicine?
Genetic medicine is the study of biological variation that is associated with illness, death, and drug response. Genetic medicine can take many different forms, from the early stages of research investigating biological variation with a certain disease or drug response all the way to the incorporation of this information into patient care.

History of Genetics
There is some evidence that the first documentation of the concepts related to human genetics dates back to Greek civilization in 500 BC, when Pythagoras stated: "Avoid fava beans." Although it is unclear why Pythagoras made this statement, some believe he recognized that some individuals who ate fava beans became sick, whereas others had no adverse effects. However, many consider the timeline of modern genetics to have begun in the 18th or 19th century with Pierre-Louis Moreau de Maupertuis or Charles Darwin. In 1745, Maupertuis proposed an adaptationist account of functional design in his publication *The Earthly Venus,* suggesting that species that "found themselves constructed in such a manner that the parts of the animal were able to satisfy its needs" had a better chance of surviving. Then, almost a century later in 1858, Darwin and Alfred Russel Wallace made a joint announcement of the theory of natural selection, stating that members of a population who are better adapted to the environment will survive and pass on their traits. This was followed in 1859 with Darwin's publication of *On the Origin of Species*, which furthered the theory that populations evolve over the course of generations through the process of natural selection.

Helpful Web Sites

1. The International HapMap Project:

    http://hapmap.ncbi.nlm.nih.gov/

2. The 1000 Genomes Project:

    www.1000genomes.org/

3. The ENCODE Project:

    www.genome.gov/Encode/

    www.nature.com/encode/#/threads

4. GTEx:

    www.broadinstitute.org/gtex/

    http://commonfund.nih.gov/GTEx/index

5. PharmGKB:

    www.pharmgkb.org/

6. NIH National Human Genome Research Institute Talking Glossary of Genetic Terms:

    www.genome.gov/Glossary/

Mendelian Genetics

During the time of Darwin's book and theories on natural selection, Gregor Mendel, who is known as the father of modern genetics, was conducting his famous experiments on heredity in pea plants. Between 1856 and 1863, he studied the inheritance patterns of various traits in pea plants and published the results in 1866. His research allowed him to deduce what he called the first and second laws of heredity. Today, these are known as Mendel's laws of inheritance: the law of segregation and the law of independent assortment. Mendel also proposed that the traits he observed were determined by discrete "factors," which are now called genes. When Mendel's work was published, the findings were all but ignored. It was not until 1902 that three scientists (Hugo de Vries, Carl Correns, and Erich von Tschermak), working independently, discovered and verified Mendel's work. This marked the beginning of a new branch of biology known as genetics.

Makeup of the Human Genome

What Is a Genome?

The first step in studying genetics is to gain an understanding about where genetic material is stored. A genome is the entire genetic makeup of an organism. In *H. sapiens* and all other eukaryotes, most genetic material is located in the nucleus of the cell and is packaged in structures called chromosomes. Humans also have a small mitochondrial genome; however, this chapter will focus on the genome located in the nucleus of the cell, which is contained in the chromosomes.

The Human Genome

Chromosomes

Each chromosome contains a portion of the genome or genetic material, which is also called DNA. Figure 20.1 depicts how a single DNA double helix is wound together. This helix forms complexes with proteins known as histones, as well as with other proteins, which enables it to wrap tightly together to form the structure of the chromosome.

Karyotype

Human beings are diploid organisms, meaning that they have two copies of every chromosome. In total, humans have 46 chromosomes, representing 23 pairs. Twenty-two chromosome pairs are called autosomes, or non–sex chromosomes, and one chromosome pair represents the sex chromosomes. Figure 20.2 shows a normal male karyotype with 22 pairs of autosomes, numbered 1–22,

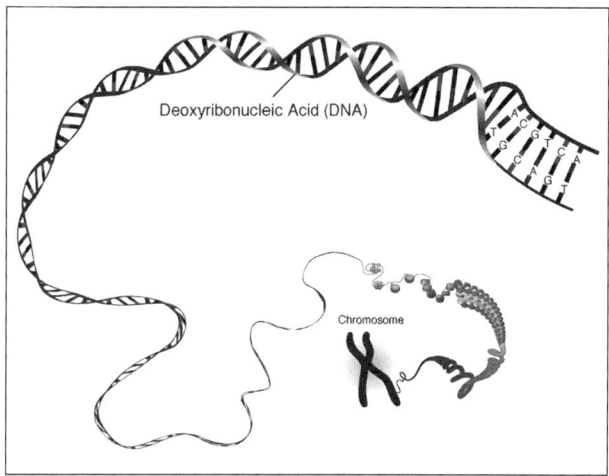

Figure 20.1. Normal chromosome structure.

Reprinted with permission from: National Human Genome Research Institute. Original image by Darryl Leja.

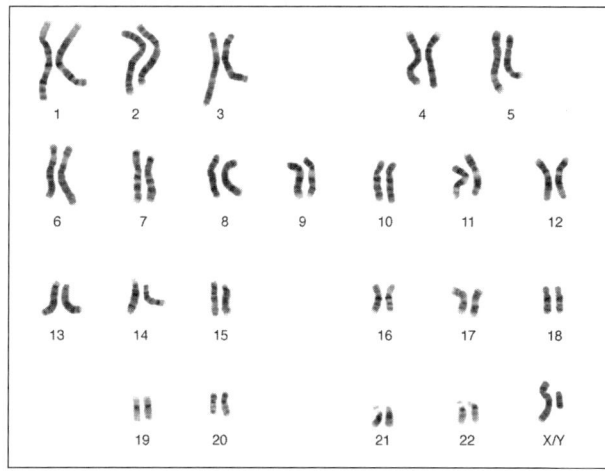

Figure 20.2. Normal male karyotype.

Reprinted with permission from: National Human Genome Research Institute. Original image by Darryl Leja.

and one pair of sex chromosomes, X/Y. As Figure 20.2 shows, males have one X and one Y chromosome, whereas a female karyotype has two X chromosomes.

DNA Characteristics

Structure—Watson and Crick 1953

In a landmark publication in *Nature* on April 25, 1953, James Watson and Francis Crick suggested that the structure of DNA was "two helical chains each coiled around the same axis," or a double helix. They also stated that "this structure has novel features which are of considerable biological interest" and determined the configuration of the DNA components (Watson 1953). The double-helix structure of DNA is

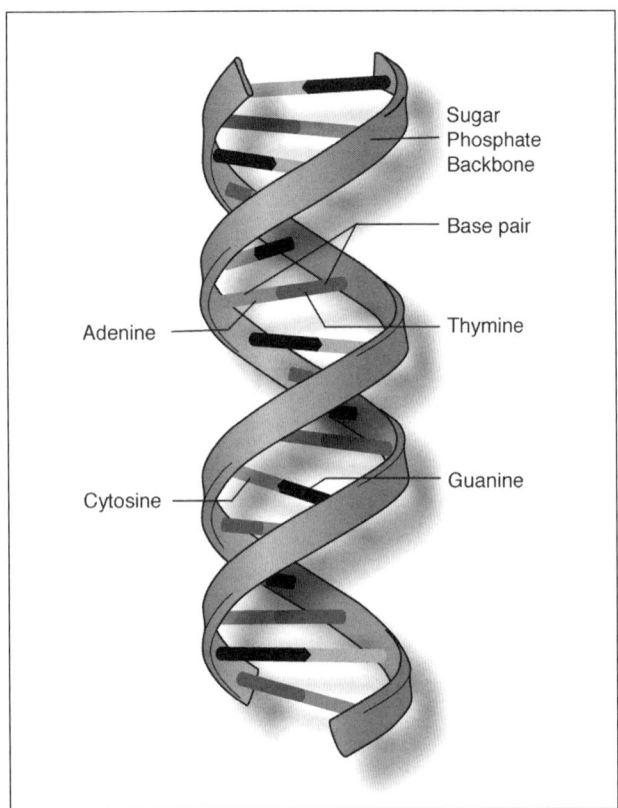

Figure 20.3. Double-helix structure of DNA.

Reprinted with permission from: National Human Genome Research Institute. Original image by Darryl Leja.

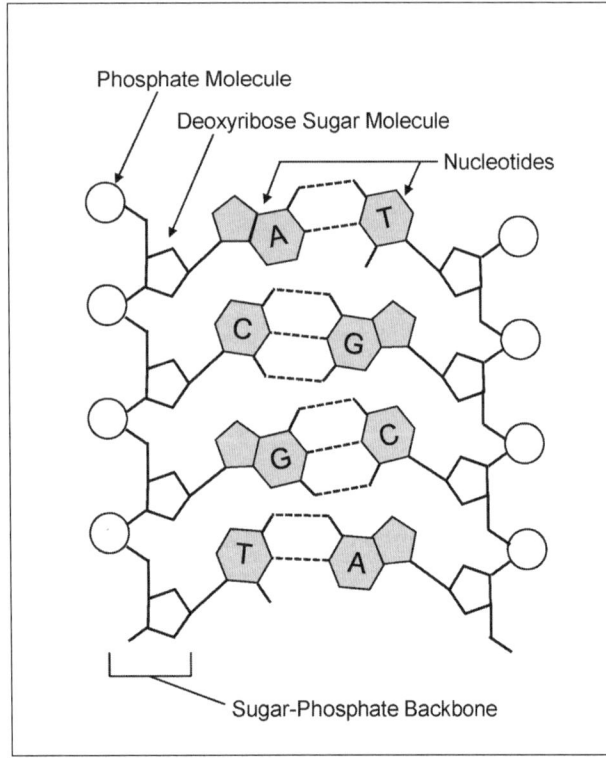

Figure 20.4. The components of DNA.

shown in Figure 20.3. In this structure, there is a nucleotide residue every 0.34 nm, and assuming an angle of 36 degrees between the adjacent residues, the double-helix structure repeats itself after every 10 nucleotide residues, or every 3.4 nm (Watson 1953).

Components

Sugar-Phosphate Backbone

The backbone of DNA is composed of 5-carbon sugars and phosphates. These strands run in the opposite direction, with the 5′ carbon of the sugar exposed on one side of the double helix and the 3′ carbon of the sugar exposed on the other side. The sugar-phosphate backbone is held together through phosphodiester bonds and is negatively charged and hydrophilic, which allows DNA to form bonds with water. Figure 20.3 and Figure 20.4 show the sugar-phosphate backbone.

Four Nucleotides

DNA has only four nucleotides: adenine (A), thymine (T), cytosine (C), and guanine (G). These nucleotides pair together through hydrogen bonds in a very specific manner; A and T always pair together through the formation of two double bonds, and C and G always pair together through the formation of three double bonds. Adenine and G are purine nucleotides, with two aromatic rings fused together, whereas C and T are pyrimidine nucleotides, with a single aromatic ring. The nucleotides pair together such that a purine will always bond with a pyrimidine. This pairing is also shown in Figure 20.3 and Figure 20.4.

The Human Genome Project

Background

The Human Genome Project was a collaboration involving 20 groups from the United States, the United Kingdom, Japan, France, Germany, and China with the ultimate goal of sequencing the entire human genome (Lander 2001). The idea was first proposed during discussions at scientific meetings between 1984 and 1986. The project began in 1990 and was coordinated by the U.S. Department of Energy and the National Institutes of Health (NIH). The draft of the human genome, published in 2001, covered about 94% of the human genome (Lander 2001). The project was completed after 13 years in 2003 (IHGSC 2004). This was ahead of the 15 years the project had been estimated to take. The Human Genome Project was swiftly completed partly because of the development of new technology and partly because of a competing effort in the private sector by Craig Venter and his company

Celera Genomics. The data from Celera Genomics were published during the same week in 2001 as the first draft of the human genome by the Human Genome Project (Venter 2001).

Project Goals
The Human Genome Project had many goals: (1) identify and map all the genes in human DNA, (2) determine the sequences of the 3 billion chemical base pairs that make up human DNA, (3) store this information in databases, (4) improve the current tools for data analysis, (5) transfer related technologies to the private sector, and (6) address the ethical, legal, and social issues that may arise from the project. The project's overall driving force was the global view that the genome could greatly accelerate biomedical research (Lander 2001).

What We Learned from the Human Genome Project
The Human Genome Project provided many lessons. For instance, it confirmed that the human genome contains about 3 billion nucleotide bases. In addition, the data from the project showed that the average gene consists of 3000 bases; however, the sizes of genes vary greatly. At the end of the Human Genome Project, many scientists were surprised to learn that less than 2% of the human genome codes for protein and that the number of genes contained in the human genome is much lower than previous estimates. The project also showed that almost all (99.9%) nucleotide bases are the same in all humans, and the variation that makes each of us different is because of a very small percentage of the genome (Lander 2001).

When the draft of the human genome was published, scientists estimated there were 30,000 protein-coding genes (IHGSC 2004), a number that has continued to decline, with the current estimate at 20,000–21,000 (Pennisi 2012; Clamp 2007). The rest of the genome was deemed "nonfunctional," and although some hypothesized it might play a role in regulation, it was termed *junk DNA*. This turned out to be a huge error, with present data suggesting that 80% of the genome is functional (Pennisi 2012) (see the ENCODE section of this chapter).

Human Genome vs. Other Genomes
The human genome is complex, but genome size does not directly translate to the complexity of the organism. The human genome contains around 3 billion bp; by comparison, the genome of the plant *Paris japonica* contains around 149 billion bp, and the genome of a marbled lungfish that lives in the Nile River contains around 133 billion bp. Furthermore, genome sizes can be very small. The genome of the bacteria *E. coli* (*Escherichia coli*) is

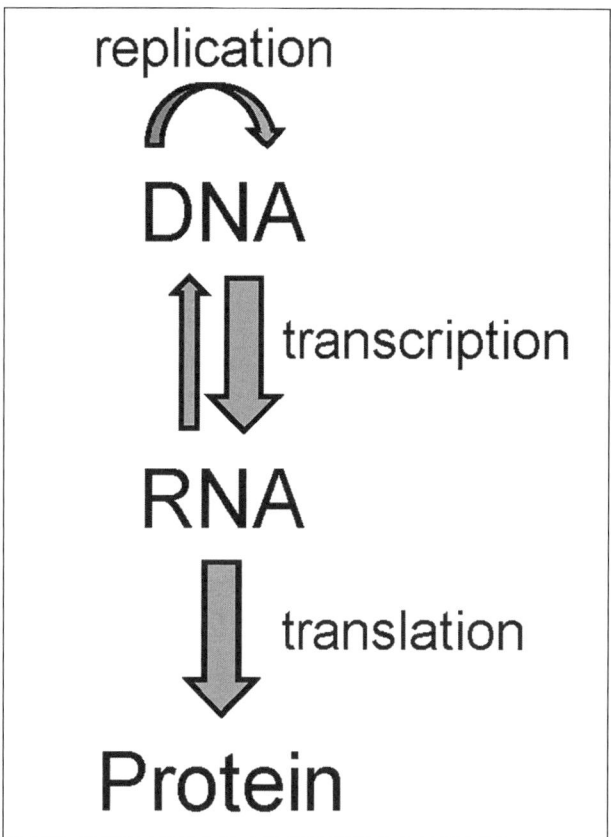

Figure 20.5. The central dogma of molecular biology.

only 4.6 million bp, and the genome of HIV (human immunodeficiency virus) is only 9100 bp and contains only nine protein-coding genes. Overall, it is best not to judge an organism by the size of its genome.

The Central Dogma of Molecular Biology
The central dogma of molecular biology states that (1) DNA is copied to make more DNA through a process known as replication, (2) DNA is converted to RNA through a process known as transcription, and (3) RNA is converted to protein through a process called translation. This pathway is depicted in Figure 20.5. RNA can also be converted back to DNA through a process known as reverse transcription. This process was first discovered in retroviruses (Temin 1970), or viruses that store their genome as a single strand of RNA, and has since been used in laboratories all over the world to convert RNA to DNA.

It is important to note the differences between DNA and RNA. DNA is double-stranded, whereas RNA is single-stranded. DNA contains four nucleotides: A, T, C, and G. Although RNA also contains four nucleotides, uracil (U) replaces T as the base complementary to A. In addition, the 5-carbon sugar is different in DNA

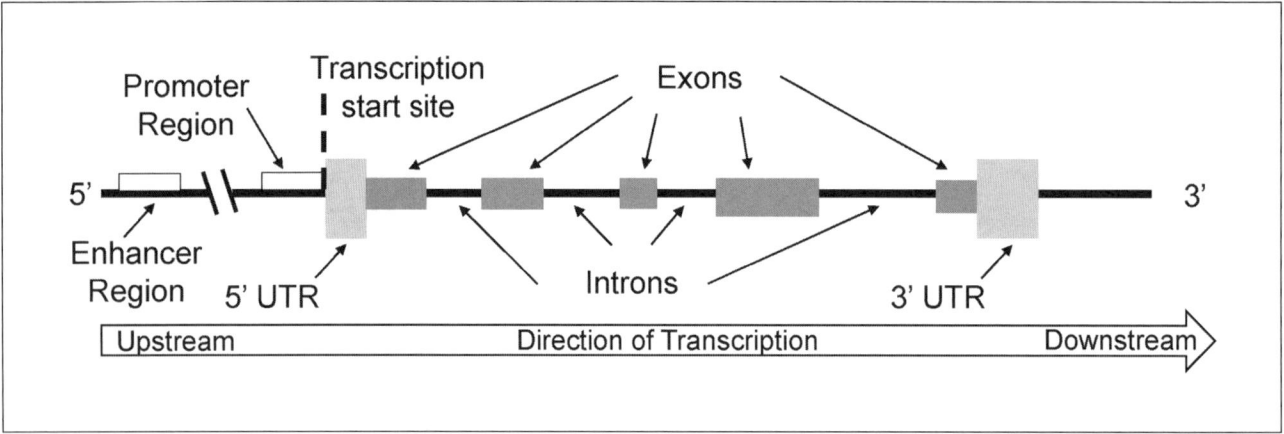

Figure 20.6. The basic structure of the gene. Although a single enhancer region is shown upstream of the coding region for simplicity, several enhancers commonly exist for a single gene, some of which can be quite far from the promoter region, including downstream of the gene.

and RNA. In DNA, the sugar is deoxyribose, whereas in RNA, the sugar is ribose.

Increasing Complexity

It was previously thought that one stretch of DNA, or one gene, was transcribed to make one RNA molecule, which was translated to make one protein. However, after the Human Genome Project was completed and the discovery had been made that there are many fewer genes than expected, scientists realized that the central dogma of molecular biology is in fact more complex. First, many factors can regulate transcription such as epigenetic changes, RNA-DNA interactions, and protein-DNA interactions. Second, there are many ways a single RNA molecule can be processed into messenger RNA (mRNA), which is the form of RNA that brings the genetic code out of the nucleus and into the cytoplasm for translation to occur. This processing is termed *splicing* and is discussed further in the next two sections (Gene Structure and How Genes Become Proteins). Finally, one single protein may be modified to form multiple mature proteins through a process called post-translational modification. Through these complex mechanisms, it is possible for a single stretch of DNA, or a single gene, to have not one but many protein products (Feero 2010).

Gene Structure

Every gene in the genome follows the same basic structure. This structure is composed of exons, the parts of the gene that are ultimately translated to protein, and introns, the parts of the gene that are transcribed from DNA to RNA but then are removed during RNA splicing. Figure 20.6 shows the basic structure of the gene.

Other important features in gene structure include the 5' and 3' untranslated regions (5' UTR and 3' UTR) of the gene. These sections at each end of the gene are transcribed to RNA but are not translated to protein. The transcription start site is at the start of the 5' UTR; however, the translation start site is at the start of exon 1. Other features that are important in gene structure are the promoter and enhancer sites. The promoter site is immediately 5' of the gene and interacts with specific proteins called transcription factors to regulate transcription, or turn it on and off at specific times, in specific cell types. The enhancer region is another type of activation sequence. This region can be a considerable distance away from the transcription start site and can be 5' (shown in Figure 20.6) or 3' of the gene. Enhancer regions work to "enhance" the expression of a specific gene or group of genes and may even work together with other enhancer sites to do so.

How Genes Become Proteins

The first step in the path from gene to protein is the transcription of DNA to RNA. This occurs within the nucleus. The double-stranded DNA molecule opens up to allow one strand to be copied to RNA, by RNA polymerase. After the RNA is transcribed, additional modifications occur; a polyA sequence, or polyA tail, is added at the 3' end of the RNA and splicing occurs. During RNA splicing, the introns are removed, and the exons are joined to form mRNA.

The next step is the translation of RNA to protein. The mRNA travels out of the nucleus into the cytoplasm to the ribosomes, where translation occurs. Within the ribosome, the translation complex reads the mRNA. A very specific code is followed for translation: every three nucleotides in the mRNA correspond to a single codon, which in turn codes for a single amino acid. This code is shown in Figure 20.7; remember that the U in RNA corresponds to a T in DNA.

As the mRNA is read, tRNAs (transfer RNAs) come in to match the mRNA code with the proper amino acid. A specific code, the start codon AUG, codes for the start of translation and the amino acid methionine. In addition, there are three codes—stop codons UAA, UAG, and UGA—that code for the stop of translation. As the mRNA continues to be read, the amino acids join, and the amino acid chain grows, eventually containing all of the amino acids for the mature protein. This process occurs for each gene in the human genome.

Genetic Polymorphisms

Types of Genetic Polymorphisms

The human genome contains variations in the DNA sequence that are different from the reference, or normal sequence at that location. These variations are called polymorphisms. Polymorphisms are likely to play a role in genetic medicine. The four main types of polymorphisms are single nucleotide polymorphisms (SNPs), insertions/deletions (indels), variable number tandem repeats (VNTRs), and CNVs (copy number variants). Figure 20.8 shows an example of each of these types of polymorphisms.

SNPs, pronounced "snip(s)," are the most common sequence variation in the human genome, accounting for about 90% of all known sequence variation. SNPs occur when a change occurs in a single base pair or nucleotide in the genomic sequence. For instance, as Figure 20.8 shows, the sequence GATCTGA would change to GATTTGA. The different base pairs of a SNP are called alleles (discussed more in the next section). On average, SNPs occur every 100–300 bp (Chorley 2008).

Indels are the presence or absence of one or more nucleotides at a specific location. Determining whether

First Base	Second Base				Third Base
	U	C	A	G	
U	UUU - Phe	UCU - Ser	UAU - Tyr	UGU - Cys	U
	UUC - Phe	UCC - Ser	UAC - Tyr	UGC - Cys	C
	UUA - Leu	UCA - Ser	**UAA - STOP**	**UGA - STOP**	A
	UUG - Leu	UCG - Ser	**UAG - STOP**	UGG - Trp	G
C	CUU - Leu	CCU - Pro	CAU - His	CGU - Arg	U
	CUC - Leu	CCC - Pro	CAC - His	CGC - Arg	C
	CUA - Leu	CCA - Pro	CAA - Gln	CGA - Arg	A
	CUG - Leu	CCG - Pro	CAG - Gln	CGG - Arg	G
A	AUU - Ile	ACU - Thr	AAU - Asn	AGU - Ser	U
	AUC - Ile	ACC - Thr	AAC - Asn	AGC - Ser	C
	AUA - Ile	ACA - Thr	AAA - Lys	AGA - Arg	A
	AUG - Met	ACG - Thr	AAG - Lys	AGG - Arg	G
G	GUU - Val	GCU - Ala	GAU - Asp	GGU - Gly	U
	GUC - Val	GCC - Ala	GAC - Asp	GGC - Gly	C
	GUA - Val	GCA - Ala	GAA - Glu	GGA - Gly	A
	GUG - Val	GCG - Ala	GAG - Glu	GGG - Gly	G

Figure 20.7. The amino acid triplet code, written in RNA bases.

A = adenine; C = cytosine; G = guanine; U = uracil.

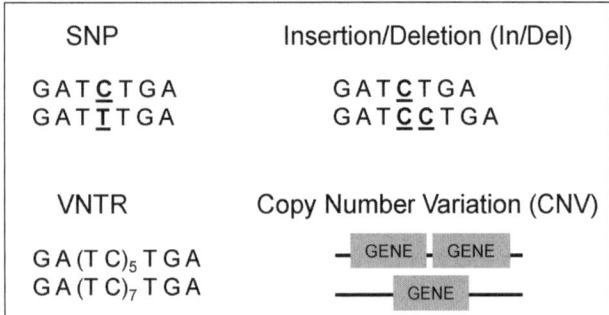

Figure 20.8. Types of genetic polymorphisms. SNP = single nucleotide polymorphism; VNTR = variable number tandem repeat.

the polymorphism is an insertion or a deletion depends on the direction from which the two genomes are being compared. VNTRs are consecutive base pair groups that are differentially repetitive. These can occur in two types: (1) microsatellites, also called simple sequence repeats or short tandem repeats, which consist of 2–6 bp that are repeated a variable number of times; and (2) minisatellites, which are 10–60 bp repeated a variable number of times.

Copy number variants are the fourth type of genetic polymorphism. They refer to the phenomenon when whole genes are deleted, duplicated, triplicated, or repeated at variable numbers. The example in Figure 20.8 shows one chromosome that has a single copy of the gene and another chromosome that has two copies of the same gene.

How Polymorphisms Affect DNA and Proteins

Why are those in the scientific community actively interested in genetic polymorphisms? They are interested because changes in DNA can affect the protein products that are ultimately transcribed. In the simplest form, a change in DNA that occurs in an exon can change the amino acid that is coded for that stretch of DNA. Figure 20.9 depicts three ways a polymorphism in the exon of a gene can affect the protein product. Row A in Figure 20.9 shows the normal DNA sequence at some point in a gene and the way in which that matches to the normal amino acid sequence in the corresponding protein product. In row B, the sixth base pair has been changed from a G to a C; however, this change in base pair does not correspond with a change in the amino acid sequence. This type of SNP is often called a silent mutation or a synonymous SNP because it frequently does not affect the protein product. By contrast, in row C, the fourth base pair was changed from a C to an A. This change, which results in a different amino acid at this position, may change the protein product, depending on where in the amino acid chain this change lies (e.g., if the amino acid is in a critical binding pocket, the change might alter the protein's function). These types of SNPs are called missense SNPs or non-synonymous SNPs because they change the amino acid code. Finally, row D depicts the insertion of 2 bp at positions 4 and 5. This type of polymorphism causes the reading frame, or the way the mRNA is read, to shift, making all amino acids different after the shift. This is called a frameshift polymorphism. In addition, this insertion causes the creation of a premature stop codon, also called a nonsense polymorphism.

Changes in the exons, or regions of DNA that code for protein, are not the only changes that can affect proteins. Polymorphisms in the introns can affect RNA transcription and RNA splicing and processing. In addition, polymorphisms in the promoter and enhancer regions of a gene can affect RNA transcription levels, or gene expression. In turn, this can lead to more or less of the protein product that is translated from the RNA. Other polymorphisms throughout the genome can also affect gene expression; these are commonly called regulatory polymorphisms. Recent studies have shown that 80% of the genome is functional, mostly acting through the regulation of gene expression (Pennisi 2012). Noncoding RNAs, RNA that is transcribed but not translated, are some of the features that can cause this regulation. These include miRNAs (microRNAs), which are RNA molecules that are about 21–24 nucleotides long with sequences that

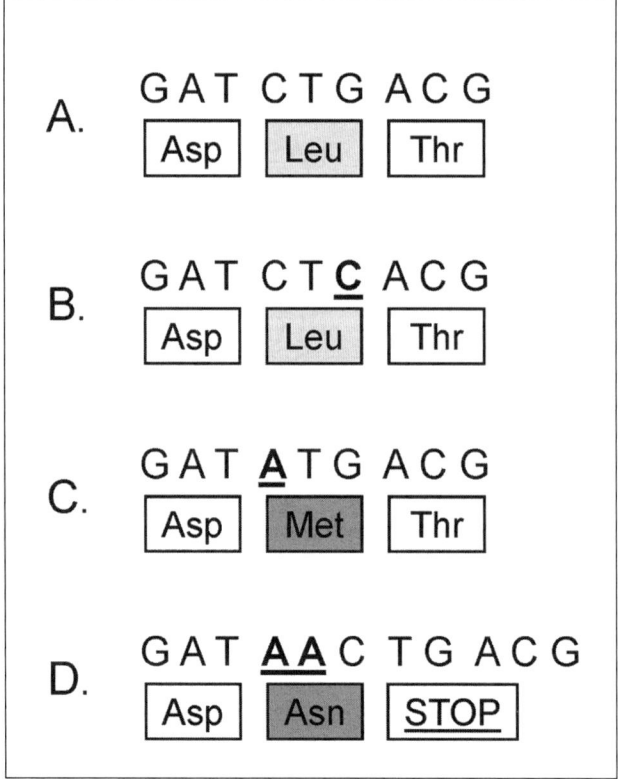

Figure 20.9. How polymorphisms in DNA can affect proteins.

are complementary to mRNA. MicroRNAs function by binding to their corresponding mRNA, which essentially silences the mRNA because it can no longer be translated to protein (Flynt 2008). Another type of noncoding RNAs is the long noncoding RNAs, or noncoding transcripts that are greater than 200 nucleotides long (Rinn 2012). Long noncoding RNAs have the ability to play a large role in gene expression and the organization of the nucleus. Long noncoding may play many other roles as well, acting in complexes with proteins, genes, and even whole chromosomes (Batista 2013). Polymorphisms in these noncoding RNAs, or in regions of the genome that would affect the transcription of noncoding RNAs, could also affect the regulatory roles of noncoding RNAs on the genome.

Genetic Nomenclature

Allele, Genotype, and Phenotype

Allele
An allele is one of two or more versions of a genetic sequence at a particular location in the genome. In the example shown in Figure 20.10, the two versions of the genetic sequence at the location indicated could be "a" and "A." In this case, the allele shown is "a," which is one version of the possible sequence.

Genotype
The term *genotype* can have two definitions in genetics. The first definition, an individual's collection of genes, is the more broad definition. The second definition, two alleles inherited for a particular gene or variation at a particular location in the genome, is the more widely used definition and is how genotype will be used in the rest of this chapter. In the examples shown in Figure 20.10, the genotype is "a/a" at the chromosomal location shown. When there are two alleles at a particular location ("a" and "A"), three genotypes are possible: aa, Aa, and AA. Two of the genotypes are called homozygous (aa and AA) because they are composed of the same alleles, and the third genotype is called heterozygous (Aa) because it is composed of one of each allele.

Phenotype
A phenotype is an individual's observable traits. The key word in this definition is observable. A phenotype is the observable property of an organism. Examples of phenotypes include height, weight, eye color, a medical condition, disease status, and drug response. Eye color is shown as the example phenotype in Figure 20.10.

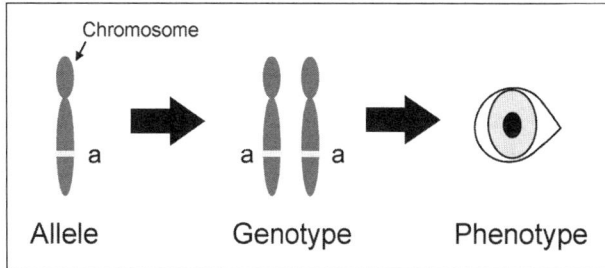

Figure 20.10. The flow from allele to genotype to phenotype.

Allele to Genotype to Phenotype
An easier way to think about all three of these terms is to think about how the three concepts connect. Figure 20.10 shows how a genotype is made of up two alleles and how a genotype at a particular location of the genome can lead to a certain phenotype. For traits and diseases that are caused by a single gene or a single genotype, this is a fairly simple concept. For example, eye color is a trait that follows a relatively easy inheritance pattern, and blue eyes is a recessive phenotype, meaning that to have the blue-eye phenotype, individuals must inherit the "blue-eye allele" from both of their parents (Figure 20.10). The line from allele to genotype to phenotype becomes more complex when scientists are studying complex traits and complex diseases that are influenced by many alleles, in many genes. However, each allele and genotype still contributes to some portion of the phenotype that is displayed. Drug response is one example of a complex phenotype. The intricacies of complex phenotypes and complex (multifactorial) diseases are further discussed in the next section.

SNP Nomenclature
When scientists discuss SNPs, special types of nomenclature are followed. The first type is the basic SNP nomenclature, which refers to a SNP using the following formula: gene position allele 1 > 2. An example of this nomenclature is *CYP2C19* 681 G>A, which indicates a G to A SNP at nucleotide position 681 in the gene *CYP2C19*. Another way SNPs are named is the reference SNP (rs) nomenclature. This naming system is used by the NCBI (National Center for Biotechnology Information) SNP database, where each SNP is assigned an rs number. For instance, *CYP2C19* 681 G>A is also known as rs4244285. The rs nomenclature is commonly used in scientific research and publications.

Allele and Genotype Nomenclature
Similarly, when scientists are discussing a particular allele or genotype, special types of nomenclature are used. Alleles can be referred to through the basic nomenclature system, stating the gene position and allele: *CYP2C19* 681 A. Alleles

can also be specified using the rs nomenclature: rs4244285 A. In pharmacogenomics, another type of allele nomenclature is known as the "star" allele nomenclature. Using the star nomenclature system, a scientist would state the gene*allele number to refer to a specific allele, where the asterisk symbol is pronounced "star." Under this system, CYP2C19 681 A is also known as CYP2C19*2. Of note, star allele nomenclature can look the same for different genes, but the functional effects associated with the star alleles may not be the same. Also of note is that some star alleles describe haplotypes, or combinations of alleles along the same chromosome. Haplotypes are discussed more in the LD and the Structure of the Human Genome section. Because of the complexity of the star allele nomenclature system and the potential confusion and misunderstandings that use of this nomenclature can create, there is a large continuing effort to eliminate the star allele nomenclature in pharmacogenomics and to use only the rs nomenclature.

Genotype nomenclature follows the same rules as allele nomenclature, except that both alleles are included. Under the basic nomenclature, a genotype of G/A at position 681 in CYP2C19 would be denoted CYP2C19 681 G/A. Along the same lines and using the rs nomenclature, the genotype would be rs4244285 G/A, and using the star genotype nomenclature, it would be CYP2C19*1/*2.

Traits, Monogenic Diseases, Complex Diseases, and Inheritance

Mendelian Genetics and Single-Gene Inheritance

Mendel's work during the mid-19th century established the basic principles of heredity through the law of segregation and law of independent assortment. The law of segregation states that in diploid organisms, each parent passes down one randomly selected copy (allele) of a gene to its offspring for any given trait. The law of independent assortment states that separate genes for separate traits are passed down independently from one another, from parent to offspring. These laws make the assumptions that a trait is determined by a single gene and that the traits being studied are on separate chromosomes with the ability to undergo independent assortment. These laws apply to most traits, diseases, and drug-response phenotypes that are attributed to a single gene. In addition, traits that are attributed to a single gene often follow either an autosomal dominant inheritance pattern or an autosomal recessive inheritance pattern.

Autosomal Dominant Inheritance

In an autosomal dominant inheritance pattern, the trait in question is displayed in both the homozygous state (AA) and the heterozygote state (Aa). This means that only one copy of the allele associated with the trait is required for the trait to be displayed. Figure 20.11 shows a pedigree that is affected by a trait that has an autosomal dominant inheritance pattern. Here, members of the pedigree affected by the trait are shaded in gray. In generation I, the father (square) is affected and has a 50:50 chance of passing the trait to each of his offspring. Two of three of the offspring in generation II inherit the trait. The generation II offspring who inherited the trait also have a 50:50 chance of passing the trait to each of their own offspring in generation III. However, the generation II offspring who did not inherit the trait have no chance of passing it on. As the pedigree suggests, the prediction of risk of autosomal dominant traits is analogous to tossing a coin (50:50); however, because each coin toss is an independent event, four tosses in a row may result in four "heads" in a row. One disease that has an autosomal dominant inheritance pattern is Huntington disease.

Autosomal Recessive Inheritance

In an autosomal recessive inheritance pattern, the trait in question is displayed only in the homozygous state (AA). This means that both copies of the allele associated with the trait are required for the trait to be displayed. Figure 20.12 shows a pedigree with an autosomal recessive inheritance pattern. Here, in generation I, both the mother (circle) and the father (square) are carriers of the trait, or heterozygotes (half-shaded); however, because the trait is displayed only in the homozygous state, neither is affected. If each parent carries the trait in an autosomal recessive inheritance pattern, the probabilities of passing the trait to their offspring are 25:50:25 for homozygous unaffected, heterozygous carrier, and homozygous affected, respectively. With these probabilities, there is a 1 in 4 chance that two heterozygous carrier parents will have an affected child. In the pedigree in Figure 20.12, generation II shows three offspring: one homozygous affected, one heterozygous carrier, and one homozygous unaffected. Then, generation III shows that the affected member of generation II has a 100% chance of passing one affected allele to her offspring, whereas the

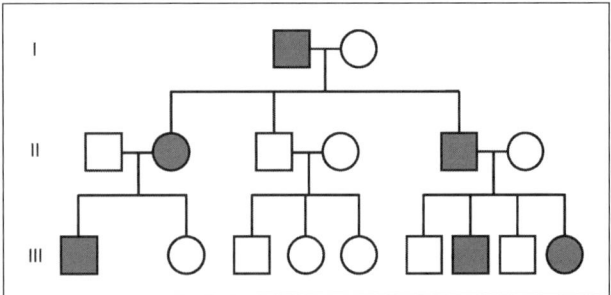

Figure 20.11. Autosomal dominant pedigree.

heterozygous carrier in generation II has a 50% chance of passing the affected allele to his offspring. Cystic fibrosis, sickle cell anemia, and Tay-Sachs disease are examples of diseases with autosomal recessive inheritance patterns.

Complex (Multifactorial) Disease

Most human traits, diseases, and drug responses are complex, or multifactorial, meaning that they arise from the interaction between genetic and non-genetic factors. Usually, the genetic factors occur because of contributions from multiple genes, and the non-genetic factors occur because of contributions from lifestyle (e.g., diet, exercise habits, smoking status) and environment (e.g., toxins, infectious diseases). The inheritance pattern in complex diseases is often not as clear as that in single-gene diseases; however, familial aggregation of complex diseases and traits is often seen. Some examples of complex diseases are hypertension, most types of cancer, type 1 and type 2 diabetes mellitus, and psychiatric disorders.

Although no clear inheritance patterns can be discerned in complex diseases and traits, inheritance patterns can still be observed with single SNPs that are associated with complex diseases and traits. Figure 20.13 depicts the relationship between the genotype at a SNP and the phenotype under three different genetic models. Under an additive model, a classic gene-dose effect occurs, in which there is an increase in the phenotype for each copy of the associated allele. Under a dominant model, only one copy of the associated allele is needed for the full phenotype to be displayed, and under the recessive model, two copies of the associated allele are needed for the phenotype to be displayed.

Single-Gene Inheritance vs. Complex (Multifactorial) Inheritance

Although the classic heredity laws laid out by Mendel are the easiest to understand, they mainly apply to traits and diseases that are caused by a single gene, and most human traits and diseases fall into the complex (multifactorial) category, as stated earlier. One of the main distinctions between the two inheritance patterns lies in the segregation patterns. In both single-gene and complex diseases, the gene causing the disease will segregate in a pedigree. However, in a single-gene disease, the disorder will segregate together with the gene in an autosomal dominant or autosomal recessive inheritance pattern, whereas in complex diseases, the disorders will most likely aggregate in the pedigree without a clear inheritance pattern because other genetic and non-genetic factors could be contributing.

Other factors that are different between single-gene diseases and complex diseases are the numbers of gene involved, role of the environment, age at onset, and risk of disease for other relatives. Single-gene diseases are primarily caused by one gene, as the name implies. In addition, the role of the

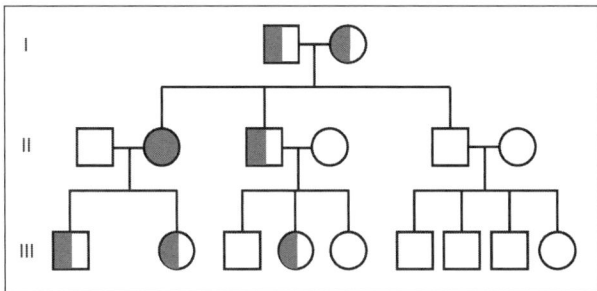

Figure 20.12. Autosomal recessive pedigree.

environment is often small in single-gene disorders, the age at onset is usually younger, and the risk of the disease in other relatives is more predictable. Conversely, complex diseases are usually caused by multiple genes; moreover, those with these diseases are older in age at disease onset, and the risk of the disease in other relatives is less predictable.

Some of the other differences between traits with single-gene inheritance and those with a complex inheritance pattern are the methods that have been successful for studying them. Overall, many methods have been used to study the genetics of disease and drug response. Methods currently being used are discussed in the Technologies and Databases Used in Genomic Medicine section; however, other methods have historically been used to identify genes associated with disease and drug response.

Candidate Gene Analysis

One of the most common methods used historically to identify genes involved in both single-gene and complex

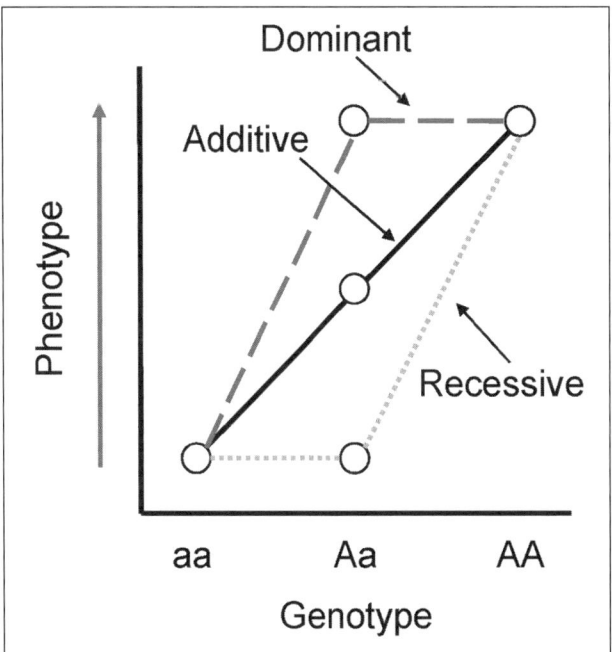

Figure 20.13. SNP (single nucleotide polymorphism) inheritance models.

diseases was the candidate gene approach. In these types of studies, specific genes were chosen for investigation because of a function that could be closely related to the phenotype of interest. Polymorphisms in and/or near the gene that may alter the protein or its expression were chosen; these variants were then genotyped in a population—often, a case-control population—and the allele frequencies were compared (Tabor 2002).

Overall, this approach has been much more successful for single-gene diseases. Although this method has been somewhat successful for complex diseases, it has some drawbacks. First, many candidate genes that have been significantly associated with disease in one population have failed to replicate in other populations. This problem, the failure to replicate, is not unique to candidate gene studies. However, the main reason behind replication failure in candidate gene studies is unique and is the other major drawback. Second, because these types of studies rely on our understanding of the disease and the pathway by which it acts, some believe that existing knowledge is inadequate to formulate proper hypotheses (Tabor 2002). Hence, our current understanding of most disease processes and drug response pathways is incomplete. Thus, when researchers select genes that are thought to be involved with a particular trait, disease, or drug response, they are probably looking at only a very small portion of the genes that are actually involved with the trait, disease, or drug response.

Genome-Wide Linkage Analysis

Genome-wide linkage scans are a type of genetic mapping, with the advantage over candidate gene analysis of uncovering genes and/or genomic regions not previously known to be related to the disease in question. In a genome-wide linkage scan, 200–400 polymorphic markers that are fairly evenly spaced across the genome are selected. These markers are genotyped in families with several cases of the disease. Statistical analysis and linkage analysis are used to locate regions of the genome that are being inherited with, or cosegregating with, the disease. This results in a locus that may contain genes contributing to the disease (Stein 2009; Altshuler 2008). These types of studies have been successfully applied in both single-gene and complex diseases; however, again, they have been more successful at identifying the locus for single-gene diseases.

Hardy-Weinberg Equilibrium

Definition and Derivation

HWE describes the distribution of genotypes for a selected polymorphism within a population. It states that gene frequencies and genotype frequencies will remain constant from generation to generation in an infinitely large interbreeding population in which mating is at random and in which no selection, migration, or mutation occurs. The equation that defines HWE is

$$p^2 + 2pq + q^2 = 1$$

where p is the frequency (Freq) of (A) (one allele), q is the frequency of (a) (the other allele), and $p + q = 1$. Thus, HWE states that from the allele frequencies p and q, it is expected that the genotype frequencies will be as follows: p^2 = Freq (AA), $2pq$ = Freq (Aa), and q^2 = Freq (aa) (Stern 1943; Hardy 1908).

When scientists test to see whether a polymorphism is in HWE, they compare the genotype frequencies they observed from their experiment with the genotype frequencies they would expect, calculated using the HWE equation and the observed allele frequencies. Scientists often compare the observed and expected genotype frequencies using either a chi-square test or an exact test (Wigginton 2005). If the observed genotype frequencies are significantly different from the expected genotype frequencies, the polymorphism is said to deviate from HWE.

What Causes a Polymorphism to Deviate from HWE?

Many things can cause a polymorphism to be out of HWE, including mutation, nonrandom mating, natural selection, limited population size, genetic drift, gene flow, or laboratory error. The first six explanations for a polymorphism deviating from HWE are caused by things a scientist cannot control. However, the last explanation for a polymorphism deviating from HWE, laboratory error, is what scientists are most concerned about. With the current genotyping and sequencing technologies and processes, there is always a chance that an error occurred during genotyping or that a sample contamination or labeling error is causing the polymorphism to deviate from HWE. Thus, when a polymorphism deviates from HWE, it is often assumed that a genotyping error has occurred and that the genotyping data are incorrect. This often causes scientists, and their knowledgeable peers, to question any relationship or association with the disease or drug response that has been observed with the polymorphism that deviates from HWE in the study.

LD and the Structure of the Human Genome

LD and Haplotype Block Structure of the Human Genome

It was previously thought that recombination could occur at any point in the human genome. However, this theory is incorrect. In fact, blocks of genomic DNA are inherited together, with little or no recombination in human history having occurring within the blocks and regions of recombination located between the blocks (Ardlie 2002; Reich 2001a). Figure 20.14 shows a stretch of DNA that is divided into four blocks, depicted by the gray triangles. Historically, the regions of DNA within each block have been inherited together.

Linkage disequilibrium is the tendency for pairs of alleles at nearby loci to be associated with each other more often than expected by chance if the loci were segregating independently in the population (Ardlie 2002; Reich 2001a). Linkage disequilibrium can be formed between two loci when a new mutation occurs at one locus on a chromosome that carries a certain allele at a locus nearby. The combination of alleles that are in LD with each other on a single chromosome is called a haplotype. The degree of LD in a population decays for two main reasons: distance and time. As the recombination distance between two polymorphisms increases, the extent of LD between the two polymorphisms decreases. Similarly, as there are more generations and more possibilities for recombination to occur between two polymorphisms over time, the extent of LD between the two polymorphisms decreases (Ardlie 2002).

Measuring LD

The two most commonly used measures of LD are the absolute value of D' and r^2 (also denoted by Δ^2). Both of these measures of LD are based on Lewontin's D (Lewontin 1964); however, each measures different things and has different properties (Ardlie 2002; Devlin 1995). Lewontin's D is calculated as

$$D = P_{AB} - P_A \times P_B$$

where P_{AB} is the observed frequency of a two-locus haplotype, and P_A and P_B are the frequencies expected if the alleles are segregating randomly (Ardlie 2002; Lewontin 1964). D' is then calculated as

$$D' = D/D_{max}$$

where D_{max} is the absolute maximum D that can be achieved given the allele frequencies at the two loci (Ardlie 2002; Devlin 1995; Lewontin 1964). D' values can range from 0 to 1, with a D' of 1 known as complete LD meaning the two loci have not been separated by recombination. When D' is equal to 1, there are at most only three of the possible four two-locus haplotypes (Ardlie 2002). The measure r^2 is considered by many the measure of choice for examining LD (Ardlie 2002; Devlin 1995; Hill 1994). r^2 is calculated as

$$r^2 = D/(P_{A1}P_{A2}P_{B1}P_{B2})^{1/2}$$

where P_{A1} and P_{A2} are the frequency of alleles 1 and 2 at the first locus, and P_{B1} and P_{B2} are the frequency of alleles

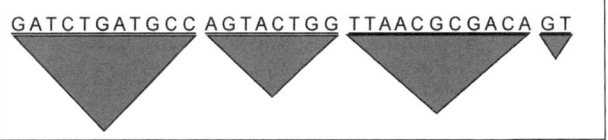

Figure 20.14. Linkage disequilibrium block structure.

1 and 2 at the second locus (Ardlie 2002; Devlin 1995). r^2 values can also range from 0 to 1. When r^2 equals 1, it is known as perfect LD. In this case, the markers have not been separated by recombination, and they have the same allele frequencies. In addition, when r^2 equals 1, only two of the possible four two-locus haplotypes are observed (Ardlie 2002).

When scientists measure LD, they use D' to ask, "Are these two polymorphisms inherited together?" or "Has recombination occurred between these polymorphisms?" However, when scientists use r^2, they are asking two questions: first, "Are these two polymorphisms inherited together?" or "Has recombination occurred between these polymorphisms?" and second, "Can the allele frequencies at one polymorphism be used to predict the allele frequencies at the other polymorphism?"

Structure of the Human Genome

The extent of LD and haplotype block size varies between different populations around the world (Gabriel 2002; Reich 2001a; Clark 1998). The estimated mean block size in Africans and African Americans ranges from 5 kb to 11 kb. Linkage disequilibrium extends farther in other populations, with an estimated mean block size of 22 kb in European and Asian samples and up to 60 kb in a U.S. population of northern European decent (Gabriel 2002; Reich 2001a). However, many larger blocks exist. A more accurate calculation of block size is possible by determining the N50 size—or the length x, where 50% of the genome lies in blocks of x or longer. Africans and African American populations have an N50 block size of 22 kb, whereas European and Asian populations have an N50 block size of 44 kb (Gabriel 2002).

Linkage disequilibrium has been reported to span longer distances, with many haplotype blocks spanning up to 100 kb (Daly 2001). Long-distance LD has also been described. One study found strong LD between markers more than 500 kb apart in regions of chromosome Xq25 and Xq28 (Taillon-Miller 2000). However, LD on a smaller scale (less than 10 kb) is not always as straightforward. When the haplotype structure was determined for a 9.7-kb region through DNA sequencing in 71 subjects, there was strong evidence for recombination in the region; however, extensive LD was also observed (Clark 1998).

The structuring of LD into distinct haplotype blocks may be the result of recombination hot spots and a breakdown of LD (Ardlie 2002; Daly 2001; Jeffreys 2001). This is consistent with other studies that have shown regions of strong LD next to regions where markers are in linkage equilibrium (Ardlie 2002; Taillon-Miller 2000). Together, these results suggest that LD can be influenced by genetic drift, population structure, population growth, temporary reductions in population size or bottlenecks, selection, recombination, and gene conversion (Ardlie 2002; Pritchard 2001; Clark 1998).

LD as a Tool in Genetics

Linkage disequilibrium provides the genetic basis for most linkage and association study strategies. These types of studies rely on the coinheritance of adjacent DNA polymorphisms. Linkage disequilibrium makes it possible to study a select number of polymorphisms and gain information on both the polymorphisms included in the study and the unstudied polymorphisms that are inherited with those being studied (Cardon 2001). This concept is often called tagging, whereby scientists use a genotyped SNP to "tag" an un-genotyped SNP that it is in LD with. The SNPs are often classified as "tagging" SNPs for one another if their pairwise r^2 value is greater than 0.80.

Genomic Medicine

What Is Genomic Medicine?

Genomic medicine is the study of the biological variation, at the whole-genome level, associated with illness, death, and drug response. The terms *genetic medicine* and *genomic medicine* are often used interchangeably, but the word genomic implies that the entirety or majority of the human genome is being studied in relation to illness, death, and drug response; in contrast, the word genetic implies a more focused study of a single polymorphism, gene, or pathway. Genomic medicine also incorporates the concepts of genomic testing, gene therapy, and pharmacogenomics.

Applications of Genomic Medicine

The many applications and potential benefits of genomic medicine are listed in Box 20.1. These applications fall into four broad categories: molecular medicine, microbial genomics, DNA identification, and agriculture. Many of the techniques and technologies that are used in genomic medicine can apply to many other applications, and it is important to remember that many potential benefits from genomic medicine lie outside the medicine realm.

Technologies and Databases Used in Genomic Medicine

Technologies in Genomic Medicine

Polymerase Chain Reaction

Polymerase chain reaction (PCR) is a technique used to amplify specific regions of a DNA strand (Figure 20.15). Developed by Kary Mullis in the 1980s, PCR uses the ability of DNA polymerase to synthesize a new complementary DNA strand, provided the DNA polymerase has a "primer," or starter, to work off. The overall process of PCR is shown in Figure 20.15. It starts with a single strand of DNA, with the region that needs to be amplified. A scientist adds DNA polymerase and primers to the specific region of DNA. One primer should bind upstream or 5′ of the region to be amplified, and the other should be downstream or 3′ of the region. With each cycle of DNA denaturation to separate the strands, hybridization of the primers, and DNA synthesis, the number of copies of the desired sequence doubles. Today, this process has been automated, and within just a few hours, billions of copies of the target DNA can be made.

Polymerase chain reaction, or the amplification of specific regions of a DNA strand, was a very important discovery, and it has many applications in genomic medicine. Many of the experiments conducted today are based on PCR and are reliant on multiple copies of a specific region of DNA. Polymerase chain reaction played a central role in the Human Genome Project, and it is currently used in laboratories around the world in genotyping and sequencing studies. In addition, PCR can be used for DNA fingerprinting in forensics, paternity testing, detection of bacteria or viruses, and diagnosis of genetic disorders (Vosberg 1989).

Reverse Transcriptase

The central dogma of molecular biology states that genetic information flows from DNA to RNA to protein. However, as mentioned previously, RNA can also flow back to DNA through a process known as reverse transcription. This method uses the enzyme reverse transcriptase, originally discovered in retroviruses (Temin 1970), to convert RNA to DNA. In genomic medicine research, this procedure is commonly used to convert mRNA to complementary DNA, which is DNA that is complementary to mRNA and contains only the exons of a gene. This technology allows researchers to study RNA more easily, given that double-stranded DNA is more stable than singled-stranded RNA.

Genome-Wide Association Study

The CD/CV (common disease/common variant) hypothesis states that common diseases in the population are likely influenced by common variants that are shared among members of the population (Reich 2001b). This is in line with evidence that association studies are more powerful than linkage studies in identifying genes that contribute to the risk of common, complex diseases (Risch 1996). In the past decade, advances in high-throughput genotyping have allowed investigators to conduct genome-wide association studies (GWASs), looking at up to 5 million SNPs per individual (Manolio 2010; Stein 2009; Altshuler 2008). Genome-wide association studies allow scientists to study the entire genome at once in an unbiased fashion. These studies are not hypothesis driven and do not rely on our current knowledge of the disease or trait being studied. In this way, scientists can discover novel genes and novel regions of the genome that are associated with the disease or trait of interest. The success of these studies is based on the premise that the markers tested either must be the causal variants or be highly correlated, in LD, with the causal variants. Using this premise as the foundation for GWASs, markers are selected according to the LD patterns observed in the human genome (Hirschhorn 2005). One of the first GWASs identified complement factor H as a novel gene strongly associated with age-related macular degeneration (Klein 2005).

The general study design of a case-control GWAS is shown in Figure 20.16. The first step in a GWAS is to collect DNA from the study participants. For many complex diseases and traits, a case-control study design is used, in which cases with the disease or trait are matched to controls without it. Next, SNPs that are spaced throughout the genome are genotyped. Today, the commonly available GWAS platforms genotype anywhere between 650,000 and 5 million SNPs. Third, statistical analyses are conducted, comparing the allele frequencies at SNPs across the genome with the allele frequencies in controls. These results are commonly shown in a Manhattan plot. In this type of plot, the x-axis represents the chromosomes, 1–22, and the y-axis is usually the negative log (p-value), meaning that the higher the point on the plot, the lower, and more significant, the p-value. Each point plotted on a Manhattan plot represents one SNP, and the dotted line represents the significance level, where SNPs above the line would be considered statistically significant. One issue in GWASs that is commonly addressed is the statistical issue of conducting so many tests (multiple comparisons). Thus, a genome-wide significance level of a p-value less than 5×10^{-8} has been established in the field (Pe'er 2008).

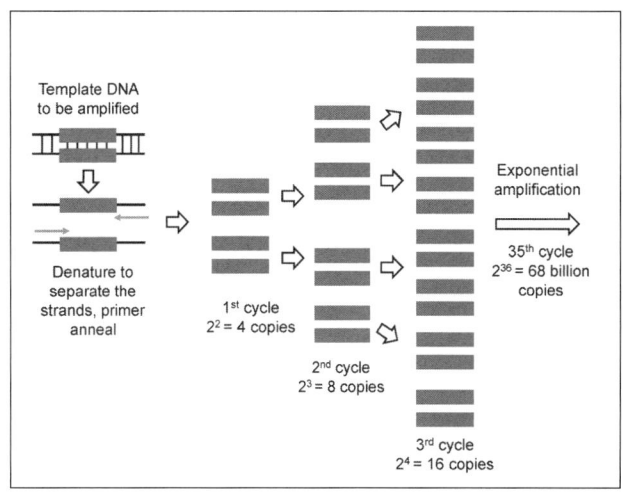

Figure 20.15. Polymerase chain reaction (PCR).

Box 20.1. Applications and Potential Benefits of Genomic Medicine

Molecular Medicine
  Disease diagnosis
  Disease risk prediction
  Drug design
  New drug targets
  Gene therapy
  Pharmacogenomics
  Explain family history of disease and/or drug response

Microbial Genomics
  Detection and treatment of pathogens
  New energy sources—biofuels
  Protection from chemical warfare

DNA Identification—Forensics
  Identify potential suspects
  Exonerate wrongly accused
  Establish paternity
  Match organ donors with recipients
  Authenticate consumables
  Genealogy

Agriculture and Bioprocessing
  Disease; insect- and drought-resistant crops
  Biopesticides
  Edible vaccines incorporated into food products

Another important concept that is used with GWASs is imputation (Li 2009). This method is commonly used in genomic medicine when scientists wish to combine their data sets, but they have used different genotyping platforms and have genotyped different SNPs. Imputation relies on the haplotype block structure of the human genome, in which blocks of DNA are inherited together. Imputation also relies on the data that are available in public databases (see the International HapMap Project and the 1000 Genomes Project sections). Phased haplotypes, which are haplotypes in which the alleles are "phased" or ordered into their positions along one chromosome, are created from the study sample and then are compared with reference phased haplotypes from the public databases, which contain many more SNPs. Using an algorithm, the haplotypes, or blocks of DNA that are inherited together along a chromosome, are matched from the study and the database. Then, the information from the additional SNPs in the database is added to the study (Li 2009). This allows scientists to study more SNPs than they have genotyped and to have a common panel to use to combine their data with those of other scientists.

Next-Generation Sequencing

Although GWASs have been very successful in identifying genomic regions that are associated with many complex diseases, traits, and drug phenotypes, GWASs have not been able to identify many causal variants. Many reasons may account for this, including the lack of casual variants on the GWAS chips, the theory that many causal variants may be rare variants, and the concept of epistasis or the interaction of genes. In addition, GWASs have not been able to identify all of the variants that account for the predicted heritability of a trait. This concept is called "missing heritability" (Manolio 2009; Maher 2008), and many think this heritability may be located in rare variants that contribute to common diseases (Cirulli 2010).

Of interest, while many scientists were wondering where the missing heritability was and how to identify the causal, and possibly rare, variants associated with various diseases and traits, the cost of sequencing was falling rapidly, faster than Moore's law. The Human Genome Project cost around $3 billion, and it took 13 years to complete a single genome. Now, the $1000 genome, generated within 1 day, is in sight (Mardis 2010).

This sudden drop in price was mainly driven by the introduction of new high-throughput sequencing machines, which use a technology called next-generation sequencing (NGS). Next-generation sequencing is a fundamentally different approach to sequencing compared with the classic Sanger sequencing method (also called "first generation"). The NGS technologies use a variety of strategies that all rely on a combination of template preparation, sequencing and imaging, and genome alignment and assembly methods (Metzker 2010). Because these methods rely on the genome being sequenced in hundreds of millions of fragments, it is important that each sample be sequenced a certain number of times to ensure that all regions of the genome are sequenced at least once. This concept is called coverage, and often, at least 30 times coverage is recommended (Schatz 2010). These technologies allow an enormous amount of data to be produced both quickly and cheaply. Next-generation sequencing has the potential to detect and identify lower-frequency variants, or rare variants, which are associated with human disease and drug response, and could possibly locate additional variants to account for the missing heritability in GWASs. Currently, NGS has many applications, with the technology being used to investigate both DNA and RNA.

DNA Sequencing

In genomic medicine, DNA NGS is being used to study both human disease and drug response. These studies use either whole-genome sequencing, in which the entire genome is sequenced, or exome sequencing, in which only the exons that code for protein are sequenced. In one study, exome sequencing revealed that a *CYP3A4* genotype was associated with severely diminished tacrolimus clearance

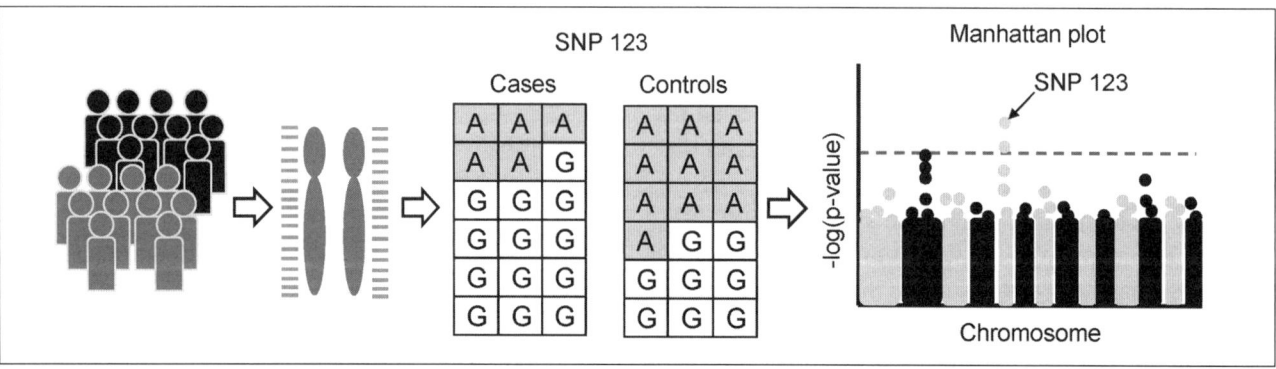

Figure 20.16. Genome-wide association study (GWAS).

in a kidney transplant patient (Werk 2014). In another study, exome sequencing was conducted in individuals with sporadic autism spectrum disorders and their parents. This study identified 21 de novo mutations that could possibly contribute to the cause of the disease (O'Roak 2011). In addition, NGS has been used to identify the previously unknown causes of some rare single-gene disorders, such as Miller syndrome. In a study that conducted whole-exome sequencing of four Miller syndrome cases from three families and eight controls, researchers found the causal gene (*DHODH*) by identifying genes that had coding indels, non-synonymous variants, or splice-site variants in all of the cases, but in none of the controls (Ng 2010).

Next-generation DNA sequencing has been used in other arenas of genomic medicine as well. One example is the use of NGS for the noninvasive assessment of the fetal genome. This process uses cell-free fetal DNA isolated from a maternal blood sample and can screen for fetal aneuploidy (Gregg 2013). Another example is the use of NGS to assess gut microbes and their association with obesity and inflammatory bowel disease (Qiu 2010). In addition, it has been suggested that in the near future, personal genomes will be part of the medical record (Collins 2009).

RNA Sequencing

Next-generation RNA sequencing (also called RNAseq) provides a quantitative method to determine the expression levels of genes in various tissues. These studies fall into the category of "transcriptomics" because they deal with the whole RNA transcript, or transcriptome. Unlike the genome, which is static and unchanging over time, the transcriptome is dynamic and continually changing. The study of the transcriptome may provide information about human disease and drug response beyond what the genome can provide (Pickrell 2010).

The RNAseq method also allows further understanding of the transcriptome through quantifying the splice variants of various genes, identifying transcripts that are coexpressed, and in combination with GWAS SNP data, identifying expression quantitative trait loci, or SNPs, that affect the expression of a gene (Pickrell 2010; Wang 2009). The use of RNAseq in genomic medicine should provide a more extensive understanding of human disease and drug response.

Other "Omic" Technologies

Other technologies are currently available that assess the flow of genetic information at other points along the central dogma of molecular biology. The study of the genome (DNA) and transcriptome (RNA) has already been discussed. In addition, methods are available to study epigenetic influences, such as DNA methylation; the proteome, or the entire set of proteins that are expressed; and the metabolome, which is the complete set of metabolites (small molecules) in an organism. Epigenetic mechanisms involve the modification of gene expression through changes that do not affect the DNA code. These changes almost certainly play a role in human disease and drug response (Egger 2004). Also of great interest is study of the proteome and the metabolome because this may provide a view that assesses both genetic and environmental influences on drug response and human disease (Kaddurah-Daouk 2008; Jungblut 1999).

Databases for Genomic Medicine

The International HapMap Project

The International HapMap Project (called HapMap) was launched in 2002, with the goal of producing a haplotype map of the human genome that would describe the common patterns of human genetic variation. Where the Human Genome Project sought to describe the ways human beings are similar, the HapMap project sought to describe the ways human beings are different. The project was designed to create a public, genome-wide database of these patterns and of the frequency and LD among common SNPs so that this information could be used by scientists around the world in genetic association studies (International HapMap Consortium 2003).

The HapMap project was conducted in phases. The first two phases (HapMap I and HapMap II) included 270 samples from four population panels: CEU, CEPH samples from Utah residents with ancestry from northern and western Europe; YRI samples from Yoruba in Ibadan, Nigeria; JPT samples from the Japanese in Tokyo, Japan; and CHB samples from the Han Chinese in Beijing, China. When phase I was complete, there were data on 1.1 million unique SNPs, with a coverage of around one SNP per 5 kb (International HapMap Consortium 2005). Then, when phase II was completed, there were data on more than 3.1 million SNPs between phase I and phase II, with a coverage of around one SNP per 1 kb (International HapMap Consortium 2007). Finally, HapMap III included 1184 samples and provided data on around 1.6 million SNPs, which represented data from the Affymetrix 6.0 GWAS chip and the Illumina 1M GWAS chip. In addition, HapMap III was extended to 11 population panels, which included the original four panels plus African Americans from the U.S. Southwest (ASW); Chinese from Denver, Colorado (CHD); Gujarati Indians in Houston, Texas (GIH); additional African samples from Kenya (LWK and MKK); individuals of Mexican ancestry in Los Angeles, California (MEL); and Tuscans (those from Tuscany, Italy) (TSI) (International HapMap Consortium 2010).

The International HapMap Project provided much new information. Overall, the four main populations showed generally similar patterns of LD. However, the YRI population had less LD overall and generally shorter haplotype blocks. In addition, on average, more haplotypes occurred per LD block (5.6) in the YRI population than in the CEU, CHB, and JPT populations (4.0). Data from HapMap showed that tagging SNPs (tag SNPs), or SNPs that represent a region of the genome that is in high LD, are generally transferrable between populations as long as the tag SNPs are not rare and are not being transferred to a population with recent African ancestry (Manolio 2008).

Information from HapMap has contributed to genomic medicine. The data have assisted in the design of GWAS chips and been used to build analytic methods and validate laboratory assays. HapMap data have also been used by scientists around the world to select tag SNPs for candidate gene studies and for imputation in GWASs. In this way, HapMap has greatly influenced genetic association studies and genomic medicine (Manolio 2008).

The 1000 Genomes Project

Whereas the HapMap project sought to describe the common patterns of human genetic variation, the goal of the 1000 Genomes Project was to find most genetic variants with a frequency of at least 1% in the populations that were included in the study. To do this, the project, as the name implies, planned to sequence 1000 genomes and sequence them "lightly" using next-generation sequencing.

The project was conducted in phases, with a pilot project and the main project. The pilot project consisted of three studies. The first aimed to assess strategies of sharing data across samples and included low-coverage (4X) whole-genome sequencing of 180 samples. The second sought to assess various coverage, platforms, and sequencing centers and included deeper-coverage (20–60X) whole-genome sequencing of two mother-father-child trios. The last study in the pilot project assessed the methods for gene-region capture by sequencing 1000 gene regions at 50 times (50X) coverage in 900 samples (1000 Genomes Project Consortium 2010).

With the success of the pilot project, the main project continued and will combine whole-genome sequences at 4 times (4x) coverage, array-based genotyping, and deep-targeted exome sequencing in about 2500 samples (1000 Geonomes Project Consortium 2010). The samples included in the 1000 Genomes Project represent five large regions of the world, with samples from East Asia (n=523), South Asia (n=494), the Americas (n=355), Africa (n=691), and Europe (n=514). Data released to date have already been used extensively in genomic medicine for the imputation of GWAS data and for variant calling and novel variant identification in sequencing data.

The ENCODE Project

The Encyclopedia of DNA Elements (ENCODE) project was launched in 2003 after the Human Genome Project had discovered that less than 2% of the human genome codes for protein. The goal of the ENCODE project was to identify all functional elements in the human genome sequence. This project was also conducted in multiple phases. The pilot phase examined the functional elements of around 1% of the genome and had great success, providing evidence that most of the genome is transcribed (ENCODE Project Consortium 2007). With the success of the pilot project, the production phase of ENCODE was funded in 2007, with the task of examining the functional elements of the entire genome and enabling all the data generated by the project to be rapidly released into public databases.

The initial results from ENCODE were published in 30 papers in September 2012, with the resounding message that noncoding DNA is not "junk" (ENCODE Project Consortium 2012; Pennisi 2012). Members of the ENCODE project were able to assign biochemical function to more than 80% of the genome. This portion of the genome has become known as the "regulome," or the DNA elements that modulate or regulate the expression of genes (ENCODE Project Consortium 2012). Data from ENCODE have been used in genomic medicine to assist in understanding the mechanism through which many strong genomic associations may act.

The Genotype-Tissue Expression Project

The Genotype-Tissue Expression (GTEx) project aims to study human gene expression and regulation in multiple tissues, which in turn will provide valuable insights into the mechanism of gene regulation and how these mechanisms are related to disease. In addition, the GTEx portal aims to be a database, with an associated tissue bank, to allow the scientific community to study the relationship between genetic variation and gene expression in human tissues (Genotype-Tissue Expression Project Consortium 2013). This type of database will be very useful in genomic medicine and will help identify genetic variations that are expression quantitative trait loci.

The Pharmacogenomics Knowledge Base

The Pharmacogenomics Knowledge Base (PharmGKB) is a resource that aims to assist scientists in understanding how genetic variation among individuals influences differences in drug response and reactions to drugs (Whirl-Carrillo 2012). A Web-based knowledge base that collects, curates, and disseminates pharmacogenomic information, PharmGKB provides data in various forms: gene variant annotations, drug-centered pathways, and VIP (very important pharmacogene) summaries. The PharmGKB also provides clinical annotations that combine multiple variant annotations into a single annotation that summarizes the relevant information for a drug-phenotype association. Each clinical annotation is assigned a "level-of-evidence" score by the PharmGKB curators. This score is a measure of the confidence in the pharmacogenomic associations that are reported in the clinical annotation (Whirl-Carrillo 2012). Because of the broad types of data that are curated at PharmGKB, it is an excellent resource for both scientists and health care professionals who are interested in pharmacogenomics.

References

1000 Genomes Project Consortium. A map of human genome variation from population-scale sequencing. Nature 2010;467:1061-73.

Altshuler D, Daly MJ, Lander ES. Genetic mapping in human disease. Science 2008;322:881-8.

Ardlie KG, Kruglyak L, Seielstad M. Patterns of linkage disequilibrium in the human genome. Nat Rev Genet 2002;3:299-309.

Batista PJ, Chang HY. Long noncoding RNAs: cellular address codes in development and disease. Cell 2013;152:1298-307.

Cardon LR, Bell JI. Association study designs for complex diseases. Nat Rev Genet 2001;2:91-9.

Chorley BN, Wang X, Campbell MR, et al. Discovery and verification of functional single nucleotide polymorphisms in regulatory genomic regions: current and developing technologies. Mutat Res 2008;659:147-57.

Cirulli ET, Goldstein DB. Uncovering the roles of rare variants in common disease through whole-genome sequencing. Nat Rev Genet 2010;11:415-25.

Clamp M, Fry B, Kamal M, et al. Distinguishing protein-coding and noncoding genes in the human genome. Proc Natl Acad Sci U S A 2007;104:19428-33.

Clark AG, Weiss KM, Nickerson DA, et al. Haplotype structure and population genetic inferences from nucleotide-sequence variation in human lipoprotein lipase. Am J Hum Genet 1998;63:595-612.

Collins F. Opportunities and challenges for the NIH—an interview with Francis Collins. Interview by Robert Steinbrook. N Engl J Med 2009;361:1321-3.

Daly MJ, Rioux JD, Schaffner SF, et al. High-resolution haplotype structure in the human genome. Nat Genet 2001;29:229-32.

Devlin B, Risch N. A comparison of linkage disequilibrium measures for fine-scale mapping. Genomics 1995;29:311-22.

Egger G, Liang G, Aparicio A, et al. Epigenetics in human disease and prospects for epigenetic therapy. Nature 2004;429:457-63.

ENCODE Project Consortium. Identification and analysis of functional elements in 1% of the human genome by the ENCODE pilot project. Nature 2007;447:799-816.

ENCODE Project Consortium. An integrated encyclopedia of DNA elements in the human genome. Nature 2012;489:57-74.

Feero WG, Guttmacher AE, Collins FS. Genomic medicine—an updated primer. N Engl J Med 2010;362:2001-11.

Flynt AS, Lai EC. Biological principles of microRNA-mediated regulation: shared themes amid diversity. Nat Rev Genet 2008;9:831-42.

Gabriel SB, Schaffner SF, Nguyen H, et al. The structure of haplotype blocks in the human genome. Science 2002;296:2225-9.

Genotype-Tissue Expression Project Consortium. The Genotype-Tissue Expression (GTEx) project. Nat Genet 2013;45:580-5.

Gregg AR, Gross SJ, Best RG, et al. ACMG statement on noninvasive prenatal screening for fetal aneuploidy. Genet Med 2013;15:395-8.

Hardy GH. Mendelian proportions in a mixed population. Science 1908;28:49-50.

Hill WG, Weir BS. Maximum-likelihood estimation of gene location by linkage disequilibrium. Am J Hum Genet 1994;54:705-14.

Hirschhorn JN, Daly MJ. Genome-wide association studies for common diseases and complex traits. Nat Rev Genet 2005;6:95-108.

International HapMap Consortium. The International HapMap Project. Nature 2003;426:789-96.

International HapMap Consortium. A haplotype map of the human genome. Nature 2005;437:1299-320.

International HapMap Consortium. A second generation human haplotype map of over 3.1 million SNPs. Nature 2007;449:851-61.

International HapMap Consortium. Altshuler DM, Gibbs RA, Peltonen L, et al. Integrating common and rare genetic variation in diverse human populations. Nature 2010;467:52-8.

International Human Genome Sequencing Consortium (IHGSC). Finishing the euchromatic sequence of the human genome. Nature 2004;431:931-45.

Jeffreys AJ, Kauppi L, Neumann R. Intensely punctate meiotic recombination in the class II region of the major histocompatibility complex. Nat Genet 2001;29:217-22.

Jungblut PR, Zimny-Arndt U, Zeindl-Eberhart E, et al. Proteomics in human disease: cancer, heart and infectious diseases. Electrophoresis 1999;20:2100-10.

Kaddurah-Daouk R, Kristal BS, Weinshilboum RM. Metabolomics: a global biochemical approach to drug response and disease. Annu Rev Pharmacol Toxicol 2008;48:653-83.

Klein RJ, Zeiss C, Chew EY, et al. Complement factor H polymorphism in age-related macular degeneration. Science 2005;308:385-9.

Lander ES, Linton LM, Birren B, et al. Initial sequencing and analysis of the human genome. Nature 2001;409:860-921.

Lewontin RC. The interaction of selection and linkage. I. General considerations; heterotic models. Genetics 1964;49:49-67.

Li Y, Willer C, Sanna S, et al. Genotype imputation. Annu Rev Genomics Hum Genet 2009;10:387-406.

Maher B. Personal genomes: the case of the missing heritability. Nature 2008;456:18-21.

Manolio TA. Genomewide association studies and assessment of the risk of disease. N Engl J Med 2010;363:166-76.

Manolio TA, Brooks LD, Collins FS. A HapMap harvest of insights into the genetics of common disease. J Clin Invest 2008;118:1590-605.

Manolio TA, Collins FS, Cox NJ, et al. Finding the missing heritability of complex diseases. Nature 2009;461:747-53.

Mardis ER. The $1,000 genome, the $100,000 analysis? Genome Med 2010;2:84.

Metzker ML. Sequencing technologies - the next generation. Nat Rev Genet 2010;11:31-46.

Ng SB, Buckingham KJ, Lee C, et al. Exome sequencing identifies the cause of a mendelian disorder. Nat Genet 2010;42:30-5.

O'Roak BJ, Deriziotis P, Lee C, et al. Exome sequencing in sporadic autism spectrum disorders identifies severe de novo mutations. Nat Genet 2011;43:585-9.

Pe'er I, Yelensky R, Altshuler D, et al. Estimation of the multiple testing burden for genomewide association studies of nearly all common variants. Genet Epidemiol 2008;32:381-5.

Pennisi E. Genomics. ENCODE project writes eulogy for junk DNA. Science 2012;337:1159, 61.

Pickrell JK, Marioni JC, Pai AA, et al. Understanding mechanisms underlying human gene expression variation with RNA sequencing. Nature 2010;464:768-72.

Pritchard JK, Przeworski M. Linkage disequilibrium in humans: models and data. Am J Hum Genet 2001;69:1-14.

Qin J, Li R, Raes J, et al. A human gut microbial gene catalogue established by metagenomic sequencing. Nature 2010;464:59-65.

Reich DE, Cargill M, Bolk S, et al. Linkage disequilibrium in the human genome. Nature 2001a;411:199-204.

Reich DE, Lander ES. On the allelic spectrum of human disease. Trends Genet 2001b;17:502-10.

Rinn JL, Chang HY. Genome regulation by long noncoding RNAs. Annu Rev Biochem 2012;81:145-66.

Risch N, Merikangas K. The future of genetic studies of complex human diseases. Science 1996;273:1516-7.

Schatz MC, Delcher AL, Salzberg SL. Assembly of large genomes using second-generation sequencing. Genome Res 2010;20:1165-73.

Stein CM, Elston RC. Finding genes underlying human disease. Clin Genet 2009;75:101-6.

Stern C. The Hardy-Weinberg law. Science 1943;97:137-8.

Tabor HK, Risch NJ, Myers RM. Candidate-gene approaches for studying complex genetic traits: practical considerations. Nat Rev Genet 2002;3:391-7.

Taillon-Miller P, Bauer-Sardina I, Saccone NL, et al. Juxtaposed regions of extensive and minimal linkage disequilibrium in human Xq25 and Xq28. Nat Genet 2000;25:324-8.

Temin HM, Mizutani S. RNA-dependent DNA polymerase in virions of Rous sarcoma virus. Nature 1970;226:1211-3.

Venter JC, Adams MD, Myers EW, et al. The sequence of the human genome. Science 2001;291:1304-51.

Vosberg HP. The polymerase chain reaction: an improved method for the analysis of nucleic acids. Hum Genet 1989;83:1-15.

Wang Z, Gerstein M, Snyder M. RNA-Seq: a revolutionary tool for transcriptomics. Nat Rev Genet 2009;10:57-63.

Watson JD, Crick FH. Molecular structure of nucleic acids; a structure for deoxyribose nucleic acid. Nature 1953;171:737-8.

Werk AN, Lefeldt S, Bruckmueller H, et al. Identification and characterization of a defective *CYP3A4* genotype in a kidney transplant patient with severely diminished tacrolimus clearance. Clin Pharmacol Ther 2014;95:416-22.

Whirl-Carrillo M, McDonagh EM, Hebert JM, et al. Pharmacogenomics knowledge for personalized medicine. Clin Pharmacol Ther 2012;92:414-7.

Wigginton JE, Cutler DJ, Abecasis GR. A note on exact tests of Hardy-Weinberg equilibrium. Am J Hum Genet 2005;76:887-93.

Chapter 21
APPLIED MOLECULAR AND CELLULAR BIOLOGY

Taimour Langaee, Ph.D.; and Issam Hamadeh, Pharm.D.

Learning Objectives
1. Define DNA replication, transcription, and translation processes.
2. Describe the effects of regulatory sequences and proteins on transcription.
3. Synthesize a timeline of the events in RNA processing and messenger RNA translation.
4. Describe the role of microRNAs, siRNAs (short interfering RNAs), and long noncoding RNAs.
5. Understand the role and importance of cell signaling pathways in cell signal transduction.
6. Describe epigenetics and the effects of DNA methylation and histone deacetylation on gene transcription and their implications in therapy.

Keywords: Deoxyribonucleic acid, ribonucleic acid, DNA replication, DNA polymerase, RNA polymerase, transcription, protein synthesis, promoter, transcriptional activator and repressor, gene regulation, noncoding RNAs, miRNA (microRNA), siRNA (short interfering RNA), cell signaling, G protein–coupled receptors, enzyme-linked receptors, epigenetics, DNA methylation, CpG (cytosine-phosphate-guanine) islands, histone acetyltransferases, histone deacetylases, DNA methyltransferases

Abbreviations in This Chapter

A	Adenine	IRE	Iron-responsive element
AC	Adenylate cyclase	IRE-BP	Iron-responsive element binding protein
AG	Adenine-guanine		
ATP	Adenine triphosphate	LBD	Ligand-binding domain
AUG	Adenosine-uracil-guanine	lncRNA	Long noncoding RNA
cAMP	Cyclic adenosine monophosphate	MAPK	Mitogen-activated protein kinase
C	Cytosine	MEK	Mitogen-activated protein kinase kinase
CDK	Cyclin-dependent kinase	miRNA	MicroRNA
CREB	Cyclic adenosine monophosphate responsive element-binding protein	mRNA	Messenger RNA
		ncRNA	Noncoding RNA
CpG	Cytosine-phosphate-guanine	NR	Nuclear receptor
C-terminal	Carboxy terminal	N-terminal	Amino terminal
CU	Cytosine-uracil	PLCβ	Phospholipase C beta
DBD	DNA-binding domain	Pol	Polymerase
DNMT	DNA methyltransferase	rRNA	Ribosomal RNA
EGFR	Epidermal growth factor receptor	RTK	Receptor tyrosine kinase
EIF	Eukaryotic initiation factor	siRNA	Short interfering RNA
G	Guanine	STAT	Signal transducer and activator of transcription
GDP	Guanosine diphosphate		
GPCR	G protein–coupled receptor	T	Thymine
GTP	Guanosine triphosphate	TBP	TATA box-binding protein
GU	Guanine-uracil	TF	Transcription factor
HAT	Histone acetyltransferase	TfR	Transferrin receptor
HDAC	Histone deacetylase	tRNA	Transfer RNA
HRE	Hormone response element	UTR	Untranslated region

Abstract

Ongoing research in molecular and cellular biology has resulted in tremendous advances in the basic, translational, and clinical sciences. Specifically, application of molecular and cell biology principles has furthered human health through improved diagnosis, prevention, and treatment of diseases. Molecular and cellular biology has also played an important role in pharmacogenomics, a fairly new discipline aimed at novel drug development and rational therapeutics based on human genetic variability and epigenetics. As this discipline continues to grow, it becomes increasingly important for clinician-scientists to become familiar with eukaryotic cell structure and function, gene processing and regulation, cell signaling processes, and epigenetics. This chapter highlights the fundamentals for practical application.

Introduction

Molecular and cellular biology is the cornerstone of all biological sciences and has wide applications in different scientific disciplines including medicine, drug development, and pharmacotherapy. New methods for diagnosing, preventing, and treating human diseases have been made possible by understanding and applying the principles of this science. The importance of molecular and cellular biology in pharmacogenomics, for example, cannot be understated. Many drugs used in clinical practice show significant differences in therapeutic efficacy and toxicity, likely partly mediated through genetic, molecular, and cellular variability. Pharmacogenomics applies molecular and cellular biology to the development of tests that can improve the efficacy and safety of drugs to be used in patients. Considering a polymorphism's potential impact on cellular processes may help translational scientists and clinicians alike in determining the likelihood for efficacy and/or toxicity of a given drug in a given patient. Current examples include genotype/protein expression–guided therapy with azathioprine (thiopurine methyltransferase gene) in leukemia and trastuzumab (HER2 [human epidermal growth factor receptor 2] protein) in metastatic breast cancer.

It is a formidable task to present the totality of molecular biology in a single chapter. The first part of this chapter provides the most essential information on eukaryotic gene processing and regulation, starting with DNA replication (regulating the initiation of DNA transcription by cell cycle machinery, close association of DNA replication with cell division). Transcription (transfer of genetic information from DNA to messenger RNA [mRNA], RNA processing, and regulation of eukaryotic gene expression by transcription) and translation or protein synthesis (processes of translation, regulation of translation, and the posttranslational modification of proteins) are subsequently described. The second part of the chapter discusses noncoding RNAs (ncRNAs), including small ncRNAs and long noncoding RNAs (lncRNAs) and their functions. Next, cell signaling processes, together with several drugs that target the cell signaling, are described. In this section, different cell signaling molecules, receptors, and cell signaling pathways are covered. The last part of the chapter describes epigenetics and some important and well-studied components of epigenetics such as histone modifications (acetylation, methylation, phosphorylation, and ubiquitination) and DNA methylation. Throughout, whenever appropriate, we discuss molecular-based drug therapy used in the treatment of diseases associated with the eukaryotic cell systems. In the end, we hope this chapter will enhance the reader's knowledge to a point that it can be used as a background for practical applications.

Background

Composed of subunits called nucleotides, DNA exists as a condensed and compact double-helical structure in the nucleus. Each nucleotide consists of a sugar, a phosphate, and either a purine (adenine [A] or guanine [G]) or a pyrimidine (cytosine [C] or thymine [T]) nitrogenous base. The deoxyribose sugar of DNA is more resistant to chemical degradation than the ribose sugar of RNA. This stabilizes the sugar-phosphate backbone of DNA, making the DNA molecule more durable than that of RNA.

Each strand of the DNA double helix has a 5′ and a 3′ end. At the 5′ end of a DNA strand, a phosphate group is attached to carbon 5 of deoxyribose. At the 3′ end, DNA has a hydroxyl group on carbon 3 of deoxyribose that is free to bind to other molecules (Cooper 1997a). The double-helix DNA strands are bound together by adenine-thymine (A-T) and cytosine-guanine (C-G) base pairing in a head-to-tail or antiparallel form (5′→3′ and 3′→5′). The bases of DNA molecules are concealed within the DNA helix, which protects them from chemical attack.

Gene Processing and Regulation

DNA Replication

The transfer of genetic information from parent to daughter cells requires the accurate and strict replication of genomic DNA (Stillman 1996). DNA replication is semiconservative given that one parent strand always serves as the template for the new complementary daughter strand (Dutta 1997). In a newly synthesized DNA molecule, the double helix is composed of an old (parent) and a new (daughter) strand. For DNA replication, several proteins are required to help facilitate the unwinding and separation of the double-stranded DNA molecule. These proteins are essential because DNA replication can occur only when DNA is single-stranded (Ogawa 1980).

Proteins Involved in DNA Replication

Topoisomerases

Topoisomerases are enzymes that act on DNA to isomerize, or convert, one topological form of DNA into another. Topoisomerases decrease the degree of DNA supercoiling (the form that DNA naturally takes in vivo) by cutting DNA strands and are involved in primary cellular functions such as replication, transcription, chromosome condensation, and maintenance of genomic stability. The two types of topoisomerases in eukaryotic cells are designated type I (IA and IB) and II. Type I topoisomerase cuts and rejoins one strand of duplex DNA in a reversible and non–adenosine triphosphate (ATP)-dependent process, resulting in relaxation of supercoiled DNA. Type II topoisomerase cleaves and rejoins the two strands of double-helix DNA in a reversible, ATP-dependent process (Pulleyblank 1997; Wang 1996).

Because these enzymes can relieve the torsional stress in DNA molecules, they are involved in both DNA replication and transcription. Topoisomerase initiates the unwinding of the DNA double helix (which is normally kept in a coiled and supercoiled state) by cleaving a phosphodiester bond in one strand. This creates a gap in the strand, relieving the conformational tension intrinsic to double-stranded DNA. This is a crucial first step in replication. These broken DNA strands can be rejoined as the phosphodiester bond forms again (Pulleyblank 1997; Wang 1996). Without topoisomerase enzymes, DNA helical unwinding and subsequent DNA replication could not occur. In fact, many commonly used antineoplastic agents, such as daunorubicin, doxorubicin, etoposide, irinotecan, and others, exploit the necessity of topoisomerases and work by interfering with the actions of topoisomerases, preventing DNA replication of malignant cells.

DNA Helicases

DNA helicases are responsible for unwinding double-stranded DNA into single-stranded DNA, a perquisite for processes such as RNA synthesis (discussed later) and homologous DNA recombination. After uncoiling the DNA supercoil by topoisomerase, helicases separate the double-stranded DNA by an energy-dependent process. To inhibit the separated strands from reannealing, single-stranded DNA-binding proteins bind to both separated strands of DNA and stabilize the single-stranded structure generated by helicases (Waga 1998; Bambara 1997; Wold 1997).

DNA Polymerases

DNA polymerases (pols) are enzymes that catalyze the synthesis of new DNA strands. DNA pols bind to the DNA template (parent strand) and synthesize DNA in the $5'\rightarrow 3'$ direction by adding a deoxyribonucleoside 5'-triphosphate (dNTP)—a molecule composed of a sugar, three phosphates, and a deoxygenated purine or pyrimidine—to the 3' hydroxyl group of the growing chain. In addition to DNA replication, DNA pols are involved in cell cycle regulation, DNA repair, DNA recombination, and telomere maintenance (Hubscher 2002; Takemura 2002).

Five DNA pols (α, β, γ, δ, and ε) in eukaryotic cells are designated classical pols that share a similar active site. DNA pols α, δ, and ε are believed to be responsible for nuclear genomic DNA replication. The classical pols are grouped into five different families (A, B, C, X, and Y) according to their structural similarities and sequence homology, with the replicative pols (α, δ, and ε) grouped in family B (Hubscher 2002; Livneh 2001).

New pols (ζ, η, θ, ι, κ, λ, μ, σ, φ, and rev1) that have been recently discovered are called novel or lesion-replicating pol enzymes. Whereas the classical pols show high accuracy and fidelity during DNA synthesis, the novel pols are not as accurate in their performance. The presence of lesions in DNA during the replication process can interfere with DNA synthesis (Hubscher 2002; Yavuz 2002; Washington 2001).

Many pols have an associated proofreading exonuclease responsibility for excising mismatched nucleotides that, if uncorrected, may manifest as clinical disorders. For example, xeroderma pigmentosum (XP) is a rare autosomal recessive genetic disorder that renders a patient highly prone to UV-induced skin cancer and is associated with defective DNA repair (deficiency in UV-specific exonuclease, which removes pyrimidine dimers during repair of UV-damaged DNA). In contrast to many patients with XP who have defective nucleotide excision repair, patients with the XP-variant have normal nucleotide excision repair but, because of a mutation in their DNA pol η, replicate UV-damaged DNA (Hubscher 2002; Yavuz 2002; Livneh 2001; Washington 2001; Tanaka 1994).

Origins and Initiation of Replication

In eukaryotic cells, the replication of DNA begins at a unique sequence called the origin of replication. To replicate DNA in a biologically timely fashion, multiple origins of replication are necessary. For example, for mammalian cells to replicate their genome within a few hours, multiple replication origins in these cells are essential. It is estimated that the human genome may contain 30,000 origins of replication (Twyman 1998; Cooper 1997a; Dutta 1997).

Recent studies suggest that the initiation of DNA replication is regulated by cell cycle machinery. Even though DNA replication occurs in the S phase of the cell cycle, the initiation complex may form at the origin of replication during the transition from M (mitosis) to G1 (G stands for gap in DNA synthesis) and await the signal to start replication. When cells progress from the G1 to the S phase, the initiation complex is activated and replication begins (Dutta 1997). One of the main regulators of this transition is the level of cyclin-dependent kinase (CDK) activity in the cell. Cells in the G1 phase show low CDK activity, which is essential for the assembly of the prereplication complex at the origins of DNA replication. The activation of CDK results in the transition from the G1 phase to the S phase, initiation of DNA replication, and disassembly of the prereplication complex at the origin of replication. The inhibition of CDK activity resets the cell cycle to the G1 phase (Hubscher 2002; Ogawa 1980). Although the details of eukaryotic DNA replication are not as well understood as those of DNA transcription and protein synthesis, it is clear that multiple origins of replication are necessary and that the process of DNA replication is closely related to the progression of the cell cycle.

DNA Replication Fork

The DNA replication fork is another component of the DNA replication machinery, containing several proteins with different functions. Helicase proteins unwind the DNA double helix to generate the replication fork. Accessory proteins (e.g., replication factor C, proliferating cell nuclear antigen, and replication factor A) regulate the interaction of pols with DNA. Finally, DNA pols synthesize the new DNA strands, and exonucleases have 3′→5′ proofreading activity (Waga 1998; Stillman 1994).

DNA replication starts at the specific chromosomal origin of replication and continues to the terminus. Because the two strands of DNA double helix are antiparallel, one strand needs to be synthesized in the 5′→3′direction and the other in then 3′→5′ direction. Because all DNA pols catalyze the polymerization of dNTPs only in the 5′→3′ orientation, one strand, designated the leading strand, can be synthesized continuously in the direction of the replication fork. The other strand, or lagging strand, is synthesized backward in relation to the movement of the replication fork (Figure 21.1). Because DNA pol cannot initiate the synthesis of a new DNA chain de novo and requires a 3′-OH terminus of an oligonucleotide primer, a short fragment of RNA-DNA hybrid (10 RNA bases, followed by 20- to 30-bp-long DNA) is synthesized by pol α/primase (an RNA primase enzyme that catalyzes the synthesis of short RNA primer molecules). Subsequently, this RNA-DNA hybrid oligonucleotide is taken over by DNA pols δ and ε for elongation on both the leading and lagging strands. On the lagging strand, short discontinuous segments of DNA (about 200 bp), called Okazaki fragments, are synthesized from the RNA primers created by primase. These RNA primers are removed by the exonuclease activity of DNA pol I/α, and the gaps between Okazaki fragments are then filled by DNA pol I/α pol activity. Finally, the completed Okazaki fragments are joined by DNA ligase (Baker 1998; Waga 1998; Bambara 1997; Yuzhakov 1996; Ogawa 1980).

Transcription

For cells to carry out their essential life processes such as reproduction, growth, and metabolism, they must continuously synthesize proteins. Protein synthesis requires precise transfer of information from DNA strands into amino acid sequences in proteins. The first step in this process of gene expression and protein synthesis is called transcription. Transcription is the process of synthesizing RNA from the information in DNA nucleotide sequences. RNA is synthesized in the 5′→3′ direction from a DNA strand that runs in the 3′→5′ direction. In a eukaryotic gene, only one strand of DNA (the complementary or antisense strand) is used for DNA replication and mRNA synthesis. The sequence of the RNA transcript is the same as for the top (sense) strand in double-stranded DNA. Transcription can be either constitutive, in which genes are transcribed in a continuous permanent manner, or regulated, in which many different extracellular and intracellular signals control RNA synthesis (Conaway 2000; Tjian 1995).

RNA Pols

RNA pols are huge multi-subunit protein complexes that catalyze the assembly of RNA from their DNA templates. Three different nuclear DNA-dependent RNA pols in eukaryotic cells transcribe various classes of genes. RNA pol I transcribes ribosomal RNA [rRNA] genes for the precursors of the 28S, 18S, and 5.8S molecules, which represent most of the RNA synthesized in transcription. RNA pol II is a megadalton-sized holoenzyme complex whose exact composition and structure have not been fully elucidated. RNA pol II transcribes all the protein-coding mRNA and several small nuclear RNA genes, but it cannot initiate transcription without the presence of several transcriptional factors. RNA pol III transcribes the 5S rRNA and all the transfer RNA (tRNA) genes involved in translation (Asturias 1999; Myer 1998; Young 1991).

Eukaryotic Promoters

Promoters are sequences on DNA that determine the site of transcription initiation and the direction of transcription. Moreover, they are sites on DNA that are upstream (5′) of the coding region to which RNA pol (and other molecules such as transcription factors [TFs]) binds to initiate transcription. Eukaryotic promoters are very diverse and may have regulatory elements that are several thousand bases away from the transcriptional start site.

Cis- and Trans-Acting Regulatory Elements

The promoter elements or DNA sequences that reside on the same molecule of DNA as the genes being transcribed are called *cis*-acting elements and serve as binding sites for general TFs. The TATA box, GC box, and CAAT box are all examples of *cis*-acting elements that are recognized by transcriptional factors. Transcriptional factors are considered *trans*-acting elements and, in contrast to *cis*-acting elements, are encoded by genes that are located on different chromosomes that migrate to their site of action. These *trans*-acting elements can either stimulate or inhibit the transcription of a specific gene.

There are two classes of basal, or core, promoter. The first important class consists only of TATA box, a 7-base sequence (5′-TATWAW-3′, where W = A or T), which is located about –25 bases from the transcription start point (+1). This sequence binds to the multiprotein transcription factor IID (TFIID), which is the TF complex that must bind for transcription to begin. The second class of basal or core promoters do not have a TATA box (TATA-less promoters), and the precise position of the transcriptional start site is determined by the initiator sequence. The consensus initiator sequence is 5′-PyPyA^{+1}NA/TPyPyPy-3′, where A is the transcription start point (+1), Py is pyrimidine (C or T), and N is any base. The TATA box is usually surrounded by guanidine (G) and cytidine (C) nucleotides. All protein-encoding genes have a basal or core promoter (Wu 2001; Pedersen 1999; Nikolov 1996; Roeder 1996; Tjian 1995).

The GC box (GGGCGG) located upstream of the transcription start site is found in a variety of genes,

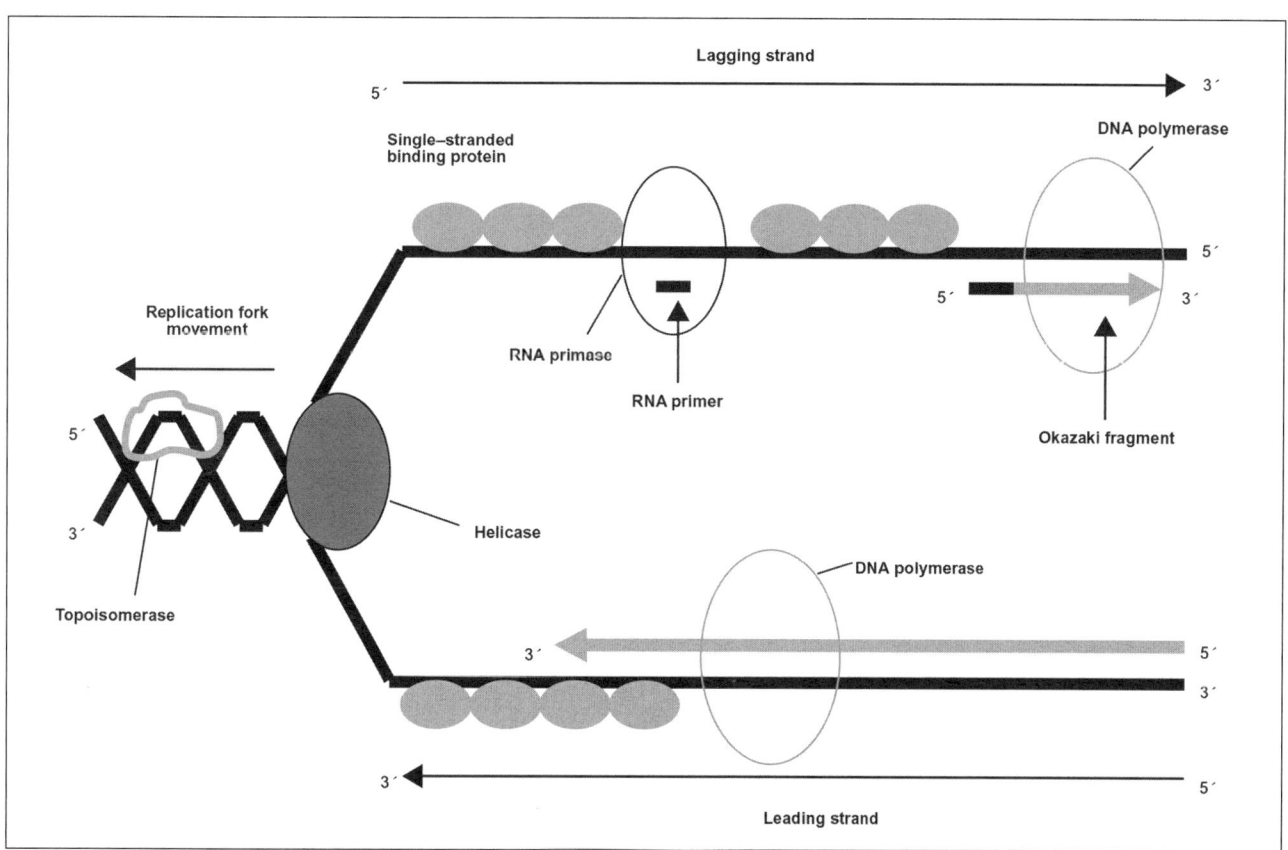

Figure 21.1. DNA replication in eukaryotic cells. DNA replication occurs in the S phase of the cell cycle. Topoisomerase relieves the stress intrinsic to the DNA double helix. Part of the double helix is unwound by helicase. DNA polymerase binds to one template strand and moves in the 3′→5′ direction, creating the complementary leading strand and reforming a double helix. Because DNA synthesis occurs only in the 5′→3′ direction, a second DNA polymerase binds to the other template strand and creates discontinuous polynucleotide strands (Okazaki fragments). DNA ligase then joins these fragments to the lagging strand.

many lacking a TATA box (e.g., "housekeeping" genes [constantly expressed genes]). The GC box is usually associated with SpI TF and may function in either orientation (forward or reverse).

The CAAT box (CCAAT), located about 50 bp upstream from the TATA box (or –80 bp from the transcription start site [+1]), usually determines the promoter efficiency. The CAAT box is associated with different TFs and, like the GC box, may function in either orientation (Pedersen 1999; Werner 1999; Tjian 1995).

Variations in the promoter region play important roles in pharmacogenomics, and many examples are reported in the literature. Some examples include C-514T in hepatic lipase (*LIPC*), where the CC genotype is associated with increased response to statins; (TA)n repeats (*28) in the TATA box of uridine diphosphate-glucuronosyltransferase 1A (*UGT1A1*) have been associated with a higher risk of diarrhea and leukopenia during irinotecan therapy in patients with cancer; promoter insertion/deletion (long/short) variations in the serotonin transporter (*SLC6A4*), with the long/long genotypes associated with a better response to fluoxetine or paroxetine (Goldstein 2003).

Stages of Transcription

The process of transcription in eukaryotes can be divided into initiation (preinitiation complex assembly, promoter clearance), elongation, and termination. Initiation of transcription is complex and takes place at the promoter. Because RNA pol II cannot recognize the promoter or initiate transcription by itself, several general TFs—TFIIA, TFIIB, TFIID, TFIIE, TFIIF, and TFIIH—assist RNA pol II in recognizing its target. The TFs first form a complex with DNA and then recruit RNA pol II to the DNA at the transcription initiation site (Dvir 2001; Werner 1999; Tjian 1995).

In the preinitiation step, the TATA box-binding protein (TBP) of TFIID recognizes and binds to the TATA element of the promoter. This interaction may be regulated by another transcription factor, TFIIA, which interacts with the TBP and other TBP-associated factors in TFIID. These TBP-associated factors, or TAFs, are cofactors that associate with TBP and help in DNA unwinding and transition from transcription initiation to elongation of the RNA transcript. At first, different TFs bind to the upstream promoter and form a multiprotein complex with DNA. Then, RNA pol II is recruited to the DNA at the transcriptional initiation site. Transcription factor IIB promotes the binding of the TFIID-TBP complex to the promoter, subsequently establishing a site for RNA pol II association with the DNA template. Binding of TFIIF to the latter complex results in recruiting both TFIIE and TFIIH into the complex, which are necessary for RNA elongation (Figure 21.2) (Dvir 2001; Roeder 1996; Tjian 1995; Buratowski 1994).

Elongation

The synthesis of the RNA transcript is not a nonstop process in which nucleotides are added at a constant rate. There are different transcription elongation blocks that RNA pol must overcome for efficient RNA synthesis. After forming the preinitiation complex (promoter, RNA pol II, and TFs), the DNA helicase activity of TFIIH unwinds the DNA template at the transcriptional start site through an ATP-dependent process. The DNA melting begins at about –10 bp upstream from the first nucleotide to be transcribed. The preinitiation complex is believed to convert the closed DNA configuration to an open complex before the initiation of RNA synthesis by RNA pol II. The C-terminal domain of the large subunit of RNA pol II, which is rich in proline, serine, and threonine residues, plays an important role in elongation. Regions rich in these amino acids are found in transcriptional activation domains (Bentley 2002; Conaway 2000; Roeder 1996).

Transcription factor IIF is bound to unphosphorylated RNA pol II, allowing RNA pol II to recognize the TFII complex. During the initiation step, the unphosphorylated C-terminal domain becomes phosphorylated by TFIIH, resulting in the release of the RNA pol II from the initiation complex, clearance from the promoter, and initiation of transcription. RNA pol II uses nucleoside triphosphates (NTPs) for RNA synthesis. These NTPs are assembled onto the strand by complementary base pairing. Because there is no thymine in RNA, each A on the DNA guides the insertion of the pyrimidine uracil (U, from uridine triphosphate, UTP). RNA pol II, in conjunction with transcription elongation factors, continues down the DNA template strand in the 3'→5' direction at about 30 bp/second, continually assembling the strand of RNA in a 5'→3' fashion. RNA pol II transcription continues until a termination signal on the template DNA strand is read, resulting in the release of both the enzyme and the RNA transcript from the DNA strand (Korzheva 2001; Conaway 2000; Lodish 2000; Cooper 1997b; Zawel 1995).

Termination

The processed eukaryotic mRNA contains a polyA addition signal (AAUAAA) that is located downstream of the last exon at the 3' end, followed by a series of adenines. The polyA signal is responsible for polyadenylation (addition of adenylate residues during RNA processing), termination of transcription, and release of RNA pol II from the DNA template molecule. Although the upstream RNA transcript undergoes posttranscriptional modification after cleavage at the polyA site, the unstable downstream RNA transcript is degraded soon after synthesis (Bentley 2002; Korzheva 2001; Burley 1996).

RNA Processing

In eukaryotes, all nuclear pre-mRNA transcripts must undergo processing and change to functional mRNA before being exported to the cytoplasm. The primary precursor of mRNA transcripts is called heterogeneous nuclear RNA (hnRNA), which contains exons and introns (coding and noncoding regions, respectively). The maturation process of hnRNA takes place in three steps (5′ capping, 3′-polyadenylation, and mRNA splicing) in the nucleus (Figure 21.3). After RNA processing, the mature and functional mRNAs are transported to the cytoplasm to be translated into proteins by ribosomes (Bentley 2002; Herbert 1999; McKeown 1992).

5′ Capping

In all eukaryotes, mRNA synthesized by RNA pol II has a cap structure at its 5′ end, which consists of a 7-methlyguanosine linked with the first nucleotide of the mRNA. The 7-methylguanosine is added to the 5′ end when the length of the RNA transcript reaches 25–30 nucleotides by RNA pol II. A capping enzyme associated with the phosphorylated C-terminal domain of RNA pol

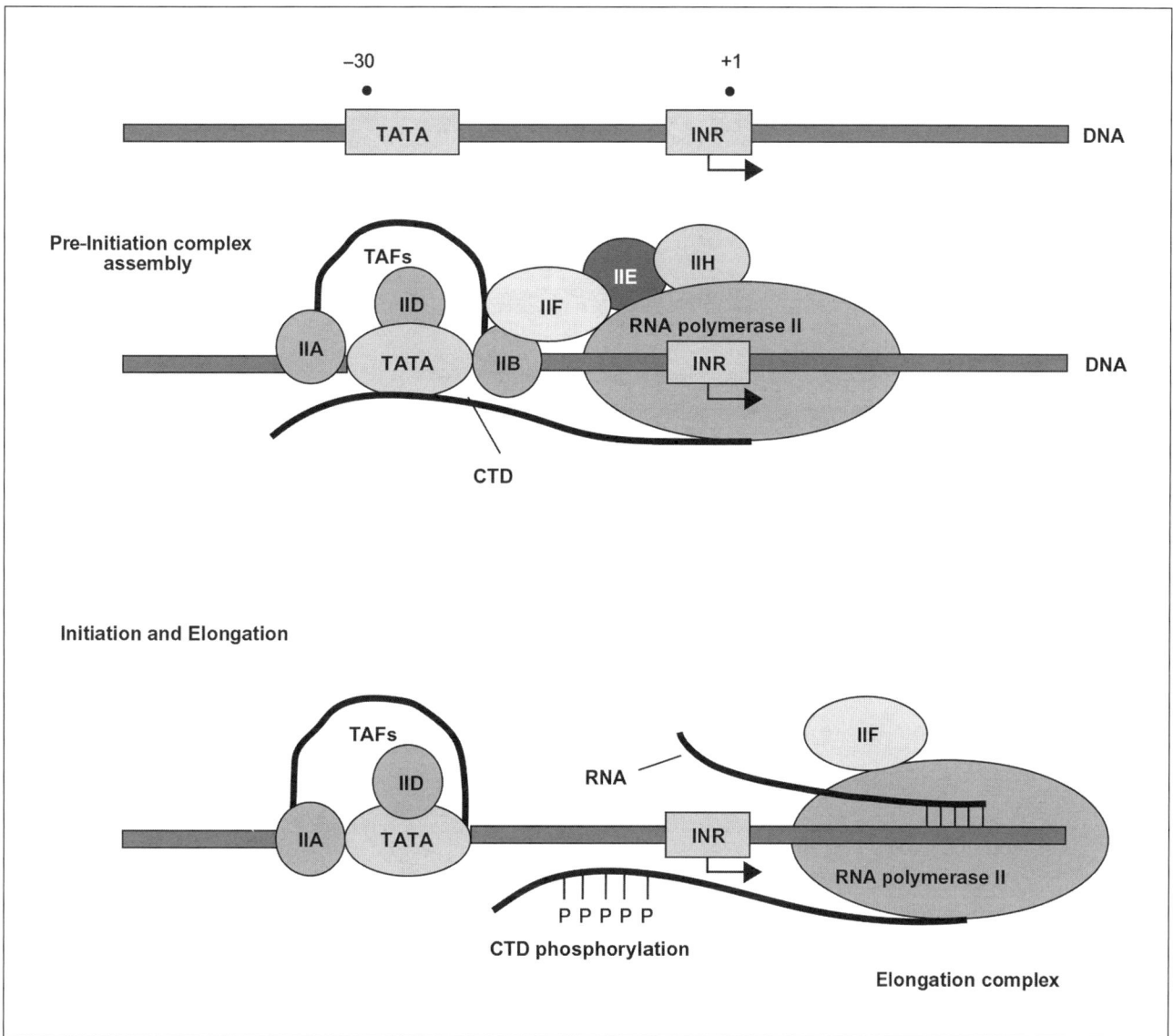

Figure 21.2. Transcription by RNA pol II. In eukaryotes, protein-encoding genes are transcribed by RNA pol II to produce mRNA. The promoter is the site of transcription initiation. However, RNA pol II has no affinity for DNA and can recognize and bind to the promoter only after TFs have bound to the promoter. The TFIID complex recognizes and binds to the TATA box of the promoter. Other TFs subsequently bind. Transcription factor IIF is bound to RNA pol II, allowing RNA pol II to recognize the TFII complex. Unphosphorylated RNA pol II enters the initiation complex. Its carboxy-terminal domain is then phosphorylated by TFIIH, signaling RNA pol II to dissociate from the initiation complex and begin transcription.

+1 = first nucleotide to be transcribed.
INR = initiator region; P = phosphorylated residue; pol = polymerase; TAFs = TATA box-binding protein-associated factors.

II catalyzes this process. 5′ capping is specific only for RNA transcripts synthesized by RNA pol II. The 5′ cap is believed to be important in mRNA stability, initiation of translation (protein synthesis), and protection of the end of mRNA from exonuclease attack (Bentley 2002; Rosenthal 1994).

3′-Polyadenylation

The 3′ ends of almost all eukaryotic mRNAs are modified by adding a long chain of adenosine residues called a polyA tail. The polyA tail is not coded for in DNA, and its length may vary from 30 to 200 adenine nucleotides. The AAUAAA sequence found near the 3′ end of most eukaryotic mRNAs may act as a signal for the site of 3′ trimming and polyA tail addition. Whereas most mRNAs have a polyA tail, some normal mRNAs without polyA tails can also be found in the cytoplasm. Because polyA tails do not exist in histone mRNA, it is assumed that they are not essential for translation. The presence of a polyA tail at the 3′ end of mRNA, like the 5′ capping of mRNA, protects the mRNA molecules from degradation by nucleases (Bentley 2002; Rosenthal 1994).

Splicing

One of the most important steps in RNA processing is RNA splicing. Most eukaryotic genes are divided into exons and introns. The parts of DNA that are transcribed into RNA but not translated into protein are known as introns (noncoding regions), whereas the stretches of

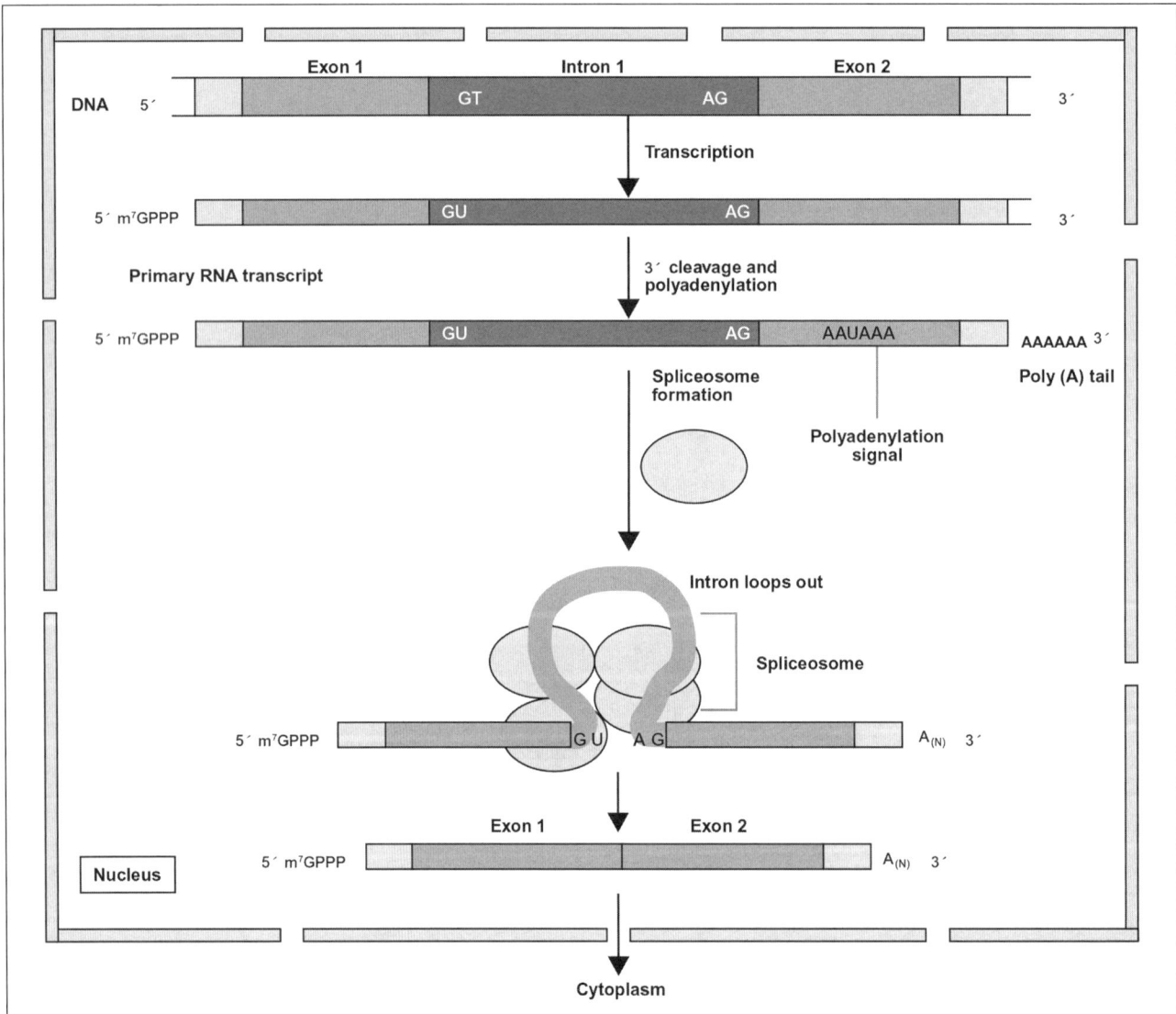

Figure 21.3. RNA processing in eukaryotes. Primary RNA transcripts undergo processing to become functional RNA before export to the cytoplasm. A guanine cap is attached to the 5′ end of pre-mRNA as it emerges from RNA pol II during transcription, protecting the pre-mRNA from degradative enzymes. A series of adenine (A) nucleotides (the polyA tail) are attached to the 3′ end of the RNA transcript when transcription is complete. Introns are excised from the pre-mRNA, and exons are spliced together by spliceosomes. The completed mRNA is transported to the cytoplasm, where it will interact with ribosomes during protein synthesis.

A(n) = polyA tail; exon = coding region; 5′m⁷GPPP = 5′-guanine cap; intron = noncoding region.

DNA that code for amino acids in the protein are called exons (coding regions). In most genes, introns are interspersed among exons. The process by which introns are removed and exons are joined (spliced) is called RNA splicing. RNA splicing is carried out on pre-mRNA by large ribonucleoproteins called spliceosomes.

In a protein-coding gene, the intron-exon junctions (splice sites) have consensus sequences (Figure 21.4). Most introns start with GT (or guanine, uracil [GU] at the RNA level) and end with adenine-guanine (AG) sequences in the 5′→3′ direction, called splice donor and acceptor sites, respectively. In addition to GU (the splice donor site) and AG (the splice acceptor site), there is another important sequence A (the branch site), which is usually located very close to the end of an intron, about 20–50 bases before the terminal AG dinucleotide (the splice acceptor site). The cytosine-uracil (CU) (A/G) A (C/U) is the consensus sequence, and "A" is conserved in all genes (Figure 21.4).

Cis-acting elements or short sequences (a few bases long) are located within intronic and exonic regions that can either enhance intronic or exonic splicing (intronic splicing enhancer and exonic splicing enhancer) or silence the splicing of the pre-mRNA (intronic splicing silencer and exonic splicing silencer). Mutations on these *cis*-acting elements can result in inappropriate inclusion or skipping of an exon. Mutations are considered gain-of-splicing function mutations when a splicing element is enhanced or created (creation of cryptic splice sites, or exonic splicing enhancer, exonic splicing silencer, intronic splicing enhancer, and intronic splicing silencer elements) and loss of function when the splicing element is weakened or destroyed (Garcia-Blanco 2004; Bentley 2002; Herbert 1999; Sharp 1994; McKeown 1992).

Alternative Splicing

Variation in splice sites can result in different isoforms of a polypeptide from a single gene. Alternative splicing (recombination of different exons) is a process whereby the introns of certain pre-mRNA from genes can be spliced in more than one way to produce several structurally and functionally distinct mRNA and protein variants. This is believed to occur in about 40%–60% of all human genes. Alternative splicing is considered a primary source of protein diversity in humans and is implicated in disease and therapy. Excision and splicing of mRNA requires great precision because the removal of a nucleotide from an exon or the carryover of a nucleotide from an intron may cause a shift in the reading frame (the codons read during translation) and result in a premature stop codon and other alterations (Figure 21.4). Many genetic diseases that result from defective proteins in a specific tissue system involve the misreading of a splice signal because of a mutation (Bentley 2002; Herbert 1999; Sharp 1994; McKeown 1992).

Dysfunction of splicing mechanisms can result in protein abnormalities with detrimental clinical consequences. For example, hemoglobin consists of two molecules of α- and β-globin polypeptides. The β-globin gene has three exons and two introns. The introns start with GT sequences that ensure the removal of introns and correct exon splicing during β-globin mRNA synthesis. The resulting mRNA is translated into β-globin polypeptide. In some individuals, a mutation from GT to AT in DNA at the beginning of the first or second intron sequences of the β-globin gene results in an alternative splice site. Consequently, these individuals are incapable of producing β-globin polypeptides and have $β^0$ thalassemia (a severe form of anemia) (Divoky 1992). Alternative splicing has implications in drug therapy as well. For example, the target of the analgesic/antipyretic acetaminophen is an alternatively spliced isoform of COX-1 (cyclooxygenase 1) (Garcia-Blanco 2004; Modrek 2003).

Transcriptional Gene Regulation

Regulation of eukaryotic gene expression originally takes place at the transcription initiation step. The transcriptional activity of single genes (e.g., through altering the efficiency of initiation) is regulated by the transcriptional regulatory components in response to environmental changes. Protein-coding genes are usually regulated by *cis*-acting transcription control elements that are located either near the start site (basal promoter) or farther upstream or downstream from the promoter (enhancer). Regulation of the expression of many genes is based on the interactions between the *cis*-acting transcription control elements and the *trans*-acting regulatory proteins. *Cis*-acting elements are gene sequences that directly affect the gene, whereas *trans*-acting elements act on a gene but are encoded by other genes. Most regulatory proteins that directly attach to DNA contain at least two functional domains. Whereas one domain is responsible for recognizing and binding to the *cis* elements in the DNA, the other domain is involved in the activation of transcription (Latchman 1996; Tjian 1995; Rosenthal 1994).

Transcriptional Activators and Repressors

Almost all eukaryotic genes are in the inactive state and require transcriptional activators for expression. The change from an inactive to an active state results from a response to an external stimulus. Short sequence motifs upstream and downstream of promoters are used as binding sites for transcriptional factors (transcriptional activators and repressors). Regulatory transcriptional proteins may function constitutively, act positively as transcriptional activators and increase the rate of transcription, or function as repressors of transcription (Kornberg 1999; Hanna-Rose 1996; Tjian 1995; Cowell 1994).

Transcriptional activators are proteins that bind regulatory elements in the upstream promoter and enhancer regions to regulate gene expression. Transcription activators have a single DNA-binding domain (DBD) and one or several activation domains. Transcriptional repressors resemble activators in having a single DBD and one or several repression domains. Repressors and activators can regulate transcription by binding to a site that is hundreds to thousands of nucleotides away from the start site. Repressors may simply interfere with the function of activators by competing for the activator-binding site. These transcriptional regulatory proteins have different structural forms or domains. We describe some of the DBDs: helix-turn-helix, zinc-finger, and leucine-zipper in Figure 21.5 (Hanna-Rose 1996; Tjian 1995).

The helix-turn-helix motif has two α-helical fragments bound to each other by a short segment of amino acids forming a bend or turn (Figure 21.5). One of the helices, designated the recognition helix, locates itself in the major or minor DNA groove, and the amino acid side chains facing the groove bind to the specific sequence in the DNA. Then, the second helix, called the stabilization helix, supports and stabilizes the recognition helix. The DBDs usually form dimers (two helix-turn-helix molecules) and fit into the major DNA groove (Harrison 1991, 1990).

The zinc-finger proteins are formed by the interaction of one or two zinc atoms with two cysteines and two histidines or, in some cases, with four cysteines. The cysteine and histidine residues are bound to a zinc atom in a way that forms a fingerlike loop pointing into the major DNA groove (Figure 21.5). There are many regulatory proteins (e.g., TFIIIA that activates 5S rRNA genes and steroid hormone receptors) that have zinc-finger motifs (Wolfe 2000; Harrison 1991).

The third DNA-binding motif is called the leucine-zipper. Because leucine amino acids are hydrophobic, they are attracted when facing each other on the outer surface of two opposing α-helices and form the teeth of the zipper that joins the two helices (Figure 21.5). The leucine-zipper maintains the α-helices in a position that fits exactly into the major DNA groove (Harrison 1991; McKnight 1991).

Enhancers, Silencers, and Insulators

Enhancers are transcriptional regulatory nucleotide sequences ranging from 50 to 200 bp in length to which TFs bind and increase gene expression. They are not part of the promoter and could be located from 200 bp to 10 kb upstream or downstream, or within an intron of a gene. Many genes require enhancers for differential expression. In addition to the DNA-binding site, the enhancer-binding proteins have other sites that bind to TFs at the promoter of the gene (Figure 21.6). The binding of an enhancer to the promoter makes the DNA form a loop. Enhancers interact even in the opposite orientation and can still be functional. In contrast to enhancers, promoters are both position- and orientation-dependent in their activity (Kornberg 1999; Tjian 1995; Rosenthal 1994).

Silencers are DNA sequences or control regions that repress the expression of genes when bound to TFs. Silencers, like enhancers, can be located thousands of base pairs away from the transcribed gene and still be effective. In contrast to transcriptional activators (not in enhancers or silencers), which can control only one gene, enhancers and silencers can control the expression of more than one gene (Burgess-Beusse 2002; Tjian 1995).

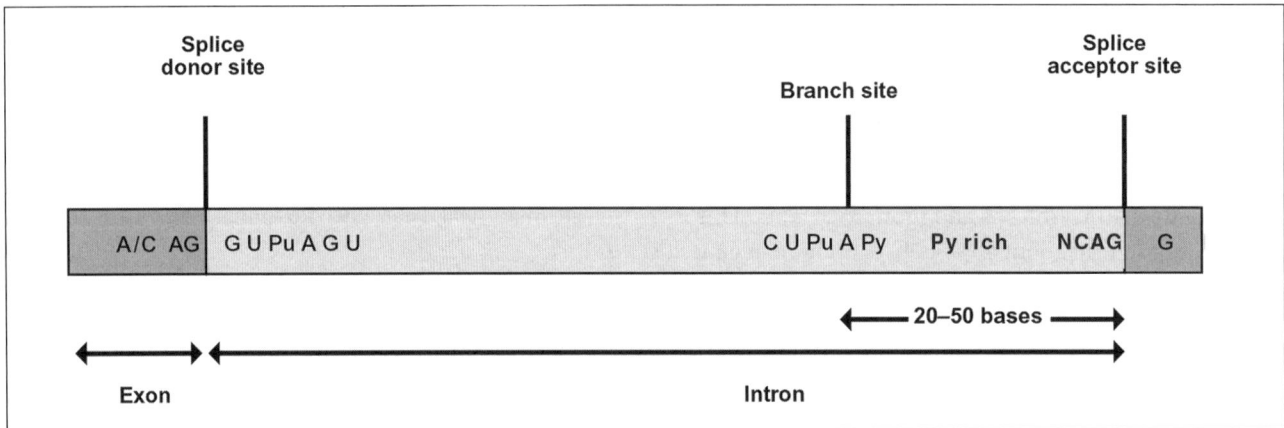

Figure 21.4. The consensus sequence for splicing. In addition to GU (the splice donor site) and AG (the splice acceptor site), there is an important sequence A (the branch site), which is usually located very close to the terminal AG dinucleotide (the splice acceptor site). The CU (A/G) A (C/U) is the consensus sequence, and "A" is conserved in all genes (Pu = A or G; Py = C or U).

A = adenine; AG = adenine-guanine; C = cytosine; CU = cytosine-uracil; G = guanine; GU = guanine-uracil; Pu = purine; Py = pyrimidine; U = uracil.

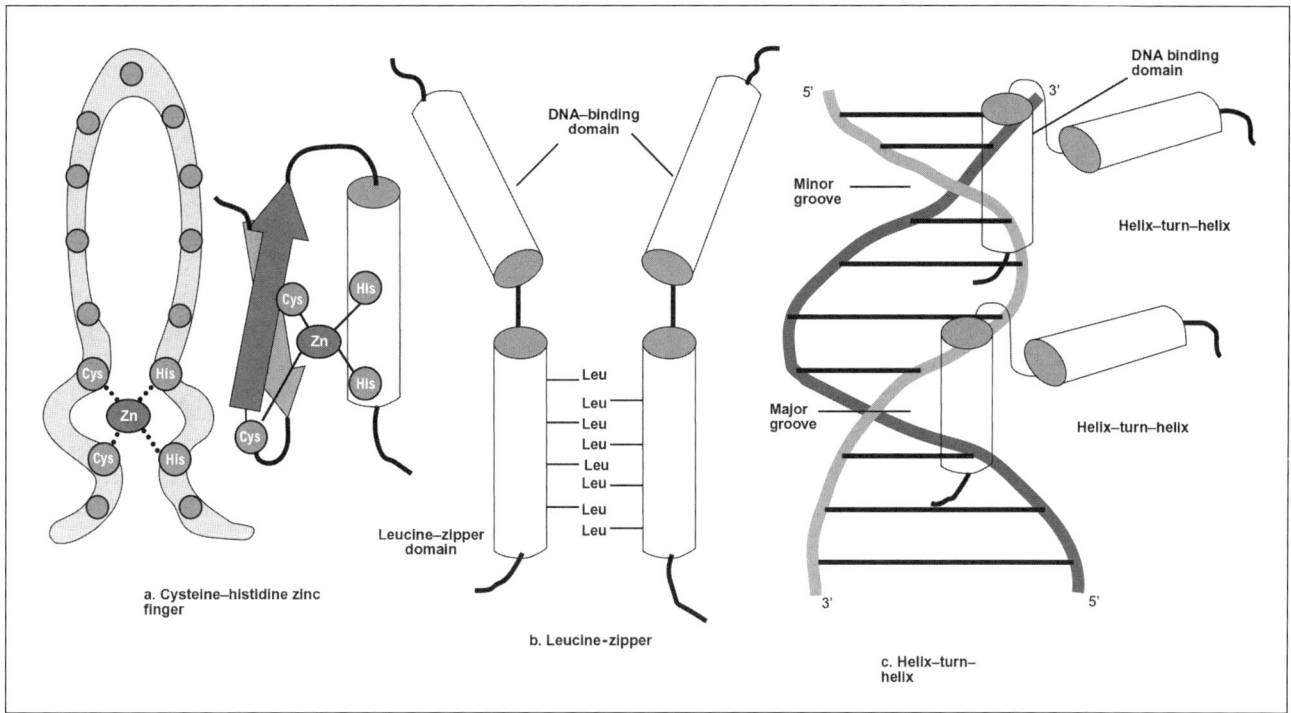

Figure 21.5. DNA-binding domains (DBDs). (a) Zinc-finger, (b) leucine-zipper, and (c) helix-turn-helix bound to the major groove of DNA. DNA-binding domains are regions of transcriptional activators or repressors that bind regulatory elements in the upstream promoter and enhancer regions to regulate gene expression.

Cys = cysteine; His = histidine; Leu = leucine; Zn = zinc.

Insulators are short stretches of DNA sequences that are located between enhancers and promoters, or silencers and promoters, that function as blocking elements to prevent the activation or repression effects of adjacent genes. Insulators need to bind a protein called CTCF (an 11 zinc-finger DNA-binding protein) to become functional (Burgess-Beusse 2002).

Steroid Hormone Receptors and Transcription Regulation

Steroid hormones (e.g., glucocorticoids, estrogen, progesterone, testosterone) are small lipid molecules that cross the plasma and nuclear membranes of the cell and bind to steroid hormone receptors. Hormones act as cell signaling molecules and control the activity of many TFs. Steroid hormones (mostly located in the nucleus) selectively regulate the transcriptional activity of a specific group of genes by binding to steroid hormone receptors and associating with hormone response elements (HREs) on the gene. The hormone receptors in the steroid, thyroid, and retinoid supergene family act as TFs that bind to target sequences in the regulatory regions of hormonally regulated genes to enhance or suppress their transcription. Eukaryotic cells may have several thousand receptor proteins for a single hormone. These receptors are members of the zinc-finger DNA-binding protein family (Khorasanizadeh 2001; Torchia 1998; Funder 1997; Tenbaum 1997).

Nuclear receptors (NRs) bind small lipophilic hormones that are produced by the endocrine system. In general, NRs are classified according to the type of hormones they bind to: steroids (glucocorticoids, mineralocorticoids, androgens, progestins, and estrogens), steroid derivates (vitamin D3), nonsteroids (thyroid hormone, prostaglandins, and retinoids), and receptors without a known ligand (orphan receptors). Through evolution, these receptors have conserved structural similarities such as a ligand-binding domain (LBD), a DBD, a dimerization domain, and one or more transactivation domains (Khorasanizadeh 2001; Kornberg 1999; Issa 1998; Torchia 1998; Funder 1997; Tenbaum 1997).

In steroid hormone receptors, the amino terminal (N-terminal) domain is involved in transcription activation, and the carboxy terminal (C-terminal) region is the hormone-binding domain. The center part of the receptor contains a cysteine-cysteine zinc motif and is responsible for recognizing and binding to HREs (Torchia 1998; Tenbaum 1997).

The binding of a hormone receptor to the HRE activates mRNA transcription or inhibits the transcription of previously activated genes. Some hormone receptors, such as glucocorticoid receptors, can bind two different types of response elements and activate or repress the transcription of a gene by inhibiting the binding of other transcriptional factors to the promoter site. Hormone response elements, like other DNA sequences that are identified by regulatory proteins, are palindromes or consensus sequences that consist of two repeats (e.g., 5'-AGAACANNNTGTTCT-3'3'-TCTTGNNNACAAGA-5') (Khorasanizadeh 2001; Torchia 1998; Tenbaum 1997).

All hormone receptors, except for androgen receptors, have recessive patterns. Mutations in hormone receptor genes affect DNA-binding affinity, ligand-binding capacity, homo/heterodimer formation, and transactivation functions. These mutations are associated with various human diseases such as breast and prostate cancer, osteoporosis, and Kennedy disease (X-linked recessive spinal and bulbar muscular atrophy). Further understanding of receptors in the hormone superfamily may promote the development of agonists or antagonists for the treatment of diseases associated with hormone receptor dysfunction (Issa 1998; Funder 1997; Tenbaum 1997).

Translation (Protein Synthesis)

Translation is the process of decoding the genetic information on mRNA and synthesizing a protein based on that sequence information. The protein synthesis machinery includes mRNA as the template containing the genetic codes for translation into protein; ribosomes, the large ribonucleoproteins that are the sites of protein synthesis; tRNA, the adaptor molecules that carry the correct amino acids to the ribosome for synthesis of the polypeptide chain; and the accessory proteins that are involved in the initiation, elongation, and termination of protein synthesis (Bolsover 1997; Cooper 1997c).

Ribosomal RNA

Ribosomal RNAs associate with certain proteins to form small and large ribosomal subunits that serve as the site for protein synthesis. There are four types of rRNA in eukaryotic cells. One molecule of 18S rRNA, together with 50 different protein molecules, make the small subunit of the ribosome. One molecule each of 28S, 5.8S, and 5S rRNA with more than 50 different protein molecules make the large subunit of the ribosome (Lodish 2000; Cooper 1997b).

Transfer RNA

The tRNAs are located in the cytoplasm, where they pick up and transfer activated amino acids to mRNA for protein synthesis. A typical eukaryotic cell has as many as 50–100 different types of tRNA. Transfer RNAs are small and contain between 70 and 90 nucleotides. The paired bases in tRNA form a section of double helix, whereas the unpaired bases form three loops. Transfer RNAs exist for each of the 20 amino acids. Some of the amino acids have more than one tRNA designated to them. The three unpaired bases at one loop form the anticodon, which are the complementary base pairs of the codon on an mRNA molecule. This consequently ensures that the correct amino acid is added to the growing polypeptide chain coded for by mRNA (Figure 21.7) (Arnez 1997; Bolsover 1997; Cooper 1997c).

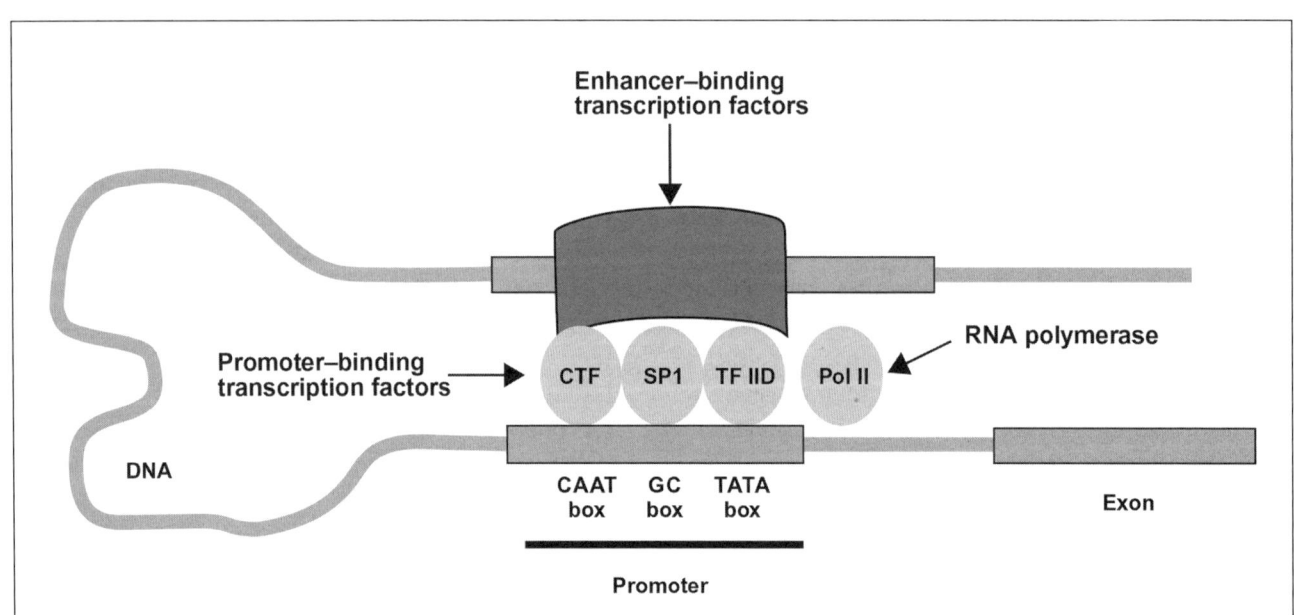

Figure 21.6. Diagram of an enhancer. Enhancers are just one of several control mechanisms for gene transcription. Enhancers bind to *trans*-acting promoter-binding transcriptional factors (CTF, SP1, and TFIID), which in turn bind to their corresponding *cis*-acting promoter elements (CAAT box, GC box, and TATA box) and stimulate the rate of transcription initiation. PolII = RNA polymerase II.

Initiation of Translation (formation of the initiation complex)

Initiation of protein synthesis starts with separation of the two ribosomal subunits (40S and 60S). Then, a ternary complex, the preinitiation complex, is formed, which consists of initiator tRNA (Met-tRNA; the initiator tRNA that carries and incorporates the initiator methionine in all proteins), guanosine triphosphate (GTP), eukaryotic initiation factor 2 (EIF-2), and the 40S ribosomal subunit. First, GTP binds to EIF-2, which is composed of α, β, and γ subunits. This binary complex then binds to the initiator tRNA, forming the ternary complex, which subsequently binds to the 40S ribosomal subunit, ultimately forming the 43S preinitiation complex. Then, the preinitiation complex is stabilized by the association of EIF-3 and EIF-1 to the 40S subunit. Binding of the 5′ cap on mRNA to the preinitiation complex is accomplished by the EIF-4F initiation factor, which is made of three proteins: EIF-4A, EIF-4E, and EIF-4G. The EIF-4A protein hydrolyzes ATP and has RNA helicase activity. The EIF-4G protein assists in binding the mRNA to the 43S preinitiation complex.

After the preinitiation complex binds to the 5′ end of mRNA, the complex scans mRNA until it reaches an initiator adenosine-uracil-guanine (AUG) codon. Binding of initiator tRNA to the initiator AUG codon is facilitated by EIF-1. Hydrolysis of EIF-2–bound GTP by EIF-5 results in the release of EIF-2 from the 40S complex and leaves the initiator tRNA in the P (peptidyl)-site of the 40S ribosomal subunit. The 60S ribosomal subunit then attaches to the 40S subunit and forms the 80S initiation complex. The energy for forming the 80S complex is provided by hydrolysis of the GTP bound to EIF-2. Now that the initiator tRNA is bound to the mRNA in the P-site of the ribosome, protein synthesis begins (Rodnina 2001; Pestova 2000, 1998; Lee 1999; Chaudhuri 1997; Kozak 1997; Sachs 1997; Thach 1992).

Elongation of Polypeptide Chain (New Protein)

In protein synthesis, tRNA acts as a translator between mRNA and the new protein by bringing the correct amino acid to mRNA. Each tRNA has an amino acid acceptor or attachment site and a nucleotide triplet (anticodon) that binds to the complementary sequence (codon) on mRNA. The codon is composed of three nucleotides and starts near the 5′ end of mRNA. The tRNA carries the amino acid attached to its 3′-terminal OH group (Figure 21.7). There are 61 codons for 20 different amino acids, and the codons for some amino acids vary only in the third position of the codon (e.g., phenylalanine; UUU and UUC). The base pairing at the first and second positions in the codon and anticodon follows standard rules (A-U and C-G), but in the third position, both the U and the C of the codon can form hydrogen bonds with a G of the anticodon. In this case, only one tRNA is needed for two codon sequences, indicating a redundancy in some tRNA functioning.

Binding of an amino acid to its correct tRNA occurs through interaction with the aminoacyl-tRNA synthetase enzyme. First, the enzyme attaches the amino acid to the α-phosphate of ATP, resulting in the release of pyrophosphate. Then, tRNA synthetase catalyzes the transfer of the amino acid to the 3′-terminal adenosine residue of tRNA, which generates activated aminoacyl-tRNA. For each amino acid and its tRNA, there is at least one aminoacyl-tRNA synthetase enzyme (Rodnina 2001; Bolsover 1997; Cooper 1997c).

Translation continues along the mRNA in the 5′→3′ direction that corresponds to the N-terminal to C-terminal direction of amino acid sequences in proteins. In the large ribosomal subunit, there are two sites: the A-site, which accepts the new tRNA carrying an amino acid, and the P-site, which sustains the tRNA bound to the growing chain. After the AUG start codon and Met-tRNA are positioned in the P-site of the ribosome, the new aminoacyl-tRNA carrying a second amino acid enters the A (aminoacyl) site or decoding site of the ribosome. Then, its anticodon pairs with the mRNA codon, and their recognition is mediated by eukaryotic elongation factor 1 associated with GTP (Rodnina 2001; Bolsover 1997; Cooper 1997c).

The mRNA codon designates which aminoacyl-tRNA should come next to the A-site. The covalent bond between the amino acid Met and initiator-tRNA in the P-site is broken, and a peptide bond is formed by the peptidyl transferase enzyme between the methionine and the second amino acid in the A-site. This process is called transpeptidation. The empty tRNA in the P-site dissociates from the ribosome, and the peptidyl tRNA (the tRNA carrying two amino acids) in the A-site is translocated into the P-site. This process of translocation is catalyzed by EIF-2 coupled with GTP hydrolysis. Another aminoacyl-tRNA carrying the third amino acid enters the A-site, and the addition of new amino acids to the growing polypeptide chain continues in this fashion until the ribosome reaches the stop codon (Rodnina 2001; Bolsover 1997; Cooper 1997c).

Termination of Protein Synthesis

In general, the termination of protein synthesis is similar to the elongation process because the stop (nonsense) codon is decoded at the ribosomal A-site. There are three stop codons (UAA, UAG, and UGA) that have no complementary anticodon or aminoacyl-tRNA to bind to mRNA. The termination

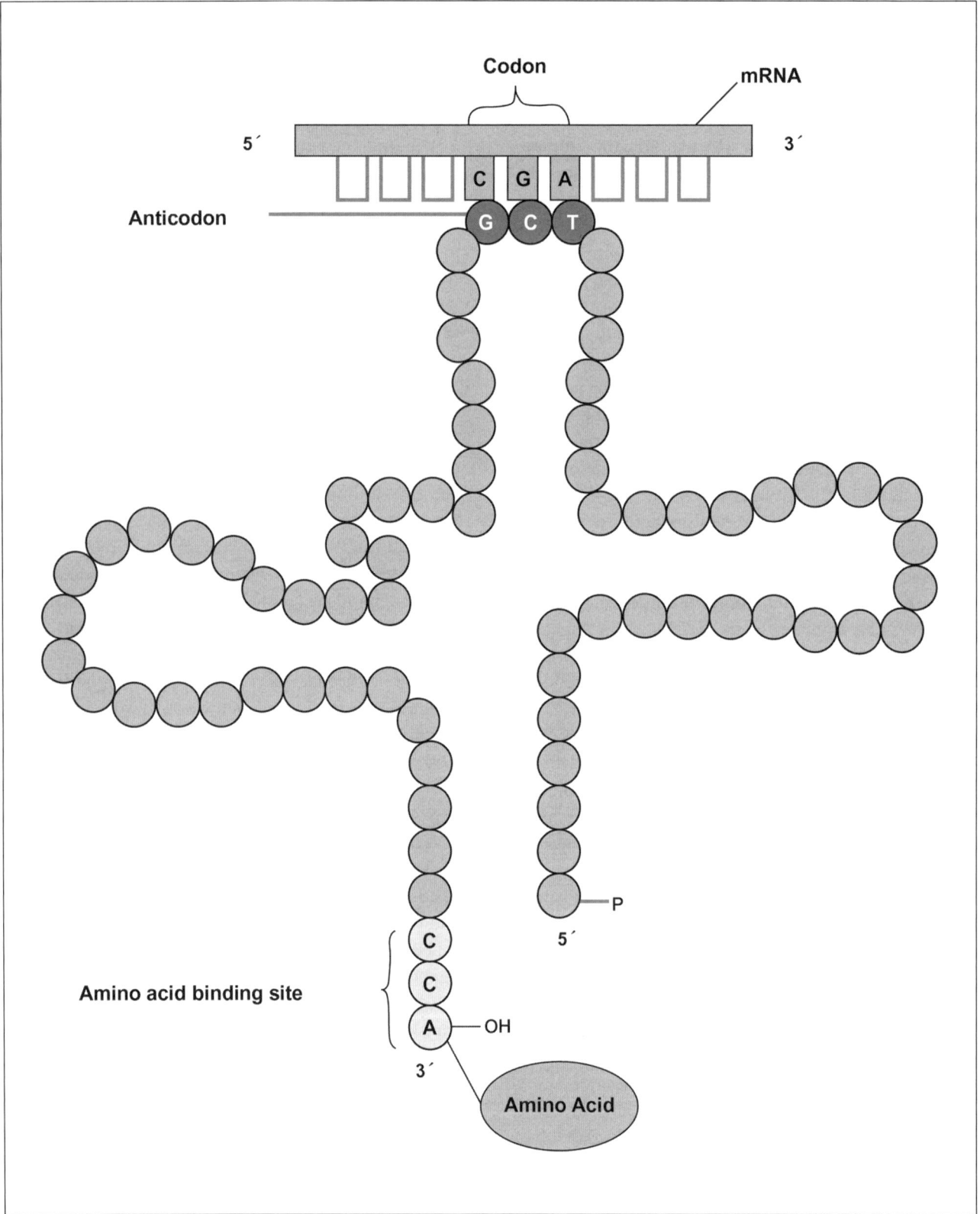

Figure 21.7. Diagram of transfer RNA (tRNA). In translation (protein synthesis), tRNA is responsible for matching the appropriate amino acid to the nucleic acid codons on mRNA. At one end, tRNA possesses an anticodon that recognizes the codon and, at the other end, carries the amino acid for that codon. The appropriate amino acid is added to the tRNA by the actions of the enzyme aminoacyl-tRNA synthetase. Amino acids are added to the growing peptide chain at the ribosome.

A = adenine; C = cytosine; G = guanine; U = uracil.

of translation occurs on the ribosomes and requires two polypeptide release factors designated eRF-1 (a codon-specific release factor) and eRF-3 (a non–codon-specific release factor) (Heurgue-Hamard 1998; Bolsover 1997; Cooper 1997c; Nakamura 1996).

The release factor recognizes and binds to the stop codon at the A-site and begins the process of translation termination. The stop codon at the A-site causes the release factor to bind to the A-site with GTP instead of aminoacyl-tRNA. After the binding of the release factor to the stop codon, the bond that holds the polypeptide chain to the tRNA at the P-site is hydrolyzed. Because there is no amino acid at the A-site, the hydrolysis permits the newly synthesized polypeptide chain to be released from the ribosome (Yokoi 2013).

After the release of the polypeptide chain, the tRNA at the P-site, together with the release factor from the A-site, is expelled. The large and small ribosomal subunits separate but can reassemble with mRNA and Met-tRNA to make a new initiation complex and continue with the translation process to produce more copies of the protein (Heurgue-Hamard 1998; Nakamura 1996).

Regulation of Translation

The second level in the regulation of gene expression involves the regulation of translation. Translational regulation can occur by control of individual initiation factors, alteration of the activity of translational factors, or interaction between the *cis*-acting sequences on mRNA and the *trans*-acting factors. One of the most common mechanisms of translation regulation is the binding of repressors to specific mRNA sites, thereby inhibiting protein synthesis by the direct blockade of translation. In eukaryotic cells, translational repression is well described in the regulation of the synthesis of ferritin, the ubiquitous iron-storage protein that chelates iron in the cytosol. The quantity of iron present in the cell controls the translation of ferritin mRNA (Goessling 1992; Kozak 1992).

The regulatory system of ferritin mRNA translation consists of the iron-responsive element (IRE), the ferritin repressor protein or IRE-binding protein (IRE-BP), and an inducer (iron). The IRE that is located at the 5′ untranslated region (5′ UTR) of ferritin mRNA contains a 28-nucleotide fragment responsible for the stimulation of ferritin synthesis by iron. The translation of ferritin mRNA is controlled by the binding of IRE-BP to the IRE sequence at the 5′ end of ferritin mRNA, which prevents the mRNA from forming an initiation complex with the ribosomal subunits. In the presence of low iron concentrations, IRE-BP binds to the IRE, and the translation of ferritin mRNA is inhibited. In the presence of sufficient amounts of iron, the IRE-BP does not bind to IRE, and the translation of ferritin continues (Figure 21.8a) (Goessling 1992; Kozak 1992).

Interference with translation and regulation is a main mode of action of several drug therapies. For example, many antibiotics block protein synthesis in both prokaryotes and eukaryotes. The inhibition of bacterial protein synthesis by antibiotics has been an effective means of fighting infectious diseases. This strategy is effective because of the structural differences between prokaryotic and eukaryotic ribosomes, initiation factors, elongation factors, and release factors and because protein translation plays an important role in the overall metabolism of the cell (Cocito 1997).

Stabilization of mRNA

Stabilization and selective degradation of mRNA play important roles in regulating gene expression at the translational level. Rapid degradation of mRNA after it enters the cytoplasm would result in the synthesis of less protein, whereas stabilization of mRNA increases its half-life and results in the production of more protein. The stability of different mRNA transcripts within a cell varies to a certain extent according to the cell's activities, differentiation, and development. Whereas mRNAs for some growth factors have half-lives of less than 30 minutes, other mRNAs (e.g., β-globin) can be stable for more than 15 hours (Jacobson 1996; Decker 1994).

Messenger RNA decay can be triggered by different events such as polyA tail shortening, translational arrest caused by a premature stop codon, and endonucleolytic cleavage. The 3′ untranslated region (3′ UTR) of mRNA plays a crucial role in the stabilization of mRNA. The 3′UTR may promote rapid deadenylation of the 3′ polyA tail, which results in the decapping of the 5′ end of mRNA and 5′→3′ degradation (Jacobson 1996; Decker 1994).

Expanding on the example of iron homeostasis, transferrin is a serum protein that transports iron to the cells that need it. Transferrin plays an important role by binding to iron and preventing the possible damage caused by free iron. Regulation of iron uptake and metabolism is mediated by a plasma membrane protein receptor called the transferrin receptor (TfR). The 3′ UTR of TfR mRNA contains a few sets of IREs that are rich in AU sequences. It has been shown that the AU-rich elements promote the degradation of mRNA (Jacobson 1996; Decker 1994).

The regulation of TfR is controlled by the 3′ UTR of TfR mRNA and the cytoplasmic IRE-BP. In the presence of low concentrations of iron, the IRE-BP binds to the IREs in the 3′ UTR and inhibits the degradation of TfR mRNA. This results in the accumulation of TfR, and more iron is transferred to the cell. At high iron concentrations, the IRE-BP does not bind to the IREs, and the AU-rich sequences stimulate the degradation of TfR mRNA (Figure 21.8b) (Jacobson 1996; Decker 1994).

Posttranslational Modification

To become active proteins, newly synthesized polypeptides often undergo posttranslational modification. For polypeptides to become functional proteins, they must fold correctly in a three-dimensional conformation. Proteins acquire the necessary information for folding from their amino acid sequences. For proper folding, newly synthesized proteins require the assistance of other proteins called molecular chaperones. Chaperones stabilize and support the partly folded polypeptide into a correct and stable three-dimensional protein (Agard 1993; Craig 1993; Hartl 1996).

Misfolded and unfolded protein molecules tend to attach to each other and form insoluble aggregates. These aggregates resemble the amyloid protein deposits found in several diseases such as Alzheimer disease. A complete understanding of the protein-folding process may help clinicians develop therapies that inhibit protein aggregation and may be useful in diseases such as Alzheimer disease (Hartl 1996; Taubes 1996).

Proteins may be cleaved by proteolytic (protein cutting) enzymes at a specific amino acid. Proteolysis is an irreversible process that regulates and controls enzyme activation. For example, the posttranslational cleavage of initiator methionine from the N-terminal of many polypeptides, which may be followed by the addition of fatty acid chains or acetyl groups, plays an important role in the translocation of proteins to different targets such as lysosomes, mitochondria, and the plasma membrane (Gereben 2000; Jentsch 1996).

The life spans of different proteins vary greatly. In contrast to long-lived structural proteins, regulatory proteins are short-lived and quickly degraded. Misfolded and abnormal proteins are eliminated in a selective protein degradation process by the ubiquitin/proteasome pathway (nonlysosomal proteolytic system). Ubiquitin is a small and stable protein that binds to the internal lysine residues

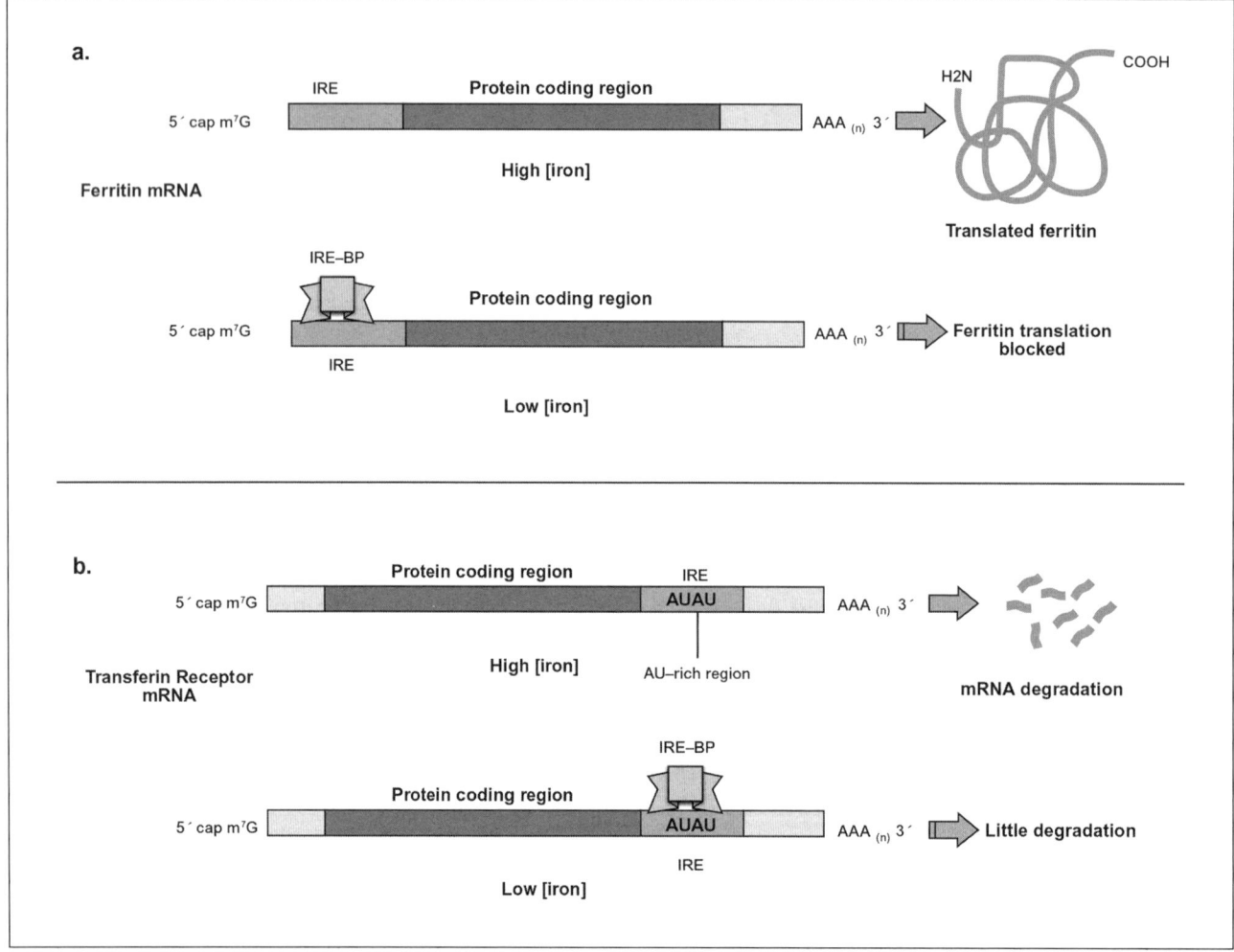

Figure 21.8. Translational regulation of ferritin (a) and transferrin (b). See text for description.

IRE = iron-responsive element; IRE-BP = IRE-binding protein.

of the substrate through its C-terminus. Ubiquitins (multiubiquitin chains) are used for tagging proteins that are to be degraded by proteasomes (Jentsch 1995).

Some amino acids of the polypeptide chain can be altered by phosphorylation and dephosphorylation. Protein phosphorylation is a reversible process catalyzed by protein kinases. Phosphorylation occurs through the transfer of phosphate groups from ATP to the -OH groups of the side chains of serine, threonine, and tyrosine. Dephosphorylation of proteins is carried out by protein phosphatases that are specific for serine, threonine, and tyrosine. The processes of phosphorylation and dephosphorylation play important roles in the activation and deactivation of proteins involved in signal transduction pathways in eukaryotic cells (Bjorklof 2002; Miranda 2002).

One of the most significant posttranslational modifications that occur in eukaryotic secretory and membrane-bound proteins is the attachment of sugar chains to an amino acid–containing hydroxyl group, a process called glycosylation. Sugars are naturally added to many proteins during and after protein synthesis. The importance of glycosylation is so great that it caused the creation of a new field in biology called "glycobiology." Serine, threonine, asparagine, tyrosine, hydroxyproline, and hydroxylysine are amino acids that contain a hydroxyl functional group and can be involved in glycosylation. The two main types of protein glycosylation are designated *N*- and *O*-glycosylation. The *N*-linked glycans are linked to asparagine, and the *O*-linked glycans are attached to serine and threonine residues. The *N*-linked glycans are involved in the folding of glycoproteins by acting as mediators in interactions between the endoplasmic reticulum chaperone proteins calnexin and calreticulin and nascent glycoproteins (Bjorklof 2002; Spiro 2002; Huby 2000; Zitzmann 1999). Glycosylation can change stability, uptake, solubility, hydrophobicity, immunological properties, and electrical charge of proteins, thus changing the protein confirmation and its biological activity (Bjorklof 2002; Huby 2000).

Defects in glycosylation are responsible for many human diseases. In the Alzheimer disease brain, the abnormal glycosylation of tau protein (one of the major microtubule-associated proteins in neurons) results in an early abnormality of neurofibrillary degeneration. The progress in elucidating the role of proteins and lipid-linked carbohydrates in various biological processes has shifted the attention toward drugs that target the enzymes involved in glycosylation. Drugs such as glycosidase inhibitors are effective in the treatment of viral infections (e.g., hepatitis B and C) in animal models (Bjorklof 2002; Liu 2002; Spiro 2002; Mehta 2001; Zitzmann 1999).

Noncoding RNAs

More than 80% of the human genome is reported to be actively transcribed to different groups of RNA transcripts known as noncoding RNAs (ncRNAs). The ncRNAs vary from having only a few nucleotides to having several kilobases and do not code for proteins but act as regulators of chromatin structure and gene expression. The ncRNAs are classified as lncRNAs and small ncRNAs according to their size (Djebali 2012; ENCODE Project Consortium 2012).

Small Noncoding RNAs

The small ncRNAs have been relatively studied more than the lncRNAs and are classified according to their known structural characteristics and biological functions. There are several different small ncRNAs, but the microRNAs (miRNAs) are the best example of known and well-studied small RNAs. We further describe the miRNA and the other small noncoding RNAs (short interfering RNAs [siRNAs]).

MicroRNAs

The miRNAs are small ncRNAs (about 20–22 nucleotides long) that are involved in posttranscriptional gene expression and act as negative posttranscriptional regulators. They regulate genes by guiding the RNA-induced silencing complex of proteins to specific targets within mRNAs, which results in immediate induction of mRNA cleavage and translational inhibition. The sites where miRNAs target are thought to be located mainly in the 3′ UTR of mRNAs (Kloosterman 2006).

The miRNAs are initially processed from primary miRNA transcripts into pre-miRNAs and then to mature miRNAs. The primary miRNAs that are usually several kilobases long are transcribed by RNA pol II in the nucleus. The primary miRNAs are then processed and cleaved by the Drosha/DGCR8 microprocessor into a 70- to 100-nucleotide hairpin structure called pre-miRNAs before being exported into cytoplasm. In the cytoplasm, the hairpin pre-miRNA structure is cleaved by an RNA pol III enzyme, into a 20- to 22-nucleotide double-stranded or duplex RNA molecule. Either strand may be functional, but usually only one strand is incorporated into the RNA-induced silencing complex, which recognizes specific miRNA targets and induces posttranscriptional gene silencing by binding mainly to the complementary sequences of the 3′ UTR) of target mRNA transcripts (Van Rooij 2011; Kim 2006).

It is estimated that miRNAs target 60% of the human mRNAs, and they are involved in different biological processes such as cell growth, cell differentiation, and apoptosis. Through their gene regulation functions, miRNAs may also provide insight into the understanding of complex traits, including drug response (Yokoi 2013).

Short Interfering RNAs
The siRNAs are usually derived from double-stranded RNAs, which play an important role in regulating gene expression in many eukaryotes. They initiate different types of gene silencing called RNA silencing or RNA interference. Double-stranded RNAs are quickly processed into short RNA duplexes (about 21–28 nucleotides long) that bind the complementary single-stranded RNA (e.g., mRNA), resulting in mRNA degradation and translational repression. Since their discovery, both miRNA and siRNA have been extensively used in biomedical research, drug discovery, and therapy, resulting in thousands of published papers and underscoring the importance of these molecules in gene regulation.

Long Noncoding RNAs
The advances made in next-generation sequencing technology and bioinformatics tools have greatly affected and accelerated the depth of RNA sequencing, analysis, and annotation. As a result, several small and long RNA species have been discovered and annotated in the human genome. Long noncoding RNAs are polyadenylated mRNA-like RNAs longer than 200 nucleotides that do not code for proteins but cause transcriptional changes in processes such as X-chromosome inactivation (a process in which one of the X chromosomes in female mammals is inactivated so that females will have the same level of gene expression as males from the X chromosome) and gene imprinting (alteration or inactivation of an allele at a given locus that is inherited from either the mother or the father, resulting in one functional allele) (Mercer 2009).

The GENCODE project, which is a subproject of ENCODE (the Encyclopedia of DNA Elements), manually annotated the GENCODE human gene annotation catalog (Harrow et al. 2012; www.genecodegenes.org/) consisting of 15,512 transcripts grouped into 9640 gene loci. Today, the GENCODE lncRNA annotation is considered the largest manually annotated catalog of human lncRNAs. After further analysis and removal of the transcripts shorter than 200 nucleotides as well as removal of the transcripts with at least one exon intersecting a protein-coding exon on the same strand, the final number of lncRNA transcripts was reported to be 14,880 consisting of 9277 gene loci. This is still the most complete annotated human lncRNA catalog (Derrien 2012).

Long noncoding RNAs are predominantly found in the chromatin and nucleus and are generated through pathways similar to those of protein-coding genes. Long noncoding RNAs have exon/intron lengths, splicing signals, and histone modifications similar to those of protein-coding genes. About 98% of lncRNAs are spliced, and of interest, 42% of the transcripts have only two exons compared with 6% of the protein-coding genes. In general, the lncRNA transcripts are shorter than the protein-coding transcripts. Most lncRNA introns share the same GT/AG splice sites, and no difference has been observed in splicing signal usage compared with protein-coding genes. More than 25% of alternative splicing events with at least two different transcript isoforms per gene occurred among lncRNAs. An expression study of several human organs showed that lncRNA expression is generally more tissue-specific and less strongly expressed compared with protein-coding genes (Derrien 2012).

Functions of the lncRNAs
In contrast to the small ncRNAs such as miRNAs, the lncRNAs are mainly categorized according to their location with respect to protein-coding transcripts because they lack a known function. Although lncRNAs have different known functions with more to be discovered, we briefly describe three of them.

Chromatin Modifications
The lncRNAs have been reported to be involved in mediating epigenetic changes through the recruitment of chromatin remodeling complexes to the specific genomic loci. For instance, hundreds of lncRNAs are expressed along the axes of human homebox (Hox) loci. Among these lncRNAs, the Hox transcript antisense RNA (HOTAIR) originated from the HOXC locus and silences the transcription across 40 kb of the HOXD locus in trans by inducing a repressive chromatin state (Rinn 2007).

Transcriptional Regulation
The lncRNAs can act as cofactors to modulate the activity of TFs or to regulate RNA pol II activity by interacting with initiation complexes that can influence promoter choice. The mechanisms by which lncRNAs control promoter usage exist in thousands of triplex structures in eukaryotic chromosomes. The lncRNAs can also interact with basal components of RNA pol II–dependent transcription machinery, leading to a decoupling of their expression (Mercer 2009).

Posttranscriptional Regulation
The lncRNAs can recognize complementary sequences, allowing specific interactions that can result in the regulation of different steps in the posttranscriptional processing of mRNAs such as mRNA splicing, editing, transport, translation, and degradation. Many mammalian genes are capable of expressing a class of ncRNAs called

antisense ncRNAs that can mask the key *cis* elements in mRNA by forming RNA duplexes that, for instance, can result in alternative splicing of mRNA transcript isoforms (He 2008).

Cell Signaling

Cells in multicellular organisms are constantly interacting with their surrounding environment. The reaction of cells to environmental stimuli is often determined by the receptors shown on the cell's surface. Cells respond to changes in their environment by altering patterns of gene expression, regulating the activity of proteins. Cell signaling refers to the mechanism by which an external change (usually in the form of ligand-receptor binding) initiates an intracellular cascade that results in a specific cellular response.

Cell Signaling Molecules

Cells respond to changes in their environment and facilitate appropriate adaptations to these changes through extracellular signals. Cell signaling molecules are incredibly diverse, ranging from light particles to odorants and pheromones to ions and peptides, among others. Binding of these molecules to their respective receptors initiates the signaling cascade that transduces the extracellular signals into intracellular biochemical reactions, eliciting the specific cellular response. Not only are the signaling molecules varied, but so, too, are their receptors, including cell surface and intracellular receptors. Extracellular signaling molecules can propagate their signals in various ways, including through plasma membrane diffusion, ion channels, G protein–coupled receptors (GPCRs), and enzyme-linked receptors.

Plasma Membrane Diffusion

Although most signaling molecules initiate the signaling process through binding with a plasma membrane–associated receptor, this is not true for all molecules. Several hydrophobic molecules can diffuse through the lipid bilayer of the plasma membrane and bind to the intracellular receptors in the cytoplasm or on the nucleus. Examples of these signaling molecules include steroids, nitric oxide, and arachidonic acid.

Steroids such as estrogen, progesterone, and testosterone and the glucocorticoids and mineralocorticoids are derived from cholesterol. They initiate cellular changes by diffusing across the plasma membrane and binding to intracellular receptors (Beato 2000). These receptors serve as TFs. Once steroids bind to their respective receptors and associate with regulatory DNA sequences (previously described HREs) in the nucleus, transcription activation occurs. Although important in a wide array of homeostatic functions, hormone-hormone receptor binding can be pathological, like in metastatic breast cancer, where estrogen-mediated transcription results in tumor proliferation. In fact, several pharmacologic agents used in current practice interrupt estrogen's effects at the receptor level. For example, tamoxifen directly binds to the estrogen receptor, causing a conformational change that blocks the transcription of the estrogen-dependent genes implicated in breast cancer.

Nitric oxide is another signaling molecule that exerts its effects by diffusing across the cell membrane. Nitric oxide, which is produced from l-arginine through nitric oxide synthase, diffuses from the cell in which it was produced and across the plasma membrane of neighboring cells to activate guanylyl cyclase, which increases the production of cyclic guanosine monophosphate (cGMP) (Ignarro 1989). The actions of cGMP are diverse, resulting in smooth muscle relaxation and vasodilation. In clinical practice, organic nitrates like nitroglycerin are converted to nitric oxide, resulting in vasodilation, which is a beneficial treatment modality for acute and chronic angina. Other drugs like sildenafil, vardenafil, and tadalafil potentiate the vasodilatory effects of nitric oxide by inhibiting phosphodiesterase type 5–mediated breakdown of cGMP in the corpus cavernosum, thereby making them commonly prescribed agents for the treatment of erectile dysfunction.

Ion Channels

Cell signaling molecules also include small ions such as Ca^{++}, K^+, and Na^+. These molecules affect cellular changes through ion channels in the plasma membrane. Ion channels are ubiquitous proteins and are crucial in many signaling processes that regulate electrolyte homeostasis, smooth muscle contraction, cell and intravascular volume, insulin release, neuronal activity, and cardiac function (Goldstein 1996). In general, ion channels are specific to their ionic molecules and can be described as either voltage-gated or ligand-gated (Sherwood 1993). These terms refer to the mechanism by which the ion channels are opened to allow passage of the ions into the intracellular space. In voltage-gated ion channels, the membrane potential of the cell determines whether the ion channel is open or closed. In ligand-gated ion channels, binding of a molecule to a receptor that is associated with the channel results in conformational changes in the ion channel that regulate the flux of ions into the cell. Given their physiologic importance, ion channels serve as target sites for several pharmacotherapies that are used clinically to treat a variety of diseases. Although covering all drugs acting on the voltage channels is beyond the scope of this chapter, we list are some important examples:

antiepileptic drugs (phenytoin and carbamazepine), which block voltage-gated Na^+ channels; antianginal drugs (verapamil and diltiazem), which block Ca^{++} channels; and antiarrhythmic drugs (dofetilide and ibutilide), which block K^+ channels.

G Protein–Coupled Receptors

In eukaryotes, GPCRs represent the largest and most diverse family of membrane receptors involved in signal transduction. They regulate an array of biological functions such as physiological homeostasis, metabolism, growth, proliferation, and differentiation of multiple cell types. That said, it is not surprising that more than 50% of the drugs used in clinical practice interfere with the GPCR signaling pathway by mimicking endogenous GPCR ligands (agonists), blocking ligand access to the receptor (antagonists), or modulating ligand production (Table 21.1).

GPCRs and G Protein Structure

Studies have shown that the human genome contains more than 800 genes that encode GPCRs (Fredriksson 2003). These genes give rise to more than 1000 distinct receptors secondary to alternative splicing. Human GPCRs can be classified into five main families according to their sequence homology: rhodopsin, secretin, glutamate, adhesion, and frizzled. The rhodopsin family of GPCRs is the largest, with more than 700 subfamilies identified thus far, of which 460 belong to the odorant receptor subfamily, with the rest being receptors for hormones (glucagon, vasopressin, angiotensin, calcitonin, bradykinin), neuropeptides (enkephalins, endorphins), eicosanoids (prostaglandins, leukotrienes), and neurotransmitters (dopamine, acetylcholine, histamine, epinephrine, serotonin).

Our initial understanding of the GPCRs dates back to the early 1980s, when rhodopsin was first sequenced and subsequently cloned. Cloning of the β2-adrenergic receptors as well as others followed in 1986. Although GPCRs are activated by a variety of agonists, they all share a common structure. The x-ray crystallographic studies showed that GPCRs are made up of seven transmembrane α helices (Palczewski 2000), an extracellular N-terminal, and an intracellular C-terminal bound to a G protein. The transmembrane domain is highly conserved among members of the GPCR family, whereas the N- and C-domains have a high degree of variability and complexity. The extracellular N-terminal plays a role in ligand binding, whereas the intracellular C-terminal is involved in signal transduction because it serves as a binding site for G protein.

G proteins act as molecular switches that have the capability to convey the changes occurring at the cell surface to the cell interior by modulating the activity of downstream effector proteins, which could potentially be enzymes or ion channels. In the inactive state, G proteins exist as a heterotrimer, where an α subunit binds to a dimer of the β and γ subunits. In humans, there are 23 Gα subunits encoded by 17 genes, 7 β subunits encoded by 5 genes, and finally 12 Gγ subunits (Downes 1999). The G proteins can be divided into four main classes according to the primary sequence homology of their α subunits: G_s, G_i, G_q, and $G_{12/13}$ (Simon 1991. The size of Gα proteins ranges from 35 to 52 kDa, and their helical and GTPase domains are involved in pivotal functions such as hydrolyzing GTP into guanosine diphosphate (GDP) plus phosphate and binding to the βγ dimer as well as to downstream effector molecules.

GPCR Signaling

When an agonist binds to the extracellular domain of a GPCR, it induces conformational changes in the structure of the receptor that cause the Gα subunit to dissociate itself from the βγ complex and exchange its bound GDP for GTP (Figure 21.9). This step is considered the rate-limiting step in the G protein activation and signal transduction processes. The three G protein–mediated signaling pathways are the adenylate cyclase (AC) pathway, the phospholipase C beta (PLCβ), and finally the RhoGTPase nucleotide exchange factor (RhoGEF) pathway. Depending on the type of activated Gα subunit, one of the three downstream effector molecules is affected. To further illustrate this, the $Gα_s$ subunit activates the AC signaling cascade (Beavo 2002), whereas the $Gα_q$ and the $G_{12/13}$ subunits activate the PLCβ and the RhoGEF pathways, respectively. By contrast, stimulation of the $Gα_i$ subunit is associated with inhibition of the AC signaling pathway. The G protein–signaling cascade remains active until the GTP bound to the Gα subunit is hydrolyzed. After hydrolysis, the βγ complex reassociates with the α subunit, and the heterotrimeric G protein returns to the receptor, where it can be reactivated upon agonist or ligand binding to the receptor.

AC Pathway

Adenylate cyclase is a membrane-bound enzyme that, on activation, catalyzes the conversion of ATP into cyclic adenosine monophosphate (cAMP) using Mg^{2+} as a cofactor for this particular reaction. The main target for cAMP is the holoenzyme, protein kinase A, which consists of two catalytic C subunits bound to two regulatory R subunits forming a tetramer (R2C2). Once the intracellular concentration of cAMP is increased, it binds to protein kinase A. Subsequently, the structure of protein kinase A undergoes substantial conformational changes that reduce the affinity of the 2C subunits for the 2R subunits. The 2C subunits can now phosphorylate a diverse array of targets (e.g., metabolic enzymes, transport proteins, and, most importantly, TFs known as cAMP responsive element-binding proteins [CREBs]). After phosphorylation, CREBs can recruit CREB-binding protein and thereby trigger the expression of several genes (tyrosine hydroxylase, iNOS [inducible macrophage-type nitric oxide synthase], angiotensinogen, glucocorticoid receptor) by virtue of the histone acetyltransferase (HAT) activity of CREB-binding protein (Mayr 2001).

PLCβ Pathway

The PLCs are cytosolic enzymes that migrate to the cytoplasm upon activation by the $G\alpha_q$ subunit. Their main function is to hydrolyze membrane phospholipids: phosphatidylinositol 1,4,5-triphosphate into diacylglycerol and 1,4,5-inositol triphosphate (IP3) (Patterson 2004). Each of these molecules has separate targets. Diacylglycerol activates the protein kinase C family, whereas IP3 diffuses to the endoplasmic reticulum to activate its own receptors and thereby stimulate the release of Ca^{2+}. After its release from the endoplasmic reticulum, Ca^{2+} binds to calmodulin, a Ca^{2+}-binding messenger protein, and then the complex activates a set of calmodulin-sensitive enzymes such as myosin light-chain kinase and phosphorylase kinase.

Clinical Implications

G protein–coupled receptors are considered the most exploited drug targets, given the diverse biological functions controlled by these receptors. As noted earlier, more than 50% of all drugs available on the market have GPCRs as their target (Table 21.1). Increased knowledge of the GPCRs and their ligands, as well as the advances in structure-based drug design, will undoubtedly pave the way for the development of novel and more selective GPCR drugs.

Receptor Tyrosine Kinases

The beginning of receptor tyrosine kinase (RTK) research goes back to 1952. Nevertheless, it was not until the early 1980s (Ushiro 1980) when the concept of signal transduction by tyrosine phosphorylation gained widespread support because of three reports showing that the epidermal growth factor receptor (EGFR), the insulin receptor, and the platelet-derived growth factor receptor

Table 21.1. Examples of Drugs That Target the GPCRs

Drug	Receptor	Examples of Approved Indications
Lorcaserin	Serotonin-2C (antagonist)	Obesity
Mirabegron	Adrenergic β_3 (agonist)	Overactive bladder
Aclidinium	M_3 (antagonist)	Chronic obstructive lung disease
Bosentan	Endothelin receptors A and B (antagonist)	Pulmonary artery hypertension
Sumatriptan, zolmitriptan	Serotonin-1D (agonists)	Migraine headaches
Atenolol, metoprolol, bisoprolol	Adrenergic β_1 (antagonists)	Heart failure, coronary artery diseases, hypertension
Diphenhydramine, cetirizine	H_1 (antagonists)	Allergic rhinitis, utricaria
Ranitidine, famotidine, nizatidine	H_2 (antagonists)	Duodenal and gastric ulcers, gastroesophageal reflux (GERD)
Alfuzocin, tamsulosin	Adrenergic α_1 (antagonists)	Benign prostate hypertrophy
Haloperidol	D_2 (antagonist)	Psychosis, delirium
Valsartan, losartan, telmisartan	AT_1 (antagonists)	Hypertension, heart failure

AT = angiotensin; D = dopamine; H = histamine; M = muscarinic.

are principally tyrosine kinases that require activation by their respective ligands.

Mammalian cells express a variety of membrane receptors characterized by an extracellular LBD and an intrinsic enzymatic activity residing within the cytoplasmic domain. These receptors include the RTKs, which have emerged as the key regulators of critical cellular processes such as proliferation, differentiation, cell survival, cell migration, and cell cycle control (Figure 21.9) (Blume-Jensen 2001). In the inactive state, the RTKs are monomeric. Binding of a signaling molecule causes the two neighboring RTKs to associate with each other and form cross-linked dimers (Ullrich 1990). Dimerization activates the intrinsic tyrosine activity of each RTK in the dimer, leading to autophosphorylation of the tyrosine residues in the receptor cytoplasmic domain. The phosphorylated tyrosine residues serve as docking sites for

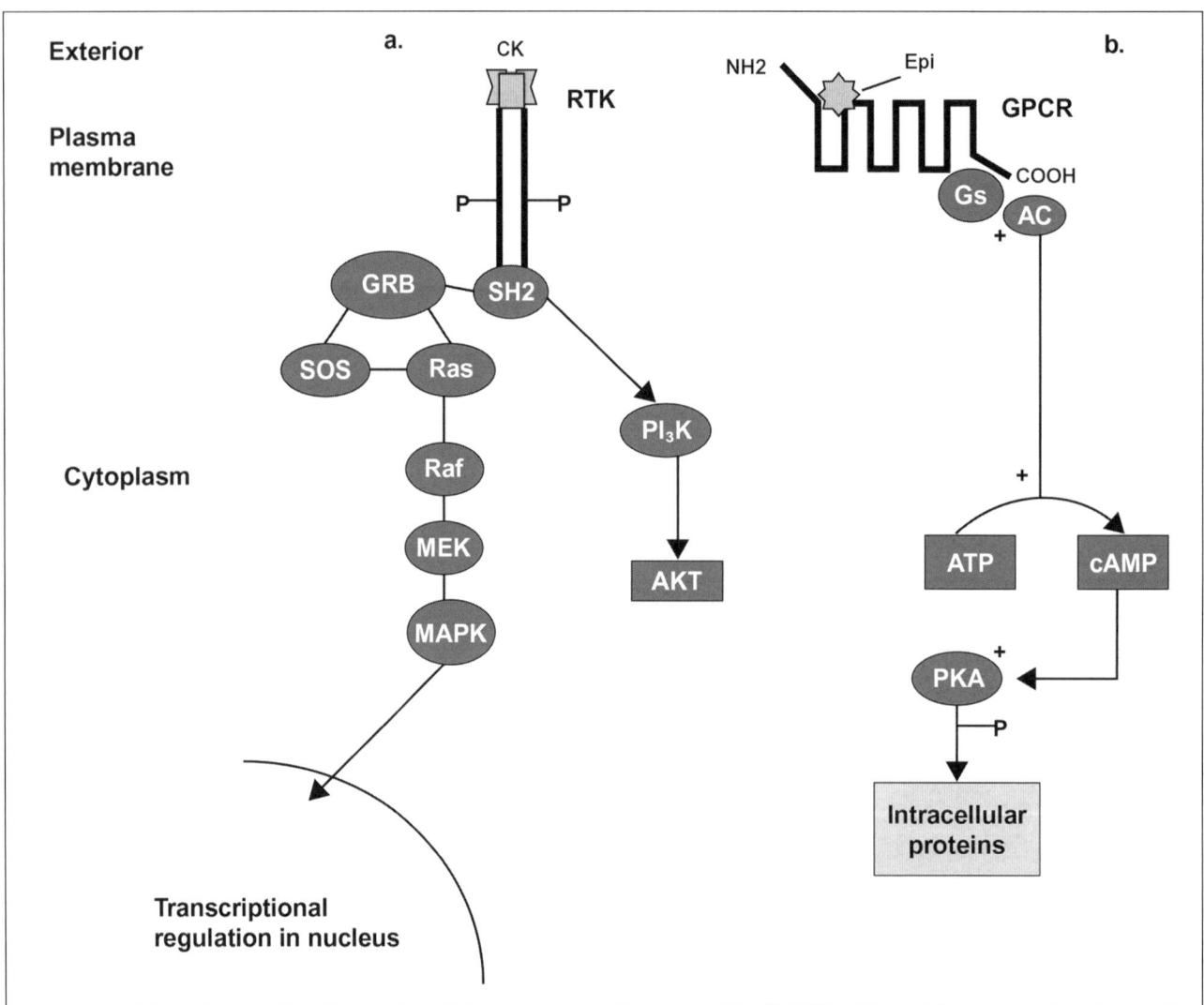

Figure 21.9. Receptor tyrosine kinase and GPCR signaling pathways. (a) RTKs are bound by an extracellular ligand such as insulin or growth factors, resulting in RTK dimerization. This allows the receptor to activate itself through autophosphorylation. The phosphorylated tyrosine residues are recognized by downstream effector molecules such as PI3K (phosphatidylinositol 3-kinase) and growth factor receptor–binding protein (GRB). Ras, a major protein in RTK signaling, is associated with GRB and SOS, a Ras-activating protein. Ras, which is then activated, further phosphorylates MEK, which in turn activates MAPK. Mitogen-activated protein kinase then phosphorylates various TFs to alter gene expression and elicit a particular biological response. (b) In this case, epinephrine binds to the β-adrenergic receptor, which is coupled with the Gs protein. The Gs protein is activated and subsequently stimulates AC. Adenylate cyclase converts ATP to cAMP, and cAMP activates protein kinase A, which phosphorylates various intracellular proteins resulting in vasodilation, bronchodilation, increased heart rate, and other effects.

AC = adenylate cyclase; ATP = adenosine triphosphate; cAMP = cyclic adenosine monophosphate; CK = cytokine; Epi = epinephrine; GPCR = G protein–coupled receptor; GRB = growth factor receptor–binding protein; MAPK = mitogen-activated protein kinase; MEK = MAPK/ERK kinase; P = phosphorylated residue; PKA = protein kinase A; RTK = receptor tyrosine kinase; SOS = son of sevenless protein; TF = transcription factor.

proteins that contain src homology-2 (SH-2) domains (Ferguson 2008) such as PI3K (phosphatidylinositol 3-kinase) α and β, which in turn increase the intracellular level of phosphatidylinositol 3,4,5-trisphosphate (PIP3). In turn, PIP3 perpetuates the RTK signaling cascade by acting as a docking site at the plasma membrane for other signaling molecules such as protein kinase B, also known as Akt, which subsequently regulates the activity of mTOR (mammalian target of rapamycin), one of the main drivers of cell growth, proliferation, and survival. In addition to recruiting enzymes with SH-2 domains, phosphorylated RTKs can attract adaptor molecules such as Grb2 that are devoid of any enzymatic activity but that can primarily attract guanine nucleotide exchange factors (GEFs) (Etienne-Manneville 2002). The son of sevenless protein is one of the GEFs involved in replacing GDP bound to Ras with GTP, thereby activating the Ras signaling pathway. The Ras pathway will be explained further in the subsequent sections.

Receptor-Associated Tyrosine Kinases

On their surface, mammalian cells express a family of receptors that can turn on a set of genes in a more rapid and direct manner relative to tyrosine kinases. Examples of such receptors are the cytokine (interleukin [IL]-2, IL-6, IL-7, IL-10 and interferon [IFN]-γ, IFN-α, and IFN-β), growth hormone, leptin, erythropoietin, thrombopoietin, G-CSF (granulocyte colony-stimulating factor), and GM-CSF (granulocyte-macrophage colony-stimulating factor) receptors, which are collectively termed *receptor-associated tyrosine kinases*. These receptors are involved in regulating hematopoiesis, inflammation, immune response, cell development, and survival. Unlike RTKs, receptor-associated tyrosine kinases do not possess an intrinsic tyrosine kinase activity. Instead, the cytoplasmic domain of the receptor forms a stable association with enzymes that have tyrosine kinase activity such as Janus kinases, or JAKs (JAK1, JAK2, JAK3, and Tyk2) (Gough 2008). The JAKs were named after Janus, the mythologic Roman god of gates with two faces, for having both true kinase and pseudokinase domains. Upon binding of a cytokine to its receptor, these kinases selectively phosphorylate and activate signal transducer and activator of transcription (STAT) proteins, which are ultimately responsible for regulating gene expression and bringing about cellular responses to stimuli (Hao 2002; Lower 2002; Zwick 2001; Robertson 2000).

The STAT proteins are seven structurally and functionally related proteins that range from 750 to 850 amino acids in length. Members of the STAT family of TFs share five conserved domains that include the N-terminal domain, the coiled-coil domain, the DBD, the linker domain, and the SH-2 tyrosine activation domain. Signal transducer and activator of transcription specificity is determined by an SH-2 domain whose sequence is quite divergent among STAT proteins and recognizes different phosphorylated motifs (Hao 2002; Lower 2002; Zwick 2001; Robertson 2000).

In the absence of specific receptor activation, STAT proteins exist as inactive transcriptional factors in the cytoplasm of the target cells. The binding of a cytokine to its cognate receptor induces receptor tyrosine phosphorylation by JAKs that specifically bind to the intracellular domains of cytokine receptors. The phosphorylated tyrosines then serve as docking sites for STAT proteins. The STAT proteins can then homodimerize or heterodimerize after being phosphorylated and are released from the receptor. The dimerized STATs are then rapidly translocated from the cytoplasm to the nucleus, where they bind to specific sequence elements (usually γ-activated sequence, or GAS, enhancers) on DNA and modulate the expression of target genes (Hao 2002; Lower 2002; Zwick 2001; Robertson 2000).

Clinical Implications

As previously mentioned, RTKs and receptor-associated tyrosine kinases regulate several key processes pertinent to cellular growth, survival, and neovascularization. In normal cells, activation of these receptors is tightly regulated by a wide range of signaling molecules or ligands. However, when their activation is abnormal because of a gain-of-function mutation, gene amplification, or gene rearrangement, they are causally involved in the development of several types of human cancers such as breast cancer, gastrointestinal stromal tumors, and lung cancers. Thus, blocking the RTKs and receptor-associated tyrosine kinases or their associated ligands using humanized antibodies or small molecules has emerged as a promising approach for the treatment of human cancers (Table 21.2).

Because of the complexity of the pathogenic alterations that occur in cancer, inhibiting a single target is less likely to be associated with therapeutic effectiveness. Hence, using a combination or a cocktail of agents or single targeted agents that inhibit multiple targets seems more promising because it offers a greater opportunity to optimize the response to therapy.

Ras, Raf, MEK, MAPK Signaling Pathways

The Ras proteins belong to a large family of GTP-binding proteins, which can be further classified into several families according to the extent of sequence homology. In general, the Ras family plays a crucial role in controlling normal cell growth or proliferation. Studies have shown that most human tumors harbor an activating point mutation in the genes that code for Ras proteins such as H-Ras, K-Ras, and N-Ras or their downstream signaling effectors (Lowy 1993) (Table 21.3).

The Ras proteins can undertake their normal function only if they undergo a posttranslational modification process whereby a 15-carbon isoprenoid chain is transferred to a cysteine residue close to the C-terminal of a Ras protein. This process is mediated by farnesyltransferase, and its sole purpose is to localize Ras proteins to their precise cellular compartment, which is the inner layer of the cytoplasmic membrane. Hence, mislocalization of Ras proteins inevitably abolishes their biological activity because they can no longer recruit their target enzymes or effectors (Seabra 1998).

The activity of Ras proteins is regulated by the ratio of bound GTP to GDP (the GDP-bound Ras is inactive, whereas the GTP-bound Ras is active) (Campbell 1998). Activation of RTKs (e.g., EGFRs) through the autophosphorylation of the tyrosine residues prompts the recruitment of SOS (son of sevenless protein) to the plasma membrane, thus allowing Ras-bound GDP to be replaced with GTP (Figure 21.9). The GTP-bound Ras can now activate its downstream signaling enzymes. The first enzyme to be activated in the Ras signaling cascade is the protein/threonine kinase Raf, which subsequently activates mitogen-activated protein kinase kinases 1 and 2 (MEK1 and MEK2). Phosphorylated MEK1 or MEK2 can now activate mitogen-activated protein kinases (MAPKs), extracellular signal-regulated kinases 1 and 2 (ERK1 and ERK2). The ERK1 and ERK2 substrates include nuclear proteins such as c-Jun, c-Myc, and c-Fos. Consequently, cell cycle regulatory proteins such as cyclins are expressed, which allow the cells to divide (Pruitt 2001).

The understanding of the Ras pathways and their downstream effector molecules has led to the discovery of new pharmacotherapies, namely vemurafenib and dabrafenib, which inhibit v-Raf murine sarcoma viral oncogene homolog B1 (*BRAF*). As more knowledge about this pathway becomes available, it will be possible to develop better and more effective drugs that will not only disrupt this pathway, which is heavily implicated in cancer

Table 21.2. Drugs Targeting RTKs

Drug Type	Drug	Molecular Target	Indication
Monoclonal antibody	Trastuzumab	HER2	Breast cancer
	Bevacizumab	VEGFR	Metastatic colorectal carcinoma Glioblastoma Metastatic non–small cell lung cancer
	Cetuximab	EGFR	Wild-type expressing metastatic colorectal carcinoma Head and neck cancer
	Panitumumab	EGFR	Wild-type expressing metastatic colorectal carcinoma
Small molecules	Erlotinib	EGFR	Metastatic non–small cell lung cancer with exon 19 deletions or exon 21 substitution mutations
	Gefitinib	EGFR	Metastatic non–small cell lung cancer with exon 19 deletions or exon 21 substitution mutations
	Sorafenib	VEGFR, PDGFR, c-kit, c-Raf	Hepatocellular carcinoma Advanced renal cell carcinoma
	Sunitinib	VEGFR, PDGFR, c-kit, flt-3	Advanced renal cell carcinoma Gastrointestinal stromal tumors (GISTs)
	Lapatinib	EGFR, HER2	HER2-positive breast cancer
	Axitinib	VEGFR	Advanced renal cell carcinoma

EGFR = epidermal growth factor receptor; HER2 = human epidermal growth factor receptor 2; PDGFR = platelet-derived growth factor receptor; RTK = receptor tyrosine kinase; VEGFR = vascular endothelial growth factor receptor.

pathogenesis, but also circumvent the resistance mechanisms that occur after *BRAF* inhibition such as *BRAF* amplification, mutations in downstream signaling pathways (MEK1), or activation of alternative signaling pathways such as EGFRs.

Nuclear Receptors

The human genome contains 48 genes encoding NRs that share a common structure as well as similar mechanisms of activation (Sonoda 2008). In general, the structure of an NR consists of a conserved N-terminal DBD and a C-terminal LBD. As the name implies, the LBD is responsible for ligand recognition and interaction with coactivators or corepressors. The NRs are often described as ligand-regulated TFs (i.e., they modulate the expression of numerous genes involved in a variety of biological processes such as cell proliferation, development, metabolism, and reproduction after ligand binding). Unlike other ligands that act by stimulating cell surface receptors, NR ligands such as estrogens, androgens, and glucocorticoids are capable of crossing the cytoplasmic membrane and interacting directly with the receptors inside the cell, which could be present either in the nucleus or in the cytoplasm.

According to DNA-binding characteristics, NRs can be broadly divided into three groups. The first group, which consists of glucocorticoid, mineralocorticoid, estrogen, and androgen receptors, binds to DNA as homodimers. They are present in the cytoplasm complexed to chaperone proteins (Echeverria 2010). Upon ligand binding, the receptor is released from the chaperone, allowing homodimerization and entry into the nucleus, where the activated receptor recognizes certain DNA sequences known as HREs (Glass 2000). The second group refers to receptors that can form heterodimers with the retinoid X receptors and includes the thyroid, retinoic acid receptor, vitamin D receptor, and peroxisome proliferator-activated receptors. In the absence of a ligand, the receptors associate with corepressor or histone deacetylase (HDAC) complexes (Watson 2012). The binding of ligands to these receptors results in corepressor dissociation and coactivator replacement, with the latter having enzymatic functions (HAT) to open up the compact chromatin. The last group functions in a manner similar to the first group. Nonetheless, they preferentially bind to DNA as monomers and not as homodimers (Mangelsdorf 1995).

Understanding the regulation of NR function has been the focus of intense research. Accordingly, the drug development process has evolved from designing synthetic drugs that can mimic the full function of the endogenous ligand to designing those that can modulate the function of these receptors (e.g., tamoxifen and raloxifene, which act as SERMs [selective estrogen receptor modulators]) (Jordan 2007). This characteristic distinguishes them from estrogen because their action varies from one type of tissue to another.

Epigenetics

Epigenetics is a relatively new and rapidly evolving field of biology that has the potential to become an integral component of modern medicine. Numerous studies have ascertained that epigenetic abnormalities are causative factors for the pathogenesis of several disease states such as cancer, pediatric syndromes, and autoimmune disorders. Although a universal definition of epigenetics remains somewhat elusive, the term can be simply defined as the study of heritable changes in gene (activity or function) expression that occur without any change in DNA sequence. Literally, the term *epigenetics* means "above" or "on top of" genetics. Hence, it refers to external modifications to DNA that turn genes "on" or "off." These modifications do not change the DNA sequence but, instead, affect how cells "read" genes.

Table 21.3. Human Cancers Associated with Mutations in the Ras Signaling Pathway

Defect	Tumor type	Frequency (%)
Ras mutation	Pancreas	90
	Non–small cell lung adenocarcinoma	35
	Colorectal	45
	Follicular thyroid	55
	Melanoma	15
	Bladder	10
	Liver	30
	Kidney	10
	Myelodysplastic syndrome	40
	Acute myeloid leukemia	30
BRAF mutation		
	Melanoma	66
	Colorectal	12

BRAF = v-Raf murine sarcoma viral oncogene homolog B1.

The genetic information is virtually the same in all cells of a multicellular organism. However, not all genes are expressed simultaneously in all cells. The culprit for this diverse gene expression profile in each cell is epigenetic modifications, which can be grouped into two main categories: DNA methylation and histone modifications. It is important to remember that an interplay or cross talk exists between these processes. In this chapter, we will provide or introduce a fundamental and practical framework for understanding the principles of epigenetic mechanisms and discuss how these principles pertain to human diseases and contribute to the development of novel targeted therapies. Epigenetic modification is divided into two major categories: histone modification and DNA methylation.

Histone Modification

In eukaryotes, DNA is wrapped around a protein core of eight histones (two copies of H2A, H2B, H3, and H4) to form nucleosomes, the smallest structural unit of chromatin. Histones are a family of basic proteins with an overall positive charge that allows them to associate with DNA. X-ray crystallographic studies have shown that each histone within a nucleosome consists of a globular domain and a highly dynamic N-terminal tail protruding from the globular domain. The extent to which nucleosomes are packed or condensed is a critical determinant of the transcriptional activity of the associated DNA, and this is partly mediated by chemical modifications of the N-terminal tail of histone proteins.

The core histones and their N-terminal tails contain many amino acids that can be modified. There are many different types of histone modifications: acetylation, methylation, phosphorylation, ubiquitinylation, sumoylation, deimination, and proline isomerization (Kouzarides 2007). Among these modifications or marks, acetylation, methylation, phosphorylation, and ubiquitinylation are the most common and well understood. Each of these modifications affects the interactions between DNA and histones, which ultimately alter gene transcriptions. In the subsequent sections, we will discuss acetylation, methylation, phosphorylation, and ubiquitinylation processes.

Histone Acetylation

Histone acetylation was first reported in 1964 (Allfrey 1964). Histone acetylation plays an important role in the regulation of transcription. In eukaryotic cells, chromosomes are packaged such that DNA is wrapped around histones about every 200 bp to form complexes called nucleosomes. One hundred forty-six base pairs of DNA are wound around a histone octamer, with the remaining 54 bp intervening between each nucleosome.

This organization greatly affects levels of transcription. The association between histone and DNA blocks the access of TFs and RNA pol II to the promoter (Cress 2000; Kouzarides 2000; Sterner 2000; Bjorklund 1999; Kornberg 1999).

Whereas histone acetylation often increases transcription, the lack of acetylation or histone deacetylation represses transcription. Histone acetylation is a reversible process catalyzed by HATs, and histone deacetylation is catalyzed by histone deacetylases (HDACs). The two main classes of HATs are type A and type B. The type B HATs reside predominantly in the cytoplasm, where they acetylate free histones but not those that are already deposited in chromatin. Type A HATs are located in the nucleus, and they can be further classified into at least three separate groups according to amino acid sequence homology as well as conformational structure: GNAT, MYST, and CREB-binding protein/p300. Histone acetyltransferases function by transferring an acetyl group from acetyl CoA (coenzyme A) to the amino group of specific lysine side chains in the N-terminal region of the histone. This removes positive charges and results in reduced affinity between the histone and DNA and generates more open DNA conformation. After relieving the bond between the histone and DNA, RNA pol II and TFs can easily access the promoter on the DNA, initiating the expression of the corresponding genes (Cress 2000; Kouzarides 2000; Sterner 2000; Bjorklund 1999; Kornberg 1999).

Histone deacetylases, in contrast, have an opposite effect on histones. These enzymes restore the positive charge of histones by removing the acetyl groups from the lysine residues. This action enables the histones to bind to DNA and consequently shield it from expression. Histone acetylation plays an important regulatory role during development, proliferation, differentiation, and gene expression. Abnormal acetylation or deacetylation results in various disorders such as leukemia, epithelial cancers, and genetic diseases associated with physical and cognitive abnormalities. Recent studies show that inhibition of HATs may be the primary cause of cellular pathogenesis in polyglutamine diseases such as Huntington disease (Marks 2001; Murata 2001; Cress 2000).

The association between histone acetylation and human diseases has raised the possibility of using pharmacologic manipulation of histone acetylation (e.g., with HDAC inhibitors) as a treatment strategy for several disorders. Several different classes of HDAC inhibitors have been identified: short-chain fatty acids (e.g., butyrates), hydroxamic acids (e.g., trichostatin A, oxamflatin, suberoylanilide hydroxamic acid [SAHA]), cyclic peptides (e.g., trapoxin A, apicidin), and benzamides. Some of these inhibitors like phenylbutyrate, pyroxamine, and SAHA are currently being investigated in clinical trials (Marks 2001; Cress 2000; Kouzarides 2000; Sterner 2000).

Histone Phosphorylation
Like histone acetylation, histone phosphorylation is a dynamic process, but it takes place on different amino acid residues in the N-terminal tail: serine, threonine, and tyrosine (Goto 2002). The extent of histone phosphorylation is controlled by kinases and phosphatases (Oki 2007). Histone kinases are responsible for the transfer of a phosphate moiety from ATP to the target amino acid. By imparting a negative charge to the histone tail, phosphorylation plays a role similar to histone acetylation with respect to modulating nucleosome dynamics. The electrostatic repulsion between DNA and the phosphorylated histone results in a loosening and opening of the histone/DNA complex in a way that renders it accessible to TFs. It is still unclear how kinases are accurately recruited to their site of action. The kinase may possess an intrinsic DBD or associate with a chromatin-bound factor to ligate it to DNA.

Histone Methylation
The targets for histone methylation are the side chains of the amino acids lysine and arginine. Unlike phosphorylation or acetylation, histone modification through methylation does not alter the overall charge of the histone proteins. Methyltransferases for both arginine and lysine catalyze the transfer of a methyl group from S-adenyl methionine (SAM) (Copeland 2009) to either the ω-guanidine group of arginine or the ε-amino group of lysine. Accordingly, lysine residues undergo mono-, di-, or trimethylation, whereas arginines undergo mono-, symmetric, or symmetric dimethylation. In general, histone methylation is associated with either transcription activation or repression.

Histone Ubiquitination
Histone ubiquitination was discovered more than 3 decades ago. Nevertheless, its exact functions are still not fully understood. In contrast to other posttranslational histone modification processes in which a small molecule is covalently attached, ubiquitination involves the addition of a 76-amino acid polypeptide, ubiquitin, to histone lysines. Ubiquitination is brought about through the sequential action of three enzymes. The first step is mediated by ubiquitin-activating enzyme (E1), which activates ubiquitin by an ATP-dependent reaction. This is followed by conjugating ubiquitin through a thioester bond to a cysteine residue in ubiquitin-conjugating enzyme (E2); finally, ubiquitin is transferred from E2 to a target lysine residue in a histone protein by the action of ubiquitin protein isopeptide ligase (E3) (Hershko 1998).

Of note, monoubiquitination of histones is the most dominant form of ubiquitination (i.e., histone proteins rarely undergo polyubiquitination, which in general targets them for degradation through the 28S proteasome complex). Monoubiquitination of histones, namely H2A and H2B, has clear implications on regulating gene transcription. Ubiquitination of H2A correlates with gene silencing, whereas ubiquitination of H2B correlates with activation of gene transcription through promoting other histone modifications and stimulating RNA pol II. Like other epigenetic processes, monoubiquitination of histones is reversible. The histone modification is removed by an isopeptidase enzyme known as the deubiquiting enzyme.

DNA Methylation
Historically, the discovery of DNA methylation occurred in about 1948 when Rollin Hotchkiss first noticed that a fraction of cytosine, which he called "modified cytosine," separated from cytosine in a manner similar to thymine (methyluracil) separation from uracil by paper chromatography. Furthermore, he suggested that this modified cytosine existed naturally in DNA. However, not until the 1980s did several studies show that DNA methylation was involved in numerous cellular processes such as gene regulation and cell differentiation. It is now well established that DNA methylation, in addition to several other regulatory processes, is a major epigenetic mechanism influencing gene activities. DNA methylation is one of the most common forms of DNA modification in eukaryotic cells that alters gene expression without changing the nucleotide sequencing. DNA methylation usually occurs in cytosine-phosphate-guanine (CpG) dinucleotides that cluster in regions called CpG islands with (G + C) and CpG content (Bird 2002).

DNA methylation is mediated by a class of enzymes, collectively known as DNA methyltransferases (DNMTs) in reference to the enzymatic reaction they catalyze, and involves the transfer of a methyl group from SAM to the fifth carbon of a cytosine residue in a CpG dinucleotide to form 5-methylcytosine. The methylation of cytosine usually takes place next to a guanine in a 5′-CG-3′ sequence. The CpG dinucleotides are not distributed randomly across the genome (i.e., they tend to cluster in regions called CpG islands, which range from 200 bp to several kilobases with a CpG content of 55%). The CpG islands, which are often unmethylated, exist in the promoter regions of actively transcribed housekeeping and tumor suppressor genes. In contrast, CpG islands located elsewhere are predominantly methylated and thus incapable of transcription activation.

DNA methylation can be reproduced during DNA replication and transferred to the next generation. In eukaryotes, the cytosines of many genes in the inactive state of transcription are highly methylated (modified to 5′-methylcytosine). Change from the inactive to the active state of transcription occurs through the removal of a methyl group from the cytosine specially located in the 5′-flanking region of the gene. It is proposed that DNA methylation suppresses gene transcription by keeping the chromatin in the condensed 30-nm fiber form that protects the DNA from being exposed to RNA pol and TFs. Studies have shown that DNA methylation plays a vital role in the transcription repression, neoplasia, and silencing of specific genes during development and cell differentiation. De novo methylation may also serve as a cellular defense mechanism to inactivate integrated foreign DNA. A complete understanding of methylation and demethylation may make these processes potential targets for the development of drugs that prevent and treat malignancies and viral infection (Kornberg 1999; Singal 1999).

Five DNMTs have been identified in mammals thus far: DNMT1, DNMT2, DNMT3a, DNMT3b, and DNMT3L. Of note, only DNMT1, DNMT3a, and DNMT3b possess methyltransferase activity (Okano 1998). Although these enzymes share a similar structure, they have distinctive functions as well as expression levels. From this perspective, DNMTs can be broadly classified as two main types: de novo DNMTs and maintenance DNMTs. DNMT1 is the most abundant DNMT, and it is mostly transcribed during the S phase of the cell cycle for the sole purpose of methylating hemimethylated sites generated during DNA replication (Gidekel 2002). DNMT1 locates to the replication fork, where it binds to the newly formed hemimethylated DNA strand and then methylates it in a manner that accurately resembles the methylation patterns in the original or the parent strand. For this main reason, DNMT1 is termed *a maintenance DNMT* (i.e., it maintains the original methylation pattern of DNA in a cell lineage). Both DNMT3a and DNMT3b constitute the other type of DNMTs, known as "de novo DNA methyltransferases" because they introduce methylation into naked DNA. DNMT3a and DNMT3b have no preference for hemimethylated DNA. When these enzymes are expressed, they catalyze the methylation of native as well as newly synthesized DNA. What distinguishes DNMT3a from DNMT3b is the expression pattern. DNMT3a is expressed ubiquitously, whereas DNMT3b is poorly expressed in tissues, except for the thyroid gland and bone marrow. It remains unclear, however, how de novo DNMTs target specific DNA regions for methylation. One of the proposed hypotheses suggests that TFs play a role in regulating DNA methylation by binding to a specific DNA sequence and thereby protect it from methylation. According to this hypothesis, only CpG sites across the genome that are unprotected by bound TFs undergo methylation by de novo DNMTs.

The intriguing question that arises is the functional consequences of DNA methylation. Studies have shown that, in most cases, DNA methylation through DNMTs triggers complete silencing or suppression of the transcription of the associated gene (i.e., this epigenetic process shuts off the gene function). Of note, the impact or the consequences of DNA methylation are location- or site-dependent. Suppression of gene transcription occurs only if the CpG islands located in the promoter regions are methylated (noncoding regions upstream from transcription start site). Bisulfite sequencing is the gold standard for DNA methylation analysis. In this method, DNA is first treated with sodium bisulfite to convert unmethylated cytosines to uracils and is then followed by conventional sequencing reaction. The unmethylated cytosine is read as thymine, whereas methylated cytosine is read as cytosine.

Cross Talk Between DNA Methylation and Histone Modification

Although DNA methylation and histone modification are carried out by different enzymes, accumulating evidence suggests a cross talk or interplay between these processes that modulate gene transcription. This cross talk is mediated by a group of proteins with methyl DNA-binding activity: methyl CpG-binding protein 2 (MeCP2), methyl CpG–binding domain protein 1 (MBD1), and KAISO, also known as ZBTB33 (zinc-finger and BTB domain containing protein 33). These proteins localize to the methylated promoter regions and recruit a protein complex that contains HDACs and histone methyltransferases to further repress gene expression (Geiman 2004).

Clinical Implications

A growing body of evidence points to the deregulation of the epigenetic machinery as the underlying etiology for the incidence of several disease states such as cancer. That epigenetic modifications are reversible, unlike genetic changes, has aroused much interest among investigators to develop new therapeutic options with the potential to disrupt these epigenetic processes/modifications by blocking the activity of the implicated enzymes. Such agents may be better termed *epigenetic therapy* because they can restore normal epigenetic patterns as well as prevent cells from acquiring further modifications to their DNA. Only a handful of these agents have already been approved by the U.S. Food and Drug Administration (FDA) (Table 21.4), but several others are being evaluated in clinical trials (Table 21.5).

As stated in the previous sections, DNA methylation is associated with the silencing of gene expression, particularly when the epigenetic modifications occurred as a result of pathological conditions. Azacitidine and decitabine, both of which inhibit DNMTs, are considered the prototypical epigenetic therapies. They underwent extensive investigation before their approval and use in clinical practice for the treatment of myelodysplastic syndromes. Initially, azacitidine and decitabine were developed as cytotoxic agents (Sorm 1964), but later, they were discovered to be powerful inhibitors of DNMTs; they thereby induce gene expression as well as cell differentiation. Of note, azacitidine and decitabine are active only when cells are in the S phase, where they serve as powerful inhibitors of DNA methylation (Constantinides 1977). One of the main concerns with using these agents is their lack of specificity attributable to hypomethylation of the whole genome (decrease in the level of methylation), the implications of which may lead to the activation of genes that are potentially deleterious. This clearly suggests that room still exists for improving the efficacy of such agents.

Because epigenetic silencing can also result from histone deacetylation, two HDAC inhibitors, vorinostat and romidepsin, were label approved by the FDA in the late 2000s for the treatment of T-cell lymphoma. Nevertheless, the exact mechanism through which HDAC inhibitors halt tumor progression has not yet been fully elucidated. However, several suggestions have been put forth. For example, HDAC inhibitors may induce CDK inhibitors, such as p21, which cause cell cycle arrest. They may also trigger the activation of apoptotic pathways in which several factors become engaged, such as NFκB (nuclear factor kappa beta), JNK (c-Jun N-terminal kinase), or BCL-2 (B-cell lymphoma 2). Because enthusiasm is continually building for this class of therapies, more agents are being developed (Marks 2003). For example, panobinostat is an HDAC inhibitor that is currently being investigated for the treatment of several hematologic malignancies.

Further work is needed to understand the substantial contributions of epigenetics to the incidence of diseases. Although deciphering the intricacies of the epigenetic modifications may sound quite challenging, it is worthwhile pursuing this path because epigenetics has the potential to guide the development of new therapeutic agents in the near future.

Summary

Molecular and cellular biology not only serve as the foundations of biological science, but are also dynamic disciplines, evolving in ways that result in technological and pharmacologic advances in the diagnosis and management of patients with a spectrum of clinical diseases. By understanding gene processing and regulation, cellular signaling, and epigenetics, practitioners can begin to narrow the gap between the basic and clinical sciences.

Table 21.4. Selected Epigenetic Therapies Label Approved by the FDA

Agent (brand name)	Class	Approval Date	Approved Indication	Main Adverse Effects
Azacitidine (Vidaza)	DNMT inhibitor	2004	MDS	Myelosuppression (thrombocytopenia, neutropenia, anemia)
Belinostat (Beleodaq)	HDAC inhibitor	2014	Relapsed or refractory peripheral T-cell lymphoma (PTCL)	Fatigue, anemia, thrombocytopenia, diarrhea, nausea
Decitabine (Dacogen)	DNMT inhibitor	2006	MDS	Myelosuppression (thrombocytopenia, neutropenia, anemia)
Vorinostat (Zolinza)	HDAC inhibitor	2006	Cutaneous T-cell lymphoma	Diarrhea, fatigue, nausea, anorexia
Romidepsin (Istodax)	HDAC inhibitor	2009	Cutaneous T-cell lymphoma	Nausea, fatigue, anemia, thrombocytopenia, ECG T-wave changes

DNMT = DNA methyltransferase; ECG = electrocardiogram; HDAC = histone deacetylase; MDS = myelodysplastic syndromes.

Table 21.5. Selected Epigenetic Therapies in Clinical Development

Agent (brand name)	Class	Development Stage
Panobinostat (LBH589)	HDAC inhibitor	Phase III trial for treatment of Hodgkin lymphoma and multiple myeloma; phase II/III trial of cutaneous T-cell lymphoma
Entinostat (MS-275)	HDAC inhibitor	Phase I and II trials for several indications such as Hodgkin lymphoma, kidney cancer, and breast cancer
Prancinostat (SB939)	HDAC inhibitor	Phase II trials for treatment of myelodysplastic syndromes (MDS), acute myeloid leukemia (AML), metastatic/recurrent sarcomas
Givinostat	HDAC inhibitor	Phase II study of myeloproliferative neoplasms
Phenelzine sulfate	HDM inhibitor	Phase II study of prostate cancer
Epigallocatechin gallate (green tea extract)	DNMT inhibitor	Phase II trial for treatment of multiple myeloma
Valproic acid	HDAC inhibitor	Phase II trial for treatment of breast cancer

DNMT = DNA methyltransferase; HDAC = histone deacetylase; HDM = histone demethylase.

With the advancement of molecular-based drug therapies through genomics research, the clinician will increasingly be called on to translate approaches to patient care from the scientific bench to the patient's bedside. A functional understanding of molecular and cellular biology principles is imperative in order to identify new drug targets, optimize currently available therapies, and, most importantly, improve patient outcomes.

References

Agard DA. To fold or not to fold. Science 1993;260:1903-4.

Allfrey VG, Faulkner R, Mirsky AE. Acetylation and methylation of histones and their possible role in the regulation of RNA synthesis. Proc Natl Acad Sci U S A 1964;51:786-94.

Arnez JG, Moras D. Structural and functional considerations of the aminoacylation reaction. Trends Biochem Sci 1997;22:211-6.

Asturias FJ, Kornberg RD. Protein crystallization on lipid layers and structure determination of the RNA polymerase II transcription initiation complex. J Biol Chem 1999;274:6813-6.

Baker TA, Bell SP. Polymerases and the replisome: machines within machines. Cell 1998;92:295-305.

Bambara RA, Murante RS, Henricksen LA. Enzymes and reactions at the eukaryotic DNA replication fork. J Biol Chem 1997;272:4647-50.

Beato M, Klug J. Steroid hormone receptors: an update. Hum Reprod Update 2000;6:225-36.

Beavo JA, Brunton LL. Cyclic nucleotide research still expanding after half a century. Nat Rev Mol Cell Biol 2002;3:710-8.

Bentley D. The mRNA assembly line: transcription and processing machines in the same factory. Curr Opin Cell Biol 2002;14:336-42.

Bird A. DNA methylation patterns and epigenetic memory. Genes Dev 2002;16.6-21.

Bjorklof K, Lundstrom K, Abuin L, et al. Co- and posttranslational modification of the alpha(1B)-adrenergic receptor: effects on receptor expression and function. Biochemistry 2002;41:4281-91.

Bjorklund S, Almouzni G, Davidson I, et al. Global transcription regulators of eukaryotes. Cell 1999;96:759-67.

Blume-Jensen P, Hunter T. Oncogenic kinase signaling. Nature 2001;411:355-65.

Bolsover SR, Hyams JS, Jones S, et al. Translation and protein targeting. In: Bolsover SL, Hyams JS, Jones S, et al., eds. From Genes to Cells. New York: John Wiley & Sons, 1997:189-208.

Buratowski S. The basics of basal transcription by RNA polymerase II. Cell 1994;77:1-3.

Burgess-Beusse B, Farrell C, Gaszner M, et al. The insulation of genes from external enhancers and silencing chromatin. Proc Natl Acad Sci U S A 2002;1:1.

Burley SK, Roeder RG. Biochemistry and structural biology of transcription factor IID (TFIID). Annu Rev Biochem 1996;65:769-99.

Campbell SL, Khosravi-Far R, Rossman KL, et al. Increasing the complexity of Ras signaling. Oncogene 1998;17:1395-413.

Chaudhuri J, Si K, Maitra U. Function of eukaryotic translation initiation factor 1A (eIF1A) (formerly called eIF-4C) in initiation of protein synthesis. J Biol Chem 1997;272:7883-91.

Cocito C, Di Giambattista M, Nyssen E, et al. Inhibition of protein synthesis by streptogramins and related antibiotics. J Antimicrob Chemother 1997;39(suppl A):7-13.

Conaway JW, Shilatifard A, Dvir A, et al. Control of elongation by RNA polymerase II. Trends Biochem Sci 2000;25:375-80.

Constantinides PG, Jones PA, Gevers W. Functional striated muscle cells from non myoblast precursors following 5-azacytidine treatment. Nature 1977;267:364-6.

Cooper GM. Replication, maintenance, and rearrangements of genomic DNA. In: Cooper GM, ed. The Cell: A Molecular Approach. Sunderland, MA: Sinauer Associates, 1997a:175-224.

Cooper GM. RNA synthesis and processing. In: Cooper GM, ed. The Cell: A Molecular Approach. Sunderland, MA: Sinauer Associates, 1997b:255-72.

Cooper GM. Protein synthesis, processing, and regulation. In: Cooper GM, ed. The Cell: A Molecular Approach. Sunderland, MA: Sinauer Associates, 1997c:273-311.

Copeland RA, Solomon ME, Richon VM. Protein methyltransferases as a target class for drug discovery. Nat Rev Drug Discov 2009;8:724-32.

Cowell IG. Repression versus activation in the control of gene transcription. Trends Biochem Sci 1994;19:38-42.

Craig EA. Chaperones: helpers along the pathways to protein folding. Science 1993;260:1902-3.

Cress WD, Seto E. Histone deacetylases, transcriptional control, and cancer. J Cell Physiol 2000;184:1-16.

Decker CJ, Parker R. Mechanisms of mRNA degradation in eukaryotes. Trends Biochem Sci 1994;19:336-40.

Derrien T, Johnson R, Bussotti G, et al. GENCODE v7 catalog of human long noncoding RNAs. Genome Res 2012;22:1775-89.

Divoky V, Bisse E, Wilson JB, et al. Heterozygosity for the IVS-I-5 (G-->C) mutation with a G-->A change at codon 18 (Val-->Met; Hb Baden) in cis and a T-->G mutation at codon 126 (Val-->Gly; Hb Dhonburi) in trans resulting in a thalassemia intermedia. Biochim Biophys Acta 1992;1180:173-9.

Djebali S, Davis CA, Merkel A, et al. Landscape of transcription in human cells. Nature 2012;489:101-8.

Downes GB, Gautam N. The G protein subunit gene families. Genomics 1999;62:544-52.

Dutta A, Bell SP. Initiation of DNA replication in eukaryotic cells. Annu Rev Cell Dev Biol 1997;13:293-332.

Dvir A, Conaway JW, Conaway RC. Mechanism of transcription initiation and promoter escape by RNA polymerase II. Curr Opin Genet Dev 2001;11:209-14.

Echeverria PC, Picard D. Molecular chaperones, essential partners of steroid hormone receptors for activity and mobility. Biochim Biophys Acta 2010;1803:641-9.

ENCODE Project Consortium, Bernstein BE, Birney E, et al. An integrated encyclopedia of DNA elements in the human genome. Nature 2012;489:57-74.

Etienne-Manneville S, Hall A. Rho GTPases in cell biology. Nature 2002;420:629-35.

Ferguson KM. Structure based view of epidermal growth factor receptor regulation. Annu Rev Biophys 2008;37:353-73.

Fredriksson R, Lagerstrom MC, Lundin LG, et al. The G protein coupled receptors in the human genome from five main families. Phylogenetic analysis, paralogon groups, and fingerprints. Mol Pharmacol 2003;63:1256-72.

Funder JW. Glucocorticoid and mineralocorticoid receptors: biology and clinical relevance. Annu Rev Med 1997;48:231-40.

Garcia-Blanco MA, Baraniak AP, Lasda EL, et al. Alternative splicing in disease and therapy. Nat Biotechnol 2004;22:535-46.

Geiman TM, Sankpai UT, Robertson AK, et al. DNMT3B interacts with hSNF2H chromatin remodeling enzyme, HDACs 1 and 2 and components of the histone methylation system. Biochem Biophys Res Commun 2004;318:544-55.

Gereben B, Goncalves C, Harney JW, et al. Selective proteolysis of human type 2 deiodinase: a novel ubiquitin-proteasomal mediated mechanism for regulation of hormone activation. Mol Endocrinol 2000;14:1697-708.

Gidekel S, Bergman Y. A unique developmental pattern of Oct-3/4 DNA methylation is controlled by a *cis* demodification element. J Biol Chem 2002;277:34521-30.

Glass CK, Rosenfeld MG. The coregulator exchange in transcriptional functions of nuclear receptors. Genes Dev 2000;14:121-41.

Goessling LS, Daniels-McQueen S, Bhattacharyya-Pakrasi M, et al. Enhanced degradation of the ferritin repressor protein during induction of ferritin messenger RNA translation. Science 1992;256:670-3.

Goldstein DB, Tate SK, Sisodiya SM. Pharmacogenetics goes genomic. Nat Rev Genet 2003;4:937-47.

Goldstein SA. Ion channels: structural basis for function and disease. Semin Perinatol 1996;20:520-30.

Goto H, Yasui Y, Nigg EA, et al. Aurora B phosphorylates histone H3 at serine 28 with regard to the mitotic chromosome condensation. Genes Cells 2002;7:11-7.

Gough DJ, Levy DE, Johnstone RW, et al. IFNγ signaling does it mean JAK-STAT? Cytokine Growth Factor Rev 2008;19:383-94.

Hanna-Rose W, Hansen U. Active repression mechanisms of eukaryotic transcription repressors. Trends Genet 1996;12:229-34.

Hao D, Rowinsky EK. Inhibiting signal transduction: recent advances in the development of receptor tyrosine kinase and Ras inhibitors. Cancer Invest 2002;20:387-404.

Harrison SC. A structural taxonomy of DNA-binding domains. Nature 1991;353:715-9.

Harrison SC, Aggarwal AK. DNA recognition by proteins with the helix-turn-helix motif. Annu Rev Biochem 1990;59:933-69.

Hartl FU. Molecular chaperones in cellular protein folding. Nature 1996;381:571-9.

He Y, Vogelstein B, Velculescu VE, et al. The antisense transcriptomes of human cells. Science 2008;322:1855-7.

Herbert A, Rich A. RNA processing and the evolution of eukaryotes. Nat Genet 1999;21:265-9.

Hershko A, Ciechanover A. The ubiquitin system. Annu Rev Biochem 1998;67:425-79.

Heurgue-Hamard V, Karimi R, Mora L, et al. Ribosome release factor RF4 and termination factor RF3 are involved in dissociation of peptidyl-tRNA from the ribosome. Embo J 1998;17:808-16.

Hubscher U, Maga G, Spadari S. Eukaryotic DNA polymerases. Annu Rev Biochem 2002;71:133-63.

Huby RD, Dearman RJ, Kimber I. Why are some proteins allergens? Toxicol Sci 2000;55:235-46.

Ignarro LJ. Endothelium-derived nitric oxide: pharmacology and relationship to the actions of organic nitrate esters. Pharm Res 1989;6:651-9.

Issa LL, Leong GM, Eisman JA. Molecular mechanism of vitamin D receptor action. Inflamm Res 1998;47:451-75.

Jacobson A, Peltz SW. Interrelationships of the pathways of mRNA decay and translation in eukaryotic cells. Annu Rev Biochem 1996;65:693-739.

Jentsch S. When proteins receive deadly messages at birth. Science 1996;271:955-6.

Jentsch S, Schlenker S. Selective protein degradation: a journey's end within the proteasome. Cell 1995;82:881-4.

Jordan VC, O'Malley BW. Selective estrogen receptor modulators and antihormonal resistance in breast cancer. J Clin Oncol 2007;25:5815-24.

Khorasanizadeh S, Rastinejad F. Nuclear-receptor interactions on DNA-response elements. Trends Biochem Sci 2001;26:384-90.

Kim VN, Nam JW. Genomics of microRNA. Trends Genet 2006;22:165-73.

Kloosterman WP, Plasterk RH. The diverse functions of microRNAs in animal development and disease. Dev Cell 2006;11:441-50.

Kornberg RD. Eukaryotic transcriptional control. Trends Cell Biol 1999;9:M46-49.

Korzheva N, Mustaev A. Transcription elongation complex: structure and function. Curr Opin Microbiol 2001;4:119-25.

Kouzarides T. SnapShot: histone-modifying enzymes. Cell 2007;131:822.

Kouzarides T. Acetylation: a regulatory modification to rival phosphorylation? EMBO J 2000;19:1176-9.

Kozak M. Regulation of translation in eukaryotic systems. Annu Rev Cell Biol 1992;8:197-225.

Kozak M. Recognition of AUG and alternative initiator codons is augmented by G in position +4 but is not generally affected by the nucleotides in positions +5 and +6. EMBO J 1997;16:2482-92.

Latchman DS. Transcription-factor mutations and disease. N Engl J Med 1996;334:28-33.

Lee JH, Choi SK, Roll-Mecak A, et al. Universal conservation in translation initiation revealed by human and archaeal homologs of bacterial translation initiation factor IF2. Proc Natl Acad Sci U S A 1999;96:4342-7.

Liu F, Zaidi T, Iqbal K, et al. Role of glycosylation in hyperphosphorylation of tau in Alzheimer's disease. FEBS Lett 2002;512:101-6.

Livneh Z. DNA damage control by novel DNA polymerases: translesion replication and mutagenesis. J Biol Chem 2001;276:25639-42.

Lodish H, Berk A, Zipursky SL, et al. RNA processing, nuclear transport, and post-transcription control. In: Losish H, Berk A, Zipursky SL, et al., eds. Molecular Cell Biology. New York: W.H. Freeman, 2000:404-94.

Lowes VL, Ip NY, Wong YH. Integration of signals from receptor tyrosine kinases and g protein-coupled receptors. Neurosignals 2002;11:5-19.

Lowy DR, William BM. Function and regulation of Ras. Rev Biochem 1993;62:851-91.

Mangelsdorf DJ, Thummel C, Beato M, et al. The nuclear receptor superfamily: the second decade. Cell 1995;83:835-9.

Marks PA, Miller T, Richon VM. Histone deacetylases. Curr Opin Pharmacol 2003;3:344-51.

Marks PA, Richon VM, Breslow R, et al. Histone deacetylase inhibitors as new cancer drugs. Curr Opin Oncol 2001;13:477-83.

Mayr B, Montminy M. Transcriptional regulation by the phosphorylation dependent factor CREB. Nat Rev Mol Cell Biol 2001;2:599-609.

McKeown M. Alternative mRNA splicing. Annu Rev Cell Biol 1992;8:133-55.

McKnight SL. Molecular zippers in gene regulation. Sci Am 1991;264:54-64.

Mehta A, Carrouee S, Conyers B, et al. Inhibition of hepatitis B virus DNA replication by imino sugars without the inhibition of the DNA polymerase: therapeutic implications. Hepatology 2001;33:1488-95.

Mercer TR, Dinger ME, Mattick JS. Long non-coding RNAs: insights into functions. Nat Rev Genet 2009;10:155-9.

Miranda FF, Teigen K, Thorolfsson M, et al. Phosphorylation and mutations of Ser16 in human phenylalanine hydroxylase. Kinetic and structural effects. J Biol Chem 2002;15:15.

Modrek B, Lee CJ. Alternative splicing in the human, mouse and rat genomes is associated with an increased frequency of exon creation and/or loss. Nat Genet 2003;34:177-80.

Murata T, Kurokawa R, Krones A, et al. Defect of histone acetyltransferase activity of the nuclear transcriptional co-activator CBP in Rubinstein-Taybi syndrome. Hum Mol Genet 2001;10:1071-6.

Myer VE, Young RA. RNA polymerase II holoenzymes and subcomplexes. J Biol Chem 1998;273:27757-60.

Nakamura Y, Ito K, Isaksson LA. Emerging understanding of translation termination. Cell 1996;87:147-50.

Nikolov DB, Chen H, Halay ED, et al. Crystal structure of a human TATA box-binding protein/TATA element complex. Proc Natl Acad Sci U S A 1996;93:4862-7.

Ogawa T, Okazaki T. Discontinuous DNA replication. Annu Rev Biochem 1980;49:421-57.

Okano M, Xie S, Li E, et al. Cloning and characterization of a family of novel mammalian DNA (cytosine-5) methyltransferases. Nat Genet 1998;19:219-20.

Oki M, Aihara H, Ito T. The role of histone phosphorylation in chromatin dynamics and its implications in diseases. Subcell Biochem 2007;41:319-36.

Palczewski K, Kumasaka T, Hori T, et al. Crystal structure of rhodopsin: a G protein coupled receptor. Science 2000;289:739-45.

Patterson RL, Boehning D, Snyder SH. Inositol 1,4,5-triphosphate receptors as signal integrators. Annu Rev Biochem 2004;73:437-65.

Pedersen AG, Baldi P, Chauvin Y, et al. The biology of eukaryotic promoter prediction—a review. Comput Chem 1999;23:191-207.

Pestova TV, Borukhov SI, Hellen CU. Eukaryotic ribosomes require initiation factors 1 and 1A to locate initiation codons. Nature 1998;394:854-9.

Pestova TV, Lomakin IB, Lee JH, et al. The joining of ribosomal subunits in eukaryotes requires eIF5B. Nature 2000;403:332-5.

Pruitt K, Der CJ. Ras and Rho regulation of cell cycle and oncogenes. Cancer Lett 2001;171:1-10.

Pulleyblank DE. Of topo and Maxwell's dream. Science 1997;277:648-9.

Rinn JL, Kertesz M, Wang JK, et al. Functional demarcation of active and silent chromatin domains in human HOX loci by noncoding RNAs. Cell 2007;129:1311-23.

Robertson SC, Tynan J, Donoghue DJ. RTK mutations and human syndromes: when good receptors turn bad. Trends Genet 2000;16:368.

Rodnina MV, Wintermeyer W. Ribosome fidelity: tRNA discrimination, proofreading and induced fit. Trends Biochem Sci 2001;26:124-30.

Roeder RG. The role of general initiation factors in transcription by RNA polymerase II. Trends Biochem Sci 1996;21:327-35.

Rosenthal N. Regulation of gene expression. N Engl J Med 1994;331:931-3.

Sachs AB, Sarnow P, Hentze MW. Starting at the beginning, middle, and end: translation initiation in eukaryotes. Cell 1997;89:831-8.

Seabra MC. Membrane association and targeting of prenylated Ras like GTPases. Cell Signal 1998;10:167-72.

Sharp PA. Split genes and RNA splicing. Cell 1994;77:805-15.

Sherwood L. Neuronal physiology. In: Sherwood L, ed. Human Physiology. St. Paul, MN: West Publishing, 1993:79-103.

Simon MI, Strathmann MP, Gautam N. Diversity of G proteins in signal transduction. Science 1991;252:802-8.

Singal R, Ginder GD. DNA methylation. Blood 1999;93:4059-70.

Sonoda J, Pei L, Evans RM. Nuclear receptors: decoding metabolic disease. FEBS Lett 2008;582:2-9.

Sorm F, Piskala A, Cibak A, et al. 5-azacytidine a new highly effective cancerostatic. Experienta 1964;20:202-3.

Spiro RG. Protein glycosylation: nature, distribution, enzymatic formation, and disease implications of glycopeptide bonds. Glycobiology 2002;12:43R-56R.

Sterner DE, Berger SL. Acetylation of histones and transcription-related factors. Microbiol Mol Biol Rev 2000;64:435-59.

Stillman B. Smart machines at the DNA replication fork. Cell 1994;78:725-8.

Stillman B. Cell cycle control of DNA replication. Science 1996;274:1659-64.

Takemura M. Evolution and degeneration of eukaryotic DNA replication system. Biosystems 2002;65:139-45.

Tanaka K, Wood RD. Xeroderma pigmentosum and nucleotide excision repair of DNA. Trends Biochem Sci 1994;19:83-6.

Taubes G. Misfolding the way to disease. Science 1996;271:1493-5.

Tenbaum S, Baniahmad A. Nuclear receptors: structure, function and involvement in disease. Int J Biochem Cell Biol 1997;29:1325-41.

Thach RE. Cap recap: the involvement of eIF-4F in regulating gene expression. Cell 1992;68:177-80.

Tjian R. Molecular machines that control genes. Sci Am 1995;272:54-61.

Torchia J, Glass C, Rosenfeld MG. Co-activators and co-repressors in the integration of transcriptional responses. Curr Opin Cell Biol 1998;10:373-83.

Twyman RM, Wisden W. Replication. In: Twyman RM, ed. Advanced Molecular Biology: A Concise Reference. New York: Bios Scientific Publishers, 1998:389-409.

Ullrich A, Schlessinger J. Signal transduction by receptors with tyrosine kinase activity. Cell 1990;61:203-12.

Ushiro H, Cohen S. Identification of phosphotyrosine as a product of epidermal growth factor activated protein kinase in A-431 cell membranes. J Biol Chem 1980;255:8363-5.

Van Rooij E. The art of microRNA research. Circ Res 2011;108:219-34.

Waga S, Stillman B. The DNA replication fork in eukaryotic cells. Annu Rev Biochem 1998;67:721-51.

Wang JC. DNA topoisomerases. Annu Rev Biochem 1996;65:635-92.

Washington MT, Johnson RE, Prakash L, et al. Accuracy of lesion bypass by yeast and human DNA polymerase eta. Proc Natl Acad Sci U S A 2001;98:8355-60.

Watson PJ, Fairall L, Schwabe JW, et al. Nuclear hormone receptor corepressor: structure and function. Mol Cell Endocrinol 2012;348:440-9.

Werner T. Models for prediction and recognition of eukaryotic promoters. Mamm Genome 1999;10:168-75.

Wold MS. Replication protein A: a heterotrimeric, single-stranded DNA-binding protein required for eukaryotic DNA metabolism. Annu Rev Biochem 1997;66:61-92.

Wolfe SA, Nekludova L, Pabo CO. DNA recognition by Cys2His2 zinc finger proteins. Annu Rev Biophys Biomol Struct 2000;29:183-212.

Wu J, Parkhurst KM, Powell RM, et al. DNA bends in TATA-binding protein-TATA complexes in solution are DNA sequence-dependent. J Biol Chem 2001;276:14614-22.

Yavuz S, Yavuz AS, Kraemer KH, et al. The role of polymerase eta in somatic hypermutation determined by analysis of mutations in a patient with xeroderma pigmentosum variant. J Immunol 2002;169:3825-30.

Yokoi T, Nakajima M. microRNAs as mediators of drug toxicity. Annu Rev Pharmacol Toxicol 2013;53:377-400.

Young RA. RNA polymerase II. Annu Rev Biochem 1991;60:689-715.

Yuzhakov A, Turner J, O'Donnell M. Replisome assembly reveals the basis for asymmetric function in leading and lagging strand replication. Cell 1996;86:877-86.

Zawel L, Reinberg D. Common themes in assembly and function of eukaryotic transcription complexes. Annu Rev Biochem 1995;64:533-61.

Zitzmann N, Mehta AS, Carrouee S, et al. Imino sugars inhibit the formation and secretion of bovine viral diarrhea virus, a pestivirus model of hepatitis C virus: implications for the development of broad spectrum anti-hepatitis virus agents. Proc Natl Acad Sci U S A 1999;96:11878-82.

Zwick E, Bange J, Ullrich A. Receptor tyrosine kinase signalling as a target for cancer intervention strategies. Endocr Relat Cancer 2001;8:161-73.

Meister G, Tuschl T. Mechanisms of gene silencing by double-stranded RNA. Nature 2004;431:343-9.

Lamba V, Ghodke Y, Guan W, et al. microRNA-34a is associated with expression of key hepatic transcription factors and cytochromes P450. Biochem Biophys Res Commun 2014;445:404-11.

Swart M, Dandara C. Genetic variation in the 3'-UTR of CYP1A2, CYP2B6, CYP2D6, CYP3A4, NR1I2, and UGT2B7: potential effects on regulation by microRNA and pharmacogenomics relevance. Front Genet 2014;5:167.

Martens-Uzunova ES, Böttcher R, Croce CM, et al. Long noncoding RNA in prostate, bladder, and kidney cancer. Eur Urol 2014;65:1140-51.

Chorawala MR, Oza PM, Shah GB. Mechanisms of anticancer drugs resistance: an overview. Int J Pharm Sci Drug Res 2012;4:1-9.

Zhang W, Dolan ME. The emerging role of microRNAs in drug responses [review]. Curr Opin Mol Ther 2010;12:695-702.

Chapter 22

STUDY DESIGN AND ANALYSIS APPROACHES IN PHARMACOGENOMICS RESEARCH

MARYLYN D. RITCHIE, PH.D.

Learning Objectives
1. Understand genetic association studies and how they can be used to look at pharmacogenomic traits.
2. Describe the importance of defining pharmacogenomic phenotypes.
3. Describe different study designs for pharmacogenomic studies.
4. Understand the different molecular technologies available to assay DNA samples for genetic variation.
5. Understand the different types of statistical analysis methods available for pharmacogenomics studies.
6. Consider the analysis plan for a pharmacogenomic study including quality control, statistical methods, and dealing with multiple testing.
7. Understand challenges with interpreting the results of association studies.

Keywords: Genetic association studies, genome-wide association studies (GWAS), study design, statistical analysis mehods, multiple testing correction, replication, validation

Abbreviations in This Chapter

ACTG	AIDS Clinical Trials Group	INR	International Normalized Range
ADME	Absorption, Distribution, Metabolism, and Elimination	KEGG	Kyoto Encyclopedia of Genes and Genomes
ADR	Adverse Drug Reaction	LD	Linkage Disequilbrium
AIDS	Acquired Immunodeficiency Syndrome	MAF	Minor Allele Frequency
AIM	Ancestry Informative Marker	MB-MDR	Model-Based Multifactor Dimensionality Reduction
AMD	Age-related Macular Degeneration		
ANOVA	Analysis of Variance	MDR	Multifactor Dimensionality Reduction
CDCV	Common Disease, Common Variant	MI	Myocardial Infarction
CDRV	Common Disease, Rare Variant	NCI	National Cancer Institute
CI	Confidence Interval	NHGRI	National Human Genome Research Institute
CNV	Copy Number Variant		
CYP	Cytochrome P450	NNT	Number Needed to Test
DALY	Disability-Adjusted Life-Year	NPV	Negative Predictive Value
DMET	Drug Metabolizing Enzymes and Transporters	OR	Odds Ratio
		QALY	Quality-Adjusted Life-Year
eMERGE	Electronic Medical Records and Genomics	QC	Quality Control
		Q-Q	Quantile-Quantile
FDR	False Discovery Rate	RCT	Randomized Clinical Trial
GENEVA	Gene, Environment Association Studies	RFLP	Restriction Fragment Length Polymorphism
GWAS	Genome Wide Association Study		
HIV	Human Immunodeficiency Virus	SNP	Single Nucleotide Polymorphism
HSR	Hypersensitivity Reaction	SURF&TuRF	Spatially Uniform ReliefF and Tuned ReliefF
HWE	Hardy Weinberg Equilibrium		
IBD	Identity By Descent		

Abstract

Pharmacogenomic studies have had great success in recent years with the identification of many genetic variants that are responsible for drug treatment response. An important component to any successful study is the overall design and analysis strategy. There are many considerations that are important for designing a powerful study including phenotype definition, epidemiologic study design, molecular data generation method, quality control procedures, statistical analysis technique, and interpretation/validation of results. In this chapter, these topics are reviewed with references to additional material with more detail. Careful selection of all these study design elements is essential for ensuring well-powered, successful pharmacogenomic studies.

Introduction

Pharmacogenomic studies are designed with the goal of finding genetic variants that account for the interindividual variability in drug treatment response including efficacy of the treatment, adverse drug reactions (ADR), or circulating blood drug levels. Prevention of serious adverse reactions is critical for improving treatment outcomes, especially with the large rate of hospital admissions due to ADRs (Pirmohamed 2004) as well as hospital fatalities estimated around 5% (Lazarou 1998; Wester 2008). Pharmacogenomics has the potential of changing health care by implementing personalized or precision (Garay 2012; Khoury 2012; Mirnezami 2012) medicine where medical decisions of treatment are based on the unique set of variants present within an individual's genome. By identifying the DNA variation responsible for drug treatment response, safe and effective treatment decisions can be made without the necessity of trial and error, which can lead to potential adverse outcomes. Although the discovery of new markers of genetic variation has been paramount to advancing the field of human genetics, the true catalysts have come in the form of large-scale collaborative research projects such as the Human Genome Project (Consortium 2004; Lander 2001) and the International HapMap project (Gibbs 2003). The HapMap project introduced two important concepts essential to establishing the currently used large-scale genetic association studies by documenting the patterns of variation across multiple populations. First, they confirmed that there are long stretches of linkage disequilibrium (LD) distributed across the genome and second, that these blocks of LD vary between populations (Gibbs 2003). Linkage disequilibrium is a concept whereby there is a nonrandom association of alleles at multiple genetic loci (Slatkin 2008). Due to the patterns of LD within the genome, it was found that all common variation in a population of European descent could be assayed by genotyping approximately 500,000 carefully selected SNPs known as tag SNPs (Gibbs 2003). These tag SNPs became the key for the creation of genome-wide association studies (GWAS), which relies on the common disease common variant (CD-CV) hypothesis. The CD-CV hypothesis states that the majority of the heritability of common disease is due to risk alleles that are found to be common in the population (Lander 1996). Heritability is defined as the proportion of the variance in a trait or disease that can be explained by genetics. In many common diseases, our current estimates of trait heritability were derived from family studies, and more specifically twin studies where the concordance in phenotype between monozygotic (identical) twins is compared with that in dizygotic (fraternal) twins (Haines 2006). If the trait or disease has a significant genetic component, we should observe a higher concordance between monozygotic twins. GWAS rely on LD between SNPs that are genotyped on one of the commercial genotyping platforms and the true susceptibility loci that modulate the trait of interest. Because of the LD, we can detect indirect association to the tagSNPs, which allows for the identification of regions of interest that can then be followed up to look for functional variants. Although it has been argued that GWAS has been extremely successful (Hindorff 2009), GWAS has not led to as much novel biology and understanding of heritability as was hoped (Maher 2008; Manolio 2009). Most of the genetic variants detected to be associated with disease outcomes have very small effects (i.e., explain little of the heritability) (Maher 2008). As a result, the majority of associated variants discovered by GWAS have little value for use in disease prediction. There are, however, a few examples of common diseases for which large genetic effects (i.e., explain much of the heritability) have been found, although it has also been shown that pharmacogenomic traits often have larger effects than general common, complex diseases (Ritchie 2012). Regardless of studying a common disease in the population or a pharmacogenomic trait, the general principles of study design and analysis are preserved. In the following sections, genetic association studies will be described in general terms followed by an in depth discussion of study designs and analysis strategies specifically for pharmacogenomics.

Genetic Association Studies

Discoveries of genetic variants with strong effects, such as the case of *APOE,* for which the ε4 allele is associated with a 4-fold increase in risk for late-onset Alzheimer's disease (Bertram 2010), and *CFH*, where individuals homozygous for the Y402H polymorphism have a 7.4-fold increase in risk for AMD (Klein 2005), seem to be the exception rather than the rule when considering genetic variations predisposing risk to common disease. There are several reasons that may explain why such a low proportion of the heritability for common diseases can be explained by the genetic variants identified thus far (Maher 2008; Manolio 2009). For example, if the CD-CV hypothesis is wrong then we should instead be considering the common disease, rare variant (CD-RV) hypothesis (Schork 2009). The CD-RV hypothesis states that the genetic burden of common diseases is most likely due to multiple rare variants accumulating in common genes or pathways. This hypothesis cannot be tested in the GWAS framework because of the lack of coverage of low frequency/rare variants genotyped on the GWAS genotyping platforms. However, whole-exome, whole-genome, and targeted-capture sequencing technologies are currently experiencing expanded use (Via 2010) and will provide a platform to explore the CD-RV hypothesis.

Another explanation for the lack of heritability accounted for thus far is that GWAS have failed to fully consider the complex genetic architecture, which is likely to underlie common, complex diseases (Maher 2008; Manolio 2009). Most genetic association studies, including GWAS, have predominantly emphasized the analysis on single variant associations or monogenic risk. Meanwhile, the literature has significant evidence that polygenic inheritance is important for many traits (Chhibber 2014; Purcell 2014). By limiting association analysis to one genetic variant at a time, these association studies cannot detect interaction between variants, also known as epistasis or gene-gene interaction, which could be essential for explaining the heritability of common disease (Maher 2008; Manolio 2009; Moore 2005; Moore 2003; Zuk 2012). Additional explanations are also possible for the unclaimed heritability in common diseases including genetic heterogeneity (where rather than a single genetic model there are multiple models that confer disease risk), gene-environment interactions (where the genetic variation is associated with disease in the context of a particular environment), as well as clinical heterogeneity (where perhaps what is being considered the disease outcome or phenotype is not simply one disease) (Thornton-Wells 2004).

Examples in Pharmacogenomics

While many genetic association studies of common disease have failed to discover genetic variants with large effects explaining a significant proportion of heritability, pharmacogenomic association studies have been comparatively much more successful (Ritchie 2012). Many influential variants in drug response phenotypes have been discovered through association studies. For example, multiple strong effect associations have been found between genetic variants in CYP (cytochrome P450) genes and pharmacogenomic outcomes. Of particular note are associations discovered among SNPs in *CYP3A5*, *CYP2B6*, and *CYP2C19* and response to the drugs tacrolimus (Ekbal 2008), efavirenz (Haas 2004), and clopidogrel (Taubert 2009; Mega 2009), respectively. Possibly the most influential genetic associations exist between SNPs in the *CYP2C9* and *VKORC1* genes and warfarin pharmacokinetics—known by many as the poster child for pharmacogenomics. Warfarin is a commonly prescribed anticoagulant that has high interindividual variability in effective dose (James 1992) and a narrow therapeutic range. Individuals with concentrations of warfarin above the target international normalized range (INR) have an increased risk of major bleeding events while warfarin levels below the target INR will not be effective in treating the thromboembolism and systemic embolism conditions for which warfarin is prescribed (Tan 2010). The *CYP2C9*2* and *CYP2C9*3* alleles that result in amino acid change in the enzyme are associated with a reduced rate of warfarin metabolism (Aithal 1999) and this association explains as much as 10% (Limdi 2010) of the variance in dose response. The *VKORC1* gene encodes vitamin K epoxide reductase complex 1, which functions to activate vitamin K (Oldenburg 2006) and then modulate proteins involved in blood clotting. Genetic variation in *VKORC1* is responsible for up to 25% of the variance in dose response to warfarin (Limdi 2010). Effects of this size are not typically observed in standard common disease association studies. However, in pharmacogenomics, there are many such examples (Ritchie 2012).

Although we have many examples of successful association studies in pharmacogenomics, still there are many drug response outcomes where we do not understand the underlying genetic etiology leading to this variation. Thus, many pharmacogenomic studies continue to be pursued. As one considers a new pharmacogenomic study, considerations of study design are important. A properly designed pharmacogenomic study has potential to discover new knowledge about the treatment response in question; however, a poorly designed study can waste time and money as well as lead to future research in the wrong direction. In the following sections of this chapter, study design considerations will be described as well as the molecular and statistical analysis components. All of these elements can result in successful pharmacogenomic studies.

Study Design

Defining the Phenotype

The first and most important element of study design is defining the phenotype. Broadly speaking, there are two possible types of traits to study in pharmacogenomics: pharmacokinetics and pharmacodynamics. Pharmacokinetics describes drug processing which includes absorption, distribution, metabolism and excretion (ADME) (Merck Manuals Online Medical Library 2010). An example of a pharmacokinetic trait that is often studied in genetic association studies is the concentration-dose ratio (Dahlin 2006; Ohara 2003; Singh 2009) (which is the plasma concentration of a drug normalized to the dosage given and often also corrected by the study participant's weight). Two other commonly used pharmacokinetic outcomes include drug clearance or drug excretion rates (Pharmacokinetics Working Group 2004). Pharmacokinetic outcomes provide the capability to assay the function of the ADME processes. Where pharmacokinetics describes the actions that the human body performs on a drug, pharmacodynamics refers to effects the drug has on the body (Merck Manuals Online Medical Library 2010). Since exploring ligand-receptor interaction dynamics are difficult, most pharmacodynamic studies look at treatment side effects and drug efficacy.

The broad classification of phenotype is important to consider, however, more importantly the form of its measurement determines other aspects of study design. There are four primary types of measurement for the phenotype/outcome typically used in genetic association studies including binary/dichotomous, continuously distributed, ordinal Poisson-distributed, and time-to-event. Binary, or case-control dichotomization, is the most commonly used phenotype in genetic association studies. Common binary phenotypes in pharmacogenomics include adverse drug reactions (ADR). While binary traits are convenient for association studies, they often have reduced statistical power in comparison to other outcomes (Majumder 2005). An ADR defined as a binary variable ignores differences that may exist in the severity of the reaction between individuals—which could also be influenced by genetic variation. As a result, a continuously distributed trait is often preferable to a binary outcome when available (Ghosh 2009; Turner 2009). As an example, consider looking for association in change in blood lipid levels in the study of statin effect, rather than identifying individuals with hyperlipidemia. Using a continuous trait may also reduce phenotypic heterogeneity and misclassification (Fredrickson 1965). It is conceivable that individuals who sit on the threshold used to define the binary outcome are not very different and by dichotomizing them, statistical power could be reduced. As opposed to binary or continuous distribution defined outcomes, event counts and rates, which typically follow a Poisson distribution, are another useful study design in pharmacogenomics. Outcomes defined by a rate attempt to look at an ordinal measurement such as a count normalized over a period of time. Utilizing a rate as the phenotypic measure in pharmacogenomic association studies also has an advantage over binary classification when possible, as it tends to be more informative and more powerful. The rate of occurrence of ADRs or even hospitalizations due to ADRs would be more descriptive than an indicator of the presence/absence of an ADR. While the use of rate-based outcomes in pharmacogenomic association studies is still somewhat rare, survival/time-to-event measures are much more commonly assessed. This is particularly true when considering long-term treatment effects using genetics guided therapy (Ginsburg 2010; Kiyotani 2010). An example study could be time to drug resistance in HIV-infected individuals on ART (Kuritzkes 2004). The dimension of time lends valuable information, all of which would be lost if the binary/dichotomous clinical phenotype were used at the end of follow-up and also alleviates issues of loss to follow-up through the use of censoring (Jiang 2007).

Treatment-Dependent and Treatment-Differential Study Designs

Pharmacogenomic association studies can be conducted in two different ways with respect to how the treatment regimen is performed: treatment-dependent and treatment-differential. A treatment-dependent study explores only the individuals on a drug of interest and looks at variation among these individuals with respect to the treatment outcome in question. In contrast, a treatment-differential study assesses two groups of patients who differ in their treatment—which can be more useful in exploring gene-treatment interactions. Therefore, some outcomes such as ADR require the use of the drug or treatment being studied and therefore must be performed as part of a treatment-dependent study. Other clinical endpoints may be pursued in either a treatment-dependent or treatment-differential manner. For example, associations between HIV antiretroviral drug efficacy and genetics could be tested either by looking across drug treatment regimens, as is done in the AIDS Clinical Trials Group (ACTG) trials through assessment of mechanisms by which genetics and regimen interact to modulate a decrease in HIV viral load. Subsequently, the treatment effects can be explored by scrutinizing patients on a single regimen and testing for association of genetic variation with efficacy. Examples of treatment-dependent and treatment-differential study strategies are provided in (Grady 2011).

Epidemiologic Study Designs

Three standard epidemiology study designs are utilized for most pharmacogenomic association studies: the case-control study (retrospective), the cohort study (prospective), and the randomized clinical trial (RCT) (prospective). The most prevalent study design used in genetic association studies is the retrospective case-control design. A case-control study involves the ascertainment of participants based on their phenotype or outcome status followed by a retrospective look at exposures of interest (including genetic variation) (Schlesselman 1982). The term exposure can refer to environmental factors including smoking, alcohol use, and diet or to genetic factors. In a pharmacogenomic case-control study, the frequency of a particular genotype (or allele) is evaluated between cases and controls with the goal of elucidating genetic markers predisposing individuals to the outcome of interest (Balding 2006). The risk estimate derived from a case-control study is the odds ratio (OR). The OR describes the odds of an exposure in cases compared to that in controls. In most situations, cases are defined by the presence of an ADR or other drug-response outcome and the genotype is referred to as the exposure. For treatment-differential studies, the exposure is often a treatment-genotype interaction. This would be the case in a study of myocardial infarction (MI) prevention through the use of statins compared with aspirin. The presence of an MI event would define cases and the exposure to test for association would be the interaction between statin use and genetic variation. Though the OR is not equivalent to the risk ratio (RR), an estimate of disease risk in individuals with the exposure as compared to those without, it is a reasonable approximation if the outcome is rare. The benefits of a case-control study design include the potential to study rare diseases, for which other designs would not be able to collect sufficient sample sizes; additionally, there is substantial cost-efficiency in case-control designs, as it alleviates the tremendous expense of patient follow-up over time (Schlesselman 1982). Unfortunately there are also several weaknesses to the case-control study design (Schlesselman 1982). First, the most concerning issue is that of potential for selection bias attributed to the ascertainment process of case-control studies. Selection bias can result from differential survival if the cases are not all newly incident, meaning that the cases ascertained represent a subset of those who have survived at the time when enrollment became possible or a subset still capable of enrolling. Selection bias could also arise from a failure to select cases and controls representative of the same underlying population. Second, in addition to selection bias, case-control studies can suffer from information bias as a consequence of differential ascertainment of study variables. Information bias refers to exposure variables that could be collected asymmetrically by researchers possibly due to knowledge about case-control status. Sometimes information bias is completely unintentional, as is recall bias. Recall bias results from cases or controls (perhaps with family history of the outcome) being *more* or *less* likely to correctly recall details of drug treatment or environmental exposures.

While the retrospective case-control study collects participants for a study based on phenotype or outcome, a prospective *cohort* study enrolls participants based on their exposures and then follows them over time to evaluate study endpoint(s) and clinical outcomes (Manolio 2006). Participants are recruited to a study based upon their drug treatment in a pharmacogenomic cohort study either as part of a single treatment cohort or as part of multiple treatment cohorts when the interest of exploring gene-treatment interactions is present. Traditionally, cohort studies are conducted prospectively, with all of the information about exposures collected prior to the appearance or measure of the study endpoints. Although it is also possible to perform the study retrospectively, this is less common (Thomas 2004). The key defining feature of a cohort study is that participant ascertainment proceeds on the basis of the exposure instead of the endpoint. Cohort studies are often selected as they have several advantages over case-control studies. First, because individuals are recruited directly based upon their exposure, the relative risk associated with an exposure can be directly estimated (Thomas 2004). The enrollment of study participants does not typically rely on genetic information, therefore cohort studies are still capable of deriving an accurate estimate of the relative risk of genetic variants. In addition, the longitudinal nature of a cohort study provides the ability to ask research questions regarding survival time. The issues from case-controls studies, including recall bias, are diminished in prospective cohort studies as are those of selection bias, although information bias can exist with respect to treatment status (Little 2005). For example, researchers might collect more complete information on individuals from one of the treatment groups because of implicit assumptions regarding differential values of the study endpoint within that group compared with the other group(s). While many biases are reduced with the cohort design, the challenge of loss during follow-up is introduced. To avoid one issue of reduced follow-up, by collecting DNA samples early in the study researchers can avoid bias in the form of a nonrepresentative participant subset with DNA available for pharmacogenomic association studies (Little 2005). Two additional areas in which cohort studies have a distinct disadvantage compared with case-control studies are (1) cost and (2) the inability to study rare outcomes (Manolio 2006). Following individuals over long periods of time, usually for years, is extremely costly. However, it is important to remember that cohort studies are the best equipped to answer questions regarding longitudinal outcomes due to their prospective nature, provided that the outcome is sufficiently common. If the outcome is not common and only occurs in 1% of the population, 1000 study participants would be required on average to observe 10 who experience such an outcome, making a well-powered study impossible.

It comes down to considering many elements of the research project to determine whether a prospective or retrospective design is preferred. While prospective cohort studies are beneficial for studying events with a dimension of time, continuously distributed traits benefit from the use of a retrospective cohort study. For a continuous trait, it is often optimal to enroll a cohort of individuals based on their prior drug exposure status and then collect information on the trait. Through the use of electronic medical records, it should also be possible to study the changes that occur in a trait over time, in particular from the initiation of drug treatment through the time of ascertainment into the research study.

The randomized clinical trial (RCT) is typically considered the gold standard of study designs in treatment-related research (Ritchie 2012). An RCT randomizes each individual enrolled to receive one treatment arm from multiple possible treatment arms (Stolberg 2004). This randomization process helps to reduce selection bias and confounding (Jadad 1998). Confounding refers to the situation where a factor is associated with both the phenotype/outcome as well as the exposure; however is not part of the causal pathway (Smith 1984). Failure to account for confounding in association studies can lead to both false positive and false negative associations, depending on the situation. The issue of confounding can be exemplified by an epidemiology study looking at the relationship between alcohol consumption and lung cancer (Zang 2001). When looking for an association between lung cancer and alcohol, if one does not control for cigarette smoking, it appears that an association exists. After adjusting for cigarette smoking, however, the association disappears. It is clear that the association between alcohol and lung cancer is confounded by cigarette smoking. The most common confounder in genetic association studies is that of genetic ancestry, which is referred to as population stratification (Hoggart 2003). Population stratification is likely to be a significant problem in pharmacogenomic studies, when the drug response has a strong ethnic disparity. Thus, this issue will be described in depth during the section on quality control (QC). By randomizing individuals to treatment status, one can reduce the likelihood that there will be an excess of individuals from one particular genetic ancestry within one treatment group and so the potential for confounding from this factor is reduced. Under the same rationale, the randomization aspect of the RCT design alleviates confounding due to other factors as well.

Another advantage of the RCT is that such studies are usually conducted in a double-blinded manner, which means that neither the researcher collecting data nor the study participant is aware of the treatment arm that the participant has been randomized to receive (Jadad 1998). The benefit of this blinding process is that it prevents information bias during data collection, as it is unlikely to observe disparities across treatment groups. The assumption behind the use of a blinding strategy is that a participant's knowledge of the study drug they are receiving might change their response and/or adherence and subsequently complicate comparability between treatment groups. Double-blinding is also used to prevent the same types of effect on the part of the clinician/research nurse, assuming that un-blinded knowledge might result in more careful monitoring of participants on a particular treatment regimen. Thus, the double blinding procedure eliminates the heterogeneity and bias that these factors could contribute to a study. There are, as with all study designs, inherent problems with the RCT design. For example, in a RCT, the treatment designation for an individual typically refers to the treatment that the participant was randomized to in the study, which may not be the optimal treatment for that participant to receive for the majority of the study due to treatment-related complications (Little 2005). Participants can be modified during the study due to adverse events, for example. The RCT design also experiences the issues of loss to follow-up, cost, and is ineffective for the study of rare outcomes. It is also much more difficult to study long-term outcomes with an RCT due to the large costs of maintaining an RCT (Black 1996; Sanson-Fisher 2007). An additional concern when performing a genetic association study using RCT data is that of selection bias related to the collection of DNA samples. Providing DNA within an RCT is encouraged but not required; therefore only a subset of participants are likely to give consent for collection and analysis of DNA. As a result, it is possible for bias to enter the study at this point due to the systematic exclusion of one or more treatment groups. Therefore, it may not be reasonable to assume that randomization still holds. A standard approach to evaluate this is to compare the demographics of the subset providing DNA samples to those of all study participants (regardless of DNA inclusion). This can help one determine the severity of the bias. Another limitation in RCTs is that of sample size. Single-center trials are often much too small to have ample statistical power for a genetic association study. Thus, the inclusion of either multiple centers or multiple studies is often necessary to achieve sufficient sample size/power. Combining multiple centers and/or studies has the potential to introduce heterogeneity with respect to both the clinical outcomes/phenotype as well as genetic ancestry (Little 2005). That said, a benefit of RCTs with respect to DNA is that a centralized repository can be a tremendous resource for future research use. The AIDS Clinical Trials Group (ACTG), for example, has collected DNA in a sample repository for many years in the interest of implementing pharmacogenomic association studies (Haas 2006). Similar efforts have been undertaken in oncology trial groups (Innocenti 2012).

Regardless of study design, specialized analysis schemas are available to answer targeted scientific questions. For example, a binary outcome (such as responder/nonresponder) can be analyzed as part of a RCT or in a cohort study through the use of either a nested case-control or nested case-cohort design. The nested case-control study (Little 2005) defines cases as individuals from the larger cohort or RCT study who developed the outcome of interest, whereas controls are defined as a random sample of the participants from the same cohort or RCT who did not develop the outcome. During the analysis, age is used to adjust for time-dependent differences between the cases and controls. Another variation of the nested case-control study, the nested case-cohort study, matches one or more controls to each case on the basis of age or other time-related variables (Little 2005). Another option, the case-only study, can be performed as part of a case-control or nested case-control dataset and is useful for exploring gene-treatment interactions. In the case-only study, a measure of association between genotype and treatment group is explored (Little 2005; Khoury 1996; VanderWeele 2010). The resulting risk estimate is a measure of the interaction between the genotype and the drug treatment. An important assumption of the case-only study is that the genotype and treatment group are independent. Due to the stringency of the assumption of independence, it is ideal to perform a case-only study within cases taken from an RCT.

Genotyping

Genome-Wide Association Studies
Molecular technologies for genotyping have continually expanded the toolbox of pharmacogenomic association studies. Four standard genotyping schemes are typically used in pharmacogenomics: (1) candidate gene genotyping, (2) specialized/targeted chip platform genotyping (candidate chip), (3) genome-wide association study (GWAS) genotyping, and (4) whole exome/whole genome/targeted sequencing. Each approach has advantages, disadvantages, and specific applications where it would be most appropriate (for more details see Grady 2011). Genome-wide genotyping arrays employing GWAS have been the standard for exploring the genetic etiology of complex human disease for the past several years (McCarthy 2008). These arrays are designed to genotype 500,000, 1 million, or up to 5 million tag SNPs (single nucleotide polymorphisms), which make it possible to capture nearly all of the common variation in the human genome of European populations due to LD (International Hapmap Consortium 2007). LD is essentially correlations among SNPs that make it possible to genotype a subset, but infer the association with a greater number of SNPs. It should be noted that LD patterns vary greatly between ancestry groups (Slatkin 2008) and therefore a tag SNP derived from a European population should not necessarily be expected to tag the same or similar region in an African-derived population. The LD in African-derived populations tends to extend across much smaller genomic regions in comparison to European populations (Tishkoff 2003) and therefore genotyping more SNPs is required to achieve comparable genome-wide coverage.

By some measures, GWAS has experienced great success in discovering genetic variation associated with risk for disease; approximately 13,750 polymorphisms from 1942 publications are reported in the NHGRI GWAS Catalog as of July 2014 (Hindorff 2010). However, the majority of these associated variants explain only a modest proportion of the outcome. It is possible that these issues with small effect size could be due to phenotypic and genetic heterogeneity. Both genetic and phenotypic heterogeneity in a study population will tend to reduce statistical power and push risk estimates towards the null hypothesis of no association (Thornton-Wells 2004). Pharmacogenomic association studies have an advantage where phenotypic heterogeneity is concerned, because the imposition of clinical standards for diagnosis of adverse drug reactions (ADRs) and the measurement of pharmacokinetics will diminish a significant proportion of the possible heterogeneity in the outcome.

There are important considerations to be made prior to performing a GWAS. In 2007, the NCI-NHGRI Working Group on Replication in Association Studies laid out the replication requirements for researchers prior to reporting GWAS results (Chanock 2007).

Replication will be discussed in detail in the post-analysis section of this chapter. One of the most significant challenges associated with GWAS is that correcting for multiple testing. Generally in a GWAS, an uncorrected p-value of 5×10^{-8} is required for the result to be considered genome-wide significant. This stringent p-value cutoff requires a very large sample size, often prohibitively large for pharmacogenomics, to detect a modest genetic effect. The issue of multiple testing and p-value corrections is covered in more depth in the post-analysis section of this chapter as well.

Candidate Chip Genotyping

An alternative to using a genome-wide genotyping platform is that of a specialized genotyping assays and arrays containing a large set of polymorphisms specific to a given trait. These custom arrays have become quite popular, especially in pharmacogenomics. Examples of these specialized genotyping chips include the Affymetrix DMET Plus Premier Pack (Burmester 2010) and the Illumina ADME BeadChip (Talmud 2009). The Affymetrix DMET microarray chip is designed to genotype approximately 2000 markers in over 200 ADME genes, many of which are not well represented on the commercial genome-wide genotyping arrays. The population allele frequency of the polymorphisms on the chip range greatly and there is a specific focus on those mutations/polymorphisms that are rare in the population. The Illumina ADME Core Panel is a similar platform that consist of 184 SNPs from 34 ADME genes (Illumina 2010). Both of these platforms allow for exploration of a large set of SNPs in candidate genes for many pharmacogenomic outcomes of interest (J 2009; Oetjens 2013; Sissung 2010).

In addition to the widely available targeted chips already created, many genotyping companies allow for the specification of custom-made chips. The Illumina ImmunoChip is an example of a genotyping array that was tailor-made for the purpose of looking at candidate genes linked to auto-immune disorders such as Lupus, Multiple Sclerosis, and Rheumatoid Arthritis (Parkes 2013). The chip provides the capability to assess variation at approximately 200,000 markers distributed across the genes and genomic regions concerned (PSC Scientific and Medical Advisory Committee 2010). Illumina provides a service, iSelect, which allows researchers to customize genotyping arrays with a number of variants ranging from either 3000 to 68000 or 68001 to 200,000 (Illumina 2010). There are multiple advantages to the application of these custom assays. First, using prior biological knowledge to focus genotyping efforts yields a higher probability of discovering biologically relevant associated polymorphisms (Roeder 2006), especially in comparison with genome-wide genotyping platforms that might not adequately assess the variation in regions of particular interest (Saccone 2009). The biological knowledge used to decide the variants added to the platforms is a primary concern, though. In terms of the currently available arrays, extensive research was done to select which genes should be included and to ensure that these genes are sufficiently covered by SNPs on the chip (Talmud 2009; Illumina 2010; Sissung 2010). Another advantage that is specific to custom arrays is the ability for the user to add variants of their choosing, including those variants which have been discovered subsequent to the release of standardized genotyping platforms. There are several methods which are used in the selection of genes and polymorphisms to add to a genotyping array (Chuang 2010; Han 2008; Li 2008; Pico 2009). The major disadvantages of the customized genotyping chips are the increased per-genotype costs as well as a great deal of initial work that is necessary to select markers for a new array. In addition, genotyping efficiency is likely to decline in custom-made chips designed by researchers when compared with arrays and assays designed by genotyping companies that contain SNPs validated for genotyping accuracy and efficiency. Finally, sharing results for meta-analysis may be more difficult if custom arrays are used, as there will be fewer other datasets in the research community using the same array.

Candidate Gene Genotyping

One of the benefits of studying pharmacogenomics is the abundance of prior biological knowledge regarding important pathways of drug metabolism, transport and elimination. Most currently known pharmacogenomic associations are the direct result of candidate gene genotyping based on knowledge of these. The candidate gene study uses small scale, targeted genotyping to assess variation at only a handful of genes (Tabor 2002). In place of a high-throughput genotyping array, a low-throughput technology such as Sequenom (Allegue 2010), TaqMan (Oberst 1998), or Illumina GoldenGate (Steemers 2005) is used. Although the era of GWAS has resulted in a decrease in the use of candidate gene association studies, the approach has had great success discovering variants with substantial effect sizes in pharmacogenomics as well as complex disease (Haas 2004; Haines 2005; Macphee 2004; Rebbeck 1998; St George-Hyslop 2000). The same general principles regarding the biological knowledge required for the candidate chip apply here to the candidate gene study. The correct biological knowledge is essential to properly focus the creation of the genotyping assay. The researcher can directly assay markers with proven functional impact at the transcript, protein or trait level or those which cause non-synonymous coding changes instead of relying on tag SNPs to detect indirect associations. One of the most significant advantages of a candidate gene study is the alleviation of the issue of multiple testing, as the study is usually small scale and thus, the multiple testing burden is limited.

Sequencing

Genotyping technologies have progressed rapidly over the last several years as the cost per genotype has steadily fallen. Now, we stand at the forefront of affordable whole-genome sequencing technology, where the costs have nearly reached $1000 per genome and continue to decrease. There are many factors driving the progress in genome sequencing, including the 1000 Genomes project (Via 2010), an initiative to uncover 95% of all genetic variants in the "accessible genome" with a population frequency of 1% or greater (Durbin 2010). The accessible genome is the portion of the genome for which sequencing reads can be unambiguously mapped back to the NCBI reference genome and this constitutes approximately 85% of the entire genomic sequence in the current build (Durbin 2010). The eventual goal has been to lower the cost of whole-genome sequencing down to $1000 per individual (Mardis 2006). For most researchers, this cost is still prohibitive to perform on a large sample of individuals. As such, large-scale whole-genome sequencing in genetic association studies has not yet become a reality. In lieu of whole-genome sequencing, a cheaper and more focused alternative is whole-exome sequencing (Choi 2009; Ng 2009; Shendure 2008). The exome describes the set of all exons in the human genome; all the regions of the genome that will be expressed. Exon-capture technology can be used to selectively sequence all protein-coding regions in the genome (Bonetta 2010).

One of the most challenging aspects of DNA sequencing at the whole-exome or whole-genome level is data management. The space required to store the data from a single sequencing run can be up to 1 terabyte (1024 gigabytes) or more (Hosseini 2010). In addition, after the sequencing data are generated, the bioinformatics infrastructure necessary to extract genetic variants from the raw sequence is another series of challenges and computational complexities. Because these variants are comingled with sequencing errors, this is a non-trivial task (Koboldt 2010). Allele frequency thresholds are required to distinguish a true genetic variant from sequencing errors. Assessing structural variation such as copy number variants (CNVs) adds an additional level of complexity, as the "break points" (genomic base-pair positions where the repeat begins and ends) of a CNV can vary between individuals (Korbel 2007). If the challenges and expense associated with sequencing technologies can be overcome, this technology allows the possibility of capturing both rare and novel genetic variants, along with common variants, which cannot be discerned through the use of genome-wide genotyping arrays. As the price of sequencing continues to fall and the data management procedures improve, sequencing will become a more viable option to capture nearly all coding and noncoding information in the human genome. The challenge at that point will not be designing the study or generating the data. Instead, the challenge will be applying powerful analytical tools to determine how this genetic information maps onto complex pharmacodynamic and pharmacokinetic traits.

Quality Control

Prior to the analysis stage of a pharmacogenomic association study, it is critical to ensure that the molecular data generated for analysis have been standardized under a set of appropriate quality control (QC) standards. The genetic data, in particular, require several checks of quality, especially high-throughput genotyping data. Many papers have been published describing QC of large scale genetic data, including Weale (2010). Further information about conducting genotype data QC is given by Laurie et al. (2010) on behalf of the GENEVA consortium or by Turner et al. (2010) or Zuvich et al. (2011) on behalf of the eMERGE network. Almost all of the QC discussed below can be conducted in the infrastructure of PLINK (Purcell 2007), developed at the Broad institute or in PLATO (Grady 2009), developed at the Pennsylvania State University (http://ritchielab.psu.edu/software/plato-download). The rationale for conducting quality checks is to remove systematic errors from the data that could otherwise bias the results of any downstream analyses. Genotype QC standards and procedures can be divided into those performed on samples and those performed on genetic markers/SNPs (Laurie 2010; Turner 2011; Zuvich 2011). If a stratified analysis will be conducted, QC should be conducted separately within each stratum especially if the stratifying variable is likely to be related to genetic variation, as is the case with ancestry (or race/ethnicity). The order by which sample and marker QC will proceed is an important consideration. Performing SNP QC first will maintain the maximum number of participants in the study; in contrast, placing sample QC first will maintain the maximum variants rather than samples. A recently developed program, genABEL (Aulchenko 2007), allows QC on markers and samples to be done iteratively to balance low-quality losses from each. The type of genotyping performed should be a factor when deciding which route of QC to pursue. If a candidate gene study has been conducted, it would be prudent to do QC on samples prior to markers to ensure the minimal discard of functionally relevant SNPs. Performing QC on SNPs first is reasonable for GWAS, as it is hypothesized that the vast majority of variants genotyped do not participate in the genetic etiology of the outcome. It is also possible with GWAS that multiple markers will be in LD with the functional variant that they are tagging; thus the discard of one will not preclude the identification of a genomic region associated with the outcome.

There are many additional QC checks and exclusion standards that could be applied to either samples or markers (Weale 2010); the ones which will ultimately be used must depend on the data available. The first step is to check for sex errors and examine relatedness in the dataset, regardless of whether the focus is on maintaining samples or markers. If sex chromosome data is available, sex errors can be identified by checking the heterozygosity of the X chromosome in comparison to the sex classification for each individual. For potential errors, it is desirable to determine whether the cause is actual misclassification or an aneuploidy (abnormality in chromosomal number) such as Kleinfelter syndrome. Through examination of raw genotype calls for X and Y chromosome marker intensities, this can usually be determined. If sex errors cannot be resolved as misclassification due to error in sample handling or data entry, it is advised to exclude these individuals from analysis. If the sex in the phenotype file does not match the genotype data, it is not known what other data are incorrect. Relatedness between individuals in the study is addressed next because interrelatedness among study participants can result in spurious associations when using statistical tests which assume independent observations. If participant relationships are known, as in pedigree data, software such as genABEL (Aulchenko 2007) or MQLS (Thornton 2007) can be used to adjust by utilizing a kinship matrix describing the relationships between all of the individuals. It is also possible to derive a kinship matrix using genABEL if relationships are unknown (Aulchenko 2010), as could be the case in association studies of unrelated individuals where cryptic relatedness, or relatedness between individuals that is not known to the researchers (Voight 2005), could be an issue. An alternative to using statistics to correct for relatedness is to remove one individual at random from each pair where there is a high estimated identity by descent (IBD), which can be done with PLINK (Purcell 2007). Identity by descent is a term that is used to describe the presence of the same allele at a genomic location as a result of inheritance from a common ancestor on the same chromosome/haplotype. Note that estimating the inbreeding coefficient and IBD or kinship coefficient can only be performed on samples for which there is sufficient genotype data and that these techniques are designed to be used with genome-wide SNP data.

To exclude low quality genotype data, the next step is the removal of samples and SNPs with a low genotyping completion rate or, put another way, a high rate of missing genotype data. Each sample and marker is passed through a filter to remove those that fall below a certain threshold; a commonly used threshold is 95%, indicating that those samples and markers for which at least 5% of the genotype data is missing are removed. For GWAS, this threshold is often more stringently placed at 98%–99%. The following step is the exclusion of samples with levels of heterozygosity well outside of that expected due to chance (determined by estimation of the inbreeding coefficient above or below certain thresholds). Low heterozygosity can be the result of inbreeding that drives down the level of individual genetic variation; whereas high heterozygosity may indicate contamination of the sample by foreign DNA, which would lead to excessively high levels of individual genetic variation (Weale 2010).

For SNP data, it is recommended to remove variants with a low minor allele frequency (MAF), as there is significantly reduced statistical power to detect associations for these markers. The minor allele frequency threshold for removing variants will depend upon the sample size for the study. For example, it has been suggested that a cutoff of $10/N$ is appropriate, where N is the sample size (Weale 2010). Filtering SNPs based on deviation from Hardy Weinberg equilibrium (HWE) is sometimes performed as well. SNPs where the test of HWE yields a p-value below a certain value are often removed from analysis. There are debates as to whether this check of HWE should result in marker exclusion or whether it would be more appropriate to flag markers with a strong deviation from HWE so that they can be followed up if they are found to be associated with the outcome (Wittke-Thompson 2005). The argument for removing SNPs that fall below some HWE p-value threshold is that they are likely to be the result of genotyping errors and might cause spurious associations. On the other hand, one of the assumptions of HWE is that there is no selection acting on the population (Hardy 1908). If a SNP is under positive or negative selection, it will violate HWE but should not be removed as that selection could indicate a role of the variant in disease etiology. These deviations from HWE may be the functionally relevant SNPs that are being searched for. The final recommended QC check is the assessment of "plate effects," also known as "batch effects," across genotyping chips (Edenberg 2009).

This can be done by comparing the MAF (Pluzhnikov 2010) or genotyping completion rate for each plate to that of all others using the genetic data that has already been filtered by MAF and sample genotyping completion rate. Plate effects constitute a particular type of batch effect, which are discussed in depth by Leek et al. (2010). Identifying plate effects is important, as they can indicate incorrect handling of DNA samples or errors during the genotyping process. The issue of batch effects recently garnered general attention in the scientific community when associations discovered in a study of human longevity (Sebastiani 2010) were demonstrated to likely be the result of batch effect and QC problems (Alberts 2010; 23andMe 2010). As such, these types of QC steps are essential to maintain high data integrity

After performing general QC exclusions to filter low-quality data, it is imperative to examine the presence of population substructure in the data. Population substructure refers to the phenomenon whereby multiple distinct subpopulations are present within the overall sample population/dataset (Tian 2008). A frequent population substructure issue in genetic association studies is population stratification, in which the distribution of subpopulations differs systematically with the phenotype or outcome (Pritchard 1999). This should be distinguished from the concept of admixture (Montana 2004), which refers to the ancestral mixing of two or more populations, as is present in populations such as Mexican Americans and African Americans. Population substructure is important to consider prior to analysis because it can confound the analysis and cause spurious associations under specific conditions. The conditions under which substructure is confounding are that the prevalence of the phenotype/outcome is different between the subpopulations in the dataset and these subpopulations also differ in allele frequency at markers unlinked to the trait of interest (Turner 2009). Due to frequent differences in drug response between ethnicities (Ribaudo 2010; Mancinelli 2001; Cavallari 2010), discerning the presence of population stratification is particularly relevant for pharmacogenomic studies. An extreme example of population stratification would be looking at risk for a severe ADR in a mixed sample of European Americans and African Americans for which 90% of the cases with the ADR are African American and 90% of the controls are European American. In this case, even though the true prevalence of the ADR might be equal between the ethnic groups, an artificial difference is introduced as a result of the poorly selected cases and controls. Any SNP that differs significantly in allele frequency between European and African Americans would be associated with the presence of an ADR, even though it might be that none of these markers are actually associated with risk of the ADR. Fortunately, many tools have been developed to mitigate the effects caused by population stratification (Hoggart 2003; Epstein 2007; Devlin 1999; Price 2006; Pritchard 2000; Satten 2001).

Historically, epidemiology studies have relied on stratification to eliminate confounding by race and other factors. Although this alleviates stratification to a degree, there are multiple disadvantages to race stratification. The most significant limitation is the drop in statistical power that results from sub-dividing the data set prior to analysis resulting in smaller sample sizes. In addition, self-described race/ancestry may not be as accurate or precise as desired (Barnholtz-Sloan 2005), particularly with individuals from an admixed population where the degree of mixed genetic ancestry can vary greatly between individuals (Galanter 2014; Reiner 2005; Salari 2007). Even in a strictly European population, where ancestry is often assumed to be primarily homogeneous, there can be local genetic differences between individuals of southern and northern or eastern and western Europe (Tian 2009). Although race stratification has some serious disadvantages, it is often the only choice for the correction of population stratification, especially in small-scale genetic association studies if ancestry-informative markers (AIMs) are not also typed. A viable alternative to stratifying by ancestry is to adjust for the presence of the type of systematic differences between individuals in the study. The Genomic Control method (Devlin 1999) was proposed as a manner of performing this adjustment. Genomic Control attempts to identify an inflation factor, λ, using markers that are unlinked to the trait of interest. This inflation factor, which describes the inflation of genetic association test statistics as a result of the systematic bias, is then used to correct the test statistics of all genetic association tests. A more recently introduced correction for population stratification involves principal components analysis (PCA), which can be conducted through use of the EIGENSTRAT software (Price 2006). EIGENSTRAT uses principal component vectors describing the overall population level genetic variation between individuals to provide an ancestry-corrected association statistic. In order to appropriately use EIGENSTRAT without correcting out the effect of truly associated genetic markers, it is recommended that EIGENSTRAT be run with AIMs or with a set of at least 100,000 genetic variants. Whether correcting for substructure by stratifying or through the use of software such as EIGENSTRAT, it is recommended that population outliers, or individuals who do not cluster cleanly with any of the identified populations in the PCA, are removed (Weale 2010). Population outliers can be visualized either by plotting principal components vectors (Weale 2010) or by using the STRUCTURE program (Pritchard 2000). Principal components are used to describe systematic variation within data as attributable to a particular factor. When applied to genetic data in the attempt to account for ancestry, one or more principal component vectors are generated where each vector places each study participant along a line for which the ends of the line describe the study participants differing most systematically by genetics related to race. If a principal component vector successfully describes systematic genetic differences attributed to race, plotting that vector should separate individuals out into two distinct clusters corresponding to a genetic split between ancestry groups. Visualization of population outliers can be enhanced by the inclusion of data from relevant HapMap Phase 3 populations (Weale 2010) or 1000 Genomes populations (1000 Genomes Project Consortium 2010), which can clarify the axes of variation in a plot of principal components vectors by labeling individuals within the plot according to ancestry. Including HapMap or 1000 Genomes samples can also be used to map study participants more robustly to an ancestral population.

Performing appropriate QC on high throughput genotype data prior to association analyses is essential to providing accurate and robust results in a genetic association study. Failure to remove systematic errors and biases can result in a dramatic increase in false positive rate, or type I errors. Noise in the data can be an issue due to the small effect sizes seen in many genetic association studies to date. It only takes a modest amount of noise to overpower the true associations in the data if they are of small effect size. The use of quantile-quantile (Q-Q) plots can help determine whether systematic error is present in the data (Weale 2010). Q-Q plots are generated by performing a test of association for each marker in the data and then graphing the resulting distribution of p-values (often using a negative log-transformation) against a set of the same number of p-values generated from a uniform distribution under the null hypothesis. If there is a systematic error in the data, it will often appear as a strong deviation from the diagonal line which would be expected under the null hypothesis of no association. After QC, it can be seen that no true association exists and systematic error was driving inflation in test statistics and p-values smaller than expected. Such plots act as a useful diagnostic and illustrate the need for thorough QC.

As mentioned above, there are many resources with detailed explanations of the QC process and important parameters to consider. We would refer readers to those resources as a guide for performing QC on large-scale pharmacogenomics data.

Analysis

Analysis of pharmacogenomic data has great potential for improved understanding of individual variability in drug responses. One of the most critical elements for a successful pharmacogenomics study is a well-designed analysis that sets up the project to answer the question of interest while minimizing false positive associations and missed true biological signals. If enough statistical tests are performed on the data many statistically significant results will be found—many of which will be due to chance alone. Due to issues with multiple testing and the nature of any "fishing expedition" in a data set, the primary objective of the analysis plan is to focus statistical testing toward answering a specific scientific question. Important considerations when developing an analysis plan include:

- Are only monogenic associations expected to be significant predictors of the outcome?
- Are gene-gene, gene-environment, or gene-treatment interactions expected?
- Should a filtering strategy be employed prior to performing statistical tests of association?
- Is there prior biological knowledge that would be useful to include in the analysis?
- Does a replication dataset exist that can be used to validate significant associations?
- Are there important covariates or confounders to consider?
- How will ancestry/race be handled in the analysis?

The answers to these questions will provide a guide to design the analyses. As the study design varies according to the phenotype/outcome of interest and the QC process varies according to the type of genotype data available, the analysis plan should take both of these into consideration and be tailored to address the primary questions of interest. In this section, a brief discussion on statistical tests for exploring monogenic associations, advanced methodology for identifying complex genetic effects, and methods for filtering the marker set will be discussed. More detail can be found in other sources including Grady and Ritchie (2011), Grady et al. (2010), Motsinger-Reif et al. (2008), and Sun et al. (2014).

Single SNP Analysis Methods

Association analyses in pharmacogenomics can typically use any of a large array of statistical methods used for any type of association analysis (Cantor 2010) such as Chi-square test (Greenwood 1996; Zheng 2004), Armitage trend test (Armitage 1955; Cree 2010), Kaplan-Meier survival curves (Kaplan 1958; Huang 2009), Bayesian statistics (Stephens 2009), or data mining methods (Coassin 2010) but most commonly, analyses are performed in the framework of regression (Woodahl 2008; Fu 2008; Epstein 2010; Aidoo 2001). In this chapter, the focus will be primarily on the use of regression techniques for single-SNP analysis. The phenotype or clinical outcome of interest will determine the type of regression to perform. Most regression analyses are based on the backbone of generalized linear models (Nelder 2010) but use different link functions. A link function describes the relationship between the dependent variable (i.e., the phenotype or outcome) and the independent variable(s) (i.e., SNPs or covariates).

When the outcome variable is continuously distributed, linear regression is usually the method of choice. Linear regression uses the identity link function, which is a linear relationship with no transformation of the outcome, to assess the effect of the variant on the mean of the outcome/trait value. The parameter, beta, provided by linear regression is an estimate of the mean difference in the continuous trait across groups differing by one genotype level/unit (Kirkwood 2001). A level with respect to genotype will depend on the coding scheme used in the regression equation. Most commonly, an additive encoding is used to statistically test each SNP for association. Coding the three genotypes of a SNP with two alleles in an additive manner involves defining one genotype as the reference genotype (typically the homozygote major allele will be coded 0) and then the heterozygote has one minor allele (coded 1) and the homozygote for the minor allele has two copies (coded 2). An additive coding assumes increased risk of each additional variant allele and is used primarily because it yields the decent power for all models except for recessive (which is least common in complex traits) (Lettre 2007). Although the additive encoding is most commonly used, it is possible that dominant, recessive or custom genotype encodings might be advantageous based on available knowledge of the genetic etiology. Each independent variable in the regression equation has a parameter, beta, which describes the strength of its effect on the study outcome.

For a binary pharmacogenomic outcome such as responder/nonresponder, logistic regression should be used. The link function for a logistic regression equation is the logit function. The logit function transforms the outcome variable to be the natural log of the odds of the outcome, where the odds are found by taking the probability of observing the outcome divided by that probability subtracted from one. The beta parameter derived from a logistic model describes the effect of the variant as the change in log odds between groups of participants differing by one unit in the genotype (Agresti 1990). When the parameter estimate is exponentiated by e, it becomes the ratio of the odds of the outcome between the genotype groups (also known as the odds ratio—OR).

With prospective cohort studies and randomized clinical trials, it is possible to collect rates of events, time-to-event, or survival data. Any associations that use a rate measure can be examined through Poisson regression. A Poisson regression uses a log link that constitutes a natural log transformation of the rate. The parameter reported by Poisson regression, when exponentiated, is the rate ratio comparing the rate of an event between groups (Agresti 1990). To analyze time-to-event outcomes, a proportional hazards regression (PHR) is the analytical approach often used. While the PHR equation can be reported in a form analogous to the other forms of regression discussed with respect to link function, PHR is not directly derived from generalized linear models. Instead, PHR uses what is known as a hazard function to perform the transformation of the outcome. The parameter estimate associated with PHR is the hazard ratio, the ratio of "instantaneous" risk (hazard) of an event over time between genotype groups (Kirkwood 2001). A benefit of the hazard regression approach is that it provides information on significant differences in long-term outcome within treatment or genotype groups.

Regression is usually the standard approach for analyzing data in genetic association studies due to many advantages provided by the framework of the regression equation. First, effects of secondary independent variables, such as confounders, can be adjusted for through their inclusion as covariates in a multivariate regression equation. Including covariates examines the effect of differences in the genotype on the outcome among individuals who have equivalent or similar values for the covariates (Agresti 1990). A nested case-control study design, for example, would be analyzed with a logistic regression equation including participant age as a covariate. Another benefit of regression is the flexibility of modeling provided. The use of different coding schemes for the genotypic effects is an important feature as different coding conventions change the scientific interpretation of the risk parameter estimates. Additive encoding, for example, assumes a linear increase in the effect of genotype on the outcome with the addition of each risk allele. Alternatively, two of the three genotypes could be coded with dummy variables (i.e., 0/1 indicator variables) (Agresti 1990), using the third genotype as the referent genotype to simulate an analysis of variance (ANOVA) model where each genotype is tested against the other two with no assumption of a linear trend in effect across genotype groups (Slinker 1988). This encoding is often termed the genotypic encoding (Grady 2009) and has the benefit of being able to detect nontraditional genotype risk models such as the interference model (Li 2000), where the heterozygote genotype is associated with either the largest or smallest effect. The problem with the genotypic encoding is that it has less statistical power than the additive encoding if there truly is a linear increase in the size of the effect with each additional variant allele. It is important to be cognizant of the effects that the coding method has on a regression analysis. Errors can result if a statistical package does not recognize the coding used for missing values of a variable, as this will cause the missing coding to be interpreted as an additional level of the independent variable and the resulting parameter estimates will be inaccurate. In addition to the potential to model a single genetic effect in multiple ways, non-additive interactions between multiple variables can be explored through the addition of one or more interaction terms. A significant gene-gene interaction would indicate that the effect of one variant differs according to the genotype of the other variant and thus the risk from one SNP cannot be assessed without knowing the genotype at the other (Hosmer 2000).

While regression offers the capability of performing most statistical analyses for a pharmacogenomic association study, it has the disadvantage of assumptions including independence of the samples, the outcome following a normal distribution, and homogeneity of variance. Small deviations from assumptions can be overcome with large sample sizes and the use of robust statistics, but nonparametric statistical methods are sometimes necessary in the instance of severe deviations. One such nonparametric test is the Kruskal-Wallis test (Maxwell 1990), which can be used to circumvent the distributional assumptions of linear regression in the analysis of a continuously distributed phenotype. The Kruskal-Wallis test is the nonparametric form of the ANOVA, which tests for a difference in mean trait values across groups of a nominal independent variable such as a SNP. Neither the ANOVA or Kruskal-Wallis test assumes linear trend in the genotypes of the variant. Nonparametric regression methods are also available (Cohen 2003; Ohno-Machado 2001; Smith 1996). Other nonparametric methods are covered at length by Hastie, Tibshirani, and Friedman (2001).

Epistasis Analysis Methods

While there are examples of tremendous success in monogenic associations predicting a pharmacogenomic outcome with accuracy, as is the case with the *HLA-B*5701* association to abacavir HSR (Mallal 2008), it is likely that in many cases the consideration of more complex genetic models could improve accuracy (Motsinger 2006). The idea that gene-gene interactions may play a large role in pharmacogenomics is supported by the knowledge of extensive and interconnected drug metabolism networks (Zanella 2010). In many cases, multiple enzymes can metabolize the same drug (Anderson 2004; Iyer 1998; Levy 2001); complementation among components in the pathways of drug action indicate that genetic variation at multiple steps in the pathway could act in synchrony to modulate drug response. The presence of non-additive gene-gene interactions (i.e., statistical epistasis) can be investigated through the use of many alternative statistical methods (Cordell 2009; Motsinger-Reif 2008; Motsinger 2007). Parametric methods such as regression can be useful in particular cases, but they tend to suffer from the "curse of dimensionality" (Motsinger 2006), a phenomenon whereby the data become increasingly sparse as higher dimensions are considered. This in turn can cause biased estimates for the size of the interaction effects. One approach that has been developed to overcome the curse of dimensionality is Multifactor Dimensionality Reduction (MDR) (Coffey 2004; Ritchie 2001). The MDR algorithm takes a set of SNPs and creates multilocus models based on the proportion of cases to controls for each genotype combination. The multilocus genotypes with more cases than would be expected are labeled as "high-risk" and the others as "low-risk." The interaction between these variants are then collapsed to a single dimension by pooling "high-risk" and "low-risk" cells separately, an approach called constructive induction (Moore 2006). A measure of accuracy is calculated to determine how well the model fits the data at hand. To prevent overfitting, a phenomenon where increasing the number of variables used for prediction improves accuracy only within the data set used to build the model, the data are divided into multiple partitions for cross validation. In cross validation, the algorithm is repeated once for each partition of the data using all but one of the partitions to build the model and the final partition for evaluation of the model by prediction accuracy. This process is performed until each partition has served as the basis for model evaluation. The best model is that which is most accurate over all iterations of the cross-validation process. Finally, permutation testing is used to generate a distribution of test statistics under the null hypothesis of no association which the statistic of the best interaction model found in the data can be compared back to in order to determine a p-value. MDR has been used to search for interactions between estrogen metabolizing genes in breast cancer patients (Ritchie 2001) and drug metabolism enzymes in response to efavirenz treatment in HIV-infected individuals (Motsinger 2006) as well as other traits (Cho 2004; Coffey 2004). The MDR interaction analysis paradigm does have disadvantages. The primary challenge is that MDR imposes a significant computational burden when considered in genome-wide data due to the exhaustive nature of its exploration of interaction effects.

In addition to dimensionality reduction methods, there are also tree-based (Breiman 1984; Schwarz 2010), evolutionary (Motsinger 2006A, 2006B; Ritchie 2003; Turner 2010), and modified regression methods (Friedman 1991; Kooperberg 2001, 2005) designed to handle complex trait etiology. Tree-based methods such as Classification and Regression Trees (CART) (Breiman 1984) and Random Forests (Breiman 2001) use an iterative algorithm that splits the data based upon the best predictor in order to build a tree that can be used for prediction of the trait value in future individuals. Evolutionary methods seek to find the optimal predictive model using an algorithm which co-opts the biological ideas of genetic mutation, recombination and reproductive fitness. Examples of evolutionary algorithms designed for application to genetic data are Grammatical Evolution Neural Networks (GENN) (Turner 2010) and Genetic Programming Neural Networks (GPNN) (Ritchie 2004). Due to the popularity of regression, many modifications have been made to allow modeling of complex processes. A couple examples of complex regression methods designed with genetic association studies in mind are Lasso regression (Ayers 2010) and Logic regression (Kooperberg 2005). The use of a wide variety of methods designed for gene-gene interaction analysis in pharmacogenomics studies is reviewed by Motsinger (2007).

Filtering

When using data from high-throughput genotyping, such as in a genome-wide pharmacogenomic association study, it is sometimes useful to reduce the data set to a more manageable subset of SNPs likely to be relevant to the outcome under research. This consideration is particularly relevant if there is intent to explore complex genetic etiology such as gene-gene and/or gene-environment interactions as the computational burden is too large to consider the entire GWAS platform. One of the challenges opposing widespread gene-gene interaction analysis in pharmacogenomic association studies is that of complexity. If a genome-wide genotyping array assaying 1 million markers is used, exhaustively exploring two-way interactions would require approximately 5×10^{11} statistical tests. Not only does this create multiple comparison issues, but it also presents an enormous computational task. Due to both computational and multiple testing issues, it is advantageous to focus or target the search for interactions. It is important to note that filtering GWAS data presents a viable alternative to performing analysis on all variants due to the inherent assumption that the majority of SNPs genotyped are not associated with the outcome.

Multiple methods are available by which to filter the initial set of genotyped markers (Bush 2009; Fong 2010; Greene 2009; Pendergrass 2013). One of the most standard filtering approaches is using a statistical test to condition upon single-locus associations and then perform the test for interactions within the subset of markers which display a marginal (i.e., single-locus) effect (Marchini 2005). The argument against this technique is that it ignores the possibility of interactions between markers with subtle or nonexistent marginal effects. Simulation studies have shown that while analysis of one study population might reveal a significant main effect, even a small shift in allele frequency in other data sets would cause detectable gene-gene interactions with no significant single-locus effects, also known as a purely epistatic effect (Greene 2009). Therefore, filtering by main effects could be problematic, though this strategy can work well if interactions are present among loci with main effects.

Another alternative for reducing the search space, the total set of variables used for analysis, uses a biological based filtering strategy, such as the Biofilter (Bush 2009). Biofilter focuses multilocus analysis using biological knowledge in the form of public databases containing information about genes and how they are related to one another such as the Kyoto Encyclopedia of Genes and Genomes (KEGG) and Reactome. SNPs in interacting genes or in genes sharing a common metabolic or biological pathway are then paired together to form biologically plausible multilocus interaction models. Biofilter is capable of reducing the search space of gene-gene interactions in a GWAS from hundreds of billions of models to several million or fewer, a reduction of about four orders of magnitude. Filtering genetic data prior to conducting gene-gene interaction analysis is pertinent to any study generating large quantities of genotypes, although it is particularly relevant for GWAS.

Other types of data driven filtering methods are also available. Spatially Uniform ReliefF and Tuned ReliefF (Greene 2009) (SURF&TuRF) is an approach that uses measures of genetic distance to uncover markers associated with the phenotype. The concept is to focus on groups of genetically similar individuals and determine whether deviations of genotype between these individuals are correlated with deviations of phenotype. For each single SNP or two-locus pair of SNPs, a set of individuals who are within a certain threshold of genetic similarity are selected by SURF and then compared for differences in phenotype. Individuals who differ in genotype at the SNP or pair of SNPs being examined and also differ in outcome cause the marker to be up-weighted while those with the same outcome and different genotypes cause the marker or pair of markers to be down-weighted. The process is performed iteratively, where the lowest scoring markers are removed subsequent to each iteration.

Post-Analysis Issues

Multiple Comparison Correction

Possibly the greatest challenge to consider when evaluating the results of a genetic association study is that of multiple testing. It is expected for a GWAS that approximately 500,000 to 1 million or more statistical tests will be performed when excluding the possibility of complex genetic effects. This results in an enormous number of significant results. When using the standard significance threshold of 0.05 for each statistical test, on average 5% of the tests, 50,000 genetic variants for a 1 million SNP GWAS, will surpass this threshold purely by chance (Fisher 1956). This exemplifies the problem of multiple testing. The most common procedure used to adjust for multiple testing is using a Bonferroni correction (Abdi 2007). The Bonferroni multiple test correction divides the p-value threshold required to consider a single test significant by the total number of statistical tests to determine the new threshold. For a GWAS with 1 million genetic markers, a p-value of 5×10^{-8} is required to be considered genome-wide significant. The reality is that the Bonferroni correction is much too conservative for the analysis of most genetic marker data (Rice 2008), as it assumes that all statistical tests performed are independent. Because of LD patterns, we know that there is significant correlation structure in the genome and thus testing all markers for association will not yield completely independent test statistics. The issue of nonindependent tests is particularly relevant when conducting gene-gene interaction analyses, as the tests containing overlapping markers will undoubtedly be correlated. MDR solves this problem utilizing permutation testing (Hahn 2003), although this can be computationally challenging for large data sets. Another option for multiple testing correction is controlling the false discovery rate (FDR), suggested by Benjamini and Hochberg (Benjamini 1995). The FDR adjustment finds a p-value cutoff to be considered significant based on the distribution of p-values and the user's definition of an acceptable proportion of false discoveries. All results with p-values below this threshold are taken as significant with the caveat that a proportion of the significant results are assumed to be false positive associations. Correcting for multiple testing requires walking the fine line between accepting false positives and discarding false negatives, a balance that is dependent on the investigator's biases and beliefs. Prior problems with false positive associations which failed to replicate in follow-up research (Ioannidis 2001, 2007) have led to an established requirement for replication studies to prove the validity of the results in genetic association studies. Whenever multiple tests are performed, the false positive rate will increase and thus the number of tests must be accounted for in order to consider the results statistically valid.

Interpretation of Results

The fact that a SNP shows evidence of genome-wide significance says nothing about the causative nature of the genetic variant that may be implicated by the association. Particularly in GWAS, the associations detected are nearly always indirect associations with tag SNPs (International Hapmap Consortium 2007), meaning that the actual causative mutation(s) will lie within a genomic region of LD which is tagged by the SNP on the GWAS platform. When the associated tag SNP is in a gene or near a gene where there is biological relevance, interpretation is likely to be more straightforward. Many times in GWAS, however, associated markers are found in intergenic regions (Hindorff 2010) such as gene deserts, in which there are no genes present or anywhere nearby. In these situations, it is useful to examine nearby genes to see if there could be some explanation for an association to that gene. It could be that the associated marker is in a long-range regulatory element for a gene of importance (Wasserman 2010). In addition, the knowledge gained by the ENCODE project have led to much more biological plausibility of GWAS signals. Many GWAS signals may not be in or near genes, but they have functional data in ENCODE that indicate that they are regulatory in some way. By changing the action of regulatory elements, gene expression and protein levels of components such as the enzymes tied to metabolism of the research drug could be modulated. Recent work has shown that some intergenic associations could also affect noncoding RNA elements (Glinskii 2009; Glinsky 2008). A particular class of noncoding RNA that has received considerable attention recently is the microRNA (Garzon 2010). A microRNA is an RNA element that is transcribed from DNA into RNA but not translated to protein. Its function is to target the transcripts of particular genes for degradation and thus act as an additional level of regulation in gene expression estimated to affect up to 30% of genes (Bader 2010). The importance of microRNAs in relation to pharmacogenomics is shown by Mishra et al. in a 2007 paper that identifies a microRNA binding site mutation located in the gene coding for dihydrofolate reductase, which results in resistance to the chemotherapy drug methotrexate (Mishra 2007). It is even hypothesized that targeting microRNAs could serve as a therapeutic intervention for diseases such as cancer for which changes in gene expression are proposed to play a significant role (Bader 2010). While genetic associations in intergenic regions can be difficult to account for, sometimes the associated marker is in a gene for which the function is not currently understood (Hindorff 2010) or for which there is no functional relevance to the outcome under study (McClellan 2010). At this point, interpretation depends on the power of the study as well as the effect size and significance level of the variant. Replication can also be a decisive factor in revealing false positives in the original study if the replication study is well powered to detect the associated variant. If the effect of the variant in an unrelated gene is large and predicts the outcome well particularly in individuals not present in the original study, or even if the effect size is small but the study was well powered to discover a variant of that effect size it could represent novel biology. For these reasons GWAS is useful for scanning the genome for novel common variation contributing to a trait but does suffer from interpretation issues. Awareness of these issues, however, can mitigate the challenge. Results of small candidate gene studies are often much easier to interpret in this respect. Candidate gene studies have the benefit of searching for direct instead of indirect association so a strong statistical signal can be more easily linked to causation. It should be noted, however, that to prove the predictive value of an associated variant from any study design validation will be required.

Replication

In 2007, the NHGRI Working Group on Replication in Association Studies laid out requirements for replication in genetic association studies to avoid the false positive associations observed in candidate gene studies (Chanock 2007). Replication was defined by the working group as significance of an effect of the same genetic variant, or one that is highly correlated with the original, with the same direction of effect (e.g., same risk allele in both studies) and in a comparable population. This working group came in response to the inability to replicate many early GWAS findings (Ioannidis 2001, 2007). Recommendations were made that replication of a finding within a population comparable to that in the original association study should be *required* for the publication of significant GWAS findings, citing the need to differentiate true positive results. While there is merit to this perspective and it has been adopted by the field, replication should take different forms under diverse circumstances. For example, the issues surrounding replication for models of gene-gene and/or gene-environment interactions have not been resolved.

The underlying rationale for replication is to strengthen the evidence that a genetic variant is truly associated with the trait under study. Ideally, the replication study should find precisely the same SNP to have a significant effect in the same direction (i.e., increased risk of disease) although the indirect nature of GWAS almost necessitates searching within the entire genomic region defined by LD structure around the original variant (Ioannidis 2009). In addition, would the same tag SNPs be expected to show the strongest interaction in multiple datasets? It is more likely that the same genes show evidence of interaction; however, the specific SNPs could differ. Another consideration for replication in pharmacogenomics specifically is the challenge of identifying a replication dataset. For many serious adverse drug reactions, there are limited patients with this reaction in the world, as many of these reactions are somewhat rare. Therefore, in some cases, all of the patients with this ADR are in the discovery analysis. This leaves no data to collect for a replication dataset. This prohibits a replication dataset from being identified. Fortunately, many early pharmacogenomic association studies give us hope that the effect sizes to be discovered in drug response phenotypes are significantly larger than that of complex diseases. Therefore they can be identified in smaller sample sizes. This means that perhaps some samples can be left out for a replication dataset. With the increasing trend toward biobanks linked to electronic health records (Gottesman 2013; McCarty 2011), there may be greater potential to find replication datasets for adverse drug reactions. As these resources continue to grow, so might the numbers of available samples with adverse drug reactions for replicating genomic signals. Alternatively, pharmacogenomic associations are sometimes taken back to a biology lab for validation.

Validation and Translation

A consideration for all pharmacogenomic association studies is the manner in which to proceed subsequent to finding significant associations. Clearly, the ideal endpoint of a pharmacogenomic study is the translation of the result into a genetic test for use of predicting drug response and guiding clinical treatment or identification of a new drug target. In order to translate a finding into a test, there are several intermediate steps that must be satisfied. First, the effect of the association must be validated (Coffey 2004) and estimates of the size of the effect refined. The latter is particularly important as it will allow cost-effectiveness estimates to be performed (Grosse 2008). It has been observed that initial association studies have a tendency to overstate estimates of risk in a phenomenon known as the "winner's curse" (Ioannidis 2009). A retrospective cohort design in a large study population would be a cost-effective technique for validating and refining the estimate of a SNP's predictive ability in a population setting if the trait is either continuously distributed or fairly common or if the goal is to show that patients with a particular genotype experience better outcomes over a long-term period (Taniguchi 2007). A retrospective cohort study would not work as well for a trait that might suffer from survival bias, however (Little 2005). Individuals at one end of the spectrum of trait values could be under-represented due to inability to participate in the study and risk estimates from the study would be biased. Given circumstances of survival bias, conducting a prospective cohort study or making use of RCT data would be more appropriate. Probably the most accurate genetic risk estimates would be gained by enrolling based on genotype as the exposure. While this would be costly and time-consuming, it has been used to look at long-term outcomes in individuals for whose clinical treatment has been determined as a result of genotyping (Epstein 2010). For rare outcomes, it would be necessary to rely on a case-control study for a risk estimate. Using either design is acceptable as long as sample size is sufficiently large and selection bias is minimized during ascertainment. A stable risk estimate will enable cost-effectiveness analysis. If a genetic test would not be cost-effective, either by a purely monetary definition or a definition integrating quality-adjusted life-years (QALYs) or disability-adjusted life-years (DALYs) (Grosse 2008), it is extremely unlikely to be accepted by the clinic or investors. Cost-effectiveness has been a factor restricting the implementation of genetic testing for warfarin dosing despite the strong genetic associations (Tan 2010). Factors determining cost-effectiveness include severity of the outcome and number needed to test (NNT) to prevent one event. Treatment setting should also be a factor in evaluating cost-effectiveness, as a different scale will be necessary when evaluating treatments in a resource-limited setting.

Conclusion

In the current era of low-cost DNA assays, genotyping and sequencing, and the availability of electronic medical records, we are getting closer to making personalized medicine a reality in more health care facilities. The success of pharmacogenomic association studies to understand the interplay between genetics and drug response phenotypes is an integral part of this success. In order to add even more true pharmacogenomic associations to the list of what would be useful in personalized medicine, however, it is paramount that these association studies are designed carefully. Ascertainment must be conducted in a manner that reduces the magnitude of information and selection bias. When possible, it can be statistically advantageous to take advantage of nonbinary phenotypes. Statistical testing should be carefully planned and focused to answer the study question in the interest of minimizing false positive results. Due to the costs of an association study and inherent publication bias, it can be tempting to pursue alternative phenotypes in attempts to discover a significant result. While it is important to not search the data too extensively, it is also important to properly account for the complexity of the genetic architecture underlying the outcome. Few association studies explore the possibility of complex genetic effects such as gene-gene or gene-environment interactions, though this is being done more and more. Due to the complex nature of drug metabolism and transport networks, it is probable that these higher order genetic effects play a significant role in determining phenotype between individuals and will be critical to understand some drug response phenotypes.

Association studies are the first step in a long road to implementing genetic tests in clinical practice. The Abacavir story has demonstrated that extensive validation efforts are necessary to prove the effects of a polymorphism and illustrate its predictive ability in practice. Although many genetic variants identified are likely to have tremendous potential to affect improvements in treatment on an individual level, the burden of proving cost-effectiveness falls on the researcher. As association studies continue to probe the genetic basis of drug response and our biological knowledge grows, it may be possible to overcome the cost-effectiveness hurdle by focusing the tests toward specific populations.

The reality is that the practice of genetic testing will not be feasible on a large-scale until significant changes are made in availability of tests. Current turn-around times on some genetic testing make it infeasible for testing data to guide initial treatment and dosing. One option that is being explored to solve this problem is the use of portable genetic tests that can be run with faster turn-around. Genetic testing to inform the prescription of warfarin is moving toward this end (Langley 2009). There are several "portable" assays that can be used to measure variation in *VKORC1* and *CYP2C9* with a considerable decrease in assay time. The cost of hardware required for testing is prohibitive to making these tests truly widespread at this point, particularly in resource-limited settings. Once we can develop truly portable assays without a decrease in accuracy, such testing could be a panacea for pharmacogenomics. This is especially true in developing countries where the density of high-quality genotyping labs is much lower and the importance of choosing an appropriate initial treatment regimen is maximized due to fewer opportunities for follow-up visits. Another option is the preemptive testing of a suite of important variants to reduce the per-genotype cost, keeping all results in the patient's medical records (Relling 2010). Having genetic information on file at the time of treatment initiation would be the ideal manifestation of pharmacogenomics. As the cost of sequencing continues to fall, it is possible that genome-wide sequence data will eventually be available to refer to at the time of treatment initiation. Even this scenario, however, relies on the use of well-designed pharmacogenomic association studies to determine upon which genetic data the treatment should be determined.

Acknowledgments

This work was supported by LM010040, AI077505, HG004608, and HL065962.

References

Agresti A. Categorical Data Analysis. New York: John Wiley & Sons, 1990.

Aidoo M, McElroy PD, Kolczak MS, et al. Tumor necrosis factor-alpha promoter variant 2 (TNF2) is associated with preterm delivery, infant mortality, and malaria morbidity in western Kenya: Asembo Bay Cohort Project IX. Genet Epidemiol 2001;21:201-11.

Aithal GP, Day CP, Kesteven PJ, et al. Association of polymorphisms in the cytochrome P450 CYP2C9 with warfarin dose requirement and risk of bleeding complications. Lancet 1999;353:717-9.

Alberts B. Editorial expression of concern. Science 2010;330:912.

Allegue C, Gil R, Sanchez-Diz P, et al. A new approach to long QT syndrome mutation detection by Sequenom MassARRAY system. Electrophoresis 2010;31:1648-55.

Anderson GD. Pharmacogenetics and enzyme induction/inhibition properties of antiepileptic drugs. Neurology 2004;63:S3-S8.

Armitage P. Tests for linear trends in proportions and frequencies. Biometrics 1955;11:375-86.

Aulchenko YS, Ripke S, Isaacs A, et al. GenABEL: an R library for genome-wide association analysis. Bioinformatics 2007;23:1294-6.

Aulchenko YS, Struchalin MV, van Duijn CM. ProbABEL package for genome-wide association analysis of imputed data. BMC Bioinformatics 2010;11:134.

Ayers KL, Cordell HJ. SNP Selection in genome-wide and candidate gene studies via penalized logistic regression. Genet Epidemiol 2010;34:879-91.

Bader AG, Brown D, Winkler M. The promise of microRNA replacement therapy. Cancer Res 2010;70:7027-30.

Balding DJ. A tutorial on statistical methods for population association studies. Nat Rev Genet 2006;7:781-791.

Barnholtz-Sloan JS, Chakraborty R, Sellers TA, et al. Examining population stratification via individual ancestry estimates versus self-reported race. Cancer Epidemiol Biomarkers Prev 2005;14:1545-51.

Benjamini Y, Hochberg Y. Controlling the false discovery rate: a practical and powerful approach to multiple testing. Journal of the Royal Statistical Society 1995;57:289-300.

Bertram L, Lill CM, Tanzi RE. The genetics of Alzheimer disease: back to the future. Neuron 2010;68:270-81.

Black N. Why we need observational studies to evaluate the effectiveness of health care. BMJ 1996;312:1215-18.

Bonetta L. Whole-genome sequencing breaks the cost barrier. Cell 2010;141:917-9.

Breiman L. Random forests. Machine Learning 2001;45:5-32.

Breiman L, Friedman JH, Olshen RA, et al. Classification and Regression Trees. New York: Chapman & Hall, 1984.

Burmester JK, Sedova M, Shapero MH, et al. DMET microarray technology for pharmacogenomics-based personalized medicine. Methods Mol Biol 2010;632:99-124.

Bush WS, Dudek SM, Ritchie MD. Biofilter: a knowledge-integration system for the multi-locus analysis of genome-wide association studies. Pac Symp Biocomput 2009;368-379.

Cantor RM, Lange K, Sinsheimer JS. Prioritizing GWAS results: A review of statistical methods and recommendations for their application. Am J Hum Genet 2010;86:6-22.

Cavallari LH, Langaee TY, Momary KM, et al. Genetic and clinical predictors of warfarin dose requirements in African Americans. Clin Pharmacol Ther 2010;87:459-64.

Chan EK, Hawken R, Reverter A. The combined effect of SNP-marker and phenotype attributes in genome-wide association studies. Anim Genet 2009;40:149-56.

Chanock SJ, Manolio T, Boehnke M, et al. Replicating genotype-phenotype associations. Nature 2007;447:655-60.

Chhibber A, Mefford J, Stahl EA, et al. Polygenic inheritance of paclitaxel-induced sensory peripheral neuropathy driven by axon outgrowth gene sets in CALGB 40101 (Alliance). Pharmacogenomics J 2014;14:336-42.

Cho YM, Ritchie MD, Moore JH, et al. Multifactor-dimensionality reduction shows a two-locus interaction associated with Type 2 diabetes mellitus. Diabetologia 2004;47:549-54.

Choi M, Scholl UI, Ji W, et al. Genetic diagnosis by whole exome capture and massively parallel DNA sequencing. Proc Natl Acad Sci U S A 2009;106:19096-101.

Chuang LY, Yang CS, Ho CH, et al. Tag SNP selection using particle swarm optimization. Biotechnol Prog 2010;26:580-88.

Coassin S, Brandstatter A, Kronenberg F. Lost in the space of bioinformatic tools: a constantly updated survival guide for genetic epidemiology. The GenEpi Toolbox. Atherosclerosis 2010;209:321-35.

Coffey CS, Hebert PR, Krumholz HM, et al. Reporting of model validation procedures in human studies of genetic interactions. Nutrition 2004;20:69-73.

Coffey CS, Hebert PR, Ritchie MD, et al. An application of conditional logistic regression and multifactor dimensionality reduction for detecting gene-gene interactions on risk of myocardial infarction: the importance of model validation. BMC Bioinformatics 2004;5:49.

Cohen J, Cohen P, West SG, et al. Applied Multiple Regression and Correlation Analysis for the Behavioral Sciences. New Jersey: Lawrence Earlbaum Associates, 2003.

Cordell HJ. Detecting gene-gene interactions that underlie human diseases. Nat Rev Genet 2009;10:392-404.

Cree BA, Rioux JD, McCauley JL, et al. A major histocompatibility Class I locus contributes to multiple sclerosis susceptibility independently from HLA-DRB1*15:01. PLoS One 2010;5:e11296.

Dahlin MG, Beck OM, Amark PE. Plasma levels of antiepileptic drugs in children on the ketogenic diet. Pediatr Neurol 2006;35:6-10.

Devlin B, Roeder K. Genomic control for association studies. Biometrics 1999;55:997-1004.

Durbin RM, Abecasis GR, Altshuler DL, et al. A map of human genome variation from population-scale sequencing. Nature 2010;467:1061-73.

Edenberg HJ, Liu Y. Laboratory methods for high-throughput genotyping. Cold Spring Harb Protoc 2009:db.

Ekbal NJ, Holt DW, Macphee IA. Pharmacogenetics of immunosuppressive drugs: prospect of individual therapy for transplant patients. Pharmacogenomics 2008;9:585-96.

Epstein MP, Allen AS, Satten GA. A simple and improved correction for population stratification in case-control studies. Am J Hum Genet 2007;80:921-930.

Epstein RS, Moyer TP, Aubert RE, et al. Warfarin genotyping reduces hospitalization rates results from the MM-WES (Medco-Mayo Warfarin Effectiveness study). J Am Coll Cardiol 2010;55:2804-12.

Fisher RA. Statistical Methods and Scientific Inference. New York: Hafner, 1956.

Fong C, Ko DC, Wasnick M, et al. GWAS analyzer: integrating genotype, phenotype and public annotation data for genome-wide association study analysis. Bioinformatics 2010;26:560-4.

Fredrickson DS, Lees RS. A system for phenotyping hyperlipoproteinemia. Circulation 1965;31:321-7.

Friedman JH. Multivariate adaptive regression splines. The Annals of Statistics 1991;19:1-67.

Fu SJ, Wang YB, Yu LX, et al. [Factors responsible for interindividual variations in dosage/concentration of tacrolimus in renal transplant recipients]. Nan Fang Yi Ke Da Xue Xue Bao 2008;28:2161-4.

Galanter JM, Gignoux CR, Torgerson DG, et al. Genome-wide association study and admixture mapping identify different asthma-associated loci in Latinos: the Genes-environments and Admixture in Latino Americans study. J Allergy Clin Immunol 2014;134:295-305.

Garay JP, Gray JW. Omics and therapy—a basis for precision medicine. Mol Oncol 2012;6:128-139.

Garzon R, Marcucci G, Croce CM. Targeting microRNAs in cancer: rationale, strategies and challenges. Nat Rev Drug Discov 2010;9:775-89.

Ghosh S. Genome-wide association analyses of quantitative traits: the GAW16 experience. Genet Epidemiol 2009;33 Suppl 1:S13-S18.

Gibbs RA, Belmont JW, Hardenbol P, et al. The International HapMap Project. Nature 2003;426:789-96.

Ginsburg GS, Voora D. The long and winding road to warfarin pharmacogenetic testing. J Am Coll Cardiol 2010;55:2813-5.

Glinskii AB, Ma J, Ma S, et al. Identification of intergenic trans-regulatory RNAs containing a disease-linked SNP sequence and targeting cell cycle progression/differentiation pathways in multiple common human disorders. Cell Cycle 2009;8:3925-42.

Glinsky GV. SNP-guided microRNA maps (MirMaps) of 16 common human disorders identify a clinically accessible therapy reversing transcriptional aberrations of nuclear import and inflammasome pathways. Cell Cycle 2008;7:3564-76.

Gottesman O, Kuivaniemi H, Tromp G, et al. The Electronic Medical Records and Genomics (eMERGE) Network: past, present, and future. Genet Med Off J Am Coll Med Genet 2013;15:761-771.

Grady BJ, Ritchie MD. Statistical optimization of pharmacogenomics association studies: key considerations from study design to analysis. Curr Pharmacogenomics Pers Med 2011;9:41-66.

Grady BJ, Torstenson E, Dudek SM, et al. Finding unique filter sets in plato: a precursor to efficient interaction analysis in GWAS data. Pac Symp Biocomput Pac Symp Biocomput 2010;315-26.

Greene CS, Penrod NM, Kiralis J, et al. Spatially Uniform ReliefF (SURF) for computationally-efficient filtering of gene-gene interactions. BioData Min 2009;2:5.

Greene CS, Penrod NM, Williams SM, et al. Failure to replicate a genetic association may provide important clues about genetic architecture. PLoS One 2009;4:e5639.

Greenwood PE, Nikulin MS. A Guide to Chi-Squared Testing. New York: Wiley, 1996.

Grosse SD, Wordsworth S, Payne K. Economic methods for valuing the outcomes of genetic testing: beyond cost-effectiveness analysis. Genet Med Off J Am Coll Med Genet 2008;10:648-54.

Hahn LW, Ritchie MD, Moore JH. Multifactor dimensionality reduction software for detecting gene-gene and gene-environment interactions. Bioinformatics 2003;19:376-82.

Haines JL, Hauser MA, Schmidt S, et al. Complement factor H variant increases the risk of age-related macular degeneration. Science 2005;308:419-21.

Haines JL, Pericak-Vance MA. Genetic Analysis of Complex Disease. Hoboken, NJ: John Wiley & Sons, 2006.

Haas DW. Human genetic variability and HIV treatment response. Curr HIV/AIDS Rep 2006;3:53-58.

Haas DW, Ribaudo HJ, Kim RB, et al. Pharmacogenetics of efavirenz and central nervous system side effects: an Adult AIDS Clinical Trials Group study. AIDS 2004;18:2391-400.

Han B, Kang HM, Seo MS, et al. Efficient association study design via power-optimized tag SNP selection. Ann Hum Genet 2008;72:834-47.

Hardy GH. Mendelian proportions in a mixed population. Science 1908;28:49-50.

Hastie T, Tibshirani R, Friedman J. The Elements of Statistical Learning: Data Mining, Inference, and Prediction. New York: Springer-Verlag, 2001.

Hindorff LA, Junkins HA, Hall PN, et al. A Catalog of Published Genome-Wide Association Studies, 2010.

Hindorff LA, Sethupathy P, Junkins HA, et al. Potential etiologic and functional implications of genome-wide association loci for human diseases and traits. Proc Natl Acad Sci USA 2009;106:9362-7.

Hoggart CJ, Parra EJ, Shriver MD, et al. Control of confounding of genetic associations in stratified populations. Am J Hum Genet 2003;72:1492-1504.

Hosmer DW, Lemeshow S. Applied Logistic Regression. New York: John Wiley & Sons Inc, 2000.

Hosseini P, Tremblay A, Matthews BF, et al. An efficient annotation and gene-expression derivation tool for Illumina Solexa datasets. BMC Res Notes 2010;3:183.

Huang SW, Chen HS, Wang XQ, et al. Validation of VKORC1 and CYP2C9 genotypes on interindividual warfarin maintenance dose: a prospective study in Chinese patients. Pharmacogenet Genomics 2009;19:226-34.

Illumina. Custom Genotyping. 2010. San Diego: Illumina, Inc.

Illumina. HumanCVD Genotyping BeadChip. 2010. San Diego: Illumina, Inc.

Illumina. The VeraCode ADME Core Panel, 2010.

Innocenti F, Owzar K, Cox NL, et al. A genome-wide association study of overall survival in pancreatic cancer patients treated with gemcitabine in CALGB 80303. Clin Cancer Res Off J Am Assoc Cancer Res 2012;18:577-84.

International Hapmap Consortium. A second generation human haplotype map of over 3.1 million SNPs. Nature 2007;449:851-61.

International Human Genome Sequencing Consortium. Finishing the euchromatic sequence of the human genome. Nature 2004;431:931-45.

Ioannidis JP. Non-replication and inconsistency in the genome-wide association setting. Hum Hered 2007;64:203-13.

Ioannidis JP, Ntzani EE, Trikalinos TA, et al. Replication validity of genetic association studies. Nat Genet 2001;29:306-9.

Ioannidis JP, Thomas G, Daly MJ. Validating, augmenting and refining genome-wide association signals. Nat Rev Genet 2009;10:318-29.

Iyer L, King CD, Whitington PF, et al. Genetic predisposition to the metabolism of irinotecan (CPT-11). Role of uridine diphosphate glucuronosyltransferase isoform 1A1 in the glucuronidation of its active metabolite (SN-38) in human liver microsomes. J Clin Invest 1998;101:847-54.

J D. The Affymetrix DMET platform and pharmacogenetics in drug development. Curr Opin Mol Ther 2009;11:260-8.

Jadad, AR. Randomised Controlled Trials: A User's Guide. London: BMJ Books, 1998.

James AH, Britt RP, Raskino CL, et al. Factors affecting the maintenance dose of warfarin. J Clin Pathol 1992;45:704-6.

Jiang H, Fine JP. Survival analysis. Methods Mol Biol 2007;404:303-18.

Kaplan EL, Meier P. Nonparametric estimation from incomplete observations. J Amer Statist Assn 1958;53:457-81.

Khoury MJ, Flanders WD. Nontraditional epidemiologic approaches in the analysis of gene-environment interaction: case-control studies with no controls! Am J Epidemiol 1996;144:207-13.

Khoury MJ, Gwinn ML, Glasgow RE, et al. A population approach to precision medicine. Am J Prev Med 2012;42:639-45.

Kirkwood B, Sterne J. Essentials of Medical Statistics, 2nd ed. Malden, MA: Wiley-Blackwell, 2001.

Kiyotani K, Mushiroda T, Hosono N, et al. Lessons for pharmacogenomics studies: association study between CYP2D6 genotype and tamoxifen response. Pharmacogenet Genomics 2010;20:565-8.

Klein RJ, Zeiss C, Chew EY, et al. Complement factor H polymorphism in age-related macular degeneration. Science 2005;308:385-9.

Koboldt DC. Challenges of sequencing human genomes. Brief Bioinform 2010.

Kooperberg C, Ruczinski I. Identifying interacting SNPs using Monte Carlo logic regression. Genet Epidemiol 2005;28:157-70.

Kooperberg C, Ruczinski I, LeBlanc ML, et al. Sequence analysis using logic regression. Genet Epidemiol 2001;21 Suppl 1:S626-S631.

Korbel JO, Urban AE, Grubert F, et al. Systematic prediction and validation of breakpoints associated with copy-number variants in the human genome. Proc Natl Acad Sci U S A 2007;104:10110-5.

Kuritzkes DR. Preventing and managing antiretroviral drug resistance. AIDS Patient Care STDS 2004;18:259-73.

Lander ES. The new genomics: global views of biology. Science 1996;274:536-9.

Lander ES, Linton LM, Birren B, et al. Initial sequencing and analysis of the human genome. Nature 2001;409:860-921.

Langley MR, Booker JK, Evans JP, et al. Validation of clinical testing for warfarin sensitivity: comparison of CYP2C9-VKORC1 genotyping assays and warfarin-dosing algorithms. J Mol Diagn 2009;11:216-25.

Laurie CC, Doheny KF, Mirel DB, et al. Quality control and quality assurance in genotypic data for genome-wide association studies. Genet Epidemiol 2010;34:591-602.

Lazarou J, Pomeranz BH, Corey PN. Incidence of adverse drug reactions in hospitalized patients: a meta-analysis of prospective studies. JAMA J Am Med Assoc 1998;279:1200-5.

Leek JT, Scharpf RB, Bravo HC, et al. Tackling the widespread and critical impact of batch effects in high-throughput data. Nat Rev Genet 2010.

Lettre G, Lange C, Hirschhorn JN. Genetic model testing and statistical power in population-based association studies of quantitative traits. Genet Epidemiol 2007;31:358-362.

Levy GN, Weber WW. Interindividual Variability in Human Drug Metabolism. Pacifici, CM and Pelkonen, O. 333-357. New York: Taylor and Francis, 2001.

Li J. Prioritize and select SNPs for association studies with multi-stage designs. J Comput Biol 2008;15:241-57.

Li W, Reich J. A complete enumeration and classification of two-locus disease models. Hum.Hered. 2000;50:334-49.

Limdi NA, Veenstra DL. Expectations, validity, and reality in pharmacogenetics. J Clin Epidemiol 2010;63:960-69.

Little J, Sharp L, Khoury MJ, et al. The epidemiologic approach to pharmacogenomics. Am J Pharmacogenomics 2005;5:1-20.

Macphee IA, Fredericks S, Tai T, et al. The influence of pharmacogenetics on the time to achieve target tacrolimus concentrations after kidney transplantation. Am J Transplant 2004;4:914-9.

Maher B. Personal genomes: The case of the missing heritability. Nature 2008;456:18-21.

Majumder PP, Ghosh S. Mapping quantitative trait loci in humans: achievements and limitations. J Clin Invest 2005;115:1419-24.

Mallal S, Phillips E, Carosi G, et al. HLA-B*5701 screening for hypersensitivity to abacavir. N Engl J Med 2008;358:568-79.

Mancinelli LM, Frassetto L, Floren LC, et al. The pharmacokinetics and metabolic disposition of tacrolimus: a comparison across ethnic groups. Clin Pharmacol Ther 2001;69:24-31.

Manolio TA, Bailey-Wilson JE, Collins FS. Genes, environment and the value of prospective cohort studies. Nat Rev Genet 2006;7:812-20.

Manolio TA, Collins FS, Cox NJ, et al. Finding the missing heritability of complex diseases. Nature 2009;461:747-53.

Marchini J, Donnelly P, Cardon LR. Genome-wide strategies for detecting multiple loci that influence complex diseases. Nat Genet 2005;37:413-7.

Mardis ER. Anticipating the 1,000 dollar genome. Genome Biol 2006;7:112.

Maxwell SE, Delaney HD. Designing Experiments and Analyzing Data. (Lawrence Erlbaum Associates). New Jersey: Routledge, 1990.

McCarthy MI, Abecasis GR, Cardon LR, et al. Genome-wide association studies for complex traits: consensus, uncertainty and challenges. Nat Rev Genet 2008;9:356-69.

McCarty CA, Chisholm RL, Chute CG, et al. The eMERGE Network: a consortium of biorepositories linked to electronic medical records data for conducting genomic studies. BMC Med Genomics 2011;4:13.

McClellan J, King MC. Genetic heterogeneity in human disease. Cell 2010;141:210-7.

Mega JL, Close SL, Wiviott SD, et al. Cytochrome p-450 polymorphisms and response to clopidogrel. N Engl J Med 2009;360:354-62.

Merck Manuals Online Medical Library. Pharmacokinetics. 2010.

Merck Manuals Online Medical Library. Pharmacodynamics. 2010.

Mirnezami R, Nicholson J, Darzi A. Preparing for Precision Medicine. N Engl J Med 2012;366:489-91.

Mishra PJ, Humeniuk R, Mishra PJ, et al. A miR-24 microRNA binding-site polymorphism in dihydrofolate reductase gene leads to methotrexate resistance. Proc Natl Acad Sci U S A 2007;104:13513-8.

Montana G, Pritchard JK. Statistical tests for admixture mapping with case-control and cases-only data. Am J Hum Genet 2004;75:771-89.

Moore JH. The ubiquitous nature of epistasis in determining susceptibility to common human diseases. Hum Hered 2003;56:73-82.

Moore JH, Gilbert JC, Tsai CT, et al. A flexible computational framework for detecting, characterizing, and interpreting statistical patterns of epistasis in genetic studies of human disease susceptibility. J Theor Biol 2006;241:252-61.

Moore JH, Williams SM. Traversing the conceptual divide between biological and statistical epistasis: systems biology and a more modern synthesis. Bioessays 2005;27:637-46.

Motsinger AA, Ritchie MD, Reif DM. Novel methods for detecting epistasis in pharmacogenomics studies. Pharmacogenomics 2007;8:1229-41.

Motsinger AA, Ritchie MD. Multifactor dimensionality reduction: an analysis strategy for modelling and detecting gene-gene interactions in human genetics and pharmacogenomics studies. Hum Genomics 2006;2:318-28.

Motsinger AA, Ritchie MD, Shafer RW, et al. Multilocus genetic interactions and response to efavirenz-containing regimens: an adult AIDS clinical trials group study. Pharmacogenet Genomics 2006;16:837-45.

Motsinger AA, Lee SL, Mellick G, et al. GPNN: power studies and applications of a neural network method for detecting gene-gene interactions in studies of human disease. BMC Bioinformatics 2006;7:39.

Motsinger AA, Dudek SM, Hahn LW, et al. Grammatical evolution for the optimization of neural networks for genetic association studies. Bioinformatics. In submission.

Motsinger-Reif AA, Reif DM, Fanelli TJ, et al. A comparison of analytical methods for genetic association studies. Genet Epidemiol 2008;32:767-78.

Nelder J, Wedderburn R. Generalized Linear Models. Journal of the Royal Statistical Society 2010;135:370-84.

Ng SB, Turner EH, Robertson PD, et al. Targeted capture and massively parallel sequencing of 12 human exomes. Nature 2009;461:272-6.

Oberst RD, Hays MP, Bohra LK, et al. PCR-based DNA amplification and presumptive detection of Escherichia coli O157:H7 with an internal fluorogenic probe and the 5' nuclease (TaqMan) assay. Appl Environ Microbiol 1998;64:3389-96.

Oetjens MT, Denny JC, Ritchie MD, et al. Assessment of a pharmacogenomic marker panel in a polypharmacy population identified from electronic medical records. Pharmacogenomics 2013;14:735-44.

Ohara K, Tanabu S, Ishibashi K, et al. CYP2D6*10 alleles do not determine plasma fluvoxamine concentration/dose ratio in Japanese subjects. Eur J Clin Pharmacol 2003;58:659-61.

Ohno Machado L. Modeling medical prognosis: survival analysis techniques. J Biomed Inform 2001;34:428-39.

Oldenburg J, Bevans CG, Muller CR, et al. Vitamin K epoxide reductase complex subunit 1 (VKORC1): the key protein of the vitamin K cycle. Antioxid Redox Signal 2006;8:347-53.

1000 Genomes Project Consortium. A map of human genome variation from population-scale sequencing. Nature 2010;467:1061-1073.

Parkes M, Cortes A, van Heel DA, et al. Genetic insights into common pathways and complex relationships among immune-mediated diseases. Nat Rev Genet 2013;14:661-73.

Pendergrass SA, Frase AT, Wallace JR, et al. Genomic analyses with biofilter 2.0: knowledge driven filtering, annotation, and model development. BioData Min 2013;6:25.

Pharmacokinetics Working Group of the AGAH. Collection of terms, symbols, equations, and explanations of common pharmacokinetic and pharmacodynamic parameters and some statistical functions. 2004. Available at www.agah.eu/fileadmin/_migrated/content_uploads/PK-glossary_PK_working_group_2004.pdf.

Pico AR, Smirnov IV, Chang JS, et al. SNPLogic: an interactive single nucleotide polymorphism selection, annotation, and prioritization system. Nucleic Acids Res 2009;37(Database issue):D803-9.

Pirmohamed M, James S, Meakin S, et al. Adverse drug reactions as cause of admission to hospital: prospective analysis of 18 820 patients. BMJ 2004;329:15-19.

Pluzhnikov A, Below JE, Konkashbaev A, et al. Spoiling the whole bunch: quality control aimed at preserving the integrity of high-throughput genotyping. Am J Hum Genet 2010;87:123-8.

Price AL, Patterson NJ, Plenge RM, et al. Principal components analysis corrects for stratification in genome-wide association studies. Nat Genet 2006;38:904-9.

Pritchard JK, Rosenberg NA. Use of unlinked genetic markers to detect population stratification in association studies. Am J Hum Genet 1999;65:220-8.

Pritchard JK, Stephens M, Donnelly P. Inference of population structure using multilocus genotype data. Genetics 2000;155:945-59.

Pritchard JK, Stephens M, Rosenberg NA, et al. Association mapping in structured populations. Am J Hum Genet 2000;67:170-81.

PSC Scientific and Medical Advisory Committee. PSC Awards Update. 2010.

Purcell SM, Moran JL, Fromer M, et al. A polygenic burden of rare disruptive mutations in schizophrenia. Nature 2014;506:185-90.

Purcell S, Neale B, Todd-Brown K, et al. PLINK: a tool set for whole-genome association and population-based linkage analyses. Am J Hum Genet 2007;81:559-75.

Rebbeck TR, Jaffe JM, Walker AH, et al. Modification of clinical presentation of prostate tumors by a novel genetic variant in CYP3A4. J Natl Cancer Inst 1998;90:1225-9.

Reiner AP, Ziv E, Lind DL, et al. Population structure, admixture, and aging-related phenotypes in African American adults: the cardiovascular health study. Am J Hum Genet 2005;76:463-77.

Relling MV, Altman RB, Goetz MP, et al. Clinical implementation of pharmacogenomics: overcoming genetic exceptionalism. Lancet Oncol 2010;11:507-9.

Ribaudo HJ, Liu H, Schwab M, et al. Effect of CYP2B6, ABCB1, and CYP3A5 Polymorphisms on Efavirenz Pharmacokinetics and Treatment Response: An AIDS Clinical Trials Group Study. J Infect Dis 2010; 202:717-22.

Rice TK, Schork NJ, Rao DC. Methods for handling multiple testing. Adv Genet 2008;60:293-308.

Ritchie MD. The success of pharmacogenomics in moving genetic association studies from bench to bedside: study design and implementation of precision medicine in the post-GWAS era. Hum Genet 2012;131:1615-26.

Ritchie MD, Coffey CSMJH. Genetic programming neural networks: A bioinformatics tool for human genetics. Lecture Notes in Computer Science 2004;3102:438-48.

Ritchie MD, Hahn LW, Roodi N, et al. Multifactor-dimensionality reduction reveals high-order interactions among estrogen-metabolism genes in sporadic breast cancer. Am J Hum Genet 2001;69:138-47.

Ritchie MD, White BC, Parker JS, et al. Optimization of neural network architecture using genetic programming improves detection and modeling of gene-gene interactions in studies of human diseases. BMC Bioinformatics 2003;4:28.

Roeder K, Bacanu SA, Wasserman L, et al. Using linkage genome scans to improve power of association in genome scans. Am J Hum Genet 2006;78:243-52.

Saccone SF, Bierut LJ, Chesler EJ, et al. Supplementing high-density SNP microarrays for additional coverage of disease-related genes: addiction as a paradigm. PLoS One 2009;4:e5225.

Salari K, Burchard EG. Latino populations: a unique opportunity for epidemiological research of asthma. Paediatr. Perinat. Epidemiol 2007;21 Suppl 3:15-22.

Salkind NJ. Encyclopedia of Measurement and Statistics. Thousand Oaks, CA: Sage, 2007.

Sanson-Fisher RW, Bonevski B, Green LW, et al. Limitations of the randomized controlled trial in evaluating population-based health interventions. Am J Prev Med 2007;33:155-61.

Satten GA, Flanders WD, Yang Q. Accounting for unmeasured population substructure in case-control studies of genetic association using a novel latent-class model. Am J Hum Genet 2001;68:466-77.

Schlesselman JJ. Case-Control Studies. Design, Conduct, Analysis. New York: Oxford University Press, 1982.

Schork NJ, Murray SS, Frazer KA, et al. Common vs. rare allele hypotheses for complex diseases. Curr Opin Genet Dev 2009;19:212-9.

Schwarz DF, Konig IR, Ziegler A. On safari to Random Jungle: a fast implementation of Random Forests for high-dimensional data. Bioinformatics 2010;26:1752-8.

Sebastiani P, Solovieff N, Puca A, et al. Genetic Signatures of Exceptional Longevity in Humans. Science 2010.

Shendure J, Ji H. Next-generation DNA sequencing. Nat Biotechnol 2008;26:1135-45.

Slatkin M. Linkage disequilibrium--understanding the evolutionary past and mapping the medical future. Nat Rev Genet 2008;9:477-85.

Singh R, Srivastava A, Kapoor R, et al. Impact of CYP3A5 and CYP3A4 gene polymorphisms on dose requirement of calcineurin inhibitors, cyclosporine and tacrolimus, in renal allograft recipients of North India. Naunyn Schmiedebergs Arch Pharmacol 2009;380:169-77.

Sissung TM, English BC, Venzon D, et al. Clinical pharmacology and pharmacogenetics in a genomics era: the DMET platform. Pharmacogenomics 2010;11:89-103.

Slatkin M. Linkage disequilibrium—understanding the evolutionary past and mapping the medical future. Nat Rev Genet 2008;9:477-85.

Slinker BK, Glantz SA. Multiple linear regression is a useful alternative to traditional analyses of variance. Am J Physiol 1988;255:R353-67.

Smith M, Kohn R. Nonparametric regression using Bayesian variable selection. Journal of Econometrics 1996;75:317-43.

Smith PG, Day NE. The design of case-control studies: the influence of confounding and interaction effects. International Journal of Epidemiology 1984;13:356-65.

St George-Hyslop P. Molecular genetics of Alzheimer's disease. Biol Psychiatry 2000;47:183-199.

Steemers FJ, Gunderson KL. Illumina, Inc. Pharmacogenomics 2005;6:777-82.

Stephens M, Balding DJ. Bayesian statistical methods for genetic association studies. Nat Rev Genet 2009;10:681-90.

Stolberg HO, Norman G, Trop I. Randomized controlled trials. AJR Am J Roentgenol 2004;183:1539-44.

Sun X, Lu Q, Mukheerjee S, et al. Analysis pipeline for the epistasis search—statistical versus biological filtering. Front Genet 2014;5:106.

Tabor HK, Risch NJ, Myers RM. Candidate-gene approaches for studying complex genetic traits: practical considerations. Nat Rev Genet 2002;3:391-7.

Talmud PJ, Drenos F, Shah S, et al. Gene-centric association signals for lipids and apolipoproteins identified via the HumanCVD BeadChip. Am J Hum Genet 2009;85:628-42.

Tan GM, Wu E, Lam YY, et al. Role of warfarin pharmacogenetic testing in clinical practice. Pharmacogenomics 2010;11:439-48.

Taniguchi A, Urano W, Tanaka E, et al. Validation of the associations between single nucleotide polymorphisms or haplotypes and responses to disease-modifying antirheumatic drugs in patients with rheumatoid arthritis: a proposal for prospective pharmacogenomic study in clinical practice. Pharmacogenet Genomics 2007;17:383-90.

Taubert D, Bouman HJ, van Werkum JW. Cytochrome P-450 polymorphisms and response to clopidogrel. N Engl J Med 2009;360:2249-50.

Thomas DC. Statistical Methods in Genetic Epidemiology. New York: Oxford University Press, Inc., 2004.

Thornton T, McPeek MS. Case-control association testing with related individuals: a more powerful quasi-likelihood score test. Am J Hum Genet 2007;81:321-37.

Thornton-Wells TA, Moore JH, Haines JL. Genetics, statistics and human disease: analytical retooling for complexity. Trends Genet 2004;20:640-47.

Tian C, Gregersen PK, Seldin MF. Accounting for ancestry: population substructure and genome-wide association studies. Hum Mol Genet 2008;17:R143-50.

Tian C, Kosoy R, Nassir R, et al. European population genetic substructure: further definition of ancestry informative markers for distinguishing among diverse European ethnic groups. Mol Med 2009;15:371-83.

Tishkoff SA, Verrelli BC. Patterns of human genetic diversity: implications for human evolutionary history and disease. Annu Rev Genomics Hum Genet 2003;4:293-340.

Turner S, Armstrong LL, Bradford Y, et al. Quality control procedures for genome-wide association studies. Curr Protoc Hum Genet Editor Board Jonathan Haines Al Chapter 1, Unit1.19. 2011.

Turner SD, Crawford DC, Ritchie MD. Methods for optimizing statistical analyses in pharmacogenomics research. Expert Rev Clin Pharmacol 2009;2:559-70.

Turner SD, Dudek SM, Ritchie MD. Grammatical evolution of neural networks for discovering epistasis among quantitative trait loci. Lect Notes Comput Sci 2010;6023:86-97.

23andMe. SNPwatch: Uncertainty Surrounds Longevity GWAS. 2010.

VanderWeele TJ, Hernandez-Diaz S, Hernan MA. Case-only gene-environment interaction studies: when does association imply mechanistic interaction? Genet Epidemiol 2010;34:327-34.

Via M, Gignoux C, Burchard EG. The 1000 Genomes Project: new opportunities for research and social challenges. Genome Med 2010;2:3.

Voight BF, Pritchard JK. Confounding from cryptic relatedness in case-control association studies. PLoS Genet 2005;1:e32.

Wasserman NF, Aneas I, Nobrega MA. An 8q24 gene desert variant associated with prostate cancer risk confers differential in vivo activity to a MYC enhancer. Genome Res 2010.

Weale ME. Quality control for genome-wide association studies. Methods Mol Biol 2010;628:341-72.

Wester K, Jönsson AK, Spigset O, Druid H, Hägg S. Incidence of fatal adverse drug reactions: a population based study. Br J Clin Pharmacol 2008;65:573-9.

Wittke-Thompson JK, Pluzhnikov A, Cox NJ. Rational inferences about departures from Hardy-Weinberg equilibrium. Am J Hum Genet 2005;76:967-86.

Woodahl EL, Hingorani SR, Wang J, et al. Pharmacogenomic associations in ABCB1 and CYP3A5 with acute kidney injury and chronic kidney disease after myeloablative hematopoietic cell transplantation. Pharmacogenomics J 2008;8:248-55.

Zanella F, Lorens JB, Link W. High content screening: seeing is believing. Trends Biotechnol 2010;28:237-45.

Zang EA, Wynder EL. Reevaluation of the confounding effect of cigarette smoking on the relationship between alcohol use and lung cancer risk, with larynx cancer used as a positive control. Prev Med 2001;32:359-70.

Zheng HX, Webber SA, Zeevi A, et al. The impact of pharmacogenomic factors on steroid dependency in pediatric heart transplant patients using logistic regression analysis. Pediatr Transplant 2004;8:551-7.

Zuk O, Hechter E, Sunyaev SR, et al. The mystery of missing heritability: Genetic interactions create phantom heritability. Proc Natl Acad Sci 2011;109:1193-8.

Zuvich RL, Armstrong LL, Bielinski SJ, et al. Pitfalls of merging GWAS data: lessons learned in the eMERGE network and quality control procedures to maintain high data quality. Genet. Epidemiol. 2011;35:887-98.

Index

Page numbers followed by *b, f,* or *t* indicate material in boxes, figures, or tables, respectively.

abacavir, 23, 38, 200–202, 238–239, 244*t*
ABCB1 gene. *See* ATP-binding cassette transporter gene
ABCG2 gene, 110
Academy of Medical Sciences, 283
ACCE framework, 20
access, 260–261
acetylation, histone, 360
acetylsalicylic acid. *See* aspirin
ACS. *See* acute coronary syndromes
activity scores, 5
acute coronary syndromes (ACS), 65
acute myelogenous leukemia (AML), 157
adenine, 318, 321*f*
adenosine diphosphate (ADP), 66–67
adenotonsillectomy, 296*t*
adenylate cyclase pathway, 354–355
ADHD. *See* attention deficit/hyperactivity disorder
ADME research, 301, 303
administrative pricing, 272
administrators, 49*t*–50*t*
ado-trastuzumab emtansine, 152
ADP. *See* adenosine diphosphate
ADP antagonists, 66
ADRA2A gene. *See* alpha-2 adrenergic receptor gene
adverse drug reaction, 2, 136, 196, 370. *See also* severe cutaneous adverse drug reactions
Adverse Event Reporting System (AERS), 296
afatinib, 154–155, 295
Affymetrix DMET Plus Premier Pack, 376
African Americans, 95–96, 181–182
age, 104, 257
agranulocytosis, 124
alcohol, 103, 374
ALK gene. *See* anaplastic lymphoma kinase gene
allele nomenclature, 3–4, 323–324
alleles, 323, 323*f*. *See also specific gene*
    dysfunctional, 5
    loss-of-function, 5
    minor allele frequency and, 378
    wild-type, 25
allopurinol, 205–206, 206*t*–207*t*, 208
alpha-2 adrenergic receptor gene (*ADRA2A* gene), 128
altered peptide repertoire concept, 209
alternative splicing, 342–343
Alzheimer disease, 40, 351, 370
American Association for Clinical Chemistry, 17
American College of Cardiology Foundation, 69
American Heart Association, 69
American Psychiatric Association, 117
American Society of Health-System Pharmacists, 56–57
American Society of Human Genetics, 257
amino acid triplet code, 321*f*
amino acids, 221*t*
aminoglycoside antibiotics, 268
amitriptyline, 118*t*
AML. *See* acute myelogenous leukemia
amphetamines, 127–128
AMPK, 21–22
AmpliChip® CYP450 test, 16, 21
analysis
    budget impact, 277–278
    candidate gene, 325–326
    cost-benefit, 278, 279*t*
    cost-effectiveness, 278–280, 279*t*
    cost-minimization, 278, 279*t*
    cost-utility, 54, 279*t*, 280–281
    CYP2C19 gene, 68
    decision, 275
    epistasis, 382–383
    filtering in, 383–384
    genome-wide linkage, 326
    pharmacogenomics research, 380–384
    principle components, 379
    retrospective, 23, 309
    risk benefit, 275–276
    single SNP, 380–382
analysis of variance (ANOVA), 382
analytic validity, 20–21
anaplastic lymphoma kinase gene (*ALK* gene), 155, 294, 300
ANOVA. *See* analysis of variance
antibiotics, 11, 197*t*, 268
anticoagulants, 83, 97. *See also* warfarin
antidepressants, 7
    ABCB1 gene and, 120
    CYP2C19 gene and, 117, 119
    CYP2D6 gene and, 117, 119
    HTR2A gene and, 120–121
    pharmacogenomics of, 117, 118*t*–119*t*
    pharmacokinetics of, 117, 119–120
    SLC6A4 gene and, 120–121
antiepileptics, 7, 197*t*, 202
antifungals, 190
antiglaucoma drugs, 197*t*
antigout drugs, 197*t*
anti-HIV drugs, 197*t*
antiplatelet drugs, 65–67, 73
antiplatelet therapy, 69–71, 70*t*
antipsychotics, 25, 121–124
antiretroviral therapy (ART), 233–234, 243–245, 244*t*
antithrombotic therapy, 277*f*
antivirals, 11, 217, 222–224
APOE. *See* apolipoprotein E
APOE gene, 40, 370
apolipoprotein E (APOE), 111
arachidonic acid, 9
aripiprazole, 118*t*, 290*t*
aromatic antiepileptic drug, 202
array-based hybridization assays, 16
arsenic trioxide, 157, 290*t*
ART. *See* antiretroviral therapy
Asians, 202, 204
aspirin, 65–66, 73
assays, 15–16, 91, 376, 387
asunaprevir, 228
ataxia-telangiectasia mutated gene (*ATM* gene), 21
atazanavir, 241–242, 244*t*
atherosclerosis, 66
ATM gene. *See* ataxia-telangiectasia mutated gene
atomoxetine, 118*t*, 127–128
atorvastatin, 105, 106*t*, 110
ATP1A2 gene, 137
ATP-binding cassette transporter gene (*ABCB1* gene), 120, 143, 183, 185
atrasentan, 303, 304*f*
attention deficit/hyperactivity disorder (ADHD), 127–129
autosomal dominant inheritance, 324, 324*f*
autosomal recessive inheritance, 324–325, 325*f*
azacitidine, 363*t*
azathioprine, 165*f*, 185–187, 185*t*, 188*f*
azole antifungals, 190

basal promoters, 339
BCR-ABL1 fusion gene. *See* breakpoint cluster region-Abelson murine leukemia viral oncogene homolog 1
belinostat, 363*t*
BiDil, 253
bilirubin, 241
billing, 54–55
bioinformatics, 49*t*
biologic variation, 316
biological sample banking, 309
biological science, 336
biology
    cellular, 336, 363–364
    molecular, 319–321, 319*f*, 336, 363–364
    of platelets, 66

biomarkers, 21
    in clinical trial design, 292f
    drug response, 149–150
    efficacy prediction with, 25
    fecal, 40
    predictive, 150
    in targeted therapy development, 292–293
bipolar disorder, 124–127
bleeding, warfarin-related, 83, 307f
boceprevir, 221
body mass index, 104
bosutinib, 158
BRAF gene. See v-Raf murine sarcoma viral oncogene homolog B
BRCA1 gene. See breast and ovarian cancer susceptibility gene 1
BRCA2 gene. See breast and ovarian cancer susceptibility gene 2
BRCA test, 33–34, 41
breakpoint cluster region-Abelson murine leukemia viral oncogene homolog 1 (*BCR-ABL1* fusion gene), 37, 150, 157–159
breast and ovarian cancer susceptibility gene 1 (*BRCA1* gene), 33
breast and ovarian cancer susceptibility gene 2 (*BRCA2* gene), 33
breast cancer, 41
    estrogen and progestin receptors and, 151
    HER2 oncogene and, 152
    pharmacogenetic testing, 152t
    prognosis of, 37–38
    risk stratification of, 33–34
    treatment of, 151–153
brentuximab vedotin, 290t
budget impact analysis, 277–278

Ca^{++}, 353–354
CAAT box, 340
CACNA1A gene, 137
CAD. See coronary artery disease
calcium channel blockers, 11, 190
cAMP. See cyclic adenosine monophosphate
Canadian Pharmacogenomics Network for Drug Safety, 16
cancer, 149–151, 359t. See also breast cancer; colon cancer; lung cancer; melanoma; prostate cancer
candidate chip genotyping, 376
candidate gene analysis, 325–326
candidate gene association studies, 67, 307–308
candidate gene genotyping, 376
capecitabine, 167f, 290t
capping. See 5' capping
carbamazepine, 22, 118t, 124
    chemical structure of, 204f
    HLA gene and, 202, 203t, 204, 207t, 208
    HSS induced by, 207t, 208
    mechanisms of action of, 126
    metabolism of, 125
    SJS/TEN induced by, 202, 203t, 204, 207t
cardiac transplants, 39
CART. See Classification and Regression Trees
case-control study, 372–373
catecholamines, 128
catechol-o-methyltransferase (COMT), 123, 128, 141–142
CDC. See Centers for Disease Control and Prevention
CD-CV hypothesis. See common disease common variant hypothesis
CDK. See cyclin-dependent kinase
CD-RV hypothesis. See common disease rare variant hypothesis
celecoxib, 290t
cell cycle machinery, 338
cell signaling, 353–363
cellular biology, 336, 363–364
Centers for Disease Control and Prevention (CDC), 26–27
Centers for Medicare & Medicaid Services (CMS), 55
cerebrovascular disease, 66
ceritinib, 155–156
CES1 gene, 97
cetuximab, 153–154, 290t, 309
CFTR gene, 36, 38, 293–294
children
    efavirenz dosing for, 240t
    with hepatitis C virus, 226–227
    pharmacogenetic testing in, 257
    warfarin and, 96
cholesterol, 106, 353
chromatin modifications, 352
chromosomes, 150, 317, 317f

chronic myeloid leukemia (CML), 37, 150, 157–159
cigarette smoking, 374
CIP. See congenital indifference to pain
CIPA. See congenital insensitivity to pain with anhidrosis
circulating cell-free DNA, 151
cirrhosis, 225–226
cis-acting regulatory elements, 339–340, 343
citalopram, 118t, 290t
Clarification of Anticoagulation through Genetics (COAG), 84–87, 86t, 92
Classification and Regression Trees (CART), 383
CLIA. See Clinical Laboratory Improvement Amendments
clinical decision support, 28
    in clinical pharmacogenetics, 52–53, 53f
    language of, 54
    for *SLCO1B1* genotyping, 108–109, 111f
Clinical Laboratory Improvement Amendments (CLIA), 20–21, 35
clinical pharmacogenetics, 45–46
    barriers to, 48
    billing and reimbursement in, 54–55
    characteristics of, 47
    clinical decision support in, 52–53, 53f
    drug-gene associations and, 50
    education in, 55–56, 59–60
    evaluation of, 56, 57
    evidence review body in, 50–51
    health care professionals in, 56–59
    pharmacogenetic testing and, 51–53
    planning, 47, 302, 302t
    return of results in, 53–54
    stakeholder engagement in, 48, 49t–50t
Clinical Pharmacogenetics Implementation Consortium (CPIC), 27–28, 70, 71f, 87–88, 208
clinical practice guidelines, 28, 45–46
clinical roles, 258–259
clinical trial design, 292f
clinical utility, 22–23, 71–72, 92–94
clinical validity, 21
clinicians, 49t–50t
clomipramine, 118t
clopidogrel, 24, 66
    activation of, 303–304
    clinical outcomes with, 68
    CYP2C19 gene and, 65, 67, 274–275, 274t, 304–305, 371
    decision models for, 274t, 276
    dose escalation studies, 68–69
    metabolism of, 66f
    pharmacodynamics of, 68
    pharmacokinetics of, 68
    response, 67–69
clozapine, 118t, 124
CML. See chronic myeloid leukemia
CMS. See Centers for Medicare & Medicaid Services
CNV. See copy number variant
COAG. See Clarification of Anticoagulation through Genetics
cochleotoxicity, 268
codeine, 6, 140–141, 296–297, 296t
co-enzyme Q10 (COQ10 enzyme), 110–111
cohort study, 372–374
colon cancer, 34, 152t, 153–154
common disease common variant hypothesis (CD-CV hypothesis), 370
common disease rare variant hypothesis (CD-RV hypothesis), 370–371
communication, 253–254
companion diagnostics, 293
competencies, 56–59, 58f
complex (multifactorial) disease, 325
COMT. See catechol-o-methyltransferase
COMT gene, 141–142
concomitant drug therapy, 55
congenital indifference to pain (CIP), 139–140
congenital insensitivity to pain with anhidrosis (CIPA), 139
congenital sensory neuropathy, 138
copy number variant (CNV), 321, 377
COQ10 enzyme. See co-enzyme Q10
coreceptor tropism, 237–238
coronary artery disease (CAD), 37, 40–41, 66
cost-benefit analysis, 278, 279t
cost-effectiveness, 386
    of pharmacogenetic testing, 273–276, 276t, 278–280
    qualitative framework, 273–276, 276t
    of warfarin genotype-guided dosing, 94–95

cost-effectiveness analysis, 278–280, 279t
cost-effectiveness study, 25–26
cost-minimization analysis, 278, 279t
cost-utility analysis, 54, 279t, 280–281
COUMA-GEN II study, 84
covariates, 381
COX-1. *See* cyclooxygenase-1
CPIC. *See* Clinical Pharmacogenetics Implementation Consortium
Crick, Francis, 317
crizotinib, 155–156, 290t, 294–295, 300
"curse of dimensionality," 382
custom assays, 376
cyclic adenosine monophosphate (cAMP), 354
cyclin-dependent kinase (CDK), 338
cyclooxygenase-1 (COX-1), 66
cyclosporine, 178f, 184–185
CYP. *See* cytochrome P450
CYP2A4 enzyme, 110
CYP2A6 enzyme, 12–13, 12t
CYP2A6 gene, 13t, 14t
CYP2B6 enzyme, 13, 14t
CYP2B6 gene
    alleles, 15t
    efavirenz and, 239–241, 243, 371
    ethnic differences, 15t
    genetic variation, 13
    genotyping, 241
    nevirapine metabolism and, 239–240
CYP2C9 enzyme
    drugs, 8–9
    inducers, inhibitors, and substrates, 10t
    statins and, 110
CYP2C9 gene
    alleles, 10t
    ethnic differences in, 10t
    genetic variation, 9–10
    warfarin and, 80–81
CYP2C19 enzyme
    clopidogrel pharmacokinetics and, 68
    drugs, 7–8
    inducers, inhibitors, and substrates, 8t
CYP2C19 gene
    alleles, 9t
    antidepressants and, 117, 119
    antiplatelet therapy guidelines, 69–71, 70t
    clinical outcomes and, 68
    clopidogrel and, 65, 67, 274–275, 274t, 304–305, 371
    decision models for, 276
    ethnic differences in, 9t
    genetic testing, 69–73
    genetic variation, 8
    genotyping, 24, 70, 71t
    meta-analysis, 68
CYP2D6 enzyme, 3
    drugs, 6
    inducers, inhibitors, and substrates, 6t
    opioid metabolism and, 140–141
    statins and, 110
CYP2D6 gene
    activity score, 5
    alleles, 7t, 8t
    antidepressants and, 117, 119
    codeine and, 296–297
    ethnic differences in, 8t
    genetic testing, 171t
    genetic variation, 6–7
    pain treatment and, 140–141
    pimozide and, 296
    stimulants and, 128
    tamoxifen and, 170, 172
CYP3A4 enzyme, 11
CYP3A4 gene
    alleles, 11t
    cyclosporine and, 184
    ethnic differences, 11t, 12t
    genetic variation in, 11
    tacrolimus and, 183
CYP3A5 enzyme, 5, 11–12, 110
CYP3A5 gene
    alleles, 12
    cyclosporine and, 184
    genetic variation in, 11–12
    tacrolimus and, 178–181, 187f, 371
CYP3A enzymes, 10–11, 11t
CYP4F2 gene, 82
CYP alleles, 3–5. *See also specific CYP gene*
CYP-allele Web site. *See* Human Cytochrome P450 Allele Nomenclature Database
cystic fibrosis, 24, 36, 293–294
cytochrome P450 (CYP), 1, 3, 121. *See also specific CYP enzyme or gene*
cytochrome P450 oxidoreductase, 183
cytosine, 318, 321f

DAAs. *See* direct-acting antivirals
dabigatran, 97
dabrafenib, 156
daclatasvir, 228
Daiichi Sankyo, 304, 306
Darwin, Charles, 316
dasabuvir, 228–229
dasatinib, 158, 290t
DAT1 gene. *See* dopamine transporter gene
DBD. *See* DNA-binding domain
death, 2
"death panel," 270
decision analysis, 275
decision modeling, 274t, 275–276
decision support. *See* clinical decision support
decision tree, 277f, 278f, 280f
decision-making, oncology, 172
decitabine, 363t
deoxyribonucleic acid (DNA)
    circulating cell-free, 151
    genetic polymorphisms and, 322–323
    junk, 319
    RNA differences from, 319
    structure and composition of, 317–318, 318f, 336
    supercoiling, 337
dephosphorylation, 351
depression, 117. *See also* antidepressants
design-defect liability, 256
desipramine, 118t
developing countries, 387
dexamethasone, 190
diabetes mellitus, 21
diagnostic manufacturers, 272
diagnostic tests, 22–23
diagnostics, companion, 293
digoxin, 40
dimensionality, 382–383
direct-acting antivirals (DAAs), 217, 222–224
discrimination, 252–253
disease. *See also specific disease or disorder*
    complex (multifactorial), 325
    diagnosis, 35–37, 36f
    heritability of, 370–371
    monitoring, 39
    monogenic, 324
    prognosis, 37–38, 37f
    research, 300
    subtypes, 300
disease risk stratification, 33–35, 34f, 35f
DNA. *See* deoxyribonucleic acid
DNA genome-wide scans, 35
DNA helicases, 337
DNA methylation, 331, 361–362
DNA methyltransferases (DNMTs), 361–362
DNA polymerase, 337
DNA sample banking, 302–303
DNA sequencing, 330–331
DNA single-gene/single-syndrome tests. *See* single-gene/single-syndrome tests
DNA-based technologies, 33
DNA-based tests, 39
DNA-binding domain (DBD), 344, 345f
DNMTs. *See* DNA methyltransferases
dopamine, 122, 128
dopamine-2 receptor gene (*DRD2* gene), 122
dopamine-4 receptor gene (*DRD4* gene), 128
dopamine transporter gene (*DAT1* gene), 137

double-blinding, 374
double-helix structure, 317–318, 318f
doxepin, 118t
DPYD gene, 167–168, 168t, 171t
DRD2 gene. See dopamine-2 receptor gene
DRD4 gene. See dopamine-4 receptor gene
drug development
    pharmacogenomics and, 289, 299, 303–306
    regulation of, 289, 291
    targeted, 289
drug interactions. See also gene-drug interactions
    in HIV therapy, 243–244
    organ transplantation and, 190
    statins and, 104
drug labeling, 289–291, 291f
drug manufacturers. See pharmaceutical manufacturers
drug monitoring, 46
drug reaction. See adverse drug reaction; severe cutaneous adverse drug reactions
drug resistance testing, 234–237, 235f, 244t
drug response
    age and, 257
    biomarkers, 149–150
    variability in, 1–2, 19, 301
drug-gene associations, 50, 267–268
drug-gene pairs, targeted, 47
drug-metabolizing enzyme genes, 5
drug-metabolizing enzymes, 2
drugs. See also specific drug; specific gene
    concomitant drug therapy, 55
    pharmacogenetic tests codeveloped with, 22
    race-based, 253
Duchenne muscular dystrophy, 38
dysfunctional alleles, 5

economic evaluations, 270, 279t
edoxaban, 306, 307f
education, 54–56, 59–60
efavirenz, 244t
    CYP2B6 gene and, 239–241, 243, 371
    dosing for children, 240t
    metabolism of, 239–240, 243–244
    plasma concentrations, 243f
EGAPP. See Evaluation of Genomic Applications in Practice and Prevention
EGAPP Working Group (EWG), 26–27
EGFR. See epidermal growth factor receptor
EGFR gene, 154
EHR. See electronic health record
EIGENSTRAT, 379
electronic health record (EHR), 51–52, 210
Eli Lilly, 304
elongation, 340, 347
EMA. See European Medicines Agency
emerging technologies, 308–309
EML4-ALK gene, 155, 300
Employee Retirement Income Security Act (ERISA), 256
ENCODE Project, 316, 332
enhancers, 344–345, 346f
enrichment studies, 24
enteric microbiome, 40
entinostat, 364t
environmental compounds, 9
enzymes, 2. See also specific enzyme
epidemiology studies, 372–375, 379
epidermal growth factor receptor (EGFR), 37, 295
epigallocatechin gallate, 364t
epigenetic therapy, 362–363, 363t–364t
epigenetics, 22, 26, 331, 359–360
epistasis analysis methods, 382–383
eQ-PCR LC Warfarin Genotyping Kit, 91
equity, 260–261
ERISA. See Employee Retirement Income Security Act
erlotinib, 154–155
eSensor platform, 91
estrogen, 353
estrogen receptor, 37, 151
ethical considerations, 23
ethnicity
    African American, 95–96, 181–182
    CYP2B6 gene and, 15t
    CYP2C9 gene and, 10t
    CYP2C19 gene and, 9t
    CYP2D6 gene and, 8t
    CYP3A4 gene and, 11t, 12t
    Hispanic, 95–96
    HLA gene and, 203t
    Southeast Asian, 202, 204
    tacrolimus and, 180f
    white, 182–183
eukaryotic cells, 339f
eukaryotic promoters, 339
EU-PACT trials. See European Pharmacogenetics of AntiCoagulation Therapy trials
European Medicines Agency (EMA), 309
European Pharmacogenetics of AntiCoagulation Therapy trials (EU-PACT trials), 86–87, 86t, 92
European Science Foundation–University of Barcelona, 69
Evaluation of Genomic Applications in Practice and Prevention (EGAPP), 26
everolimus, 151–152, 290t
evidence
    challenges with, 22–26
    with pharmacogenetic testing, 268
    review of, 26–28, 50–51
    thresholds of, 23
evidence-based guidelines, 26–29
evidence-based medicine, 26
evolutionary methods, 383
EWG. See EGAPP Working Group
exemestane, 290t
exercise, 103
exon skipping, 38
exons, 320, 322, 343
expressed genome, 37

familial dysautonomia (FD), 138
familial hemiplegic migraine (FHM), 137
family health history (FHH), 33–34, 41
FD. See familial dysautonomia
FDA. See Food and Drug Administration
fecal biomarker, 40
female gender, 104
ferritin mRNA, 349, 350f
FHH. See family health history
FHM. See familial hemiplegic migraine
fibromyalgia, 137–138
filtering, 383–384
5' capping, 341–342
5' untranslated region (5' UTR), 320
FK-binding proteins, 177
fluorouracil, 290t
fluvastatin, 106, 106t
fluvoxamine, 118t
Food and Drug Administration (FDA), 16, 21, 46, 288
    companion diagnostics regulation by, 293
    drug development regulation by, 289, 291
    package inserts approved by, 267
    purine antimetabolites and, 166
    warfarin genotyping platforms and, 91, 95t
fulvestrant, 290t

G protein-coupled receptors, 303–304, 354–355, 355t, 356f
gamma-aminobutyric acid (GABA), 126
GC box, 339
gefitinib, 154–155
genABEL, 378
gene-drug interactions, 50, 267–268
gene-gene interactions, 382
genes. See also specific gene
    basic structure of, 320, 320f
    drug-metabolizing enzyme, 5
    processing and regulation, 336–353
    protein-coding, 319–321
genetic and genomic testing, 26–27. See also pharmacogenetic testing; single-gene/single-syndrome tests; specific gene
    analytic validity of, 20–21
    clinical utility of, 22
    clinical validity of, 21
    turn-around time on, 387

genetic discrimination, 252–253
Genetic Information Nondiscrimination Act (GINA), 253
genetic medicine, 315–316
genetic professionals, 259–260
Genetic Programming Neural Networks (GPNN), 383
Genetic Testing Registry, 15, 28, 71
genetics, 315–316
    epigenetics, 22, 26, 331, 359–360
    germline, 164f
    linkage disequilibrium in, 328
    Mendelian, 317, 324
Genetics and Genomics Competency Center for Education, 56
Genetics Information Trial (GIFT), 92
GENN. See Grammatical Evolution Neural Networks
genome, 37, 221f. See also human genome
genome-wide assays, 16
genome-wide association study (GWAS), 21–22, 308, 329–330, 330f
    of clopidogrel response, 67–68
    genotyping, 375
    limitations of, 370–371
    of warfarin dose requirements, 82–83
genome-wide linkage analysis, 326
genome-wide scans, 35
genome-wide testing, 36
genomic applications, 29
genomic data, 24–26, 40
genomic medicine, 33
    applications of, 328, 329
    benefits of, 329
    databases for, 331–333
    evolution of, 39–40
    technologies in, 328–331
genotype, 323f. See also warfarin genotype-guided dosing
    nomenclature, 323–324
    resistance test, 236
    statin therapy and, 107–110, 109t
    tacrolimus dosage equation, 181
genotype-phenotype associations, 21–22
Genotype-Tissue Expression Project (GTEx), 316, 332
genotyping. See also specific gene
    candidate chip, 376
    candidate gene, 376
    GWAS, 375
    sequencing, 377
    technologies, 15–16
    warfarin, 91, 95t
    whole genome, 35
germline genetics, 164f
germline mutations, 149
germline pharmacogenetics, 163, 171t
GIFT. See Genetics Information Trial
GINA. See Genetic Information Nondiscrimination Act
givinostat, 364t
glutamate, 123
glycosylation, 351
GPNN. See Genetic Programming Neural Networks
Grammatical Evolution Neural Networks (GENN), 383
grapefruit juice, 10
green tea extract, 364t
GTEx. See Genotype-Tissue Expression Project
guanine, 318, 321f
GWAS. See genome-wide association study

haloperidol, 118t
Han Chinese, 22
haplotype block, 326–328, 327f
haplotypes, 330
hapten concept, 209
Hardy-Weinberg equilibrium, 326
HBOC syndrome, 41
HCV. See hepatitis C virus
health care, 271–273
health care professionals, 56–59, 272
health economist, 270
health outcome data, 24–26
"healthy ills," 252
helix-turn-helix motif, 344, 345f
hematologic malignancies, 157–159, 158t
hepatitis B virus, 224
hepatitis C virus (HCV)
    children with, 226–227
    cirrhosis and, 225–226
    genome structure, 221f
    hepatitis B virus co-infection with, 224
    HIV co-infection with, 224–225
    IFNL3 gene and, 220–221
    liver transplantation and, 225
    pharmacotherapy, 217–218, 228–229
    prevalence of, 217
    resistance of, 219, 221t
    RNA of, 218–220
    sustained virologic response rates, 223f–225f
    treatment guidelines, 219f–220f
HER2. See human epidermal growth factor receptor 2
HER2 oncogene, 37–38, 152, 254
hereditary disorders insensitive to pain (HSAN), 138–140
heredity, 317, 370–371
heterogenous nuclear RNA (hnRNA), 341
HFE gene, 33
high on-treatment platelet reactivity (HTPR), 66, 68
HIPAA Privacy Rule, 258
Hispanics, 95–96
histone acetylation, 360
histone methylation, 361
histone modification, 360, 362
histone phosphorylation, 361
histone ubiquitination, 361
HIV. See human immunodeficiency virus
HIV antivirals, 11
HIV therapy. See also antiretroviral therapy
    anti-HIV drugs for, 197t
    drug resistance testing in, 234–237, 235f, 244t
    drug transporter pharmacogenetics, 242–243
    drug-drug interactions in, 243–244
    genetic testing in, 238–242
    genotype resistance test in, 236
    maraviroc for, 237–238
    pharmacogenetic testing in, 244–245, 244t
    phenotypic resistance testing in, 236
    tropism testing in, 238
HLA. See human leukocyte antigen
HLA gene
    abacavir and, 200–202, 238–239
    alleles, 201t, 207t
    allopurinol and, 205–206, 206t–207t, 208
    carbamazepine and, 202, 203t, 204, 207t, 208
    ethnicity and, 203t
    genotyping, 38
    pharmacogenetics, 200
    SCAR and, 201t, 208–209
    screening, 200–202, 204–205
    Southeast Asians and, 202, 204
HMG-CoA. See 3-hydroxy 3-methylglutaryl coenzyme A
HMG-CoA reductase (HMGCR), 111–112
HMG-CoA reductase inhibitors. See statins
HMGCR. See HMG-CoA reductase
hnRNA. See heterogenous nuclear RNA
"home brews," 268
homeostasis, 353
horizon scanning, 28–29
hormone response element (HRE), 345
Hotchkiss, Rollin, 361
HRE. See hormone response element
HSAN. See hereditary disorders insensitive to pain
HSS. See hypersensitivity syndrome
HTPR. See high on-treatment platelet reactivity
HTR2A gene, 120–123
Human Cytochrome P450 Allele Nomenclature Database (CYP-allele Web site), 3, 14
human epidermal growth factor receptor 2 (HER2), 37
human genome, 19
    characteristics of, 317, 319
    haplotype block structure of, 326–328
Human Genome Project, 318–319
human immunodeficiency virus (HIV), 233–234. See also HIV therapy
    coreceptor tropism and, 237–238
    hepatitis C virus co-infection with, 224–225
    reverse transcriptase inhibitor resistance mutations and, 235f
    in sanctuary sites, 243
human leukocyte antigen (HLA), 38, 198–199, 198f
hydralazine, 253
hydrocodone, 140

hydrolysis probe assays, 15–16
3-hydroxy 3-methylglutaryl coenzyme A (HMG-CoA), 111
hypersensitivity syndrome (HSS), 197–198, 197t
    abacavir-induced, 200–202, 238–239
    carbamazepine-induced, 207t, 208

IBD. *See* identity by descent
ICER. *See* incremental cost-effectiveness ratio
identity by descent (IBD), 378
IEM. *See* inherited erythromelalgia
IFNL3 gene
    DAAs and, 222–224
    hepatitis C virus and, 220–221
    PEG/RBV response and, 221–222
    polymorphisms, 226f
    testing, 227
IFNL4 gene, 227
IKBAP gene, 139
IL28B gene. *See IFNL3* gene
Illumina ADME BeadChip, 376
iloperidone, 118t, 290t
imatinib, 157–158, 290t
imipramine, 118t
immunophilins, 177
immunosuppressants, 176–177, 191
imputation, 330
incidental information, 254–255
incremental cost-effectiveness ratio (ICER), 279–280
indels. *See* insertions/deletions
independent assortment, law of, 324
Industry Pharmacogenomics Working Group (I-PWG), 299
INFINITI Warfarin Assay, 91
informatics, 28, 49t
informed consent, 253–254, 256
inheritance
    autosomal dominant, 324, 324f
    autosomal recessive, 324–325, 325f
    complex (multifactorial), 325
    laws of, 317
    single-gene, 325
    SNP and, 325f
inherited erythromelalgia (IEM), 139–140
initiation, 337–338, 347
initiation complex, 347
INR monitoring, 90
insertions/deletions (indels), 321
insulators, 344–345
insurance
    genetic discrimination and, 252–253
    pharmacogenetic testing and, 55, 261
    provider liability, 256
integrase strand transfer inhibitors, 242
International HapMap Project, 316, 331–332
intron-exon junctures, 343
ion channels, 353–354
I-PWG. *See* Industry Pharmacogenomics Working Group
irinotecan, 168–169, 169f
isoniazid, 2
isosorbide dinitrate, 253
ITPA gene, 227–228
ivacaftor, 24, 290t, 293–294

Janus tyrosine kinase 2 gene (*JAK2* gene), 35–36
junk DNA, 319

K^+, 353–354
Kalow, Werner, 2
karyotype, 317, 317f
kinship coefficient, 378
Kirsten RAS viral oncogene homolog (*KRAS* gene), 37, 153–154, 309
Kruskal-Wallis test, 382

laboratory processing, 51–52
laboratory staff, 49t
laboratory-developed test (LDT), 268
lamotrigine, 124–126, 204f
lapatinib, 153, 290t
law of independent assortment, 324
law of segregation, 324

laws of inheritance, 317
LBD. *See* ligand-binding domain
LDD. *See* lumbar disc degeneration
LDL. *See* low density lipoprotein
LDT. *See* laboratory-developed test
ledipasvir, 228
lenalidomide, 159, 290t
letrozole, 290t
leucine-zipper motif, 344, 345f
liability, 255–257
ligand-binding domain (LBD), 345
linear regression, 381
linkage disequilibrium, 326–328, 327f, 370
lithium, 124, 126
liver, 2
liver transplantation, 177, 179, 225
long noncoding RNA (lncRNA), 352
loss-of-function alleles, 5
lovastatin, 110
low back pain, 136–137
low density lipoprotein (LDL), 103
lumbar disc degeneration (LDD), 136–137
lung cancer
    EGFR gene and, 154
    EML4-ALK gene and, 155
    non-small cell, 154, 294–295
    pharmacogenetic testing, 152t
    treatment of, 154–156
Lyme disease, 256
LYMErix, 256
Lynch syndrome, 34, 36, 41–42

macrolide antibiotics, 11
MAF. *See* minor allele frequency
major histocompatibility complex (MHC), 198, 199f
malaria, 2
MammaPrint, 38
MAPK. *See* mitogen-activated protein kinase
maraviroc, 237–238, 244t, 290t
Maupertuis, Pierre-Louis Moreau de, 316
MC1R gene, 142–143
MC4R. *See* melanocortin 4 receptor
MDR. *See* Multifactor Dimensionality Reduction
MDR1 gene. *See* multidrug resistance gene
MEDCO-MAYO Warfarin Effectiveness study, 84
Medicaid. *See* Centers for Medicare & Medicaid Services
medical liability, 255–257
medical malpractice, 256
Medicare. *See* Centers for Medicare & Medicaid Services
medicine. *See also* genomic medicine
    evidence-based, 26
    genetic, 315–316
    personalized, 19
    "stratified," 283
MEFV gene, 137
MEK. *See* mitogen-activated protein kinase kinase
melanocortin 1 receptor, 142
melanocortin 4 receptor (MC4R), 123
melanoma, 152t, 156, 295
Mendel, Gregor, 317, 324
Mendelian genetics, 317, 324
mental illness, 116–117
mercaptopurine, 164, 165f
messenger RNA (mRNA), 320–321, 343, 349, 350f
metabolizer phenotypes, 4, 4t
metabolome, 331
metformin, 21
methotrexate, 169–170
methylation, 32, 331, 361–362
methylenetetrahydrofolate reductase (MTHFR), 123
MHC. *See* major histocompatibility complex
microRNA (miRNA), 322, 351, 385
migraine disorder, 137
minor allele frequency (MAF), 378
minority populations, 95–96. *See also* ethnicity
miRNA. *See* microRNA
mitogen-activated protein kinase (MAPK), 358
mitogen-activated protein kinase kinase (MEK), 357–359
modified regression methods, 383
molecular biology, 319–321, 319f, 336, 363–364
molecular target, 289

molecules, cell signaling, 353
monogenic disease, 324
mood stabilizers, 124–127
morphine, 6
mRNA. *See* messenger RNA
MRP2. *See* multidrug resistance protein 2
MTHFR. *See* methylenetetrahydrofolate reductase
Mullis, Kary, 328
multidrug resistance gene (*MDR1* gene), 143
multidrug resistance protein 2 (MRP2), 189–190
Multifactor Dimensionality Reduction (MDR), 382–383
multiple comparison correction, 384
multiple testing, 384
multiplexed testing, 25
mutations, 149–150, 235*f*
myalgia, 103
mycophenolic acid, 186*f*, 187–190
myelodysplastic syndromes, 159
myocardial infarction, 66
myopathy, 103. *See also* statin-induced myopathy
myositis, 103

Na$^+$, 353
natural selection, 316
ncRNA. *See* noncoding RNA
nerve growth factor (NGF), 139
neuroblastoma RAS viral oncogene homolog (*NRAS* gene), 153
nevirapine, 239–241
next-generation sequencing, 16, 35–36, 39, 330–331
NGF. *See* nerve growth factor
NGFB gene, 139
NIH National Human Genome Research Institute, 316
nilotinib, 158, 290*t*
nitric oxide, 353
NNT. *See* number needed to test
noise, in data, 380
noncoding RNA (ncRNA), 322–323, 351
nonparametric statistical methods, 382
non-small cell lung cancer, 154, 294–295
nonsteroidal anti-inflammatory drugs (NSAIDs), 8, 197*t*
norepinephrine, 128
nortriptyline, 118*t*
NRAS gene. *See* neuroblastoma RAS viral oncogene homolog
NSAIDs. *See* nonsteroidal anti-inflammatory drugs
NTRK1 gene, 139
nuclear receptors, 345, 359
nucleotides, 318
number needed to test (NNT), 386

OATP1B1. *See* organic anion-transporting polypeptide 1B1
Office of Public Health Genomics, 28–29
omacetaxine, 158–159
ombitasvir, 228–229
Omnibus Budget Reconciliation Act of 1990, 256
On the Origin of Species (Darwin), 316
oncology
    decision-making, 172
    germline pharmacogenetics in, 163, 171*t*
    pharmacogenetic testing in, 159
    pharmacogenomics research, 309–310
Oncotype DX, 38–39, 149–150
opioids, 140–141
OPRM1 gene, 142
optimal therapy, 38–39
organ transplantation, 177, 190
organic anion-transporting polypeptide 1B1 (OATP1B1), 104, 107, 303, 304*f*
outcome variable, 381
oxcarbazepine, 204*f*
oxycodone, 140–141

P2Y$_{12}$ receptor inhibitors, 66–67
package insert, 267
pain
    with CIP, 139–140
    with CIPA, 139
    with HSAN, 138–140
    with low back pain, 136–137
    with PEPD, 139–140
    pharmacogenetics and, 144–145, 144*t*

receptors involved in, 142–144
treatment of, 140–142
panitumumab, 153–154, 290*t*, 309
panobinostat, 364*t*
PANSS. *See* Positive and Negative Symptom Scale
paritaprevir, 228–229
PARK2 gene, 137
paroxysmal extreme pain disorder (PEPD), 139–140
patient confidentiality, 257–258
Patient Protection and Affordable Care Act (PPACA), 253, 261
patients
    education of, 54
    provider communication with, 253–254
    as stakeholders, 273
PCA. *See* principle components analysis
PCI. *See* percutaneous coronary intervention
PCR. *See* polymerase chain reaction
PCR-based assays, 15
PCSK9. *See* proprotein convertase subtilisin/kexin type 9
pedigree, 324–325, 324*f*–325*f*, 378
peginterferon alfa-2a (PEG), 217–218, 221–222
PEPD. *See* paroxysmal extreme pain disorder
percutaneous coronary intervention (PCI), 65
perphenazine, 118*t*
personalized medicine, 19
pertuzumab, 152, 290*t*
Pfizer, 308
pharmaceutical manufacturers, 24, 271, 299
pharmacists, 48, 50
    competencies of, 56–59, 58*f*
    liability of, 256–257
    role of, 260
pharmacodynamics, 303, 371. *See also* specific drug
pharmacogenetic associations, 24
pharmacogenetic information storage, 255
pharmacogenetic studies, 2–3. *See also* pharmacogenomics research
pharmacogenetic testing, 20, 251
    adoption of, 108
    for breast cancer, 152*t*
    budget impact analysis, 277–278
    in children, 257
    clinical pharmacogenetics and, 51–53
    for colon cancer, 152*t*
    cost-effectiveness of, 273–275, 275*t*, 278–280
    cost-utility analysis and, 280–281
    decision modeling, 275–276
    drugs codeveloped with, 22
    economic incentives for, 283
    evidence levels with, 268
    in HIV therapy, 244–245, 244*t*
    incidental information and, 254–255
    informed consent for, 253–254
    insurance and, 55, 261
    laboratory processing, 51–52
    lung cancer, 152*t*
    market size of, 268
    medical liability and, 255–257
    in oncology, 159
    ordering, 51–52
    in patient care process, 51–53
    preemptive, 24–26, 51–52, 72, 108–109, 387
    reactive, 51–52
    reimbursement for, 261
    risk classification in, 108, 110
    stigmatization and, 252
    targeted, 24–26
pharmacogenetics, 3, 251–252. *See also* clinical pharmacogenetics; *specific drug; specific gene*
    of drug-metabolizing enzyme genes, 5
    evaluation of, 269–270
    evolution of, 1, 282
    familial implications of, 257–258
    germline, 163, 171*t*
    pain and, 144–145, 144*t*
    reimbursement barriers and reforms, 282–284
    translational pathways for, 23–24
pharmacogenomic associations, 21–22
pharmacogenomics, 3. *See also specific drug*
    ADME research and, 303
    drug development and, 289, 299, 303–306
    drug labeling and, 289–291, 291*f*

guidelines for, 16–17
optimal therapy selection and, 38
pharmacokinetics and pharmacodynamics research and, 303
targeted therapies and, 291–293
Pharmacogenomics Knowledge Base (PharmGKB), 3, 14, 27, 46, 316, 332–333
pharmacogenomics research, 370. *See also* study design
analysis, 380–384
clinical, 301–310
multiple comparison correction in, 384
in oncology research, 309–310
post-analysis issues in, 384–386
preclinical, 299–301
quality control, 377–380
replication in, 309, 385
results interpretation in, 384–385
techniques, 306–307
validation and translation in, 386
pharmacokinetics, 2, 303, 371. *See also specific drug*
pharmacy informatics, 49t
PharmGKB. *See* Pharmacogenomics Knowledge Base
phase I enzymes, 2
phase II enzymes, 2
phased haplotypes, 330
phenelzine sulfate, 364t
phenotype, 323, 323f
genotype associations with, 21–22
metabolizer, 4, 4t
study design and, 371–372
phenotypic resistance testing, 236
phenytoin, 204f
Philadelphia chromosome, 150
phospholipase C beta (PLCβ), 345–355
phosphorylation, 351, 361
physician alert, 94f
physician assistants, 56
physicians, 56
p-i concept, 209
pimozide, 290t, 296
plasma membrane diffusion, 353
Platelet Inhibition and Patient Outcomes (PLATO), 73, 377
platelets, 66, 303
PLATO. *See* Platelet Inhibition and Patient Outcomes
PLCβ. *See* phospholipase C beta
Plexxikon, 310
PLINK, 377–378
point-of-care testing, 51, 72
Poisson regression, 381
polyadenylation. *See* 3' polyadenylation
polymerase chain reaction (PCR), 15, 40, 328, 329f
polymerases, 320, 337–338
polymorphisms, 321–323, 322f, 326. *See also* single nucleotide polymorphism; *specific gene*
polypeptides, 347, 350–351
ponatinib, 158–159
population substructure, 378–379
portable assays, 387
Positive and Negative Symptom Scale (PANSS), 123
post traumatic stress disorder (PTSD), 137–138
posttranscriptional regulation, 352
PPACA. *See* Patient Protection and Affordable Care Act
pracinostat, 364t
prasugrel, 65, 67, 73, 303
pravastatin, 105, 106t
predicted activity scores, 5
predicted metabolizer phenotypes, 4, 4t
predictive biomarkers, 150
preemptive pharmacogenetic testing, 24–26, 51–52, 72, 108–109, 387
pregnancy, 6
preinitiation complex, 347
pricing, 272
primaquine, 2–3
primary care providers, 259
principle components analysis (PCA), 379
product liability, 256
professional guidelines, 259
professional readiness, 258
progesterone, 353
progesterone receptor, 37
progestin receptor, 151
proprotein convertase subtilisin/kexin type 9 (PCSK9), 111–112

prospective cohort study. *See* cohort study
prostate cancer, 34
protease inhibitors, 241–242
protein synthesis. *See* translation (protein synthesis)
protein-coding genes, 319–321
proteins. *See also specific protein or enzyme*
in DNA replication, 337–338
genetic polymorphisms and, 322–323, 322f
life spans of, 350
production of, 320–321
proteome, 331
proton pump inhibitors, 7
PTSD. *See* post traumatic stress disorder
public interest, 252
PubMed Central repository, 28
purine antimetabolites, 164–166
pyrimidine antimetabolites, 166–168, 168t

QALY. *See* quality-adjusted life-year
Q-Q plots. *See* quantile-quantile plots
quality control, 377–380
quality-adjusted life-year (QALY), 280–281, 282f
quantile-quantile plots (Q-Q plots), 380

race stratification, 379. *See also* ethnicity
race-based drugs, 253
Raf signaling pathway, 357–359
Random Forests, 383
randomized controlled trials (RCTs), 23–24, 71–72, 372, 374–375
RAS oncogene, 37
Ras signaling pathway, 357–359, 359t
rasburicase, 290t
rates of events, 372, 381
RBV. *See* ribavirin
RCTs. *See* randomized controlled trials
reactive pharmacogenetic testing, 51–52
receptor tyrosine kinase (RTK), 355–357, 356f, 358t
receptor-associated tyrosine kinases, 357
regression, 381–383
regulation
of companion diagnostics, 293
of drug development, 289, 291
stakeholders in, 272–273
of targeted therapies, 291–293
of warfarin pharmacogenetics, 87
reimbursement
barriers and reforms, 282–284
in clinical pharmacogenetics, 54–55
for *CYP2C19* genetic testing, 72
models of, 272
for pharmacogenetic testing, 261
value-based, 283
with warfarin, 95
relatedness, 378
release factor, 349
replication (DNA), 319, 336
in eukaryotic cells, 339f
fork, 338
origins and initiation of, 337–338
proteins involved in, 337–338
replication (research), 309, 385
respiratory depression, 296–297
results, 53–54, 384–385
retrospective analysis, 23, 309
retrospective case-control study. *See* case-control study
reverse transcriptase, 328
reverse transcriptase inhibitor resistance mutations, 235f
reverse transcription, 328
rhabdomyolysis, 103
ribavirin (RBV), 217–218, 221–222
ribonucleic acid (RNA)
DNA differences from, 319
of hepatitis C virus, 218–220
hnRNA, 341
lncRNA, 352
miRNA, 322, 351, 385
mRNA, 320–321, 343, 349, 350f
ncRNA, 322–323, 351
rRNA, 346
siRNA, 352
tRNA, 320, 346, 348f

ribosomal RNA (rRNA), 346
Riley-Day syndrome, 138
risk, 23
risk benefit analysis, 275–276
risk classification, 108, 110
risk stratification, 33–35, 34f, 35f
risperidone, 118t
ritonavir, 228–229, 241
rivaroxaban, 97
RNA. *See* ribonucleic acid
RNA gene expression profiling, 36–39
RNA polymerase, 320, 338
RNA processing, 341–343, 342f
RNA sequencing, 331
RNA-based technologies, 33
Roche, 310
romidepsin, 363t
rosuvastatin, 106, 106t
Royal Dutch Association for the Advancement of Pharmacy, 16, 69–70
rRNA. *See* ribosomal RNA
RTK. *See* receptor tyrosine kinase

sample banking, 302–303, 309
SCAR. *See* severe cutaneous adverse drug reactions
schizophrenia, 121
SCN1A gene, 137
SCN9A gene, 139–140
SEARCH. *See* Study of Effectiveness of Additional Reductions in Cholesterol and Homocysteine
segregation, law of, 324
segregation patterns, 325
selective serotonin reuptake inhibitor (SSRI), 117, 119–120
sequencing
    consensus, for splicing, 344f
    DNA, 330–331
    genotyping, 377
    next-generation, 16, 35–36, 39, 330–331
    RNA, 331
    whole exome, 33, 35–36
    whole genome, 33, 35–36, 377
serine palmitoyltransferase, 138
serotonin and norepinephrine reuptake inhibitor (SNRI), 117, 119–120
severe cutaneous adverse drug reactions (SCAR), 196–198
    clinical practice and, 209–210
    genetic studies of, 199–200, 209
    HLA gene and, 201t, 208–209
    models of, 208–209
sex errors, 378
short interfering RNA (siRNA), 352
signal transducer and activator of transcription (STAT), 357
silencers, 344–345
simeprevir (SMV), 218
simvastatin, 104, 106t, 110, 111f
single gene assays, 15–16
single nucleotide polymorphism (SNP), 32, 321, 323
    analysis method, 380–382
    genome-wide scans for, 35
    inheritance models, 325f
    quality control and, 377–378
single-gene/single-syndrome tests, 34–39
siRNA. *See* short interfering RNA
SJS. *See* Stevens-Johnson syndrome
SLC6A4 gene, 120–121, 128
SLCO1B1 gene, 102
    genotype implementation, 107–110, 109t
    methotrexate and, 169–170
    mycophenolic acid and, 189
    statin-induced myopathy and, 106–107, 107t
    statins and, 104–110, 109t
    testing, 171t
    variants, 104–107, 105t, 107t
SLCO1B1 genotyping
    clinical decision support for, 108–109, 111f
    implementation of, 107–110, 109t
    preemptive, 108–109
SLCO1B3 gene, 188–189
smoking, 374
SMV. *See* simeprevir
SNP. *See* single nucleotide polymorphism
SNRI. *See* serotonin and norepinephrine reuptake inhibitor
sofosbuvir (SOF), 218–219, 223, 228

solid tumors, 151–156, 152t
somatic mutations, 149–150
Southeast Asians, 202, 204
splicing, 320, 342–343, 344f
SPTLC1 gene, 138
SSRI. *See* selective serotonin reuptake inhibitor
stakeholders, 48, 49t–50t, 271–273
standardized evidence review, 26–28
star allele nomenclature, 3–4, 323–324
STAT. *See* signal transducer and activator of transcription
Statin Response Examined by Genetic Haplotype Markers (STRENGTH), 106
statin-induced myopathy
    COQ10 enzyme and, 110–111
    incidence of, 103
    risk classification, 108
    risk factors for, 103–104
    SLCO1B1 gene and, 106–107, 107t
statins, 11, 102–103
    ABCG2 gene and, 110
    candidate genes involved with, 308
    CYP enzymes and, 110
    drug interactions and, 104
    efficacy of, 111–112
    pharmacokinetics of, 104–106
    SLCO1B1 gene and, 104–110, 109t
    systemic exposure to, 110–111
statistics, 329, 382. *See also* regression
STEP-BD. *See* Systematic Treatment Enhancement Program for Bipolar Disorder
steroid hormone receptors, 345–346
steroids, 11, 353
Stevens-Johnson syndrome (SJS), 197, 197t
    allopurinol-induced, 205–206, 206t–207t, 208
    carbamazepine-induced, 202, 203t, 204, 207t
stigmatization, 252
stimulants, 127–128
stop codon, 347
"stratified medicine," 283
STRENGTH. *See* Statin Response Examined by Genetic Haplotype Markers
stroke, 66
STRUCTURE, 379
study design
    biomarkers in, 292f
    case-control, 372–373
    cohort study, 372–374
    epidemiologic, 372–375
    phenotype and, 371–372
    rate measurement in, 372
    RCT, 372, 374–375
    treatment-dependent, 372
    treatment-differential, 372
Study of Effectiveness of Additional Reductions in Cholesterol and Homocysteine (SEARCH), 106
succinylcholine, 2
sugar-phosphate backbone, 318
sulfotransferase-4A1 gene (*SULT4A1* gene), 123–124
supplier pricing, 272
surgery, 103
survival bias, 386
survival data, 381
systematic evidence review, 26–28
Systematic Treatment Enhancement Program for Bipolar Disorder (STEP-BD), 124

tacrolimus, 12
    ABCB1 gene and, 183
    African Americans and, 181–182
    azole antifungals interaction with, 190
    calcium channel blockers interaction with, 190
    CYP3A4 gene and, 183
    CYP3A5 gene and, 178–181, 187f, 371
    cytochrome P450 oxidoreductase and, 183
    dexamethasone interaction with, 190
    dose of, 179–181, 180
    ethnicity and, 180f
    genotype dosage equation, 181
    metabolism of, 177–178, 178f
    pharmacodynamics of, 183–184
    pharmacogenomics of, 178
    pharmacokinetics of, 177–178, 179t

troughs of, 180, 180f
whites and, 182–183
tamoxifen, 170, 172
TaqMan® assays, 15–16
tardive dyskinesia, 25
targeted chips, 376
targeted drug development, 289
targeted drug-gene pairs, 47
targeted pharmacogenetic testing, 24–26
targeted therapies, 290t, 291–293
TATA box, 339
TCA. *See* tricyclic antidepressants
technical expert panel (TEP), 27
telaprevir, 222
TEN. *See* toxic epidermal necrolysis
TEP. *See* technical expert panel
termination, 340, 347, 349
test developers and manufacturers, 271–272
testosterone, 353
tetrabenazine, 290t
TGMP. *See* thioguanine monophosphate
therapeutic index, 2, 79–80
therapeutics, 22–23
thienopyridines, 303
thioguanine, 164, 165f
thioguanine monophosphate (TGMP), 164
thiopurine methyltransferase (TPMT), 164–166
thioridazine, 290t
1000 Genomes Project, 316, 332
3' polyadenylation, 342
3' untranslated region (3' UTR), 320
thromboxane A_2 (TXA_2), 66
thymine, 318
ticagrelor, 65, 67, 73
time-to-event, 381
tolbutamide, 24
topoisomerases, 337
tositumomab, 290t
toxic epidermal necrolysis (TEN), 197, 197t
allopurinol-induced, 205–206, 206t–207t, 208
carbamazepine-induced, 202, 203t, 204, 207t
toxicogenomics, 300–301
TPMT. *See* thiopurine methyltransferase
TPMT gene, 164–166
azathioprine dose recommendations and, 185t, 187
testing, 171t
variants, 165t–166t
TPP. *See* Translational Pharmacogenetics Project
tramadol, 140–141
trametinib, 156
trans-acting regulatory elements, 339–340
transcription, 319–320, 338, 341f
elongation in, 340
regulation, 345–346, 352
reverse, 328
stages of, 340
termination in, 340
transcriptional activators, 343–344
transcriptional gene regulation, 343–346
transcriptional repressors, 343–344
"transcriptomics," 331
transfer RNA (tRNA), 320, 346, 348f
translation (protein synthesis), 319–321, 346
initiation of, 347
polypeptide chain elongation in, 347
posttranslational modification after, 350–351
regulation of, 349, 350f
termination of, 347, 349
translation (research), 386
Translational Pharmacogenetics Project (TPP), 28
trastuzumab, 152, 254, 290t
treatment-dependent study design, 372
treatment-differential study design, 372
tree-based methods, 383
tretinoin, 157, 290t
tricyclic antidepressants (TCA), 117, 119–120
trimipramine, 119t
tRNA. *See* transfer RNA
tropism testing, 238
tuberculosis, 2
tumors, 39, 151–156, 152t

twins, 370
TXA_2. *See* thromboxane A_2
TXA_2 inhibitors, 66
tyrosine kinase inhibitors, 168–169

ubiquitination, 361
UGT1A1 gene, 168–169, 171t, 241–242
UGT1A4 gene, 125
UGT1A8 gene, 188–189
UGT1A9 gene, 188–189
UGT2B7 gene, 125, 189
UGT enzymes. *See* uridine 5'-diphosphate glucuronosyl-transferase enzymes
unipolar depression, 117
untranslated regions, 320
uracil, 321f
uridine 5'-diphosphate glucuronosyl-transferase enzymes (UGT enzymes), 125, 168, 241

validation, 386
valproic acid, 119t, 124–126, 290t, 364t
variable number tandem repeat (VNTR), 321
vasoconstriction, 66
vemurafenib, 156, 290t, 295–296, 309–310
Verigene System, 91
vitamin K epoxide reductase complex subunit 1 (VKORC1 enzyme), 80
VKORC1 gene, 81–82
VNTR. *See* variable number tandem repeat
vorinostat, 363t
vortioxetine, 119t
v-Raf murine sarcoma viral oncogene homolog B (*BRAF* gene), 154, 156, 309–310, 359t

Wallace, Alfred Russel, 316
warfarin, 10, 22, 24
bleeding with, 83, 307f
children and, 96
CPIC and, 87–88
CYP2C9 gene and, 80–81
CYP4F2 gene and, 82
dose requirements, 80, 82–83
genes involved with, 80, 81f, 82t
genotyping platforms, 91, 95t
GWAS with, 82–83
minority populations and, 95–96
pharmacodynamics of, 80, 81f
pharmacogenetics, 87, 90–94, 96
pharmacokinetics of, 80, 81f
physician alert, 94f
reimbursement with, 95
therapeutic index of, 79–80
VKORC1 gene and, 81–82
warfarin genotype-guided dosing, 83–87, 85t, 93f
cost-effectiveness of, 94–95
dosing algorithms, 88–90, 88t
dosing prediction accuracy, 89
inter-individual dose variability and, 79–80, 306
Watson, James, 317
WES. *See* whole exome sequencing
WGS. *See* whole genome sequencing
whites, 182–183
whole exome sequencing (WES), 33, 35–36
whole genome genotyping, 35
whole genome sequencing (WGS), 33, 35–36, 377
wild-type alleles, 25
"winner's curse," 386
WNK1 gene, 138

zinc-finger motif, 344, 345f

DISCLOSURE OF POTENTIAL CONFLICTS OF INTEREST

Company relationships: Deepak Voora (RenaissanceRx); Sook Wah Yee (Apricity Therapeutics, Inc.).

Consultancies: Jeffery R. Bishop (Physician's Choice Laboratory Services); William J. Canestaro (Genetech); Vicki L. Ellingrod (ACCP Research Institute, Lexicomp); Andrea Gaedigk (Pfizer; Translational Software); Deanna L. Kroetz (Millenium); Grace M. Kuo (NIH); Jeannine S. McCune (Saladex; Pharmalink); Jeremiah Momper (Omnitura Therapeutics, Inc., Epocrates, Inc., Pfizer); Stuart A. Scott (USDS, Inc.).

Equity: Deanna L. Kroetz (Amgen, Bristol Myers Squibb, Incyte, Exelexis, Merck, Gilead, Sunesis Pharmaceuticals, Celgene, Talon Therapeutics); Jeremiah Momper (Illumina).

Grants: Amber Beitelshees (NIH); Jeffery R. Bishop (NIMH); Kyle J. Burghardt (NIH); Vicki L. Ellingrod (NIMH); Andrea Gaedigk (NIH); Deanna L. Kroetz (NIH, California Breast Cancer Research, CALGB Foundation/Genentech); Grace M. Kuo (Centers for Disease Control and Prevention); Caitrin W. McDonough (NIH/NHLBI); Jeremiah Momper (USDA); Lori A. Orlando (NIH); Jamie Park (Roche Laboratories, Inc.); Stuart A. Scott (NIH/NIGMS); Sharon Seifert (Institute of Medicine, ASHP Research and Education Foundation); Deepak Voora (NIH, U.S. Air Force).

Lecture services: James Hoffman (American Society of Health System Pharmacists); Andrea Gaedigk (NIH); Peter O'Donnell (University of Florida; Genetic Task Force of Illinois).